DUALITY SYSTEM IN APPLIED MECHANICS AND OPTIMAL CONTROL

Advances in Mechanics and Mathematics

Volume 5

DUALITY SYSTEM IN APPLIED MECHANICS AND OPTIMAL CONTROL

by

Wan-Xie Zhong
Dalian University of Technology

SPRINGER SCIENCE+BUSINESS MEDIA, LLC

Library of Congress Cataloging-in-Publication

Zhong, Wanxie
Duality System in Applied Mechanics and Optimal Control
ISBN 978-1-4757-7917-2 ISBN 978-1-4020-7881-1 (eBook)
DOI 10.1007/978-1-4020-7881-1

Originally published by Kluwer Academic Publishers in 2004
Softcover reprint of the hardcover 1st edition 2004

Printed on acid-free paper.

Contents

Preface

Applied mechanics, as the foundation of various engineering disciplines, is greatly promoted in aerospace, naval architecture, machine building, civil, power, chemical, and material engineering etc. The texts and references of applied mechanics have developed several series of books, such as the text books by S.P. Timoshenco and others. Since the computer impact, the development of computational mechanics changes the traditional methodology of problem solving. However, the theoretical basis is still fundamentally the same as before. That is the same PDEs to be solved, the same kind of variational principles used, etc.

Analytical dynamics is a fundamental part in applied mechanics, however in fact, it is less taught in applied mechanics, because the curriculums such as Structural mechanics, Theory of elasticity, Fluid dynamics, Structural vibration, Structural stability, are not so much related to analytical dynamics. Control theory was originated from mechanics but is less taught in applied mechanics curriculums. The theory and methodology in these curriculums are self-contained such as in the books about theory of elasticity, it is hardly seen to have close connections with analytical dynamics.

The analogy between structural mechanics and optimal control manifests that they have the same mathematical basis, *i.e.* the state space approach. Along this way of consideration, the various curriculums in applied mechanics are closely interrelated and they have a common theoretical basis, so that if one studies one curriculum among them, then it will be easier for him to understand the others. The common theoretical basis is the duality system theory. It is one of the purposes of the present book, to ease the studying of applied mechanics.

Classical analytical dynamics proposes the most fundamental system in applied mechanics. The Lagrange equation, the Least action principle, the Hamilton canonical equations, Canonical transformation, and Hamilton-Jacobi theory, etc., compose a graceful theoretical system, and develops as the foundation of statistical mechanics, electro-dynamics, quantum mechanics etc. On the contrary, its appearance in the curriculums of applied mechanics is far from enough. The starting point of state space method, *i.e.* the basis of modern control theory, should at least trace back to the Hamilton canonical equation system. The Hamilton canonical system is really a duality system, it is composed of dual variables, dual equations etc. Further, the linear programming, quadratic programming and non-linear programming theory are also closely related to the duality system. The applied mathematics develops toward the duality system too, see [39]. Based on the above observation, the duality system method should also be consciously and systematically used in various parts of applied mechanics, so as to provide the powerful mathematical methodology, and also to ease the readers with a common mathematical foundation.

Based on the analogy relationship between structural mechanics and optimal control, the dual variables and the related theoretical system are introduced into theory of elasticity, which changes the traditional solution methodology. Under the traditional approach, the solution methodology is by the try-and-error technique called semi-inverse method. However, the duality system methodology derives the basic equation into Hamiltonian dual variables and the respective dual equation form

(Canonical variables and equations), and the mathematics related becomes symplectic geometric rather than the traditional Euclidean geometric. The new solution system for elasticity changes the traditional solution technique to become rational, which far extends the solvable problems, that a number of problems formerly cannot be solved by the semi-inverse method can now be solved by the symplectic eigen-function expansion approach. The same methodology can also be applied to vibration problems for gyroscopic system and wave propagation. The modern control, LQG and also robust control H_∞ theory, has been described by the dual variable system, when the basic equations in applied mechanics are transformed to duality system, then the solution methodology such as eigen-function expansion etc. can be transplanted to the related problems in control theory. The inter-disciplinary development is quite fruitful for both sides. The duality system methodology is unified to various curriculums, which eases both teaching and research.

Computational mechanics is one of the most active parts in applied mechanics, and also is the bridge from applied mechanics toward engineering applications. Combining with the duality system description, numerical methods and algorithms will be emphasized in the text. The precise integration method can be used both for initial value problems and also for two point boundary value problems (TPBVP). For the differential equations in dynamics and the matrix Riccati differential equations in control theory, the precise integration method can give precise solution on the computer. The precise integration in control theory not only solves numerical precise solution for Riccati differential equation, but also provides the numerical precise solution for both the time-variant state and filter differential equations, for which real-time computation is needed, that the related algorithm is also given in the book. Various precise integration algorithms and the symplectic eigen-solution method are the feature of this book on computations. Thus as a preliminary step, the precise integration method is given in the introduction.

Chapter 1, Provides the preliminary of analytical dynamics, Lagrange and Hamilton system, Legendre transformation, dual variables, canonical transformation, symplectic system, the Hamilton-Jacobi method and separation of variables etc., which manifests that the common basis for the later parts of contents is fundamentally from analytical dynamics.

Chapter 2, Heavily describes the structural vibration theory, eigen-problem solution, especially the symplectic eigen-problem for gyroscopic system and its algorithm. The eigenvalue count is also described.

Chapter 3, Preliminary of probability and stochastic process theory, the analysis object of applied mechanics, its parameters, external forces and the instrument measurements are all stochastic. Taking these stochastic factors into consideration is a trend today.

Chapter 4, Random vibration, this is very important for structural aseismatic design, turbulent induced vibrations etc. Although the basic theory of linear random vibration had been established, but its application in engineering practice still has tremendous difficulty, mainly computational. A recently proposed highly efficient algorithm, the Pseudo-Excitation-Method is provided, which speeds up 2~5 orders in computational expense with comparison to the usual approach proposed before.

Chapter 5, Single continuous coordinate elasticity problem, such as the semi-analytical method and wave propagation problems in elasticity, these problems can be solved by the duality system theory. The analogy between structural mechanics and optimal control is based on these developments. The precise integration algorithm solves the Two Point Boundary Value Problem and the matrix Riccati differential equation, which is quite useful also in optimal control problems.

Chapter 6, Optimal control theory, its derivation uses the methodology parallel to that in applied mechanics, which is based on the analogy between structural mechanics and optimal control. The precise integration algorithm solves various matrix Riccati differential equations precisely, which not only solves the Riccati equation but also both the filter and state differential equations. It is found that the critical parameters γ_{cr}^{-2} in both robust H_∞-control or else in H_∞-filter are just the extended Rayleigh quotient in applied mechanics. The analogy relationship between structural mechanics and optimal control can ease the reader from getting familiar with the control subjects.

So far the above contents are for finite degrees of freedom system, which are comparable with that in analytical dynamics. However, the duality methodology can also be used for continuums, *i.e.* infinite degrees of freedom systems, such as elasticity. Based on the analogy relationship between structural mechanics and optimal control, the new solution system for theory of elasticity can be promoted as follows,

1) State space approach is developed and the elasticity problem is derived to be in Hamiltonian system, then
2) Method of separation of variables is applied, and the symplectic eigen-solution problem follows. The eigen-solutions are mutually adjoint symplectic orthogonal to each other, and span the whole state space.
3) Expansion solution method can be used to get the solutions, which can far extend the solvable problems in elasticity.
4) The same methodology is further applied to plate bending, which far extends the solvable problems too. All the above mentioned is completely a new solution methodology for elasticity.

However, because of the limitation of size, the infinite degrees of freedom problems will not be described in this book, a later book will provide such development.

This book is supported by the National natural science foundation of China #19732020, the China NKBRSF project #G1999032805, the foundation for Doctoral program of Education ministry of China, and part of the work was carried out in my visiting to the City University of Hong Kong. My sincere gratitude is for all these supports and the helps from my colleagues and friends.

ZHONG Wan-Xie

中行独复，以从道也
--易经，复卦六四

Introduction

Applied mechanics, as an engineering foundation, has greatly contributed to the progress of various engineering disciplines, such as astro/aeronautics, mechanical, civil, material, chemical, power, electrical and electronic engineering etc. Meanwhile, applied mechanics has also been promoted from applications, developed various theories and methods. From mathematical point of view, when the basic differential equations are established, the problem has been clear and remains only how to solve the equations. The requirement from application asks numerical results but may not cease at theoretical stage. It is quite often that the basic equations have long been established, however, it is very difficult to solve. Such as the theory of elasticity, its basic equation has well been established one more century ago, but its solution is far from completed till now.

Facing with the difficulty of strict solution, various application theories were developed, such as structural mechanics, theory of plates and shells, thin-walled structures, engineering vibrations, structural stability, soil mechanics, fluid mechanics etc. The system of applied mechanics is thus composed. Even though, the differential equations for these simplified theories are still difficult to solve analytically. Mathematicians contributed big efforts to develop mathematical methods for solving applied mechanics problems. The typical books are such as 《Methods of mathematical physics》 by R. Courant and D. Hilbert [1]; the series of textbooks by S.P. Timoshenco & others: 《Theory of elasticity》, 《Theory of elastic stability》, 《Plates and shells》, 《Advanced strength of material》, 《Vibration problems in engineering》 [2~6], and also a number of other related books. These analytical solution systems are quite fruitful classical methodologies, and greatly influenced the following developments.

Since 195* computer and programming language emerged, the finite element method (FEM) first appeared in computational structural mechanics, see [7~9], the situation abruptly changed. Based on the established theoretical system of applied mechanics, and also backed-up with the powerful computational technology, the FEM program systems developed. The problems of structural mechanics, solid mechanics, field problem etc., for which the problem can be described by the linear algebraic simultaneous equations, can be solved very quickly, and then the computer numerical method becomes the powerful tool on engineers' hand. The FEM propagated very quickly to various parts of applied mechanics, scientific and engineering computations, and have made great success.

The success of FEM does not reduce the importance of analytical approach. The reason is, firstly FEM is a kind of approximate method and its theoretical basis is still analytical, secondly there are a number of problems, such as the crack tip element, infinite element etc. their features are naturally analytic. Further, for boundary layer effect, localization, the composite material boundary effect etc., the analytical approach will still be interested. The precise integration method maps the influence of analytical approach in numerical method. The boundary element method also needs analytical solution.

A scan of the book 《Theory of elasticity》 by S.P. Timoshenco & J.N. Goodier can find that the majority of the text is devoted to problem solving, and the solution methodology applies heavily the semi-inverse approach. However, the semi-inverse method is really of try-and-error solution approach, being problem dependent but not a general approach. The semi-inverse method does can find some solutions but cannot make sure that all the solutions have been found. The readers always have the question, how to assume the appropriate form of solution so that the problem on hand can be solved, it is quite a puzzle.

The reason for applying the semi-inverse method is the complexity of the PDE-s. The traditional solution methodology for PDE set solving is first to eliminate the unknown functions as possible, prefer to incur a higher order PDE but with only one unknown function, and then try to solve this unknown function. It often derives to a complicated high order PDE, for which the usual effective solution method, such as the method of separation of variables and the eigen-function expansion method etc., cannot be applied.

A question is raised now, is it the sole way to use such traditional elimination procedure for solution? In fact, such traditional approach is not unique, the duality methodology, state space approach is right the answer.

In analytical dynamics, after the derivation of Euler-Lagrange equation, W.R. Hamilton proposed the canonical equation system [10~15], which is the beginning of the state space approach. Note that the basic theory of ordinary differential equation (ODE) is also settled on the system of first order ODEs. However, in the classical theory of automatic control, the typical formulation was also based on high order ODE with single-input—single-output (SISO) system description. The control theory changed its course under the impact of computing technique, that the modern control theory developed [16~19]. The modern control theory did not simply extend its theory along the original theoretic frame of classical control theory, but made dramatic changes that the fundamental theoretic system was also updated, *i.e.* the state space approach is used instead of the SISO description, completely a different consideration. Applied mechanics can share the useful experience.

The control theory updated its theoretical system description along its own evolution, at a first glance it must be much depart from the theory in applied mechanics. However, the situation is just the contrast, that the mathematical problem of modern control theory is analogous to a class of problems in structural mechanics, having a one-one correspondence relationship to each other [20~22]. This book is written based on the analogy relationship between structural mechanics and optimal control. Looking from mathematical theory, the analogy relationship is established based on the duality system theory. Since the control theory constructed a completely new systematic theory based on shifting from the traditional approach to the state space method, the applied mechanics can also develop its state space method and duality system theory toward success. Such development seems natural looking from the analogy theory between structural mechanics and optimal control.

Examining the traditional solution system, the typical textbook is 《Theory of Elasticity》 by S.P. Timoshenco & J.N. Goodier [2], all the solution methods are first derived to the fundamental equation with one kind of variables, then looking for solutions. From mathematical point of view, one kind of variables solution methodology is classified as Lagrange system theory. Hence deriving to the higher order partial differential equation (PDE) is the definite consequence, and then the powerful methods such as method of separation of variables cannot be applied, the eigen-function expansion method, the canonical transformation method etc. cannot be applied, and the semi-inverse solution method must be applied then. However, after changing the solution methodology systematically, introducing the state space, which is composed of the original variables and the dual variables, into theory of elasticity, the Saint-Venant problem for prismatic domain derives to a new set of fundamental equations, and the method of separation of variables applies smoothly. The solutions formerly found via the semi-inverse method can be solved by the method of separation of variables directly in the state space. And the solutions in a prismatic domain, which relate to the boundary conditions at the two ends, were quite difficult to solve and were covered by the Saint-Venant principle, can now be solved by the direct method, see [23]. The direct method, contrast versus the semi-inverse method, derives the solution rationally that the reader can follow the typical steps to solve problems and easy to understand.

Recent information technology development stimulates intelligence, smart material, smart structures, smart system, smart device, etc. which demonstrates the potential of control technology. The structural control is under ever increasing concern [24,25]. The teaching of engineering mechanics should not ignore such trend. The combined structure-control design is appealing. The present book's intention is to combine the applied mechanics and control theory with a **unified systematic approach**, so as to expose the intrinsic theoretic interrelationship of applied mechanics and optimal control, which will be beneficial to the new generation of engineers.

The transition from Lagrange to Hamilton system implies that the geometry correspondingly shifts from Euclidean to symplectic. It breaks the traditional consideration and brings the duality system theory into the vast areas of applied mechanics. The present book gives a number of different fields such as vibration, structural mechanics, wave propagation and also the LQG and H_∞ control theory, and their precise integration. All these subjects are described with the same solution system. After the reader studied one of them, then the other fields will be easier to understand as the mathematical methodology is the same. A unified approach will also be beneficial to teaching.

Scientific ***computation*** combined with ***theory*** and ***experiment*** have been the three major supports for modern sciences, which states that only describing from the theoretical aspect is not enough. Carrying out computation giving numerical results is necessary. Hence in this book, algorithm is also emphasized, especially the numerical solution of differential equations. The proposed "precise integration method" can be applied both for evolutionary time history integration and for two point boundary value problem (TPBVP) with its induced Riccati differential equation, the numerical results will be approaching the computer precision. In contrast to the traditional numerical integration algorithm, which always uses finite difference approximation, the precise integration method combines the techniques of 2^N ***algorithm*** and ***keeping track of the incremental part*** to reach high precision. The mathematics needed for precise integration is simple. Early understanding such featured new algorithm will be beneficial in future applications. Because of the importance of numerical computations, especially after the system description derived to state space and the fundamental equations become a system of ODEs, numerical integration of the ODEs becomes a fundamental step toward applications. To ease the application of duality system methodology, the precise integration method of ODE set is introduced first here.

§0.1, Introduction to Precise Integration method

The precise integration method [22,23] applies to first order ordinary differential equations (ODEs). In fact, the mathematical theory of ODE treats the normal form as a set of first order equations. Both the state space method and the Hamiltonian system theory derive the differential equations into first order ODEs to solve. The numerical integration for ODEs can be classified into two classes of boundary value problems:

1) Initial value problem—Dynamical systems or evolutionary type problems need to integrate the equations with given initial state, [26,27].
2) Two point boundary value problem (TPBVP)—Elasticity, structural mechanics, wave-guide, optimal control and filter problems need integration with given boundary conditions at the two ends [23,97].

In this introductory section, the precise integration of ODEs with initial value is introduced first. Let a set of ordinary differential equations be given in matrix/vector form as

$$\dot{\mathbf{v}} = \mathbf{A}\mathbf{v} + \mathbf{f}, \qquad \mathbf{v}(0) = \mathbf{v}_0 = \text{given vector} \tag{0.1}$$

where a dot above $(\dot{\ })$ means the differentiation with respect to time t, $\mathbf{v}(t)$ is the n dimensional vector function to be determined, $\mathbf{A}$ is a $n \times n$ given constant matrix, and $\mathbf{f}(t)$ is a given external force vector, n dimensional.

§0.1.1, Homogeneous equation, algorithm for exponential matrix

According to the solution theory for ODEs, the homogeneous equation should be solved first.

$$\dot{\mathbf{v}} = \mathbf{A}\mathbf{v} \tag{0.2}$$

Because $\mathbf{A}$ is a time-invariant matrix, its general solution can be given as

$$\mathbf{v} = \exp(\mathbf{A}t) \cdot \mathbf{v}_0 \tag{0.3}$$

The exponential matrix is defined as usual, see [1], vol.I,p.9

$$\exp(\mathbf{A}t) = \mathbf{I}_n + \mathbf{A}t + (\mathbf{A}t)^2/2 + (\mathbf{A}t)^3/3! + \cdots \tag{0.4}$$

Now the problem is its numerical computation, as precise as possible. A time step, denoted as η, is necessary for numerical integration, and a series of equally duration instants are

$$t_0 = 0, \quad t_1 = \eta, \quad \cdots, \quad t_k = k\eta, \; \cdots \tag{0.5}$$

for which

$$\mathbf{v}_1 = \mathbf{v}(\eta) = \mathbf{T}\mathbf{v}_0, \qquad \mathbf{T} = \exp(\mathbf{A}\eta) \tag{0.6}$$

After the matrix $\mathbf{T}$ is computed, time step integration becomes the following recurrence

$$\mathbf{v}_1 = \mathbf{T}\mathbf{v}_0,\ \mathbf{v}_2 = \mathbf{T}\mathbf{v}_1,\ \cdots,\ \mathbf{v}_{k+1} = \mathbf{T}\mathbf{v}_k, \cdots \tag{0.7}$$

i.e. a series of matrix-vector multiplications. Therefore, the problem is reduced to the computation of exponential matrix $\mathbf{T}$ in equation (0.6). The precise computation of exponential matrix has two cruxes, namely

1) ***The additional theorem of exponential function is used, i.e. the*** 2^N ***algorithm*** [28];
2) ***Keeping track of the incremental part of the exponential matrix, rather than the total value.***

The additional theorem of exponential function gives

$$\exp(\mathbf{A}\eta) \equiv [\exp(\mathbf{A}\eta/m)]^m \tag{0.8}$$

where m is an arbitrary integer. It is suggested to select

$$m = 2^N, \quad \text{such as} \quad N = 20, \quad m = 1048576 \tag{0.9}$$

Because η should be a small time interval, so $\tau = \eta/m$ is an extremely small time interval. Hence for the τ interval, the truncated Taylor expansion can be applied

$$\exp(\mathbf{A}\tau) \approx \mathbf{I}_n + (\mathbf{A}\tau) + (\mathbf{A}\tau)^2/2 + (\mathbf{A}\tau)^3/3! + (\mathbf{A}\tau)^4/4! \tag{0.10}$$

Because τ is extremely small, the first five term series expansion is good enough. The exponential matrix $\mathbf{T}$ departs the unit matrix $\mathbf{I}_n$ also extremely small, hence it must be disintegrated as

$$\exp(\mathbf{A}\tau) \approx \mathbf{I}_n + \mathbf{T}_a, \quad \mathbf{T}_a = (\mathbf{A}\tau) + (\mathbf{A}\tau)^2[\mathbf{I}_n + (\mathbf{A}\tau)/3 + (\mathbf{A}\tau)^2/12]/2 \tag{0.11}$$

where the ***matrix*** $\mathbf{T}_a$ ***is very small.***

In computations, one of the most important cruxes is that only the additional matrix $\mathbf{T}_a$ of (0.11) is kept in the memory rather than the matrix $\mathbf{T} = (\mathbf{I}_n + \mathbf{T}_a)$. Because $\mathbf{T}_a$ is extremely small, if it is added to the unit matrix $\mathbf{I}_n$, $\mathbf{T}_a$ will become an appended part and its precision will be greatly dropped in the round-off operations in computer arithmetic. In fact, $\mathbf{T}_a$ is an incremental, the second crux mentioned above.

For computing the matrix $\mathbf{T}$, the equation (0.8) should be factored as

$$\mathbf{T} = (\mathbf{I} + \mathbf{T}_a)^{2^N} = (\mathbf{I} + \mathbf{T}_a)^{2^{(N-1)}} \times (\mathbf{I} + \mathbf{T}_a)^{2^{(N-1)}} \tag{0.12}$$

Such factorization is carried out continuously for N times. Next, for arbitrary matrices $\mathbf{T}_b, \mathbf{T}_c$, the following identity holds

$$(\mathbf{I} + \mathbf{T}_b) \times (\mathbf{I} + \mathbf{T}_c) \equiv \mathbf{I} + \mathbf{T}_b + \mathbf{T}_c + \mathbf{T}_b \times \mathbf{T}_c \tag{0.13}$$

and when $\mathbf{T}_b, \mathbf{T}_c$ are extremely small, the multiplication must not be carried out after the addition of unit matrix $\mathbf{I}$. Treating the matrices $\mathbf{T}_b, \mathbf{T}_c$ as $\mathbf{T}_a$, the N times multiplication of equation (0.12) correspond to the following instruction

$$\text{for}\,(iter = 0; iter < N; iter++)\quad \mathbf{T}_a = 2\mathbf{T}_a + \mathbf{T}_a \times \mathbf{T}_a \tag{0.14}$$

After the execution of this instruction, the addition

$$\mathbf{T} = \mathbf{I} + \mathbf{T}_a \tag{0.15}$$

is finally executed. After N times multiplication, $\mathbf{T}_a$ is no longer an extremely small matrix, and this addition will have no serious numerical round-off error any more. The algorithm given above is called as the ***precise computation of exponential matrix***.

Exponential matrix is widely used and is one of the most frequently computed matrix functions. Quite a number of algorithms had been proposed before, however, still not be so satisfied. Reference [29] reviewed nineteen **dubious** algorithms among them, but in their later book [30] pointed out again that the problem needs further investigation. It should be mentioned that the eigenvector expansion method is effective in case of no Jordan form nearly to appear. However, the precise computation method proposed above **always** works perfectly even if the Jordan form really appears for the matrix $\mathbf{A}$, that it is **never a dubious algorithm**.

§0.1.2, Solution of inhomogeneous equation

Let us go back to the equation (0.1), the external force $\mathbf{f}(t)$ should be considered now. According to the theory of linear differential equation, after found the ***impulse response matrix*** $\mathbf{\Phi}(t,t_1)$, where $t_1 < t$ is an arbitrary time instant, the external force $\mathbf{f}(t)$ induced response can be computed by the Duhamel's integration

$$\mathbf{v}(t) = \mathbf{\Phi}(t,t_0)\mathbf{v}_0 + \int_{t_0}^{t} \mathbf{\Phi}(t,t_1)\mathbf{f}(t_1)dt_1 \tag{0.16}$$

where the matrix $\mathbf{\Phi}(t,t_1)$ has the following characteristics:

1) $\mathbf{\Phi}(t,t) = \mathbf{I}$ (0.17)

2) $\mathbf{\Phi}(t,t_1) = \mathbf{\Phi}(t,t_2)\mathbf{\Phi}(t_2,t_1)$ (0.18)

3) Satisfies the differential equation $\dot{\mathbf{\Phi}}(t,t_1) = \mathbf{A}(t)\mathbf{\Phi}(t,t_1)$ (0.19)

In above equation, the time-variant matrix $\mathbf{A}(t)$ means that the equation (0.16) applies also to time-variant system. For the special case of time-invariant system,

$$\mathbf{\Phi}(t,t_1) = \mathbf{\Phi}(t-t_1) = \exp[\mathbf{A}\cdot(t-t_1)] \tag{0.20}$$

is an exponential matrix, and obviously $\mathbf{\Phi}(\eta) = \mathbf{T}$.

In numerical computations, the numerical results at equally distant instants are needed only. The integration need not be carried out from the initial instant t_0, that it can be from t_k to t_{k+1}. Thus the equation (0.16) can be updated as

$$\begin{aligned}\mathbf{v}_{k+1} &= \mathbf{T}\mathbf{v}_k + \int_{t_k}^{t_{k+1}} \mathbf{\Phi}(t_{k-1}-t)\mathbf{f}(t)dt \\ &= \mathbf{T}\mathbf{v}_k + \int_0^{\eta} \exp[\mathbf{A}\cdot(\eta-\xi)]\mathbf{f}(t_k+\xi)d\xi\end{aligned} \tag{0.21}$$

The analytical expression of the external force $\mathbf{f}(t_k+\xi)$ is required, but it is not definitely available. If linear interpolation approximation is used in $t_k \sim t_{k+1}$,

$$\mathbf{f}(t_k + \xi) \approx \mathbf{r}_0 + \mathbf{r}_1 \cdot \xi \tag{0.22}$$

the equation (0.21) can be integrated as

$$\mathbf{v}_{k+1} = \mathbf{T}[\mathbf{v}_k + \mathbf{A}^{-1}(\mathbf{r}_0 + \mathbf{A}^{-1}\mathbf{r}_1)] - \mathbf{A}^{-1}[\mathbf{r}_0 + \mathbf{A}^{-1}\mathbf{r}_1 + \eta\mathbf{r}_1] \tag{0.23}$$

However, linear interpolation is a rough approximation, but there are a number of different approximate expressions. If $\mathbf{f}(t_k + \xi)$ is approximated by the following functions,

1) Polynomials;
2) Exponential functions;
3) Trigonometric functions;
4) The product of the above functions, etc.

the integration in equation (0.21) can be carried out analytically, see ref.[31].

To check the effect of precise integration method, a numerical example is given as follows:

Example 0.1, The numerical integration of the ODEs is needed up to $t = 20$, see [43]

$$\dot{u}_1 = -2000u_1 + 999.75u_2 + 1000.25,$$

$$\dot{u}_2 = u_1 - u_2, \qquad u_1(0) = 0,\ u_2(0) = -2$$

Solution: The eigenvalues of the matrix $\mathbf{A}$ are $\lambda_1 = -2000.5,\ \lambda_2 = -0.5$. The eigenvalues depart so large to each other, which means that the equation set is ***stiff***, and the stiff ratio is 4000. The analytical solution is

$$u_1(t) = -1.499875e^{-0.5t} + 0.49975e^{-2000.5t} + 1$$

$$u_2(t) = -2.99975e^{-0.5t} - 0.00025e^{-2000.5t} + 1$$

If the 4-th order Runge-Kutta algorithm is used in computation, the numerical stability requirement confines the time step-size being smaller than 0.00138, it requires 14493 steps to reach $t = 20$. A big amount of computational expense, and also there is numerical errors accumulation. However, using the precise integration method, no matter how many time-steps subdivided in the time interval, it always gives $u_1(20) = 0.9999329$, $u_2(20) = 0.9998638$, the precise numerical result, as checked by the analytical solution. ##

§0.1.3, Precision analysis

The main step in precise integration algorithm is the computation of exponential matrix $\mathbf{T} = \exp(\mathbf{A}\eta)$. Except the round-off errors in the usual computer arithmetic in matrix multiplication, error can only be induced from the power series expansion truncation in equation (0.10). In the 2^N algorithm, the main part of matrix $\mathbf{T}_a$ is the first term $\mathbf{A}\cdot\tau$, hence the truncation error should be compared with it. The first term truncated in equation (0.10) is $(\mathbf{A}\tau)^5/5!$, so the relative error is estimated as

$$(\mathbf{A}\tau)^4/120 \tag{0.24}$$

Now suppose all the eigen-solutions are solved for matrix $\mathbf{A}$,

$$\mathbf{AY} = \mathbf{Y}\mathrm{diag}[\mu_i], \quad or \quad \mathbf{A} = \mathbf{Y}\mathrm{diag}[\mu_i]\mathbf{Y}^{-1} \tag{0.25}$$

where $\mathbf{Y}$ is a $n \times n$ matrix composed of all the eigenvectors of $\mathbf{A}$ as its columns, and μ_i represents all the eigenvalues, $\mathrm{diag}[\cdots]$ means diagonal matrix. Then

$$\exp(\mathbf{A}\tau) = \mathbf{Y}\exp(\mathrm{diag}[\mu_i]\tau)\mathbf{Y}^{-1} = \mathbf{Y}\mathrm{diag}[\exp(\mu_i\tau)]\mathbf{Y}^{-1}$$

is derived. Thus, the truncation in equation (0.11) corresponds to the truncation in

$$\exp(\mu\tau) \approx 1 + \mu\tau + (\mu\tau)^2/2 + (\mu\tau)^3/3! + (\mu\tau)^4/4!$$

The above analysis disintegrates the errors come from each eigenvalues. The relative errors in equation (0.10) for each eigen-solution being of the order of $(\mu\tau)^4/120$, the absolute value is $(\mathrm{abs}(\mu)\tau)^4/120$. Note the double precision of real type number in present day computer has 16 decimal digits. Therefore within the computer double precision, it requires

$$[\mathrm{abs}(\mu)\cdot\eta/2^N]^4/120 < 10^{-16}$$

Let $N = 20, 2^N \approx 10^6$, it derives

$$\mathrm{abs}(\mu)\cdot\eta < 300 \tag{0.26}$$

For natural vibration with no damping, $\mu = i\omega$, where ω is the circular frequency, which means, even the integration step size η being as large as 50 cycles, the numerical result still feels no truncation error in the expansion. Certainly, ω should be the highest frequency, however, for practical problems, vibration always accompanies with damping. After a number of cycles, the influence of high frequency component decays to be negligible. It means that the estimation (0.26) for high frequency tends to be quite conservative.

Based on the above analysis, the high precision of the method is understandable. The numerical result of precise integration method reaches the computer precision.

Discussion: The success of precise computation of the exponential matrix $\mathbf{T}$ is based on further subdividing the time step η into 1048576 fine steps. But solely subdividing η does not bring precise result home. The second crux of precise computation is always ***keeping track of the incremental part***, so as to avoid the numerical ill-conditioned arithmetic operations. Such as using the Runge-Kutta integration method, subdivides one time step η into 1048576 fine steps, and integrates the dynamic equation with the initial vector $\mathbf{x}_0$, the numerical result still cannot reach the precision of the precise integration. Because the Runge-Kutta stepwise integration algorithm uses the full value of vector in computation, but not uses its increments.

Comparing with precise integration, the time step integration algorithm proposed before are all finite difference approximation, so the numerical results never approach the computer precision. Some kind of numerical problem will appear in practical computations, such as numerical stability problem or stiff problem etc. These problems were brought with the finite difference approximation. The finite difference method (FDM) is executed with the full vector, so that if the step size selected too small then it may bring another kind of numerical difficulty as mentioned before. The precise integration method truncates as given in equation (0.10), but the truncation error has been beyond the real number double precision, which implies that with reasonable integration step size selection of η, no stability or stiff problem may appear. Certainly, this assertion applies only to time-invariant ODEs and exponential matrix computation.

When considering the numerical integration based on precise integration method for time-variant ODEs or non-linear dynamical system equations, some other approximation must be introduced, for which some problems may appear, and further investigations are needed.

§0.1.4, Discussions on time-variant system or non-linear system

The differential equations derived from application problems are usually non-linear or time-variant. Numerical integration cannot avoid these equations. The precise integration method proposed above is only for time-invariant ODEs, however, it also brings a basis for these difficult equations [32~34]. The ODE can be rewritten as

$$\dot{\mathbf{v}} = (\mathbf{A}_0 + \mathbf{A}_1)\mathbf{v} + \mathbf{f} \tag{0.27}$$

where $\mathbf{A}_0$ is a time-invariant matrix, and $\mathbf{A}_1$ is time-variant or is related to the unknown vector $\mathbf{v}$, *i.e.* non-linear equation.

The analytical solution of time-variant equation or non-linear equation is usually very difficult. Rewriting the equation (0.27) as

$$\dot{\mathbf{v}} = \mathbf{A}_0\mathbf{v} + (\mathbf{f} + \mathbf{A}_1\mathbf{v})$$

the terms within the parenthesis above can be treated as some 'external force', then the equation will be of the form of equation (0.1) and the precise integration method can be applied again. The impulse response matrix for time-invariant matrix $\mathbf{A}_0$ can be computed first, $\mathbf{T} = \exp(\mathbf{A}_0\eta)$. Afterwards, using equation (0.21) gives

$$\mathbf{v}_{k+1} = \mathbf{T}\mathbf{v}_k + \int_0^{\eta}\exp[\mathbf{A}_0 \cdot (\eta - \xi)]\mathbf{f}_c(t_k + \xi)\mathrm{d}\xi \ , \quad \mathbf{f}_c = \mathbf{f} + \mathbf{A}_1\mathbf{v} \tag{0.28}$$

where the expression of the force vector $\mathbf{f}_c$ involves the unknown vector $\mathbf{v}$. Hence, the above equation (0.28) becomes an integral equation of Volterra type. Generally speaking, to solve the integral equation requires numerical method. One factor should be mentioned at this point that in numerical computation, the approximation of integration is usually better than the finite difference.

Making approximation for the integral equation has a number of approaches. In general, making use of analytical functions such as polynomial, exponential, trigonometric functions etc. to approximate $\mathbf{f}_c$, such that the integration can be analytically carried out. Thereafter one may select similar techniques used in finite difference, such as single step or multiple steps, explicit or implicit, predictor-corrector methods etc., to use for the integral equation (0.28), see references [33,34].

Finally, the precise integration method can not only be used for the integration of initial value problems but also can be used for the integration of two point boundary value problems (TPBVP), which will be given in relevant chapters later.

Chapter 1, Introduction to analytical dynamics

Sir Isaac Newton proposed the fundamental equations of dynamics, which is considered the mile stone of sciences. With the advancement of applications, the dynamical systems under constraints were necessary to be considered, such as in mechanical engineering. In 1788, Lagrange published the book "Analytical Mechanics", the mathematical modeling for a system of bodies under constraints was developed analytically, and the generalized displacement method was proposed. Hamilton in 1834 introduced the dual variables from the Lagrange system, and proposed the canonical equation system, called the Hamilton system. With the continuous contributions from a number of brilliant great mathematicians, the whole classical system of analytical dynamics was well established. It composes a cornerstone for mechanics and physics. However, only some basic topics are described here based on the requirements of applications.

The method of analytical dynamics mainly treats discrete systems, such as for the system composed of finite number of particles or rigid bodies. The configuration of the system can be described with finite number of independent parameters, so that it is called ***discrete system***. For flexible body composed system, such as elastic body or else fluid, its configuration needs infinite degrees of freedom to describe, so that the flexible body composed system is called as continuous system, continuum mechanics. When using finite element method (FEM) or other discretization approximation, it reduces to discrete system. For majority mechanical systems in nature or in engineering, the discrete model often describes the system dynamical behavior quite satisfactorily, such as the FEM. For ease of mathematical theory and solution, the discrete system is first selected in the analytical dynamics rather than the system composed of continuum, hence the dynamics for discrete system is widely accepted.

Analytical dynamics and the respective variational principles are very important foundation for applied mechanics. The basic part is concisely introduced in this chapter. Interested readers can find more contents from such as the references [10~14].

§1.1, Holonomic and nonholonomic constraints

Treating the mechanical system as a group of N particles, then the system configuration can be described with $3N$ values of the coordinates. If the system is free from constraint, then the $3N$ coordinate values are all independent. Now let the system configuration be restrained with l constraints, if these constraints can be expressed with only the functions of coordinates $u_1, u_2, \cdots, u_{3N}$, (but not their time derivative)

$$f_r(u_1, u_2, \cdots, u_{3N}) = 0, \qquad r = 1,2,\cdots l \tag{1.1.1a}$$

or

$$f_r(u_1, u_2, \cdots, u_{3N}, t) = 0, \qquad r = 1,2,\cdots l \tag{1.1.1b}$$

then they are called as holonomic constraints. Note in equation (1.1.1a) there is no time t, and is called time-invariant (scleronomous) constraints, however equation (1.1.1b) is time dependent and is called time-variant (rheonomous) constraints. Only equality constraints are considered here.

Example 1.1, A crankshaft mechanism is shown in figure 1.1, the displacement parameters are $u_1 = \theta,\ u_2 = \varphi,\ u_3 = x$, obviously, u_1, u_2, u_3 are not independent, the constraint equations are

$$x = r\cos\theta + l\cos\varphi\ ,$$

and $\quad r\sin\theta = l\sin\varphi$.

Because the two constraint equations do not explicitly involve the time t, hence the constraints are time-invariant and holonomic. ##

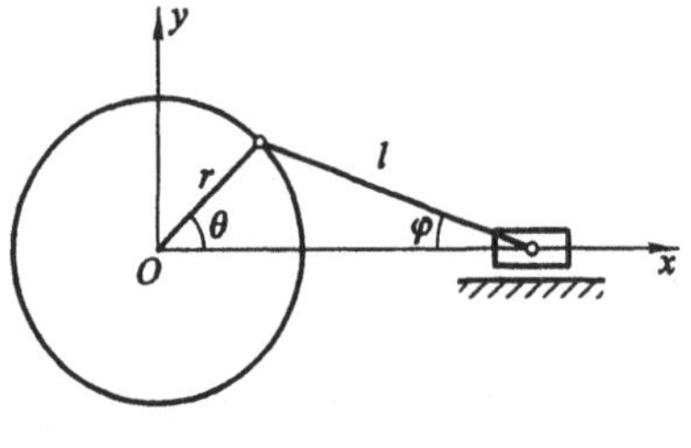

Figure 1.1, *A crankshaft*

Example 1.2, A rotating disc has a groove, in which a key slides, figure 1.2. Select the relative sliding distance x' as variable. The inertia coordinate of this key particle is the fixed frame Oxy, and the displacements of the particle are

$$u_1 = x = a\cos\omega t + x'\sin\omega t\ ,$$

$$u_2 = y = a\sin\omega t - x'\cos\omega t$$

the constraint equation is

$$(u_1 - a\cos\omega t)\cos\omega t$$

$$+ (u_2 - a\sin\omega t)\sin\omega t = 0$$

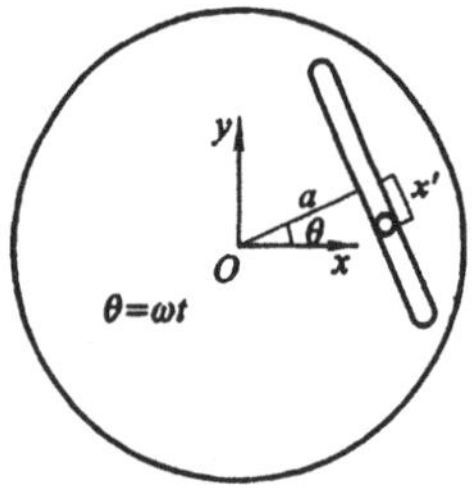

Figure 1.2, *A key slides in a rotating disk groove*

The constraint equation for the inertia coordinates u_1, u_2 is time variant, so that it is a rheonomous holonomic constraint. The x' plays a freedom in the relative coordinate system, that the problem has one degree of freedom. ##

If the first partial differential of the functions f_r with respect to all the variables u_i in the equations (1.1.1a) and (1.1.1b) exist and are continuous, then the complete time derivative gives

$$\sum_{s=1}^{3N} \frac{\partial f_r}{\partial u_s}\dot{u}_s = 0, \qquad r = 1,2,\cdots,l \tag{1.1.2a}$$

or

$$\sum_{s=1}^{3N} \frac{\partial f_r}{\partial u_s}\dot{u}_s + \frac{\partial f_r}{\partial t} = 0, \qquad r = 1,2,\cdots,l \tag{1.1.2b}$$

Obviously, the above two equations can be integrated into the form of equations (1.1.1a) and (1.1.1b), the difference is only a constant. This is the characteristic of holonomic constraint. Equation (1.1.2) can be written in differential form as

$$\sum_{s=1}^{3N} \frac{\partial f_r}{\partial u_s}\mathrm{d}u_s = 0, \qquad \text{or} \quad \mathrm{d}f_r = 0, \qquad r = 1,2,\cdots,l \tag{1.1.3a}$$

or $$\sum_{s=1}^{3N}\frac{\partial f_r}{\partial u_s}\mathrm{d}u_s+\frac{\partial f_r}{\partial t}dt=0,\quad \text{or}\quad \mathrm{d}f_r=0,\quad r=1,2,\cdots,l \tag{1.1.3b}$$

Integrability means that they are a complete differential. When all the constraints of a system are holonomic, it is called a holonomic system.

In the constraint equation (1.1.1) there are only coordinate values. In applications, however, there are constraints involving velocities of system variables. A classical example of such non-holonomic constraint system is a vertical disc of radius r rolling on a horizontal plane x,y, see figure 1.3.

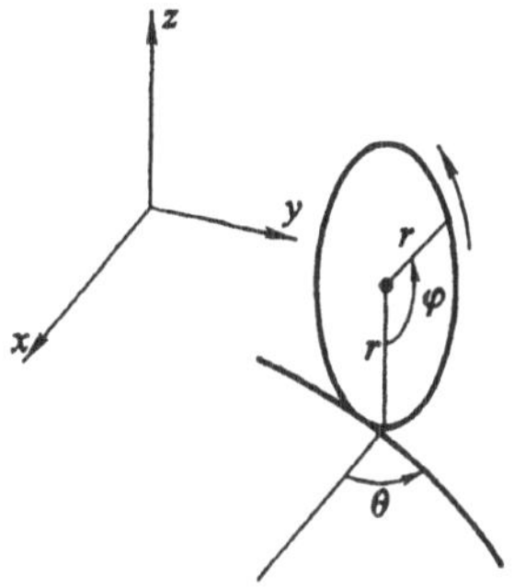

Figure 1.3, *A rolling vertical disc*

The center of disc is denoted as x, y and z that $z=r$ and $v=r\dot{\varphi}$ where φ is the angle of rotation. Let the rolling direction be denoted by the angle θ and the configuration is described by (x,y,φ,θ), then the motion is $\dot{x}=v\cos\theta$, $\dot{y}=v\sin\theta$, from which derives the constraint with time derivatives

$$\dot{y}\cos\theta-\dot{x}\sin\theta=0$$

This constraint equation cannot be integrated as a complete differential hence non-holonomic. A non-holonomic constraint equation involves velocity (time derivative), but it cannot be integrated to be a complete differential of a displacement function. On the other hand, any given configuration is accessible.

Let the system have m non-holonomic constraints denoted as

$$g_r(u_1,u_2,\cdots,u_{3N};\dot{u}_1,\dot{u}_2,\cdots,\dot{u}_{3N};t)=0,\quad r=1,2,\cdots,m \tag{1.1.4}$$

usually the constraint equations are linear with respect to the velocity and called the linear non-holonomic constraints, expressed as

$$\sum_{s=1}^{3N}A_{rs}\,\dot{u}_s+A_r=0,\quad r=1,2,\cdots,m \tag{1.1.5}$$

where A_{rs} and A_r are functions of u_s and t. These equations are often written in Pfaff form

$$\sum_{s=1}^{3N}A_{rs}\,\mathrm{d}\dot{u}_s+A_r\mathrm{d}t=0,\quad r=1,2,\cdots,m \tag{1.1.6}$$

If $A_r\equiv 0$ the above constraint is linear homogeneous with respect to velocity, otherwise it is non-homogeneous.

A system having one non-holonomic constraint is a non-holonomic system.

Comparing the equation (1.1.2b) with equation (1.1.6) gives

$$A_{rs}=\partial f_r/\partial u_s,\quad A_r=\partial f_r/\partial t,\quad r=1,2,\cdots,l\quad s=1,2,\cdots,3N \tag{1.1.7}$$

which means that the constraint equation (1.1.5) can still be holonomic. According to equation (1.1.7)

$$\frac{\partial A_{rs}}{\partial u_j}=\frac{\partial A_{rj}}{\partial u_s}\left(=\frac{\partial^2 f_r}{\partial u_s\partial u_j}\right),\quad s,j=1,2,\cdots,3N \tag{1.1.8}$$

which gives the necessary condition of integrability as equation (1.1.8). According to theory of differential equations [1,36], it is also the sufficient condition.

§1.2, Generalized displacement, degrees of freedom, virtual displacement

A system composed of N particles uses $3N$ inertial Cartesian coordinates denoted by $u_1, u_2, \cdots, u_{3N}$ to describe its space configuration. If the system is subjected to l holonomic constraints as in equation (1.1.1), then these coordinate values are not independent. A set of ***independent*** parameters $q_1, q_2, \cdots, q_n$, which gives an unambiguous representation of the configuration of the system, will serve as a system coordinates and is known as the ***generalized coordinates*** (or ***generalized displacements***). It can also be written in vector form $\mathbf{q}$. Obviously

$$n = 3N - l$$

Generalized coordinates usually have a readily visualized geometrical significance, and are often chosen on this basis. If the generalized coordinates can vary independently without violating the constraints, then the number of generalized coordinates is equal to the number of ***degrees of freedom***.

If except the l holonomic constraints, the system has another m non-holonomic constraints (1.1.5). Because the non-holonomic constraints cannot be integrated that they can only constrain velocity but cannot influence the number of independent parameters specifying the configuration of the system. Hence the number n of generalized coordinates is unchanged. After selected the generalized coordinates $q_i, i = 1, \cdots, n$, the position vectors $\mathbf{r}_i$ of various particles can be expressed as the functions (transformation) of the generalized displacements

$$\mathbf{r}_i = \mathbf{r}_i(q_1, q_2 \cdots q_n; t), \quad i = 1, 2, \cdots, N \tag{1.2.1}$$

Using generalized displacement vector $\mathbf{q}$ to describe the configuration of system, the holonomic constraints have been satisfied automatically. The another m non-holonomic constraints can be expressed with the generalized displacement and velocity vectors $\mathbf{q}$ and $\dot{\mathbf{q}}$ as

$$\sum_{i=1}^{n} a_{ri}\, \dot{q}_i + a_r = 0, \qquad r = 1, 2, \cdots, m \tag{1.2.2}$$

where a_{ri} and a_r are the functions of generalized displacements q_i and time t. When $a_r = 0$, then the constraint equations are homogeneous. Quite often the equations are written in differential form

$$\sum_{i=1}^{n} a_{ri}\, \mathrm{d}q_i + a_r \mathrm{d}t = 0, \quad r = 1, 2, \cdots, m \tag{1.2.3}$$

The idea of ***virtual displacement*** is introduced now. Suppose the configuration of a system with N particles is given by the $3N$ Cartesian coordinates $u_1, u_2, \cdots, u_{3N}$, which are measured relative to an inertia frame, but subjected to constraints. At any given time, let us assume that the coordinates move through infinitesimal displacements $\delta u_1, \delta u_2, \cdots, \delta u_{3N}$, which are virtual in the sense that they are assumed to ***occur without the passage of time***. Such small change of $\delta u_1, \delta u_2, \cdots, \delta u_{3N}$ in the configuration of the system is known as a ***virtual displacement***.

Note that a virtual displacement conforms to the ***instantaneous constraints***, that is, any moving constraints are looked stopped during the virtual displacement. Suppose the system is subjected to l holonomic constraints of equation (1.1.1b) then the virtual displacement should satisfy

$$\sum_{i=1}^{3N} \frac{\partial f_r}{\partial u_i} \delta u_i = 0, \qquad r = 1,2,\cdots l$$

note that the term δt does not present in the above equation even for holonomic constraint.

For m non-holonomic constraints the virtual displacement should satisfy the equation

$$\sum_{i=1}^{n} a_{ri}\, \delta q_i = 0, \quad r = 1,2,\cdots,m \tag{1.2.4}$$

comparing with equation (1.2.3), for which the variation $\delta t = 0$ exists. It means the virtual displacements regard the time variation equals to zero. The holonomic constraints can be considered in combination with non-holonomic constraints, *i.e.* combining (1.1.3) and (1.1.6), that the virtual displacements $\delta u_1, \delta u_2, \cdots, \delta u_{3N}$ satisfy the set of equations

$$\sum_{s=1}^{3N} A_{rs}\, \delta u_s = 0, \qquad r = 1,2,\cdots,(l+m) \tag{1.2.5}$$

For a holonomic system the particle configuration is described by the generalized displacements q_i via the equation (1.2.1). The virtual displacements in Cartesian coordinate satisfy

$$\delta \mathbf{r}_i = \sum_{j=1}^{n} \frac{\partial \mathbf{r}_i}{\partial q_j} \delta q_j, \qquad i = 1,2,\cdots,N \tag{1.2.6}$$

The various δq_i are independent, hence there are n linearly independent virtual displacements satisfying the constraints.

However, if there are another m non-holonomic constraints

$$\sum_{s=1}^{3N} A_{rs}\, \mathrm{d}u_s + A_r \mathrm{d}t = 0, \qquad r = 1,2,\cdots,m \tag{1.2.7}$$

Although the generalized displacement expression (1.2.1) is used for $\mathbf{r}_i$, but only the holonomic constraints are considered. Because there are non-holonomic constraints, so $\delta q_1, \cdots, \delta q_n$ are no longer independent. Note that $u_1, u_2, \cdots, u_{3N}$ are the components of position vectors $\mathbf{r}_i$. Substituting expression (1.2.1) into equation (1.2.7) gives

$$\begin{gathered} \sum_{j=1}^{n} a_{rj}\, \dot{q}_j + a_r = 0, \qquad r = 1,2,\cdots,m \\ a_{rj}(\mathbf{q},t) = \sum_{s=1}^{3N} A_{rs} \frac{\partial u_s}{\partial q_j}, \quad a_r(\mathbf{q},t) = \sum_{s=1}^{3N} A_{rs} \frac{\partial u_s}{\partial t} + A_r \end{gathered} \tag{1.2.8}$$

hence the virtual generalized displacements satisfy

$$\sum_{j=1}^{n} a_{rj}\,\delta q_j = 0, \qquad r = 1,2,\cdots,m \tag{1.2.9}$$

Therefore the number of variations of independent generalized displacements is $(n-m)$, *i.e.* the number n of generalized displacements minus number m of non-holonomic constraints.

The ***degrees of freedom*** are defined as the ***number of independent variations of displacements***. Hence, for a holonomic system, degrees of freedom equal the number of generalized displacements; however, for a non-holonomic system the number of degrees of freedom equals the difference between the number of generalized displacements and the number of non-holonomic constraints.

§1.3, Principle of virtual displacement and the D'Alembert principle

An equilibrium system of forces does no work on arbitrary virtual displacements; on the contrary, if a system of forces does no work on all the virtual displacements, then it is an equilibrium system of forces. This is the ***principle of virtual displacement***.

The D'Alembert principle is also called the dynamic equilibrium method. Denoting the masses of all the particles as m_i, the inertia forces of the particles are $-m_i\ddot{\mathbf{r}}_i, i = 1,2,\cdots,N$. Imagining these inertia forces in combination with the active forces $\mathbf{F}_i$ and the constraint forces $\mathbf{R}_i$ acting on these particles, compose a system of equilibrium forces, where $\mathbf{r}_i$ is the vector of the i-th particle in an inertia frame. Therefore the principle of virtual displacement is used to the dynamic equilibrium force system to give

$$\sum_{i=1}^{N} (\mathbf{F}_i + \mathbf{R}_i - m_i\ddot{\mathbf{r}}_i)^T \cdot \delta\mathbf{r}_i = 0 \tag{1.3.1}$$

where the superscript ^{T} denotes transpose. The expression above transposing the column vector of combined forces becoming a row vector then multiplying the virtual displacement column vector gives a scalar, which equals zero. All the constraint forces $\mathbf{R}_i$ do no work on any virtual displacements, hence

$$\sum_{i=1}^{N} \mathbf{R}_i^T \cdot \delta\mathbf{r}_i = 0 \tag{1.3.2}$$

Such constraints are the ***ideal***, for example, the constraint of a rigid connection, or smooth surface constraint and/or perfectly rolling with no friction etc. Hence

$$\sum_{i=1}^{N} (\mathbf{F}_i - m_i\ddot{\mathbf{r}}_i)^T \cdot \delta\mathbf{r}_i = 0 \tag{1.3.3}$$

This equation is called as D'Alembert-Lagrange principle. This principle requires only that the constraints are ideal, hence can be applied to holonomic as well as to non-holonomic systems.

Equation (1.3.3) is given with the displacements in Cartesian coordinate. However, using the generalized displacement description for a constraint system has a number of conveniences. The complete time t derivative of the generalized displacement expression (1.2.1) gives

$$\dot{\mathbf{r}}_i = \sum_{j=1}^{n}(\frac{\partial \mathbf{r}_i}{\partial q_j})\dot{q}_j + \frac{\partial \mathbf{r}_i}{\partial t}, \qquad i = 1,2,\cdots,N \tag{1.3.4}$$

where $\mathbf{r}_i$ are originally functions of the generalized displacements q_j. However, its time derivative $\dot{\mathbf{r}}_i$ can be considered as the functions of q_j, $\dot{q}_j$ and time t. Temporarily, consider holonomic constraint system first, that $\dot{q}_j$ and q_j can be regarded as ***independent variables***. Taking partial derivative of equation (1.3.4) with respect to $\dot{q}_j$ gives

$$\frac{\partial \dot{\mathbf{r}}_i}{\partial \dot{q}_j} = \frac{\partial \mathbf{r}_i}{\partial q_j} \qquad i = 1,2,\cdots,N; j = 1,2,\cdots,n \tag{1.3.5}$$

Partial differentiating the expression (1.3.4) with respect to q_j gives

$$\frac{\partial \dot{\mathbf{r}}_i}{\partial q_j} = \sum_{s=1}^{n}(\frac{\partial^2 \mathbf{r}_i}{\partial q_s \partial q_j})\dot{q}_s + \frac{\partial^2 \mathbf{r}_i}{\partial t \partial q_j}, \qquad i = 1,2,\cdots,N; j = 1,2.\cdots,n$$

On the other hand, taking complete time derivative of $\partial \mathbf{r}_i / \partial q_j$ gives

$$\frac{\mathrm{d}}{\mathrm{d}t}(\frac{\partial \mathbf{r}_i}{\partial q_j}) = \sum_{s=1}^{n}(\frac{\partial^2 \mathbf{r}_i}{\partial q_j \partial q_s})\dot{q}_s + \frac{\partial^2 \mathbf{r}_i}{\partial q_j \partial t}$$

Comparing the two expressions gives

$$\frac{\mathrm{d}}{\mathrm{d}t}(\frac{\partial \mathbf{r}_i}{\partial q_j}) = \frac{\partial \dot{\mathbf{r}}_i}{\partial q_j}, \qquad i = 1,2,\cdots,N; \quad j = 1,2,\cdots,n \tag{1.3.6}$$

Let the virtual displacement in equation (1.3.3) satisfies the holonomic constraints first

$$\delta \mathbf{r}_i = \sum_{j=1}^{n}\frac{\partial \mathbf{r}_i}{\partial q_j}\delta q_j$$

Therefore equation (1.3.3) derives

$$\sum_{j=1}^{n}\left[\sum_{i=1}^{N}\mathbf{F}_i^T \cdot \frac{\partial \mathbf{r}_i}{\partial q_j} - \sum_{i=1}^{N}(m_i \ddot{\mathbf{r}}_i^T \cdot \frac{\partial \mathbf{r}_i}{\partial q_j})\right]\delta q_j = 0 \tag{1.3.7}$$

Defining the ***generalized forces*** as $\mathbf{Q}_j$

$$\mathbf{Q}_j = \sum_{i=1}^{N}\mathbf{F}_i^T \cdot \frac{\partial \mathbf{r}_i}{\partial q_j} \tag{1.3.8}$$

and the second summation in (1.3.7) can be rewritten as

$$\begin{aligned}\sum_{i=1}^{N}(m_i \ddot{\mathbf{r}}_i^T \cdot \frac{\partial \mathbf{r}_i}{\partial q_j}) &= \sum_{i=1}^{n} m_i\left[\frac{\mathrm{d}}{\mathrm{d}t}(\dot{\mathbf{r}}_i^T \cdot \frac{\partial \mathbf{r}_i}{\partial q_j}) - \dot{\mathbf{r}}_i^T \cdot \frac{\mathrm{d}}{\mathrm{d}t}(\frac{\partial \mathbf{r}_i}{\partial q_j})\right] \\ &= \sum_{i=1}^{N} m_i\left[\frac{\mathrm{d}}{\mathrm{d}t}(\dot{\mathbf{r}}_i^T \cdot \frac{\partial \dot{\mathbf{r}}_i}{\partial \dot{q}_j}) - \dot{\mathbf{r}}_i^T \cdot \frac{\partial \dot{\mathbf{r}}_i}{\partial q_j}\right] \\ &= \frac{\mathrm{d}}{\mathrm{d}t}\frac{\partial}{\partial \dot{q}_j}(\sum_{i=1}^{N} m_i \dot{\mathbf{r}}_i^T \cdot \dot{\mathbf{r}}_i / 2) - \frac{\partial}{\partial q_j}(\sum_{i=1}^{N} m_i \dot{\mathbf{r}}_i^T \cdot \dot{\mathbf{r}}_i / 2) = \frac{\mathrm{d}}{\mathrm{d}t}(\frac{\partial T}{\partial \dot{q}_j}) - \frac{\partial T}{\partial q_j}\end{aligned} \tag{1.3.9}$$

using the equations (1.3.5) and (1.3.6), where

$$T = \sum_{i=1}^{N} \frac{1}{2} m_i \dot{\mathbf{r}}_i^T \cdot \dot{\mathbf{r}}_i \tag{1.3.10}$$

is the kinetic energy of the system. Then the variational equation (1.3.7) is derived as

$$\sum_{j=1}^{n} \left[-\frac{\mathrm{d}}{\mathrm{d}t}\left(\frac{\partial T}{\partial \dot{q}_j}\right) + \frac{\partial T}{\partial q_j} + Q_j \right] \delta q_j = 0 \tag{1.3.11}$$

called the D'Alembert-Lagrange equation.

§1.4, Lagrange equation

After derived the D'Alembert-Lagrange equation (1.3.11) the dynamic differential equations under the generalized displacements can be given as follows.

For a holonomic system, there is no other non-holonomic constraint, so that the n variations δq_j in equation (1.3.11) are all independent, hence the dynamic equations can easily be written as

$$\frac{\mathrm{d}}{\mathrm{d}t}\left(\frac{\partial T}{\partial \dot{q}_j}\right) - \frac{\partial T}{\partial q_j} = Q_j \quad , \qquad j = 1, 2, \cdots, n \tag{1.4.1}$$

The kinetic energy T is a function of $\mathbf{q}, \dot{\mathbf{q}}, t$, where $\mathbf{q} = \{q_1, q_2, \cdots, q_n\}^T$ is the ***generalized displacement vector***. The generalized external forces Q_j should be considered further, which can be distinguished according to the ***potential forces*** and ***general external forces***. The general external forces can be denoted as Q'_j. The potential forces are written as

$$Q_{jp} = -\partial \Pi / \partial q_j , \qquad \Pi = \Pi(q_1, q_2, \cdots, q_n, t) \tag{1.4.2}$$

where Π is a function of ***potential energy***. Introducing the Lagrange function as

$$\mathrm{L}(\mathbf{q}, \dot{\mathbf{q}}, t) = T - \Pi \tag{1.4.3}$$

then the dynamic equation (1.4.1) becomes

$$\frac{\mathrm{d}}{\mathrm{d}t}\left(\frac{\partial \mathrm{L}}{\partial \dot{q}_j}\right) - \frac{\partial \mathrm{L}}{\partial q_j} = Q'_j \quad , \qquad j = 1, 2, \cdots, n \tag{1.4.4}$$

which is called the ***Lagrange equation*** or ***Euler-Lagrange equation***. This equation is derived for a holonomic system. This book mainly considers holonomic system.

The composition of the Lagrange function is ***Kinetic energy minus Potential energy***. The derivation of Lagrange equation involves inertial and potential forces, but has no damping force. In order to include the damping force in Lagrange equation, Rayleigh proposed a dissipative function

$$R = \dot{\mathbf{q}}^T \mathbf{C} \dot{\mathbf{q}} / 2 = \sum_{i=1}^{n} \sum_{j=1}^{n} C_{ij} \dot{q}_i \dot{q}_j / 2 , \text{ damping force: } \; Q_i = \partial R / \partial \dot{q}_i \tag{1.4.5}$$

Therefore, equation (1.4.4) becomes

$$\frac{\mathrm{d}}{\mathrm{d}t}\left(\frac{\partial \mathrm{L}}{\partial \dot{q}_j}\right) - \frac{\partial \mathrm{L}}{\partial q_j} + \frac{\partial R}{\partial \dot{q}_j} = Q'_j \quad , \qquad j = 1, 2, \cdots, n \tag{1.4.6}$$

The dissipative forces always expense energy, hence the dissipative matrix $\mathbf{C}$ is positive definite or at least non-negative definite.

However, the dynamic equation of a non-holonomic system should also be considered. Except the holonomic constraints, which have been considered by means of the generalized displacements, there are another m non-holonomic constraints

$$\sum_{j=1}^{n} a_{rj}\dot{q}_j + a_r = 0, \qquad r = 1,2,\cdots,m \tag{1.4.7}$$

and the virtual displacements δq_j should satisfy the conditions

$$\sum_{j=1}^{n} a_{rj}\delta q_j = 0, \qquad r = 1,2,\cdots,m \tag{1.4.8}$$

Hence the variations δq_j $(j = 1,\cdots,n)$ in equation (1.2.9) are not completely independent, for which the Lagrange multiplier method can be applied. Introducing the multipliers

$$\lambda_r, \quad r = 1,2,\cdots,m \tag{1.4.9}$$

which is to be determined. Multiplying λ_r to the non-holonomic variational condition (1.4.8) and adding into the equation (1.3.7) gives

$$\sum_{j=1}^{n}\left(\frac{\mathrm{d}}{\mathrm{d}t}\frac{\partial T}{\partial \dot{q}_j} - \frac{\partial T}{\partial q_j} - Q_j - \sum_{r=1}^{m}\lambda_r a_{rj}\right)\delta q_j = 0 \tag{1.4.10}$$

There are $(n-m)$ independent variations of δq_j, so suppose $\delta q_{m+1},\cdots,\delta q_n$ are independent, then selecting λ_r so that the former m terms in the parenthesis are zero. Thereafter because of the independent variations, the latter $(n-m)$ terms in the parenthesis should also be zero, therefore

$$\frac{\mathrm{d}}{\mathrm{d}t}\left(\frac{\partial T}{\partial \dot{q}_j}\right) - \frac{\partial T}{\partial q_j} = Q_j + \sum_{r=1}^{m}\lambda_r a_{rj} \tag{1.4.11}$$

The unknown functions are $q_j (j = 1,\cdots,n)$ and the Lagrange multipliers $\lambda_r (r = 1,\cdots,m)$ totally $(n+m)$ variables to be solved. Corresponding equations are (1.4.11) and constraints (1.4.7) totally also $(n+m)$ equations, and so the number of variables and of equations are compatible.

§1.5, Hamilton variational principle

From what follows, the description is only for holonomic systems, and the system configuration is given with generalized displacement vector $\mathbf{q}$ only.

In dynamic equation (1.4.4) when the external forces are only potential forces, then $Q_j' = 0$, it derives the dynamic equation

$$(\mathrm{d}/\mathrm{d}t)(\partial \mathrm{L}/\partial \dot{q}_j) - \partial \mathrm{L}/\partial q_j = 0, \quad j = 1,2,\cdots,n, \quad \mathrm{L} = T - \Pi \tag{1.5.1}$$

However, this equation can be derived from a variational principle, which is called the ***Hamilton principle***

$$\delta S = 0, \quad S = \int_{t_0}^{t_f}(T - \Pi)\mathrm{d}t, \quad \text{or} \quad \delta\int_{t_0}^{t_f}\mathrm{L}(\mathbf{q},\dot{\mathbf{q}}.t)\mathrm{d}t = 0 \tag{1.5.2}$$

where t_0 is the beginning time, t_f is the finish time, given instances. The displacement vectors $\mathbf{q}_0$ and $\mathbf{q}_f$ are given at the two instants and having no variations. L is the Lagrange function and is composed of ***kinetic energy minus potential energy***.

There are various paths (orbits) from the generalized displacement $\mathbf{q}_0$ at initial instant t_0 to reach the finish point, but generally speaking there is only one path, which satisfies the differential equation (1.5.1), reaching the generalized displacement vector $\mathbf{q}_f$ at the finish time t_f, which is called the real orbit. The variational principle says that the integration S along the real orbit takes stationary value. In other words, the ***real orbit in the configuration space of a holonomic system during the fixed interval*** t_0 ***to*** t_f ***is that the integration*** S ***is stationary with respect to orbit variations, which vanish at the two end points.*** The orbit variations are first order small values, which need not satisfy the dynamic equation (1.5.1), but satisfy the (holonomic) constraints. Introducing the ***action*** integral

$$S = \int_{t_0}^{t_f} \mathrm{L}(\mathbf{q}, \dot{\mathbf{q}}, t)\mathrm{d}t \ , \qquad \delta S = 0 \tag{1.5.3}$$

and the principle of that its variation equals zero is called the ***Hamilton principle***. The variable S is a function of the generalized displacement $\mathbf{q}(t)$, but $\mathbf{q}(t)$ itself is also a function of time. Hence S is a function of functions, and is termed as a ***functional***. Thus the action S is a functional of generalized displacement $\mathbf{q}(t)$. For a holonomic system, every component of $\mathbf{q}(t)$ can vary independently.

Denoting the solution (*i.e.* the real orbit) as $\mathbf{q}_*(t)$, nearby the solution there is another orbit $\mathbf{q}(t)$ from which the functional S can also be computed which will not coincide with the real action S_*. The difference $(S - S_*)$ is the variation of the functional and is computed in first variation as

$$\begin{aligned}
\delta S &= S(\mathbf{q}) - S(\mathbf{q}_*) = \int_{t_0}^{t_f} [\mathrm{L}(\mathbf{q}, \dot{\mathbf{q}}, t) - \mathrm{L}(\mathbf{q}_*, \dot{\mathbf{q}}_*, t)]\mathrm{d}t \\
&= \int_{t_0}^{t_f} [(\partial \mathrm{L}/\partial \dot{\mathbf{q}})^T \cdot \delta \dot{\mathbf{q}} + (\partial \mathrm{L}/\partial \mathbf{q})^T \cdot \delta \mathbf{q}]\mathrm{d}t \\
&= \int_{t_0}^{t_f} [-(\mathrm{d}/\mathrm{d}t)(\partial \mathrm{L}/\partial \dot{\mathbf{q}}) + \partial \mathrm{L}/\partial \mathbf{q}]^T \cdot \delta \mathbf{q}\mathrm{d}t + [(\partial \mathrm{L}/\partial \dot{\mathbf{q}})^T \cdot \delta \mathbf{q}]_{t_0}^{t_f} \\
&= \int_{t_0}^{t_f} [-(\mathrm{d}/\mathrm{d}t)(\partial \mathrm{L}/\partial \dot{\mathbf{q}}) + \partial \mathrm{L}/\partial \mathbf{q}]^T \cdot \delta \mathbf{q}\mathrm{d}t
\end{aligned} \tag{1.5.4}$$

The vectorial notation is used in the derivation. A scalar function taking derivative with respect to a vector gives a column vector, for which the geometric meaning is the gradient. Hence $\partial \mathrm{L}/\partial \dot{\mathbf{q}}$ is a vector. The second equal sign gives the first variation, the third equal sign is integration by parts, and the fourth equal sign is because of the variations of generalized displacement at two ends equal zero. Because for a holonomic system, the variation $\delta \mathbf{q}$ of generalize displacement in time interval (t_0, t_f) is arbitrary, so that the term in the bracket in the last line of (1.5.4) equals zero, from which the dynamic equation (1.5.1) is obtained in vectorial form as

$$\frac{\mathrm{d}}{\mathrm{d}t}\left(\frac{\partial \mathrm{L}}{\partial \dot{\mathbf{q}}}\right)-\frac{\partial \mathrm{L}}{\partial \mathbf{q}}=\mathbf{0} \tag{1.5.5}$$

called the ***Euler-Lagrange equation***.

The above paragraph defines the ***action*** S and gives the Hamilton principle. The ***action functional*** S holds a stationary value at the real orbit $\mathbf{q}_*(t)$. For analytical dynamics, the Lagrange function in the action integration $\mathrm{L}=T-\Pi$ is composed of (***kinetic energy*** – ***potential energy***). However, variational method, Hamilton principle and so on are not necessarily limited to be applied only for analytical dynamics. These methods can also be applied in eletro-dynamics, in quantum mechanics, etc. In the present book, these methods will be applied also in elasticity, in structural mechanics, wave propagation and optimal control problems. In various field applications there are also the Lagrange function $\mathrm{L}(\mathbf{q},\dot{\mathbf{q}},t)$, but the argument vector $\mathbf{q}(t)$ is not necessarily generalized displacement vector, the coordinate t is not necessarily the time, and the Lagrange function is no longer the composition of (kinetic energy – potential energy). For all these applications, the variational equation (1.5.3) always holds. Generally speaking, the composition rule of the Lagrange function $\mathrm{L}(\mathbf{q},\dot{\mathbf{q}},t)$ can be quite versatile, that it is only necessary a function of $\mathbf{q},\dot{\mathbf{q}},t$. For analytical dynamics, the dynamic equation is evolutionary, initial value problem. But for other problems, it can also be a two point boundary value problem (TPBVP).

The Hamilton principle does not limit the number n of degrees of freedom of the generalized displacement $\mathbf{q}(t)$ (the argument function). Hence this variational principle not only can be used to discrete system but also can be use to continuous system. Thereafter, it can develop to discrete as well as to continuous mixed variable systems. Such versatility is quite beneficial to theory of elasticity, complex structures, electro-magnetic fields and wave-guide problems etc.

Example 1.3, A system composed of two masses m_1 and m_2 connected with two thin hanging rods for which the mass is neglected, see figure 1.4(a), a given horizontal force $f(t)$ acts on the mass m_2. It is required to derive the dynamic equation with the Lagrange method.

Solution: The system has two degrees of freedom, and θ_1 and θ_2 are selected as the generalized displacement. The system kinetic energy is

$$T=\frac{1}{2}m_1(l_1\dot{\theta}_1)^2+\frac{1}{2}m_2[(l_1\dot{\theta}_1\cos\theta_1+l_2\dot{\theta}_2\cos\theta_2)^2+(l_1\dot{\theta}_1\sin\theta_1+l_2\dot{\theta}_2\sin\theta_2)^2]$$
$$=m_1(l_1\dot{\theta}_1)^2/2+m_2[(l_1\dot{\theta}_1)^2+(l_2\dot{\theta}_2)^2+2l_1l_2\dot{\theta}_1\dot{\theta}_2\cos(\theta_2-\theta_1)]/2$$

In the external forces, the gravity m_1g and m_2g are potential forces, for which the potential expression is

$$\Pi=m_1gl_1(1-\cos\theta_1)+m_2g[l_1(1-\cos\theta_1)+l_2(1-\cos\theta_2)]$$

The horizontal force $f(t)$ is non-potential. To compute the generalized forces induced by $f(t)$, first let the virtual generalized displacement be $\delta\theta_1=1$ and $\delta\theta_2=0$, see figure 1.4(b). According to virtual work equivalence gives

$$Q_1'\delta\theta_1 = f(t)l_1\delta\theta_1\cos\theta_1, \qquad \text{hence} \quad Q_1' = fl_1\cos\theta_1$$

Taking $\delta\theta_1 = 0$ and $\delta\theta_2 \neq 0$, as shown in figure 1.4(c), and the virtual work equivalence gives

$$Q_2'\delta\theta_2 = fl_2\delta\theta_2\cos\theta_2, \qquad \text{hence} \quad Q_2' = fl_2\cos\theta_2$$

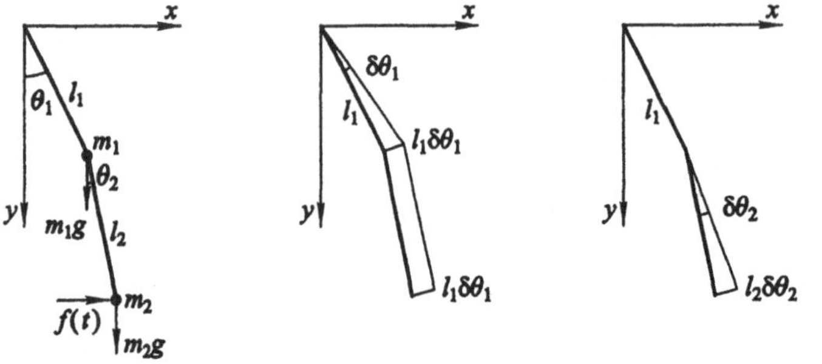

(a) Double pendulum system of two particles, **(b) The virtual displacement of $\delta\theta_1$,** **(c) The virtual displacement of $\delta\theta_2$**

Figure 1.4, *Double pendulum*

According to $L = T - \Pi$, substituting into the Euler-Lagrange equation (1.5.1) gives

$$(m_1 + m_2)l_1\ddot{\theta}_1 + m_2l_2\ddot{\theta}_2\cos(\theta_2 - \theta_1) - m_2l_2\dot{\theta}_2^2\sin(\theta_2 - \theta_1) + (m_1 + m_2)g\sin\theta_1$$
$$= f(t)\cos\theta_1$$

$$m_2l_2\ddot{\theta}_2 + m_2l_1\ddot{\theta}_1\cos(\theta_2 - \theta_1) + m_2l_1\dot{\theta}_1^2\sin(\theta_2 - \theta_1) + m_2g\sin\theta_2 = f(t)\cos\theta_2$$

These two ODEs are the dynamic equations required. However, they are second order non-linear differential equations, to find their analytical solution is not easy.

When the disturbance force $f(t)$ is small, and the initial displacements and velocities are all small, the system will vibrate slightly nearby the equilibrium position $\theta_1 = \theta_2 = 0$. Neglecting the higher order small quantities, let $\cos(\theta_2 - \theta_1) \approx 1$, $\sin(\theta_2 - \theta_1) \approx \theta_2 - \theta_1$, $\cos\theta_1 \approx 1$, $\cos\theta_2 \approx 1$, $\sin\theta_1 \approx \theta_1$, and $\sin\theta_2 \approx \theta_2$, the dynamic equations are simplified as

$$(m_1 + m_2)l_1\ddot{\theta}_1 + m_2l_2\ddot{\theta}_2 + (m_1 + m_2)g\theta_1 = f(t)$$

$$m_2l_1\ddot{\theta}_1 + m_2l_2\ddot{\theta}_2 + m_2g\theta_2 = f(t)$$

They are a set of ODEs and can be easily solved by the linear vibration theory, see next chapter. ##

The Lagrange method is often used in the vibration analysis of discrete system. In the above derivation, the linearization step is carried out after the dynamic equations are derived. This linearization step can be executed in generating T and Π, in this linearization process the second order displacement terms must be kept in the functional. Hence

$$T \approx m_1l_1^2\dot{\theta}_1^2/2 + m_2[l_1^2\dot{\theta}_1^2 + l_2^2\dot{\theta}_2^2 + 2l_1l_2\dot{\theta}_1\dot{\theta}_2]/2$$

$$\Pi \approx m_1 g l_1 \theta_1^2/2 + m_2 g[l_1\theta_1^2 + l_2\theta_2^2]/2$$

Afterwards, from the Euler-Lagrange equation derives

$$(m_1+m_2)l_1^2\ddot{\theta}_1 + m_2 l_1 l_2 \ddot{\theta}_2 + (m_1+m_2) g l_1 \theta_1 = l_1 f(t)$$

$$m_2 l_1 l_2 \ddot{\theta}_1 + m_2 l_2^2 \ddot{\theta}_2 + m_2 g l_2 \theta_2 = l_2 f(t)$$

which are completely in agreement with those derived above.

The characteristic of these dynamic equations is that there is no $\dot{\theta}$ term, which implies that the system has neither damping nor gyroscopic term. The pendulums are described in an inertia frame and also time invariant, hence there is no gyroscopic term. ##

The expression of the kinetic energy T is worth to explore below. Using the expression (1.3.4) of $\dot{\mathbf{r}}_i$ gives

$$\begin{aligned} T &= \sum_{i=1}^{N} m_i \dot{\mathbf{r}}_i^T \cdot \dot{\mathbf{r}}_i/2 \\ &= \frac{1}{2}\sum_{i=1}^{N} m_i \left[\sum_{r=1}^{n}\sum_{s=1}^{n} \left(\frac{\partial \mathbf{r}_i}{\partial q_r}\right)^T \cdot \frac{\partial \mathbf{r}_i}{\partial q_s} \dot{q}_r \dot{q}_s + 2\left(\frac{\partial \mathbf{r}_i}{\partial t}\right)^T \cdot \sum_{r=1}^{n} \frac{\partial \mathbf{r}_i}{\partial q_r} \dot{q}_r + \left(\frac{\partial \mathbf{r}_i}{\partial t}\right)^T \cdot \frac{\partial \mathbf{r}_i}{\partial t} \right] \end{aligned} \tag{1.5.6}$$

There are second order, first order and zero order terms of $\dot{\mathbf{q}}$ in the above expression. The condition for the first and zero order terms to exist is that, the expression of $\mathbf{r}_i$ is time variant, so that the partial derivative with respect to time t is not zero (rheonomic system). Classifying the kinetic energy expression T as the summation of quadratic $\dot{\mathbf{q}}$, linear $\dot{\mathbf{q}}$ and $\dot{\mathbf{q}}$ independent terms gives

$$T = T_2 + T_1 + T_0 \tag{1.5.7}$$

where

$$\begin{aligned} T_2 &= \sum_{r=1}^{n}\sum_{s=1}^{n} m_{rs}(\mathbf{q},t)\dot{q}_r\dot{q}_s, \qquad m_{rs} = \sum_{i=1}^{N} m_i \left(\frac{\partial \mathbf{r}_i}{\partial q_r}\right)^T \cdot \frac{\partial \mathbf{r}_i}{\partial q_s}, \\ T_1 &= \sum_{r=1}^{n} \mu_r(\mathbf{q},t)\dot{q}_r, \qquad \mu_r = \sum_{i=1}^{N} m_i \left(\frac{\partial \mathbf{r}_i}{\partial t}\right)^T \cdot \frac{\partial \mathbf{r}_i}{\partial q_r}, \\ T_0 &= \gamma(\mathbf{q},t), \qquad \gamma = \frac{1}{2}\sum_{i=1}^{N} m_i \left(\frac{\partial \mathbf{r}_i}{\partial t}\right)^T \cdot \frac{\partial \mathbf{r}_i}{\partial t} \end{aligned} \tag{1.5.8}$$

For a time invariant system (scleronomic system), μ_r and γ are both zero, hence the kinetic energy T becomes a homogeneous quadratic form of the generalized displacement vector $\dot{\mathbf{q}}$. According to the physical meaning of kinetic energy, T_2 is positive definite or at least non-negative definite.

There is some explanation for the generalized forces Q_j, the definition of which is

$$Q_j = \sum_{i=1}^{N} \mathbf{F}_i^T \cdot \frac{\partial \mathbf{r}_i}{\partial q_j}, \quad j = 1,2,\cdots,n \tag{1.5.9}$$

from the expression of virtual work of the external forces

$$\sum_{i=1}^{N}\mathbf{F}_i^T\cdot\delta\mathbf{r}_i=\sum_{i=1}^{N}\mathbf{F}_i^T\cdot\sum_{j=1}^{n}\frac{\partial\mathbf{r}_i}{\partial q_j}\delta q_j=\sum_{j=1}^{n}(\sum_{i=1}^{N}\mathbf{F}_i^T\cdot\frac{\partial\mathbf{r}_i}{\partial q_j})\delta q_j=\sum_{j=1}^{n}Q_j\delta q_j \tag{1.5.10}$$

which is also the virtual work of generalized external force Q_j. Hence the equation for finding the external generalized forces is thus derived, which is again the method of virtual displacement principle. The double pendulum example given above uses such method.

§1.6, Hamiltonian canonical equations

The Hamiltonian system was first given systematically in 1834 (W. R. Hamilton). Before Hamilton, the French mathematicians Poisson, Lagrange, Pfaff, Cauchy also contributed the development as mentioned in [10], p.264. The importance of Hamilton system lies in providing a framework for theoretical extensions in many areas of physics. Within analytical dynamics it forms the basis for further developments, such as the canonical transformation, the Hamilton-Jacobi theory and perturbation approaches etc. The Hamiltonian formulation provides much of the language with which present day statistical mechanics, quantum mechanics are constructed.

The importance of Hamiltonian formulation in applied mechanics is not limited only in analytical dynamics, that it also provides a foundation for optimal control, elasticity, vibration theory, wave propagation, multi-body dynamics etc. Hence, although the Hamiltonian formulation was introduced in the extent of analytical dynamics, its theoretical implication is never only limited in analytical dynamics.

§1.6.1, Legendre transformation and Hamiltonian canonical equations

The unknown functions in Lagrange system formulation are expressed with the generalized displacement vector $\mathbf{q}(t)$, which describes the displacements of particles, *i.e.* a point in the displacement space. Such description is termed the ***configuration*** and the dynamic equation is

$$\frac{\mathrm{d}}{\mathrm{d}t}(\frac{\partial \mathrm{L}}{\partial \dot{q}_i})-\frac{\partial \mathrm{L}}{\partial q_i}=0,\quad i=1,2,\cdots,n \tag{1.6.1}$$

which is a second order ODE of the vector $\mathbf{q}(t)$. The number of boundary conditions needed to fix the solution is $2n$, which can be either giving the initial n generalized displacements and velocities at time t_0 or giving the generalized displacements q_i, $i=1,2,\cdots.n$ at t_0 and t_f, etc. In Lagrange system, the generalized displacement vector $\mathbf{q}(t)$ of configuration description is the main unknown, whereas the velocity $\dot{\mathbf{q}}$ is considered only the time derivative of $\mathbf{q}$.

In Hamiltonian system formulation, the consideration is completely different. The governing differential equations are first order, but the number of equations is doubled becoming $2n$. The number of fundamental unknowns is also $2n$, which coincides with the number of boundary conditions. In classical mechanics and in physics it is termed as ***phase space***, however, it is also termed as ***state space method***.

The first n unknowns are naturally selected as q_i, $i=1,2,\cdots.n$, and the other half, n unknowns, are selected as the dual variables—***generalized momentum***

$$p_i = \partial \mathrm{L}(q_i, \dot{q}_i, t)/\partial \dot{q}_i \tag{1.6.2}$$

The dual vector variables $(\mathbf{q},\mathbf{p})$ are termed the ***canonical variables*** in analytical dynamics or in physics. However, in other applications it is often termed as the ***state variables***. The vectors $\mathbf{q}$ and $\mathbf{p}$ are dual to each other, hence they are often called as ***dual variables***.

Mathematically, the partial differentials in the Lagrange equations have practically treated the components of vectors $\mathbf{q}$ and $\dot{\mathbf{q}}$ as independent to each other. Such as $\partial \mathrm{L}/\partial q_i$, which means only the component q_i varies but all the components of $\dot{\mathbf{q}}$ and the other components of $\mathbf{q}$ are regarded unchanged. The transition from Lagrange system to Hamilton system corresponds to, from the variables $(\mathbf{q},\dot{\mathbf{q}},t)$ transformed to the variables $(\mathbf{q},\mathbf{p},t)$, where the relation of $\mathbf{p}$ to $\mathbf{q},\dot{\mathbf{q}}$ is given by (1.6.2). Such variable transformation is described with the Legendre transformation. The geometric interpretation of Legendre transformation can be found from [1]-vol.2, §6.

The Hamilton description uses two kinds of variables, *i.e.* the dual variables of $\mathbf{q}$ and $\mathbf{p}$. The corresponding variational principle is derived as follows. Introducing the variable vector $\mathbf{s}$ to denote the generalized velocity $\dot{\mathbf{q}}$, the Lagrange function becomes $\mathrm{L}(\mathbf{q},\mathbf{s},t)$, and $\mathbf{s}=\dot{\mathbf{q}}$ becomes a prerequisite condition of the vairational principle. Correspondingly, introducing the vector $\mathbf{p}$ as a Lagrange multiplier vector of the prerequisite condition, hence the variational principle (1.5.2) becomes

$$\delta\int_{t_0}^{t_f} [\mathbf{p}^T(\dot{\mathbf{q}}-\mathbf{s}) + \mathrm{L}(\mathbf{q},\mathbf{s},t)]\mathrm{d}t = 0 \tag{1.6.3}$$

where the three kinds of variables $\mathbf{q},\mathbf{p}$ and $\mathbf{s}$ are considered varying independently. Performing the variational derivation gives

$$\delta\mathbf{s}: \quad \mathbf{p} = \partial \mathrm{L}(\mathbf{q},\mathbf{s},t)/\partial\mathbf{s}, \qquad [\text{from which solves } \mathbf{s}=\mathbf{s}(\mathbf{q},\mathbf{p},t)] \tag{1.6.4a}$$

$$\delta\mathbf{p}: \quad \dot{\mathbf{q}} = \mathbf{s} \tag{1.6.4b}$$

$$\delta\mathbf{q}: \quad \dot{\mathbf{p}} = \partial \mathrm{L}(\mathbf{q},\mathbf{s},t)/\partial\mathbf{q} \tag{1.6.4c}$$

Substituting (1.6.4b) into (1.6.4a) gives (1.6.2). Introducing into equation (1.6.3) the function

$$\mathrm{H}(\mathbf{q},\mathbf{p},t) = \mathbf{p}^T\mathbf{s} - \mathrm{L}(\mathbf{q},\mathbf{s},t) \quad [= \mathbf{p}^T\dot{\mathbf{q}} - \mathrm{L}(\mathbf{q},\dot{\mathbf{q}},t)] \tag{1.6.5}$$

where the variable $\mathbf{s}$ is regarded a function of $\mathbf{q},\mathbf{p},t$. The function H is termed as the ***Hamilton function***. Based on the chain rule of differentiation gives

$$\mathrm{dH} = (\partial\mathrm{H}/\partial\mathbf{q})^T \cdot \mathbf{dq} + (\partial\mathrm{H}/\partial\mathbf{p})^T \cdot \mathbf{dp} + (\partial\mathrm{H}/\partial t)\cdot \mathrm{d}t \tag{1.6.6a}$$

However, from equation (1.6.5) the complete differential derives

$$\mathrm{dH} = (\mathbf{dp})^T \cdot \mathbf{s} + \mathbf{p}^T\mathbf{ds} - (\partial\mathrm{L}/\partial\mathbf{q})^T \cdot \mathbf{dq} - (\partial\mathrm{L}/\partial\mathbf{s})^T \cdot \mathbf{ds} - (\partial\mathrm{L}/\partial t)\cdot \mathrm{d}t$$

Substituting equation (1.6.4a), the $\mathbf{ds}$ term is cancelled, then using equations (1.6.4b,c) the above equation derives to

$$\mathrm{dH} = \dot{\mathbf{q}}^T \cdot \mathrm{d}\mathbf{p} - \dot{\mathbf{p}}^T \cdot \mathrm{d}\mathbf{q} - (\partial \mathrm{L}/\partial t) \cdot \mathrm{d}t \tag{1.6.6b}$$

Comparing with (1.6.6a) gives

$$\dot{\mathbf{q}} = \partial \mathrm{H}/\partial \mathbf{p} \tag{1.6.7a}$$

$$\dot{\mathbf{p}} = -\partial \mathrm{H}/\partial \mathbf{q} \tag{1.6.7b}$$

$$-\partial \mathrm{L}/\partial t = \partial \mathrm{H}/\partial t \tag{1.6.8}$$

Equations (1.6.7a,b) are called the Hamiltonian ***canonical equations.*** Then the $2n$ first order ODE substitutes the Lagrange equation.

The equation (1.6.7a) corresponds to the inverse of equation (1.6.2), which can be regarded as the constitutive relation, and equation (1.6.7b) is the dynamic equation. The formulation of Hamilton function $\mathrm{H}(\mathbf{q},\mathbf{p},t)$ explicitly expresses that it is a function of the dual variables $\mathbf{q}$, $\mathbf{p}$ and time t. Eliminating the vector $\mathbf{s}$ by substituting relation (1.6.5) into the variational principle (1.6.3) gives

$$\delta \int_{t_0}^{t_f} [\mathbf{p}^T \dot{\mathbf{q}} - \mathrm{H}(\mathbf{q},\mathbf{p})] \mathrm{d}t = 0 \tag{1.6.9}$$

It is the variational principle corresponding to the set of Hamiltonian canonical equation (1.6.7), a variational principle with two kinds of dual variables $\mathbf{q}$, $\mathbf{p}$.

Hamilton function is very important, that a number of basic equations and fundamental theorems are derived from it. The main steps to derive the Hamilton function from the Lagrange function are outlined as follows:

1) Select the generalized displacement $\mathbf{q}$, then compose the Lagrange function $\mathrm{L}(\mathbf{q},\dot{\mathbf{q}},t)$.
2) Substituting $\dot{\mathbf{q}}$ with $\mathbf{s}$, introduce the generalized momentum by means of equation (1.6.4a).
3) Solve $\mathbf{s} = \mathbf{s}(\mathbf{q},\mathbf{p},t)$ from the equation $\mathbf{p} = \partial \mathrm{L}(\mathbf{q},\mathbf{s},t)/\partial \mathbf{s}$.
4) Substitute $\mathbf{s}$ into (1.6.5) to obtain the Hamilton function, which is a function of $(\mathbf{q},\mathbf{p},t)$.

The steps given above are quite general. For a dynamical system $\mathrm{L} = T - \Pi$ where the kinetic energy T is composed of quadratic, linear and zero-th order terms of $\dot{\mathbf{q}}$, see equation (1.5.7). Suppose

$$\mathrm{L}(\mathbf{q},\dot{\mathbf{q}},t) = \mathrm{L}_0(\mathbf{q},t) + \dot{\mathbf{q}}^T \cdot \mathbf{a}(\mathbf{q},t) + \dot{\mathbf{q}}^T \mathbf{M} \dot{\mathbf{q}}/2 \tag{1.6.10}$$

where $\mathbf{a}$ is an n-dimensional vector, $\mathbf{M}(\mathbf{q},t)$ is an $n \times n$ symmetric positive definite matrix. Hence the dual variable vector (generalized momentum) is

$$\mathbf{p} = \mathbf{M}\dot{\mathbf{q}} + \mathbf{a} \quad (= \mathbf{M}\mathbf{s} + \mathbf{a}) \tag{1.6.11}$$

Solving $\dot{\mathbf{q}}$ gives

$$\dot{\mathbf{q}} = \mathbf{M}^{-1}(\mathbf{p} - \mathbf{a}) \tag{1.6.12}$$

which gives the relation between momentum and velocity and can be regarded as a constitutive relation. Hence the Hamilton function is derived as

$$\begin{aligned} \mathrm{H}(\mathbf{q},\mathbf{p},t) &= \mathbf{p}^T \dot{\mathbf{q}} - [\mathrm{L}_0 + \dot{\mathbf{q}}^T \mathbf{a} + \dot{\mathbf{q}}^T \mathbf{M} \dot{\mathbf{q}}/2] = \dot{\mathbf{q}}^T \mathbf{M} \dot{\mathbf{q}}/2 - \mathrm{L}_0 \\ &= (\mathbf{p} - \mathbf{a})^T \mathbf{M}^{-1} (\mathbf{p} - \mathbf{a})/2 - \mathrm{L}_0(\mathbf{q},t) \end{aligned} \tag{1.6.13}$$

It is noticed, that the Hamiltonian canonical equations (1.6.7a,b) are not entirely symmetric with respect to the dual variable vectors $\mathbf{q}$ and $\mathbf{p}$, because the negative sign in equation (1.6.7b). Compose the ***state vector*** using the dual vectors $\mathbf{q}$ and $\mathbf{p}$ as components

$$\mathbf{v} = \begin{Bmatrix} \mathbf{q} \\ \mathbf{p} \end{Bmatrix} \tag{1.6.14}$$

Hence the Hamilton function can be written as $\mathrm{H}(\mathbf{q},\mathbf{p},t) = \mathrm{H}(\mathbf{v},t)$ and $\partial\mathrm{H}/\partial\mathbf{v}$ becomes a $2n$-dimensioned vector, where

$$(\partial\mathrm{H}/\partial\mathbf{v})_i = \partial\mathrm{H}/\partial q_i, \quad (\partial\mathrm{H}/\partial\mathbf{v})_{n+i} = \partial\mathrm{H}/\partial p_i, \quad i \le n \tag{1.6.15}$$

For combining the two canonical equations as one, introducing a $2n \times 2n$ matrix as

$$\mathbf{J} = \begin{bmatrix} \mathbf{0} & \mathbf{I}_n \\ -\mathbf{I}_n & \mathbf{0} \end{bmatrix} \begin{matrix} n \\ n \end{matrix} \tag{1.6.16}$$

then the Hamilton canonical equations are given in combined form

$$\dot{\mathbf{v}} = \mathbf{J} \cdot (\partial\mathrm{H}/\partial\mathbf{v}) \tag{1.6.17}$$

This form of canonical equation is called the ***symplectic*** expression of Hamilton canonical equation. The term ***symplectic*** comes from Greek, stands for the meaning of 'intertwined'. H. Weyl first introduced the term in 1939, see [15]. The matrix $\mathbf{J}$ is of special importance in symplectic formulations. Some easily verified properties of $\mathbf{J}$ are listed as follows:

$$\mathbf{J}^2 = -\mathbf{I}_{2n}, \quad \text{Comment: similar the imaginary number}$$

$$\mathbf{J}^T\mathbf{J} = \mathbf{I}_{2n}, \quad \text{Comment: an orthogonal matrix} \tag{1.6.18}$$

thus $\mathbf{J}^T = \mathbf{J}^{-1} = -\mathbf{J}$, and $\det(\mathbf{J}) = 1$

For arbitrary vector $\mathbf{v}_a$ of $2n$-dimensions, it always holds $\mathbf{v}_a^T\mathbf{J}\mathbf{v}_a \equiv 0$, because $\mathbf{J}$ is a skew-symmetric matrix.

Here, it is seen that the ***symplectic*** behaviour and Hamiltonian system are closely interrelated. All conservative systems can be described by a Hamiltonian formulation, hence has the symplectic property. There are quite a number of conservative systems in applied mechanics and modern control theory. Using the duality variable system and deriving the governing equations into Hamiltonian system description, it is hopefully to establish the ***unified methodology under the symplectic frame***, which is beneficial to the exchange among various disciplines.

Go back to the equation (1.6.3), a variational principle with three kinds of variables $\mathbf{q},\mathbf{p},\mathbf{s}$. The variational principle (1.6.9) of two kinds of independent variables corresponding to the Hamiltonian canonical equations is obtained by eliminating $\mathbf{s}$ from (1.6.3). Another approach is to eliminate $\mathbf{p}$ but keeps the variables $\mathbf{q}$ and $\mathbf{s}$, so as to obtain another variational principle with two kinds of variables. Substituting $\mathbf{p} = \partial\mathrm{L}/\partial\mathbf{s}$ into equation (1.6.3) gives

$$\delta\int_{t_0}^{t_f}[(\partial\mathrm{L}/\partial\mathbf{s})^T(\dot{\mathbf{q}} - \mathbf{s}) + \mathrm{L}(\mathbf{q},\mathbf{s},t)]\mathrm{d}t = 0 \tag{1.6.19}$$

where $\mathbf{q}$ and $\mathbf{s}$, the generalized displacement and generalized velocity, respectively, are the two kinds of independent variables. The differential equations derived from which are ([10] p.247)

$$\dot{\mathbf{q}} - \mathbf{s} = \mathbf{0}, \qquad -\frac{\mathrm{d}}{\mathrm{d}t}\left(\frac{\partial \mathrm{L}}{\partial \mathbf{s}}\right) + \frac{\partial \mathrm{L}}{\partial \mathbf{q}} = \mathbf{0} \tag{1.6.20}$$

Such form of variational principle (1.6.19) is sometime useful in computations.

§1.6.2, Cyclic coordinate and conservation

A generalized coordinate q_i, which does not appear in the Lagrange function L, is called a cyclic coordinate, however, $\dot{q}_i$ can still appear in the function L. From equations (1.6.4c) and (1.6.7b) it derives

$$\dot{p}_i = \partial \mathrm{L}/\partial q_i = -\partial \mathrm{H}/\partial q_i = 0$$

Therefore p_i is a constant, and will absent in the Hamilton function. Conversely, if a generalized coordinate does not occur in H, the dual momentum is conserved. The cyclic coordinate is often used in solving the problem of momentum conservation or angular momentum conservation etc.

The problem of how the Hamilton function H changing with time is also very important. From canonical equations the complete differential of H is given as

$$\frac{\mathrm{dH}}{\mathrm{d}t} = \left(\frac{\partial \mathrm{H}}{\partial \mathbf{q}}\right)^T \dot{\mathbf{q}} + \left(\frac{\partial \mathrm{H}}{\partial \mathbf{p}}\right)^T \cdot \dot{\mathbf{p}} + \frac{\partial \mathrm{H}}{\partial t} = \frac{\partial \mathrm{H}}{\partial t} = -\frac{\partial \mathrm{L}}{\partial t} \tag{1.6.21}$$

Therefore if L is not an explicit function of time t, then the Hamilton function H is a constant of motion.

The physical meaning of Hamilton function is of great concern. If there is only $T = T_2$ in kinetic energy expression (1.5.7), then $\mathrm{H} = \dot{\mathbf{q}}^T \mathbf{M}\dot{\mathbf{q}}/2 + \Pi = T + \Pi$, which is (***kinetic energy +potential energy***). The equation $\mathrm{H} = \text{constant}$ gives the ***mechanical energy conservation principle.*** Only when the generalized displacement vector $\mathbf{q}$ is expressed in an inertia coordinate system, then $T = T_2$, the principle of energy conservation is arrived.

Note that such interpretation applies only to analytical dynamics. In other fields, the interpretation for conservation theorem is different, that the interpretation is problem dependent.

A cyclic coordinate q_i implies a conservative rule, for which the respective dual variable p_i takes a constant denoted as α_i. In such case the Routh method applies. From the Lagrange function, selecting only these cyclic coordinates introducing the dual variables p_i, for other generalized coordinates no transformation is made. Without loss of generality, assuming the i-th coordinate $s+1 \le i \le n$ being cyclic, then the generalized coordinates with subscript $1 \sim s$ do not transform, and $p_{s+1} \sim p_n$ take constant values $\alpha_{s+1}, \cdots, \alpha_n$. After Legendre transformation for these cyclic coordinates is carried out, the Routh function is obtained as

$$\mathrm{R}(q_1, \cdots, q_s; p_1, p_2, \cdots, p_s; \alpha_{s+1}, \cdots, \alpha_n, t) = \sum_{i=s+1}^{n} p_i \dot{q}_i - \mathrm{L} \tag{1.6.22}$$

From the Routh function derives the dynamic equations for the former s generalized coordinates

$$\frac{\mathrm{d}}{\mathrm{d}t}(\frac{\partial \mathrm{R}}{\partial \dot{q}_i}) - \frac{\partial \mathrm{R}}{\partial q_i} = 0, \quad i = 1,2,\cdots,s \tag{1.6.23}$$

where R is regarded as a Lagrange function, however there are only the generalized coordinates $q_i, 1 \le i \le s$. The values of constants $\alpha_i, s+1 \le i \le n$ can be determined by the boundary conditions.

Using purely analytical method to find solutions, less unknown functions are generally easier. However solution by numerical method, transforming to state space may be easier, so that all generalized coordinate should be transformed to Hamiltonian formulation.

§1.7, Canonical transformation

The above expression for canonical dual equations extends the variables from generalized displacements to the state space (phase space), *i.e.* expressed in the combination of $\mathbf{q}$ and $\mathbf{p}$, which has been completely different from only the configuration (generalized displacement) space $\mathbf{q}$ for Lagrange system. In Lagrange system description, the coordinate transformation is from the original configuration $\mathbf{q}$ transforming to another configuration $\mathbf{Q}$, and such kind of transformation is termed as the ***point transformation***. Configuration is only a point in the generalized displacement space, that each component of $\mathbf{q}$ or $\mathbf{Q}$ is only a position parameter for this point, which is different from the state of the system. The state of a system should also involve momentum $\mathbf{p}$.

For Hamilton system, the description is the state of system. So the transformation for Hamilton system should be about the state vectors, *i.e.* from $\mathbf{q}$ and $\mathbf{p}$ to $\mathbf{Q}$ and $\mathbf{P}$, it is a $2n$ dimensional state space (invertible) transformation and is termed as the ***canonical transformation.*** An arbitrary transformation in the $2n$ dimension may not be called as canonical, only when after the transformation, the system dual equations formulated with the dual vectors $\mathbf{Q}$ and $\mathbf{P}$ still have the canonical form then the transformation is called canonical. Here the notation is made that the bold lowercase characters are usually regarded as vectors, whereas the bold capital characters are matrices. However in describing the canonical transformation, $\mathbf{Q}$ and $\mathbf{P}$ are still vectors. To formulate the transformations

Point transformation: $\mathbf{Q} = \mathbf{Q}(\mathbf{q}, t)$ (1.7.1)

Canonical transformation: $\mathbf{Q} = \mathbf{Q}(\mathbf{q}, \mathbf{p}, t), \quad \mathbf{P} = \mathbf{P}(\mathbf{q}, \mathbf{p}, t)$ (1.7.2a,b)

These equations are for time-variant transformations. The simpler case is time-invariant, or termed as steady (time-invariant) transformations

Steady point transformation: $\mathbf{Q} = \mathbf{Q}(\mathbf{q})$ (1.7.3)

Steady canonical transformation: $\mathbf{Q} = \mathbf{Q}(\mathbf{q}, \mathbf{p}), \quad \mathbf{P} = \mathbf{P}(\mathbf{q}, \mathbf{p})$ (1.7.4a,b)

In the application for vibration theory later, the usual method of eigenvector expansion solution in the inertia coordinate (structural dynamics) belongs to the steady point transformation, whereas the eigenvector expansion solution for gyroscopic system belongs to the steady canonical transformation. Steady transformation is easier to deal with.

The canonical transformation requires that the dual differential equations under the dual coordinates $\mathbf{Q}$ and $\mathbf{P}$ after transformation still keep the canonical duality form. This requirement will be satisfied provided that a new function $\mathrm{K}(\mathbf{Q},\mathbf{P},t)$ exists in the transformed dual coordinate system $\mathbf{Q},\mathbf{P}$ such that the dual equations are given as

$$\dot{\mathbf{Q}} = \partial \mathrm{K}/\partial \mathbf{P}, \qquad \dot{\mathbf{P}} = -\partial \mathrm{K}/\partial \mathbf{Q} \tag{1.7.5}$$

The function $\mathrm{K}(\mathbf{Q},\mathbf{P},t)$ plays the role of a Hamilton function in $\mathbf{Q},\mathbf{P}$ coordinate system. Correspondingly, the variational principle appears as

$$\delta \int_{t_0}^{t_f} [\mathbf{P}^T \dot{\mathbf{Q}} - \mathrm{K}(\mathbf{Q},\mathbf{P},t)] \mathrm{d}t = 0 \tag{1.7.6}$$

At the same time, the variational principle under the original coordinate system is

$$\delta \int_{t_0}^{t_f} [\mathbf{p}^T \dot{\mathbf{q}} - \mathrm{H}(\mathbf{q},\mathbf{p},t)] \mathrm{d}t = 0 \tag{1.7.7}$$

The two variational principles describe the same problem, however it is by no means to say that the integrands in both variational principles must be equal. Because at the two ends t_0 and t_f, it is considered no variation for the state variables, hence except a scale transformation, which is not so important [11], the difference between the two integrands is a complete differential, *i.e.*

$$\mathbf{p}^T \dot{\mathbf{q}} - \mathrm{H}(\mathbf{q},\mathbf{p},t) = \mathbf{P}^T \dot{\mathbf{Q}} - \mathrm{K}(\mathbf{Q},\mathbf{P},t) + \mathrm{d}F/\mathrm{d}t \tag{1.7.8}$$

where F is any function of the state space coordinates with second order derivatives, and is useful only when it is a function of $2n$ argument variables, in which ***half of the arguments are the original coordinates and another half are the transformed coordinates***, termed the ***generating function***. The ***first class of generating function*** uses two displacement vectors $\mathbf{q}$ and $\mathbf{Q}$ as arguments

$$F = F_1(\mathbf{q},\mathbf{Q},t) \tag{1.7.9}$$

where F_1 is a function to be selected. Substituting into equation (1.7.8) derives

$$\mathbf{p}^T \dot{\mathbf{q}} - \mathrm{H} = \mathbf{P}^T \dot{\mathbf{Q}} - \mathrm{K} + (\partial F_1/\partial \mathbf{q})^T \dot{\mathbf{q}} + (\partial F_1/\partial \mathbf{Q})^T \dot{\mathbf{Q}} + \partial F_1/\partial t$$

Note that before transformation, $\mathbf{q},\mathbf{p}$ are independent vectors, so there are totally $2n$ independent variables. After transformation, there are still $2n$ independent variables, but there have been $4n$ variables namely $\mathbf{q},\mathbf{p}$ and $\mathbf{Q},\mathbf{P}$. Transformation means that there are $2n$ relations among them. Now treat $\mathbf{q}$ and $\mathbf{Q}$ as independent vectors, so that there are still $2n$ independent variables. Hence the above equation can hold identically zero only if both the coefficients of $\dot{\mathbf{q}}$ and $\dot{\mathbf{Q}}$ vanish, (which supplies $2n$ conditions), *i.e.*

$$\mathbf{p} = \partial F_1/\partial \mathbf{q}, \quad \mathbf{P} = -\partial F_1/\partial \mathbf{Q}, \quad \mathrm{K} = \mathrm{H} + \partial F_1/\partial t \tag{1.7.10a~c}$$

where equation (1.7.10a) implies that $\mathbf{p}$ is a function of $\mathbf{q},\mathbf{Q}$ and t, solving it with respect to $\mathbf{Q}$ derives the equation (1.7.2a). Then substituting the solved $\mathbf{Q}$ into the equation (1.7.10b) gives equation (1.7.2b). The canonical transformation (1.7.2a,b) is thus derived. This derivation explains that ***the canonical transformation depends only on the generating function*** $F_1(\mathbf{q},\mathbf{Q},t)$ ***but not***

relates to the function H. Hence the canonical transformation (1.7.10) can be used to arbitrary Hamilton functions. After an individual Hamilton function $H(\mathbf{q},\mathbf{p},t)$ is selected, the dual variables $\mathbf{q},\mathbf{p}$ are expressed as functions of the new dual variables $\mathbf{Q},\mathbf{P}$. Then substituting into equation (1.7.10c) gives the new Hamilton function $K(\mathbf{Q},\mathbf{P},t)$.

The above canonical transformation is derived from $F_1(\mathbf{q},\mathbf{Q},t)$, called the ***first class of generating function***. If F_1 does not depend on time t explicitly, the transformation is called as a steady canonical transformation and the function K equals to H numerically. Though K equals H numerically, but the expression is completely different, for the arguments are different.

Next, if using the generating function (1.7.9) is inconvenient, the ***second class of generating function*** $F_2(\mathbf{q},\mathbf{P},t)$ can be applied. Let the function F in equation (1.7.8) be

$$F = F_2(\mathbf{q},\mathbf{P},t) - \mathbf{P}^T\mathbf{Q} \tag{1.7.11}$$

substituting into (1.7.8) gives

$$\mathbf{p}^T\dot{\mathbf{q}} - H = -\dot{\mathbf{P}}^T\mathbf{Q} - K + (\partial F_2/\partial\mathbf{q})^T\dot{\mathbf{q}} + (\partial F_2/\partial\mathbf{P})^T\dot{\mathbf{P}} + \partial F_2/\partial t$$

in which both the coefficients of variables $\dot{\mathbf{q}}$ and $\dot{\mathbf{P}}$ should be zero, hence derives

$$\mathbf{p} = \partial F_2/\partial\mathbf{q}, \quad \mathbf{Q} = \partial F_2/\partial\mathbf{P}, \quad K = H + \partial F_2/\partial t \tag{1.7.12a~c}$$

From equation (1.7.12a), $\mathbf{P}$ should be solved as functions of $\mathbf{q},\mathbf{p},t$ which gives equation (1.7.2); then substituting the solved $\mathbf{P}$ into (1.7.12b) and $\mathbf{Q}(\mathbf{q},\mathbf{p},t)$ is obtained. Again, this ***canonical transformation depends only on the function*** $F_2(\mathbf{q},\mathbf{P},t)$ ***but not depends on the function*** H. Finally, for the given Hamilton function $H(\mathbf{q},\mathbf{p},t)$, the new Hamilton function $K(\mathbf{Q},\mathbf{P},t)$ is found from equation (1.7.12c).

The ***third class generating function*** is denoted as $F_3(\mathbf{p},\mathbf{Q},t)$. Let the function F be

$$F = \mathbf{p}^T\mathbf{q} + F_3(\mathbf{p},\mathbf{Q},t) \tag{1.7.13}$$

and substituting it into equation (1.7.8) gives

$$-H = \mathbf{P}^T\dot{\mathbf{Q}} - K + (\partial F_3/\partial\mathbf{p})^T\dot{\mathbf{p}} + (\partial F_3/\partial\mathbf{Q})^T\dot{\mathbf{Q}} + \partial F_3/\partial t + \dot{\mathbf{p}}^T\mathbf{q}$$

The coefficients of variables $\dot{\mathbf{p}}$ and $\dot{\mathbf{Q}}$ should be zero, which derives to

$$\mathbf{q} = -\partial F_3/\partial\mathbf{p}, \quad \mathbf{P} = -\partial F_3/\partial\mathbf{Q}, \quad K = H + \partial F_3/\partial t \tag{1.7.14a~c}$$

The former two in the above equations are the implicit canonical transformation, which is independent on the Hamilton function. Then, $K(\mathbf{Q},\mathbf{P},t)$ is solved for the selected Hamilton function $H(\mathbf{q},\mathbf{p},t)$.

The ***fourth class of generating function*** $F_4(\mathbf{p},\mathbf{P},t)$ is expressed as

$$F = \mathbf{p}^T\mathbf{q} - \mathbf{P}^T\mathbf{Q} + F_4(\mathbf{p},\mathbf{P},t) \tag{1.7.15}$$

similarly

$$\mathbf{q} = -\partial F_4/\partial\mathbf{p}, \quad \mathbf{Q} = \partial F_4/\partial\mathbf{P}, \quad K = H + \partial F_4/\partial t \tag{1.7.16a~c}$$

The four classes of generating function can be used interchangeably [11,12].

As an example, if select the second class of generating function as

$$F_2(\mathbf{q},\mathbf{P},t) = \mathbf{P}^T\mathbf{f}(\mathbf{q},t) + g(\mathbf{q},t) \tag{1.7.17}$$

According to equation (1.7.12b) the new displacement vector becomes

$$\mathbf{Q} = \partial F_2/\partial \mathbf{P} = \mathbf{f}(\mathbf{q},t) \tag{1.7.18}$$

which looked like a point transformation (1.7.1). However according to (1.7.12a)

$$\mathbf{p} = \partial F_2/\partial \mathbf{q} = (\partial \mathbf{f}/\partial \mathbf{q})\mathbf{P} + \partial g/\partial \mathbf{q}$$

from which solves $\mathbf{P} = (\partial \mathbf{f}/\partial \mathbf{q})^{-1}(\mathbf{p} - \partial g/\partial \mathbf{q})$ (1.7.19)

where the differential of vector function $\mathbf{f}$ with respect to the argument vector $\mathbf{q}$ is defined as

$$\frac{\partial \mathbf{f}}{\partial \mathbf{q}} \underset{\text{def}}{=} \begin{bmatrix} \partial f_1/\partial q_1 & \partial f_2/\partial q_1 & \cdots & \partial f_m/\partial q_1 \\ \partial f_1/\partial q_2 & \cdots & \cdots & \partial f_m/\partial q_2 \\ \vdots & & & \vdots \\ \partial f_1/\partial q_n & \partial f_2/\partial q_n & \cdots & \partial f_m/\partial q_n \end{bmatrix} \tag{1.7.20}$$

Here, $\mathbf{f}$ is written as a m-dimensioned vector for generality, as a special case it is $m = n$ for canonical transformation. Since an arbitrary function $g(\mathbf{q},t)$ appears in equation (1.7.19), hence it is not a point transformation, but still canonical.

§1.8, Symplectic description of the canonical transformation

In last section, the canonical transformation is derived from generating function, however, the generating function method is not unique. The canonical transformation can also be described by the symplectic method. The two methods are equivalent, however, are quite different in form. For simplicity, only the time invariant canonical transformation (1.7.4) is described.

The time-invariant canonical transformation does not change the value of Hamilton function, *i.e.* $\mathrm{H}(\mathbf{q},\mathbf{p}) = \mathrm{K}(\mathbf{Q},\mathbf{P})$ as given in the last section. The symplectic form of canonical equations is

$$\dot{\mathbf{v}} = \mathbf{J}\cdot(\partial \mathrm{H}/\partial \mathbf{v}), \quad \text{where } \mathbf{v} = \{\mathbf{q}^T, \mathbf{p}^T\}^T \tag{1.6.17}$$

and is a $2n$-dimensional state vector. The time-invariant canonical transformation (1.7.4) can be expressed as a $2n$-dimensional state vector transformation

$$\boldsymbol{\zeta} = \boldsymbol{\zeta}(\mathbf{v}), \quad \text{where } \boldsymbol{\zeta} = \{\mathbf{Q}^T, \mathbf{P}^T\}^T \tag{1.8.1}$$

The reverse canonical transformation can be expressed as the inverse function, which is certainly time-invariant too

$$\mathbf{v} = \mathbf{v}(\boldsymbol{\zeta}) \tag{1.8.2}$$

Substituting the above equation into (1.6.17) gives

$$(\partial \mathbf{v}/\partial \boldsymbol{\zeta})^T \dot{\boldsymbol{\zeta}} = \mathbf{J}(\partial \boldsymbol{\zeta}/\partial \mathbf{v})\cdot(\partial \mathrm{K}/\partial \boldsymbol{\zeta}) \tag{1.8.3}$$

On the other hand, substituting the vector $\boldsymbol{\zeta}$ in (1.8.1) into equation (1.8.2) gives identical transformation, *i.e.* $\mathbf{v}(\boldsymbol{\zeta}(\mathbf{v})) = \mathbf{v}$, for which taking partial differential with respect to $\mathbf{v}$ gives

$$(\partial\mathbf{v}/\partial\boldsymbol{\zeta})^T(\partial\boldsymbol{\zeta}/\partial\mathbf{v})^T = \mathbf{I}_{2n}, \quad \text{or} \quad (\partial\boldsymbol{\zeta}/\partial\mathbf{v})\cdot(\partial\mathbf{v}/\partial\boldsymbol{\zeta}) = \mathbf{I}_{2n}$$

and the equation (1.8.3) becomes

$$\dot{\boldsymbol{\zeta}} = (\partial\boldsymbol{\zeta}/\partial\mathbf{v})^T\mathbf{J}(\partial\boldsymbol{\zeta}/\partial\mathbf{v})\cdot(\partial\mathrm{K}/\partial\boldsymbol{\zeta}) = \mathbf{S}^T\mathbf{J}\mathbf{S}\cdot(\partial\mathrm{K}/\partial\boldsymbol{\zeta})$$

where the matrix $\mathbf{S}$ is defined as

$$\mathbf{S} \underset{\text{def}}{=} \partial\boldsymbol{\zeta}/\partial\mathbf{v} \tag{1.8.4}$$

Canonical transformation requires that after transformation, the symplectic expression for canonical equations are still of the form (1.6.17), *i.e.*

$$\dot{\boldsymbol{\zeta}} = \mathbf{J}\cdot(\partial\mathrm{K}/\partial\boldsymbol{\zeta}) \tag{1.8.5}$$

From which derives the condition for canonical transformation (1.8.1)

$$\mathbf{S}^T\mathbf{J}\mathbf{S} = \mathbf{J} \tag{1.8.6}$$

It is defined that ***any matrix satisfying the condition (1.8.6) is said to be a symplectic matrix.***

The derivation above has not mentioned generating function, but directly proposed the condition for canonical transformation that the partial differential of the transformation matrix $\mathbf{S}$ in equation (1.8.4) should be a symplectic matrix. The two methods for deriving canonical transformation, *i.e.* 1) deriving by generating function or 2) using symplectic matrix, get the equivalent results in fact. However, only time-invariant canonical transformation is derived above with the symplectic matrix method. For time-variant canonical transformation, please confer references [11,14] and the papers cited there. The time-invariant canonical transformation can solve quite a number of problems.

The symplectic matrices will be used quite frequently in what follows, hence a list of its behavior is necessary. First taking the determinant to its definition equation (1.8.6), because the determinant value of matrix $\mathbf{J}$ is 1, and the determinants of both the original and transpose matrices are equal, hence $\det(\mathbf{S}) = \pm 1$. Therefore symplectic matrices can be classified into **two classes**, the first class is $\det(\mathbf{S}) = 1$ and the other class is -1. Recall that the orthogonal matrices can also be classified as two classes according to the determinant being 1 or -1, which is the similar situation. Because its determinant not equal to zero, the symplectic matrix must have its inverse matrix.

It is easy to verify that $\mathbf{J}$ and $\mathbf{I}_{2n}$ are all symplectic matrices.

The transpose of a symplectic matrix is again symplectic. To prove the statement, taking inversion for the equation (1.8.6) gives $\mathbf{S}^{-1}\mathbf{J}\mathbf{S}^{-T} = \mathbf{J}$ then left multiplying with $\mathbf{S}$ and right multiplying with $\mathbf{S}^T$ derives $\mathbf{J} = \mathbf{S}\mathbf{J}\mathbf{S}^T$, so $\mathbf{S}^T$ is a symplectic matrix.

The inverse matrix of a symplectic matrix is symplectic.

The product of two symplectic matrices is symplectic, and $\mathbf{I}_{2n}$ is a unit element.

The multiplication of symplectic matrices is the usual matrix multiplication, so the associative rule applies

$$(\mathbf{S}_1\mathbf{S}_2)\mathbf{S}_3 = \mathbf{S}_1(\mathbf{S}_2\mathbf{S}_3) = \mathbf{S}_1\mathbf{S}_2\mathbf{S}_3$$

Hence the ***symplectic matrices compose a group***. **All the symplectic matrices with determinant being 1 compose a normal subgroup**. Also all the symplectic matrices with determinant being -1 compose another normal subgroup. The situation is very similar to the orthogonal matrices.

§1.9, Poisson bracket

Definition: for two arbitrary functions of the canonical dual vectors $\mathbf{q}$ and $\mathbf{p}$, *i.e.* $u_1(\mathbf{q},\mathbf{p})$ and $u_2(\mathbf{q},\mathbf{p})$, their Poisson bracket is defined as a bilinear expression

$$[u_1,u_2]_{\mathbf{q},\mathbf{p}} \underset{\text{def}}{=} \left(\frac{\partial u_1}{\partial \mathbf{q}}\right)^T \frac{\partial u_2}{\partial \mathbf{p}} - \left(\frac{\partial u_1}{\partial \mathbf{p}}\right)^T \frac{\partial u_2}{\partial \mathbf{q}} \tag{1.9.1}$$

The value of a ***Poisson bracket*** is a scalar, where u_1,u_2 can also be functions of time, the argument t here is only a parameter. Using symplectic notation is perhaps simpler that

$$[u_1,u_2]_{\mathbf{v}} = (\partial u_1/\partial \mathbf{v})^T \cdot \mathbf{J} \cdot (\partial u_2/\partial \mathbf{v}), \quad \text{where} \quad \mathbf{v} = \{\mathbf{q}^T,\mathbf{p}^T\}^T \tag{1.9.2}$$

If the functions u_1,u_2 are selected directly from the components of the canonical variables, it is easily verified that

$$\begin{aligned} &[q_i,q_j]_{\mathbf{v}} = 0, \qquad [p_i,p_j]_{\mathbf{v}} = 0, \quad i,j = 1,\cdots,n \\ &[q_j,p_k]_{\mathbf{v}} = \delta_{jk}, \quad [p_j,q_k]_{\mathbf{v}} = -\delta_{jk}, \quad j,k = 1,\cdots,n \end{aligned} \tag{1.9.3}$$

These canonical variables (arguments), $\mathbf{q}$ and $\mathbf{p}$, are all components of the state vector $\mathbf{v}$. Composing the Poisson bracket equation (1.9.3) as a $2n \times 2n$ matrix gives

$$[\mathbf{v},\mathbf{v}]_{\mathbf{v}} = \mathbf{J} \tag{1.9.3'}$$

Now selecting the functions u_1,u_2 from the components of canonical transformed dual vectors $\mathbf{Q}$ and $\mathbf{P}$, or from $\zeta = \{\mathbf{Q}^T,\mathbf{P}^T\}^T$, then the Poisson bracket matrix can be computed as

$$[\zeta,\zeta]_{\mathbf{v}} = (\partial\zeta/\partial\mathbf{v})^T \mathbf{J} (\partial\zeta/\partial\mathbf{v}) = \mathbf{S}^T\mathbf{J}\mathbf{S} = \mathbf{J} \tag{1.9.4}$$

because of the equation (1.8.4) and that $\mathbf{S}$ is a symplectic matrix. Conversely, if equation (1.9.4) is valid then the transformation is canonical. The Poisson bracket matrix for the canonical variables is called as the fundamental Poisson bracket matrix. Equation (1.9.4) explains that the fundamental Poisson bracket matrix keeps unchanged under canonical transformations. In other words, the fundamental Poisson brackets are invariant under canonical transformation, and the invariance is thus in all ways equivalent to the symplectic condition for a canonical transformation.

It is easy to show that the Poisson bracket is invariant under any canonical transformation. Let $u_1(\mathbf{q},\mathbf{p})$ and $u_2(\mathbf{q},\mathbf{p})$ be two functions, because

$$(\partial u/\partial \mathbf{v}) = (\partial\zeta/\partial\mathbf{v})\partial u/\partial\zeta = \mathbf{S}\,\partial u/\partial\zeta$$

then $[u_1,u_2]_{\mathbf{v}} = (\partial u_1/\partial \mathbf{v})^T \cdot \mathbf{J} \cdot (\partial u_2/\partial \mathbf{v}) = (\partial u_1/\partial\zeta)^T \mathbf{S}^T\mathbf{J}\mathbf{S}(\partial u_2/\partial\zeta)$

Because the transformation from $\mathbf{v}$ to ζ is canonical, so that $\mathbf{S}$ is a symplectic matrix satisfying (1.8.6), so

$$[u_1,u_2]_{\mathbf{v}} = (\partial u_1/\partial \zeta)^T \cdot \mathbf{J} \cdot (\partial u_2/\partial \zeta) = [u_1,u_2]_\zeta \tag{1.9.5}$$

Thus the Poisson bracket has the same value when evaluated with respect to any canonical set of variables, that all Poisson brackets are canonical invariant. Hence the indication subscript $\mathbf{v}$ to ζ of the Poisson bracket is immaterial and can be taken off thereafter.

The essence of Hamilton canonical equations is that its form is unchanged under canonical transformations. Similarly the Poisson bracket is also unchanged under canonical transformation, which implies that the canonical equations can be expressed with Poisson bracket as

$$\dot{\mathbf{q}} = [\mathbf{q},\mathrm{H}], \quad \dot{\mathbf{p}} = [\mathbf{p},\mathrm{H}], \quad \text{or} \quad \dot{\mathbf{v}} = [\mathbf{v},\mathrm{H}] \tag{1.9.6}$$

where inside Poisson bracket, the former is a vector and the latter is a scalar of Hamilton function H, and the result is again a vector. So (1.9.6) is a $2n$-vector equation. Using the symplectic expression of Poisson bracket gives

$$[\mathbf{v},\mathrm{H}] = \mathbf{J}\,\partial\mathrm{H}/\partial\mathbf{v} \tag{1.9.7}$$

In dynamic equation (1.9.6) and (1.9.7), the time derivatives of canonical variables are expressed using Poisson bracket. For arbitrary function $u(\mathbf{q},\mathbf{p},t)$, its differential with respect to time t is

$$\begin{aligned} \dot{u} = \mathrm{d}u/\mathrm{d}t &= (\partial u/\partial\mathbf{q})^T\dot{\mathbf{q}} + (\partial u/\partial\mathbf{p})^T\dot{\mathbf{p}} + \partial u/\partial t \\ &= (\partial u/\partial\mathbf{q})^T(\partial\mathrm{H}/\partial\mathbf{p}) - (\partial u/\partial\mathbf{p})^T(\partial\mathrm{H}/\partial\mathbf{q})\partial u/\partial t \\ &= [u,\mathrm{H}] + \partial u/\partial t \end{aligned} \tag{1.9.8}$$

or

$$\dot{u} = (\partial u/\partial\mathbf{v})^T\dot{\mathbf{v}} + \partial u/\partial t = (\partial u/\partial\mathbf{v})^T\mathbf{J}(\partial\mathrm{H}/\partial\mathbf{v}) + \partial u/\partial t \tag{1.9.8'}$$

which is the symplectic expression of complete time derivative. If the function $u(\mathbf{q},\mathbf{p},t)$ is selected as the Hamilton function $\mathrm{H}(\mathbf{q},\mathbf{p},t)$, it derives

$$\mathrm{dH}/\mathrm{d}t = \partial\mathrm{H}/\partial t$$

Because $\mathbf{v}_a^T\mathbf{J}\mathbf{v}_a \equiv 0$ is valid for arbitrary vector $\mathbf{v}_a$.

The dual coordinates $\mathbf{q},\mathbf{p}$ are used to describe motion, whereas Hamilton function $\mathrm{H}(\mathbf{q},\mathbf{p},t)$ is for an individual motion. It can be interpreted as, that the Hamilton function $\mathrm{H}(\mathbf{q},\mathbf{p},t)$ generates a motion.

§1.9.1, Algebra of Poisson bracket

The importance of Poisson bracket has been shown in the previous section, hence the algebra for Poisson bracket is very interested. First, it is skew-symmetric, as

$$[u_1,u_2] = -[u_2,u_1], \quad [u,u] = 0 \tag{1.9.9}$$

Next, it is linear and distributive, that

$$[au_1 + bu_2, u_3] = a[u_1,u_3] + b[u_2,u_3] \tag{1.9.10}$$

where a,b are arbitrary constants and u_1,u_2,u_3 are arbitrary functions of $\mathbf{q},\mathbf{p},t$. Thirdly

$$[u_1 \cdot u_2, u_3] = [u_1, u_3]u_2 + u_1[u_2, u_3] \tag{1.9.11}$$

In addition, Poisson bracket has another behavior, the ***Jacobi identity***

$$[u,[v,w]] + [v,[w,u]] + [w,[u,v]] \equiv 0 \tag{1.9.12}$$

where a Poisson bracket appears as an argument in the Poisson bracket, a double Poisson bracket. For functions $u(\mathbf{q},\mathbf{p},t), v(\mathbf{q},\mathbf{p},t), w(\mathbf{q},\mathbf{p},t)$ having continuous second derivatives, the sum of their cyclic permutations of the double Poisson bracket is zero. This proposition is proved below.

The first term in (1.9.12) has only first-order partial derivatives for u, the second-order partial derivatives for u only appear in the second and third terms in equation (1.9.12). Hence in the expanded expression of (1.9.12), all terms are the product of two first-order partial derivative factors and one second-order partial derivative. Combining according to the second-order partial derivative, if all the coefficients are zero then the equation equals zero.

Expanding the third term in (1.9.12) and denoting ζ as canonical arguments, the equation (1.9.2) becomes

$$[u,v] = (\partial u/\partial \zeta)^T \mathbf{J}(\partial v/\partial \zeta)$$

again a function of ζ. The third double Poisson bracket in (1.9.12) is derived as

$$[w,[u,v]] = (\frac{\partial w}{\partial \zeta})^T \mathbf{J}\frac{\partial}{\partial \zeta}\left[\left(\frac{\partial u}{\partial \zeta}\right)^T \mathbf{J}\frac{\partial v}{\partial \zeta}\right] = \left(\frac{\partial w}{\partial \zeta}\right)^T \mathbf{J}\left(\frac{\partial^2 u}{\partial \zeta \partial \zeta}\mathbf{J}\frac{\partial v}{\partial \zeta} - \frac{\partial^2 v}{\partial \zeta \partial \zeta}\mathbf{J}\frac{\partial u}{\partial \zeta}\right)$$

where only the first term in bracket appears the second-order partial derivative of u. Note $\partial^2 u/\partial \zeta \partial \zeta$ is a symmetric $2n \times 2n$ matrix. Cyclic permuting the functions u, v, w gives

$$[v,[w,u]] = (\partial v/\partial \zeta)^T \mathbf{J}\big((\partial^2 w/\partial \zeta \partial \zeta)\mathbf{J}(\partial u/\partial \zeta) - (\partial^2 u/\partial \zeta \partial \zeta)\mathbf{J}(\partial w/\partial \zeta)\big)$$

where second-order partial derivative of u has only one term, which is a scalar, taking transpose gives

$$-(\partial v/\partial \zeta)^T \mathbf{J}(\partial^2 u/\partial \zeta \partial \zeta)\mathbf{J}(\partial w/\partial \zeta) = -(\partial w/\partial \zeta)^T \mathbf{J}(\partial^2 u/\partial \zeta \partial \zeta)\mathbf{J}(\partial v/\partial \zeta)$$

Hence in the equation (1.9.2) two terms of second-order partial derivative for u cancelled each other. With the same reason, the second-order partial derivative terms for v and for w cancel too. Therefore the Jacobi identity is proved.

If the Poisson bracket of two functions u_1 and u_2 is defined as a 'product' of the two functions, then the Jacobi identity is the replacement for the associative law of multiplication. It is remembered that the ordinary multiplication is associative, such as for matrix multiplication $(\mathbf{AB})\mathbf{C} = \mathbf{A}(\mathbf{BC})$, which says that the order of multiplication is immaterial. However, if Poisson bracket operation is regarded as the multiplication operation, the corresponding associative rule no longer applies, that $[u,[v,w]] \neq [[u,v],w] = -[w,[u,v]]$. The ***Jacobi identity applies instead of the associative rule***.

Therefore the equations (1.9.9~12) proposed a kind of non-associative algebra, called the ***Lie algebra***. The Poisson bracket is not the unique Lie algebra, if the cross multiplication of matrices $\mathbf{A}$ and $\mathbf{B}$

$$[\mathbf{A},\mathbf{B}] \underset{\text{def}}{=} \mathbf{AB}-\mathbf{BA}$$

is treated as the multiplication operation, it gives another Lie algebra.

§1.10, Action

In § 1.5, the Hamilton principle is proposed based on the ***action*** integral

$$S=\int_{t_0}^{t_f}\mathrm{L}(\mathbf{q},\dot{\mathbf{q}},t)\mathrm{d}t \tag{1.10.1}$$

The Hamilton principle stated that as the configuration vector $\mathbf{q}$ varies nearby the real solution $\mathbf{q}_*$ in the interval $t_0<t<t_f$ with $\mathbf{q}$ given at both ends t_0,t_f, the first variation of action S being zero derives the Euler-Lagrange dynamic equation.

However, the orbit $\mathbf{q}$ discussed below is a real one, *i.e.* $\mathbf{q}_*=\mathbf{q}$. The subscript $_*$ will be cancelled for ease of notation. Now let the displacement vector $\mathbf{q}_f$ at the end point t_f varies, such that $\mathbf{q}(t)$ varies correspondingly in the time interval (t_0,t_f), the variation of S is to analyze. Further, the ***action*** of an arbitrary interval (t_1,t_2) within $[t_0,t_f]$ is to consider, that the action S is considered as a function of the two end generalized displacements $\mathbf{q}_1$ and $\mathbf{q}_2$, *i.e.* the function $S(\mathbf{q}_1,\mathbf{q}_2)$ is to be analyzed.

Carrying out the variational derivation

$$\delta S=[(\partial\mathrm{L}/\partial\dot{\mathbf{q}})^T\cdot\delta\mathbf{q}]_{t_1}^{t_2}+\int_{t_1}^{t_2}[\partial\mathrm{L}/\partial\mathbf{q}-(\mathrm{d}/\mathrm{d}t)(\partial\mathrm{L}/\partial\dot{\mathbf{q}})]\mathrm{d}t$$

since the real solution $\mathbf{q}(t)$ is used within the interval (t_1,t_2), the integrand in the above equation equals zero. If the initial displacement vector $\mathbf{q}_1$ is given, only $\delta\mathbf{q}_2$ varies, then

$$\delta S(\mathbf{q}_2)=[(\partial\mathrm{L}/\partial\dot{\mathbf{q}})^T\delta\mathbf{q}]_{t_2}=\mathbf{p}_2^T\delta\mathbf{q}_2 \tag{1.10.2}$$

The action S is now a function of the end time displacement $\mathbf{q}$ (dropping the subscript 2), then

$$\partial S/\partial\mathbf{q}=\mathbf{p} \tag{1.10.3}$$

The above derivation assumes t_2 unchanged and the variation applies only to the displacement $\mathbf{q}_2$ at t_2 for the action $S(\mathbf{q}_2,t_2)$. Next, let t_2 varies to $t_2+\delta t$ and let $\mathbf{q}_2$ extends smoothly, *i.e.*

$$\delta\mathbf{q}_2=\dot{\mathbf{q}}_2\delta t$$

which implies that the orbit internal to time interval is unchanged, so $\delta S=\mathrm{L}\delta t$, or

$$\dot{S}=\mathrm{d}S/\mathrm{d}t=\mathrm{L} \tag{1.10.4}$$

However, if the end time t_2 varies δt_2 with $\mathbf{q}_2$ keeps unchanged, then the internal orbit varies too, such variation is $\partial S/\partial t$. Combining gives

$$\dot{S}=\partial S/\partial t+(\partial S/\partial\mathbf{q})^T\dot{\mathbf{q}}=\partial S/\partial t+\mathbf{p}^T\dot{\mathbf{q}} \tag{1.10.5}$$

Hence

$$\partial S/\partial t = \mathrm{L} - \mathbf{p}^T\dot{\mathbf{q}} = -\mathrm{H}(\mathbf{q},\mathbf{p},t) \tag{1.10.6}$$

The action S can be treated as a both ends t_1,t_2 varied function $S(\mathbf{q}_1,t_1;\mathbf{q}_2,t_2)$, that

$$\mathrm{d}S = \mathbf{p}_2^T\mathrm{d}\mathbf{q}_2 - \mathrm{H}_2\mathrm{d}t - \mathbf{p}_1^T\mathrm{d}\mathbf{q}_1 + \mathrm{H}_1\mathrm{d}t \tag{1.10.7}$$

The action function is very important in applications and has different terms in different fields.

§1.11, Hamilton-Jacobi equation

The equation (1.10.6) is practically the Hamilton-Jacobi equation, where S is a function of right end arguments $\mathbf{q},t$. However, in the Hamilton function there is $\mathbf{p}$, which is the momentum at the right end and can be substituted with equation (1.10.3), which derives

$$\partial S/\partial t + \mathrm{H}(\mathbf{q},\partial S/\partial \mathbf{q},t) = 0 \tag{1.11.1}$$

This is a first order PDE, called the ***Hamilton-Jacobi equation***, to be satisfied by the action function $S(\mathbf{q},t)$.

Mathematically, equation (1.11.1) has $n+1$ independent variables (arguments), composed of a n-dimensional generalized displacement vector $\mathbf{q}$ and 1 time variable t. The ***complete integral*** of this PDE should have $n+1$ ***independent arbitrary constants***, see [1,11~13]. Because the unknown function S appears in equation (1.11.1) only with its partial differential, and the function S can be added with an arbitrary constant A, so its complete integral has the following form

$$S = S(\mathbf{q},\alpha_1,\alpha_2,\cdots,\alpha_n,t) + A = S(\mathbf{q},\boldsymbol{\alpha},t) + A \tag{1.11.2}$$

where $\alpha_1,\alpha_2,\cdots,\alpha_n,A$ are arbitrary constants (parameters). Evidently, A is irrelevant.

After found the complete integral (1.11.2), how to derive the integration of the dynamic equation is of concern. The canonical transformation with generating function can serve for this purpose. The initial conditions at $t=t_0$ are $\mathbf{q}_0$ and $\mathbf{p}_0$ given. The action $S(\mathbf{q},t,\alpha_1,\alpha_2,\cdots,\alpha_n)$, *i.e.* the ***complete integral*** of Hamilton-Jacobi equation, is regarded as the first class of generating function $F_1(\mathbf{q},\mathbf{Q},t)$, where the arbitrary constants α_i in S are treated as Q_i

$$Q_i = \alpha_i, \qquad i = 1,2,\cdots,n \tag{1.11.3}$$

Then the three transformation equations (1.7.10a~c) are applied. The first equation (1.7.10a) is

$$\mathbf{p} = \partial S/\partial \mathbf{q} \tag{1.11.4a}$$

which is valid for arbitrary time t. Based on the initial conditions at $t=t_0$ with $\mathbf{p}_0,\mathbf{q}_0$ given, the n equations of (1.11.4a) at $t=t_0$ can solve all the parameters $Q_i \equiv \alpha_i(\mathbf{q}_0,\mathbf{p}_0,t_0)$. Next, the equation (1.7.10b) becomes

$$P_i = -\partial S/\partial Q_i = -\partial S/\partial \alpha_i \tag{1.11.4b}$$

As the Hamilton function $\mathrm{K}(\mathbf{Q},\mathbf{P},t)$ after canonical transformation, since F_1 is selected as S, the right hand side of equation (1.7.10c) is just the Hamilton-Jacobi equation (1.11.1), so $\mathrm{K}=0$ and then $\dot{\mathbf{P}}=\mathbf{0}$. This result is due to the selection of complete integral of Hamilton-Jacobi equation as the first class of generating function. Hence, the generalized momentum $\mathbf{P}$ after transformation is also a constant vector, denoted as $\boldsymbol{\beta}$

$$\mathbf{P}=-\partial S(\mathbf{q},\boldsymbol{\alpha},t)/\partial\boldsymbol{\alpha}=\boldsymbol{\beta} \tag{1.11.5}$$

Let $t=t_0$, substituting $\mathbf{q}_0$ and the constant vector $\boldsymbol{\alpha}(\mathbf{q}_0,\mathbf{p}_0,t_0)$ just solved into the above equation, the constant $\boldsymbol{\beta}$ is computed as a vector function of $(\mathbf{q}_0,\boldsymbol{\alpha},t)$. Thereafter, solving the vector $\mathbf{q}$ from equation (1.11.5) for arbitrary time t gives the orbit $\mathbf{q}(\boldsymbol{\alpha},\boldsymbol{\beta},t)$. Afterwards, substituting $\mathbf{q}(\boldsymbol{\alpha},\boldsymbol{\beta},t)$ into (1.11.4a) gives $\mathbf{p}(\boldsymbol{\alpha},\boldsymbol{\beta},t)$. The n-dimensional constant vector $\alpha_i(\mathbf{q}_0,\mathbf{p}_0,t_0)$, which has been expressed with the initial condition and $\boldsymbol{\beta}$, can also be computed from the initial condition, therefore the arbitrary vectors $\boldsymbol{\alpha}$ and $\boldsymbol{\beta}$ can be used instead of the initial condition. Hence, $S(\mathbf{q},\boldsymbol{\alpha},t)$ is called as a ***complete integral***.

The above solution is based on the ***complete integral***, however sometimes, only incomplete integral of the Hamilton-Jacobi equation is found, *i.e.* the integral involves only $m<n$ arbitrary constants. In such case, although the general integration of the motion cannot be found readily, but the integration problem can be simplified. Let the integration function S involve arbitrary constants $\alpha_i, i=1,2,\cdots,m$, from the equation

$$\partial S/\partial\boldsymbol{\alpha}_m=\boldsymbol{\beta}_m \tag{1.11.6}$$

Both $\boldsymbol{\alpha}_m$ and $\boldsymbol{\beta}_m$ are m-dimensional arbitrary constant vectors, and the above relation gives the equations connecting $q_1,\cdots,q_n$ and t. The problem is simplified but not solved yet.

§1.11.1, A simple harmonic oscillator

A one-dimensional oscillator is used as an example. Its Hamilton function is

$$\mathrm{H}(q,p)=(p^2+m^2\omega^2q^2)/(2m), \qquad \omega=\sqrt{k/m} \tag{1.11.7}$$

where m,k are mass and spring constants respectively, and ω is the circular frequency. Substituting the p with $\partial S/\partial q$ derives the Hamilton-Jacobi equation as

$$\partial S/\partial t+[(\partial S/\partial q)^2+m^2\omega^2q^2]/(2m)=0 \tag{1.11.8}$$

The time t appears only in $\partial S/\partial t$, so the solution must have the form

$$S(q,\alpha,t)=W(q,\alpha)-\alpha\, t \tag{1.11.9}$$

where α is an arbitrary integration constant. The equation for W is

$$[(\partial W/\partial q)^2+m^2\omega^2q^2]/(2m)=\alpha \tag{1.11.10}$$

and α is interpreted as conserved mechanical energy. Integration for W gives

$$W = \sqrt{2m\alpha}\int\sqrt{1 - m\omega^2 q^2/(2\alpha)}\mathrm{d}q$$
$$= \frac{\alpha}{\omega}\left[\arcsin\left(q\sqrt{\frac{m\omega^2}{2\alpha}}\right) + \sqrt{\frac{m\omega^2}{2\alpha}}q\sqrt{1 - \frac{m\omega^2}{2\alpha}q^2}\right] \qquad (1.11.10')$$

and
$$S = \sqrt{2m\alpha}\int\sqrt{1 - m\omega^2 q^2/(2\alpha)}\mathrm{d}q - \alpha t$$

Then from equation (1.11.5) derives

$$\beta = -\partial S/\partial\alpha = t - \sqrt{2m/\alpha}\int \mathrm{d}q\Big/\sqrt{1 - m\omega^2 q^2/(2\alpha)} = t - \arcsin(q\sqrt{m\omega^2/2\alpha})/\omega$$

from which, q is solved as a function of time t with integration constants α, β

$$q = \sqrt{2\alpha/(m\omega^2)}\sin\omega(t - \beta) \qquad (1.11.11)$$

the well-known solution of simple harmonic oscillator. From equation (1.11.4) gives the momentum

$$p = \partial S/\partial q = \partial W/\partial q = \sqrt{2m\alpha - m^2\omega^2 q^2}$$

Substituting equation (1.11.11) into the above equation gives

$$p = \sqrt{2m\alpha}\cos\omega(t - \beta)$$

This momentum p coincides with $m\dot{q}$.

The constants α, β should be determined with the initial condition q_0, p_0. Firstly, it derives

$$p_0^2/2m\alpha + m\omega^2 q_0^2/2\alpha = 1$$

then $\mathrm{tg}\omega\beta = -(q_0/\sqrt{2\alpha/m\omega^2})/(p_0/\sqrt{2m\alpha}) = -m\omega\, q_0/p_0$ gives the phase angle β_0.

One dimensional simple harmonic oscillator is a simplest problem. The expressions for W and S looked complicated. Direct integrating the dynamic equation is simpler. However, the Hamilton-Jacobi theory gives deeper insight. Such canonical transformation method can be applied to the solution of non-linear differential equations, which is given in the next chapter.

§1.11.2, Time invariant system

When the Hamilton function $\mathrm{H}(\mathbf{q},\mathbf{p})$ does not involve the time t explicitly, the system is time-invariant. Quite a number of application problems are time-invariant, for which equation (1.11.1) reduces to

$$\partial S/\partial t + \mathrm{H}(\mathbf{q}, \partial S/\partial\mathbf{q}) = 0$$

The time appears explicitly only in the first term, so that the action function has the form

$$S(\mathbf{q},\boldsymbol{\alpha},t) = W(\mathbf{q},\boldsymbol{\alpha}) - \alpha_1 t \qquad (1.11.12)$$

Substituting into (1.11.1) derives the Hamilton-Jacobi ***characteristic equation***

$$\mathrm{H}(\mathbf{q}, \partial W/\partial\mathbf{q}) = \alpha_1 \qquad (1.11.13)$$

In this equation time t does not appear explicitly, which indicates conservation of the Hamilton function. The vector $\boldsymbol{\alpha}$ is a n-dimensional constant vector and its first component α_1 is the conserved energy. The function W is a part of action, termed as the Hamilton ***characteristic function***. Let $W(\mathbf{q},\boldsymbol{\alpha})$ be treated as the second class of generating function $F_2(\mathbf{q},\mathbf{P})$ and the constant vector $\boldsymbol{\alpha}$ be regarded as the transformed canonical momentum $\mathbf{P}$. Hence

$$\mathbf{p} = \partial W/\partial \mathbf{q}, \quad \mathbf{Q} = \partial W/\partial \mathbf{P} = \partial W/\partial \boldsymbol{\alpha} \quad , \quad \mathrm{K} = \alpha_1 \tag{1.11.14}$$

which are just equations (1.7.12a~c). The dual equations after canonical transformation are

$$\dot{\mathbf{P}} = -\partial \mathrm{K}/\partial \mathbf{Q} = \mathbf{0}, \quad \dot{\mathbf{Q}} = \partial \mathrm{K}/\partial \mathbf{P} = \partial \mathrm{K}/\partial \boldsymbol{\alpha} = \{1,0,0,\cdots 0\}^T$$

The first equation coincides $\mathbf{P} = \boldsymbol{\alpha}$, and the integration of the second equation gives

$$Q_1 = t + \beta_1 = \partial W/\partial \alpha_1, \quad Q_i = \beta_i = \partial W/\partial \alpha_i, \quad (i > 1) \tag{1.11.15}$$

where β_i is the integration constants. Therefore after transformation, except the generalized displacement Q_1 being linear function of t, all dual variables are constants. The original displacement function $\mathbf{q}$ can be solved from the equation set (1.11.15) and $\mathbf{p}$ can be obtained from substituting the solved $\mathbf{q}$ into the equation $\mathbf{p} = \partial W/\partial \mathbf{q}$.

The feature of characteristic function is independent on time t explicitly, its complete differential is

$$\mathrm{d}W/\mathrm{d}t = (\partial W/\partial \mathbf{q})^T \dot{\mathbf{q}} = \mathbf{p}^T \dot{\mathbf{q}}, \quad W = \int \mathbf{p}^T \mathrm{d}\mathbf{q} \tag{1.11.16}$$

But for the action function S,

$$\dot{S} = \mathbf{p}^T \dot{\mathbf{q}} - \mathrm{H}(\mathbf{q},\mathbf{p},t) \ , \quad S = \int [\mathbf{p}^T \dot{\mathbf{q}} - \mathrm{H}]\mathrm{d}t$$

Hence W is also called as the abbreviated action function.

The Hamilton-Jacobi theory appears quite graceful however, the difficulty is *how to find the complete integral of the action function S or the characteristic function W*. The constant valued vector $\boldsymbol{\alpha}$ in S of (1.11.2) or in W of equation (1.11.12) is not only a given vector, but also a vector of parametric variables that differentiation with respect of $\boldsymbol{\alpha}$ is necessary. Solving non-linear PDE is far more difficult than solving ODE-s.

However, if the problem can separate variables, the Hamilton-Jacobi theory will be quite helpful.

§1.11.3, Separation of variables

If in the Hamilton-Jacobi equation there is a coordinate, say q_1, and its respective partial differential $\partial S/\partial q_1$ appear only with the combination form of $\varphi(q_1, \partial S/\partial q_1)$, and in the combination function φ, there is no other coordinate (or time) or derivative, *i.e.* the Hamilton-Jacobi equation has the form:

$$\Phi\left\{ q_i, t, \frac{\partial S}{\partial q_i}, \frac{\partial S}{\partial t}; \varphi(q_1, \frac{\partial S}{\partial q_1}) \right\} = 0 \tag{1.11.17}$$

where the q_i means all the other coordinates except q_1. In such case, the coordinate q_1 is separable, [13]. The solution can be found in the form

$$S = S'(q_i, t) + S_1(q_1) \tag{1.11.18}$$

where in the function S_1 there is only q_1 and possibly there are other integration constants. Therefore the variable q_1 is separated with other q_i. Substituting (1.11.18) into (1.11.17) gives

$$\Phi\left\{q_i, t, \frac{\partial S'}{\partial q_i}, \frac{\partial S'}{\partial t}; \varphi(q_1, \frac{\mathrm{d}S_1}{\mathrm{d}q_1})\right\} = 0 \tag{1.11.19}$$

This equation is valid for arbitrary value of q_1, however when q_1 varies only the function φ can be varied, hence it must be

$$\varphi\left(q_1, \mathrm{d}S_1/\mathrm{d}q_1\right) = \alpha_1 \tag{1.11.20}$$

$$\Phi\left\{q_i, t, \frac{\partial S'}{\partial q_i}, \frac{\partial S'}{\partial t}, \alpha_1\right\} = 0 \tag{1.11.21}$$

where α_1 is an arbitrary constant. The above separated equation (1.11.20) has been an ordinary differential equation of S_1 with respect to the argument q_1. Solving the function $S_1(q_1)$ excludes the argument q_1 in the partial differential equation (1.11.21).

If all the generalized coordinates $q_i, i = 1, \cdots, n$ and time t are separated, finding the complete solution for the Hamilton-Jacobi equation is reduced to solve n separated ordinary differential equations. For a conservative system, the time t can be separated first, that

$$S(\mathbf{q}, \alpha, t) = W(\mathbf{q}, \alpha) + S_0(t, \alpha) \tag{1.11.22}$$

Because the time t does not appear explicitly in H, the Hamilton-Jacobi equation becomes

$$\mathrm{H}(\mathbf{q}, \partial W/\partial \mathbf{q}) + \partial S_0/\partial t = 0$$

Since S_0 is independent to $\mathbf{q}$ so that

$$\partial S_0/\partial t = -\alpha_0 \quad , \qquad S_0 = -\alpha_0 t \tag{1.11.23a}$$

$$\mathrm{H}(\mathbf{q}, \partial W/\partial \mathbf{q}) = \alpha_0 \tag{1.11.23b}$$

the time t is thus separated. Certainly, under appropriate conditions other variables can also be separated.

After the separation of variables, the ODE (1.11.20) is obtained, however its integration is still not easily obtained, because the n arbitrary constants must be selected arbitrarily. The variable separation is only a rare case. Only for some special problems, the separation of variable can be realized in the properly selected generalized coordinates, such as for problems of cyclic coordinate, central potential force field etc.

§1.11.4, Separation of variables for linear systems

Theoretically, Hamilton-Jacobi theory is certainly applicable both for linear and non-linear problems. However, the solution for a PDE is very difficult. A general non-linear problem solution is too complicated in practice. In applications, the first step is to solve based on the linear theory. For a non-linear problem, which is very difficult to solve, the approximate solution is usually to find based on the respective linearized solution. After solved the linear problem, the perturbation method, averaging method etc. can be used to find a better approximate solution. Only for time-invariant linear systems, the Hamilton function is quadratic, for which the method of separation of variables always works.

Let us consider first the small vibration problem in an inertia coordinate system. The kinetic energy is a homogeneous quadratic function of the generalized displacement $\dot{\mathbf{q}}$ with only the time-invariant constraints that the kinetic energy is expressed as

$$T = \dot{\mathbf{q}}^T \mathbf{M} \dot{\mathbf{q}} / 2 \tag{1.11.24a}$$

where $\mathbf{M}$ is an $n \times n$ symmetric and positive definite mass matrix. After neglecting the higher order small terms the potential energy nearby the equilibrium point becomes

$$\Pi = \mathbf{q}^T \mathbf{K} \mathbf{q} / 2, \qquad \mathrm{L} = T - \Pi \tag{1.11.24b,c}$$

where $\mathbf{K}$ is a symmetric stiffness matrix. The generalized momentum $\mathbf{p}$ and the Hamilton function are

$$\mathbf{p} = \mathbf{M}\dot{\mathbf{q}} \ , \qquad \mathrm{H} = \mathbf{p}^T \mathbf{M}^{-1} \mathbf{p} / 2 + \mathbf{q}^T \mathbf{K} \mathbf{q} / 2 \tag{1.11.25a,b}$$

For this time-invariant system, the characteristic equation is

$$\mathbf{q}^T \mathbf{K} \mathbf{q} / 2 + (\partial W / \partial \mathbf{q})^T \mathbf{M}^{-1} (\partial W / \partial \mathbf{q}) / 2 = E \tag{1.11.26}$$

where E is the mechanical energy. Directly finding the W solution of n-dimensional problem is still difficult hence a linear transformation for $\mathbf{q}$ is necessary. In the present case, a point transformation is proposed as

$$\mathbf{q} = \mathbf{U} \mathbf{q}_a , \qquad \dot{\mathbf{q}} = \mathbf{U} \dot{\mathbf{q}}_a \tag{1.11.27}$$

where $\mathbf{U}$ is a $n \times n$ matrix. The kinetic and potential energy are transformed as

$$T = \dot{\mathbf{q}}_a^T \mathbf{M}_a \dot{\mathbf{q}}_a / 2, \quad \mathbf{M}_a = \mathbf{U}^T \mathbf{M} \mathbf{U}; \tag{1.11.28a}$$

$$\Pi = \mathbf{q}_a^T \mathbf{K}_a \mathbf{q}_a / 2, \quad \mathbf{K}_a = \mathbf{U}^T \mathbf{K} \mathbf{U}; \tag{1.11.28b}$$

and the characteristic equation is transformed simultaneously as

$$\mathbf{q}_a^T \mathbf{K}_a \mathbf{q}_a / 2 + (\partial W / \partial \mathbf{q}_a)^T \mathbf{M}_a^{-1} (\partial W / \partial \mathbf{q}_a) / 2 = E \tag{1.11.29}$$

Hence the problem becomes to ***find the transformation matrix*** $\mathbf{U}$ ***so as to diagonalize both the symmetric matrices*** $\mathbf{K}_a$ ***and*** $\mathbf{M}_a$ ***simultaneously***, so that all the variables are separated. Such statement and its related algorithms are the main subject in the ***theory of vibrations*** and can be found in chapter 2 and various textbooks.

Next, the small vibration problem in a rotating coordinate system is considered. The kinetic energy is a quadratic function of generalized velocities and displacements and is given as

$$T = \dot{\mathbf{q}}^T \mathbf{M} \dot{\mathbf{q}} / 2 + \dot{\mathbf{q}}^T \mathbf{G} \mathbf{q} / 2 + \mathbf{q}^T \mathbf{K}_t \mathbf{q} / 2 \tag{1.11.30}$$

Hence the Lagrange function is

$$\mathrm{L}(\mathbf{q}, \dot{\mathbf{q}}) = \dot{\mathbf{q}}^T \mathbf{M} \dot{\mathbf{q}} / 2 + \dot{\mathbf{q}}^T \mathbf{G} \mathbf{q} / 2 - \mathbf{q}^T \mathbf{K} \mathbf{q} / 2 \tag{1.11.31}$$

where the matrix $\mathbf{K}$ has superimposed the matrix $\mathbf{K}_t$ from T, hence $\mathbf{K}$ is symmetric but cannot ensure positive definite. The mass matrix $\mathbf{M}$ still keeps symmetric and positive definite, and the matrix $\mathbf{G}$, termed as ***gyroscopic matrix***, is skew-symmetric $\mathbf{G}^T = -\mathbf{G}$.

The generalized momentum vector is introduced as

$$\mathbf{p} = \partial L/\partial \dot{\mathbf{q}} = \mathbf{M}\dot{\mathbf{q}} + \mathbf{G}\mathbf{q}/2 \tag{1.11.32}$$

$$\begin{aligned} &H(\mathbf{q},\mathbf{p}) = \mathbf{p}^T\mathbf{D}\mathbf{p}/2 + \mathbf{p}^T\mathbf{A}\mathbf{q} + \mathbf{q}^T\mathbf{B}\mathbf{q}/2 \\ &\mathbf{D} = \mathbf{M}^{-1}, \quad \mathbf{A} = -\mathbf{M}^{-1}\mathbf{G}/2, \quad \mathbf{B} = \mathbf{K} + \mathbf{G}^T\mathbf{M}^{-1}\mathbf{G}/4 \end{aligned} \tag{1.11.33}$$

Note, $\mathbf{D}$ is symmetric and positive definite, $\mathbf{B}$ is symmetric but not ensured positive definite, the matrix $\mathbf{A}$ is $n \times n$ dimensioned, all being time-invariant matrices. Such system is again time-invariant, the characteristic equation (1.11.13) is

$$\mathbf{q}^T\mathbf{B}\mathbf{q}/2 + (\partial W/\partial \mathbf{q})^T\mathbf{A}\mathbf{q} + (\partial W/\partial \mathbf{q})^T\mathbf{D}(\partial W/\partial \mathbf{q})/2 = E \tag{1.11.34}$$

The key to the solution of this PDE is again the method of ***separation of variables***. A linear transformation is to look for, in order to diagonalize the Hamilton function, which is a quadratic function presently, and its arguments are the dual vectors $\mathbf{q},\mathbf{p}$. A point transformation cannot diagonalize it, so that a linear canonical transformation is looking for.

The dual canonical equations is examined first, corresponding to (1.11.33)

$$\begin{aligned} \dot{\mathbf{q}} &= \mathbf{A}\mathbf{q} + \mathbf{D}\mathbf{p} \\ \dot{\mathbf{p}} &= -\mathbf{B}\mathbf{q} - \mathbf{A}^T\mathbf{p} \end{aligned} \tag{1.11.35}$$

or combined written as

$$\dot{\mathbf{v}} = \mathbf{H}\mathbf{v}, \quad \mathbf{v} = \begin{Bmatrix} \mathbf{q} \\ \mathbf{p} \end{Bmatrix}, \quad \mathbf{H} = \begin{bmatrix} \mathbf{A} & \mathbf{D} \\ -\mathbf{B} & -\mathbf{A}^T \end{bmatrix} \tag{1.11.36}$$

where $\mathbf{v}$ is the state vector and $\mathbf{H}$ is termed a ***Hamilton matrix***, whose characteristic is

$$-\mathbf{J}\mathbf{H} = \begin{bmatrix} \mathbf{B} & \mathbf{A}^T \\ \mathbf{A} & \mathbf{D} \end{bmatrix}, \quad (\mathbf{J}\mathbf{H})^T = \mathbf{J}\mathbf{H} \quad \text{or} \quad \mathbf{J}\mathbf{H}\mathbf{J} = \mathbf{H}^T \tag{1.11.37}$$

The Hamilton matrix is asymmetric but instead, the matrix $(\mathbf{J}\mathbf{H})$ is symmetric. The Hamilton matrix has the feature of symplectic behavior, which is conceivable.

In previous sections, the canonical transformation is used to transform the dual canonical variables $\mathbf{q},\mathbf{p}$, or to transform the state vector $\mathbf{v}$. Equation (1.8.4) and (1.8.6) set up the condition for a canonical transformation. Such form of condition is for general case and certainly applies to a linear system, that for a linear time-invariant system the linear canonical transformation is expressed by a time-invariant matrix $\mathbf{S}$

$$\boldsymbol{\zeta} = \mathbf{S}^{-1}\mathbf{v}\,, \qquad \boldsymbol{\zeta} = \{\mathbf{Q}^T, \mathbf{P}^T\}^T\,, \quad \mathbf{v} = \mathbf{S}\boldsymbol{\zeta} \tag{1.11.38}$$

where $\mathbf{S}^{-1}$ should satisfy the equation (1.8.6), thus $\mathbf{S}^{-1}$ and $\mathbf{S}$ are both symplectic matrices.

In the textbooks of linear algebra, one of the main subjects is using an orthogonal transformation to diagonalize a quadratic form. As mentioned above, such method and algorithm applies to Lagrange system and point transformations. The point transformation applies to the system description with one kind of variables and the corresponding geometry is of Euclidean type. In present case, the transformation is a symplectic matrix $\mathbf{S}$ applied to two kinds of variables $\mathbf{q}$ and $\mathbf{p}$. The original function is a quadratic Hamiltonian $\mathrm{H}(\mathbf{q},\mathbf{p})$ and will be transformed to $\mathrm{K}(\mathbf{Q},\mathbf{P})$ which leads the dual equations to be diagonalized. The transformed canonical equations are

$$\dot{\mathbf{Q}} = \partial\mathrm{K}/\partial\mathbf{P}, \quad \dot{\mathbf{P}} = -\partial\mathrm{K}/\partial\mathbf{Q} \quad \text{or} \quad \dot{\zeta} = \mathbf{J}(\partial\mathrm{K}/\partial\zeta)$$

The best form is that at the right hand side of equation for $\dot{\mathbf{Q}}$ there is only $\mathbf{Q}$, and the equation for $\dot{\mathbf{P}}$ there is only $\mathbf{P}$, all being diagonalized. Hence it requires that within the new Hamiltonian $\mathrm{K}(\mathbf{Q},\mathbf{P})$, there are only terms with P_i and Q_i together, expressed in matrix notation

$$\mathrm{K}(\mathbf{Q},\mathbf{P}) = \mathbf{P}^T \times \mathrm{diag}(\mu_i) \times \mathbf{Q} \tag{1.11.39}$$

where $\mathrm{diag}(\mu_i)$ denotes a diagonal matrix with $\mu_i (i = 1,\cdots,n)$ being its (diagonal) elements, and the dual canonical equations are

$$\dot{Q}_i = \mu_i Q_i, \qquad \dot{P}_i = -\mu_i P_i \qquad (i = 1,2,\cdots,n) \tag{1.11.40}$$

the simplest form.

Therefore, the selection of the symplectic matrix $\mathbf{S}$ should transform the quadratic form $\mathrm{H}(\mathbf{q},\mathbf{p})$ to that given in equation (1.11.39). According to equation (1.11.33)

$$\mathrm{H}(\mathbf{q},\mathbf{p}) = \frac{1}{2}\begin{Bmatrix}\mathbf{q}\\ \mathbf{p}\end{Bmatrix}^T \begin{bmatrix}\mathbf{B} & \mathbf{A}^T\\ \mathbf{A} & \mathbf{D}\end{bmatrix}\begin{Bmatrix}\mathbf{q}\\ \mathbf{p}\end{Bmatrix} = \mathbf{v}^T(-\mathbf{JH})\mathbf{v}/2$$

To compute $\mathrm{K}(\mathbf{Q},\mathbf{P})$, substituting the transformation (1.11.38) into the above equation gives

$$\mathrm{K}(\mathbf{Q},\mathbf{P}) = -\zeta^T\mathbf{S}^T\mathbf{JHS}\zeta/2 = \zeta^T\begin{bmatrix}0 & \mathrm{diag}(\mu_i)\\ \mathrm{diag}(\mu_i) & 0\end{bmatrix}\zeta/2$$

The right hand side is only the rewriting of equation (1.11.39). The congruent transformation (1.11.41) with the symplectic matrix $\mathbf{S}$ requires bringing the symmetric matrix $\mathbf{JH}$ into the form

$$\mathbf{S}^T(-\mathbf{JH})\mathbf{S} = \begin{bmatrix}0 & \mathrm{diag}(\mu_i)\\ \mathrm{diag}(\mu_i) & 0\end{bmatrix} \tag{1.11.41}$$

Left multiplying the above equation with $\mathbf{SJ}$ and note that $\mathbf{S}$ is a symplectic matrix, it gives

$$\mathbf{HS} = \mathbf{S}\begin{bmatrix}\mathrm{diag}(\mu_i) & 0\\ 0 & -\mathrm{diag}(\mu_i)\end{bmatrix} \tag{1.11.42}$$

which is an eigen-equation for the Hamilton matrix $\mathbf{H}$ and each column of the symplectic matrix $\mathbf{S}$ is an eigen-vector of the Hamilton matrix. The algorithm for finding all the eigen-vectors of a Hamilton matrix will be given in chapter 2. But some theoretical problem should be explained here. Because the matrix $\mathbf{H}$ itself is asymmetric, complex valued eigen-solutions may appear and their complex conjugate must also be eigen-solutions. Combining these complex conjugate eigen-solutions a real valued symplectic matrix $\mathbf{S}_r$ can be composed. The congruent transformation by the matrix $\mathbf{S}_r$ carries the transformed Hamilton function $\mathrm{K}(\mathbf{Q},\mathbf{P})$ being real valued. However, real valued transformation cannot fully diagonalize $\mathrm{K}(\mathbf{Q},\mathbf{P})$. The numerical algorithms in later chapters require real transformation $\mathbf{S}_r$.

However, the following description still uses the fully diagonalized transformation $\mathbf{S}$.

After the diagonal transformation, in the canonical coordinate $\mathbf{Q},\mathbf{P}$ system, the dual equation (1.11.35) is transformed as given in (1.11.40) and the system matrix becomes

$$\mathbf{A}_Q = \mathrm{diag}(\mu_i); \quad \mathbf{B}_Q = \mathbf{0}; \quad \mathbf{D}_Q = \mathbf{0} \tag{1.11.43}$$

Hence the transformed characteristic equation (1.11.34) becomes

$$(\partial W/\partial \mathbf{Q})^T \mathrm{diag}(\mu_i)\mathbf{Q} = E; \quad \text{or} \quad \sum_{i=1}^{n} (\partial W(Q_i,\alpha_i)/\partial Q_i)\cdot Q_i\mu_i = E \tag{1.11.44}$$

All the generalized coordinates Q_i have been separated. According to the equations (1.11.18,20), the variable separated Hamilton-Jacobi characteristic equation is derived to

$$W = \sum_{i=1}^{n} W_i; \quad (\partial W_i/\partial Q_i)\cdot Q_i\mu_i = \alpha_i; \qquad i = 1,\cdots,n \tag{1.11.45}$$

The solution has been greatly simplified.

The canonical transformation given here is for the separation of variables, which brings the combined n-dimensional oscillator into n one-dimensional oscillators. However, the canonical transformation given in section 1.11.1 works only for one-dimensional oscillator, that this one-dimensional oscillator is transformed using the energy and phase angle as primary variables. The two canonical transformations can be applied successively, because the combination of two successive canonical transformations is again a canonical transformation. The eigen-solution of a Hamilton matrix is a fundamental subject in state space formulation, but it appears not so much in published textbooks of applied mechanics. In next chapters this subject will appear frequently.

The above discussion is only for linear problems, which is easier to understand. The importance of linear problem is to provide a starting point of such as perturbation approach, iterative method etc. so as to analyze the time-variant system and/or non-linear system problems.

Chapter 2, Vibration Theory

In mechanical, electrical and aerospace engineering and other fields, vibration problems exist everywhere. Only fundamental theory and methods can be presented in this book. Among various topics, some selected subjects are described here.

The usual solution methods for vibrations are 1) ***Direct integration method***, 2) ***Eigenvector expansion method*** and others. Quite a number of textbooks or monographs have been published. Hence, in addition to the fundamental theory, the present book introduces some relatively new contents as we can. For example, the precise integration method supplies an entirely different approach versus the usual FDM. Also, the solution method of gyroscopic system vibrations will introduce the state space method, the dual spaces, Hamilton system theory and symplectic geometry applications etc., the featured methods for vibration analysis. For gyroscopic system vibrations, the method of separation of variables and the respective eigenvector expansion method will bring the system geometry from Euclidean to Symplectic. Such methodology appears not so much in traditional applied mechanics textbooks, however, as will be seen in this book, for wave propagation, for semi-analytical method in structural mechanics, for theory of elasticity and also for modern control theory and so on, the symplectic methodology appears over and over again. Based on such new considerations, it is reasonable to develop a systematic methodology in applied mechanics.

Using dual variable state-space and symplectic geometric method, a ***unified approach*** is proposed for the solution of various field problems, which is the main purpose of this book. The precise integration method, as a numerical integration method, can be regarded as the algorithm back up for the duality system theory so as to be used in practice.

§2.1, Vibration of single degree of freedom system

Vibration of single degree of freedom system is the simplest problem. However, only linear time-invariant system vibration is simple, which provides the foundation for the vibration analysis of multi-degrees of freedom system. The analysis of time-variant single degrees of freedom is not so easy as compared to time-invariant systems. The non-linear single degree of freedom system analysis is more difficult, that there are still a number of problems to be solved.

§2.1.1, Linear vibration

Linear vibration is the simplest problem and has been studied in the curriculums of engineering mechanics. The importance of which is that for multi-degrees of freedom system after modal analysis, the system vibration is reduced to be the superposition of single degree of freedom vibrations. And in applications, a number of problems are treated as single degree of freedom vibration.

The equation for single degree of freedom system vibration is

$$m\ddot{x}+c\dot{x}+kx=f(t)\,,\quad x(0)=\text{known},\quad \dot{x}(0)=\text{known} \tag{2.1.1}$$

This equation is for forced vibration with damping. Its respective free vibration equation with no damping is $m\ddot{x}+kx=0$, and the natural vibration circular frequency is $\omega_1=\sqrt{k/m}$, its period is $T=2\pi/\omega_1=2\pi\sqrt{m/k}$. The general solution is

$$x=x_0\cos\omega_1 t+(v_0/\omega_1)\sin\omega_1 t \tag{2.1.2}$$

The equation for free vibration with damping is $m\ddot{x}+c\dot{x}+kx=0$, for which the ***critical damping constant*** C_c is

$$C_c=2\sqrt{km},\quad \text{let}\ \ \varsigma=c/C_c \tag{2.1.3}$$

where ς is the damping ratio. The characteristic equation is $ms^2+cs+k=0$ and its roots are

$$s_{1,2}=(-\varsigma\pm\sqrt{\varsigma^2-1})\omega_1 \tag{2.1.4}$$

When $\varsigma>1$, the solution behaves only decay with time; and when $\varsigma<1$, the system vibrates forth-and-back. For case of small damping $\varsigma<1$, the logarithmic damping ratio is defined as

$$\delta=\ln(x_i/x_{i+1}) \tag{2.1.5}$$

where x_i, x_{i+1} represent two successive maximum displacement, which is independent on the subscript i. The period T and logarithmic damping ratio δ of vibration are, respectively

$$T=2\pi\big/(\omega_1\sqrt{1-\varsigma^2}),\qquad \delta=2\pi\varsigma\big/\sqrt{1-\varsigma^2}\approx 2\pi\varsigma \tag{2.1.6}$$

The measurement of ς can use its amplitude half damped count n of reciprocate

$$\delta=\ln(x_0/x_n)/n=\ln 2/n\approx 2\pi\varsigma\,,\quad \text{so}\ \ n\varsigma\approx 0.110 \tag{2.1.7}$$

Forced vibration under periodical force excitation is considered next

$$m\ddot{x}+c\dot{x}+kx=F_0\sin\omega t \tag{2.1.8}$$

With longer time, the influence of initial condition damps out, only steady forced vibration remains. Let X be its amplitude and ϕ be the phase lag

$$x=X\sin(\omega t-\phi) \tag{2.1.9a}$$

Substituting into (2.1.8) solves

$$X=F_0\big/\sqrt{(k-m\omega^2)^2+(c\omega)^2}\,,\qquad \text{tg}\phi=c\omega/(k-m\omega^2) \tag{2.1.9b}$$

Let $X_0=F_0/k$—the static displacement, gives

$$X/X_0=\left[(1-\omega^2/\omega_1^2)^2+(2\varsigma\omega/\omega_1)^2\right]^{-\frac{1}{2}},\quad tg\phi=\frac{2\varsigma\omega/\omega_1}{1-(\omega/\omega_1)^2} \tag{2.1.10}$$

Generally speaking, when the circular frequency ω of exciting force approaching the damping free natural vibration frequency $\omega_1=\sqrt{k/m}$, resonance appears. According to equation (2.1.10), when $\omega=\omega_1$, $X/X_0=1/(2\varsigma)$ and phase lag $\phi=\pi/2$, the curves are shown in figure 2.1.

For non-periodic excitations, the unit step excitation is considered first. At

$t=0$ a constant force F_0 is suddenly applied on a damped spring-mass system originally at rest. The equation is

$$m\ddot{x}+c\dot{x}+kx=F_0\,,\quad \text{or}\quad \ddot{x}+2\varsigma\omega_1\dot{x}+\omega_1^2x=F_0/m\,,\quad x(0)=\dot{x}(0)=0 \tag{2.1.11}$$

The solution can be written in the form $x=F_0\cdot h_1(t)$ where

$$h_1(t)=\left[1-\left(e^{-\varsigma\omega_2 t}\Big/\sqrt{1-\varsigma^2}\right)\sin(\sqrt{1-\varsigma^2}\,\omega_1 t+\phi)\right]\Big/k\,,\ tg\phi=\sqrt{1-\varsigma^2}\Big/\varsigma \tag{2.1.12}$$

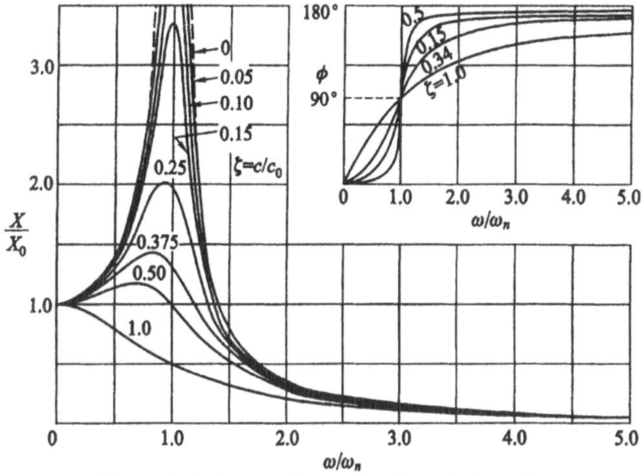

Figure 2.1, *Curves for amplitude and phase lag*

Let $F_0=1$, then the unit step force response is just $x=h_1(t)$, called as ***unit step force response function***. Based on this function the response of arbitrary external force $F(t)$ can be written as

$$x(t)=F(0)h(t)+\int_0^t\dot{F}(\zeta)h_1(t-\zeta)d\zeta \tag{2.1.13}$$

called as Duhamel integration. Integration by parts gives

$$x(t)=\int_0^t F(\zeta)\dot{h}_1(t-\zeta)d\zeta \tag{2.1.14}$$

because $h_1(0)=0$ always holds. The Duhamel integration in the form of (2.1.14) appears more frequently, where $\dot{h}_1(t-\zeta)$ is called the ***unit impulse response function***. Differentiating (2.1.12) with respect to time t gives

$$\dot{h}_1(t)=\left[e^{-\varsigma\omega_1 t}\Big/\sqrt{mk\cdot(1-\varsigma^2)}\right]\sin(\sqrt{1-\varsigma^2}\,\omega_1 t)\underset{\text{def}}{=}h(t) \tag{2.1.15}$$

which is applied much frequently.

§2.1.2, Parametric resonance

The above analysis is only for time-invariant single degree of freedom system, when the frequency of external force approaches the system natural frequency, resonance happens. However, time-variant linear system vibration has another kind of resonance, namely the ***parametric resonance,*** for which even if no external force at the right-hand side, the system still vibrates severely. Only some introductory theory of parametric resonance can be presented here. Suppose a single pendulum of length l in figure 2.2, for which the terminal mass m is acted on by gravity and also a vertical periodical force $f_0 \cos 2\omega t$. The position $\theta = 0$ is obviously an equilibrium point and the dynamic equation is given as

$$ml^2\ddot{\theta} + (mgl + lf_0 \cos 2\omega t)\theta = 0$$

or $$\ddot{\theta} + (\omega_1^2 + 2\varepsilon \cos 2\omega t)\theta = 0, \quad \omega_1^2 = g/l, \; 2\varepsilon = f_0/ml \qquad (2.1.16)$$

The equation (2.1.16) is called the ***Mathieu equation*** and is the simplest periodical coefficient linear dynamic differential equation.

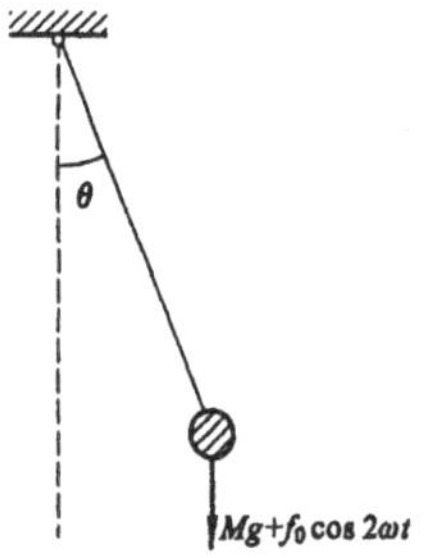

Figure 2.2, *A single pendulum*

The analysis can be given as follows. For a linear equation, the superposition principle applies. The equation (2.1.16) has two basic solutions $\Theta_1(t)$ and $\Theta_2(t)$, which satisfy the differential equation (2.1.16) with the initial conditions

$$\Theta_1(0) = 1, \; \dot{\Theta}_1(0) = 0, \quad \text{when } t = 0 \qquad (2.1.17a)$$

$$\Theta_2(0) = 0, \; \dot{\Theta}_2(0) = 1, \quad \text{when } t = 0 \qquad (2.1.17b)$$

respectively. After a whole period of the coefficient external force, the time is $t = T = \pi/\omega$, the differential equation coincides with (2.1.16) at $t = 0$, but the value of functions $\Theta_1(T)$, $\dot{\Theta}_1(T)$; and $\Theta_2(T)$, $\dot{\Theta}_2(T)$ will no longer be (2.1.17). Let the value be denoted as k_{11}, k_{12}, and k_{21}, k_{22}, respectively. According to the superposition principle gives

$$\Theta_1(t+T) = k_{11}\Theta_1(t) + k_{12}\Theta_2(t); \quad \Theta_2(t+T) = k_{21}\Theta_1(t) + k_{22}\Theta_2(t) \qquad (2.1.18)$$

These equations can be written in matrix form as

$$\mathbf{u}(t+T) = \mathbf{K}\mathbf{u}(t) \qquad (2.1.18')$$

where $$\mathbf{u} = \{\Theta_1, \Theta_2\}^T, \quad \text{and} \quad \mathbf{K} = \begin{bmatrix} k_{11} & k_{12} \\ k_{21} & k_{22} \end{bmatrix} \qquad (2.1.19)$$

After the matrix $\mathbf{K}$ is computed, the next step is to transform from the vector $\mathbf{u}$ to a vector $\mathbf{v}$ by a linear transformation matrix $\mathbf{P}$, as follows

$$\mathbf{v}(t) = \mathbf{P}\mathbf{u}(t), \qquad \mathbf{P} = \begin{bmatrix} p_{11} & p_{12} \\ p_{21} & p_{22} \end{bmatrix} \qquad (2.1.20)$$

Substituting (2.1.20) into equation (2.1.18') gives the matrix transformation

$$\mathbf{v}(t+T) = \mathbf{PKP}^{-1}\mathbf{v}(t) = \mathbf{B}\mathbf{v}(t), \qquad \mathbf{B} = \mathbf{PKP}^{-1} \qquad (2.1.21)$$

Because $\mathbf{K}$ and $\mathbf{B}$ are similar matrices to each other, they have the same eigenvalue λ

$$\det(\mathbf{B} - \lambda\mathbf{I}) = \det[\mathbf{P}(\mathbf{K} - \lambda\mathbf{I})\mathbf{P}^{-1}] = \det(\mathbf{K} - \lambda\mathbf{I})$$

The matrix $\mathbf{B}$ requires diagonalized, which means to solve the eigenvalues λ_1 and λ_2 of matrix $\mathbf{K}$ with eigenvectors $\mathbf{v}_1, \mathbf{v}_2$, respectively. Hence

$$\mathbf{v}_1(t+T)=\lambda_1\mathbf{v}_1(t),\quad \mathbf{v}_1(t+2T)=\lambda_1^2\mathbf{v}_1(t),\quad \cdots,\quad \mathbf{v}_1(t+nT)=\lambda_1^n\mathbf{v}_1(t),\quad \cdots \tag{2.1.22}$$

Similar equations also hold for $\mathbf{v}_2$. Therefore, if one of the eigenvalues appears $\text{abs}(\lambda_i)>1$, then the corresponding eigenvector $\mathbf{v}_i$ increases indefinitely with time, and the system is unstable. This theory applies to the systems of linear periodical coefficient differential equations, called the ***Floquet theory*** [36], and $\mathbf{K}$ is called the ***Floquet matrix***.

The set of periodic coefficient linear differential equations has stability problem. Solution composes of two steps. The first step is the computation of Floquet matrix. The next step is to solve the eigen-problem of Floquet matrix for stability analysis. Mathieu equation is only a second order ODE, however, Floquet theory applies to any order ODE set.

The periodical coefficients ODE-s are not necessarily unstable. Such as for equation (2.1.16) only when the parameters of circular frequency ω and the coefficient ε have some combinations, then the system may fall into instability, hence termed as ***parametric resonance***.

Therefore, the problem requires first to find the Floquet matrix $\mathbf{K}$, for which the basic initial value problem of differential equation (2.1.17) needs to be solved. Analytical method is usually difficult to find the solutions of a time-variant coefficient differential equation, so that numerical integration must be used. Fortunately, the powerful computer has been available now, numerical solution can be found via various algorithms, especially the precise integration method. As eigen-solutions, there are standard algorithms to invoke, see [42,43].

Numerical method for stability analysis of periodical coefficient ODEs is general. However, such method is not enough for physical insight. Hence the perturbation method for solving the Mathieu equation is beneficial. When ε is neglected, the natural vibration of the system is $\theta=A\cos\omega_1 t$. Based on this solution, when the small ε is taken into account, the first order approximation of differential equation can be written as

$$\ddot{\theta}+\omega_1^2\theta=-2\varepsilon\cos 2\omega t\cdot A\cos\omega_1 t=-\varepsilon A[\cos(2\omega-\omega_1)t+\cos(2\omega+\omega_1)t]$$

This equation resonates when $2\omega-\omega_1=\omega_1$, *i.e.* when $\omega=\omega_1$. Note that 2ω is the circular frequency of the external force of parametric resonance. For Mathieu equation, when 2ω is twice as the natural frequency ω_1, parametric resonance appears. Because the resonance frequency is half of the frequency of external force, it is called as sub-harmonic resonance.

The above derivation moves the small parameter term to the right hand side of the equation, which corresponds to the first order approximation of perturbation with the small parameter ε. The small parameter perturbation method is a usual analytical approximate approach for parametric resonance or for non-linear vibration analysis.

Perturbation method had published quite a number of papers, see [37~39]. However, it is noted that the algebraic derivation is quite cumbersome. Presently, the complicated algebraic derivation can be carried out via the symbolic algebraic software, for instance MATHEMATICA, MAPLE, etc.

Stable and unstable regions of Mathieu equation are sketched in figure 2.3.

The periodic solution of a non-linear differential equation proposes a limit cycle, and the stability analysis for the limit cycle should use the perturbation method near by the limit cycle, which derives a set of periodical coefficient differential equation (Hill equation). The Floquet theory can be used to solve the stability problem.

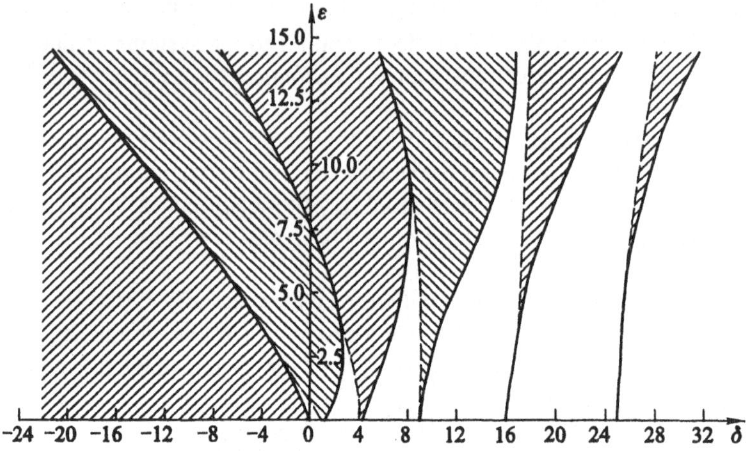

Figure 2.3, *Stable and unstable regions of Mathieu function*

§2.1.3, *Introduction to non-linear vibration*

Non-linear vibration is a vast area of analysis. Even for single degree of freedom system, the non-linear vibration has quite a number of problems to be further investigated. Only some problems are selected here to discuss.

§2.1.3.1, *Limit cycle*

A limit cycle problem with non-linear damping is introduced first. After dimensionless transformation, let $m=1, k=1$; suppose the non-linear damping term gives the vibration equation as

$$\ddot{x} - c\dot{x}[1-(x^2+\dot{x}^2)] + x = 0, \qquad c>0 \tag{2.1.23}$$

Inspection determines that there are two damping terms, where $F_{d1} = -c\dot{x}$ is the negative linear damping, whereas $F_{d2} = c\dot{x}(x^2+\dot{x}^2)$ is the non-linear damping. Evidently, the origin $x=0$ is an equilibrium point. When the vibration amplitude is small, the non-linear term plays a negligible role and the equation can be approximated as

$$\ddot{x} - c\dot{x} + x = 0, \qquad c>0$$

The eigenvalue of which is

$$s_{1,2} = (c/2) \pm \mathrm{i}(1-c^2/4)^{1/2}$$

The real parts take positive value, which mean that the original point is unstable. Physically, because of the negative damping, the work done by the negative damping is calculated as

$$W_1 = \int c\dot{x}\mathrm{d}x = \int c\dot{x}^2 \mathrm{d}t$$

which always takes positive value. Hence the negative damping force $F_{d1} = -c\dot{x}$ continuously supplies energy to the system, which causes the original equilibrium point $x = 0$ unstable.

The effect of non-linear damping term is considered next, it does work

$$W_2 = \int -c\dot{x}(x^2 + \dot{x}^2)\mathrm{d}x = -\int c\dot{x}^2(x^2 + \dot{x}^2)\mathrm{d}t$$

and always takes negative value, hence the non-linear term consumes energy for the system. Nearby the original point, $x, \dot{x}$ are small, therefore $\mathrm{abs}(W_2) << \mathrm{abs}(W_1)$, the positive damping is far less than the negative damping, the vibration will be increasing. On the other hand, if the vibration amplitude is large, such that $(x^2 + \dot{x}^2) >> 1$, the energy consumed by the positive non-linear damping far exceeds the work done by the negative damping, the vibration will be decreasing. The work done by the two kinds of damping forces opposes each other, when amplitude is small, the negative damping work W_1 dominates and the vibration amplitude increases, whereas when the amplitude is large, W_2 dominates and the amplitude decreases. Therefore in between, there should be appropriate amplitude such that the vibration reaches energy balance. To express in the state space (phase plane) $(x, \dot{x})$, there is a closed curve called ***limiting cycle.*** As the vibration moved along the limiting cycle, the negative damping work W_1 balanced with the non-linear damping work W_2 and the vibration keeps on the limit cycle, see [35].

For the equation (2.1.23), it is easy to find the limit cycle. Substituting $x(t) = \cos t$ into equation (2.1.23) gives

$$-\cos t + c\sin t[1 - (\cos^2 t + \sin^2 t)] + \cos t = 0$$

Thus the limiting cycle is a unit circle in state space. The initial state inside the unit circle will gradually approach the limiting cycle, and the state outside the unit circle will go back to this limit cycle, so that this limiting cycle is stable. Figure 2.4 sketches such curves of motion.

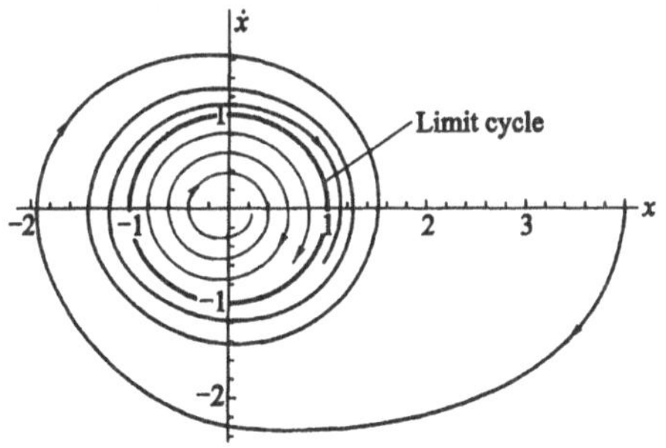

Figure 2.4, *The limiting cycle*

The stability problem is very important for limiting cycle. The original point $x = 0, \dot{x} = 0$ in state space is an equilibrium point for equation (2.1.23), but it is unstable. Even if a very small departure to the equilibrium point appears, the motion will further depart from the equilibrium point and never return. This equilibrium point is thus unstable. The stability problem also exists for limit cycle.

The limit cycle of equation (2.1.23) is stable, because for **any** point neighbor to this limit cycle, the successive motion will always turn back toward this limit cycle, which verifies the stability of the limit cycle.

Let the coefficient c in equation (2.1.23) be negative, the system will have an unstable limiting cycle. It is easy to verify that the unit cycle is still a periodic solution of the dynamic equation, *i.e.* limiting cycle. However, the linear damping term becomes positive whereas the non-linear damping term becomes negative now. From any point inside the unit circle, the successive motion tends to the original point and the original point is a stable equilibrium point. However, from a point slightly apart outside from the unit circle, its successive motion will go further depart from the unit circle and never return, hence the limit cycle is unstable.

Stability problem of motion is very important in applications. Fluid structure interaction proposes a lot of practical problems, for which vibration instability phenomenon will cause serious consequence. Such as the wind induced vibration of long span bridges, the aircraft wing flutter etc.

Equation (2.1.23) is a special example just for illustration. The proposed equation is very simple, that the limit cycle solution is easily obtained by inspection in order to explain the nature of limit cycle and stability. For non-linear differential equation, the superposition principle does not apply, and therefore the solution of general non-linear vibration problem is usually very difficult. In next section, the solution for one degree of freedom Duffing differential equation is considered, so as to illustrate the application of canonical transformation to non-linear vibrations.

§2.1.3.2, Duffing equation

Duffing differential equation is typical to non-linear vibrations. Usually, the solution applies the perturbation method, see [35,37]. In this section, the canonical transformation is applied by deriving the system into duality system to solve. Duffing equation can be given as

$$\ddot{x}+c\dot{x}+x+bx^3 = f\cos\omega_a t, \quad c\geq 0 \tag{2.1.24}$$

where b, c and f are given parameters. Because of the term bx^3, it is a non-linear differential equation, and the periodic solution (limit cycle) is to find. Note, for the solution being nearly resonant the frequency ω_a should not be far apart from unity. Obviously, the periodic solution must have the same frequency ω_a as the external driven force.

The basic conservative system for the above equation can be selected as

$$\ddot{x}+x+bx^3 = 0 \tag{2.1.25}$$

for which the Lagrange function is

$$\mathrm{L}(x,\dot{x}) = \dot{x}^2/2 - x^2/2 - bx^4/4 \tag{2.1.26}$$

The duality system can then be introduced under the Hamilton frame, let the dual variable be

$$p = \partial\mathrm{L}/\partial\dot{x} = \dot{x} \tag{2.1.27}$$

then the Legendre transformation gives

$$\mathrm{H}(x,p) = p\dot{x} - \mathrm{L}(x,\dot{x}) = p^2/2 + x^2/2 + bx^4/4 \tag{2.1.28}$$

a time invariant system. Obviously, the energy conservation gives

$$\mathrm{H}(x,p) = p^2/2 + x^2/2 + bx^4/4 = C \quad (2.1.29)$$

where C is kept constant, the conserved energy. The canonical dual equations are written as

$$\dot{x} = \partial\mathrm{H}/\partial p = p, \quad \dot{p} = -\partial\mathrm{H}/\partial x = -x - bx^3 \quad (2.1.30)$$

Up to here, the derivation is popular as usual. Corresponding to the canonical system, the respective Hamilton-Jacobi equation is given as

$$\partial S/\partial t + \mathrm{H}(x, \partial S/\partial x) = 0 \quad (2.1.31)$$

where the function $S(x,t)$ is the action function and is defined as

$$S = \int_{t_0}^{t} \mathrm{L}(x, \dot{x}, \tau)\mathrm{d}\tau \quad (2.1.32)$$

with the initial condition

$$x(t_0) = x_0 = \text{given}, \quad \text{when } \tau = t_0 \quad (2.1.33)$$

The intention now is to find a canonical transformation for equation (2.1.25), introducing a new canonical coordinate system and solving the original problem in the transformed coordinate system. Using the method given in section 1.11 that the solution of the Hamilton-Jacobi equation (2.1.31) must have the form

$$S = S(\mathbf{q}, t, \alpha_1, \alpha_2, \cdots, \alpha_n) + A \quad (2.1.34)$$

where $\alpha_1, \alpha_2, \cdots, \alpha_n, A$ are arbitrary constants. Evidently, the constant A is useless, presently $n = 1$. So

$$\partial S/\partial t + [(\partial S/\partial x)^2 + x^2 + bx^4/2]/2 = 0 \quad (2.1.35)$$

Because the equation (2.1.25) is a time-invariant conservative system, that the time t appears only in $\partial S/\partial t$, thus the solution form must be

$$S(x,C,t) = W(x,C) - C\,t \quad (2.1.36)$$

where C is the integration constant. The equation for W is

$$[(\mathrm{d}W/\mathrm{d}x)^2 + x^2 + bx^4/2]/2 = C \quad (2.1.37)$$

and C is the conserved mechanical energy. Integration for W gives

$$W(x,C) = \sqrt{2C}\int\sqrt{1 - (x^2 + bx^4/2)/(2C)}\mathrm{d}x \quad (2.1.38)$$

where $W(x,C)$ is an elliptic function with C being a parameter to be determined. Substituting $W(x,C)$ into equation (2.1.36) gives the action S. Using energy conservation $p^2/2 + x^2/2 + bx^4/4 = C$ at $t = t_0$ gives

$$p_0^2 = 2C - x_0^2 - bx_0^4/2 \quad (2.1.39)$$

that the initial momentum p_0 can be expressed with the parameter C.

Treat the complete action integral $S(x,C,t)$ as the first class generating function $F_1(\mathbf{q},\mathbf{Q},t)$, and the integration constant C in S is regarded as the new canonical coordinate Q

$$Q = C \quad (2.1.40a)$$

Then the three transformation equations (1.7.10a-c) are used. The first one gives

$$p = \partial S/\partial x = \partial W/\partial x = \sqrt{(2C) - (x^2 + bx^4/2)} \quad (2.1.41)$$

where $W(x,Q)$ is a known function. The next step is

$$P = -\partial S/\partial Q = -\partial S/\partial \alpha \tag{2.1.42}$$

and the transformed Hamilton function $K = 0$, because the complete integration of Hamilton-Jacobi equation is used as the generating function for the canonical transformation. The generalized momentum P becomes also a constant, denoted as

$$P = \beta \tag{2.1.40b}$$

While equations (2.1.40a,b) give Q, P being constants, which corresponding to the natural vibration of the system equation (2.1.25). Now the period of vibration is to determine. The maximum amplitude x_{max} can be found from the energy conservation as

$$x_{max}^2/2 + bx_{max}^4/4 = C, \quad x_{max} = \left[\left(\sqrt{1+4Cb}-1\right)/b\right]^{1/2}$$

In order to match the given frequency ω_a of external force, the period should be $T = 2\pi/\omega_a$, and can be computed as

$$T = 2\int_{-x_{max}}^{x_{max}} \left(2C - x^2 - bx^4/2\right)^{-1/2} \mathrm{d}x$$

which is a function of parameter C. According to equations (2.1.36), (2.1.41)

$$\beta = -\partial S/\partial C = t - w(x,Q), \quad w(x,Q) \underset{\text{def}}{=} \partial W(x,Q)/\partial Q \tag{2.1.43}$$

where
$$w(x,C) = \sqrt{1/2C}\int_0^x \mathrm{d}x_a \Big/ \sqrt{1-(2x_a^2 + bx_a^4)/(4C)} \tag{2.1.43a}$$

From equation (2.1.43) x is solved as a function of time t with two integration parameters C, β, and the integration from $x = 0$ only influences the constant value β.

The parameters C, β are now treated as the new canonical coordinate system according to the equations (2.1.40a,b), they are constants for the energy conservation system of equation (2.1.25). But when the transformation applied to the original equation (2.1.24), the coordinates Q, P will no longer be constants. That Q, P can be treated as the unknowns to be solved for the transformed original problem.

Rewriting the parameters C, β in (2.1.41), (2.1.43) as Q, P gives

$$p = \sqrt{(2Q) - (x^2 + bx^4/2)} \quad \text{and} \quad P = t - w(x,Q) \tag{2.1.44a,b}$$

which is a canonical transformation. From (2.1.44b), x is solved as

$$x = x(Q,P,t)\left[= x(C,\beta,t)\right] \tag{2.1.45a}$$

Afterwards, substituting the above equation into equation (2.1.44a) gives

$$p = \sqrt{(2Q) - (x^2 + bx^4/2)} = p(Q,P,t) \tag{2.1.45b}$$

Equations (2.1.45a,b) give the canonical transformation which transforms the original variables x, p to be the functions of Q, P and t.

Now, the dual equations of the original system is

$$\dot{x} = p, \quad \dot{p} = -x - bx^3 - cp + f\cos\omega_a t \tag{2.1.46a,b}$$

Substituting (2.1.45a,b) into equations (2.1.46a,b), respectively, gives

$$p = \frac{\partial x}{\partial Q}\dot{Q} + \frac{\partial x}{\partial P}\dot{P} + \frac{\partial x}{\partial t} \quad \text{or} \quad \frac{\partial x}{\partial Q}\dot{Q} + \frac{\partial x}{\partial P}\dot{P} = p - \frac{\partial x}{\partial t} \tag{a}$$

$$\dot{p} = -x - bx^3 - c\left(\frac{\partial x}{\partial Q}\dot{Q} + \frac{\partial x}{\partial P}\dot{P} + \frac{\partial x}{\partial t}\right) + f\cos\omega_a t$$

Directly differentiating equation (2,1,45b) gives $\dot{p} = \frac{\partial p}{\partial Q}\dot{Q} + \frac{\partial p}{\partial P}\dot{P} + \frac{\partial p}{\partial t}$, so

$$\left(\frac{\partial p}{\partial Q} + c\frac{\partial x}{\partial Q}\right)\dot{Q} + \left(\frac{\partial p}{\partial P} + c\frac{\partial x}{\partial P}\right)\dot{P} = -x - bx^3 - c\frac{\partial x}{\partial t} - \frac{\partial p}{\partial t} + f\cos\omega_a t \tag{b}$$

Note that $x = x(Q,P,t), p = p(Q,P,t)$ are the canonical transformation of the conservative system (2.1.25), then the equations (a) and (b) are derived as

$$\frac{\partial x}{\partial Q}\dot{Q} + \frac{\partial x}{\partial P}\dot{P} = 0 \tag{2.1.47a}$$

$$\left(\frac{\partial p}{\partial Q} + c\frac{\partial x}{\partial Q}\right)\dot{Q} + \left(\frac{\partial p}{\partial P} + c\frac{\partial x}{\partial P}\right)\dot{P} = -c\frac{\partial x}{\partial t} + f\cos\omega_a t \tag{2.1.47b}$$

Solving (2.1.47a,b) with respect to $\dot{Q}$ and $\dot{P}$ gives the new dual equations of Q, P for the original system (2.1.24), which is not a conservative system. Up to here, the derivation has not introduced any approximation yet.

In the equations (2.1.47a,b), all the terms involving x, p should be considered as functions of Q, P. The equations obtained are still non-linear certainly. To solve the non-linear system, numerical integration is necessary or some approximate analytical approaches can be used for the comparatively simple problems.

§2.1.3.3, Simplified solution method for Duffing equation

The purpose of selection of equation (2.1.25) with a conservative system is to generate a canonical transformation. It can be further simplified as that, instead of the non-linear equation (2.1.25) a simple harmonic oscillator can also be used, which corresponds to the popular approach. The canonical transformation for a simple harmonic oscillator is easy to handle

$$\ddot{x} + \omega_a^2 x = 0 \tag{2.1.25'}$$

Because a periodic solution is to find, its natural frequency is selected equal to the frequency of the external force. The Lagrange function is

$$\mathrm{L}(x,\dot{x}) = \dot{x}^2/2 - \omega_a^2 x^2/2 \tag{2.1.26'}$$

The dual variables are

$$p = \partial\mathrm{L}/\partial\dot{x} = \dot{x} \tag{2.1.27'}$$

The Legendre transformation derives the Hamilton function as

$$\mathrm{H}(x,p) = p\dot{x} - \mathrm{L}(x,\dot{x}) = p^2/2 + \omega_a^2 x^2/2 \tag{2.1.28'}$$

Energy conservation gives

$$\mathrm{H}(x,p) = p^2/2 + \omega_a^2 x^2/2 = C \tag{2.1.29'}$$

where C is a constant, the conserved energy of the simplified system. The dual equations are

$$\dot{x} = \partial\mathrm{H}/\partial p = p, \quad \dot{p} = -\partial\mathrm{H}/\partial x = -\omega_a^2 x \tag{2.1.30'}$$

Hamilton-Jacobi equation is (2.1.31), and the action $S(x,t)$ is defined as (2.1.32)

$$\partial S/\partial t+[(\partial S/\partial x)^2+\omega_a^2x^2]/2=0 \tag{2.1.35'}$$

For a time-invariant system, its solution form must be

$$S(x,C,t)=W(x,C)-Ct \tag{2.1.36'}$$

where C is an integration constant, W is a characteristic function and the differential equation is

$$[(\mathrm{d}W/\mathrm{d}x)^2+\omega_a^2x^2]/2=C \tag{2.1.37'}$$

The simple harmonic motion is so simple that the characteristic function W is integrated as

$$\begin{aligned}W(x,C)&=\sqrt{2C}\int\sqrt{1-(\omega_a^2x^2)/(2C)}\mathrm{d}x\\&=\frac{C}{\omega}\left[\arcsin\left(x\sqrt{\frac{\omega_a^2}{2C}}\right)+\sqrt{\frac{\omega_a^2}{2C}}x\sqrt{1-\frac{\omega_a^2}{2C}x^2}\right]\end{aligned} \tag{2.1.38'}$$

$$S=\sqrt{2C}\int\sqrt{1-\omega_a^2x^2/(2C)}\mathrm{d}x-Ct$$

It is important that

$$\beta=-\partial S/\partial C=t-\sqrt{2/C}\int\mathrm{d}x\Big/\sqrt{1-\omega_a^2x^2/(2C)}=t-\arcsin(x\sqrt{\omega_a^2/2C})\Big/\omega_a$$

From which, x is solved as a function of time t and two parameters C,β

$$x=\sqrt{2C/(\omega_a^2)}\sin\omega_a(t-\beta)$$

the well-known simple harmonic vibration, and the momentum is

$$p=\sqrt{2C}\cos\omega_a(t-\beta)$$

The physical meaning of parameters C,β is conserved energy and phase angle, respectively. All the above derivation is for the simplified conservative system (2.1.25'). Now change the parameters C,β to be the dual variables Q,P, the transformation is still canonical

$$x=\left(\sqrt{2Q}/\omega_a\right)\times\sin\omega_a(t-P),\quad p=\sqrt{2Q}\cos\omega_a(t-P) \tag{2.1.45'a,b}$$

The above derivation is parallel to the previous section. Now return to the original non-conservative system (2.1.24). The dual equations are

$$\dot{x}=p,\quad \dot{p}=-\omega_a^2x+(\omega_a^2-1)x-bx^3-cp+f\cos\omega_a t \tag{2.1.46'a,b}$$

Substituting transformation (2.1.45'a,b) into equations (2.1.46'a,b) gives

$$\left[\sin\omega_a(t-P)\big/\left(\omega_a\sqrt{2Q}\right)\right]\dot{Q}-\left[\sqrt{2Q}\cos\omega_a(t-P)\right]\dot{P}=0 \tag{2.1.47'a}$$

Directly differentiate (2.1.45'b) gives

$$\dot{p}=\left[\cos\omega_a(t-P)\big/\sqrt{2Q}\right]\dot{Q}+\left[\omega_a\sqrt{2Q}\sin\omega_a(t-P)\right]\dot{P}-\omega_a\sqrt{2Q}\sin\omega_a(t-P)$$

Substituting into (2.1.46'b) gives

$$\begin{aligned}&\left[\cos\omega_a(t-P)\big/\sqrt{2Q}\right]\dot{Q}+\left[\omega_a\sqrt{2Q}\sin\omega_a(t-P)\right]\dot{P}=\\&\quad(\omega_a^2-1)\left(\sqrt{2Q}/\omega_a\right)\sin\omega_a(t-P)-c\sqrt{2Q}\cos\omega_a(t-P)\\&\quad-b\left(\sqrt{2Q}/\omega_a\right)^3\sin^3\omega_a(t-P)+f\cos\omega_a t\end{aligned} \tag{2.1.47'b}$$

Solving equations (2.1.47'a,b) with respect to $\dot{Q}, \dot{P}$ gives

$$\dot{Q} = \begin{bmatrix} f\cos\omega_a t + (\omega_a^2 - 1)\left(\sqrt{2Q}/\omega_a\right)\sin\omega_a(t-P) \\ -c\sqrt{2Q}\cos\omega_a(t-P) - b\left(\sqrt{2Q}/\omega_a\right)^3\sin^3\omega_a(t-P) \end{bmatrix} \times \sqrt{2Q}\cos\omega_a(t-P)$$

$$\dot{P} = \begin{bmatrix} (\omega_a^2 - 1)\left(\sqrt{2Q}/\omega_a\right)\sin\omega_a(t-P) + f\cos\omega_a t - b\cdot \\ \left(\sqrt{2Q}/\omega_a\right)^3\sin^3\omega_a(t-P) - c\sqrt{2Q}\cos\omega_a(t-P) \end{bmatrix} \cdot \frac{\sin\omega_a(t-P)}{\left(\omega_a\sqrt{2Q}\right)} \qquad (2.1.48a,b)$$

Up to here, all the derivation is exact. However, the differential equations (2.1.48a,b) are very complicated. Numerical integration directly can solve the problem. However, analytical solution gives physical insight and so is still attractive, but some approximations are necessary. Note $Q(t)$ is the energy of the simplified oscillator (2.1.25'), a periodic solution should not change this value after one cycle of oscillation with time period $T = 2\pi/\omega_a$. Hence the integration of $\dot{Q}(t)$ in one cycle must be zero, which derives

$$\int_0^T \begin{bmatrix} f\cos\omega_a t + (\omega_a^2 - 1)\left(\sqrt{2Q}/\omega_a\right)\sin\omega_a(t-P) \\ -c\sqrt{2Q}\cos\omega_a(t-P) - b\left(\sqrt{2Q}/\omega_a\right)^3\sin^3\omega_a(t-P) \end{bmatrix} \times \sqrt{2Q}\cos\omega_a(t-P)\mathrm{d}t = 0 \qquad (2.1.49a)$$

Similarly, the phase angle does not change after one cycle of oscillation, it derives

$$\int_0^T \begin{bmatrix} f\cos\omega_a t + (\omega_a^2 - 1)\left(\sqrt{2Q}/\omega_a\right)\sin\omega_a(t-P) \\ -c\sqrt{2Q}\cos\omega_a(t-P) - b\left(\sqrt{2Q}/\omega_a\right)^3\sin^3\omega_a(t-P) \end{bmatrix} \times \sin\omega_a(t-P)/\left(\omega_a\sqrt{2Q}\right)\mathrm{d}t = 0 \qquad (2.1.49b)$$

The simultaneous equations (2.1.49a,b) are to be solved for the average unknown functions $\sqrt{2Q}$ and P, however they are under the integration sign. As a first approximation, these functions are regarded as constants $\sqrt{2Q_0}$ and P_0 within the integration sign. Hence integration directly gives the simultaneous equations for $\sqrt{2Q_0}$ and P_0 as

$$f\cos\omega_a P_0 - c\sqrt{2Q_0} = 0$$

$$-f\sin\omega_a P_0 + (\omega_a - 1/\omega_a)\sqrt{2Q_0} - b\left(\sqrt{2Q_0}/\omega_a\right)^3 3/4 = 0$$

Eliminate P_0 gives

$$(2Q_0)^3 \times 9b^2/(16\omega_a^6) + (2Q_0)^2 \times (3b/2\omega_a^4)(1-\omega_a^2) + (2Q_0)\left[c^2 + (\omega_a - 1/\omega_a)^2\right] - f^2 = 0$$

After solving the root $2Q_0$ of the triple polynomial equation then the average phase angle P_0 is solved. Note that the solved Q_0, P_0 are the average value of one period. Afterwards substituting Q_0, P_0 into right hand side of equation (2.1.48'a,b), then integration, the time functions are found as $Q_{0a}(t), P_{0a}(t)$, the first order approximate solution.

The higher order solution should substitute $Q_{0a}(t), P_{0a}(t)$ into the integration sign of (2.1.49a,b), and carrying out the integration. The two periodic conditions (2.1.49a,b) should still be satisfied. The average value should be updated as $2Q_1$ and P_1, and the periodic conditions are required. Non-linear equation solution is always cumbersome.

Non-linear vibration analysis certainly concerns multi-degrees of freedom system and the canonical transformation method can also be used. This is quite interested and will be considered later in this chapter, after the multi-degrees of freedom linear system solutions have been found.

A special case is considered now, *i.e.* the conservative system given in (2.1.25), for which analytical solution is available and the solution is ***symplectic conservative.*** But the first order approximate solution by equations (2.1.49) gives $2Q_0 = 4\omega_a^2(\omega_a^2 - 1)/3b$ with $P_0 = 0$. Then the integration gives

$$\begin{aligned} Q_{0a}(t) &= (2Q_0/\omega_a)\int_0^t [(\omega_a^2 - 1)\sin\omega_a\tau - b(2Q_0/\omega_a^2)\sin^3\omega_a\tau]\cos\omega_a\tau \mathrm{d}\tau \\ &= (Q_0/2\omega_a^2)\left((\omega_a^2 - 1) + bQ_0/2\omega_a^2\right)(1 - \cos 2\omega_a t) - (bQ_0^2/16\omega_a^4)(1 - \cos 4\omega_a t) \end{aligned}$$

$$\begin{aligned} P_{0a}(t) &= (1/\omega_a^2)\int_0^t [(\omega_a^2 - 1)\sin\omega_a\tau - b(2Q_0/\omega_a^2)\sin^3\omega_a\tau]\sin\omega_a\tau \mathrm{d}\tau \\ &= [\sin 2\omega_a t(1 - \omega_a^2 + 2bQ_0/\omega_a^2) + bQ_0 \sin 4\omega_a t/4\omega_a^2]/(4\omega_a^3) \end{aligned}$$

Because of approximation, the motion can only ensure symplectic conservation after a whole cycle of vibration, but not for any time t. Note that after transformation $\dot{Q} = \partial \mathrm{K}/\partial P$, $\dot{P} = -\partial \mathrm{K}/\partial Q$, where

$$\mathrm{K}(Q,P) = Q(\omega_a^{-2} - 1)\sin^2\omega_a(t - P) + (bQ^2/\omega_a^4)\sin^4\omega_a(t - P)$$

The approximation with symplectic conservation at all the time means to find an approximate Hamilton function $\mathrm{K}_a(Q,P)$, such that $\mathrm{K}_a(Q,P) \approx \mathrm{K}(Q,P)$ and the motion follows the dual canonical equations

$$\dot{Q} = \partial \mathrm{K}_a/\partial P, \qquad \dot{P} = -\partial \mathrm{K}_a/\partial Q$$

precisely. Suppose selecting

$$\mathrm{K}_a(Q,P) = Q_0(\omega_a^{-2} - 1)\sin^2\omega_a(t - P) + (bQ_0^2/\omega_a^4)\sin^4\omega_a(t - P) \approx \mathrm{K}(Q,P)$$

the dual equation derived are

$$\begin{aligned} \dot{Q}_{0a}(t) &= Q_0(\omega_a - \omega_a^{-1})\sin 2\omega_a(t - P_{0a}) \\ &\qquad - (2bQ_0^2/\omega_a^3)\sin^2\omega_a(t - P_{0a})\sin 2\omega_a(t - P_{0a}) \end{aligned}$$

$$\dot{P}_{0a} = 0$$

integration gives $P_{0a} = 0$, and from

$$\begin{aligned} \dot{Q}_{0a}(t) &= Q_0(\omega_a - \omega_a^{-1})\sin 2\omega_a t - (2bQ_0^2/\omega_a^3)\sin^2\omega_a t \sin 2\omega_a t \\ &= \left[Q_0(\omega_a - \omega_a^{-1}) + bQ_0^2/2\omega_a^3\right]\sin 2\omega_a t - (bQ_0^2/4\omega_a^3)\sin 4\omega_a t \end{aligned}$$

Integration gives

$$Q_{0a}(t) = \left[Q_0(1 - \omega_a^{-2})/2 + bQ_0^2/4\omega_a^4\right](1 - \cos 2\omega_a t) - (bQ_0^2/16\omega_a^4)(1 - \cos 4\omega_a t)$$

That $Q_{0a}(t)$ is the same as first approximation, but $P_{0a}(t)$ is different.

The periodical solutions imply limit cycles. Limit cycle solution has the ***stability*** problem. The stability analysis should be, first perturbs the periodical motion to derive the time-variant linear ODEs, then using the Floquet theory analyze the stability problem of the ODEs. In some cases, the non-linear vibration is very sensitive with the initial values, that a very small departure of the initial value may develope ***Chaos***. Chaotic motion means unstable. For non-linear systems, the stable band nearby the limit cycle may be very thin even does not exist, which cannot bear disturbances.

For vibration problems, the integration is along the time coordinate. For wave propagation problems in frequency domain, the time coordinate is changed to be the longitudinal coordinate, however the system is still in duality form. For non-linear wave propagation problems, the similar methodology can also be applied for solution. Especially, for non-linear state space optimal control problems, based on the analogy between structural mechanics and optimal control, the duality methodology can be used too. Anyway, the canonical transformation method supplies a new methodology for solving non-linear vibration problems. It requires first solving an approximate linear system, in order to derive a canonical transformation as described above, which can be carried out by means of the eigen-solution expansion method. This is one of the reasons of why the eigen-solution is heavily considered in this book.

§2.2, Vibration of multi-degrees of freedom system

Vibration of system having two or more degrees of freedom with masses is considered multi-degrees of freedom. In various applications there are multi-degrees of freedom vibration problems everywhere. Let us derive the vibration equations using the analytical dynamics approach. According to equation (1.5.7), the expression of kinetic energy is composed of the second order, first order and zero order terms of generalized velocity $\dot{\mathbf{q}}$. In an inertia coordinate, if the position vectors $\mathbf{r}_i$ of particles relate only to the generalized vector $\mathbf{q}$ but not relate to the time t explicitly, then $T_1 = T_0 = 0$ from equation (1.5.8), the kinetic energy remains only T_2 *i.e.* the kinetic energy is a homogeneous quadratic form of the generalized velocity $\dot{\mathbf{q}}$. For small vibration problem in inertia coordinate, the factors $m_{rs}(\mathbf{q},t)$ in equation (1.5.8) are regarded as constants. Writing in vector/matrix form, the kinetic energy is

$$T = \dot{\mathbf{q}}^T \mathbf{M} \dot{\mathbf{q}} / 2 \tag{2.2.1a}$$

where $\dot{\mathbf{q}}$ denotes the n-dimensional vector of velocity, $\mathbf{M}$ denotes an $n \times n$ symmetric positive definite matrix. Nearby the equilibrium point $\mathbf{q} = \mathbf{0}$, the potential energy Π is given as

$$\Pi = \mathbf{q}^T \mathbf{K} \mathbf{q} / 2 , \qquad \mathbf{K}^T = \mathbf{K} \tag{2.2.1b}$$

where $\mathbf{K}$ is a $n \times n$ symmetric stiffness matrix, usually non-negative definite. The general form of linear vibration equation for n degrees of freedom system is

$$\mathbf{M}_n \ddot{\mathbf{y}} + \mathbf{C}_n \dot{\mathbf{y}} + \mathbf{K}_n \mathbf{y} = \mathbf{f}(t) \tag{2.2.2}$$

where the subscript n denotes the degrees of freedom and $\mathbf{C}_n$ is a damping matrix, which is symmetric, usually positive definite, and $\mathbf{f}(t)$ is an exciting force vector. The damping free vibration problems are considered first.

§2.2.1, Free vibration with no damping, eigen-solutions

For free vibration with no damping, using Lagrange function $\mathrm{L} = T - \Pi$ the multi-degrees of freedom vibration equation is derived as

$$\mathbf{M}\ddot{\mathbf{q}} + \mathbf{K}\mathbf{q} = \mathbf{0} \tag{2.2.3}$$

There are only two terms in the dynamic equation, so that the method of separation of variables applies. Let $\mathbf{q}(t) = \psi \cdot f(t)$ where $f(t)$ is a scalar function of time, and the n-dimensional vector ψ represents how the displacement varies with the component number i, $1 \le i \le n$. Substituting $\mathbf{q}$ into equation (2.2.3) derives

$$-(\ddot{f}/f)\mathbf{M}\psi = \mathbf{K}\psi \tag{2.2.4}$$

The right-hand side relates only to the subscript i but not depends on time t, so does the left-hand side, hence it must be $-(\ddot{f}/f) = \omega^2$, where ω^2 is a constant. Usually $\mathbf{K}$ is a non-negative definite matrix, so that writing the constant as ω^2 is appropriate. It is solved as

$$f(t) = a\cos\omega t + B\sin\omega t$$

and the equation for the generalized eigenvector ψ is

$$\mathbf{K}\psi - \omega^2\mathbf{M}\psi = \mathbf{0} \tag{2.2.5}$$

which is the vibration eigen-equation for multi-degrees of freedom system. The equation for eigenvalue ω^2 is therefore

$$\det(\mathbf{K} - \omega^2\mathbf{M}) = 0 \tag{2.2.6}$$

which is an n-degrees polynomial equation for ω^2. The algebraic equation must have n roots, where the m-multiple root should be accounted for as m roots. Corresponding to each eigenroot $\omega_i^2 (i = 1,2,\cdots,n)$, there is an eigenvector ψ_i satisfying equation (2.2.5). For two eigen-solutions i, j

$$\mathbf{K}\psi_i = \omega_i^2\mathbf{M}\psi_i, \qquad \mathbf{K}\psi_j = \omega_j^2\mathbf{M}\psi_j \qquad (i,j = 1,2,\cdots,n)$$

To prove the orthogonality between the two eigenvectors, multiply ψ_j^T and ψ_i^T from left, respectively, to the above two equations and then subtract each other. Based on that both the matrices $\mathbf{K},\mathbf{M}$ are symmetric, so that $\psi_j^T\mathbf{M}\psi_i = \psi_i^T\mathbf{M}\psi_j$ and similarly for $\mathbf{K}$, it derives

$$(\omega_i^2 - \omega_j^2)\psi_j^T\mathbf{M}\psi_i = 0$$

From which, it gives the orthogonal relations

$$\text{when } \omega_i^2 \ne \omega_j^2, \qquad \psi_j^T\mathbf{M}\psi_i = 0 \tag{2.2.7a}$$

and

$$\text{when } \omega_i^2 \ne \omega_j^2, \qquad \psi_i^T\mathbf{K}\psi_j = 0 \tag{2.2.7b}$$

The eigenvectors are mutually orthogonal with respect to both the mass matrix

and the stiffness matrix. Each eigenvector has an arbitrary constant factor to select, and it is selected as

$$\boldsymbol{\psi}_i^T \mathbf{M} \boldsymbol{\psi}_i = 1 \tag{2.2.8}$$

which is the normalization condition, *i.e.* normalizing with respect to the mass matrix. Therefore, all the eigenvectors compose an ortho-normal set with respect to the mass matrix.

To show that all the eigenvalues ω_i^2 ***are real numbers.*** From eigen-equation (2.2.5), that if ω_i^2 is complex, then $\boldsymbol{\psi}_i$ is also a complex vector. Using a bar above to denote the complex conjugate value, taking complex conjugate for both sides of equation (2.2.5) gives

$$\mathbf{K}\overline{\boldsymbol{\psi}}_i = \overline{\omega}_i^2 \mathbf{M}\overline{\boldsymbol{\psi}}_i, \qquad \text{which derives to} \qquad \boldsymbol{\psi}_i^T \mathbf{K}\overline{\boldsymbol{\psi}}_i = \overline{\omega}_i^2 \boldsymbol{\psi}_i^T \mathbf{M}\overline{\boldsymbol{\psi}}_i$$

Deriving as in the above proof of orthogonality relationship leads to $\boldsymbol{\psi}_i^T \mathbf{M}\overline{\boldsymbol{\psi}}_i = 0$. Triangular factorize the mass matrix

$$\mathbf{M} = \mathbf{L}\mathbf{L}^T \tag{2.2.9}$$

called the Cholesky factorization, the equation becomes $(\mathbf{L}^T\boldsymbol{\psi}_i)^T\overline{(\mathbf{L}^T\boldsymbol{\psi}_i)} = 0$, which is valid only for $\mathbf{L}^T\boldsymbol{\psi}_i$ being a null vector. But $\boldsymbol{\psi}_i$ is not a null vector, which leads to a contradiction, thus ω_i^2 should not be a complex number. Note that equation (2.2.9) implies that $\mathbf{M}$ ***is positive definite.***

Sorting the eigenvalues ω_i^2 from smaller to larger and then making use of the eigenvectors as columns composes a $n \times n$ matrix as

$$\boldsymbol{\Psi} = [\boldsymbol{\psi}_1, \boldsymbol{\psi}_2, \cdots, \boldsymbol{\psi}_n] \tag{2.2.10}$$

According to the ortho-normality of the eigenvectors

$$\boldsymbol{\Psi}^T \mathbf{M} \boldsymbol{\Psi} = \mathbf{I}_n \tag{2.2.11}$$

Let $\mathbf{S} = \mathbf{L}^T\boldsymbol{\Psi}$, then from equation (2.2.9) determines $\mathbf{S}^T\mathbf{S} = \mathbf{I}$, or

$$\mathbf{S}^T \mathbf{I}_n \mathbf{S} = \mathbf{I}_n \tag{2.2.12}$$

which explains that $\mathbf{S}$ is an orthogonal matrix. The determinant of an orthogonal matrix equals ± 1. It can always select this determinant being 1.

The eigenvalue can be solved by the variational method. According to the Hamilton variational principle, the action integration takes stationary value

$$\delta \int_0^{t_f} [\dot{\mathbf{q}}^T \mathbf{M}\dot{\mathbf{q}}/2 - \mathbf{q}^T \mathbf{K}\mathbf{q}/2] \mathrm{d}t = 0 \tag{2.2.13}$$

where the displacements $\mathbf{q}$ at the two ends $0, t_f$ are given n vectors, see (1.5.4). Substituting

$$\mathbf{q} = \mathbf{u}\sin\omega t \ , \qquad \dot{\mathbf{q}} = \omega\mathbf{u}\cos\omega t \tag{2.2.14}$$

into the variational equation (2.2.13) and let $t_f = 2m\pi/\omega$, where m is a large integer. In this trial function, ω is treated as a parameter but not varies, that the varying argument is the vector $\mathbf{u}$. Carrying out the integration gives

$$\delta\left[\mathbf{u}^T(\omega^2\mathbf{M} - \mathbf{K})\mathbf{u}\right] = 0 \tag{2.2.15}$$

where the variation is taken only for the vector $\mathbf{u}$. Left multiplying the equation (2.2.5) with $\boldsymbol{\psi}^T$ determines that the substitution of eigenvectors $\boldsymbol{\psi}_i$ into equation

(2.2.15) gives a null value inside the bracket, therefore

$$\omega_i^2 = \psi_i^T \mathbf{K} \psi_i / \psi_i^T \mathbf{M} \psi_i \quad (= \Pi/T) \tag{2.2.16}$$

called the ***Rayleigh quotient***, and the variational equation (2.2.15) can be rewritten in Rayleigh quotient form

$$\delta\left[\mathbf{u}^T\mathbf{K}\mathbf{u}/\mathbf{u}^T\mathbf{M}\mathbf{u}\right] = 0, \qquad \text{or } \delta(\Pi/T) = 0 \tag{2.2.17}$$

For the base frequency ω_1^2, the above equation takes a minimum

$$\omega_1^2 = \min_{\mathbf{u}}(\Pi/T) \tag{2.2.17'}$$

In the variational equation for natural vibrations with no damping, the argument is only the displacement vector $\mathbf{u}$ that all its components vary independently. From the variational equation all the n eigen-solutions (eigenvalue and eigenvector)

$$[\omega_i^2, \psi_i], \qquad (i = 1, 2, \cdots, n)$$

can be solved.

One of the important applications of eigenvector expansion is that an arbitrary n-dimensional vector $\mathbf{u}$ can always be expanded as the linear combination of all eigenvectors, called the eigenvector expansion (modal expansion). From equation (2.2.11), $\mathbf{\Psi}$ is not a singular matrix and all its columns (eigenvectors) compose a linearly independent basis for the n-dimensional space. An arbitrary vector $\mathbf{u}$ can always be expanded as

$$\mathbf{u} = \mathbf{\Psi} \cdot \mathbf{a} \quad (= a_1\psi_1 + a_2\psi_2 + \cdots + a_n\psi_n) \tag{2.2.18}$$

where the coefficient a_i is to determine. Based on the ortho-normality relations (2.2.6), (2.2.8), left multiplying (2.2.18) with $\psi_i^T\mathbf{M}$ gives

$$a_i = \psi_i^T \mathbf{M}\mathbf{u}, \qquad (i = 1, 2, \cdots, n) \tag{2.2.19}$$

or in combined form

$$\mathbf{a} = \mathbf{\Psi}^T\mathbf{M}\mathbf{u} \tag{2.2.19'}$$

which is the equation of ***eigenvector expansion.*** The eigenvector expansion solution is one of the fundamental methods in vibration analysis. A simple numerical example is used to demonstrate.

Example 2.1, Giving a two degrees of freedom $n = 2$ damping free vibration system, the mass and stiffness matrices $\mathbf{M}$ and $\mathbf{K}$ are given as, respectively

$$\mathbf{M} = \begin{bmatrix} 100 & 0 \\ 0 & 1 \end{bmatrix}, \qquad \mathbf{K} = \begin{bmatrix} 900 & 1 \\ 1 & 9 \end{bmatrix}$$

The time history of vibration is required with the initial condition given as

$$\text{when } t = 0, \qquad q_1 = 1,\ q_2 = 0;\ \dot{q}_1 = 0,\ \dot{q}_2 = 0$$

Solution: First solving the eigen-problem, from $\det(\mathbf{K} - \omega^2\mathbf{M}) = 0$ it solves

$\omega_1^2 = 8.9$, eigenvector: $\psi_{11} = 1.0,\ \psi_{21} = -10.0$,

and $\omega_2^2 = 9.1$, eigenvector: $\psi_{12} = 1.0,\ \psi_{22} = 10.0$

After normalization it composes

$$\mathbf{\Psi} = \frac{1}{\sqrt{2}}\begin{bmatrix} 0.1 & 0.1 \\ -1.0 & 1.0 \end{bmatrix}$$

According to the expansion theorem, the displacements can be expanded as

$$q_1 = A_1\psi_{11}\cos\omega_1 t + A_2\psi_{12}\cos\omega_2 t, \quad q_2 = A_1\psi_{21}\cos\omega_1 t + A_2\psi_{22}\cos\omega_2 t.$$

where initial condition of $\dot{\mathbf{q}} = 0$ has been considered. The initial displacement conditions are

$$A_1 + A_2 = 1, \quad -10A_1 + 10A_2 = 0. \quad \text{It solves } A_1 = A_2 = 0.5$$

then

$$q_1 = (0.5\cos\omega_1 t + 0.5\cos\omega_2 t)$$
$$= \cos[(\omega_2 - \omega_1)t/2]\cos[(\omega_2 + \omega_1)t/2] = \cos(0.0166t)\cos(3t)$$
$$q_2 = -5\cos\omega_1 t + 5\cos\omega_2 t$$
$$= 10\sin[(\omega_2 - \omega_1)t/2]\sin[(\omega_2 + \omega_1)t/2] = 10\sin(0.0166t)\sin(3t)$$

The curves of vibration are plotted in figure 2.5, which is typically vibration with beating. A slowly vibrating envelope encloses a high frequency vibration. ##

The physical interpretation for beating is that at $t = 0$ all the mechanical energy is preserved in the large mass-spring system, but the small mass-spring system having no energy. However, the two frequencies of the mass-spring systems are very close to each other, that the coupling is slight between the two mass-spring sub-systems, since $M_{11}/K_{11} \approx M_{22}/K_{22}$, and K_{12} is small. The vibration of large mass (curve a) excites the small mass-spring subsystem (curve b) to be almost in resonance (system internal resonance), that the energy is transferred from the large mass to the small mass gradually and causing the small mass vibrates severely. Afterwards, the vibration energy of the small mass transfer back to the large mass. Such transferring of energy is carrying on forth and back, and is called the wandering of energy [44].

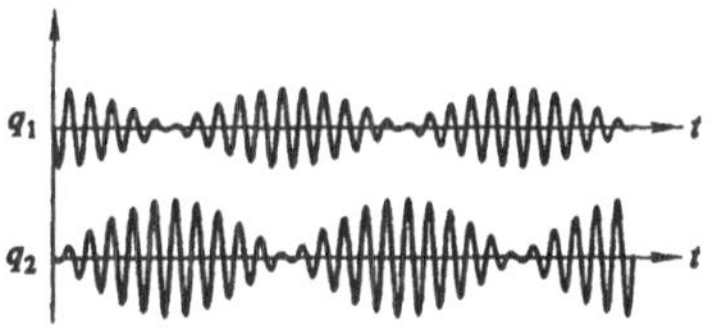

Figure 2.5, *Beating phenomenon, energy wandering*

Such kind of system ***internal resonance*** should be concerned. Anyway, all resonance should be concerned. The internal resonance can be applied to the design of dynamic absorber [4]. The effect of resonance is the exciting of the absorber mass (usually a small mass), and at the same time energy damping device is used to eliminate the vibration of the structure. However, such internal resonance may appear at different parts of a structure and may cause harmful consequence. For instance, at the top floor of a tall building some additional part may be attached, which composes a local vibration subsystem. Its local vibration frequency should be designed apart from the natural frequencies of the main part of the building. Otherwise, the whole building vibration can be imagined as a large mass M_1, and the additional part is the small mass as described in the example. Then the vibration of the local part may appear local resonance, which is called the "whip-tip effect" internal resonance [45], a severe vibration for the small mass.

The coupling between two parts in a structure is not only limited in linear systems. There are non-linear system internal couplings, which may also cause internal resonance, or non-linear parametric resonance. The same kind of internal wandering of energy is often appeared in practice. A simple model of internal parametric resonance of a cable-structure system is described later in section 2.4.

§2.2.2, Constraints, count of eigenvalues

The previous section describes eigen-solution, ortho-normality, expansion with eigenvectors etc. These behaviors are derived under the assumption that the natural vibration system (2.2.3) has no constraints. This section considers that the vibration system is under constraint, and some important theorems will be derived.

Examine the Rayleigh quotient first, if the trial vector in equation (2.2.17) is substituted with eigenvector expansion, *i.e.* the $\mathbf{u}$ in equation (2.2.18) is substituted into (2.2.17). Because

$$\boldsymbol{\Psi}^T\mathbf{M}\boldsymbol{\Psi}=\mathbf{I}_n\ ,\qquad \boldsymbol{\Psi}^T\mathbf{K}\boldsymbol{\Psi}=\mathrm{diag}(\omega_1^2,\omega_2^2,\cdots,\omega_n^2) \tag{2.2.20}$$

it derives

$$\omega^2=\mathbf{a}^T\mathrm{diag}(\omega_1^2,\cdots\omega_n^2)\mathbf{a}/(\mathbf{a}^T\mathbf{a})=\sum_{i=1}^n\omega_i^2a_i^2\Big/\sum_{i=1}^n a_i^2 \tag{2.2.21}$$

Let $\mathbf{u}$ denote the trial vector in equation (2.2.17), if the components of $\mathbf{u}$ are not all independent, *i.e.* if there are m linear constraints, then the system has only $n-m$ degrees of freedom and can have only $n-m$ eigenvalues. The eigenvalue behavior of this constraint system is of concern.

Let the linear constraints be expressed as

$$\mathbf{b}_j^T\mathbf{u}=0\ ,\qquad (j=1,2,\cdots,m),\quad m<n \tag{2.2.22}$$

where $\mathbf{b}_j$ is a given n-dimensional vector. It needs to determine the varying of eigenvalues of the constraint system, when $\mathbf{b}_j$ changes.

Using the eigenvector expansion of the unconstrained system, substituting $\mathbf{u}=\boldsymbol{\Psi}\cdot\mathbf{a}$ gives

$$(\mathbf{b}_j^T\times\boldsymbol{\Psi})\mathbf{a}=0,\quad \text{or}\quad \boldsymbol{\beta}_j^T\cdot\mathbf{a}=0,\quad \text{where}\quad \boldsymbol{\beta}_j=\boldsymbol{\Psi}^T\mathbf{b}_j,\ \ j\le m \tag{2.2.23}$$

The parameter vector $\mathbf{a}$ is varied instead of $\mathbf{u}$ in the variational principle. To find the changing of ω^2 with constraints, the equation (2.2.21) should be appended on m conditions of (2.2.23), then taking minimization for ω^2 with respect to $\mathbf{a}$. The ω^2 will depend on the constraint condition expressed with $\boldsymbol{\beta}_j$.

A special case is to consider, if the m constraints $\boldsymbol{\beta}_j$ are selected as $\boldsymbol{\beta}_1^T=\{1,0,0,\cdots,0\}$, $\boldsymbol{\beta}_2^T=\{0,1,\cdots,0\},\cdots$, then only $\mathbf{a}=\{0,\cdots,0,a_{m+1},a_{m+2},\cdots,a_n\}^T$ can be selected, then

$$\omega^2=\min_{\mathbf{a}}\ \sum_{i=m+1}^n\omega_i^2a_i^2\Big/\sum_{i=m+1}^n a_i^2=\omega_{m+1}^2$$

where the $n-m$ parameters $a_j, j=m+1,\cdots,n$ vary independently. Therefore, under this special selection of m constraints, the eigenvalue ω^2 does reach ω_{m+1}^2. Now consider the case of $\boldsymbol{\beta}_j(j=1,2,\cdots,m)$ being arbitrary vectors. Let the parameters be selected as

$$a_{m+2}=a_{m+3}=\cdots=a_n=0$$

then the arbitrary selection is limited within the range of $a_1,a_2,\cdots,a_m,a_{m+1}$, so that there are still $m+1$ variables to be selected out of the m constraint conditions (2.2.23). Limitation of the selection had been applied to the minimization, that

$$\omega^2 = \min_{\mathbf{a}} \sum_{i=1}^{m+1} \omega_i^2 a_i^2 \Big/ \sum_{i=1}^{m+1} a_i^2 \le \omega_{m+1}^2$$

from which it concludes that under m linear constraints, the minimum eigenvalue can never exceed ω_{m+1}^2. But it does can reach ω_{m+1}^2. Therefore, ω_{m+1}^2 is the true upper bound of the minimum eigenvalue under the m linear constraints. This is the well-known ***maximum-minimum principle for eigenvalues*** [1]. The characteristic of the $(m+1)$-th eigenvalue ω_{m+1}^2 is, selecting m-constraints so as to maximize the lowest eigenvalue of the constrained system. Note also, the eigenvector $\mathbf{u}$ selection of the system with given m-constraints is to minimize the eigenvalue, so that

$$\omega_{m+1}^2 = \max_{\mathbf{b}_j, j=1,\cdots,m} \left[\min_{\mathbf{u} \,|\, \mathbf{b}_j^T \cdot \mathbf{u}=0} \left(\mathbf{u}^T \mathbf{K} \mathbf{u} / \mathbf{u}^T \mathbf{M} \mathbf{u} \right) \right] \tag{2.2.24}$$

§2.2.2.1, Inclusion theorem

The maximum-minimum behavior of eigenvalue has been discussed above. Now the eigenvalue distribution for the system under one linear constraint is considered. The unconstrained system originally has n-degrees of freedom, then adding on one given constraint

$$\mathbf{b}^T \mathbf{u} = 0 \tag{2.2.25}$$

it becomes a system of $n-1$ degrees of freedom. The original unconstrained system eigenvalues are $\omega_1^2, \omega_2^2, \cdots, \omega_n^2$ sorted from small to large and the constrained system eigenvalues are $\omega_{c1}^2, \omega_{c2}^2, \cdots, \omega_{c,n-1}^2$. Then the ***inclusion theorem*** says

$$\omega_1^2 \le \omega_{c1}^2 \le \omega_2^2 \le \omega_{c2}^2 \le \cdots \le \omega_{n-1}^2 \le \omega_{c,n-1}^2 \le \omega_n^2 \tag{2.2.26}$$

which is to be proved, *i.e.* the constrained system eigenvalues are located in between the two corresponding eigenvalues of the original unconstrained system.

The proof can use the maximum-minimum behavior for eigenvalues. To prove the inequality

$$\omega_m^2 \le \omega_{c,m}^2 \le \omega_{m+1}^2, \qquad m < n \tag{2.2.26'}$$

the m-th eigenvalue $\omega_{c,m}^2$ of the original system can be considered as, the system under one given constraint (2.2.25) is further added on $m-1$ constraints and taking maximization with respect to these $m-1$ constraints. For ω_{m+1}^2, it can be considered based on the maxi-minimization principle as, the original system is added on m constraints and taking maximization with respect to these m constraints. Both the constraints for $\omega_{c,m}^2$ and for ω_{m+1}^2 are $1+(m-1)$ versus m constraints, respectively. However, for the case of $\omega_{c,m}^2$ one constraint in the m constraints is given, whereas for the case of ω_{m+1}^2 all m constraints are selected arbitrarily that the selection range is larger. Therefore it concludes that $\omega_{c,m}^2 \le \omega_{m+1}^2$. As the proof of $\omega_m^2 \le \omega_{c,m}^2$, that ω_m^2 is obtained from the original system under $m-1$ arbitrary constraints however, $\omega_{c,m}^2$ is obtained from $m-1$ arbitrary constraints and also a given constraint (2.2.25), *i.e.* $\omega_{c,m}^2$ is under one more constraints, so $\omega_{m'}^2 \le \omega_{c,m}^2$. Up to here, both inequalities of (2.2.26') have been proved. Since the m in (2.2.26') is arbitrary, so that (2.2.26) is proved also.

§2.2.2.2, Dynamic stiffness matrix and eigenvalue count

For a given ω^2, the matrix

$$\mathbf{R}(\omega^2) = \mathbf{K} - \omega^2 \mathbf{M} \tag{2.2.27}$$

is usually called as the ***dynamic stiffness matrix*** [46]. The dynamic stiffness matrix is certainly dependent on ω, and is often used in wave propagation problems, vibration problems etc. The dynamic stiffness matrix has an important index, the ***eigenvalue count***, which is given as follows. For a given $\omega_\#^2$, the n eigenvalues ω_i^2 of natural vibration of the given $\mathbf{M},\mathbf{K}$ system are subdivided into two groups of $\omega_j^2 > \omega_\#^2$ and

$$\omega_i^2 < \omega_\#^2 \,, \qquad (i = 1,2,\cdots,m) \tag{2.2.28}$$

where the number (eigenvalue count) m is needed.

The simplest consideration is, compute all the eigenvalues for the given $\mathbf{M},\mathbf{K}$ system, then check with the condition (2.2.28), the number m is readily obtained. However, solving the eigenvalue problem $\det(\mathbf{K} - \omega^2\mathbf{M})$ is not an easy problem. Especially the matrices $\mathbf{M},\mathbf{K}$ may also be frequency dependent. The requirement is that only the dynamic stiffness matrix $\mathbf{R}(\omega_\#^2)$ is calculated, and the count m is required without solving the eigen-problem for the $\mathbf{M},\mathbf{K}$ system. To find the eigenvalue count, the Sturm sequence should be considered first. ***Sturm sequence*** is defined as

$$d_0 = 1, \quad d_1 = \det[r_{11}], \quad d_2 = \det\begin{bmatrix} r_{11} & r_{12} \\ r_{21} & r_{22} \end{bmatrix}, \quad \cdots \tag{2.2.29}$$

i.e. the sequence of determinant of diagonal block of the matrix $\mathbf{R}$, see figure 2.6. The picked up main diagonal sub-matrices of $\mathbf{R}$ are still composed of the respective sub-matrices of $\mathbf{M},\mathbf{K}$ as in (2.2.27).

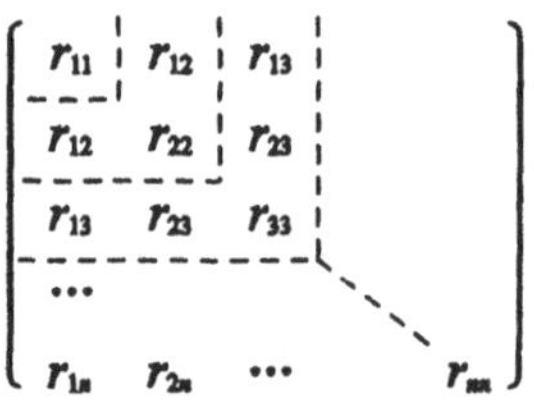

Figure 2.6, *Sturm sequence, determinant of main diagonal matrix*

Counting the sign change of the Sturm sequence $d_0, d_1, d_2, \cdots, d_n$ can reply the eigenvalue count problem (2.2.28). Because the sequence of determinant of diagonal sub-matrices, such as d_k, which corresponds to the sub-system, obtained from that the last

$(n-k)$ displacements of the original n-degrees of freedom system $\mathbf{M},\mathbf{K}$ being constrained with only the first k displacements free. Note that the dynamic stiffness matrix of this k-degrees of freedom system is just the k-th diagonal submatrix, and d_k is its determinant. Examining the sign of d_n first, that d_n is a n-degrees polynomial of $\omega_\#^2$, which can be factorized as

$$d_n \underset{\text{def}}{=} \det(\mathbf{K}-\omega_\#^2\mathbf{M}) = \det(\mathbf{K})\cdot(1-\omega_\#^2/\omega_1^2)(1-\omega_\#^2/\omega_2^2)\cdots(1-\omega_\#^2/\omega_n^2) \qquad (2.2.30)$$

Because $\mathbf{K}$ is positive definite, so that all its main diagonal determinants are positive. Hence according to (2.2.28), if m factors in (2.2.30) being negative then the sign of d_n is $(-1)^m$.

Next examining d_{n-1}, it corresponds to the determinant of the system matrix, whose last degree of freedom is constrained

$$d_{n-1} = \det(\mathbf{K}_{n-1})\cdot(1-\omega_\#^2/\omega_{c1}^2)\cdot(1-\omega_\#^2/\omega_{c2}^2)\cdots(1-\omega_\#^2/\omega_{c,n-1}^2)$$

The constrained system eigenvalue distribution is as shown in the inclusion theorem of equation (2.2.26), where $\omega_{c1}^2,\cdots,\omega_{c,m-1}^2$ is definitely less than $\omega_\#^2$; whereas $\omega_{c,m+1}^2,\cdots,\omega_{c,n-1}^2$ is definitely larger than $\omega_\#^2$; only $\omega_{c,m}^2$ is uncertain. When $\omega_{c,m}^2 < \omega_\#^2$, d_{n-1} and d_n have the same sign, but when $\omega_{c,m}^2 > \omega_\#^2$, d_{n-1} and d_n change sign. Hence, the determinant sign change means the number of eigenvalues less than $\omega_\#^2$ reduced one. Then for the $n-1$ degrees of freedom system putting on again one constraint, it becomes a $n-2$ degrees of freedom system, the same derivation applies again, etc., until all the degrees of freedom are constrained, then the determinant equals $d_0 = 1$. Therefore, *the sign change of the Sturm sequence of the dynamic stiffness matrix* $\mathbf{R}(\omega_\#^2)$ *equals the number* m *of eigenvalues* $\omega_i^2 < \omega_\#^2$ *for the original system.*

This proposition needs Sturm sequence computation, which means a number of determinant computations and is quite cumbersome. Reducing the computational expense is anticipated. Since the dynamic stiffness matrix $\mathbf{R}(\omega_\#^2)$ has been available so that the modified triangular-factorization can be executed as in [42]

$$\mathbf{R} = \mathbf{LDL}^T \qquad (2.2.31)$$

which is the commonly used factorization for symmetric matrix algorithm. The matrices $\mathbf{L},\mathbf{D}$ can be used to pick up the main diagonal matrices of $\mathbf{R}(\omega_\#^2)$. Simple verification shows that factorization (2.2.31) is still valid for the diagonal sub-matrices of $\mathbf{R}(\omega_\#^2)$. Note that $\det(\mathbf{L}) \equiv 1$, hence $\det(\mathbf{R}) = \det(\mathbf{D})$. The determinant of matrix $\mathbf{D}$ equals to the product of the diagonal elements. Hence the number of negative elements in matrix $\mathbf{D}$ equals the sign change of the Sturm sequence. Therefore, the eigenvalue count m of $\mathbf{R}(\omega_\#^2)$, as defined in (2.2.28), needs only to ***factorize the dynamic stiffness matrix*** $\mathbf{R}(\omega_\#^2)$ ***as in (2.2.31), then the number of negative elements in the diagonal matrix*** $\mathbf{D}$ ***gives the count*** m.

§2.2.3, Eigenvalue count and substructure analysis

Finite element method (FEM) has been applied in almost all the structural engineering projects. In large structural system computations, sub-structural analysis is the commonly used approach.

The vibration eigenvalue computation for the whole structure is certainly very concerned also in using substructure analysis. The structure is composed of a number of substructures and/or finite elements. The pre-request is that all the external dynamic stiffness matrices of substructures are computed and their eigen-solutions should be analyzed. Based on which, the whole structural eigen-solution analysis can be investigated, and there is the ***modal synthesis method*** to solve this problem. The sub-structural eigenvalue count is considered now. The majority FEM program systems today using the displacement method, hence the eigenvalue count for sub-structural analysis is also based on displacement method.

Looking from the elimination process of simultaneous algebraic equation, the sub-structural analysis is nothing but the sub-structural internal variables are eliminated first. Hence the previous section analysis on Sturm sequence can still be applied. In order to clarify the composition of sub-structural analysis, the case of two substructures combined together is considered first, as sketched in figure 2.7. The nodes of a substructure can be classified as

a) The ***internal nodes,*** which have no connection with external of the substructure, and the displacements on these nodes (internal displacement) can be eliminated inside the substructure;

b) The ***external nodes,*** which will further be connected to the structure, and the displacements of these external nodes (external displacement) cannot be eliminated on this substructure level, and the external stiffness matrix of the substructure is defined with respect to these external displacements.

Besides the above, there can be fixed nodes for which the displacements are given as zero (or given value), so no elimination at all for these displacements. If the composed structure is only a higher level substructure, then the external node of the higher level substructure will also be connected to the even higher level substructure (or structure).

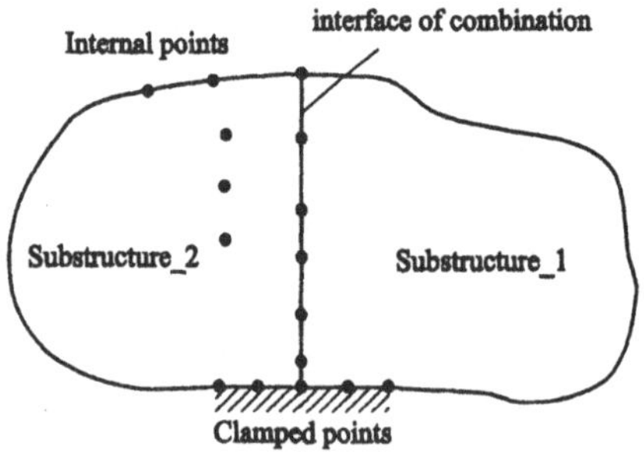

Figure 2.7, *Substructure combination*

Using displacement method implies that the dynamic stiffness matrix is used in elimination process. Using sub-structural analysis implies that the internal displacements of the substructures are eliminated first. Then at the structure level, the displacements contributed from the external displacements of the composed substructures and the displacements from the structure level itself (if any) is analyzed.

Using displacement method implies that the dynamic stiffness matrix is used in elimination process. Using sub-structural analysis implies that the internal displacements of the substructures are eliminated first. Then at the structure level, the displacements contributed from the external displacements of the composed substructures and the displacements from the structure level itself (if any) is analyzed.

The first step is the internal displacement elimination of substructure #S1. For this substructure, the internal displacements are sorted first, then the external displacements. At the substructure level, the elimination is only for the internal displacements. Let $\omega_\#$ denote the given frequency for dynamic stiffness matrix $\mathbf{R}(\omega_\#^2)$ or simply written as $\mathbf{R}$. Since displacements are classified as internal and external, the stiffness matrix is block partitioned correspondingly as

$$\mathbf{v}=\begin{Bmatrix}\mathbf{v}_i\\ \mathbf{v}_0\end{Bmatrix}\begin{matrix}n_{\mathrm{i}}\\ n_{\mathrm{o}}\end{matrix},\qquad \mathbf{R}(\omega_\#^2)=\begin{bmatrix}\mathbf{R}_{\mathrm{ii}} & \mathbf{R}_{\mathrm{io}}\\ \mathbf{R}_{\mathrm{oi}} & \mathbf{R}_{\mathrm{oo}}\end{bmatrix} \tag{2.2.32}$$

where the subscript i means internal whereas the subscript o means external, the dimension of internal displacements denotes n_{i}, respectively, etc. The elimination at the substructure level is only for the n_{i} internal displacements. After the elimination of internal displacements, the external stiffness matrix of substructure is computed as

$$\mathbf{R}'_{\mathrm{oo}}=\mathbf{R}_{\mathrm{oo}}-\mathbf{R}_{\mathrm{oi}}\mathbf{R}_{\mathrm{ii}}^{-1}\mathbf{R}_{\mathrm{io}},\quad \mathbf{R}_{\mathrm{ii}}=\mathbf{LDL}^T \tag{2.2.33}$$

These are well-known algorithms and are easily verified. The above equation shows that the inverse matrix of $\mathbf{R}_{\mathrm{ii}}$ can be executed by the modified triangular factorization method. Such as in the program system JIGFEX in China, the substructure elimination is carried out via the $\mathbf{LDL}^T$ factorization.

For static analysis, the supplying of $\mathbf{R}'_{\mathrm{oo}}$ has been enough, because static analysis does not care about natural frequency. For dynamic analysis problem, after the dynamic stiffness is computed, there is still the problem of eigenvalue count, *i.e.* the eigenvalue count $J_{\mathrm{i}}(\omega_\#^2)$ of internal natural frequencies less than the given $\omega_\#^2$, which is denoted as

$$J_{\mathrm{i}}(\omega_\#^2)\ ,\quad \text{such that}\quad \omega_{\mathrm{i}1}^2,\cdots,\omega_{\mathrm{i}J_i}^2<\omega_\#^2 \tag{2.2.34}$$

To determine this eigenvalue count $J_{\mathrm{i}}(\omega_\#^2)$, the internal stiffness matrix $\mathbf{R}_{\mathrm{ii}}$ is modified triangular factorized as in (2.2.33), and then the number of negative elements of the diagonal matrix $\mathbf{D}$ gives the $J_{\mathrm{i}}(\omega_\#^2)$. This conclusion can be seen from the sub-structural elimination of the internal variables, that all the external displacements are treated as fixed. Hence, after composing the dynamic stiffness matrix $\mathbf{R}_{\mathrm{ii}}(\omega_\#^2)$ of the internal displacements of the substructure, which is regarded as the dynamic stiffness matrix in (2.2.27), then the reason below (2.2.27) draws the conclusion, that

$$J_{\mathrm{i}}(\omega_\#^2)=\text{number of negative elements of }\mathbf{D}\text{ in (2.2.33)} \tag{2.2.35}$$

This equation can be applied to all the substructures.

After the sub-structural internal displacement elimination, the next step is the displacement elimination at the structural level. The sub-structural analysis means that all the elimination internal to the substructures had been performed for the n_i internal displacements of all the substructures, and this internal elimination gives the internal eigenvalue count $J_i(\omega_\#^2)$ of all the substructures. All the internal eigenvalue counts of the substructures should be accounted for, that is

$$\sum_{\text{subs.}} J_i(\omega_\#^2) = \sum_{\text{subs.}} (\text{number of negative elements in } \mathbf{D})$$

where the subscript 'subs' represents all the substructures. Continued process is the elimination of displacements at the structural level. This step is again a displacement elimination (constraint release) process. The dynamic stiffness matrix is composed of the external stiffness matrices of all the substructures and finite elements at the structural level. Let the dynamic stiffness matrix at the structural level be denoted as $\mathbf{R}(\omega_\#^2)$. At the structural level, its dynamic stiffness matrix can still be factorized as $\mathbf{LDL}^T$, and the method in previous section applies, *i.e.* accounting the number of negative elements for the diagonal matrix $\mathbf{D}$. Therefore the number of eigenvalue count for whole the structure (including all its substructures) is

$$J_i(\omega_\#^2) = \sum_{\text{subs.}} J_i(\omega_\#^2) + s\{\mathbf{R}(\omega_\#^2)\} \tag{2.2.36}$$

where the subscript subs. represents all the substructures, and $s\{\cdots\}$ means 'Sign count', *i.e.* the matrix in the brace is $\mathbf{LDL}^T$ factorized then the number of negative elements of the diagonal matrix $\mathbf{D}$ is accounted. The meaning is the same as in equation (2.2.35), but using a different formulation. The equation (2.2.36) is called as W-W (Wittrick-Williams) algorithm for sub-structural analysis, using the displacement method, see [40].

For multi-level sub-structural analysis, equation (2.2.36) can be applied for all the substructures level by level. The above proof can be applied to all levels of substructures.

§2.2.3.1, Mixed energy, the eigenvalue count for dual variables

The above derivation uses the displacement method. However, the elimination in purely displacement method has a problem in computations, *i.e.* when the mesh becomes extremely dense, the round off error increases very fast. Transforming to the mixed energy and dual variables method can bypass such kind of round off error problem. Correspondingly, the ***extended W-W algorithm*** for dual variable system description and ***mixed energy*** must be proposed [41].

Firstly, the mixed energy and dual (mixed) variable formulation for sub-structural analysis need to be clarified before the description of eigenvalue count. In the displacement formulation for sub-structural analysis the external variables are all displacements and then the external dynamic stiffness follows. However, in the formulation of mixed energy system, the dual variables are applied, the external nodes can be sub-divided into two classes, the first class nodes use displacements as its variables, whereas the another class nodes use the nodal force as its variables. The mixed energy formulated sub-structural analysis can be used in the 'wave front' type elimination procedure in terms of mixed variables. Formerly, the wave front method elimination method is commonly used for chain type structure. Figure 2.8

shows a sub-structural chain and the elimination from one end to the other end corresponds to the wave front progressing.

Since displacement method is the widely used model and the discussion can be led from such model. A typical $\#k$ substructure in figure.2.8 has two external surfaces k and $k+1$, or denoted as end-a and end-b of the substructure. The external displacement vector $\mathbf{q}_o$ can naturally be partitioned as the composition of $\mathbf{q}_k$ and $\mathbf{q}_{k+1}$, or denoted as $\mathbf{q}_a$ and $\mathbf{q}_b$, the respective dimensions are, respectively, n_a and n_b. The external dynamic stiffness matrix $\mathbf{R}_o(\omega_\#^2)$, which is originally of dimension $(n_a+n_b)\times(n_a+n_b)$, should be partitioned corresponding to the displacement partition as given in the equation (2.2.37). Correspondingly, the sub-structural external forces $\mathbf{p}_o$ should also be partitioned as $-\mathbf{p}_a$ and $\mathbf{p}_b$ of dimensions n_a and n_b, respectively. The relation between external forces and displacements are

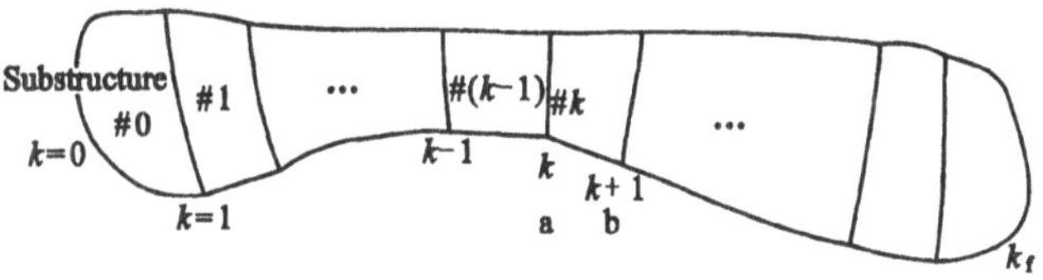

Figure 2.8, *Sub-structural chain, wave front progressing*

$$\mathbf{R}_o(\omega_\#^2)=\begin{bmatrix}\mathbf{R}_{aa} & \mathbf{R}_{ab}\\ \mathbf{R}_{ba} & \mathbf{R}_{bb}\end{bmatrix},\quad \mathbf{q}_o=\begin{Bmatrix}\mathbf{q}_a\\ \mathbf{q}_b\end{Bmatrix}\begin{matrix}n_a\\ n_b\end{matrix},\quad \mathbf{p}_o=\begin{Bmatrix}-\mathbf{p}_a\\ \mathbf{p}_b\end{Bmatrix}=\mathbf{R}_o\mathbf{q}_o \tag{2.2.37}$$

where $\mathbf{R}_o$ ***is the condensed dynamic stiffness matrix*** $\mathbf{R}'_{oo}$ of (2.2.33). It is the formulation of displacement method. The mixed formulation is a transformation from displacement to dual variables as follows.

$$\mathbf{M}_e\mathbf{w}_{ab}=\mathbf{w}_{ba},\quad \mathbf{M}_e=\begin{bmatrix}\mathbf{F} & \mathbf{G}\\ -\mathbf{Q} & \mathbf{F}^T\end{bmatrix},\quad \mathbf{w}_{ab}=\begin{Bmatrix}\mathbf{q}_a\\ \mathbf{p}_b\end{Bmatrix},\ \mathbf{w}_{ba}=\begin{Bmatrix}\mathbf{q}_b\\ \mathbf{p}_a\end{Bmatrix} \tag{2.2.38}$$

where $\mathbf{M}_e$ is termed as the ***mixed energy matrix*** of the substructure, the sub-matrices of which are derived as

$$\mathbf{Q}=\mathbf{R}_{aa}-\mathbf{R}_{ab}\mathbf{R}_{bb}^{-1}\mathbf{R}_{ba},\quad \mathbf{G}=\mathbf{R}_{bb}^{-1},\quad \mathbf{F}=-\mathbf{R}_{bb}^{-1}\mathbf{R}_{ba} \tag{2.2.39}$$

and having respective dimensions, $\mathbf{Q}$ and $\mathbf{G}$ are symmetric. Obviously, the matrix $\mathbf{M}_e$ is derived from the dynamic stiffness matrix $\mathbf{R}_o$ and the mixed energy formulation relates to the dynamic stiffness matrix formulation by a Legendre transformation. The theory is of great importance, so it is simply described as follows.

The substructure potential energy Π_o is given as

$$\Pi_o=\mathbf{q}_o^T\mathbf{R}_o\mathbf{q}_o/2=\mathbf{q}_a^T\mathbf{R}_{aa}\mathbf{q}_a/2+\mathbf{q}_b^T\mathbf{R}_{ba}\mathbf{q}_a+\mathbf{q}_b^T\mathbf{R}_{bb}\mathbf{q}_b/2$$

The nodal forces $\mathbf{p}_b$ is defined by partial differential, the Legendre transformation, then goes along the way that the mixed energy V_o is also obtained from the Legendre transformation, as follows

$$\mathbf{p}_b = \partial \Pi_o / \partial \mathbf{q}_b = \mathbf{R}_{ba}\mathbf{q}_a + \mathbf{R}_{bb}\mathbf{q}_b, \quad \mathbf{q}_b = -\mathbf{R}_{bb}^{-1}\mathbf{R}_{ba}\mathbf{q}_a + \mathbf{R}_{bb}^{-1}\mathbf{p}_b \tag{2.2.40}$$

$$\begin{aligned} V_o(\mathbf{q}_a, \mathbf{p}_b) &= \mathbf{p}_b^T\mathbf{q}_b - \Pi_o = \mathbf{q}_b^T\mathbf{R}_{bb}\mathbf{q}_b/2 - \mathbf{q}_a^T\mathbf{R}_{aa}\mathbf{q}_a/2 \\ &= \mathbf{p}_b^T\mathbf{G}\mathbf{p}_b/2 + \mathbf{p}_b^T\mathbf{F}\mathbf{q}_a - \mathbf{q}_a^T\mathbf{Q}\mathbf{q}_a/2 \end{aligned} \tag{2.2.41}$$

and the equation (2.2.38) can be expressed by

$$\mathbf{q}_b = \partial V_o / \partial \mathbf{p}_b, \qquad \mathbf{p}_a = \partial V_o / \partial \mathbf{q}_a \tag{2.2.38'}$$

For static structural analysis, $\mathbf{R}_o$ is only a usual stiffness matrix, and the eigenvalue count is unnecessary. Dynamic stiffness matrix $\mathbf{R}_o(\omega_\#^2)$ not only depends on vibration frequency but also needs the eigenvalue count. Hence the eigenvalue count $J_i(\omega_\#^2)$ of the substructure internal vibration is necessary to express its vibration characteristics. However, in mixed variable formulation, the dynamic stiffness matrix $\mathbf{R}_o(\omega_\#^2)$ is transformed to the mixed energy matrix $\mathbf{M}_e(\omega_\#^2)$, so that the eigenvalue count should also be transformed from $J_i(\omega_\#^2)$ to the mixed energy eigenvalue count $J_m(\omega_\#^2)$.

Transforming from J_i to J_m correspond to the two end conditions being that $\mathbf{q}_a$ is fixed but $\mathbf{q}_b$ is released. Using the identity

$$\mathbf{R}_o(\omega_\#^2) = \begin{bmatrix} \mathbf{R}_{aa} & \mathbf{R}_{ab} \\ \mathbf{R}_{ba} & \mathbf{R}_{bb} \end{bmatrix} = \begin{bmatrix} \mathbf{I} & \mathbf{R}_{ab}\mathbf{R}_{bb}^{-1} \\ \mathbf{0} & \mathbf{I} \end{bmatrix} \begin{bmatrix} \mathbf{R}_{aa} - \mathbf{R}_{ab}\mathbf{R}_{bb}^{-1}\mathbf{R}_{ba} & \mathbf{0} \\ \mathbf{0} & \mathbf{R}_{bb} \end{bmatrix} \begin{bmatrix} \mathbf{I} & \mathbf{0} \\ \mathbf{R}_{bb}^{-1}\mathbf{R}_{ba} & \mathbf{I} \end{bmatrix}$$

At the right hand side, both the left and right factorial matrices are right triangular and left triangular matrices, respectively, and their determinants are all unity. Hence from (2.2.34)

$$\det(\mathbf{R}_o) = \det(\mathbf{Q}) \times \det(\mathbf{R}_{bb}) \tag{2.2.42}$$

Carrying out triangular factorization for the matrices $\mathbf{R}_o, \mathbf{Q}, \mathbf{R}_{bb}$ in the form of $\mathbf{LDL}^T$, from equation (2.2.42), their diagonal matrix is the same, *i.e.* the diagonal matrix $\mathbf{D}_o$ of $\mathbf{R}_o$ is just $\mathbf{D}_o = \mathrm{diag}(\mathbf{D}_Q, \mathbf{D}_{bb})$, hence their count of negative elements have the relation

$$s\{\mathbf{R}_o\} = s\{\mathbf{Q}\} + s\{\mathbf{R}_{bb}\} \tag{2.2.43}$$

Furthermore, the matrices $\mathbf{G}$ and $\mathbf{R}_{bb}$ are mutually inverse to each other, hence $s\{\mathbf{G}\} = s\{\mathbf{R}_{bb}\}$, therefore

$$s\{\mathbf{R}_o\} = s\{\mathbf{Q}\} + s\{\mathbf{G}\} \tag{2.2.43'}$$

For the original structure (substructure is regarded as structure), the eigenvalue count equation in the displacement method formulation is given as

$$J(\omega_\#^2) = J_i(\omega_\#^2) + s\{\mathbf{R}_o\}$$

When the constraints at the right end-b are released, the left end-a stiffness matrix is $\mathbf{Q}$, therefore

$$J(\omega_\#^2) = J_m(\omega_\#^2) + s\{\mathbf{Q}\}$$

Comparison gives

$$J_{\mathrm{m}}(\omega_{\#}^2) = J_{\mathrm{i}}(\omega_{\#}^2) + s\{\mathbf{G}\} \tag{2.2.44}$$

This is the corresponding eigenvalue sign count under the mixed energy formulation.

§2.2.3.2, Substructure combination and eigenvalue count of mixed energy

The substructure combination of the displacement method makes use of the minimum potential energy. However, for dynamic stiffness matrix the potential energy will no longer be positive definite so that it does not necessarily take the minimum value, in such case minimization should be changed as taking stationary value of potential energy. The additivity of potential energy is easily understood for there is only one kind of variables.

In case of mixed energy formulation, there are two kinds of variables, so that the mixed energy should not be simply added together and the variational principle need be considered further. First, the physical background of mixed energy sub-matrices $\mathbf{Q},\mathbf{G},\mathbf{F}$ should be interpreted. Since $\mathbf{G}$ and $\mathbf{R}_{\mathrm{bb}}$ are inverse matrices to each other, hence $\mathbf{G}$ ***is the dynamic flexibility*** (admittance, compliance) matrix, under the condition that the end-a is considered fixed and the end-b is released (free) with unit forces (n unit vectors) acted on in turn. The corresponding displacements at the end-b under these unit forces compose the dynamic flexibility matrix. The two-end conditions are similar to a 'cantilever beam' and this model is used for all the submatrices $\mathbf{Q},\mathbf{G},\mathbf{F}$. Matrix $\mathbf{Q}$ uses the same 'cantilever beam' model, that the end-b is free and the end-a is given the n unit vectors of displacements in turn, the corresponding reactionary forces at the end-a compose a stiffness matrix. So $\mathbf{Q}$ ***is the stiffness matrix at end-a with end-b free***, that $\mathbf{Q}$ is a $n_{\mathrm{a}}\times n_{\mathrm{a}}$ symmetric matrix, whereas $\mathbf{G}$ is a $n_{\mathrm{b}}\times n_{\mathrm{b}}$ symmetric matrix. The matrix $\mathbf{F}^T$ is interpreted as that when end-b acted on unit forces, the reactionary force at the end-a; and the matrix $\mathbf{F}$ is a $n_{\mathrm{b}}\times n_{\mathrm{a}}$ matrix, which gives the displacement at end-b when end-a is given the unit vectors of displacement,. So, $\mathbf{F}$, $\mathbf{F}^T$ ***are displacement and force transfer matrices, respectively***.

The variational form of the mixed energy V_{ol} of substructure #Sl can make use of (2.2.38'),

$$\delta V_{\mathrm{ol}} = (\partial V_{\mathrm{ol}}/\partial \mathbf{q}_{\mathrm{a}})^T \delta\mathbf{q}_{\mathrm{a}} + (\partial V_{\mathrm{ol}}/\partial \mathbf{p}_{\mathrm{b}})\delta\,\mathbf{p}_{\mathrm{b}} = -\mathbf{p}_{\mathrm{a}}^T\delta\mathbf{q}_{\mathrm{a}} - \mathbf{q}_{\mathrm{b}}^T\delta\mathbf{p}_{\mathrm{b}} \tag{2.2.45}$$

The variational principle is quite useful to sub-structural combination. Suppose the end-b of substructure #Sl is a surface to combine another substructure #S2 , for which its external surfaces are end-b and end-c. The degrees of freedom of #S2 at end-b are also n_{b} for satisfying the condition of combination consistency, and the degrees of freedom at end-c are n_{c} , see figure 2.9. After combination, the end-b surface becomes internal to the higher level substructure, denoted as #Sc , whereas the surfaces end-a and end-c are external to the composed #Sc . Both the substructures #Sl and #S2 are expressed with mixed energy and dual variables.

The mixed energy variational equation for substructure #Sl is (2.2.45) and for substructure #S2 is

$$\delta V_{02} = -\mathbf{p}_b^T \delta \mathbf{q}_b - \mathbf{q}_c^T \delta \mathbf{p}_c$$

Then variational equation for the combined higher level substructure #Sc is

$$V_{oc}(\mathbf{q}_a, \mathbf{p}_c) = V_{01}(\mathbf{q}_a, \mathbf{p}_b) + \mathbf{p}_b^T \mathbf{q}_b + V_{02}(\mathbf{q}_b, \mathbf{p}_c) \tag{2.2.46}$$

$$\begin{aligned} \delta V_{oc} &= -\mathbf{p}_a^T \delta \mathbf{q}_a - \mathbf{q}_b^T \delta \mathbf{p}_b + \mathbf{p}_b^T \delta \mathbf{q}_b + \mathbf{p}_b^T \delta \mathbf{q}_b - \mathbf{p}_b^T \delta \mathbf{q}_b + \mathbf{q}_c^T \delta \mathbf{p}_c \\ &= -\mathbf{p}_a^T \delta \mathbf{q}_a - \mathbf{q}_c^T \delta \mathbf{p}_c \end{aligned} \tag{2.2.46'}$$

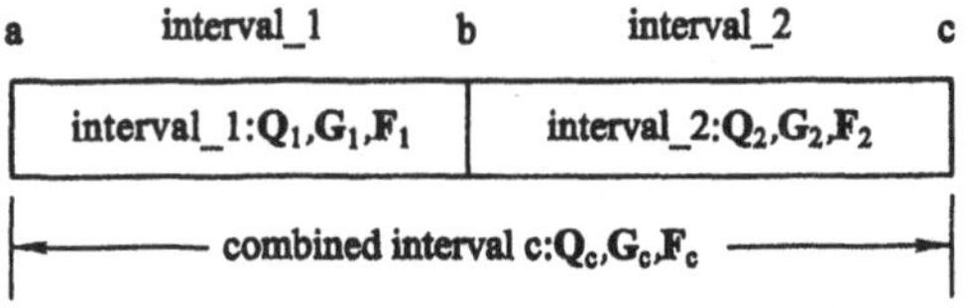

Figure 2.9, *Combination of substructures with mixed energy expressions*

Equation (2.2.46) is the mixed energy combination rule, that the elimination of internal dual variables $\mathbf{p}_b$ and $\mathbf{q}_b$ is necessary. Applying the equation (2.2.41) to both substructures #S1 and #S2 and substituting into (2.2.46) gives

$$\begin{aligned} V_{oc}(\mathbf{q}_a, \mathbf{p}_c) &= -\mathbf{q}_a^T \mathbf{Q}_1 \mathbf{q}_a/2 - \mathbf{p}_b^T \mathbf{F}_1 \mathbf{q}_a + \mathbf{p}_b^T \mathbf{G}_1 \mathbf{p}_b/2 + \mathbf{p}_b^T \mathbf{q}_b - \mathbf{q}_b^T \mathbf{Q}_2 \mathbf{q}_b/2 \\ &\quad - \mathbf{p}_c^T \mathbf{F}_2 \mathbf{q}_b + \mathbf{p}_c^T \mathbf{G}_2 \mathbf{p}_c/2 = -\mathbf{q}_a^T \mathbf{Q}_c \mathbf{q}_a/2 - \mathbf{p}_c^T \mathbf{F}_c \mathbf{q}_a + \mathbf{p}_c^T \mathbf{G}_c \mathbf{p}_c/2 \end{aligned} \tag{2.2.47}$$

To eliminate $\mathbf{q}_b$ and $\mathbf{p}_b$, the partial differentials with respect to $\mathbf{q}_b$ and $\mathbf{p}_b$ must be zero

$$\partial V_{oc}/\partial \mathbf{q}_b = 0: \quad \mathbf{p}_b - \mathbf{Q}_2 \mathbf{q}_b = \mathbf{F}_2^T \mathbf{p}_c$$

$$\partial V_{oc}/\partial \mathbf{p}_b = 0: \quad \mathbf{q}_b - \mathbf{G}_1 \mathbf{p}_b = \mathbf{F}_1 \mathbf{q}_a$$

Solving gives

$$\begin{aligned} \mathbf{q}_b &= (\mathbf{I} + \mathbf{G}_1 \mathbf{Q}_2)^{-1} (\mathbf{F}_1 \mathbf{q}_a - \mathbf{G}_1 \mathbf{F}_2^T \mathbf{p}_c) \\ \mathbf{p}_b &= (\mathbf{I} + \mathbf{Q}_2 \mathbf{G}_1)^{-1} (\mathbf{Q}_2 \mathbf{F}_1 \mathbf{q}_a + \mathbf{F}_2^T \mathbf{p}_c) \end{aligned} \tag{2.2.48}$$

Substituting $\mathbf{q}_b, \mathbf{p}_b$ back into equation (2.2.47) of V_{oc} derives

$$\left.\begin{aligned} \mathbf{Q}_c &= \mathbf{Q}_1 + \mathbf{F}_1^T (\mathbf{Q}_2^{-1} + \mathbf{G}_1)^{-1} \mathbf{F}_1, && n_a \times n_a \text{ matrix} \\ \mathbf{G}_c &= \mathbf{G}_2 + \mathbf{F}_2 (\mathbf{G}_1^{-1} + \mathbf{Q}_2)^{-1} \mathbf{F}_2^T, && n_c \times n_c \text{ matrix} \\ \mathbf{F}_c &= \mathbf{F}_2 (\mathbf{I} + \mathbf{G}_1 \mathbf{Q}_2)^{-1} \mathbf{F}_1, && n_c \times n_a \text{ matrix} \end{aligned}\right\} \tag{2.2.49}$$

These equations are the sub-structural combination equation under mixed energy formulation. Substituting these matrices into equation (2.2.47), the combined mixed energy of higher level substructure #Sc is obtained.

These combination equations are found the same as discrete time optimal control theory, see section 6.5. However in structural analysis, each combination surface can have its own dimensions, such as n_a, n_b, n_c, respectively, and is more general, see figure 2.9.

The matrices $\mathbf{Q}_1, \mathbf{Q}_2, \mathbf{G}_1, \mathbf{G}_2$ are symmetric originally. From (2.2.49) it is easily seen that $\mathbf{Q}_c, \mathbf{G}_c$ are again symmetric matrices. If $\omega_\#^2$ is small (less than the fundamental natural frequency) or for static problems $\omega_\#^2 = 0$, then $\mathbf{Q}, \mathbf{G}$ are non-negative matrices. The addition sign in the combination equation (2.2.49) determines that the mixed energy elimination equation will not appear numerical ill-conditioned problem. Such behavior is very important for localized boundary layer analysis or very dense mesh analysis.

The combination/elimination equation (2.2.49) for dynamic problem is not enough, for eigenvalue problem the eigenvalue count should be supplied further. The equation (2.2.49) can be used recursively so that the eigenvalue count should also be recursive. In addition to the equation (2.2.49), based on the known eigenvalue counts $J_{\mathrm{m1}}(\omega_{\#}^2)$ and $J_{\mathrm{m2}}(\omega_{\#}^2)$ of substructures, the combined eigenvalue count $J_{\mathrm{mc}}(\omega_{\#}^2)$ of #Sc should be determined.

To derive $J_{\mathrm{mc}}(\omega_{\#}^2)$, the first step is to transform back to the both ends given displacement form. From equation (2.2.44), it derives $J_{\mathrm{i1}}(\omega_{\#}^2) = J_{\mathrm{m1}}(\omega_{\#}^2) - s\{\mathbf{G}\}$. However the #S2 substructure is still in mixed energy formulation, *i.e.* given displacement at end-b and given force at end-c. Thus at the combination surface, end-b, both substructures #S1 and #S2 are given displacement. Then the internal stiffness matrix at end-b must be the superposition from both the sub-structural contributions $(\mathbf{G}_1^{-1} + \mathbf{Q}_2)$. Such kind of displacement elimination equation can use the W-W algorithm equation (2.2.36), which derives the eigenvalue count for the combined higher level substructure

$$J_{\mathrm{mc}}(\omega_{\#}^2) = J_{\mathrm{i1}}(\omega_{\#}^2) + J_{\mathrm{m2}}(\omega_{\#}^2) + s\{\mathbf{G}_1^{-1} + \mathbf{Q}_2\}$$

Transforming the eigenvalue count $J_{\mathrm{i1}}(\omega_{\#}^2)$ back to $J_{\mathrm{m1}}(\omega_{\#}^2)$ gives

$$J_{\mathrm{mc}}(\omega_{\#}^2) = J_{\mathrm{m1}}(\omega_{\#}^2) + J_{\mathrm{m2}}(\omega_{\#}^2) + s\{\mathbf{G}_1^{-1} + \mathbf{Q}_2\} - s\{\mathbf{G}_1\} \tag{2.2.49a}$$

which is the required equation for the ***extended W-W algorithm*** [41].

Eigenvalue count is very important for substructure vibration, stability, wave-guide and a number of other problems. For optimal control, it is again a fundamental issue. The displacement method for solving two point boundary value problems (TPBVP) may possibly have numerical ill-conditioning problem, the mixed energy and dual variables method can avoid such problem and the precise integration algorithm can effectively be used to TPBVP.

For eigenvalue problem, eigenvalue count is usually necessary.

§2.2.3.3, Essence of modal synthesis method

The generation of dynamic stiffness matrix $\mathbf{R}(\omega_{\#}^2)$ at structural level has various methods. One of sub-structural ***modal synthesis*** methods is simply described here.

Suppose the structure is composed of one substructure #S1 and other finite elements. The method is easily extended to the case of multiple substructures. The displacements of a substructure are classified as external and internal displacement vectors $\mathbf{v}_{\mathrm{o}}$ and $\mathbf{v}_{\mathrm{i}}$ of n_{o} and n_{i}-dimensioned, respectively. In practical applications of FEM, number n_{o} is finite, however, n_{i} may be very large. Note that the dynamic stiffness matrix $\mathbf{R}(\omega_{\#}^2)$ depends on the given frequency, hence solving the eigen-solution needs to compute the dynamic stiffness matrix $\mathbf{R}(\omega_\#^2)=\mathbf{K}-\omega_\#^2\mathbf{M}$ repeatedly for different $\omega_\#^2$, which means tremendous computational expense and is difficult to bear. Suppose the static stiffness and mass matrices are computed, then corresponding to the classification of displacements as internal and external, the submatrices of internal, external and mutual $\mathbf{R}_{\text{ii}}(\omega_\#^2)$, $\mathbf{R}_{\text{oo}}(\omega_\#^2)$ and $\mathbf{R}_{\text{io}}(\omega_\#^2)$ dynamic stiffness matrices, respectively, are formed. The sub-structural analysis needs to eliminate the internal displacement in order to supply the condensed external stiffness matrix $\mathbf{R}'_{\text{oo}}(\omega_\#^2)$, as in equation (2.2.33). However, if the condensation is executed according to the equation (2.2.33) for all the frequencies $\omega_\#^2$ in iteration, tremendous computational expense is resulted. Hence it is required to avoid the modified triangular factorization of the internal dynamic stiffness matrix $\mathbf{R}_{\text{ii}}$ for all the iterations of $\omega_\#^2$. In application, the required natural frequencies are usually at the low end, hence the higher frequency vibration internal to the substructure can be neglected, that computing some low frequency internal eigen-solutions of the internal displacement $\mathbf{v}_\text{i}$ is usually good enough.

$$\omega_{\text{i}j}^2,\boldsymbol{\psi}_{\text{i}j},(j=1,\cdots,n_{\text{i}d})$$

where $n_{\text{i}d} << n_\text{i}$ is the number of low frequencies of the internal eigen-solutions. These internal eigen-solutions are obtained based on the n_o external displacements being treated as fixed, $\mathbf{v}_\text{o}=\mathbf{0}$. Usually $n_{\text{i}d} << n_\text{i}$, these internal eigen-solutions $\omega_{\text{i}j}^2,\boldsymbol{\psi}_{\text{i}j},(j=1,\cdots,n_{\text{i}d})$ can be solved by means of various methods, such as subspace iteration method. These eigen-solutions are unrelated to the given frequency $\omega_\#^2$, so once computed they can be used over and over again. After these internal eigen-solutions are computed, the elimination of internal displacements can use the eigen-solution expansion method. The condensation equation can be derived as

$$\left[\mathbf{R}_{\text{ii}}(\omega_\#^2)\right]^{-1}=\sum_{j=1}^{n_{\text{i}d}}\left(\frac{\boldsymbol{\psi}_{\text{i}j}\boldsymbol{\psi}_{\text{i}j}^T}{\omega_{\text{i}j}^2-\omega_\#^2}\right)\quad\text{and}\quad\mathbf{R}_{\text{oi}}\mathbf{R}_{\text{ii}}^{-1}\mathbf{R}_{\text{io}}=\sum_{j=1}^{n_{\text{i}d}}\left(\frac{(\mathbf{R}_{\text{oi}}\boldsymbol{\psi}_{\text{i}j})(\mathbf{R}_{\text{oi}}\boldsymbol{\psi}_{\text{i}j})^T}{\omega_{\text{i}j}^2-\omega_\#^2}\right)$$

If $n_{\text{i}d}=n_\text{i}$ these equations are exact. Using these condensation equations in (2.2.33) gives the sub-structural modal synthesis approach.

However, the above equation considers only the former $n_{\text{i}d}$ eigenvectors, the effect of the latter $(n_\text{i}-n_{\text{i}d})$ eigenvectors are completely neglected. The reason is because of $\omega_{\text{i}j}^2>>\omega_\#^2$ so that the error is not large and also to save the computational expense. In order to improve the approximation, the following equation can be used instead

$$\left[\mathbf{R}_{\text{ii}}(\omega_\#^2)\right]^{-1}\cong\sum_{j=1}^{n_{\text{i}d}}\left(\boldsymbol{\psi}_{\text{i}j}\boldsymbol{\psi}_{\text{i}j}^T/(\omega_{\text{i}j}^2-\omega_\#^2)\right)+\sum_{j=n_{\text{i}d}+1}^{n_\text{i}}\left(\boldsymbol{\psi}_{\text{i}j}\boldsymbol{\psi}_{\text{i}j}^T/\omega_{\text{i}j}^2\right)$$

However, only the former $n_{\text{i}d}$ eigen-solutions are available, therefore the latter term in the above equation should be reformulated. That is

$$\sum_{j=n_{id}+1}^{n_i}\left(\boldsymbol{\psi}_{ij}\boldsymbol{\psi}_{ij}^T/\omega_{ij}^2\right)=\sum_{j=1}^{n_i}\left(\boldsymbol{\psi}_{ij}\boldsymbol{\psi}_{ij}^T/\omega_{ij}^2\right)-\sum_{j=1}^{n_{id}}\left(\boldsymbol{\psi}_{ij}\boldsymbol{\psi}_{ij}^T/\omega_{ij}^2\right)=\mathbf{K}_{\rm ii}^{-1}-\sum_{j=1}^{n_{id}}\left(\boldsymbol{\psi}_{ij}\boldsymbol{\psi}_{ij}^T/\omega_{ij}^2\right)$$

where the former summation is just $\mathbf{R}_{\rm ii}^{-1}(0)=\mathbf{K}_{\rm ii}^{-1}$, and this inverse matrix has been used in finding the internal eigen-solutions and is available. The eigenvectors in the latter summation are also available so that the above equation can be executed effectively.

§2.2.4, Subspace iteration for the eigen-solution of symmetric matrices

The eigenvalue count is discussed in some detail in the previous sections, so as to ensure no eigenvalue is dropped in numerical computation. However, the eigenvector expansion (modal analysis) is one of the most popular methods used in structural vibration analysis. Application requires numerical eigen-solutions. Theoretically, all the eigen-solutions should be found for expansion solution. In applications, the FEM analysis model often has degrees of freedom $n=$ dozens of thousands. The eigen-solutions have also same amount of n, among which the majority is high frequency vibrations. Under damping action, the high frequency vibration will quickly be damped out and the long time vibration is mainly dominated by lower frequency components. Also too many eigen-solutions brings a lot of difficulties in computation. Hence in structural engineering applications, only the number q lowest frequency eigen-solutions need to be solved, for example $q=50\sim 500$ for structural random vibrations. The eigen-solutions of q dimensional subspace with lowest frequency eigenvalues for a structural system attracted much attention, and the subspace iteration method [8] is one of the most effective algorithms. In sub-structural internal eigen-solution analysis, the subspace iteration method can also be used.

Selecting number q approximate eigenvectors of lowest frequency as basis, a q dimensional subspace is composed in the n-dimensional complete space. To find these eigen-solutions, iteration method is necessary. Suppose the first q *approximate* eigenvectors $\boldsymbol{\psi}'_j, j=1,\cdots,q$ have been selected, and according to the expansion theorem these approximate vectors can be expanded with the eigenvectors as

$$\boldsymbol{\psi}'_j=\sum_{i=1}^{n}c_{ji}\boldsymbol{\psi}_i,\quad j=1,\cdots,q \tag{2.2.50}$$

The purpose of subspace iteration is to improve the approximate eigenvectors $\boldsymbol{\psi}'_j, j=1,\cdots,q$ converging to the first q eigenvectors, *i.e.*

$$c_{ji}=\delta_{ji},\quad j\le q,\ i\le n, \text{ where } \delta_{ji}=1 \text{ when } i=j,\quad \text{otherwise } \delta_{ij}=0$$

Note that the eigenvectors are ortho-normal to each other with respect to the mass matrix, although $\boldsymbol{\psi}'_j, j=1,\cdots,q$ are not really eigenvectors, however, they still need to satisfy the ortho-normality condition. Below, the ortho-normalization algorithm is proposed as the Gram-Schmidt algorithm.

§2.2.4.1, Ortho-normalization algorithm

The ortho-normality for the eigenvectors is with respect to the mass matrix $\mathbf{M}$. The approximate vectors $\psi'_j, j=1,\cdots,q$ should also be ortho-normalized with respect to $\mathbf{M}$. The algorithm is:

for (*j=1; j<=q; j++*) { *comment: below line is orthogonalization*

for (*k=1; k<j; k++*) { $a = \psi'^T_k \mathbf{M}\psi'_j;\ \psi'_j = \psi'_j - a\psi'_k$ }

[$a = \psi'^T_j \mathbf{M}\psi'_j;\quad a = \text{sqrt}(a);\quad \psi'_j = \psi'_j / a;$] *comment: normalize*

} (2.2.51)

These ortho-normalized approximate vectors can be treated as the basis of subspace. And the mass and stiffness matrices should be projected to the subspace as follows.

§2.2.4.2, Subspace projection and its eigen-solutions

In a n-dimensional space, there are n basis vectors, and $\psi'_j, j=1,\cdots,q$ are only a part of them. The other $(n-q)$ basis vectors should be orthogonal to the q basis with respect to the mass matrix. Compose a $n \times q$ approximate eigenvector matrix using $\psi'_j, j=1,\cdots,q$ as columns

$$\mathbf{\Phi} = [\psi'_1, \psi'_2, \cdots, \psi'_q] \tag{2.2.52}$$

For ortho-normalized basis, the ***projection mass matrix*** is then $\mathbf{\Phi}^T\mathbf{M}\mathbf{\Phi} = \mathbf{I}_q$; and the ***projection stiffness matrix*** is

$$\mathbf{K}_q = \mathbf{\Phi}^T\mathbf{K}\mathbf{\Phi} \tag{2.2.53}$$

Evidently, it is still a symmetric matrix. Hence the eigen-problem in the q-dimensional projected subspace is

$$\mathbf{K}_q\xi = s^2\xi \tag{2.2.54}$$

which is a standard eigen-problem for a symmetric matrix with dimension q, and standard algorithm is available. For example, the standard procedures given in [42,43] can be invoked, so that all the eigen-solutions $s_j^2, \xi_j, j=1\sim q$ in subspace are obtained readily. Where s_j^2 are eigenvalues in q-dimensional subspace. Please do not confuse with the eigenvalues ω_j^2 in the original n-dimensional complete space.

§2.2.4.3, Subspace iteration

After all the q eigen-solutions $s_j^2, \xi_j, j=1\sim q$ are solved, these q-dimensional subspace eigen-solutions should be transformed back to the original n-dimensional space, *i.e.*

$$\boldsymbol{\varphi}_j = \boldsymbol{\Phi}\boldsymbol{\xi}_j, \qquad j = 1 \sim q \tag{2.2.55}$$

Although $\boldsymbol{\xi}_j$ is the eigenvector in the q-dimensional subspace, however, $\boldsymbol{\varphi}_j$ is not an eigenvector in the original n-dimensional space but only an approximate eigenvector. Although the eigen-solutions have been solved in the subspace and transformed back to the original space, but the linear combination of these q vectors is still within the same subspace spanned by the matrix $\boldsymbol{\Phi}$, *i.e.* the subspace does not change. To improve the subspace, these basis vectors $\boldsymbol{\varphi}_j$, $j = 1, \cdots, q$ should be updated in the original space via iteration, *i.e.* ***subspace rotation***. The new improved approximate basis are computed as

$$\boldsymbol{\psi}_j' = s_j^2 \mathbf{K}^{-1}\mathbf{M}\boldsymbol{\varphi}_j \ , \ \ j = 1, \cdots, q \tag{2.2.56}$$

The iteration equation above means that the subspace basis vectors rotate in the original n-dimensional space. This is the ***subspace iteration***.

The effect of equation (2.2.56) is to clarify. Expanding the vectors $\boldsymbol{\varphi}_j$ with all the eigenvectors $\boldsymbol{\psi}_i, i = 1, \cdots, n$ gives

$$\boldsymbol{\varphi}_j = \sum_{i=1}^{n} d_{ji}\boldsymbol{\psi}_i, \quad j = 1, \cdots, q \tag{2.2.57}$$

Substituting into equation (2.2.56) gives

$$\boldsymbol{\psi}_j' = \sum_{i=1}^{n} s_j^2 d_{ji}\boldsymbol{\psi}_i / \omega_i^2, \quad j = 1, \cdots, q \tag{2.2.58}$$

These updated approximate vectors $\boldsymbol{\psi}_j', j = 1, \cdots, q$ are no longer involved in the q-dimensional subspace before updating (spanned by $\boldsymbol{\varphi}_j$), which means that the subspace is rotated in the original n-dimensional space. Except ω_1^2 in (2.2.58) the higher the frequency ω_i^2, the larger the denominator, and after the iteration step its involvement is reduced. Therefore, the purpose of subspace rotation is to filter out the high frequency components. After quite a number of subspace iterations, only the first q eigenvectors will remain, *i.e.* $\boldsymbol{\psi}'_j, j = 1, \cdots, q$ will converge to $\boldsymbol{\psi}_j, j = 1, \cdots, q$.

It is the essence of subspace iteration. The algorithm of one iteration step is given as:

[The subspace's dimension q, and $\boldsymbol{\psi}'_j, j = 1, \cdots, q$ are selected]
[Ortho-normalize with respect to $\mathbf{M}$]
[Compose the $n \times q$ matrix $\boldsymbol{\Phi}$, see (2.2.52)]
[Computing the projected stiffness matrix, see (2.2.53)]
[Finding the subspace eigen-solutions, $s_j^2, \boldsymbol{\xi}_j, j = 1 \sim q$]
[Returning to the original space, (2.2.55)]
[Subspace rotation, (2.2.56)] (2.2.59)

To speed up convergence, the subspace rotation (2.2.56) can execute twice in one iteration rather than once.

§2.2.4.4, Transition of the subspace

To find an initial approximation is always a problem for the iteration method. The subspace dimension q can also be selected flexibly. Let the number of eigen-solutions required be denoted as p, such as in structural random vibration $p = 50 \sim 500$ is needed. However, the iterative dimension q of subspace can be volatile, such as at the beginning $q = 1$ and the initiate 'approximation' selected as $\psi'_1 = \{1,0,\cdots,0\}^T$ or any, because of no effective information available originally. After a round of iteration, the updated ψ'_1 has filtered out much higher frequency eigenvectors, therefore for the next round of iteration, the last ψ'_1 can be regarded as ψ'_2, and the dimension of subspace is updated as $q = 2$. For further iteration steps, the dimension q of subspace can be further enlarged, and the possible approximate vectors supplied are from the last subspace basis vectors.

The iteration needs a stop check, such as to check the current smallest eigenvalue with the last (several) iterations, if the change is negligible, then this eigen-solution is considered converged and can be picked up to store into the database. Then the subspace should be shifted one eigen-solution, which has been picked up and stored into the database, and one another new approximate eigenvector must be supplied. There are q approximate vectors updated in one round of iteration, one of the approximate vectors before iteration is used to substitute the picked up one, then continue the iteration. Note, in the ortho-normalization with respect to the mass matrix, these eigenvectors stored in the database must be involved.

When the number of eigenvectors filled in the database exceeds or equals to the required number p, the iteration stops.

§2.2.5, Eigen-solutions of asymmetric real matrix

The eigen-solution of an asymmetric real matrix is often encountered in applications. To solve an eigen-problem of a $n \times n$ asymmetric matrix is a key step for the eigen-solution of a $2n \times 2n$ Hamilton matrix as given later in section 2.3.3.3. The eigen-problem of a n-dimensional real matrix $\mathbf{A}$ is

$$\mathbf{Ax} = \mu\mathbf{x} \tag{2.2.60}$$

Its eigenvalue μ should satisfy the equation

$$\det(\mathbf{A} - \mu\mathbf{I}) = 0 \tag{2.2.61}$$

a n-degrees polynomial algebraic equation for roots μ. It has n eigen-roots and the m-multiple roots must be accounted as m. According to the eigen-problem in matrix theory, an asymmetric matrix may have degenerate (m-multiple) eigen-solutions and the ***Jordan normal form*** may appear. A theorem was proven as: ***Given an*** $n \times n$ ***matrix*** $\mathbf{A}$, ***there exists a non-singular*** $n \times n$ ***complex valued matrix*** $\mathbf{X}$ ***such that***

$$\mathbf{AX} = \mathbf{X} \cdot \mathrm{diag}(\mathbf{J}_1, \mathbf{J}_2, \cdots, \mathbf{J}_t) \tag{2.2.62}$$

where $m_1 + m_2 + \cdots + m_t = n$ ***, and the*** $m_i \times m_i$ ***Jordan block matrix*** $\mathbf{J}_i$ ***,*** ($i = 1, \cdots, t$) ***has the form***

$$\mathbf{J}_i = \begin{bmatrix} \mu_i & 1 & & \\ & \mu_i & 1 & \\ & & \ddots & 1 \\ & & & \mu_i \end{bmatrix} \quad m_i \tag{2.2.63}$$

where μ_i ***is the*** m_i ***-multiple eigenvalue and there are*** m_i ***corresponding columns in the matrix*** $\mathbf{X}$***, and the correspondent equations are***

$$\begin{aligned} \mathbf{A}\mathbf{x}_{i1} &= \mu_i \mathbf{x}_{i1} \\ \mathbf{A}\mathbf{x}_{i2} &= \mu_i \mathbf{x}_{i2} + \mathbf{x}_{i1} \\ &\cdots \\ \mathbf{A}\mathbf{x}_{i,m_i} &= \mu_i \mathbf{x}_{i,m_i} + \mathbf{x}_{i,(m_i-1)} \end{aligned} \tag{2.2.64}$$

for each Jordan block. Different Jordan blocks may possibly have the same eigenvalue. This is the eigen-solution theorem for the general real matrix, see [30,47,48].

For small n-dimensional matrix $\mathbf{A}$, there is standard procedure to invoke for eigen-solutions, see [42,43]. However, when Jordan form appears, the numerical result is unstable. Although the Jordan form rarely appears in applications, but cannot be excluded. The usual cases are that all eigenvalues are different and the eigenvectors are not nearly parallel. Only when some eigenvalues are very close to each other, then the corresponding eigenvectors are almost parallel.

An asymmetric matrix usually has complex conjugate eigenvalue pairs and the corresponding eigenvectors are also complex conjugate to each other.

When the dimension n is large, finding all its eigen-solutions is unnecessary then the dual subspace iteration method [49] can be used to find the number p eigen-solutions with smallest absolute eigenvalues.

§2.2.5.1, Dual subspace iteration method for asymmetric matrix

In the case of no Jordan form appears, the eigen-matrix equation becomes

$$\mathbf{AX} = \mathbf{X} \cdot \mathrm{diag}(\mu_i), \quad \mathrm{diag}(\mu_i) \underset{\mathrm{def}}{=} \mathrm{diag}(\mu_1, \mu_2, \cdots, \mu_n) \tag{2.2.65a}$$

Its transposed matrix $\mathbf{A}^T$, having the same eigenvalues as $\mathbf{A}$, also has the eigen-matrix equation

$$\mathbf{A}^T\mathbf{Y} = \mathbf{Y} \cdot \mathrm{diag}(\mu_i), \quad \text{where } \mathbf{Y} = \mathbf{X}^{-T} \tag{2.2.65b}$$

Solving the eigen-matrix $\mathbf{X}$, which is composed of all the eigenvectors, implies that the dual eigen-matrix $\mathbf{Y}$ is also solved, equation (2.2.65b) gives the relation between the two matrices $\mathbf{X}$ and $\mathbf{Y}$, the dual ortho-normality relation. However if only number q eigenvectors $\mathbf{x}_1, \mathbf{x}_2, \cdots, \mathbf{x}_q$ are known, then from the composed defective $n \times q$ matrix

$$\mathbf{X}_q = [\mathbf{x}_1 \ \ \mathbf{x}_2 \ \ \cdots \ \ \mathbf{x}_q] \tag{2.2.66a}$$

the equation (2.2.65b) cannot be used to find the dual defective eigen-matrix which is composed of the dual eigenvectors $\mathbf{y}_1, \mathbf{y}_2, \cdots, \mathbf{y}_q$. Hence the two defective eigen-matrices $\mathbf{X}_q$ and $\mathbf{Y}_q$ can only be solved simultaneously via iteration. Let

$$\mathbf{Y}_q = [\mathbf{y}_1 \quad \mathbf{y}_2 \quad \cdots \quad \mathbf{y}_q] \tag{2.2.66b}$$

$$\mathbf{Y}_q^T \mathbf{X}_q = \mathbf{I}_q \tag{2.2.67}$$

In the iteration process, the eigenvectors have not been found yet, so that all vectors are approximate. Dual subspaces are composed of using these approximate vectors as basis. The subspace projection $q \times q$ matrix is defined as

$$\mathbf{A}_q = \mathbf{Y}_q^T \mathbf{A} \mathbf{X}_q \tag{2.2.68}$$

the dimension has been reduced to be manageable, for $q << n$. Hence all the eigen-solutions of projection matrix $\mathbf{A}_q$ can be solved via invoking the standard procedure in [42]

$$\mathbf{A}_q \boldsymbol{\xi} = \boldsymbol{\xi} \cdot \mathrm{diag}(\mu_1, \mu_2, \cdots, \mu_q) \ , \quad [\text{or simply } = \boldsymbol{\xi} \cdot \mathrm{diag}(\mu_q)\,] \tag{2.2.69a}$$

and

$$\mathbf{A}_q^T \boldsymbol{\eta} = \boldsymbol{\eta} \cdot \mathrm{diag}(\mu_q) \tag{2.2.69b}$$

where $\boldsymbol{\xi}, \boldsymbol{\eta}$ are $q \times q$ dual eigen-matrices, $\boldsymbol{\eta}^T \boldsymbol{\xi} = \mathbf{I}_q$. However, these are only eigen-solutions in the project dual subspaces but not eigen-solutions in the original spaces, such situation is similar to the subspace iteration for symmetric matrix. Especially, the eigenvalues are not the eigenvalues of $\mathbf{A}$ either, so $\mathrm{diag}(\mu_q)$ is also approximate.

Having solved the dual eigen-matrices $\boldsymbol{\xi}, \boldsymbol{\eta}$ in the dual subspaces, the next step is to transform back to the original dual subspaces

$$\mathbf{X}_q := \mathbf{X}_q \cdot \boldsymbol{\xi}; \qquad \mathbf{Y}_q := \mathbf{Y}_q \cdot \boldsymbol{\eta} \tag{2.2.70a,b}$$

where the sign := is read as 'substituted by'. This step is only recomposing the dual subspaces, that the dual ortho-normality relation (2.2.67) still holds, but the dual subspaces have not been updated. In order to improve the columns of dual matrices $\mathbf{X}_q, \mathbf{Y}_q$ approaching the dual eigenvectors, the dual subspaces must be updated (rotated) in the original dual spaces. From equation (2.2.65a,b), the rotations can be

$$\mathbf{X}_q := \mathbf{A}^{-1} \mathbf{X}_q \cdot \mathrm{diag}(\mu_q)\,; \quad \mathbf{Y}_q := \mathbf{A}^{-T} \mathbf{Y}_q \cdot \mathrm{diag}(\mu_q) \tag{2.2.71a,b}$$

After this round of dual subspace rotation, the defective dual sub-matrices $\mathbf{X}_q, \mathbf{Y}_q$ should be dual ortho-normalized again, which can be executed as

```
for  (i = 1; i ≤ q; i++)  {
   for  (j = 1;  j ≤ i−1;  j++) {
      [Orthogonalize the  i-th column of  X_q  to the  j-th column of  Y_q]
      [Orthogonalize the  i-th column of  Y_q  to the  j-th column of  X_q]
   }
   [Normalize both the  i-th columns of  X_q  and  Y_q]
}
```

(2.2.72)

It is pointed out that the eigen-solution of an asymmetric matrix $\mathbf{A}_q$ possibly has complex conjugate eigenvalues. The complex conjugate eigenvalues $\alpha \pm \beta i$ appears simultaneously with the eigenvectors denoted as $\mathbf{f} \pm \mathbf{g}i$. However it is easier to compute with real arithmetic in numerical computation, for which the real equation is

$$\mathbf{A}_q \cdot [\mathbf{f}, \mathbf{g}] = [\mathbf{f}, \mathbf{g}] \cdot \begin{bmatrix} \alpha & \beta \\ -\beta & \alpha \end{bmatrix} \tag{2.2.73}$$

and the complex valued diagonal matrix $\text{diag}(\mu_q)$ is updated as block diagonal matrix with 2×2 sub-blocks as in (2.2.73). The original approximate $n \times q$ dual matrices $\mathbf{X}_q, \mathbf{Y}_q$ are also real. After the above clarification, the dual subspace iteration algorithm of a real asymmetric matrix can be given as [23,49]

1) [Selecting q and the initial dual approximate $n \times q$ matrices $\mathbf{X}_q, \mathbf{Y}_q$]
2) [Dual ortho-normalizing according to (2.2.72) for $\mathbf{X}_q, \mathbf{Y}_q$]
3) [According to (2.2.68) compute the projection submatrix $\mathbf{A}_q$]
4) [Solving all eigen-solutions of $\mathbf{A}_q$ get $\text{diag}(\mu_q)$ and $\boldsymbol{\xi}, \boldsymbol{\eta}$]
5) [Returning the original dual subspaces, (2.2.70a,b)]
6) [If the successive iterative solutions satisfying precision condition, stop]
7) [Dual subspaces rotation (2.2.71a,b), then go to the 2nd step] (2.2.74)

Some improvements in the implementation, such as transition subspaces, are similar to the symmetric matrix case and are neglected. However, asymmetric matrix may appear complex conjugate eigen-solutions. According to the above algorithm, the converged eigen-solutions are sorted according to the absolute eigenvalues in ascending order. But the fixed dual subspace dimension q may possibly separate the two complex conjugate eigen-solutions, *i.e.* the q-th approximate solution is just one of the complex conjugate but the another one is the $(q+1)$-th, not involved in the subspace. The effect of such separation of complex conjugate eigen-solutions will cause eigenvalues jumping phenomenon in successive iterations. Hence, if the case of jumping eigenvalues emerges the subspace dimension q should increase 1 and continuing the iteration. The reasoning of convergence of dual subspace for asymmetric matrix is similar to the case of subspace iteration method for symmetric matrix.

§2.2.6, Singular value decomposition

For a given $m \times n$ $(m \geq n)$ real matrix $\mathbf{A}$, the $n \times n$ product matrix $\mathbf{A}^T\mathbf{A}$ is obviously symmetric and non-negative definite, because

$$\mathbf{x}^T \mathbf{A}^T \mathbf{A} \mathbf{x} = (\mathbf{A}\mathbf{x})^T (\mathbf{A}\mathbf{x}) = \|\mathbf{A}\mathbf{x}\|^2 \geq 0$$

hence $\mathbf{A}^T\mathbf{A}$ may have no negative eigenvalue. Zero eigenvalue appears only when $\mathbf{A}\mathbf{x} = \mathbf{0}$, hence if all the column vectors of $\mathbf{A}$ are linearly independent, *i.e.*

the rank of $\mathbf{A}$ equals to n, then $\mathbf{A}^T\mathbf{A}$ is a positive definite matrix. Otherwise $\mathbf{A}^T\mathbf{A}$ is a singular matrix and has at least one zero eigenvalue. The matrices $\mathbf{A}$ and $\mathbf{A}^T\mathbf{A}$ have the same rank.

§2.2.6.1, QR *decomposition*

The decomposition of $\mathbf{A}=\mathbf{QR}$ is introduced first, which is very important. The $m\times n$ real matrix $\mathbf{A}$ is considered composed of linearly independent column vectors $\mathbf{a}_1,\cdots,\mathbf{a}_n$ $(m\geq n)$, the $(m\times n)$ matrix $\mathbf{Q}$ is composed of ortho-normalized column vectors $\mathbf{q}_1,\cdots,\mathbf{q}_n$, *i.e.* $\mathbf{q}_i^T\mathbf{q}_j=\delta_{ij}$, and the $n\times n$ matrix $\mathbf{R}$ is a right upper triangular matrix. The generation of matrix $\mathbf{Q}$ is step by step column-wise. First, select the vector $\mathbf{a}_1$, normalizing, then used as $\mathbf{q}_1$. Next, select $\mathbf{a}_2$, orthogonalizing to the vector $\mathbf{q}_1$ and then normalizing gives $\mathbf{q}_2$. Further successively selecting $\mathbf{a}_i$, orthogonalizing to $\mathbf{q}_j(j<i)$ and then normalizing gives $\mathbf{q}_i$, and so on. Then all the following identity holds

$$\mathbf{a}_i=\mathbf{q}_1\cdot(\mathbf{q}_1^T\mathbf{a}_i)+\cdots+\mathbf{q}_{i-1}\cdot(\mathbf{q}_{i-1}^T\mathbf{a}_i)+\mathbf{q}_i\cdot(\mathbf{q}_i^T\mathbf{a}_i),\quad i=1,\cdots,n \tag{2.2.75}$$

until the vector $\mathbf{q}_n$. The matrix $\mathbf{Q}$ is thus obtained.

As the matrix $\mathbf{R}$, obviously $\mathbf{a}_i$ is unrelated to the vectors $\mathbf{q}_j(j<i)$, hence $\mathbf{R}$ is a right upper triangular matrix. From (2.2.75), the elements of matrix $\mathbf{R}$ are determined easily. For example, for the case of $n=3$

$$\mathbf{A}=[\mathbf{a}_1\quad \mathbf{a}_2\quad \mathbf{a}_3]=[\mathbf{q}_1\quad \mathbf{q}_2\quad \mathbf{q}_3]\times\begin{bmatrix}\mathbf{q}_1^T\mathbf{a}_1 & \mathbf{q}_1^T\mathbf{a}_2 & \mathbf{q}_1^T\mathbf{a}_3\\ 0 & \mathbf{q}_2^T\mathbf{a}_2 & \mathbf{q}_2^T\mathbf{a}_3\\ 0 & 0 & \mathbf{q}_3^T\mathbf{a}_3\end{bmatrix}=\mathbf{QR} \tag{2.2.76}$$

§2.2.6.2, Singular value decomposition

Singular value decomposition (SVD) [30,43,48,50] means to factorize a given real matrix $\mathbf{A}$ as

$$\mathbf{A}=\mathbf{Q}_1\mathbf{D}\mathbf{Q}_2^T \tag{2.2.77}$$

where $\mathbf{A}$ is a real $m\times n$ matrix with rank r, and $\mathbf{Q}_1$, $\mathbf{Q}_2$ are, respectively, $m\times m$ and $n\times n$ orthogonal matrices and

$$\mathbf{D}=\begin{bmatrix}\mathbf{S}_r & 0\\ 0 & 0\end{bmatrix},\quad \mathbf{S}_r=\mathrm{diag}(\sigma_1,\sigma_2,\cdots,\sigma_r),\quad \sigma_1\geq\sigma_2\geq\cdots\geq\sigma_r \tag{2.2.78}$$

where $\sigma_i, i=1\sim r$ are called the singular values, which are the r positive square roots of eigenvalues of the matrix $\mathbf{A}^T\mathbf{A}$. They are certainly the r square roots of eigenvalues of the matrix $\mathbf{A}\mathbf{A}^T$. The columns of matrices $\mathbf{Q}_1$ and $\mathbf{Q}_2$ are the eigenvectors of the matrices $\mathbf{A}\mathbf{A}^T$ and $\mathbf{A}^T\mathbf{A}$, respectively.

When $\mathbf{A}$ is a $n \times n$ symmetric and positive definite matrix, $\mathbf{Q}_2$ is reduced to be the eigen-vector matrix and $\mathbf{Q}_1 = \mathbf{Q}_2^T$.

Consider from the variational principle [50]

$$\sigma_1(\mathbf{A}) = \max_{\mathbf{q}}(\|\mathbf{Aq}\|/\|\mathbf{q}\|) \underset{\text{def}}{=} \|\mathbf{A}\| \tag{2.2.79}$$

where the inner product of the vector $\mathbf{q}$ (Euclidean type norm) is used to define the norm of the matrix $\mathbf{A}$, hence termed as the ***induced norm.*** Such induced norm has the below behavior

1, $\|\mathbf{A}\| > 0$

2, $\|c\mathbf{A}\| = c\|\mathbf{A}\|$

3, $\|\mathbf{A}+\mathbf{B}\| \le \|\mathbf{A}\| + \|\mathbf{B}\|$

The proof is $\|\mathbf{A}+\mathbf{B}\| = \max\limits_{\|\mathbf{q}\|=1}\|\mathbf{Aq}+\mathbf{Bq}\| \le \max\limits_{\|\mathbf{q}\|=1}\|\mathbf{Aq}\| + \max\limits_{\|\mathbf{q}\|=1}\|\mathbf{Bq}\| = \|\mathbf{A}\| + \|\mathbf{B}\|$

The above behavior must be fulfilled by any norm (Euclidean type), but the induced norm further satisfies the ***Submultiplicative property*** [50]

4. $\|\mathbf{AB}\| \le \|\mathbf{A}\| \cdot \|\mathbf{B}\|$.

The proof is

$$\|\mathbf{AB}\| = \max_{\|\mathbf{q}\|=1}\|\mathbf{ABq}\| = \max_{\|\mathbf{q}\|=1}((\|\mathbf{ABq}\|/\|\mathbf{Bq}\|)\cdot\|\mathbf{Bq}\|)$$

$$\le \max_{\|\mathbf{p}\|=1}(\|\mathbf{Ap}\|/\|\mathbf{p}\|)\cdot\max_{\|\mathbf{q}\|=1}\|\mathbf{Bq}\| = \|\mathbf{A}\|\cdot\|\mathbf{B}\|$$

Based on the above behavior, it is further proved that

$$\sigma_1(\mathbf{A}+\mathbf{B}) \le \sigma_1(\mathbf{A}) + \sigma_1(\mathbf{B})$$

$$\sigma_1(\mathbf{AB}) \le \sigma_1(\mathbf{A})\cdot\sigma_1(\mathbf{B}), \qquad \text{etc.}$$

§2.3, Small vibration of gyroscopic system

The vibration of rotating machine or vibration in moving coordinate system often appears in applications. According to analytical dynamics, the kinetic energy equation (1.5.7) is $T = T_2 + T_1 + T_0$ for non-inertia coordinate system, where not only $T_2 = \dot{\mathbf{q}}^T\mathbf{M}\dot{\mathbf{q}}/2$, but also T_1 and T_0 exist. The first order term T_1 of generalized velocity $\dot{\mathbf{q}}$ is given by

$$T_1 = \dot{\mathbf{q}}^T\mathbf{G}\mathbf{q}/2, \qquad \mathbf{G}^T = -\mathbf{G} \tag{2.3.1}$$

Note that $\mathbf{G}$ denotes the gyroscopic matrix, its meaning is entirely different to the damping matrix $\mathbf{C}$. The T_0 term can be merged into the potential function Π. The Lagrange function is then

$$L(\mathbf{q},\dot{\mathbf{q}}) = \dot{\mathbf{q}}^T\mathbf{M}\dot{\mathbf{q}}/2 + \dot{\mathbf{q}}^T\mathbf{G}\mathbf{q}/2 - \mathbf{q}^T\mathbf{K}\mathbf{q}/2 \tag{2.3.2}$$

where $\mathbf{q}$ is the n-dimensional generalized vector, from which the damping free dynamic equation is derived as [12]

$$\mathbf{M}\ddot{\mathbf{q}} + \mathbf{G}\dot{\mathbf{q}} + \mathbf{K}\mathbf{q} = \mathbf{f}_1(t) \tag{2.3.3}$$

where $\mathbf{f}_1(t)$ is the external force. Comparing with the free vibration equation in previous section, a gyroscopic term $\mathbf{G}\dot{\mathbf{q}}$ appears, where $\mathbf{G}$ is a skew-symmetric matrix. Equation (2.3.3) is still a second order ODE. According to the theory of differential equation, the homogeneous equation should be solved first to find its impulse response function then the Duhamel integration is used to solve the inhomogeneous term. The homogeneous equation is

$$\mathbf{M}\ddot{\mathbf{q}} + \mathbf{G}\dot{\mathbf{q}} + \mathbf{K}\mathbf{q} = \mathbf{0} \tag{2.3.4}$$

which has three terms, therefore direct use the usual method of separation of variables does not work.

However, the method of separation of variables really works also to gyroscopic system. The key step is transforming to the state space, for which the dual variable, *i.e.* the momentum vector, is introduced as

$$\mathbf{p} = \partial \mathrm{L}/\partial \dot{\mathbf{q}} = \mathbf{M}\dot{\mathbf{q}} + \mathbf{G}\mathbf{q}/2 \tag{2.3.5}$$

or

$$\dot{\mathbf{q}} = -\mathbf{M}^{-1}\mathbf{G}\mathbf{q}/2 + \mathbf{M}^{-1}\mathbf{p} \tag{2.3.6}$$

and the Hamilton function is derived as

$$\begin{aligned}\mathrm{H}(\mathbf{q},\mathbf{p}) &= \mathbf{p}^T\dot{\mathbf{q}} - \mathrm{L}(\mathbf{q},\dot{\mathbf{q}}) = \dot{\mathbf{q}}^T\mathbf{M}\dot{\mathbf{q}}/2 + \mathbf{q}^T\mathbf{K}\mathbf{q}/2 \\ &= \mathbf{p}^T\mathbf{M}^{-1}\mathbf{p}/2 - \mathbf{p}^T\mathbf{M}^{-1}\mathbf{G}\mathbf{q}/2 + \mathbf{q}^T(\mathbf{K} + \mathbf{G}^{\mathrm{T}}\mathbf{M}^{-1}\mathbf{G}/4)\mathbf{q}/2\end{aligned}$$

or written as

$$\mathrm{H}(\mathbf{q},\mathbf{p}) = \mathbf{p}^T\mathbf{D}\mathbf{p}/2 + \mathbf{p}^T\mathbf{A}\mathbf{q} + \mathbf{q}^T\mathbf{B}\mathbf{q}/2 \tag{2.3.7}$$

$$\mathbf{D} = \mathbf{M}^{-1}, \qquad \mathbf{A} = -\mathbf{M}^{-1}\mathbf{G}/2, \qquad \mathbf{B} = \mathbf{K} + \mathbf{G}^T\mathbf{M}^{-1}\mathbf{G}/4 \tag{2.3.7a}$$

Note that the matrices $\mathbf{D}$ and $\mathbf{B}$ are symmetric, $\mathbf{D}$ is positive definite but $\mathbf{B}$ may not be positive definite, which is because of the T_0 term has been merged into the potential energy Π that the matrix $\mathbf{K}$ may be indefinite. The variational principle is still read as

$$\delta\int_{t_0}^{t_f}[\mathbf{p}^T\dot{\mathbf{q}} - \mathrm{H}(\mathbf{q},\mathbf{p})]\mathrm{d}t = 0 \tag{2.3.8}$$

The dual canonical equations derived from the variational principle are

$$\dot{\mathbf{q}} = \partial\mathrm{H}/\partial\mathbf{p} = \mathbf{A}\mathbf{q} + \mathbf{D}\mathbf{p} \tag{2.3.9a}$$

$$\dot{\mathbf{p}} = -\partial\mathrm{H}/\partial\mathbf{q} = -\mathbf{B}\mathbf{q} - \mathbf{A}^T\mathbf{p} \tag{2.3.9b}$$

where the former equation is just (2.3.6).

Composing the state vector $\mathbf{v}(t)$ by combining the dual vectors $\mathbf{q},\mathbf{p}$

$$\mathbf{v} = \{\mathbf{q}^T, \mathbf{p}^T\}^T \tag{2.3.10}$$

and the dual canonical equations becomes

$$\dot{\mathbf{v}} = \mathbf{H}\mathbf{v}, \qquad \mathbf{H} = \begin{bmatrix} \mathbf{A} & \mathbf{D} \\ -\mathbf{B} & -\mathbf{A}^T \end{bmatrix} \tag{2.3.11}$$

The initial condition is

$$\mathbf{q}(0) = \mathbf{q}_0 = \text{given}, \qquad \dot{\mathbf{q}}(0) = \dot{\mathbf{q}}_0 = \text{given} \tag{2.3.12}$$

The solution methods can be classified as, 1) the ***direct integration*** method, and 2) the ***eigen-solution expansion*** method. Direct integration can use the precise integration method, which has been given in the introductory, so that the eigen-solution method is described in some detail with its algorithms.

The eigen-solutions propose the basis for the method of separation of variables in the Hamilton-Jacobi theory, also for the canonical transformation, and thus are very important. Even for non-linear vibration, the canonical transformation can still be applied as seen in the solution for Duffing equation. For multi-degrees of freedom non-linear vibration, similar methodology can be applied and the eigen-solution is a necessary step.

§2.3.1, Method of separation of variables, eigen-problem

To solve the vibration equation, the method of separation of variables and eigen-solution is concentrated below. The matrix $\mathbf{K}$ is assumed only symmetric but not necessarily positive definite. The dynamic equation (2.3.11) in state space has a time coordinate, and the components of the state vector can be regarded as discrete space variables that the argument is its subscript. The intention of separation of variables is to separate the time variable t with this subscript. Let

$$\mathbf{v} = \boldsymbol{\psi} \cdot \varphi(t)$$

where $\boldsymbol{\psi}$ is a $2n$-dimensional time-invariant vector, and $\varphi(t)$ is a scalar function but not related to the subscript. Substituting into (2.3.11) derives

$$\boldsymbol{\psi} \cdot (\dot{\varphi}/\varphi) = \mathbf{H}\boldsymbol{\psi}$$

The right hand side is independent on time t, so $\dot{\varphi}/\varphi = \mu$ must be a constant and it derives to

$$\mathbf{H}\boldsymbol{\psi} = \mu\boldsymbol{\psi}, \qquad \varphi = \exp(\mu t) \tag{2.3.13}$$

which lead to an eigen-problem for the matrix $\mathbf{H}$.

It is easily verified that

$$\left.\begin{array}{l} \mathbf{JH} = \begin{bmatrix} -\mathbf{B} & -\mathbf{A}^T \\ -\mathbf{A} & -\mathbf{D} \end{bmatrix} = (\mathbf{JH})^T, \qquad \mathbf{J} = \begin{bmatrix} \mathbf{0} & \mathbf{I} \\ -\mathbf{I} & \mathbf{0} \end{bmatrix} \\ \mathbf{JJ} = -\mathbf{I}_{2n}, \quad \mathbf{J}^T = \mathbf{J}^{-1} = -\mathbf{J}, \quad \mathbf{JHJ} = \mathbf{H}^T \end{array}\right\} \tag{2.3.14}$$

Hence $\mathbf{H}$ is a Hamilton matrix. The eigen-problem of a Hamilton matrix has a number of special features. First, both μ and $-\mu$ are eigenvalues of $\mathbf{H}$ simultaneously. The proof is as follows. From (2.3.13) it derives

$$\mathbf{JHJJ}\boldsymbol{\psi} = \mu\mathbf{J}\boldsymbol{\psi}, \qquad \text{or} \qquad \mathbf{H}^T(\mathbf{J}\boldsymbol{\psi}) = -\mu(\mathbf{J}\boldsymbol{\psi})$$

which says that $(\mathbf{J}\boldsymbol{\psi})$ is an eigenvector of the matrix $\mathbf{H}^T$ with the eigenvalue $-\mu$. Because the eigenvalue of $\mathbf{H}^T$ is also an eigenvalue of $\mathbf{H}$, so the statement is proved. Therefore, the $2n$ eigenvalues of matrix $\mathbf{H}$ can be classified as the two classes:

$$\begin{array}{ll} \alpha) \quad \mu_i, \quad \mathrm{Re}(\mu_i) < 0 \text{ or } \mathrm{Re}(\mu_i) = 0 \wedge \mathrm{Im}(\mu_i) > 0, \quad i = 1,2,\cdots,n; \\ \beta) \quad \mu_{n+i}, \quad \mu_{n+i} = -\mu_i \end{array} \tag{2.3.15}$$

where the case of $\mathrm{Re}(\mu_i) = 0$ is special. The classification of eigenvalues $\mathrm{Im}(\mu_i) > 0$ as belonging to the α class has some consequences as shown in section 2.4 later. The case of eigenvalue $\mu = 0$ is not included in the above classification, that $\mu = 0$ is definitely a duplicate eigenvalue, for $0 = -0$, and the

Jordan form follows. In theory of elasticity such situation appears quite frequently, see [23]. In vibration or wave propagation problem, the $\mu = 0$ eigenvalue rarely appears. The eigen-solutions corresponding to μ_i and μ_{n+i} are termed as mutually ***symplectic adjoint***.

It is seen in analytical dynamics, that the appearance of matrix $\mathbf{J}$ implies symplectic behavior. The symplectic orthogonal relation for eigenvectors of matrix $\mathbf{H}$ is proved below. Let

$$\mathbf{H}\psi_i = \mu_i \psi_i \quad , \qquad \mathbf{H}\psi_j = \mu_j \psi_j \ , \quad \text{then}$$

$$\mathbf{H}^T(\mathbf{J}\psi_i) = -\mu_i \mathbf{J}\psi_i \quad , \qquad \mathbf{J}\mathbf{H}\psi_j = \mu_j \mathbf{J}\psi_j$$

$$\psi_j^T \mathbf{H}^T \mathbf{J}\psi_i = -\mu_i \psi_j^T \mathbf{J}\psi_i \quad , \qquad \psi_i^T \mathbf{J}\mathbf{H}\psi_j = \mu_j \psi_i^T \mathbf{J}\psi_j$$

$$\psi_i^T \mathbf{J}\mathbf{H}\psi_j = -\mu_i \psi_i^T \mathbf{J}\psi_j$$

adding together gives

$$(\mu_i + \mu_j)\psi_i^T \mathbf{J}\psi_j = 0$$

where the two columns above are all derivations. Equation (2.3.16) explains that except the case of symplectic adjoint, *i.e.* $j = n+i$ or $i = n+j$, for which $\mu_i + \mu_j = 0$, otherwise the eigenvectors ψ_i and ψ_j must be

$$\psi_i^T \mathbf{J}\psi_j = 0 \ , \quad \psi_j^T \mathbf{J}\psi_i = 0 \ , \quad \text{when } \mu_i + \mu_j \neq 0 \qquad (2.3.16)$$

such orthogonal relation is termed as ***symplectic orthogonal,*** since in between the multiplication of two eigenvectors the matrix $\mathbf{J}$ appears. The eigenvectors of symmetric matrix have also orthogonal relation, however, only the unit matrix $\mathbf{I}$ is in between the two vectors, or for the generalized eigen-solutions the positive definite mass matrix $\mathbf{M}$ is in between the two eigenvectors. Presently, the in between matrix $\mathbf{J}$ is skew-symmetric, which is the characteristic of ***symplectic geometry.*** One must know that ***any state vector must be symplectic orthogonal to itself.***

Note that from the symplectic orthogonal relation of eigenvectors, it is easy to verify that

$$\psi_i^T \mathbf{J}\mathbf{H}\psi_j = 0, \quad \text{when } \mu_i + \mu_j \neq 0 \qquad (2.3.16')$$

Each eigenvector has an arbitrary multiplier, which can be selected as

$$\psi_i^T \mathbf{J}\psi_{n+i} = 1, \quad \text{or} \quad \psi_{n+i}^T \mathbf{J}\psi_i = -1 \qquad (2.3.17)$$

and is termed as ***symplectic normalization***. The combination is called the ***adjoint symplectic ortho-normality*** relation. Note that both ψ_i and ψ_{n+i} have a constant multiplier each, so that when $\text{Re}(\mu_i) < 0$, one additional condition can be added such as $\psi_i^T \psi_i = \psi_{n+i}^T \psi_{n+i}$ or other conditions.

Using all the eigenvectors as columns composes a $2n \times 2n$ matrix $\mathbf{\Psi}$

$$\mathbf{\Psi} = [\psi_1, \psi_2, \cdots, \psi_n; \psi_{n+1}, \psi_{n+2}, \cdots, \psi_{2n}] \qquad (2.3.18)$$

Based on the adjoint symplectic ortho-normality relation, it is verified that

$$\mathbf{\Psi}^T \mathbf{J}\mathbf{\Psi} = \mathbf{J} \qquad (2.3.19)$$

Hence, the eigenvector matrix $\mathbf{\Psi}$ of a Hamilton matrix $\mathbf{H}$ is a symplectic matrix. The determinant of $\mathbf{\Psi}$ equals 1, therefore all the column vectors of $\mathbf{\Psi}$, *i.e.* the eigenvectors, span the whole $2n$ dimensional state space. Hence, an arbitrary vector $\mathbf{v}$ in the $2n$ dimensional state space can always be expressed as the linear combination of these eigenvectors

$$\mathbf{v} = \sum_{i=1}^{n} [a_i \boldsymbol{\psi}_i + b_i \boldsymbol{\psi}_{n+i}], \quad \text{where} \quad a_i = -\boldsymbol{\psi}_{n+i}^T \mathbf{J}\mathbf{v}, \quad b_i = \boldsymbol{\psi}_i^T \mathbf{J}\mathbf{v} \tag{2.3.20}$$

It is the ***expansion theorem*** with the eigenvectors of a Hamilton matrix.

It should be pointed out that the above derivation is based on the assumption, that all eigenvalues μ_i are single roots. Under this assumption, a theorem should be proved further, that the ***mutually symplectic adjoint eigenvectors*** $\boldsymbol{\psi}_i$ ***and*** $\boldsymbol{\psi}_{n+i}$ ***are never symplectic orthogonal.*** Otherwise any constant multiplier will not be able to reach the adjoint symplectic normality relation of (2.3.17). The proof needs a **fundamental algebraic theorem** for linear algebraic simultaneous equations, which is stated as:

For a set of n ***-dimensional algebraic simultaneous equations***

$$\mathbf{A}\mathbf{x} = \mathbf{y} \tag{2.3.21}$$

where $\mathbf{A}$ ***is a*** $n \times n$ ***given matrix,*** $\mathbf{y}$ ***is a*** n ***-dimensional given force vector,*** $\mathbf{x}$ ***is the vector to be solved. When the determinant*** $\det(\mathbf{A}) \neq 0$, $\mathbf{x}$ ***has unique solution, in particular the solution*** $\mathbf{x} = \mathbf{0}$ ***for*** $\mathbf{y} = \mathbf{0}$. ***Otherwise, when the right hand side vector*** $\mathbf{y} = \mathbf{0}$, ***then the homogeneous equation must has number*** $\rho > 0$ ***linearly independent non-trivial solutions*** $\mathbf{x}_1, \mathbf{x}_2, \cdots, \mathbf{x}_\rho$, ***which may be assumed normalized; and at the same time the transposed homogeneous equation*** $\mathbf{A}^T\mathbf{z} = \mathbf{0}$ ***has also exactly number*** ρ ***linearly independent non-trivial solutions*** $\mathbf{z}_1, \mathbf{z}_2, \cdots, \mathbf{z}_\rho$. ***The condition for the inhomogeneous equation (2.3.21) having solution is that the vector*** $\mathbf{y}$ ***is orthogonal to all*** $\mathbf{z}_1, \mathbf{z}_2, \cdots, \mathbf{z}_\rho$, ***and the solution*** $\mathbf{x}$ ***is composed of a special solution of the inhomogeneous equations superimposed with an arbitrary linear combination of*** $\mathbf{x}_1, \mathbf{x}_2, \cdots, \mathbf{x}_\rho$. See for example [1], page 6.

Based on this fundamental algebraic theorem, the proof of that $\boldsymbol{\psi}_i$ and $\boldsymbol{\psi}_{n+i}$ is not symplectic orthogonal can be as follows. According to the assumption that they are all single roots, hence $(\mathbf{H} - \mu_i \mathbf{I})\mathbf{x} = \boldsymbol{\psi}_i$ has no solution, otherwise it becomes a Jordan form duplicate root. According to the fundamental algebraic theorem, the right hand side vector $\boldsymbol{\psi}_i$ is not orthogonal to all solutions of the transpose equation $(\mathbf{H}^T - \mu_i \mathbf{I})\mathbf{z} = \mathbf{0}$. Because $\mathbf{H}$ is a Hamilton matrix, according to the equation (2.3.14) it derives $(\mathbf{H} + \mu_i \mathbf{I})(\mathbf{J}\mathbf{z}) = \mathbf{0}$, just the eigen-equation of the symplectic adjoint, hence $\mathbf{J}\mathbf{z} = \boldsymbol{\psi}_{n+i}$ or $\mathbf{z} = -\mathbf{J}\boldsymbol{\psi}_{n+i}$, that the statement of $\boldsymbol{\psi}_i$ is not orthogonal to all $\mathbf{z}$, turns to be $\boldsymbol{\psi}_i^T \mathbf{J} \boldsymbol{\psi}_{n+i} \neq 0$, *i.e.* not symplectic orthogonal. The statement is proved. ##

Because $\mathbf{H}$ is not a symmetric matrix, so when duplicate eigen-roots appear, the Jordan form may emergent, see [47,48]. The Jordan normal form is very important for elasticity and structural static and will be described in some detail in chapter 5.

The expansion theorem is very useful for the solution of inhomogeneous equations

$$\dot{\mathbf{v}}(t) = \mathbf{H}\mathbf{v} + \mathbf{f}, \qquad \mathbf{v}(0) = \mathbf{v}_0 = \text{given} \tag{2.3.22}$$

the expansion expression for vector $\mathbf{v}$ is (2.2.30), similar expansions exist also for $\mathbf{f}$ and for $\mathbf{v}_0$, but the coefficients should be changed as f_{ai}, f_{bi} for $\mathbf{f}$ and a_{i0}, b_{i0} for $\mathbf{v}_0$, respectively, instead of a_i, b_i in equation (2.2.30). The coefficients a_i, b_i are functions of time t and satisfying the equations

$$\dot{a}_i = \mu_i a_i + f_{ai}, \quad \dot{b}_i = -\mu_i b_i + f_{bi}, \quad a_i(0) = a_{i0}, \quad b_i(0) = b_{i0} \tag{2.3.23}$$

which are derived based on the adjoint symplectic ortho-normality relation. The impulse response functions for the coefficient functions a_i and b_i are

$$\Phi_{ai}(t,\tau) = \exp[\mu_i(t-\tau)], \quad \Phi_{bi}(t,\tau) = \exp[-\mu_i(t-\tau)] \tag{2.3.24}$$

They are only simply scalar functions, because the eigenvector expansion method completely separates all the vibration components. Using the Duhamel integration gives

$$a_i(t) = a_{i0}e^{\mu_i t} + \int_0^t \Phi_{ai}(t,\tau) f_{ai}(\tau)d\tau, \quad b_i(t) = b_{i0}e^{-\mu_i t} + \int_0^t \Phi_{bi}(t,\tau) f_{bi}(\tau)d\tau \tag{2.3.25}$$

Hence the problem is reduced to find the eigen-solutions for the Hamilton matrix $\mathbf{H}$. The way of thinking is parallel to the multi-degrees of freedom vibration for symmetric matrices. However, the analysis should distinguish the cases of positive definite or indefinite Hamilton function.

§2.3.2, Positive definite Hamilton function

The eigen-solution theory is applied to Hamilton matrix, however, the definiteness of the Hamilton function should be distinguished. This section is devoted to the case of Hamilton function $\mathrm{H}(\mathbf{q},\mathbf{p})$ being a positive definite quadratic form, see (2.3.7). Although the theory is based on the form of Hamilton function, it can also be transformed back to the representation by matrices $\mathbf{M},\mathbf{K},\mathbf{G}$. From equation (2.3.7), it is seen that the positive definiteness of Hamilton function is ensured from the positive definiteness of the matrices $\mathbf{M}$ and $\mathbf{K}$. The eigen-solution algorithm given in reference [51] is for this case.

Although in some applications the matrix $\mathbf{K}$ cannot ensure positive definiteness. However, for some problems the matrix $\mathbf{K}$ is positive definite. For the case of vibration with no gyroscopic term, the eigen-solution theory are well developed, however, the present case adding on a gyroscopic term, similar methodology may also be applied. The relation between these two cases should be clarified, so as to find successful method for the vibration problem with gyroscopic term.

The eigen-solution of a Hamilton matrix is explored first, for which the conclusion in previous section certainly applies. Now the assumption of Hamilton function $\mathrm{H}(\mathbf{q},\mathbf{p})$ being positive definite assigns other behaviors to the eigen-solutions, *i.e.* the eigenvalues of the respective Hamilton matrix (2.3.13) are ***all purely imaginary valued***, its proof is given below.

The simplest proof is to consider that the Hamilton function $\mathrm{H}(\mathbf{q},\mathbf{p})$ keeps a constant that

$$\mathrm{dH}/\mathrm{d}t = (\partial \mathrm{H}/\partial \mathbf{q})^T \dot{\mathbf{q}} + (\partial \mathrm{H}/\partial \mathbf{p})\dot{\mathbf{p}} = -\dot{\mathbf{p}}^T \dot{\mathbf{q}} + \dot{\mathbf{q}}^T \dot{\mathbf{p}} = 0$$

according to the dual equations (2.3.9a,b). If an eigenvalue μ takes a complex value then the respective complex valued state vector solution $\mathbf{v} = \boldsymbol{\psi}_i e^{\mu_i t}$ follows. The Hamilton function can be extended to complex valued vector as

$$\mathrm{H}(\mathbf{q},\mathbf{p}) = -\mathbf{v}^H \mathbf{J}\mathbf{H}\mathbf{v}/2, \qquad \text{where} \qquad \mathbf{v} = \{\mathbf{q}^T, \mathbf{p}^T\}^T \tag{2.3.7'}$$

then $\mathrm{H} = -\boldsymbol{\psi}_i^H \mathbf{J}\mathbf{H}\boldsymbol{\psi}_i/2 \cdot \exp[(\mu_i + \bar{\mu}_i)t]$, where the superscript H means taking Hermite transpose, *i.e.* the operation of ***taking complex conjugate then transpose,*** and the bar above means taking complex conjugate.

Thus the Hamilton function being positive definite implies the matrix $(-\mathbf{JH})$ being positive definite. Hence $\boldsymbol{\psi}_i^H(-\mathbf{JH})\boldsymbol{\psi}_i/2$ must have positive value. The conservation of Hamilton function $\mathrm{H}(\mathbf{q},\mathbf{p})$ requires $(\mu_i + \bar{\mu}_i) = 0$, which determines that μ_i must be purely imaginary. ##

The complex arithmetic appears here. The eigen-solutions of Hamilton matrix cannot free from complex arithmetic. Now the statement is proved again, and it is also useful later. That

$$\mathbf{H}\boldsymbol{\psi} = \mu\boldsymbol{\psi}, \qquad \text{implies} \quad -\boldsymbol{\psi}^H \mathbf{J}\mathbf{H}\boldsymbol{\psi} = \boldsymbol{\psi}^H \mathbf{J}\boldsymbol{\psi} \cdot (-\mu) = \text{positive number}$$

But $\boldsymbol{\psi}^H \mathbf{J}\boldsymbol{\psi} = \mathbf{q}^H \mathbf{p} - \mathbf{p}^H \mathbf{q}$, the two terms are complex conjugate to each other and gives purely imaginary number. Hence the eigenvalue μ must be purely imaginary to give a real number. Both the proofs are quite simple. Duplicate eigenvalues may appear, however, which are never ***Jordan form eigenvalues***. Otherwise, there is the solution with multiplier $t\exp(\mathrm{i}\omega t)$, which increases indefinitely with time t and the system cannot be energy conserved.

The characteristic of purely imaginary eigenvalue is that ***the complex conjugate of the eigenvalue equals the symplectic adjoint eigenvalue***.

A number of theories of variational principle, Rayleigh quotient for the eigenvalues, the maximum-minimum characteristics, the eigenvalues under linear constraints and eigenvalue count etc. can be extended to the case of the Hamilton function being positive definite.

§2.3.2.1, Variational principle of eigenvalues

Because the eigenvalues are purely imaginary, so the equation is written as

$$\mathbf{H}\boldsymbol{\psi} = \mu\boldsymbol{\psi}, \qquad \mu = \mathrm{i}\omega, \qquad \boldsymbol{\psi} = \boldsymbol{\psi}_\mathrm{r} + \mathrm{i}\boldsymbol{\psi}_\mathrm{i} \tag{2.3.26}$$

where ω is a real number, *i.e.* the imaginary part of the eigenvalue. The eigenvector $\boldsymbol{\psi}$ is complex valued, and is expressed with two real valued vectors $\boldsymbol{\psi}_\mathrm{r}$ and $\boldsymbol{\psi}_\mathrm{i}$, both vectors are $2n$ dimensional.

Pure imaginary eigenvalue implies that the symplectic adjoint eigenvalue $-\mathrm{i}\omega$ equals its complex conjugate. Hence the complex conjugate of the eigen-equation is just the symplectic adjoint eigen-equation

$$\mathbf{H}(\boldsymbol{\psi}_\mathrm{r} - \mathrm{i}\boldsymbol{\psi}_\mathrm{i}) = -\mathrm{i}\omega(\boldsymbol{\psi}_\mathrm{r} - \mathrm{i}\boldsymbol{\psi}_\mathrm{i}),$$

the symplectic adjoint eigenvector is $(\psi_i + i\psi_r)$. The complex conjugate eigenvector multiplying $-i$ gives the symplectic adjoint eigenvector. Subdividing the complex equation into two real one

$$\mathbf{H}\psi_r = -\omega\psi_i, \qquad \mathbf{H}\psi_i = \omega\psi_r \tag{2.3.27}$$

Left multiplying $-\psi_r^T\mathbf{J}$ and $-\psi_i^T\mathbf{J}$ to the first and second equations respectively, then adding together derives

$$\omega = [\psi_r^T(-\mathbf{JH})\psi_r + \psi_i^T(-\mathbf{JH})\psi_i]/2\psi_r^T\mathbf{J}\psi_i \tag{2.3.28}$$

which expresses that the imaginary part of the eigenvalue is computed by the real and imaginary parts of the eigenvector. The statement can be extended as a variational principle, that let the selection of ψ_i ensures the denominator positive in above equation, then the variational principle for ω is given as

$$\omega = \min_{\mathbf{v}_r^T\mathbf{J}\mathbf{v}_i > 0} [\mathbf{v}_r^T(-\mathbf{JH})\mathbf{v}_r + \mathbf{v}_i^T(-\mathbf{JH})\mathbf{v}_i]/2\mathbf{v}_r^T\mathbf{J}\mathbf{v}_i \tag{2.3.28'}$$

called as the generalized Rayleigh quotient.

Proof: This functional involves two real state vectors $\mathbf{v}_r$ and $\mathbf{v}_i$, but has only one inequality condition. In fact, if $\mathbf{v}_r^T\mathbf{J}\mathbf{v}_i < 0$ then let $\mathbf{v}_i := -\mathbf{v}_i$ so that the positive condition satisfies, however, the numerator keeps unchanged. The numerator in the functional is ensured positive definite because of the positive definiteness of Hamilton function. Hence the fractional functional is lower bounded, and the minimization is legitimate. Carrying out the variation gives

$$\delta\omega = [\delta\mathbf{v}_r^T(-\mathbf{JH}\mathbf{v}_r - \mathbf{J}\mathbf{v}_i\omega) + \delta\mathbf{v}_i^T(-\mathbf{JH}\mathbf{v}_i + \mathbf{J}\mathbf{v}_r\omega)]/\mathbf{v}_r^T\mathbf{J}\mathbf{v}_i = 0$$

both the multipliers to the variations $-\delta\mathbf{v}_r^T$ and $-\delta\mathbf{v}_i^T$, respectively, must be zero, which derives the two sets of real equations in (2.3.27). The variational formulation (2.3.28') is proved. ##

The variational principle (2.3.28') has no unique eigenvector $\psi = \psi_r + i\psi_i$, because an arbitrary complex constant factor can be multiplied.

In (2.3.28) the variation is written as a minimization, hence valid only for the minimum positive ω. If rewritten as $\delta\omega = 0$, then the derived equation can be applied to all the purely imaginary eigenvalues. However, the first order variation equal to zero is not so convenient as maxi-minimum. Hence the maxi-minimum behavior in equation (2.2.24) should be recalled. The maxi-minimum for the present eigenvalue problem is to establish below.

Because the variational vectors $\mathbf{v}_r$ and $\mathbf{v}_i$ are real, hence the symplectic orthogonal conditions (2.3.16') and (2.3.16") should also be expressed with real vectors. The subscripts r and i are used to mark the real and imaginary parts, and the subscripts j, k are used for the order of the eigenvectors. The symplectic orthogonality is given in real form as

$$\left.\begin{array}{l} \psi_{rj}^T\mathbf{J}\psi_{rk} - \psi_{ij}^T\mathbf{J}\psi_{ik} = 0,\ \psi_{rj}^T\mathbf{J}\psi_{ik} + \psi_{ij}^T\mathbf{J}\psi_{rk} = 0,\ \text{and} \\ \psi_{rj}^T\mathbf{JH}\psi_{rk} - \psi_{ij}^T\mathbf{JH}\psi_{ik} = 0,\ \psi_{rj}^T\mathbf{JH}\psi_{ik} + \psi_{ij}^T\mathbf{JH}\psi_{rk} = 0 \end{array}\right. ,\quad \mu_i + \mu_j \neq 0 \tag{2.3.29}$$

The present case is $(-\mathbf{JH})$ being positive definite, that complex conjugate is the symplectic adjoint. In the symplectic orthogonality relation (2.3.29), except the symplectic adjoint case $n+j=k$, otherwise the negative (–) and positive (+) signs can all be replaced with $\pm$. Since the symplectic adjoint eigenvectors are the complex conjugate eigenvectors, the formulation for the symplectic orthogonality and symplectic normality can be combined in real form

$$\begin{aligned} &\boldsymbol{\psi}_{rj}^T \mathbf{J} \boldsymbol{\psi}_{rk} = \boldsymbol{\psi}_{ij}^T \mathbf{J} \boldsymbol{\psi}_{ik} = 0, \quad \boldsymbol{\psi}_{rj}^T \mathbf{J} \boldsymbol{\psi}_{ik} = -\boldsymbol{\psi}_{ij}^T \mathbf{J} \boldsymbol{\psi}_{rk} = \delta_{jk}/2, \qquad j,k \le n \\ &\boldsymbol{\psi}_{rj}^T \mathbf{J} \mathbf{H} \boldsymbol{\psi}_{rk} = \boldsymbol{\psi}_{ij}^T \mathbf{J} \mathbf{H} \boldsymbol{\psi}_{ik} = -\omega_j \delta_{jk}/2, \quad \boldsymbol{\psi}_{rj}^T \mathbf{J} \mathbf{H} \boldsymbol{\psi}_{ik} = \boldsymbol{\psi}_{ij}^T \mathbf{J} \mathbf{H} \boldsymbol{\psi}_{rk} = 0 \end{aligned} \tag{2.3.30}$$

where the eigenvalues are sorted as

$$0 < \omega_1 < \omega_2 < \cdots < \omega_n \tag{2.3.31}$$

Expanding the trial functions in (2.3.28') with the eigenvectors and also using the symplectic orthogonality relation (2.3.30), the variational expression for the imaginary part of eigenvalues ω is derived in expansion form

$$\mathbf{v}_r = \sum_{j=1}^{n} (a_{rj} \boldsymbol{\psi}_{rj} + a_{ij} \boldsymbol{\psi}_{ij}), \quad \mathbf{v}_i = \sum_{j=1}^{n} (b_{rj} \boldsymbol{\psi}_{rj} + b_{ij} \boldsymbol{\psi}_{ij}) \tag{2.3.32}$$

$$\omega = \sum_{j=1}^{n} \omega_j (a_{rj}^2 + a_{ij}^2 + b_{rj}^2 + b_{ij}^2) \Big/ 2 \sum_{j=1}^{n} (a_{rj} b_{ij} - b_{rj} a_{ij}) \tag{2.3.32'}$$

When selecting $a_{r1} = b_{i1} = 1$ with other parameters being zero gives ω_1. Equation (2.3.32') corresponds to equation (2.3.28), but the variation of (2.3.28') is necessary. The inequality condition is for the denominator in (2.3.32'). Equation (2.3.32') is the extension of (2.2.21).

Let the imaginary eigenvalues for ω_j be sorted as in equation (2.3.31) then the maxi-minimum behavior for ω follows. It can be formulated as in equation (2.2.24)

$$\omega_{m+1} = \max_{\mathbf{c}_j, j=1,m} \left(\min_{\mathbf{v}, \mathbf{c}_j^T \mathbf{J} \mathbf{v} = 0, \mathbf{v}_r^T \mathbf{J} \mathbf{v}_i > 0} [\mathbf{v}_r^T (-\mathbf{J}\mathbf{H}) \mathbf{v}_r + \mathbf{v}_i^T (-\mathbf{J}\mathbf{H}) \mathbf{v}_i] / 2 \mathbf{v}_r^T \mathbf{J} \mathbf{v}_i \right) \tag{2.3.33}$$

where $\mathbf{c}_j, j = 1, \cdots, m$ is also real valued state vectors, and $\mathbf{v} = \mathbf{v}_r + i\mathbf{v}_i$, that $\mathbf{v}_r$, $\mathbf{v}_i$ are argument vectors of the functional. If the selection is $\mathbf{c}_1 = \alpha_1 \boldsymbol{\psi}_{r1} + \alpha_2 \boldsymbol{\psi}_{i1}$, where α_1, α_2 are given constants, then according to the symplectic orthogonality (2.3.30), it must be

$$\begin{aligned} -a_{r1}\alpha_2 + a_{i1}\alpha_1 &= 0 \\ -b_{r1}\alpha_2 + b_{i1}\alpha_1 &= 0 \end{aligned}$$

Because α_1, α_2 cannot be simultaneously zero, so the determinant is zero $\Delta_1 = a_{r1} b_{i1} - b_{r1} a_{i1} = 0$, which implies that the parameters $a_{r1}, a_{i1}, b_{r1}, b_{i1}$ having no contribution to the denominator in (2.3.33), but adding positive value to the numerator. Hence it must be $a_{r1} = b_{i1} = a_{i1} = b_{r1} = 0$.

The proof for the maximum-minimum variational principle (2.3.33) can apply the similar method to the variational principle (2.2.24). Transforming to eigenvectors coordinate, the variational equation is expressed by equation (2.3.32'), then m conditions are expressed with $\mathbf{c}_j$. When selecting

$$\mathbf{c}_j = \boldsymbol{\psi}_{rj}, (j = 1, 2, \cdots, m) \tag{2.3.34}$$

the summation for the denominator of (2.3.32) begins from $j = m+1$, because the minimization gives $a_{rj} = a_{ij} = b_{rj} = b_{ij} = 0$, $(j = 1,2,\cdots,m)$; and therefore the minimization for (2.3.33) give $\omega = \omega_{m+1}$ and $a_{rj} = a_{ij} = b_{rj} = b_{ij} = 0, (j > m+1)$ also. Hence the maximization with $\mathbf{c}_j$ can really reach the value ω_{m+1}.

It needs to show further that the selection of $\mathbf{c}_j$ other than (2.3.34) can only make the minimization in the large parenthesis in (2.3.33) reduced. The proof is similar to $\mathbf{M}, \mathbf{K}$ system. Firstly, let $a_{rj} = a_{ij} = b_{rj} = b_{ij} = 0, (j > m+1)$, that a special selection can only increase the value of min. However, there are $4 \times (m+1)$ parameters to be selected arbitrarily, where number m of constraints $\mathbf{c}_j, (j = 1,2,\cdots,m)$ supply $2m$ real conditions. There are still $2m+4$ parameters to select, for which $2 \times (m+1)$ conditions can be

$$b_{ij} = a_{rj},\ b_{rj} = -a_{ij}, j = 1,\cdots,m+1, \quad \text{then } \Delta_j = a_{rj}b_{ij} - b_{rj}a_{ij} = a_{rj}^2 + a_{ij}^2 \qquad (2.3.34a)$$

The remaining 2 conditions can be selected arbitrarily. In such case equation (2.3.33) gives

$$\omega = \sum_{j=1}^{m+1} \omega_j (a_{rj}^2 + a_{ij}^2) \Big/ \sum_{j=1}^{m+1} (a_{rj}^2 + a_{ij}^2) \leq \omega_{m+1} \qquad (2.3.34b)$$

Therefore the selection of $\mathbf{c}_j$ other than (2.3.34) will not give the ω larger than ω_{m+1}. This proves that the equation (2.3.33) gives the maxi-minimum theorem for ω_{m+1}. ##

Having established the maxi-minimum variational principle and the inclusion theorem, the eigenvalue count theorem etc. are the natural developments, which is the same situation as before for the case of symmetric matrices $\mathbf{M}, \mathbf{K}$. However, the former variational principle can be described in the point transformation, Lagrange system, whereas for gyroscopic system, it should use dual vectors, state space, Hamilton matrix and symplectic geometry etc., quite an extension.

§2.3.2.2, Eigenvalue count and the inclusion theorem

The maxi-minimum theorem for eigenvalues supplies a foundation, from which the ***inclusion theorem*** for the eigenvalue distribution under constraints can be derived. The derivation given below is similar to that in section 2.2.

Let the gyroscopic system before constraint (original system) has n-degrees of freedom. Now, one given constraint is added on

$$\mathbf{b}^T\mathbf{q} = 0 \qquad (2.3.35)$$

where $\mathbf{b}$ is a n-dimensional real vector. This is a displacement constraint. Then the system becomes $n-1$ degrees of freedom. Suppose the eigenvalues of the original system be $\pm i\omega_1, \pm i\omega_2, \cdots, \pm i\omega_n$, and sorted as $0 < \omega_1 < \omega_2 < \cdots < \omega_n$. The constrained system sorted eigenvalues as $\pm i\omega_{c1}, \pm i\omega_{c2}, \cdots, \pm i\omega_{c,n-1}$. For a system with positive definite Hamilton function, it needs to prove the inclusion theorem formulated as

$$\omega_1 \leq \omega_{c1} \leq \omega_2 \leq \omega_{c2} \leq \cdots \leq \omega_{n-1} \leq \omega_{c,(n-1)} \leq \omega_n \qquad (2.3.36)$$

i.e. the eigenvalues of the constrained system are located in between the corresponding two eigenvalues of the original system.

Proof: Note first that the constraint (2.3.35) can be given as

$$\mathbf{c}^T\mathbf{J}\mathbf{v}=0\,,\ \text{where}\ \ \mathbf{c}=\{\mathbf{0}^T\ \ \mathbf{b}^T\}^T \tag{2.3.35'}$$

Using the maximum-minimum behavior of the eigenvalues proves first that

$$\omega_m \le \omega_{c,m} \le \omega_{m+1}, \qquad m<n \tag{2.3.36'}$$

The m-th eigenvalue $\omega_{c,m}$ of the constrained system can be obtained as follows. The original system are under m constraints, among which one constraint is the given one, $\mathbf{c}$, and the other $m-1$ constraints are selected according to the maximum-minimum variational principle, *i.e.* the $m-1$ constraints are selected in order to maximize $\omega_{c,m}$. When comparison is made with ω_{m+1}, for which the maxi-minimum principle requires that all the m constraints be selected to maximize ω_{m+1}. Hence ω_{m+1} is obtained with one more maximized constraint selection, therefore it must be $\omega_{m+1}\ge\omega_{c,m}$ and the latter inequality in (2.3.36') is proved.

Next, the inequality for ω_m in (2.3.36') is to check with. For which, the maxi-minimum has $m-1$ constraint conditions. When comparing with $\omega_{c,m}$, the $m-1$ arbitrary constraints to be selected is along the same way, however for $\omega_{c,m}$ there is one more constraint, which can only increase $\omega_{c,m}$. Hence the former inequality in (2.3.36') is valid too. When m runs from 1 to $n-1$, the inclusion theorem (2.3.36) is proved. ##

B.E. Yang proved the inclusion theorem for a special problem of rotating axis, see [52].

The eigenvalue count is to consider next. For a given frequency $\omega_\#$, the ***dynamic stiffness matrix*** is defined as

$$\mathbf{R}(\omega_\#)=\mathbf{K}+i\omega_\#\mathbf{G}-\omega_\#^2\mathbf{M} \tag{2.3.37}$$

For a positive definite Hamilton function the matrix $\mathbf{K}$ is positive definite, and it is proved that all the eigenvalues are purely imaginary. In the $\mathbf{M},\mathbf{G},\mathbf{K}$ gyroscopic system description, which has natural vibration frequencies ω_i, according to the condition

$$\omega_i<\omega_\#\,, \qquad (i=1,2,\cdots,m)$$

The number m is to be determined. It requires directly determining m with the dynamic stiffness matrix $\mathbf{R}(\omega_\#)$. Because $\mathbf{G}$ is skew-symmetric, therefore $\mathbf{R}(\omega_\#)$ is Hermite symmetric. The determinant of a Hermite matrix is a real number and the main diagonal submatrices of $\mathbf{R}$ are all Hermite symmetric. Computing a series of determinants of main diagonal submatrices of $\mathbf{R}$, see equation (2.2.29), constitutes the Sturm sequence.

For eigenvalue count m, one can count the sign changes of the Sturm sequence $d_0,d_1,\cdots,d_n$. Since the inclusion theorem is the same as in the case of no gyroscopic term, the derivation is almost the same. The conclusion is that: ***the count of sign change for the Sturm sequence gives the number*** m ***of eigenvalues***

$\omega < \omega_{\#}$.

In order to avoid computing a series of determinants of Hermite matrices, the modified triangular factorization $\mathbf{LDL}^H$ of the Hermite matrix can be used instead. Counting the number of negative entries in the diagonal matrix $\mathbf{D}$ gives the required eigenvalue count m.

In substructure combination for Hermite dynamic stiffness matrices, the algorithm and eigenvalue count are similar to that for symmetric matrices, the difference is that the real arithmetic should be changed as complex arithmetic. The detail is neglected.

Above, the eigenvalue problem is described parallel to the gyroscopic force free case. However, it is under the assumption that the Hamilton function is positive definite, *i.e.* both $\mathbf{M}$ and $\mathbf{K}$ are positive definite.

The problem of numerical computation requires further description.

Discussion: Positive definite Hamilton function means the matrix $(-\mathbf{JH})$ is positive definite. From (2.3.27) eliminating ψ_i gives the eigen-equation $\mathbf{H}^2\psi_r = -\omega^2\psi_r$, which derives in turn

$$\mathbf{JH}^3\psi_r = \omega^2(-\mathbf{JH})\psi_r \tag{2.3.38}$$

an eigenvalue problem with both the matrix operators $(-\mathbf{JH})$ ***and*** $\mathbf{JH}^3$ ***positive definite***, since

$$\mathbf{v}^T\mathbf{JH}^3\mathbf{v} = \mathbf{v}^T(\mathbf{JH})^T\mathbf{H}\cdot(\mathbf{Hv}) = (\mathbf{Hv})^T\cdot(-\mathbf{JH})\cdot(\mathbf{Hv}) > 0$$

where $\mathbf{v}$ is an arbitrary state vector. So the above development is conceivable.

§2.3.3, Indefinite Hamilton function

The previous section requires the positive definiteness of system Hamilton function. But in applications, a number of engineering problems cannot fulfill this condition. Thus the present section discusses under only the positive definiteness of mass matrix $\mathbf{M}$. The eigen-equation

$$\mathbf{H}\psi = \mu\psi \tag{2.3.13}$$

is as before. However the matrix $-\mathbf{JH}$ is no longer positive definite although it is still symmetric. For ease of discussion, a point transformation is carried out first, which uses the matrix $\mathbf{\Psi}_n$ in equation (2.2.10), and is denoted as $\mathbf{Q}$ now, in order to avoid confusing. Let $\mathbf{q} = \mathbf{Qq}'$, then

$$\mathbf{Q}^T\mathbf{MQ} = \mathbf{I}_n, \quad \mathbf{Q}^T\mathbf{KQ} = \mathrm{diag}(k_i) = \mathbf{K}', \quad \mathbf{Q}^T\mathbf{GQ} = \mathbf{G}' \tag{2.3.39}$$

After this point transformation, the mass becomes a unit matrix, and the stiffness becomes a real diagonal matrix, for which the diagonal elements $k_i, (i = 1,\cdots,n)$ will have some negative values because of $-\mathbf{JH}$ is indefinite. The gyroscopic matrix $\mathbf{G}$ is transformed as $\mathbf{G}'$ and is still skew-symmetric. For convenience below, the point transformation is considered performed and the prime sign is removed for ease of notation.

One another form of state vector $\mathbf{w}_i$ can be used in state space approach,

$$\mathbf{w}_t = \begin{Bmatrix} \mathbf{q} \\ \dot{\mathbf{q}} \end{Bmatrix}, \quad \mathbf{p} = \mathbf{M}\dot{\mathbf{q}} + \mathbf{G}\mathbf{q}/2, \quad \text{where} \quad \mathbf{v} = \begin{Bmatrix} \mathbf{q} \\ \mathbf{p} \end{Bmatrix} = \mathbf{L}\mathbf{w}_t, \quad \dot{\mathbf{v}} = \mathbf{N}\mathbf{w}_t \tag{2.3.40}$$

where
$$\mathbf{L} = \begin{bmatrix} \mathbf{I}_n & \mathbf{0} \\ \mathbf{G}/2 & \mathbf{M} \end{bmatrix}, \quad \mathbf{N} = \begin{bmatrix} \mathbf{0} & \mathbf{I}_n \\ -\mathbf{K} & -\mathbf{G}/2 \end{bmatrix}, \quad \mathbf{H} = \mathbf{N}\mathbf{L}^{-1} \tag{2.3.41}$$

The system equation expressed with $\mathbf{w}_t$ is

$$\mathbf{L}\dot{\mathbf{w}}_t = \mathbf{N}\mathbf{w}_t \tag{2.3.42}$$

and the corresponding eigen-equation is derived as

$$\mathbf{N}\mathbf{w} = \mu\mathbf{L}\mathbf{w}, \qquad \mathbf{w}_t = \mathbf{w} \times \exp(\mu t) \tag{2.3.43}$$

where μ is the eigenvalue and $\mathbf{w}$ is the correspondent eigenvector. Evidently $\psi = \mathbf{L}\mathbf{w}$ is the respective eigenvector of the matrix $\mathbf{H}$. Theoretically no significant difference is made, however using $\mathbf{w}$ would be easier for computation.

It is easy to verify the identities

$$\mathbf{L}^T\mathbf{J}\mathbf{L} = \begin{bmatrix} \mathbf{G} & \mathbf{M} \\ -\mathbf{M} & \mathbf{0} \end{bmatrix}, \quad \mathbf{N}^T\mathbf{J}\mathbf{N} = \begin{bmatrix} \mathbf{0} & \mathbf{K} \\ -\mathbf{K} & -\mathbf{G} \end{bmatrix}, \quad \mathbf{N}^T\mathbf{J}\mathbf{L} = -\mathbf{L}^T\mathbf{J}\mathbf{N} = \begin{bmatrix} \mathbf{K} & \mathbf{0} \\ \mathbf{0} & \mathbf{M} \end{bmatrix} \tag{2.3.44}$$

then the eigen-equation (2.3.43) leads to

$$\begin{bmatrix} \mathbf{0} & \mathbf{K} \\ -\mathbf{K} & -\mathbf{G} \end{bmatrix}\mathbf{w} = \mu\begin{bmatrix} \mathbf{K} & \mathbf{0} \\ \mathbf{0} & \mathbf{M} \end{bmatrix}\mathbf{w}, \text{ and } -\mu\begin{bmatrix} \mathbf{G} & \mathbf{M} \\ -\mathbf{M} & \mathbf{0} \end{bmatrix}\mathbf{w} = \begin{bmatrix} \mathbf{K} & \mathbf{0} \\ \mathbf{0} & \mathbf{M} \end{bmatrix}\mathbf{w} \tag{2.3.45}$$

and in combined form

$$\begin{bmatrix} \mathbf{0} & -\mathbf{K} \\ \mathbf{K} & \mathbf{G} \end{bmatrix}\mathbf{w} = \mu^2\begin{bmatrix} \mathbf{G} & \mathbf{M} \\ -\mathbf{M} & \mathbf{0} \end{bmatrix}\mathbf{w} \tag{2.3.46}$$

Both sides are skew-symmetric matrices.

Solving the eigen-problem (2.3.46) is an important step for solving the original eigen-problem (2.3.43). The eigen-solution of (2.3.46) is unnecessarily the eigen-solution of (2.3.43) however, both the eigen-vectors of (2.3.43) with eigenvalues μ and $-\mu$, respectively, are the duplicate eigen-vectors of (2.3.46) with eigenvalue μ^2, as is easily verified.

Therefore, one can first solve the eigen-equation (2.3.46), whose duplicate eigen-vectors subdivide the whole space into a number of subspaces. These subspaces are composed of that each eigen-solution in class α is combined with its symplectic adjoint eigen-solution in class β. If the eigenvalue is real then this subspace is two dimensional, or if the eigenvalue is a complex number, then the complex conjugate should also be included and the dimension of the subspace is four. After finding all these subspaces, their eigen-problems should be solved in order to find the eigen-solutions of the original equation (2.3.43). Because the equation (2.3.46) is for eigenvalues of μ^2, hence it is unable to distinguish $+\mu$ and $-\mu$ solutions. Therefore, the $\pm\mu$ eigen-solutions of (2.3.43) are solved after the duplicate μ^2 eigen-solutions of (2.3.46) are found. Thus solving the eigen-problem (2.3.46) is an important intermediate step for solving the eigen-problem of the Hamilton matrix [53].

The adjoint symplectic ortho-normality relation is now

$$\mathbf{w}_i^T(\mathbf{L}^T\mathbf{J}\mathbf{L})\mathbf{w}_j = 0, \quad \mathbf{w}_i^T(\mathbf{L}^T\mathbf{J}\mathbf{N})\mathbf{w}_j = 0, \quad \text{when} \quad \mu_i + \mu_j \neq 0 \tag{2.3.47}$$

and the normality condition becomes

$$\mathbf{w}_i^T(\mathbf{L}^T\mathbf{J}\mathbf{L})\mathbf{w}_{n+i}=1 \quad \text{or} \quad \mathbf{w}_{n+i}^T(\mathbf{L}^T\mathbf{J}\mathbf{L})\mathbf{w}_i=-1 \ \text{and}\ \mathbf{w}_i^T\mathbf{w}_i=\mathbf{w}_{n+i}^T\mathbf{w}_{n+i} \tag{2.3.48}$$

Based on such weighted symplectic ortho-normality relation, the expansion theorem can be obtained readily.

§2.3.3.1, Effect of gyroscopic force to the stability of vibration

When there is no gyroscopic force, vibration stability is determined by the stiffness matrix $\mathbf{K}$, because the mass matrix $\mathbf{M}$ is positive definite. For gyroscopic system, when $\mathbf{K}$ is positive definite, according to the analysis in section 2.3.2, the eigenvalues are always purely imaginary and the system is always stable but not asymptotically stable, disregard how large is the gyroscopic force. If there is again damping forces, the system must be asymptotically stable, that when $t\to\infty$ the state vector tends to zero, because the Hamilton function, which is positive definite, decreases with time and tends to zero.

But when $\mathbf{K}$ is indefinite, its eigenvalue appears negative number, if there were no gyroscopic force term, the system is unstable. Now the problem is that, if $\mathbf{K}$ is indefinite and there is gyroscopic force, then the stability behavior of system vibration needs to be determined.

The theorem of Thomson and Tait determined that, ***if the matrix*** $\mathbf{K}$, ***when diagonalized by an orthogonal matrix, appears odd number of negative elements, then the system is always unstable regardless of any gyroscopic force applied on.***

Explain first that the gyroscopic force may stabilize a system, which is unstable if no gyroscopic force. Certainly the diagonalized stiffness matrix $\mathbf{K}$ must have even number of negative entries. A two degrees of freedom system is used as a model problem

$$\ddot{q}_1+k_1q_1=0\ , \qquad \ddot{q}_2+k_2q_2=0\ ,(k_1<0,k_2<0)$$

which is unstable if no gyroscopic force. Adding on the gyroscopic forces the system equations become

$$\ddot{q}_1+\Gamma\dot{q}_2+k_1q=0\ , \qquad \ddot{q}_2-\Gamma\dot{q}_1+k_2q=0$$

$$\mathbf{M}=\begin{bmatrix}1&0\\0&1\end{bmatrix},\quad \mathbf{K}=\begin{bmatrix}k_1&0\\0&k_2\end{bmatrix},\quad \mathbf{G}=\begin{bmatrix}0&\Gamma\\-\Gamma&0\end{bmatrix}$$

$$\mathbf{L}=\begin{bmatrix}1&0&0&0\\0&1&0&0\\0&\Gamma/2&1&0\\-\Gamma/2&0&0&1\end{bmatrix},\quad \mathbf{N}=\begin{bmatrix}0&0&1&0\\0&0&0&1\\-k_1&0&0&-\Gamma/2\\0&-k_2&\Gamma/2&0\end{bmatrix}$$

where Γ represents the gyroscopic term. The equation for eigenvalue μ is $\det(\mu\mathbf{L}-\mathbf{N})=0$ and the expanded form is

$$\mu^4+(\Gamma^2+k_1+k_2)\mu^2+k_1k_2=0 \tag{2.3.49}$$

if $\qquad k_1k_2>0,\quad \Gamma^2+k_1+k_2>0,\quad (\Gamma^2+k_1+k_2)-4k_1k_2>0$

then both the roots of μ^2 are negative, and μ is purely imaginary, thus the motion is stable, but not asymptotically stable. The conditions $k_1<0,k_2<0$ have ensured the first condition $k_1 k_2 > 0$ satisfied, therefore it is only necessary that

$$\Gamma^2 + (k_1 + k_2) > 0$$

then the system is stable and it will be satisfied if the gyroscopic term is large enough. However, it is also seen that if $k_1 > 0, k_2 < 0$ then the system is definitely unstable, regardless how large Γ is.

Proof of the Thomson-Tait theorem: The eigen-equation for μ is

$$\Delta(\mu) = \det(\mu\mathbf{L} - \mathbf{N}) = 0$$

where $\mathbf{N}, \mathbf{L}$ are all real matrices. When μ is an arbitrary real number, $\Delta(\mu)$ is also real and is a continuous function of μ. When $\mu = 0$, it gives $\Delta(0) = \det(\mathbf{JN}) = \prod_1^n k_i$, whereas as $\mu \to \infty$, $\Delta(\infty)$ approaches μ^{2n} and is definitely positive. But the sign of $\Delta(0)$ is determined from the number of negative k_i, when there are odd number of negative k_i then $\Delta(0) < 0$ and when μ increases from 0 to ∞, the continuous function $\Delta(\mu)$ changes from negative to positive, hence at certain $\mu > 0$ it must be $\Delta(\mu) = 0$. It implies a positive eigen-root $\mu > 0$ exists, hence the system is unstable. The statement is proved. ##

The damping of system needs to be considered now. The damping is usually expressed with the Rayleigh dissipation, with the dissipation function $\dot{\mathbf{q}}^T \mathbf{C} \dot{\mathbf{q}}/2$, where $\mathbf{C}$ is a symmetric positive definite matrix, or at least non-negative $n \times n$ matrix. Damping force is $\mathbf{C}\dot{\mathbf{q}}$, and the dissipation function is the work done by the damping force.

For the system composed of $\mathbf{M}, \mathbf{K}$ only, the mechanical energy is conserved. The damping means the mechanical energy is dissipated and the dissipation function gives the rate of dissipation. For gyroscopic system vibration, the effect of damping force should distinguish the two cases of Hamilton function $H(\mathbf{q}, \mathbf{p})$ being positive definite or indefinite. In treating stability of the system, if the Hamilton function is positive definite then this Hamilton function (system mechanical energy) can be selected as the Lyapunov function and from the dynamic equation it is seen that the Lyapunov function is always decreasing due to dissipation. Therefore the system is always asymptotically stable, see section 6.2.

In case of indefinite Hamilton function, which implies indefinite $\mathbf{K}$, the indefinite Hamilton function cannot be used as the Lyapunov function although it still decreases. Chetaev [54] proved that the damping force will bring the damping free system stabilized by the gyroscopic force (certainly with even degrees of freedom unstable originally) back to be unstable.

According to the Thomson-Tait theorem, if the stiffness matrix $\mathbf{K}$ have even number of unstable degrees of freedom, then gyroscopic force can stabilize the system. But their analysis is based on the assumption of no damping factor. Now if the system has damping, the system is brought back to be unstable. Section 2.5 will go back to this stability analysis again.

§2.3.3.2, Symplectic eigenvalue problem and its algorithm

Eigen-solutions are the basis for a number of applications. Only theoretical analysis is not enough, an algorithm is necessary for solving the application problem. The eigen-solutions are the basis for modal analysis and canonical transformations etc. The non-linear system equations can be numerically solved based on the canonical transformation. The eigen-solution algorithm given in reference [51] applies only to the case of positive definite Hamilton function. Hence a general eigen-solution algorithm is given here, which can be used to indefinite Hamilton function.

It has been shown at the beginning of section 2.3, that solving eigen-equation (2.3.43) should solve the eigen-equation (2.3.46) first. The eigenvectors, say $\mathbf{w}$, of (2.3.43) with roots μ and $-\mu$, are eigenvectors of eigen-equation (2.3.46). However, the reverse is not true, the eigenvectors of (2.3.46) must be duplicated, denoted as $\mathbf{w}_a$ and $\mathbf{w}_b$, which may not be the eigenvectors of (2.3.43); but the eigenvectors of (2.3.43) can be found from the linear combinations of $\mathbf{w}_a$ and $\mathbf{w}_b$. Hence, the equation (2.3.43) or (2.3.45) should be expressed in the subspace spanned by $\mathbf{w}_a$ and $\mathbf{w}_b$. The projection into subspace is a very important technique, which can be carried out under the variational principle. Because of the composition of state vector $\mathbf{w}(t)$ in (2.3.40), the variational principle (1.6.19) should be used

$$\delta\int_{t_0}^{t_f}[(\mathbf{Ms}+\mathbf{Gq}/2)^T(\dot{\mathbf{q}}-\mathbf{s})+\mathrm{L}(\mathbf{q},\mathbf{s})]\mathrm{d}t=0$$

where $\quad \mathrm{L}(\mathbf{q},\mathbf{s})=\mathbf{s}^T\mathbf{Ms}/2+\mathbf{s}^T\mathbf{Gq}/2-\mathbf{q}^T\mathbf{Kq}/2$

which gives

$$\delta\int_{t_0}^{t_f}\left\{\dot{\mathbf{w}}_t^T\begin{bmatrix}\mathbf{G} & \mathbf{M}\\ -\mathbf{M} & 0\end{bmatrix}\mathbf{w}_t/2-\mathbf{w}_t^T\begin{bmatrix}\mathbf{K} & 0\\ 0 & \mathbf{M}\end{bmatrix}\mathbf{w}_t/2\right\}\mathrm{d}t=0,\quad \mathbf{w}_t=\begin{Bmatrix}\mathbf{q}\\ \mathbf{s}\end{Bmatrix} \tag{2.3.50}$$

After substituted with $\mathbf{w}_t=\mathbf{w}e^{\mu t}$, it becomes

$$\delta\left\{\mu\mathbf{w}^T\begin{bmatrix}\mathbf{G} & \mathbf{M}\\ -\mathbf{M} & 0\end{bmatrix}\mathbf{w}/2-\mathbf{w}^T\begin{bmatrix}\mathbf{K} & 0\\ 0 & \mathbf{M}\end{bmatrix}\mathbf{w}/2\right\}=0 \tag{2.3.51}$$

Carrying out the variation derives (2.3.45). This variational principle is convenient for subspace projection.

Since the two basis state vectors are found as $\mathbf{w}_a$ and $\mathbf{w}_b$, a two-dimensional subspace of ξ is composed as

$$\mathbf{w}=[\mathbf{w}_a,\mathbf{w}_b]\cdot\xi,\qquad \xi=\{\xi_1,\xi_2\}^T \tag{2.3.52}$$

and the variational principles for the subspace becomes

$$\delta\{\mu\xi^T\mathbf{A}_2\xi/2-\xi^T\mathbf{B}_2\xi/2\}=0$$

$$\mathbf{A}_2=\begin{bmatrix}\mathbf{w}_a^T\\ \mathbf{w}_b^T\end{bmatrix}\begin{bmatrix}\mathbf{G} & \mathbf{M}\\ \mathbf{M} & 0\end{bmatrix}[\mathbf{w}_a,\mathbf{w}_b],\quad \mathbf{B}_2=\begin{bmatrix}\mathbf{w}_a^T\\ \mathbf{w}_b^T\end{bmatrix}\begin{bmatrix}\mathbf{K} & 0\\ 0 & \mathbf{M}\end{bmatrix}[\mathbf{w}_a,\mathbf{w}_b] \tag{2.3.53}$$

This gives the 2-D projection matrices $\mathbf{A}_2$ and $\mathbf{B}_2$, and the variational principle derives

$$(\mathbf{B}_2 - \mu\mathbf{A}_2)\xi = 0, \qquad \det(\mathbf{B}_2 - \mu\mathbf{A}_2) = 0$$

for which, the eigenvalue problem in two dimensions is easily solved, and also gives a check for the eigenvalue μ. However, because $\mathbf{w}_a$ and $\mathbf{w}_b$ may be complex valued vectors and the 2-D matrices are also complex valued. In such case, the complex conjugate vectors of $\mathbf{w}_a$ and $\mathbf{w}_b$ can be considered simultaneously, which corresponds to using the real and imaginary parts of the vectors $\mathbf{w}_a$ and $\mathbf{w}_b$, that a 4-D subspace basis is resulted. The computation is still not difficult. Hence, the 2-D projection subspace eigen-solution problem is regarded solved.

Therefore, the main problem is to solve equation (2.3.46), the point transformation (2.3.39) can be regarded as a performed pre-step, and thus the equation to be solved becomes

$$\begin{bmatrix} 0 & \text{diag}(-k_i) \\ \text{diag}(k_i) & 0 \end{bmatrix} \mathbf{w}' = \begin{bmatrix} \mathbf{G}' & \mathbf{I}_n \\ -\mathbf{I}_n & 0 \end{bmatrix} \mathbf{w}' \cdot \mu^2, \quad \mathbf{w} = \begin{bmatrix} \mathbf{Q} & \mathbf{0} \\ \mathbf{0} & \mathbf{Q} \end{bmatrix} \mathbf{w}' \tag{2.3.54}$$

where the eigenvalues k_i of matrix $\mathbf{K}$ can be regarded sorted as $k_i \le k_{i+1}$, and negative values may exist. For ease of computation, the vector $\mathbf{w}'$ can be reordered in the form of $\mathbf{w}_n$

$$\mathbf{w}_n = \{q'_1, \dot{q}'_1;\ q'_2, \dot{q}'_2;\ \cdots;\ q'_n, \dot{q}'_n\}^T \tag{2.3.55}$$

Correspondingly, the matrix in equation (2.3.54) should also permute its rows and columns. The case of $n = 4$ is used as an example. Originally,

$$\mathbf{G}' = \begin{bmatrix} 0 & g_{12} & g_{13} & g_{14} \\ -g_{12} & 0 & g_{23} & g_{24} \\ -g_{13} & -g_{23} & 0 & g_{34} \\ -g_{14} & -g_{24} & -g_{34} & 0 \end{bmatrix} \qquad \text{for } n = 4$$

after reordering it gives (the sign $\Rightarrow$ is read as 'derives to')

$$\mathbf{A}_k \mathbf{w}_n = \mu^2 \mathbf{B} \mathbf{w}_n \tag{2.3.54'}$$

$$\mathbf{J} = \begin{bmatrix} \mathbf{0} & \mathbf{I}_n \\ -\mathbf{I}_n & \mathbf{0} \end{bmatrix} \Rightarrow \mathbf{J}'_n, \quad \mathbf{J}'_n = \text{diag}(\mathbf{J}_1), \quad \mathbf{J}_1 = \begin{bmatrix} 0 & 1 \\ -1 & 0 \end{bmatrix} \tag{2.3.56}$$

where $\mathbf{J}'_n$ is a block diagonal matrix, obtained by repeating the 2×2 matrix $\mathbf{J}_1$ for n times at the diagonal, which derives a $2n \times 2n$ matrix. The matrix in (2.3.54') is transformed as

$$\begin{bmatrix} \mathbf{G}' & \mathbf{I}_n \\ -\mathbf{I}_n & \mathbf{0} \end{bmatrix} \Rightarrow \mathbf{B} = \begin{bmatrix} 0 & 1 & g_{12} & 0 & g_{13} & 0 & g_{14} & 0 \\ -1 & 0 & 0 & 0 & 0 & 0 & 0 & 0 \\ -g_{12} & 0 & 0 & 1 & g_{23} & 0 & g_{24} & 0 \\ 0 & 0 & -1 & 0 & 0 & 0 & 0 & 0 \\ -g_{13} & 0 & -g_{23} & 0 & 0 & 1 & g_{34} & 0 \\ 0 & 0 & 0 & 0 & -1 & 0 & 0 & 0 \\ -g_{14} & 0 & -g_{24} & 0 & -g_{34} & 0 & 0 & 1 \\ 0 & 0 & 0 & 0 & 0 & 0 & -1 & 0 \end{bmatrix} \tag{2.3.57}$$

also $\begin{bmatrix} \mathbf{0} & -\text{diag}(k_i) \\ \text{diag}(k_i) & \mathbf{G}' \end{bmatrix} \Rightarrow \mathbf{A}_k$, where

$$\mathbf{A}_k=\begin{bmatrix} 0 & -k_1 & 0 & 0 & 0 & 0 & 0 & 0 \\ k_1 & 0 & 0 & g_{12} & 0 & g_{13} & 0 & g_{14} \\ 0 & 0 & 0 & -k_2 & 0 & 0 & 0 & 0 \\ 0 & -g_{12} & k_2 & 0 & 0 & g_{23} & 0 & g_{24} \\ 0 & 0 & 0 & 0 & 0 & -k_3 & 0 & 0 \\ 0 & -g_{13} & 0 & -g_{23} & k_3 & 0 & 0 & g_{34} \\ 0 & 0 & 0 & 0 & 0 & 0 & 0 & -k_4 \\ 0 & -g_{14} & 0 & -g_{24} & 0 & -g_{34} & k_4 & 0 \end{bmatrix} \tag{2.3.58}$$

From the above equation it is seen that the skew-symmetric matrices $\mathbf{B}, \mathbf{A}_k$ etc. can be considered as composed of 2×2 matrices, called as ***cells***, *i.e.* the $2n\times 2n$ matrix is considered as a $n\times n$ ***cell*** matrix; and the 2×2 matrix $\mathbf{J}_1$ corresponds to a ***unit skew-cell***. The equation (2.3.54') of skew-symmetric matrices can be transformed to a standard form [55], using the cell skew-symmetric modified triangular factorization $\mathbf{LD}_\mathrm{J}\mathbf{L}^T$ [56]. For the matrix $\mathbf{B}$ in (2.3.57), it is factorized as

$$\mathbf{B}=\mathbf{L}_a\mathbf{J}'_n\mathbf{L}_a^T \tag{2.3.59}$$

where

$$\mathbf{L}_a=\begin{bmatrix} 1 & 0 & 0 & 0 & 0 & 0 & 0 & 0 \\ 0 & 1 & 0 & 0 & 0 & 0 & 0 & 0 \\ 0 & g_{12} & 1 & 0 & 0 & 0 & 0 & 0 \\ 0 & 0 & 0 & 1 & 0 & 0 & 0 & 0 \\ 0 & g_{13} & 0 & g_{23} & 1 & 0 & 0 & 0 \\ 0 & 0 & 0 & 0 & 0 & 1 & 0 & 0 \\ 0 & g_{14} & 0 & g_{24} & 0 & g_{34} & 1 & 0 \\ 0 & 0 & 0 & 0 & 0 & 0 & 0 & 1 \end{bmatrix} \tag{2.3.60}$$

Substituting (2.3.60) into the equation (2.3.59) verifies directly. Similarly,

$$\mathbf{L}_a^{-1}=\begin{bmatrix} 1 & 0 & 0 & 0 & 0 & 0 & 0 & 0 \\ 0 & 1 & 0 & 0 & 0 & 0 & 0 & 0 \\ 0 & -g_{12} & 1 & 0 & 0 & 0 & 0 & 0 \\ 0 & 0 & 0 & 1 & 0 & 0 & 0 & 0 \\ 0 & -g_{13} & 0 & -g_{23} & 1 & 0 & 0 & 0 \\ 0 & 0 & 0 & 0 & 0 & 1 & 0 & 0 \\ 0 & -g_{14} & 0 & -g_{24} & 0 & -g_{34} & 1 & 0 \\ 0 & 0 & 0 & 0 & 0 & 0 & 0 & 1 \end{bmatrix} \tag{2.3.60'}$$

Substituting (2.3.59) back into the eigen-equation (2.3.54') and left multiplying with $\mathbf{L}_a^{-1}$ derives

$$\mathbf{A}\mathbf{w}_\mathrm{b}=\mu^2\mathbf{J}'_n\mathbf{w}_\mathrm{b} \tag{2.3.61}$$

$$\mathbf{A}=\mathbf{L}_a^{-1}\mathbf{A}_k\mathbf{L}_a^{-T}, \quad \mathbf{w}_\mathrm{b}=\mathbf{L}_a^T\mathbf{w}_n \tag{2.3.62}$$

For the sake of simplicity, $\mathbf{w}_\mathrm{b}$ is written as $\mathbf{w}$ again. The matrix $\mathbf{A}$ is obviously still skew-symmetric. The eigen-equation (2.3.61) is considered the standard form of a skew-symmetric symplectic eigen-problem. Note that the form of eigen-problem (2.3.61) is entirely different to the usual form of $\mathbf{Aw}=\mu^2\mathbf{w}$.

There is $\mathbf{J}'_n$ at the right hand side of (2.3.61), so that it is called as ***symplectic eigen-problem***, whereas the usual form is an eigen-problem of a skew-symmetric matrix, no $\mathbf{J}'_n$ at the right hand side.

When dimension n is large for the eigen-equation (2.3.61), the ***adjoint symplectic subspace iteration method*** can be used to find the main eigen-solutions. The transformation equation is similar to that given in (2.3.53) for subspace projection. Note that (2.3.53) has only two subspace basis vectors. Similar to the subspace iteration for symmetric matrix, the subspace rotation requires multiplication of $\mathbf{A}^{-1}$ to the cell vectors hence the cell triangular factorization $\mathbf{L}\mathbf{D}_J\mathbf{L}^T$ of the skew-symmetric matrix $\mathbf{A}$ is very useful. In fact, the matrix multiplication in equation (2.3.62) is unnecessary to carry out numerically, that the matrix $\mathbf{A}_k$ should also be factorized as

$$\mathbf{A}_k = \mathbf{L}_k[\text{diag}(-k_i\mathbf{J}_1)]\mathbf{L}_k^T \tag{2.3.63}$$

where

$$\mathbf{L}_k = \begin{bmatrix} 1 & & & & & & & 0 \\ 0 & 1 & & & & & & \\ 0 & 0 & 1 & & & & & \\ g_{12}/k_1 & 0 & 0 & 1 & & & & \\ 0 & 0 & 0 & 0 & 1 & & & \\ g_{13}/k_1 & 0 & g_{23}/k_2 & 0 & 0 & 1 & & \\ 0 & 0 & 0 & 0 & 0 & 0 & 1 & \\ g_{14}/k_1 & 0 & g_{24}/k_2 & 0 & g_{34}/k_3 & 0 & 0 & 1 \end{bmatrix} \tag{2.3.64}$$

and

$$\mathbf{L}_k^{-1} = \begin{bmatrix} 1 & & & & & & & 0 \\ 0 & 1 & & & & & & \\ 0 & 0 & 1 & & & & & \\ -g_{12}/k_1 & 0 & 0 & 1 & & & & \\ 0 & 0 & 0 & 0 & 1 & & & \\ -g_{13}/k_1 & 0 & -g_{23}/k_2 & 0 & 0 & 1 & & \\ 0 & 0 & 0 & 0 & 0 & 0 & 1 & \\ -g_{14}/k_1 & 0 & -g_{24}/k_2 & 0 & -g_{34}/k_3 & 0 & 0 & 1 \end{bmatrix} \tag{2.3.64'}$$

for $n = 4$. These matrices are useful in computer programming.

The algorithm for transforming to standard form of symplectic eigen-problem is described as:

Giving $n, \mathbf{M}, \mathbf{K}, \mathbf{G}$

1) Solving the generalized eigen-problem of symmetric matrices $\mathbf{M}, \mathbf{K}$ (2.3.39) gives $\mathbf{Q}, k_i, (i = 1, \cdots, n)$.
2) Compute $\mathbf{G}' = \mathbf{Q}^T\mathbf{G}\mathbf{Q}$, then compose matrix $\mathbf{A}_k$ according to (2.3.58).
3) Compose $\mathbf{L}_a, \mathbf{L}_a^{-1}$ according to equation (2.3.60).
4) Compute matrix $\mathbf{A}$, according to equation (2.3.62).
5) Solving the symplectic eigen-solution (2.3.61) gives $\mu^2, \mathbf{w}_b$; then compute $\mathbf{w}_n = \mathbf{L}_a^{-T}\mathbf{w}_b$.

6) Back transforming from (2.3.55) gives $\mathbf{w}'$, then $\mathbf{q} = \mathbf{Q}\mathbf{q}', \dot{\mathbf{q}} = \mathbf{Q}\dot{\mathbf{q}}'$ and composing $\mathbf{w}$.
7) From μ^2 and its corresponding eigenvectors $\mathbf{w}_a, \mathbf{w}_b$, according to (2.3.52) and (2.3.53), solve the eigen-solutions of (2.3.45).

Among all of these steps, only the fifth step, the algorithm for the standard problem of ***solving the symplectic eigen-solution*** (2.3.61), needs to be further clarified, which is given in the next section.

§2.3.3.3, Symplectic eigen-solution of skew-symmetric matrix

The standard form of symplectic eigen-problem of skew-symmetric matrix is expressed as

$$\mathbf{A}\mathbf{w} = \mu^2 \mathbf{J}'_n \mathbf{w} \tag{2.3.65}$$

where $\mathbf{A}$ is a given $2n \times 2n$ skew-symmetric matrix. Treating 2×2 sub-matrix as ***cell***, then $\mathbf{A}$ is a $n \times n$ ***cell matrix*** and $\mathbf{J}'_n$ is defined in (2.3.56). The ***symplectic eigen-problem*** is different to the traditional eigen-problem $\mathbf{A}\mathbf{w} = \mu^2 \mathbf{w}$, which has no unit symplectic matrix $\mathbf{J}'_n$ at the right hand side, hence belonging to Euclidean metric, however (2.3.65) corresponds to the symplectic metric, hence called ***symplectic eigen-problem***. The problem now is to develop an effective algorithm for the solution of ***symplectic eigen-problem***.

The methodology of solution of symplectic eigen-problem can take benefit from the algorithm for the solution of symmetric and/or general eigen-problems. The solution usually is via a series of transformations to derive the original matrix to some diagonal form or other standard form, and then use the iteration method. The traditional eigen-problem uses the Euclidean metric hence the transformation matrix should use the orthogonal matrix to keep the metric unchanged. However the symplectic eigen-problem uses ***symplectic metric***. For two arbitrary $2n$-D state vectors $\mathbf{w}_a, \mathbf{w}_b$, the ***symplectic metric*** is defined as

$$d(\mathbf{w}_a, \mathbf{w}_b) = \mathbf{w}_a^T \mathbf{J}'_n \mathbf{w}_b = -\mathbf{w}_b^T \mathbf{J}'_n \mathbf{w}_a = -d(\mathbf{w}_b, \mathbf{w}_a) \tag{2.3.66}$$

which is anti-symmetric. In mathematics, the usual definition for **Euclidean metric** requires 1) positive definite; 2) symmetric; 3) the triangular inequality. However, the ***symplectic metric*** is indefinite, skew-symmetric, nor triangular inequality either. For a transformation matrix $\mathbf{S}$

$$\mathbf{w}_a = \mathbf{S}\mathbf{w}'_a, \quad \mathbf{w}_b = \mathbf{S}\mathbf{w}'_b \quad \text{then} \quad d(\mathbf{w}_a, \mathbf{w}_b) = \mathbf{w}_a'^T (\mathbf{S}^T \mathbf{J}'_n \mathbf{S}) \mathbf{w}'_b$$

If the symplectic metric keeps unchanged under the transformation, then

$$d(\mathbf{w}_a, \mathbf{w}_b) = d(\mathbf{w}'_a, \mathbf{w}'_b)$$

is valid for arbitrary two vectors $\mathbf{w}_a$ and $\mathbf{w}_b$, it must be

$$\mathbf{S}^T \mathbf{J}'_n \mathbf{S} = \mathbf{J}'_n \tag{2.3.67}$$

which means that $\mathbf{S}$ is a symplectic matrix. Note that the symplectic matrix here corresponds to the form of row and column permutation of the usual symplectic matrix, see (2.3.55), so that the dual variables $\mathbf{q}, \mathbf{p}$ are sorted in mixed order. The symplectic matrices compose a group, which has been mentioned in chapter 1.

One of the most effective eigen-solution methods for symmetric matrix is, first using the Householder orthogonal transformation to derive the matrix to tri-diagonal form, then to solve by the QL iterative algorithm. For real general matrix the method is similar, see [42] for example. For symplectic eigen-solution of skew-symmetric matrix, similar technique can be applied, and is introduced in three steps described below: see [56]

1) Symplectic-Householder transformation (*SH* transformation) and/or ***orthogonal SH transformation***
2) Transform to half tri-diagonal cell matrix;
3) The symplectic eigen-solution for the half tri-diagonal cell matrix.

The three steps are described successively below.

(1) The ***SH*** transformation and the ***orthogonal SH*** transformation:

The usual Householder transformation matrix is an orthogonal matrix and is composed of a projection vector, see [30]. For symplectic Householder transformation matrix, it is composed of a ***cell vector*** $\mathbf{u}$. The cell vector $\mathbf{u}$ is a n-dimensional vector with cells (2×2 sub-matrices) as elements. Therefore $\mathbf{u}$ is really a $2n\times2$ matrix in the usual sense. From $\mathbf{u}$, a $2n\times2n$ transformation matrix ($n\times n$ cell matrix) is composed as

$$\mathbf{S}_h = \mathbf{I}_{2n} - 2\mathbf{J}'_n\mathbf{u}(\mathbf{u}^T\mathbf{J}'_n\mathbf{u})^{-1}\mathbf{u}^T \tag{2.3.68}$$

Note that $\mathbf{J}'^T_n = -\mathbf{J}'_n$, $\mathbf{J}'^{-1}_n = -\mathbf{J}'_n$, $(\mathbf{J}'_n)^2 = -\mathbf{I}_{2n}$ and the verification of $\mathbf{S}_h$ being a symplectic matrix is given as

$$\begin{aligned}\mathbf{S}_h^T\mathbf{J}'_n\mathbf{S}_h &= \left[\mathbf{I}_{2n} - 2\mathbf{u}(\mathbf{u}^T\mathbf{J}'_n\mathbf{u})^{-1}\mathbf{u}^T\mathbf{J}'_n\right]\mathbf{J}'_n\left[\mathbf{I}_{2n} - 2\mathbf{J}'_n\mathbf{u}(\mathbf{u}^T\mathbf{J}'_n\mathbf{u})^{-1}\mathbf{u}^T\right]\\ &= \mathbf{J}'_n + 2\mathbf{u}(\mathbf{u}^T\mathbf{J}'_n\mathbf{u})^{-1}\mathbf{u}^T + 2\mathbf{u}(\mathbf{u}^T\mathbf{J}'_n\mathbf{u})^{-1}\mathbf{u}^T - 4\mathbf{u}(\mathbf{u}^T\mathbf{J}'_n\mathbf{u})^{-1}\mathbf{u}^T\mathbf{J}'_n\times(\mathbf{u}^T\mathbf{J}'_n\mathbf{u})^{-1}\mathbf{u}^T\\ &= \mathbf{J}'_n + 4\mathbf{u}(\mathbf{u}^T\mathbf{J}'_n\mathbf{u})^{-1}\mathbf{u}^T - 4\mathbf{u}(\mathbf{u}^T\mathbf{J}'_n\mathbf{u})^{-1}\mathbf{u}^T = \mathbf{J}'_n\end{aligned}$$

In the class of SH transformation matrices, the subclass of ***orthogonal SH transformation matrices*** are of great interest. Because, ***transformation with orthogonal matrix is numerically stable.*** The SH transformation given in (2.3.68) is not necessarily an orthogonal matrix. The ***orthogonal SH transformation matrix*** is still defined as in (2.3.64), however, the cell vector $\mathbf{u}$ can no longer be arbitrarily selected. The matrix $\mathbf{S}_h$ is orthogonal, if it satisfies further the condition $\mathbf{S}_h^T\mathbf{S}_h = \mathbf{I}_{2n}$, and then $\mathbf{S}_h$ is an ***orthogonal SH*** matrix. Let the selection of the cell vector $\mathbf{u}$ be, that all its cells (2×2 submatrices) have the form

$$\mathbf{u}_i = \begin{bmatrix} a_i & -b_i \\ b_i & a_i \end{bmatrix} \tag{2.3.69}$$

where a_i, b_i are arbitrary real numbers. It needs to verify that under condition (2.3.69) for cells, the SH transformation matrix composed in (2.3.68) is simultaneously an orthogonal matrix. To verify, note that for an arbitrary 2×2 matrix $\mathbf{P}$

$$\mathbf{P}^T\mathbf{J}_1\mathbf{P} \equiv \mathbf{J}_1\det(\mathbf{P})$$

Note next, that $\mathbf{u}_i$ should not all be zero cells (2×2 matrices), otherwise $\mathbf{S}_h = \mathbf{I}_{2n}$ and the transformation is unnecessary. Hence

$$(\mathbf{u}^T\mathbf{J}_n'\mathbf{u})^{-1} = -\mathbf{J}_1 d^{-1}, \quad d = \sum_i \det(\mathbf{u}_i) = \sum_i (a_i^2 + b_i^2) > 0 ;$$

and $\mathbf{u}^T\mathbf{u} = \mathbf{I}_2 d$. Therefore

$$\begin{aligned}\mathbf{S}_h^T\mathbf{S}_h &= [\mathbf{I}_{2n} - 2\mathbf{u}(\mathbf{u}^T\mathbf{J}_n'\mathbf{u})^{-1}\mathbf{u}^T\mathbf{J}_n']\cdot[\mathbf{I}_{2n} - 2\mathbf{J}_n'\mathbf{u}(\mathbf{u}^T\mathbf{J}_n'\mathbf{u})^{-1}\mathbf{u}^T] \\ &= \mathbf{I}_{2n} + 2d^{-1}\mathbf{u}\mathbf{J}_1\mathbf{u}^T\mathbf{J}_n' + 2d^{-1}\mathbf{J}_n'\mathbf{u}\mathbf{J}_1\mathbf{u}^T - 4d^{-2}\mathbf{u}\mathbf{J}_1\mathbf{u}^T\mathbf{u}\mathbf{J}_1\mathbf{u}^T \\ &= \mathbf{I}_{2n} - 2d^{-1}\mathbf{u}\mathbf{u}^T - 2d^{-1}\mathbf{u}\mathbf{u}^T + 4d^{-1}\mathbf{u}\mathbf{u}^T = \mathbf{I}_{2n}\end{aligned}$$

where the identity $\mathbf{u}^T\mathbf{J}_n' = \mathbf{J}_1\mathbf{u}^T$, which is easily verified, is applied. So that $\mathbf{S}_h$ is an orthogonal SH transformation matrix. Note that the multiplication of two orthogonal SH matrices is also an orthogonal SH matrix.

(2) Transform to ***half tri-diagonal cell matrix***

The main application of the SH transformation matrix is, transforming the skew-symmetric matrix $\mathbf{A}$ to become a tri-diagonal cell matrix; and the main application of the orthogonal SH transformation matrix is, transforming a skew-symmetric matrix $\mathbf{A}$ to become a half tri-diagonal cell matrix, which has the form

$$\begin{bmatrix} 0 & d_1 & & \text{skew -} & \text{symmetry} & & & \\ -d_1 & 0 & & & & & & \\ 0 & * & 0 & d_2 & & & & \\ * & * & -d_2 & 0 & & & & \\ 0 & * & 0 & * & 0 & d_3 & & \\ 0 & * & * & * & -d_3 & 0 & & \\ 0 & * & 0 & * & 0 & * & 0 & d_4 \\ 0 & * & 0 & * & * & * & -d_4 & 0 \end{bmatrix} \tag{2.3.70}$$

where * represents real numbers. Because the numerical stability is of great concern, hence only the solution with half tri-diagonal cell matrix is given here. Bringing the row-column permutation back to a half tri-diagonal cell matrix, *i.e.* the reverse sorting for equation (2.3.55), (the odd rows/columns are sorted separately). After reverse sorting, the matrix (2.3.70) becomes of the form

$$\begin{bmatrix} \mathbf{0} & \mathbf{M}_1^T \\ -\mathbf{M}_1 & \mathbf{C} \end{bmatrix} = \begin{bmatrix} & & & & d_1 & * & 0 & 0 \\ & \mathbf{0} & & & * & d_2 & * & 0 \\ & & & & * & * & d_3 & * \\ & & & & * & * & * & d_4 \\ -d_1 & * & * & * & 0 & \text{skew -} & & \\ * & -d_2 & * & * & * & 0 & \text{symm.} & \\ 0 & * & -d_3 & * & * & * & 0 & \\ 0 & 0 & * & -d_4 & * & * & * & 0 \end{bmatrix} \tag{2.3.71}$$

where the upper-left block becomes a $n \times n$ zero matrix. The lower-left block $n \times n$ matrix $-\mathbf{M}_1$ has been transformed to be a $n \times n$ Hessenberg matrix. Lower-right block matrix $\mathbf{C}$ is $n \times n$ skew-symmetric and upper-right block $\mathbf{M}_1^T$ is mutually negative transpose to the lower-left block. Transformed to the above form is convenient to solve the symplectic eigen-solution, and which is to be explained in step (3). Now, it is to describe how to use the ***orthogonal SH matrix*** to transform the matrix $\mathbf{A}$ into the form of (2.3.70).

It can use $n-2$ steps, for the cell columns of $\text{1st}, \text{2nd}, \cdots, (\text{n}-2)\text{-th}$ progressively carrying out the orthogonal SH transformations. The transformation is similar to that of a usual matrix, which is transformed step by step toward the Hessenberg type matrix with the Householder transformation [42]. Suppose the cell columns of $1, \cdots, r-1$ in matrix $\mathbf{A}$ have been transformed to be in half tri-diagonal cell matrix form, *i.e.* $\mathbf{A}$ had been transformed to be a matrix $\mathbf{A}_r$. It is to find an appropriate cell column $\mathbf{u}$, so as to compose an orthogonal SH matrix $\mathbf{S}_r$, such that the congruent transformation by $\mathbf{S}_r$

$$\mathbf{A}_{r+1} = \mathbf{S}_r^T \mathbf{A}_r \mathbf{S}_r \tag{2.3.72}$$

which updates the r-th cell row and cell column in $\mathbf{A}_{r+1}$ to be in the form of a half tri-diagonal cell matrix, and at the same time the former $r-1$ cell columns and rows are kept unchanged. The matrix $\mathbf{S}_r$ can be composed of a Givens rotation followed by an orthogonal SH transformation. Because, both are orthogonal SH transformations so their product $\mathbf{S}_r$ is also an orthogonal SH matrix. The effect of the Givens rotation is to eliminate the upper-left element in $\mathbf{a}_{r+1,r}^{(r)}$, which is the cell $(r+1, r)$ in the matrix $\mathbf{A}_r$. Here the superscript (r) in the cell $\mathbf{a}_{r+1,r}^{(r)}$ is the subscript of $\mathbf{A}_r$, and will be used below. Let the cell before this round of transformation be

$$\mathbf{a}_{r+1,r}^{(r)} = \begin{bmatrix} a_1 & a_2 \\ a_3 & a_4 \end{bmatrix}$$

The Givens rotation matrix $\mathbf{S}_g$ is a unit matrix $\mathbf{I}_{2n}$ except that the diagonal cell $(r+1, r+1)$ is given as

$$\mathbf{S}_{g,(r+1,r+1)} = \begin{bmatrix} c & s \\ -s & c \end{bmatrix}, \quad c = a_3 \Big/ \sqrt{a_1^2 + a_3^2}\,, \quad s = a_1 \Big/ \sqrt{a_1^2 + a_3^2}\,, \quad c^2 + s^2 = 1$$

After the congruent transformation with $\mathbf{S}_g$ (Givens transformation), the upper-left element in the $(r+1, r)$ cell $\mathbf{a}_{r+1,r}^{(rg)}$ of the matrix $\mathbf{A}_{rg} = \mathbf{S}_g^T \mathbf{A}_r \mathbf{S}_g$ has become zero, *i.e.*

$$\mathbf{a}_{r+1,r}^{(rg)} = \begin{bmatrix} 0 & \beta \\ \alpha & \gamma \end{bmatrix}, \qquad \alpha = \sqrt{a_1^2 + a_3^2}$$

the superscript $^{(rg)}$ denotes that the cell is after the Givens transformation. Afterward, the orthogonal SH matrix $\mathbf{S}_{h0}$ is to be found, such that the r-th transformation matrix $\mathbf{S}_r$ is combined as

$$\mathbf{S}_r = \mathbf{S}_g \times \mathbf{S}_{h0} \tag{2.3.72a}$$

The cell vector $\mathbf{u}$ used to compose $\mathbf{S}_{h0}$ is selected that the cells in $\mathbf{u}$ are given as

$$\mathbf{u}_i = \mathbf{0}, \qquad \text{when } i \le r, \tag{2.3.73a}$$

$$\mathbf{u}_{r+1} = -(\alpha+\sigma)\mathbf{J}_1, \quad \text{when} \quad i = r+1 \tag{2.3.73b}$$

$$\mathbf{u}_i = \begin{bmatrix} a_i & -b_i \\ b_i & a_i \end{bmatrix}, \quad \text{when} \quad i > r+1, \quad \text{where} \quad \mathbf{a}_{i,r}^{(rg)} = \begin{bmatrix} a_i & c_i \\ b_i & d_i \end{bmatrix} \tag{2.3.73c}$$

where σ is a parameter to be determined. Because the selection of (2.3.73a), the orthogonal SH transformation matrix $\mathbf{S}_{h0}$ generated by $\mathbf{u}$ differs from the unit matrix $\mathbf{I}_{2n}$ only below and right to the $(r+1,r+1)$ cell. Hence the congruent transformation with $\mathbf{S}_{h0}$ is written as

$$\mathbf{B}_r = \mathbf{S}_{h0}^T\mathbf{A}_{rg}, \qquad \text{then} \qquad \mathbf{A}_{r+1} = \mathbf{B}_r\mathbf{S}_{h0} \tag{2.3.74}$$

and it is easy to see that the r-th cell column can be affected only by the left multiplication of $\mathbf{S}_{h0}^T$, hence the r-th cell column in matrix $\mathbf{A}_{r+1}$ is the same as the matrix $\mathbf{B}_r$. Because, the half tri-diagonalization for the r-th cell column in the cell matrix $\mathbf{A}_{r+1}$ is to be carried out, so that the left column within the r-th cell column is of concern, it is shown as follows.

The r-th cell column in matrix $\mathbf{B}_r$ is multiplied explicitly as follows. Because in $\mathbf{S}_{h0}$

$$(\mathbf{u}^T\mathbf{J}_n'\mathbf{u})^{-1} = \left[\mathbf{J}_1\cdot(\sigma^2+2\sigma\alpha+\alpha^2+\sum_{i=r+2}^{n}(a_i^2+b_i^2))\right]^{-1} = -(\sigma^2+2\sigma\alpha+d)^{-1}\mathbf{J}_1$$

where
$$d = \alpha^2 + \sum_{i=r+2}^{n}(a_i^2+b_i^2) > 0 \tag{2.3.75}$$

for the r-th cell column in matrix $\mathbf{A}_{rg}$, denoted as $\mathbf{A}_r^{(rg)}$, carrying out the multiplication, gives

$$\mathbf{u}^T\mathbf{J}_n'\mathbf{A}_r^{(rg)} = -(\alpha+\sigma)\times\begin{bmatrix} 0 & \beta \\ \alpha & \gamma \end{bmatrix} + \sum_{i=r+2}^{n}\begin{bmatrix} -b_i & a_i \\ -a_i & -b_i \end{bmatrix}\times\begin{bmatrix} a_i & c_i \\ b_i & d_i \end{bmatrix}$$

$$= \begin{bmatrix} 0 & -\beta(\alpha+\sigma)+\sum_{r+2}^{n}(a_id_i-b_ic_i) \\ -\alpha^2-\alpha\sigma-\sum_{r+2}^{n}(a_i^2+b_i^2) & -\gamma(\alpha+\sigma)-\sum_{r+2}^{n}(a_ic_i+b_id_i) \end{bmatrix}$$

hence

$$2\mathbf{u}(\mathbf{u}^T\mathbf{J}_n'\mathbf{u})^{-1}\mathbf{u}^T\mathbf{J}_n'\mathbf{A}_r^{(rg)} =$$

$$2(\sigma^2+2\sigma\alpha+d)^{-1}\mathbf{u}\times\begin{bmatrix} (d+\sigma\alpha) & \gamma(\alpha+\sigma)-\sum_{r+2}^{n}(a_ic_i+b_id_i) \\ 0 & -\beta(\alpha+\sigma)+\sum_{r+2}^{n}(a_id_i-b_ic_i) \end{bmatrix}$$

as the product, the (i,r)-th cell with $i > r+1$ in matrix $\mathbf{B}_r$, denoted as $\mathbf{B}_{i,r}$, being

$$\mathbf{B}_{i,r} = \begin{bmatrix} a_i & c_i \\ b_i & d_i \end{bmatrix} - (\sigma^2+2\sigma\alpha+d)^{-1}(2d+2\sigma\alpha)\begin{bmatrix} a_i & * \\ b_i & * \end{bmatrix}, \quad i > r+1$$

In order to make the left column within $\mathbf{B}_{i,r}$ be zero, the selection must be

$$2d + 2\sigma\alpha = \sigma^2 + 2\sigma\alpha + d\ , \quad \text{it solves} \quad \sigma = \sqrt{d} \tag{2.3.75'}$$

From equation (2.3.75) $d > 0$, and $\alpha \geq 0$ is ensured. After σ is determined, the cell vector $\mathbf{u}$ is obtained. Then the ***orthogonal SH transformation matrix*** $\mathbf{S}_{h0}$ is obtained as

$$\mathbf{S}_{h0} = \mathbf{I}_{2n} - (d + \sqrt{d}\alpha)^{-1}\mathbf{u}\mathbf{u}^T \tag{2.3.76}$$

which is easy to compute. From (2.3.72a), the matrix $\mathbf{S}_r$ is obtained. From (2.3.72), the r-th step of half tri-diagonalization for the cell matrix is generated.

When r runs for $1,2,\cdots,n-2$, the cell matrix $\mathbf{A}$ is half tri-diagonalized.

(3) The symplectic eigen-solutions for skew-symmetric half tri-diagonalized cell matrix

After the cell skew-symmetric matrix $\mathbf{A}$ half tri-diagonalized, the symplectic eigen-solutions have been easier to solve. Equation (2.3.71) showed that the back permutation of columns and rows brings the symplectic eigen-problem of the skew-symmetric half tri-diagonal cell matrix (2.3.70) to the form

$$\begin{bmatrix} \mathbf{0} & \mathbf{M}_1^T \\ -\mathbf{M}_1 & \mathbf{C} \end{bmatrix} \begin{Bmatrix} \mathbf{q} \\ \mathbf{p} \end{Bmatrix} = \mu^2 \begin{bmatrix} \mathbf{0} & \mathbf{I}_n \\ -\mathbf{I}_n & \mathbf{0} \end{bmatrix} \begin{Bmatrix} \mathbf{q} \\ \mathbf{p} \end{Bmatrix} \tag{2.3.77}$$

where $\mathbf{q},\mathbf{p}$ are n-dimensional dual vectors. After a series of symplectic transformation (*i.e.* canonical transformation), they have been no longer the original displacement and momentum vectors. The equation (2.3.77) can be disintegrated as the composition of eigen-problems

$$\mathbf{M}_1^T\mathbf{p} = \mu^2\mathbf{p} \tag{2.3.77a}$$

$$\mathbf{M}_1\mathbf{q} - \mathbf{C}\mathbf{p} = \mu^2\mathbf{q} \tag{2.3.77b}$$

where equation (2.3.77a) is obviously a n-dimensional eigen-problem of a real matrix, and the matrix $\mathbf{M}_1^T$ has been of lower Hessenberg form. The equation (2.3.77b) is an inhomogeneous equation of vector $\mathbf{q}$. The corresponding homogeneous equation of (2.3.77b)

$$\mathbf{M}_1\mathbf{q}_0 = \mu^2\mathbf{q}_0$$

is just the dual to the eigen-equation (2.3.77a), which is an eigen-problem of upper Hessenberg type matrix, and coincides completely to the condition of the algorithm given in [42]-II-15. Hence the standard procedure HQR2 can be invoked to find all the eigenvalues μ_i^2 and the eigenvectors $\mathbf{q}_{0i} (i = 1,\cdots,n)$. Expressed with the matrix form as

$$\mathbf{M}_1\mathbf{Q}_0 = \mathbf{Q}_0\text{diag}(\mu_i^2) \tag{2.3.78}$$

where $\mathbf{Q}_0$ is the matrix compose of eigenvectors $\mathbf{q}_{0i}$, and $\text{diag}(\mu_i^2)$ denotes a diagonal matrix. Thereafter all the eigen-solutions of (2.3.77a) are derived as (all eigenvalues are single)

$$\mathbf{M}_1^T\mathbf{P}_0 = \mathbf{P}_0\text{diag}(\mu_i^2), \quad \mathbf{P}_0 = \mathbf{Q}_0^{-T} \tag{2.3.79}$$

Next step is to find all the solutions of (2.3.77b), which is a non-homogeneous equation for $\mathbf{q}$. Because μ_i^2 is an eigenvalue, so $\det(\mathbf{M}_1 - \mu_i^2\mathbf{I}_n) = 0$, and the equation (2.3.77b) is a singular algebraic linear simultaneous equation set. According to the ***fundamental theorem of linear algebraic equations***, the solution existence condition is that the solutions of its dual simultaneous equation (2.3.77a), *i.e.* the eigenvectors $\mathbf{p}_i$ or the columns in $\mathbf{P}_0$, should all be orthogonal to the non-homogeneous term. Left multiplying $\mathbf{p}_i^T$ to equation (2.3.77b), based on (2.3.77a) gives

$$\mathbf{p}_i^T \mathbf{C} \mathbf{p}_i = 0 \tag{2.3.80}$$

which is the condition for the existence of solutions. But the matrix $\mathbf{C}$ is skew-symmetric, hence the above equation is valid for arbitrary vector $\mathbf{p}$, an identity indeed. Hence (2.3.77b) is a consistent equation. The solution of which is to find the inhomogeneous special solution $\mathbf{q}_i'$ of the equation

$$\mathbf{M}_1 \mathbf{q}_i' - \mu_i^2 \mathbf{q}_i' = \mathbf{C} \mathbf{p}_i \tag{2.3.81}$$

Using the eigenvector expansion method

$$\mathbf{q}_i' = \sum_{j=1}^{n}{}^{*} b_{ij} \mathbf{q}_{0j}, \qquad b_{ij} \text{ are to be determined} \tag{2.3.82}$$

where the diagonal elements b_{ii} is selected as zero, which is the meaning of $\sum^{*}$, because $\mathbf{q}_{0i}$ is the solution of homogeneous equation. Substituting (2.3.82) into (2.3.81) gives

$$\sum_{j=1}^{n}{}^{*} (\mu_j^2 - \mu_i^2) b_{ij} \mathbf{q}_{0j} = \mathbf{C} \mathbf{p}_i$$

To determine the coefficients b_{ij}, left multiplying the above equation with $\mathbf{p}_k^T$, because $\mathbf{p}_k$ is the k-th column of $\mathbf{Q}_0^T$, see (2.3.79), so when $k \neq j$, $\mathbf{p}_k^T \mathbf{q}_{0j} = 0$. It derives

$$b_{ik} = \mathbf{p}_k^T \mathbf{C} \mathbf{p}_i / (\mu_k^2 - \mu_i^2) \tag{2.3.83}$$

Because the matrix $\mathbf{C}$ is skew-symmetric, so $b_{ik} = b_{ki}$.

Up to now the $2n$ symplectic eigen-solutions of (2.3.77) have all been found. They are composed of n duplicate eigen-roots $\mu_i^2, (i = 1, 2, \cdots, n)$, and the corresponding two eigenvectors are

$$\mathbf{w}_i = \begin{Bmatrix} \mathbf{q}_{0i} \\ \mathbf{0} \end{Bmatrix}, \quad \text{and} \quad \mathbf{w}_{n+i} = \begin{Bmatrix} \mathbf{q}_i' \\ \mathbf{p}_i \end{Bmatrix}, \quad i = 1, 2, \cdots, n \tag{2.3.84}$$

They are obviously independent to each other when the Jordan form does not appear. They are mutually symplectic adjoint to each other, *i.e.*

$$\mathbf{w}_i^T \mathbf{J} \mathbf{w}_{n+i} = \mathbf{q}_{0i}^T \mathbf{p}_i = 1$$

and the eigenvectors for different eigenvalues μ_i^2 are definitely symplectic orthogonal to each other mutually.

The solution algorithm of symplectic eigen-solution problem (2.3.65) for a skew-symmetric matrix $\mathbf{A}$ can be described as

1) Using the orthogonal SH transformation $\mathbf{S}$, the matrix $\mathbf{A}$ is brought to the

form of half tri-diagonalized cell matrix, denoted as $\mathbf{A}_{s3d} = \mathbf{S}^T \mathbf{A} \mathbf{S}$, see the paragraph between (2.3.68)~(2.3.76), with $\mathbf{S} = \mathbf{S}_1 \mathbf{S}_2 \cdots \mathbf{S}_{n-2}$;

2) Reverse permutation for rows and columns, the $\mathbf{A}_{s3d}$ is brought back to the form of (2.3.77), which gives $\mathbf{M}_1, \mathbf{C}$;
3) Invoking the procedure HQR2 solves all eigen-solutions of (2.3.78), to obtain the matrices $\mathbf{Q}_0$ and $\mathrm{diag}(\mu_i^2)$;
4) According to (2.3.79), compute the matrix $\mathbf{P}_0$;
5) According to (2.3.83) compute b_{ik}, and solve the eigenvectors (2.3.84), $\mathbf{w}_i, \mathbf{w}_{n+i}$;
6) Reordering the eigen-vectors: $\mathbf{w}' = \{q_1, p_1; q_2, p_2; \cdots; q_n, p_n\}^T$;
7) Equation $\mathbf{w} = \mathbf{S}\mathbf{w}'$ gives all the symplectic eigen-solutions of (2.3.65) of the given matrix $\mathbf{A}$.

§2.3.3.4, Numerical example

Example 2.2, the numerical example here is to demonstrate two facets,

1) The gyroscopic system is transformed to the standard form of symplectic eigen-problem for a skew-symmetric matrix; and
2) Solve the symplectic eigen-problem for a skew-symmetric matrix $\mathbf{A}$.

The given matrices are

$$\mathbf{M} = \begin{bmatrix} 8 & & \text{symmetry} & \\ -2 & 10 & & \\ 1 & 4 & 10 & \\ 0 & 4 & -1.2 & 8 \end{bmatrix}, \quad \mathbf{G} = \begin{bmatrix} 0 & -16 & -8 & -12 \\ 16 & 0 & -40 & -12 \\ 8 & 40 & 0 & -16 \\ 12 & 12 & -16 & 0 \end{bmatrix}, \quad \mathbf{K} = \begin{bmatrix} 4 & & \text{symmetry} & \\ -3 & -3 & & \\ 2 & 1 & -3 & \\ 0 & -3 & -2 & 4 \end{bmatrix}.$$

Solution: From $\mathbf{M}, \mathbf{K}$ the ortho-normalized matrix $\mathbf{Q}$ is found, and the matrix $\mathbf{G}'$ is given as

$$\mathbf{Q} = \begin{bmatrix} -.0201 & 0.3091 & -.0652 & 0.2137 \\ -.1526 & -.0099 & -.2186 & 0.3549 \\ -.1835 & -.0720 & -.0311 & -.3356 \\ -.1196 & -.1135 & -.3822 & -.1385 \end{bmatrix}, \quad \mathbf{G}' = \begin{bmatrix} 0 & & & \\ 0.8818 & 0 & & \\ 1.4241 & -.2528 & 0 & \\ -6.7678 & 0.9914 & 0.9546 & 0 \end{bmatrix}$$

$$\mathbf{K}' = \mathrm{diag}\begin{bmatrix} -0.2577 & 0.5498 & 0.9889 & -1.3272 \end{bmatrix}$$

Then according to (2.3.62) the skew-symmetric matrix $\mathbf{A}$ is computed as

$$\mathbf{A} = \begin{bmatrix} 0 & 0.2571 & & & & & & \\ -.2571 & 0 & & & \text{skew-} & & & \\ -.2267 & 0 & 0 & -1.3274 & & & & \\ 0 & -.8818 & 1.3274 & 0 & \text{symmtry} & & & \\ -.3661 & -.2230 & -.3356 & -1.2558 & 0 & -3.0810 & & \\ 0 & 1.4241 & 1.2558 & -.2528 & 3.0810 & 0 & & \\ -1.7397 & 2.2337 & 2.5148 & -6.2093 & -5.6951 & 93876 & 0 & -46.3700 \\ 0 & 6.7678 & 5.9680 & 0.9914 & -9.3876 & 0.9546 & 46.3700 & 0 \end{bmatrix}$$

Thus the first facet (1) is performed, *i.e.* transform to standard form of symplectic eigen-problem for a skew-symmetric matrix.

Next stage is to find the solution of a symplectic eigen-problem for a skew-symmetric matrix $\mathbf{A}$. Within which, the first step is using the Givens rotation and orthogonal SH transformation to carry out the half tri-diagonal cell matrix form for $\mathbf{A}$. After half tri-diagonalization transformation

$$\mathbf{A}_{s3d} = \mathbf{S}^T\mathbf{A}\mathbf{S} =$$

$$= \begin{bmatrix} 0 & 0.2571 & & & \text{skew -} & \text{symmetry} & & \\ -.2571 & 0 & & & & & & \\ 0 & -6.9720 & 0 & -49.1260 & & & & \\ -1.7922 & 2.1227 & 49.1260 & 0 & & & & \\ 0 & 0.6797 & 0 & 5.8466 & 0 & -.8750 & & \\ 0 & -.0670 & -1.3674 & -.6339 & 0.8750 & 0 & & \\ 0 & 0.2136 & 0 & 1.8491 & 0 & -.3880 & 0 & -.7730 \\ 0 & -.1452 & 0 & -.9635 & 0.3348 & -.2071 & 0.7730 & 0 \end{bmatrix}$$

where the accumulated symplectic transformation matrix, certainly orthogonal matrix $\mathbf{S}$ is

$$\mathbf{S} = \begin{bmatrix} 1 & 0 & 0 & 0 & 0 & 0 & 0 & 0 \\ 0 & 1 & 0 & 0 & 0 & 0 & 0 & 0 \\ 0 & 0 & 0 & 0.1256 & -.6660 & -.0030 & 0.6909 & -.2511 \\ 0 & 0 & -.1265 & 0 & 0.0030 & -.6660 & 0.2511 & 0.6909 \\ 0 & 0 & 0 & 0.2043 & -.7011 & 0.0979 & -.5983 & 0.3149 \\ 0 & 0 & -.2043 & 0 & -.0979 & -.7011 & -.3149 & -.5983 \\ 0 & 0 & 0 & 0.9707 & 0.2343 & -.0202 & 0.0359 & -.0336 \\ 0 & 0 & -.9707 & 0 & 0.0202 & 0.2343 & 0.0336 & 0.0359 \end{bmatrix}$$

It can be verified that $\mathbf{S}^T\mathbf{S} = \mathbf{I}_{2n}$ and $\mathbf{S}^T\mathbf{J}'_n\mathbf{S} = \mathbf{J}'_n$, which is checked by the computer. Thereafter, the matrix $\mathbf{A}_{s3d}$ is brought to the block matrix form or equation (2.3.77), giving

$$\mathbf{M}_1 = \begin{bmatrix} 0.2571 & -6.9720 & 0.6797 & 0.2136 \\ 1.7922 & -49.1260 & 5.8466 & 1.8491 \\ 0 & 1.3674 & -.8795 & -.3880 \\ 0 & 0 & -.3348 & -.7730 \end{bmatrix}, \quad \mathbf{C} = \begin{bmatrix} 0 & \text{skew -} & & \\ 2.1227 & 0 & \text{symmetry} & \\ -.0670 & -.6339 & 0 & \\ -.1452 & -.9635 & -.2071 & 0 \end{bmatrix}$$

where matrix $\mathbf{M}_1$ has been in upper Hessenberg form. Invoking the procedure HQR2 solves the eigen-solutions

$$\mu^2 = -0.0088041, \quad -.39988, \quad -1.07452, \quad -49.0382$$

$$\mu = \pm 0.09383\text{i}, \quad \pm 0.63236\text{i}, \quad \pm 1.03659\text{i}, \quad \pm 7.00273\text{i}$$

and the eigenvectors respectively. There are some computations further, which are as usual and are omitted. ##

The modal analysis for natural vibration is a fundamental stage for various applications. Such as random vibration problems, a pre-assigned number of modals are selected for structural response analysis. For large systems, the symplectic subspace iteration method can be used to find these modals. The symplectic subspace iteration method is similar to that given for symmetric matrices, and the algorithm given above can be used in the symplectic subspace eigen-solution.

The canonical transformation requires eigen-solutions too.

§2.4, Non-linear vibration of multi-degrees of freedom system

The analysis for linear vibration of multi-degrees of freedom systems is described in the previous sections, especially the eigen-solutions. However, many applications require taking the non-linear effect of system into consideration. Therefore non-linear vibration analysis of multi-degrees of freedom system is necessary.

Even non-linear analysis problem of single degree of freedom is difficult. The vibration of Duffing equation with sinusoidal excitation is discussed in previous section, for which only the periodical solution is considered. However, the forced vibration with arbitrary initial condition will often fall into chaotic motion, for which time step integration is necessary. Using precise integration method for time step integration is a good choice. But numerical integration for initial value problem could go somewhere uncertainly. Hence analytical method is still very interesting.

The non-linear vibration analysis for multi-degrees of freedom system is certainly much difficult than single degree of freedom system. However, the effective analysis method in single degree of freedom can be transplanted into multi-degrees of freedom. Such as the canonical transformation approach in single degree of freedom can also be applied in multi-degrees of freedom system. In present section, the canonical transformation for multi-degrees of freedom system vibration analysis is considered. Canonical transformation needs to find the complete analytical solutions for the fundamental system, which can be selected as a linear system. Thus the eigen-solutions given in previous sections can be used as the analytical solutions for the canonical transformation. For slightly non-linear system, making use of a linear system as the fundamental one is reasonable. Let the vibration equation of a multi-degrees of freedom non-linear system be given as

$$\mathbf{M}\ddot{\mathbf{q}} + (\mathbf{G}+\mathbf{C})\dot{\mathbf{q}} + \mathbf{K}\mathbf{q} + \mathbf{K}_n(\mathbf{q},\dot{\mathbf{q}})\mathbf{q} = \mathbf{f}_1(t) \tag{2.4.1}$$

where $\mathbf{M},\mathbf{G},\mathbf{K}$ are mass, gyroscopic and stiffness matrices, respectively, $\mathbf{q}$ is the n-dimensional generalized displacement vector to be solved, $\mathbf{C}$ is a symmetric positive definite $n\times n$ dimensional damping matrix, $\mathbf{K}_n(\mathbf{q},\dot{\mathbf{q}})$ is a slightly non-linear $n\times n$ matrix and is a function of the displacement $\mathbf{q}$ and velocity $\dot{\mathbf{q}}$. The external force term $\mathbf{f}_1(t)$ is usually a given simple harmonic excitation.

The fundamental approximate linear system for the non-linear vibration system is selected as

$$\mathbf{M}\ddot{\mathbf{q}} + \mathbf{G}\dot{\mathbf{q}} + \mathbf{K}\mathbf{q} = \mathbf{0} \tag{2.4.2}$$

for which all the eigen-solutions are regarded as solved using the algorithm given in previous sections. These eigen-solutions are used to carry out the canonical transformation. The equation (2.4.2) is certainly transformed into Hamilton system that the derivations in section 1.11.3 and the derivation and computation described in section 2.3 are regarded performed.

For a linear system, the corresponding Hamilton-Jacobi equation can separate all its variables. Theoretically, all the eigen-solutions, which give the complete solution, must be solved for expansion solution. However, in application, only the low frequency components are necessary, a reasonable approximation. In what follows, the canonical transformation of the fundamental system is regarded performed, that the variables are separated for the linear system, and corresponding to the dual vectors $\mathbf{Q},\mathbf{P}$ used in equation (1.11.38), the symplectic matrix $\mathbf{S}$ has been found.

$$\zeta = \mathbf{S}^{-1}\mathbf{v}\ , \qquad \zeta = \{\mathbf{Q}^{\mathrm{T}},\mathbf{P}^{T}\}^{T}\ , \qquad \mathbf{v} = \mathbf{S}\zeta \tag{1.11.38}$$

The notations $\mathbf{Q},\mathbf{P}$ will be used for further canonical transformation, so that $\mathbf{Q},\mathbf{P}$ is replaced with $\hat{\mathbf{q}},\hat{\mathbf{p}}$ below. Now after transformation (1.11.38), the Hamilton function $\mathrm{K}(\hat{\mathbf{q}},\hat{\mathbf{p}})$ has diagonalized and given in matrix form as

$$\mathrm{K}(\hat{\mathbf{q}},\hat{\mathbf{p}}) = \hat{\mathbf{p}}^{T} \times \mathrm{diag}(\mu_i) \times \hat{\mathbf{q}} \tag{1.11.39}$$

that the above canonical transformation separates the variables. However, Hamilton matrix is asymmetric so that its eigen-solutions may appear complex conjugate pairs, *i.e.* the matrix $\mathbf{S}$ can be complex valued and the corresponding vectors $\hat{\mathbf{q}},\hat{\mathbf{p}}$ are complex vectors, which are somewhat inconvenient to further investigations. The real symplectic matrix $\mathbf{S}_{\mathrm{r}}$ is then used instead of $\mathbf{S}$.

For a stable linear gyroscopic system, all its eigenvalues $\mu_i, i = 1,2,\ldots,n$ are purely imaginary. Let the corresponding eigen-solutions be denoted as

$$\mu_i = \mathrm{i}\omega_i,\ \omega_i > 0\ , \quad \psi_i = \{\mathbf{q}_i^T,\mathbf{p}_i^T\}^T \tag{2.4.3a}$$

Because of complex conjugate eigenvalues appearing in pairs, so that the respective eigenvectors ψ_i are also complex valued. The symplectic adjoint eigenvalue of μ_i is $\mu_{n+i} = -\mathrm{i}\omega_i = \bar{\mu}_i$ which means that the complex conjugate eigenvalues equals its symplectic adjoint, where a bar above represents the complex conjugate. The symplectic eigen-solution can be denoted as

$$\mu_{n+i} = -\mathrm{i}\omega_i\ , \quad \psi_{n+i} = a_i\{\bar{\mathbf{q}}_i^T,\bar{\mathbf{p}}_i^T\}^T = a_i\bar{\psi}_i \tag{2.4.3b}$$

where a_i is a complex constant multiplier with $\mathrm{abs}(a_i) = 1$. Because the variables are separated, so ***all solutions of various subscripts i are linearly independent.*** According to the adjoint symplectic ortho-normality relation, it holds $\psi_{n+i}^T\mathbf{J}\psi_i = -1$. Expressed with dual vectors $\mathbf{q}_i,\mathbf{p}_i$

$$a_i(\bar{\mathbf{q}}_i^T\mathbf{p}_i - \bar{\mathbf{p}}_i^T\mathbf{q}_i) = a_i(\bar{\mathbf{q}}_i^T\mathbf{p}_i - \mathbf{q}_i^T\bar{\mathbf{p}}_i) = -1$$

where the parenthesis $(\bar{\mathbf{q}}_i^T\mathbf{p}_i - \mathbf{q}_i^T\bar{\mathbf{p}}_i)$ is the difference between two complex conjugate numbers. Because $\mathrm{abs}(a_i) = 1$, so a_i must be a purely imaginary number $a_i = \pm\mathrm{i}$. To distinguish the two cases of $a_i = \mathrm{i}$ and/or $a_i = -\mathrm{i}$ is very important.

It is seen from equation (2.4.3b) that ψ_i and ψ_{n+i} have two real vector basis. Since these two basis separate with other basis for the fundamental system, then the two real vectors of $\psi_i^{(\mathrm{r})}$ and $\psi_i^{(\mathrm{i})}$, which are the real and imaginary parts of the eigenvector ψ_i respectively, are used instead of the complex basis vectors ψ_i and ψ_{n+i} . The normalization of these two vectors corresponds to the transformation matrices is

$$[\psi_i \ \ \psi_{n+i}] = [\psi_i^{(\mathrm{r})} \ \ \psi_i^{(\mathrm{i})}] \times \begin{bmatrix} 1 & \mathrm{i} \\ \mathrm{i} & 1 \end{bmatrix} \Big/ \sqrt{2}\ , \quad \text{when } a_i = \mathrm{i} \tag{2.4.4a}$$

and $$[\psi_i \quad \psi_{n+i}] = [\psi_i^{(r)} \quad \psi_i^{(i)}] \times \begin{bmatrix} 1 & -\mathrm{i} \\ \mathrm{i} & -1 \end{bmatrix} \Big/ \sqrt{2}, \quad \text{when } a_i = -\mathrm{i}$$

However, it should be pointed out that both $\begin{bmatrix} 1 & \mathrm{i} \\ \mathrm{i} & 1 \end{bmatrix} \Big/ \sqrt{2}$ and $\begin{bmatrix} 1 & -\mathrm{i} \\ \mathrm{i} & -1 \end{bmatrix} \Big/ \sqrt{2}$ are symplectic matrices but with determinant 1 and -1, respectively. They belong to different subgroups, but the -1 determinant is quite inconvenient. For the $a_i = -\mathrm{i}$ case, the symplectic adjoint eigenvectors ψ_i and ψ_{n+i} can be composed of the two real vectors of $\psi_i^{(r)}$ and $\psi_i^{(i)}$, instead as

$$[\psi_i \quad \psi_{n+i}] = [\psi_i^{(r)} \quad \psi_i^{(i)}] \times \begin{bmatrix} \mathrm{i} & 1 \\ -1 & -\mathrm{i} \end{bmatrix} \Big/ \sqrt{2}, \quad \text{when } a_i = -\mathrm{i} \tag{2.4.4b}$$

The use of real vector basis, instead of the pure eigenvector basis, composes

$$\mathbf{S}_r = [\psi_1^{(r)} \quad \cdots \quad \psi_n^{(r)}; \quad \psi_1^{(i)} \quad \cdots \quad \psi_n^{(i)}] \tag{2.4.5}$$

where $\mathbf{S}_r$ is also a symplectic matrix, that $\mathbf{S}_r^T \mathbf{J} \mathbf{S}_r = \mathbf{J}$ can be verified. The transformation between $\mathbf{S}$ and $\mathbf{S}_r$ is a diagonal block matrix and the sub-blocks have been given in equations (2.4.4a,b). Furthermore, from the congruent transformation given as

$$\mathbf{S}^T(-\mathbf{JH})\mathbf{S} = \begin{bmatrix} \mathbf{0} & \mathrm{diag}(\mu_i) \\ \mathrm{diag}(\mu_i) & \mathbf{0} \end{bmatrix} \tag{1.11.41}$$

the real form congruent transformation is derived as

$$\mathbf{S}_r^T(-\mathbf{JH})\mathbf{S}_r = \begin{bmatrix} \mathrm{diag}(\pm\omega_i) & \mathbf{0} \\ \mathbf{0} & \mathrm{diag}(\pm\omega_i) \end{bmatrix} \tag{2.4.6}$$

where the sign selection rule of $\pm\omega_i$ is, ***when*** $a_i = \mathrm{i}$ ***it equals*** ω_i, ***whereas*** $a_i = -\mathrm{i}$ ***gives*** $-\omega_i$. Instead of equation (1.11.38), the real canonical transformation is

$$\zeta = \mathbf{S}_r^{-1}\mathbf{v}, \qquad \zeta = \{\hat{\mathbf{q}}^T, \hat{\mathbf{p}}^T\}^T, \qquad \mathbf{v} = \mathbf{S}_r\zeta \tag{2.4.7}$$

and the Hamilton function $\mathrm{K}(\hat{\mathbf{q}}, \hat{\mathbf{p}})$ after transformation becomes

$$\begin{aligned} \mathrm{H}(\mathbf{q},\mathbf{p}) &= \mathbf{p}^T\mathbf{D}\mathbf{p}/2 + \mathbf{p}^T\mathbf{A}\mathbf{q} + \mathbf{q}^T\mathbf{B}\mathbf{q}/2 = \mathbf{v}^H(-\mathbf{JH})\mathbf{v}/2 \\ &= \mathrm{K}(\hat{\mathbf{q}},\hat{\mathbf{p}}) = [\hat{\mathbf{p}}^T\mathrm{diag}(\pm\omega_i)\hat{\mathbf{p}} + \hat{\mathbf{q}}^T\mathrm{diag}(\pm\omega_i)\hat{\mathbf{q}}]/2 = \sum_{i=1}^{n} \pm\,\omega_i(p_i^2 + q_i^2)/2 \end{aligned} \tag{2.4.8}$$

for which the canonical dual equations are $\dot{\hat{\mathbf{q}}} = \mathrm{diag}(\pm\omega_i)\hat{\mathbf{p}}$, $\dot{\hat{\mathbf{p}}} = -\mathrm{diag}(\pm\omega_i)\hat{\mathbf{q}}$. The sign $\pm$ selection is then: ***when*** $a_i = \mathrm{i}$ ***it takes*** ω_i***; and when*** $a_i = -\mathrm{i}$ ***it takes*** $-\omega_i$.

After the variables separated, the Hamilton-Jacobi characteristic equation uses $\mathrm{K}(\hat{\mathbf{q}}, \hat{\mathbf{p}})$ as the Hamilton operator, and the characteristic equation, after the variables separated, gives as

$$W = \sum_{i=1}^{n} W_i, \quad \pm\,\omega_i[(\partial W_i/\partial \hat{q}_i)^2 + \hat{q}_i^2]/2 = \alpha_i, \qquad i = 1,\cdots,n \tag{2.4.9}$$

where $\mu_i = \mathrm{i}\omega_i$ is the eigenvalue and ω_i is always positive. The two cases of $\pm$ should be considered separately. The discussion below assumes first the case of $a_i = \mathrm{i}$ all, which implies the case of **positive definite Hamilton function.** The case of $a_i = -\mathrm{i}$ for some i will be given later in section 2.5.

Based on the assumption above, it derives

$$W_i(\hat{q}_i,\alpha_i)=\sqrt{2\alpha_i/\omega_i}\int\sqrt{1-\omega_i\hat{q}_i^2/(2\alpha_i)}\mathrm{d}\hat{q}_i+A_i\,,\qquad i=1,\cdots,n \tag{2.4.10}$$

correspondingly, the action function is

$$S_i(\hat{q}_i,\alpha_i,t)=W_i(\hat{q}_i,\alpha_i)-\alpha_i t \tag{2.4.11}$$

which has been given in section 1.11.1 for one-dimensional vibrator. After separation of variables, α_i is the conservative mechanical energy of this i-th vibrator. Treating the action function $S_i(\hat{q}_i,\alpha_i,t)$ as a first class generating function derives a further canonical transformation, where α_i is one of the new canonical coordinates

$$Q_i=\alpha_i\,,\qquad i=1,\cdots,n \tag{2.4.12}$$

Using the first transformation equation (1.7.10a) gives

$$\hat{p}_i=\partial S_i/\partial\hat{q}_i=\sqrt{(2\alpha_i/\omega_i)-\hat{q}_i^2}\,,\qquad i=1,\cdots,n \tag{2.4.13}$$

This equation is valid for arbitrary time t. The second equation (1.7.10b) gives

$$P_i=-\partial S_i/\partial\alpha_i\,,\qquad i=1,\cdots,n$$

As the Hamilton function $\mathrm{K}_2(\mathbf{Q},\mathbf{P},t)$, because the generating function F_1 is selected as S, the equation (1.7.10c) coincides with the Hamilton-Jacobi equation, so $\mathrm{K}_2=0$. The generalized momentum P_i is again a constant, denoted as β_i

$$P_i=\beta_i=-\partial S_i/\partial\alpha_i=-\arcsin(\sqrt{\omega_i/(2\alpha_i)}\hat{q}_i)/\omega_i+t \tag{2.4.14}$$

Solving the above equation for arbitrary time t with respect to $\hat{q}_i$ gives

$$\hat{q}_i(\alpha_i,\beta_i,t)=\sqrt{2\alpha_i/\omega_i}\sin[\omega_i(t-\beta_i)] \tag{2.4.15a}$$

Substituting into (2.4.13) derives

$$\hat{p}_i(\alpha_i,\beta_i,t)=\sqrt{2\alpha_i/\omega_i}\cos[\omega_i(t-\beta_i)] \tag{2.4.15b}$$

The two equations (2.4.15a,b) are the transformation required, where the new dual variables are $Q_i=\alpha_i$ and $P_i=\beta_i$ all constants, because the approximate fundamental linear system is selected as (2.4.2).

But equations (2.4.15a,b) are only a coordinate transformation, using this canonical transformation gives constant solutions α_i,β_i for the conservative system (2.4.2). Under this transformation, the vibration of the original non-linear system (2.4.1) is needed, for which the dual variables Q_i, P_i are no longer constants. Hence the canonical transformation is rewritten as

$$\hat{q}_i(Q_i,P_i,t)=\sqrt{2Q_i/\omega_i}\sin[\omega_i(t-P_i)] \tag{2.4.15'a}$$

$$\hat{p}_i(Q_i,P_i,t)=\sqrt{2Q_i/\omega_i}\cos[\omega_i(t-P_i)] \tag{2.4.15'b}$$

The dual equations should be derived for the original system. The dual vector is introduced the same as the fundamental system (2.4.2)

$$\mathbf{p}=\mathbf{M}\dot{\mathbf{q}}+\mathbf{G}\mathbf{q}/2 \tag{2.3.5}$$

and the original equation (2.4.1) is transformed to the dual equations

$$\dot{\mathbf{q}} = \mathbf{A}\mathbf{q} + \mathbf{D}\mathbf{p} \tag{2.4.16a}$$

$$\dot{\mathbf{p}} = -\mathbf{B}\mathbf{q} - \mathbf{A}^T\mathbf{p} - \mathbf{C}(\mathbf{A}\mathbf{q} + \mathbf{D}\mathbf{p}) - \mathbf{K}_n\mathbf{q} + \mathbf{f}_1(t) \tag{2.4.16b}$$

where $\mathbf{D} = \mathbf{M}^{-1}, \quad \mathbf{A} = -\mathbf{M}^{-1}\mathbf{G}/2, \quad \mathbf{B} = \mathbf{K} + \mathbf{G}^T\mathbf{M}^{-1}\mathbf{G}/4$ (2.3.7a)

and denote
$$\mathbf{H} = \begin{bmatrix} \mathbf{A} & \mathbf{D} \\ -\mathbf{B} & -\mathbf{A}^T \end{bmatrix}$$

For dual vectors $\mathbf{q}$ and $\mathbf{p}$, the real form of canonical transformation (2.4.7) is expressed as

$$\mathbf{S}_r \underset{\text{def}}{=} \begin{bmatrix} \mathbf{S}_{r11} & \mathbf{S}_{r12} \\ \mathbf{S}_{r21} & \mathbf{S}_{r22} \end{bmatrix}, \qquad \mathbf{q} = \mathbf{S}_{r11}\hat{\mathbf{q}} + \mathbf{S}_{r12}\hat{\mathbf{p}}, \quad \mathbf{p} = \mathbf{S}_{r21}\hat{\mathbf{q}} + \mathbf{S}_{r22}\hat{\mathbf{p}} \tag{2.4.17}$$

where $\mathbf{S}_r$ is a time invariant matrix and is computed by the eigen-solutions of the fundamental system, known matrix. Substituting into (2.4.16a,b) gives

$$(\dot{\mathbf{q}} =)\mathbf{S}_{r11}\dot{\hat{\mathbf{q}}} + \mathbf{S}_{r12}\dot{\hat{\mathbf{p}}} = \mathbf{A}(\mathbf{S}_{r11}\hat{\mathbf{q}} + \mathbf{S}_{r12}\hat{\mathbf{p}}) + \mathbf{D}(\mathbf{S}_{r21}\hat{\mathbf{q}} + \mathbf{S}_{r22}\hat{\mathbf{p}})$$

$$(\dot{\mathbf{p}} =)\mathbf{S}_{r21}\dot{\hat{\mathbf{q}}} + \mathbf{S}_{r22}\dot{\hat{\mathbf{p}}} = -\mathbf{B}(\mathbf{S}_{r11}\hat{\mathbf{q}} + \mathbf{S}_{r12}\hat{\mathbf{p}}) - \mathbf{A}^T(\mathbf{S}_{r21}\hat{\mathbf{q}} + \mathbf{S}_{r22}\hat{\mathbf{p}})$$
$$- (\mathbf{C}\mathbf{A} + \mathbf{K}_n)(\mathbf{S}_{r11}\hat{\mathbf{q}} + \mathbf{S}_{r12}\hat{\mathbf{p}}) - \mathbf{C}\mathbf{D}(\mathbf{S}_{r21}\hat{\mathbf{q}} + \mathbf{S}_{r22}\hat{\mathbf{p}}) + \mathbf{f}_1(t)$$

Solving the simultaneous algebraic equations of $\dot{\hat{\mathbf{q}}}$ and $\dot{\hat{\mathbf{p}}}$, because $\mathbf{S}_r$ satisfies (2.4.6) and $\mathbf{S}_r$ is symplectic $\mathbf{S}_r^{-1} = -\mathbf{J}\mathbf{S}_r^T\mathbf{J}$, it gives

$$\dot{\hat{\mathbf{q}}} = \text{diag}(\pm\omega_i)\hat{\mathbf{p}} + \mathbf{S}_{r12}^T\left[(\mathbf{C}\mathbf{A} + \mathbf{K}_n)(\mathbf{S}_{r11}\hat{\mathbf{q}} + \mathbf{S}_{r12}\hat{\mathbf{p}}) + \mathbf{C}\mathbf{D}(\mathbf{S}_{r21}\hat{\mathbf{q}} + \mathbf{S}_{r22}\hat{\mathbf{p}}) - \mathbf{f}_1(t)\right]$$

$$\dot{\hat{\mathbf{p}}} = -\text{diag}(\pm\omega_i)\hat{\mathbf{q}} - \mathbf{S}_{r11}^T\left[(\mathbf{C}\mathbf{A} + \mathbf{K}_n)(\mathbf{S}_{r11}\hat{\mathbf{q}} + \mathbf{S}_{r12}\hat{\mathbf{p}}) + \mathbf{C}\mathbf{D}(\mathbf{S}_{r21}\hat{\mathbf{q}} + \mathbf{S}_{r22}\hat{\mathbf{p}}) - \mathbf{f}_1(t)\right]$$

(2.4.18a,b)

Up to here, the fundamental part has been separated the variables via the canonical transformation and becomes a number of single degree of freedom vibrators, and such separation of variables is usually called as modal analysis for $\mathbf{M}, \mathbf{K}$ system. It should be pointed out, that both the damping matrix $\mathbf{C}$ and the matrix of non-linear term $\mathbf{K}_n$ are small. In applications, how to identify the value of the damping matrix $\mathbf{C}$ is somewhat vague, their determination perhaps can also be given in the modal analysis coordinate system as usually used in the linear engineering vibration analysis, which can also be used for non-linear case.

Based on the above canonical transformation with matrix $\mathbf{S}_r$, which corresponds to the modal analysis, the next canonical transformation (2.4.15'a,b) can be applied further. For each vibrator, the mechanical energy Q_i and the phase angle P_i can be selected as the unknown dual variables, for which

$$\dot{\hat{q}}_i(Q_i, P_i, t) = \left(\dot{Q}_i\big/\sqrt{2Q_i\omega_i}\right)\sin[\pm\,\omega_i(t - P_i)] - (\dot{P}_i - 1)\sqrt{2Q_i\omega_i}\,\cos[\omega_i(t - P_i)]$$

$$\dot{\hat{p}}_i(Q_i, P_i, t) = \left(\dot{Q}_i\big/\sqrt{2Q_i\omega_i}\right)\cos[\omega_i(t - P_i)] + (\dot{P}_i - 1)\sqrt{2Q_i\omega_i}\,\sin[\pm\,\omega_i(t - P_i)]$$

(2.4.19a,b)

which is again a canonical transformation. For positive definite Hamilton function, the $\pm$ sign takes always positive value. Substituting into the dual equations (2.4.18a,b) and solving with respect to $\dot{Q}_i$ and $\dot{P}_i$ derives

$$\dot{\mathbf{Q}} = \begin{bmatrix} \mathrm{diag}\{\sqrt{2Q_i\omega_i}\sin[\pm\omega_i(t-P_i)]\}\times\mathbf{S}_{r12}^T \\ -\mathrm{diag}\{\sqrt{2Q_i\omega_i}\cos[\omega_i(t-P_i)]\}\times\mathbf{S}_{r11}^T \end{bmatrix} \times\left[(\mathbf{CA}+\mathbf{K}_n)(\mathbf{S}_{r11}\hat{\mathbf{q}}+\mathbf{S}_{r12}\hat{\mathbf{p}})+\mathbf{CD}(\mathbf{S}_{r21}\hat{\mathbf{q}}+\mathbf{S}_{r22}\hat{\mathbf{p}})-\mathbf{f}_1(t)\right] \tag{2.4.20a}$$

$$\dot{\mathbf{P}} = -\begin{bmatrix} \mathrm{diag}\{\cos[\omega_i(t-P_i)]/\sqrt{2Q_i\omega_i}\}\times\mathbf{S}_{r12}^T \\ +\mathrm{diag}\{\sin[\pm\omega_i(t-P_i)]/\sqrt{2Q_i\omega_i}\}\times\mathbf{S}_{r11}^T \end{bmatrix} \times\left[(\mathbf{CA}+\mathbf{K}_n)(\mathbf{S}_{r11}\hat{\mathbf{q}}+\mathbf{S}_{r12}\hat{\mathbf{p}})+\mathbf{CD}(\mathbf{S}_{r21}\hat{\mathbf{q}}+\mathbf{S}_{r22}\hat{\mathbf{p}})-\mathbf{f}_1(t)\right] \tag{2.4.20b}$$

where the variables $\hat{\mathbf{q}}$ and $\hat{\mathbf{p}}$ at right hand side should be substituted with equations (2.4.15'a,b) and then the differential equations for $\mathbf{Q}$ and $\mathbf{P}$ are established. The damping matrix $\mathbf{C}$ and the matrix $\mathbf{K}_n$ for non-linear term must be given at this stage. According to the equation (2.4.1) they are defined in the original coordinate system. However in applications, the determination of $\mathbf{C}$ and $\mathbf{K}_n$ should be investigated further, perhaps they should be determined in the modal coordinate system $\hat{\mathbf{q}}$ and $\hat{\mathbf{p}}$.

The integration of equations (2.4.20a,b) should be further investigated for non-linear system. A two-degrees of freedom system is considered in next section.

§2.4.1, Non-linear internal parametric resonance

To understand the essence of cable and beam coupled vibration for cable stayed bridges, a simple two-degrees of freedom non-linear vibration system is investigated as given in figure 2.10.

According to Newton's law, the differential equations for geometrically non-linear system vibration with two-degrees of freedom are derived as

$$\begin{cases} m_1\ddot{x}_1 + c\dot{x}_1 + (2T_0/L + EAx_1^2/L^3)x_1 + EAx_1x_2/L^2 = 0 \\ m_2\ddot{x}_2 + EAx_1^2/(2L^2) + EAx_2/(2L) = 0 \end{cases} \tag{2.4.21}$$

where a large mass m_2 vibrates vertically and a small mass m_1 vibrates horizontally. If the non-linear terms are neglected in the equations, then the equations are just two uncoupled single degree of freedom vibration of $x_1(t)$ and $x_2(t)$, each with natural frequencies ω_1 and ω_2, respectively. However the second order and third order terms in equation (2.4.21) claim the non-linear coupling between the two natural vibration modes. Although the system parameters do not explicitly involve time t variant terms, but the vertical vibration of large mass m_2 will cause the tension of cable be time variant, which corresponds to the small mass m_1 being subject to time variant restoration stiffness. Along the horizontal

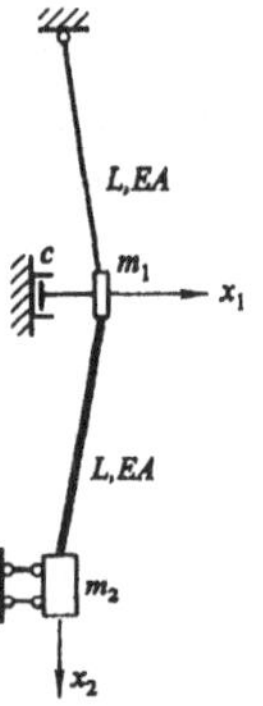

Figure 2.10, *Cable-Mass vibration model, 2-degrees of freedom system*

direction, the restoration force of the small mass m_1 is

$$(2T_0 / L + EAx_2 / L^2 + EAx_1^2 / L^3)x_1$$

where the parenthesis factor is the stiffness of restoring force, the first term in parenthesis is time invariant fundamental linear part of stiffness, the second term is the non-linear coupling term between the two-degrees of freedom, and the third term is non-linear for displacement x_1 itself. Because of the coupling term, x_2 vibrates with frequency ω_2, which supplies a time variant term for the stiffness of x_1. Hence m_1 is subjected to ***parametric resonance*** excitation correspondingly, that the time variant restoring stiffness of m_1 is due to the vertical vibration of the large mass m_2. According to the parametric resonance analysis given in section 2.1.2, the vibration of the large mass m_2 may possibly induce parametric resonance for m_1, which is the two-degrees of freedom system ***internal parametric resonance***. It is anticipated that when the natural vibration frequency ω_1 of m_1 is close to half the natural vibration frequency ω_2 of m_2, *i.e.* $\omega_1 \approx 0.5\omega_2$, the vertical vibration of m_2 will excite the mass m_1 vibrating violently along the horizontal direction.

Rewriting the equation set (2.4.21) in the form

$$\begin{cases} m_1\ddot{x}_1 + m_1\omega_1^2 x_1 = -c\dot{x}_1 + 2\kappa x_1 x_2 + \beta_1 x_1^3 \\ m_2\ddot{x}_2 + m_2\omega_2^2 x_2 = \kappa x_1^2 \end{cases} \tag{2.4.22a,b}$$

where $\omega_1^2 = 2T_0/(m_1 L)$, $\omega_2^2 = EA/(2m_2 L)$, $\beta_1 = -EA/L^3$, $\kappa = -EA/2L^2$. This is not a dimensionless form. The equations (2.4.22a,b) are solved below using the canonical transformation method.

A linear fundamental system selected for the canonical transformation is

$$m_1\ddot{x}_1 + m_1\omega_1^2 x_1 = 0, \qquad m_2\ddot{x}_2 + m_2\omega_2^2 x_2 = 0 \tag{2.4.23a,b}$$

for which the variables x_1 and x_2 are separate. Let $q_i \underset{\text{def}}{=} x_i, i = 1,2$, $p_i = m_i\dot{q}_i$, the Hamilton function is derived as

$$\mathrm{H}(\mathbf{q},\mathbf{p}) = \sum_{i=1}^{2}(m_i\omega_i^2 q_i^2 + p_i^2 / m_i)/2 \tag{2.4.24}$$

and the corresponding Hamilton-Jacobi characteristic equation is given as

$$W = \sum_{i=1}^{2} W_i, \quad [(\partial W_i / \partial \hat{q}_i)^2 / m_i + m_i\omega_i^2 \hat{q}_i^2]/2 = \alpha_i, \quad i = 1,2 \tag{2.4.25}$$

Integration of W gives

$$W_i(q_i, \alpha_i) = \sqrt{2m_i\alpha_i}\int\sqrt{1 - m_i\,\omega_i^2 q_i^2/(2\alpha_i)}\mathrm{d}q_i$$

$$= \frac{\alpha_i}{\omega_i}\left[\arcsin\left(q_i\sqrt{\frac{m_i\omega_i^2}{2\alpha_i}}\right) + \sqrt{\frac{m_i\omega_i^2}{2\alpha_i}}q_i\sqrt{1 - \frac{m_i\omega_i^2}{2\alpha_i}q_i^2}\right]$$

and
$$S_i = \sqrt{2m_i\alpha_i}\int\sqrt{1 - m_i\omega_i^2 q_i^2/(2\alpha_i)}\mathrm{d}q_i - \alpha_i t$$

It is important that

$$\beta_i = -\partial S_i/\partial\alpha_i = t - \sqrt{2m_i/\alpha_i}\int \mathrm{d}q_i/\sqrt{1 - m_i\omega_i^2 q_i^2/(2\alpha_i)}$$
$$= t - \arcsin(q_i\sqrt{m_i\omega_i^2/2\alpha_i})/\omega_i \tag{2.4.26}$$

from which solves q_i as function of t and two integration constants α_i, β_i

$$q_i(\alpha_i, \beta_i, t) = \sqrt{2\alpha_i/(m_i\omega_i^2)}\sin\omega_i(t - \beta_i) \tag{2.4.27a}$$

the well-known equation for simple harmonic vibrator. The momentum is given as

$$p_i(\alpha_i, \beta_i, t) = \partial S_i/\partial q_i = \partial W_i/\partial q_i = \sqrt{2m_i\alpha_i - m_i^2\omega_i^2 q_i^2}$$

Substituting with (2.4.27a) gives

$$p_i(\alpha_i, \beta_i, t) = \sqrt{2m_i\alpha_i}\cos\omega_i(t - \beta_i) \tag{2.4.27b}$$

For the fundamental system, α_i, β_i, $i = 1,2$ are constants. However, for the original non-linear system (2.4.22a,b) these parameters α_i, β_i are no longer constants but can be treated as dual variables, and are rewritten as Q_i, P_i, then the equations (2.4.27a,b) become the canonical transformation

$$\begin{aligned} q_i(Q_i, P_i, t) &= \sqrt{2Q_i/m_i\omega_i^2}\sin\omega_i(t - P_i) \\ p_i(Q_i, P_i, t) &= \sqrt{2m_iQ_i}\cos\omega_i(t - P_i) \end{aligned}, \quad i = 1,2 \tag{2.4.27'a,b}$$

The unknowns are transformed to be $Q_i(t)$ and $P_i(t)$, and the original dynamic equations (2.4.22a,b) should also be transformed as differential equations for $Q_i(t)$ and $P_i(t)$, respectively.

To transform the dynamic equations (2.4.22a,b) into dual variables' form, let $q_i \underset{\text{def}}{=} x_i$, $i = 1,2$ the equations become

$$\begin{aligned} &\dot{q}_1(t) = p_1/m_1, \quad \dot{p}_1(t) = -m_1\omega_1^2 q_1 - cp_1/m_1 + 2\kappa q_1 q_2 + \beta_1 q_1^3 \\ &\dot{q}_2(t) = p_2/m_2, \quad \dot{p}_2(t) = -m_2\omega_2^2 q_2 + \kappa q_1^2 \end{aligned} \tag{2.4.28a~d}$$

Which can be substituted with the equations (2.4.27'a,b). Differentiating derives

$$\dot{q}_i = \left[\dot{Q}_i/(\omega_i\sqrt{2m_iQ_i})\right]\sin[\omega_i(t - P_i)] + (1 - \dot{P}_i)\sqrt{2Q_i/m_i}\cos[\omega_i(t - P_i)]$$
$$\dot{p}_i = \dot{Q}_i\cdot\sqrt{m_i/2Q_i}\cos[\omega_i(t - P_i)] - (1 - \dot{P}_i)\omega_i\sqrt{2m_iQ_i}\sin[\omega_i(t - P_i)]$$

Substituting into (2.4.28a,b) then solving with respect to $\dot{Q}_i$ and $\dot{P}_i$ gives

$$\dot{Q}_1 = \sqrt{2Q_1/m_1}\cos[\omega_1(t - P_1)]\times[2\kappa q_1 q_2 + \beta_1 q_1^3 - cp_1/m_1] \tag{2.4.29a}$$

$$\dot{P}_1 = (\omega_1\sqrt{2m_1Q_1})^{-1}\sin[\omega_1(t - P_1)]\times[2\kappa q_1 q_2 + \beta_1 q_1^3 - cp_1/m_1] \tag{2.4.29b}$$

$$\dot{Q}_2 = \sqrt{2Q_2/m_2}\cos[\omega_2(t - P_2)]\times\kappa q_1^2 \tag{2.4.29c}$$

$$\dot{P}_2 = (\omega_2\sqrt{2m_2Q_2})^{-1}\sin[\omega_2(t - P_2)]\times\kappa q_1^2 \tag{2.4.29d}$$

where q_i, p_i should be substituted with (2.4.27'a,b). The remaining problem is the integration for these equations.

Careful analysis is necessary before integration. Assuming $c = 0$, then the system is conservative, for which the Hamilton function for (2.4.28a~d) is

$$\mathrm{H}(\mathbf{q},\mathbf{p})=\sum_{i=1,2}\left[p_i^2/m_i+m_i\omega_i^2q_i^2\right]/2-\kappa q_1^2q_2-\beta_1q_1^4/4=E \tag{2.4.30}$$

where E is the conserved energy. Substituted with the canonical transformation of (2.4.27'a,b) derives

$$\begin{aligned}\mathrm{K}(\mathbf{Q},\mathbf{P})&=\mathrm{H}(\mathbf{q}(\mathbf{Q},\mathbf{P}),\mathbf{p}(\mathbf{Q},\mathbf{P}))=E\\&=Q_1+Q_2-\beta_1(2Q_1/m_1\omega_1^2)^2\sin^4[\omega_1(t-P_1)]\\&\quad-\kappa(2Q_1/m_1\omega_1^2)\sqrt{2Q_2/m_2\omega_2^2}\sin^2[\omega_1(t-P_1)]\sin[\omega_2(t-P_2)]\end{aligned} \tag{2.4.31}$$

Substituting $c=0$ into (2.4.29) and for simplicity let $\beta_1=0$, this term is not so important for parametric resonance, then the differential equations become

$$\dot{Q}_1=\kappa[2Q_1/(m_1\omega_1\omega_2)]\sqrt{2Q_2/m_2}\sin[2\omega_1(t-P_1)]\times\sin\omega_2(t-P_2) \tag{2.4.32a}$$

$$\dot{Q}_2=\kappa[Q_1/(m_1\omega_1^2)]\sqrt{2Q_2/m_2}\cos[\omega_2(t-P_2)]\times[1-\cos2\omega_1(t-P_1)] \tag{2.4.32c}$$

$$\dot{P}_1=\kappa[\sqrt{2Q_2}/(m_1\omega_1^2\omega_2\sqrt{m_2})]\times[1-\cos2\omega_1(t-P_1)]\times\sin\omega_2(t-P_2) \tag{2.4.32b}$$

$$\dot{P}_2=\kappa[Q_1/(m_1\omega_1^2\omega_2\sqrt{2m_2Q_2})]\sin[\omega_2(t-P_2)]\times[1-\cos2\omega_1(t-P_1)] \tag{2.4.32d}$$

which coincide with the dual canonical equations $\dot{\mathbf{Q}}=\partial\mathrm{K}/\partial\mathbf{P}$, $\dot{\mathbf{P}}=-\partial\mathrm{K}/\partial\mathbf{Q}$ derived from $\mathrm{K}(\mathbf{Q},\mathbf{P})$.

The parameter κ represents the coupling between two vibrators. When amplitudes are small, the vibrators' energies Q_1,Q_2 are small too, and the conserved energy E in equation (2.4.31) is mainly composed of Q_1+Q_2. Because the parameter κ is small, so that the exchange rate of energy between the vibrators is slow, and the system vibration composes mainly from the vibrations of each vibrator individually. However because of the parameter κ, there is energy exchange between Q_1 and Q_2 of the two vibrators. The energy exchange is small within one period $T_1=2\pi/\omega_1$ or $T_2=2\pi/\omega_2$ of vibration, hence Q_1 and Q_2 can be regarded approximately as constants in one period of T_2. In equation (2.4.32a) there is the product of two sinusoidal functions

$$\begin{aligned}2\sin[2\omega_1(t-P_1)]\times\sin\omega_2(t-P_2)&=\cos[(\omega_2-2\omega_1)t+2\omega_1P_1-\omega_2P_2]\\&\quad-\cos[(\omega_2+2\omega_1)t-2\omega_1P_1-\omega_2P_2]\end{aligned}$$

The latter term is high frequency vibration and will be cancelled in the time integration, but when $\omega_1\approx0.5\omega_2$, the former term is a very low frequency vibration and is almost a constant in a short time duration, such as within a period of T_1 or T_2. Similarly

$$\begin{aligned}2\cos[\omega_2(t-P_2)]\cos[2\omega_1(t-P_1)]&=\cos[(\omega_2+2\omega_1)t-(2\omega_1P_1+\omega_2P_2)]\\&\quad+\cos[(\omega_2-2\omega_1)t+2\omega_1P_1-\omega_2P_2]\end{aligned}$$

for the time integration of $\dot{P}_1$ and $\dot{P}_2$ the same analysis applies too. Anyway, the energy Q_1,Q_2 and phase P_1,P_2 changes are mainly appearing nearby $\omega_1\approx0.5\omega_2$. If the time integration step size η is taken as $\eta=T_2=2\pi/\omega_2$, then in each step of integration, Q_1,Q_2 and P_1,P_2 at right hand sides of equations (2.4.32a~d) can be treated as constants, and the increments (decrements) $\Delta Q_1, \Delta Q_2$ and $\Delta P_1, \Delta P_2$ of each cycle are obtained. For each cycle of vibration the energy conservation (2.4.31) can be checked.

This problem expresses the non-linear ***internal parametric resonance*** for multi-degrees of freedom vibration, which may also appear in more degrees of freedom problems, such as cable stayed bridges, etc. The precise integration method was applied in numerical simulation for this problem [57], although the approximate model is not so precise, however the basic picture is clear for ***internal parametric resonance and energy wandering*** between the two parts of the system. The computation can be carried out easily for the present case.

Numerical example can be found from [57].

The analysis of conservative $c = 0$ case of parametric vibration is discussed by the canonical transformation method. For damping system, the derivation is almost the same and more terms should be considered. The purpose of this section is to show the application of dual system method and canonical transformation for non-linear system vibrations. The canonical transformation can be used to develop a featured analysis method in such kind of problems.

§2.4.2, Non-linear internal sub-harmonic resonance

Stability and resonance are very important issues for vibration problems. The internal parametric resonance is analyzed in the last section however the internal non-linear sub-harmonic resonance should also be discussed. Suppose a large mass m_1 vibrates, such as a Duffing vibrator. Further, another small mass m_2 vibrator attaches to the Duffing vibrator and the system becomes of two degrees of freedom.

For the Duffing vibrator, in addition to the basic harmonic with frequency ω_1 there is also sub-harmonic $3\omega_1$ component and others. If the natural frequency ω_2 of small mass nearby the sub-harmonic frequency $3\omega_1$ of large mass, internal resonance can also appear. Let the dynamic equations for two degrees of freedom system be

$$\begin{aligned} &m_1\ddot{x}_1 + K_1 x_1 + K_b x_1^3 + k_c x_2 = f\cos\omega_1 t, \\ &m_2\ddot{x}_2 + c\dot{x}_2 + k_2 x_2 + k_c x_1 = 0, \qquad c > 0 \end{aligned} \tag{2.4.33}$$

where k_c represents the coupling term. The non-linear system can still use the canonical transformation method to solve. The nearly periodic vibration is considered by selecting the fundamental linear system as

$$\begin{aligned} &m_1\ddot{x}_1 + k_1 x_1 = 0, \quad k_1 = m_1\omega_1^2 \\ &m_2\ddot{x}_2 + k_2 x_2 = 0 \end{aligned} \tag{2.4.34}$$

which is composed of two separately vibrating systems. Let $K_1 / m_1 = \omega_a^2 \approx \omega_1^2$ and denote $q_1 = x_1$ and $q_2 = x_2$. The Hamilton function is derived as

$$\begin{aligned} &p_1 = m_1\dot{x}_1, \quad \dot{p}_2 = m_2\dot{x}_2 \\ &\mathrm{H}(\mathbf{q},\mathbf{p}) = \mathrm{H}_1(q_1,p_1) + \mathrm{H}_2(q_2,p_2) = A, \\ &\mathrm{H}_1(q_1,p_1) = \dot{p}_1^2 / 2m_1 + k_1 x_1^2 / 2, \quad \mathrm{H}_2(q_2,p_2) = \dot{p}_2^2 / 2m_2 + k_2 x_2^2 / 2 \end{aligned} \tag{2.4.35}$$

where $A=\alpha_1+\alpha_2$ is the conserved energy of the fundamental system. The corresponding Hamilton-Jacobi characteristic functions are, respectively

$$[(\mathrm{d}W_1/\mathrm{d}q_1)^2/m_1+m_1\omega_1^2q_1^2]/2=\alpha_1,\qquad [(\mathrm{d}W_2/\mathrm{d}q_2)^2/m_2+m_2\omega_2^2q_2^2]/2=\alpha_2,\quad \omega_2^2=k_2/m_2 \tag{2.4.36a,b}$$

from which solves

$$W_i=\sqrt{2m_i\alpha_i}\int\sqrt{1-m_i\omega_i^2q_i^2/(2\alpha_i)}\mathrm{d}q_i$$
$$=\frac{\alpha_i}{\omega_i}\left[\arcsin\left(q_i\sqrt{\frac{m_i\omega_i^2}{2\alpha_i}}\right)+\sqrt{\frac{m_i\omega_i^2}{2\alpha_i}}q_i\sqrt{1-\frac{m_i\omega_i^2}{2\alpha_i}q_i^2}\right],\quad i=1,2$$
$$S_i=W_i-\alpha_it \tag{2.4.37}$$

and the important things are

$$\beta_i=-\partial S/\partial\alpha_i=t-\sqrt{2m_i/\alpha_i}\int\mathrm{d}q_i\Big/\sqrt{1-m_i\omega_i^2q_i^2/(2\alpha_i)}$$
$$=t-\arcsin(q_i\sqrt{m_i\omega_i^2/2\alpha_i})\big/\omega_i \tag{2.4.38}$$

from which q_i are solved as the functions of time t with two integration constants α_i,β_i as

$$q_i=\sqrt{2\alpha_i/(m_i\omega_i^2)}\sin\omega_i(t-\beta_i)$$

Only for the fundamental system, α_i,β_i are constants, however when the canonical transformation is applied to solve the original equations (2.4.33), the transformed coordinates α_i,β_i will no longer be constants but functions of time, hence they are replaced with Q_i,P_i and it derives

$$q_i=\sqrt{2Q_i/m_i\omega_i^2}\sin\omega_i(t-P_i),\qquad p_i=\sqrt{2m_iQ_i}\cos\omega_i(t-P_i),\quad i=1,2 \tag{2.4.39a,b}$$

where Q_i,P_i are dual variables, the physical meaning is that they are the energies and phase angles of the two vibrators of the fundamental system, respectively. Rewrite the original equation (2.4.33) as

$$\dot q_1=p_1/m_1,\quad \dot p_1=-K_1q_1-K_bq_1^3-k_cq_2+f\cos\omega_1t$$
$$\dot q_2=p_2/m_2,\quad \dot p_2=-cp_2/m_2-k_2q_2-k_cq_1 \tag{2.4.40a-d}$$

Differentiating the equations (2.4.39a,b) gives

$$\dot q_i=\dot Q_i/(\omega_i\sqrt{2m_iQ_i})\sin[\omega_i(t-P_i)]+(1-\dot P_i)\sqrt{2Q_i/m_i}\cos[\omega_i(t-P_i)]$$
$$\dot p_i=\dot Q_i\sqrt{m_i/2Q_i}\cos[\omega_i(t-P_i)]-(1-\dot P_i)\omega_i\sqrt{2m_iQ_i}\sin[\omega_i(t-P_i)]$$

substituting into (2.4.40a~d) gives the simultaneous equations for $\dot Q_i,\dot P_i$, $i=1,2$. Solving gives

$$\dot Q_1=\sqrt{2Q_1/m_1}\cos[\omega_1(t-P_1)]\times[f\cos\omega_1t-(K_1-k_1)q_1-K_bq_1^3-k_cq_2] \tag{2.4.41a}$$

$$\dot P_1=[\sin(\omega_1(t-P_1))/(\omega_1\sqrt{2m_1Q_1})]\times\begin{bmatrix}f\cos\omega_1t-(K_1-k_1)q_1\\-K_bq_1^3-k_cq_2\end{bmatrix} \tag{2.4.41b}$$

$$\dot{Q}_2 = \sqrt{2Q_2/m_2}\cos[\omega_2(t-P_2)]\times[-cp_2/m_2 - k_c q_1] \tag{2.4.41c}$$

$$\dot{P}_2 = [\sin[\omega_2(t-P_2)]/(\omega_2\sqrt{2m_2Q_2})]\times[-cp_2/m_2 - k_c q_1] \tag{2.4.41d}$$

which is the differential equations for Q_i, P_i, $i = 1,2$, where q_1, q_2 at the right hand side should be substituted with (2.4.39a), certainly.

According to the description before, the frequency of external force is nearby the natural linear vibration frequency, *i.e.* $\omega_1 \approx \omega_a$, where the external force period is $T_1 = 2\pi/\omega_1$. If the integration is carried out directly with the given initial condition, then the development will be easily fallen into chaotic motion. The periodic solution is a special solution. It is similar to the 1-D Duffing equation that the periodic conditions become

$$\int_0^{T_1} \dot{Q}_i \mathrm{d}t = 0, \quad \int_0^{T_1} \dot{P}_i \mathrm{d}t = 0, \qquad i = 1,2 \tag{2.4.42a-d}$$

The differential equations (2.4.41a~d) are complicated, that the purely analytical integration is hopeless, and the approximate analysis or just numerical integration must be applied. The first approximation is to treat Q_i, P_i, $i = 1,2$ as constants in a period of vibration, then after integration the conditions (2.4.42a~d) become 4 equations for the 4 unknown constants. Substituting the right hand sides of (2.4.41a~d) into (2.4.42a~d) derives

$$\int_0^{T_1}\begin{bmatrix} f\cos(\omega_1 t) - K_b\left(\sqrt{2Q_1/m_1}/\omega_1\right)^3 \sin^3(\omega_1(t-P_1)) \\ -(K_1 - k_1)\left(\sqrt{2Q_1/m_1}/\omega_1\right)\sin(\omega_1(t-P_1)) \\ -k_c\left(\sqrt{2Q_2/m_2}/\omega_2\right)\sin(\omega_2(t-P_2)) \end{bmatrix} \tag{2.4.43a}$$

$$\times\sqrt{\frac{2Q_1}{m_1}}\cos(\omega_1(t-P_1))\mathrm{d}t = 0$$

$$\int_0^{T_1}\begin{bmatrix} f\cos(\omega_1 t) - K_b\left(\sqrt{2Q_1/m_1}/\omega_1\right)^3 \sin^3(\omega_1(t-P_1)) \\ -(K_1 - k_1)\left(\sqrt{2Q_1/m_1}/\omega_1\right)\sin(\omega_1(t-P_1)) \\ -k_c\left(\sqrt{2Q_2/m_2}/\omega_2\right)\sin(\omega_2(t-P_2)) \end{bmatrix}\frac{\sin(\omega_1(t-P_1))}{\omega_1\sqrt{2Q_1m_1}}\mathrm{d}t = 0 \tag{2.4.43b}$$

$$\int_0^{T_1}\begin{bmatrix} -c\sqrt{2Q_2/m_2}\cos(\omega_2(t-P_2)) \\ -k_c\left(\sqrt{2Q_1/m_1}/\omega_1\right)\sin(\omega_1(t-P_1)) \end{bmatrix}\sqrt{\frac{2Q_2}{m_2}}\cos(\omega_2(t-P_2))\mathrm{d}t = 0 \tag{2.4.43c}$$

$$\int_0^{T_1}\begin{bmatrix} -c\sqrt{2Q_2/m_2}\cos(\omega_2(t-P_2)) \\ -k_c\left(\sqrt{2Q_1/m_1}/\omega_1\right)\sin(\omega_1(t-P_1)) \end{bmatrix}\frac{\sin(\omega_2(t-P_2))}{\omega_2\sqrt{2Q_2m_2}}\mathrm{d}t = 0 \tag{2.4.43d}$$

However after integration, it is found that when $\omega_2 = 3\omega_1$, Q_1, P_1 and Q_2, P_2 are still uncoupled to each other. It expresses that the first approximation can only get Q_2, P_2 as constants, and Q_1, P_1 are the first approximate solution of Duffing vibrator. The sub-harmonic resonance can appear only in the higher approximation. However, this is because the fundamental system (2.4.34) selected only involve linear terms, if the fundamental system selects

$$m_1\ddot{x}_1 + K_1 x_1 + K_b x_1^3 = 0$$
$$m_2\ddot{x}_2 + k_2 x_2 = 0 \tag{2.4.34'}$$

which is still composed of two independent 1-D sub-systems, but has been taken the non-linear effect into consideration and the higher harmonic factor appears, then sub-harmonic resonance can appear in the first approximation.

The discussion given here is quite rough, since the purpose is only to show the possible application of canonical transformation to non-linear vibration problems, no details will be given further.

§2.5, Discussion on the stability of gyroscopic system

The discussion for the equations of gyroscopic system in section 2.4 is continued here. As is found in section 2.3, the analysis of gyroscopic system must distinguish the cases of positive definite Hamilton function with the case of indefinite Hamilton function in the eigen-solution numerical computation. The stability analysis for gyroscopic system must distinguish the two cases too. Because time invariant system is considered, the Hamilton-Jacobi characteristic equation is given as

$$\mathrm{H}(\mathbf{q}, \partial W/\partial \mathbf{q}) = E$$

§2.5.1, Gyroscopic system with positive definite Hamilton function

The free vibration equation of a gyroscopic system is $\mathbf{M}\ddot{\mathbf{q}} + \mathbf{G}\dot{\mathbf{q}} + \mathbf{K}\mathbf{q} = \mathbf{0}$, and the derivation follows as

$$\mathbf{p} = \mathbf{M}\dot{\mathbf{q}} + \mathbf{G}\mathbf{q}/2, \quad \mathbf{v} = \begin{Bmatrix} \mathbf{q} \\ \mathbf{p} \end{Bmatrix} = \mathbf{L}\mathbf{w}_t, \quad \mathbf{w}_t = \begin{Bmatrix} \mathbf{q} \\ \dot{\mathbf{q}} \end{Bmatrix}, \quad \dot{\mathbf{v}} = \mathbf{N}\mathbf{w}_t \tag{2.3.40}$$

where
$$\mathbf{L} = \begin{bmatrix} \mathbf{I}_n & \mathbf{0} \\ \mathbf{G}/2 & \mathbf{M} \end{bmatrix}, \quad \mathbf{N} = \begin{bmatrix} \mathbf{0} & \mathbf{I}_n \\ -\mathbf{K} & -\mathbf{G}/2 \end{bmatrix}, \quad \mathbf{H} = \mathbf{N}\mathbf{L}^{-1} \tag{2.3.41}$$

and the system equation expressed by $\mathbf{w}_t$ is

$$\mathbf{L}\dot{\mathbf{w}}_t = \mathbf{N}\mathbf{w}_t \tag{2.3.42}$$

The corresponding eigen-equation is

$$\mathbf{N}\mathbf{w} = \mu\mathbf{L}\mathbf{w}, \quad \mathbf{w}_t = \mathbf{w} \times \exp(\mu t) \tag{2.3.43}$$

the eigenvalues μ can be found from $\det(\mu\mathbf{L} - \mathbf{N}) = 0$. The vibration stability problem had been discussed in the previous sections.

The eigenvector of the corresponding Hamilton matrix is $\psi = \mathbf{L}\mathbf{w}$. Two real vectors $\psi_i^{(r)}$ and $\psi_i^{(i)}$, which are the real and imaginary parts of the complex eigenvectors ψ_i can be used instead of ψ_i and ψ_{n+i}. When the Hamilton function is positive definite, only the case of $a_i = \mathrm{i}$ appears in equation (2.4.3b), and the symplectic matrix

$$\mathbf{S}_r = \left[\psi_1^{(r)} \quad \cdots \quad \psi_n^{(r)}; \ \psi_1^{(i)} \quad \cdots \quad \psi_n^{(i)}\right] \tag{2.4.5}$$

is composed, which is described in section 2.4. The real canonical transformation

(2.4.5) and the following derivation are preferable. The vibration computation of the original system (2.4.1) is considered under this transformation, for which Q_i and P_i are no longer constant vectors. Note further that indefinite Hamilton function is considered so that the factor $\pm\omega_i$ must be taken into consideration now. The first stage of canonical transformation is

$$\mathbf{S}_r^T(-\mathbf{JH})\mathbf{S}_r = \begin{bmatrix} \mathrm{diag}(\pm\omega_i) & \mathbf{0} \\ \mathbf{0} & \mathrm{diag}(\pm\omega_i) \end{bmatrix}, \quad \begin{Bmatrix} \mathbf{q} \\ \mathbf{p} \end{Bmatrix} = \begin{bmatrix} \mathbf{S}_{r11} & \mathbf{S}_{r12} \\ \mathbf{S}_{r21} & \mathbf{S}_{r22} \end{bmatrix} \begin{Bmatrix} \hat{\mathbf{q}} \\ \hat{\mathbf{p}} \end{Bmatrix}$$

and the differential equation (2.4.18) become

$$\dot{\hat{\mathbf{q}}} = \mathrm{diag}(\pm\omega_i)\hat{\mathbf{p}} + \mathbf{S}_{r12}^T \begin{bmatrix} (\mathbf{CA}+\mathbf{K}_n)(\mathbf{S}_{r11}\hat{\mathbf{q}} + \mathbf{S}_{r12}\hat{\mathbf{p}}) \\ +\mathbf{CD}(\mathbf{S}_{r21}\hat{\mathbf{q}} + \mathbf{S}_{r22}\hat{\mathbf{p}}) - \mathbf{f}_1(t) \end{bmatrix} \tag{2.4.18a}$$

$$\dot{\hat{\mathbf{p}}} = -\mathrm{diag}(\pm\omega_i)\hat{\mathbf{q}} - \mathbf{S}_{r11}^T \begin{bmatrix} (\mathbf{CA}+\mathbf{K}_n)(\mathbf{S}_{r11}\hat{\mathbf{q}} + \mathbf{S}_{r12}\hat{\mathbf{p}}) \\ +\mathbf{CD}(\mathbf{S}_{r21}\hat{\mathbf{q}} + \mathbf{S}_{r22}\hat{\mathbf{p}}) - \mathbf{f}_1(t) \end{bmatrix} \tag{2.4.18b}$$

Following the derivation below (2.4.1), the transformation (2.4.15) is given as

$$\hat{\mathbf{q}} = \mathrm{diag}\left(\sqrt{2Q_i/\omega_i}\,\sin[\pm\,\omega_i(t-P_i)]\right)\times\{\mathbf{1}\} \tag{2.4.15'a}$$

$$\hat{\mathbf{p}} = \mathrm{diag}\left(\sqrt{2Q_i/\omega_i}\,\cos[\omega_i(t-P_i)]\right)\times\{\mathbf{1}\} \tag{2.4.15'b}$$

where Q_i takes the same sign as $\pm\omega_i$, and $\{\mathbf{1}\} \underset{\mathrm{def}}{=} \{1,1,\cdots,1\}^T$ is a n-dimensional vector, whose components are all unity.

To derive the dual equations for the original system, the vector

$$\mathbf{p} = \mathbf{M}\dot{\mathbf{q}} + \mathbf{G}\mathbf{q}/2 \tag{2.3.5}$$

is introduced first, and the original equation is derived as the equation (2.4.16)

$$\dot{\mathbf{q}} = \mathbf{Aq} + \mathbf{Dp}, \quad \dot{\mathbf{p}} = -\mathbf{Bq} - \mathbf{A}^T\mathbf{p} - \mathbf{C}(\mathbf{Aq}+\mathbf{Dp}) - \mathbf{K}_n\mathbf{q} + \mathbf{f}_1(t)$$

For the problem of free vibration with damping $\mathbf{C}$ only, $\mathbf{K}_n = \mathbf{0}$ and $\mathbf{f}_1(t) = \mathbf{0}$, the differential equations are derived as

$$\dot{\mathbf{Q}} = \begin{bmatrix} \mathrm{diag}\{\sqrt{2\mathrm{Q}_i\omega_i}\,\sin[\pm\omega_i(t-P_i)]\}\times\mathbf{S}_{r12}^T \\ -\mathrm{diag}\{\sqrt{2\mathrm{Q}_i\omega_i}\,\cos[\omega_i(t-P_i)]\}\times\mathbf{S}_{r11}^T \end{bmatrix} \times\mathbf{C}\times \begin{bmatrix} (\mathbf{AS}_{r11}+\mathbf{DS}_{r21})\hat{\mathbf{q}} \\ +(\mathbf{AS}_{r12}+\mathbf{DS}_{r22})\hat{\mathbf{p}} \end{bmatrix} \tag{2.5.1a}$$

$$\dot{\mathbf{P}} = -\begin{bmatrix} \mathrm{diag}\{\cos[\omega_i(t-P_i)]/\sqrt{2\mathrm{Q}_i\omega_i}\}\times\mathbf{S}_{r12}^T \\ +\mathrm{diag}\{\sin[\pm\omega_i(t-P_i)]/\sqrt{2\mathrm{Q}_i\omega_i}\}\times\mathbf{S}_{r11}^T \end{bmatrix} \cdot\mathbf{C}\cdot \begin{bmatrix} (\mathbf{AS}_{r11}+\mathbf{DS}_{r21})\hat{\mathbf{q}} \\ +(\mathbf{AS}_{r12}+\mathbf{DS}_{r22})\hat{\mathbf{p}} \end{bmatrix} \tag{2.5.1b}$$

Left multiplying $\mathbf{S}_r\mathbf{J}$ to equation (2.4.4a) and noting that $\mathbf{S}_r$ is a symplectic matrix gives

$$\mathbf{HS}_r = \begin{bmatrix} \mathbf{A} & \mathbf{D} \\ -\mathbf{B} & -\mathbf{A}^T \end{bmatrix} \times \begin{bmatrix} \mathbf{S}_{r11} & \mathbf{S}_{r12} \\ \mathbf{S}_{r21} & \mathbf{S}_{r22} \end{bmatrix} = \mathbf{S}_r \begin{bmatrix} \mathbf{0} & \mathrm{diag}(\pm\omega_i) \\ -\mathrm{diag}(\pm\omega_i) & \mathbf{0} \end{bmatrix}$$

Writing in block matrix form

$$(\mathbf{AS}_{r11}+\mathbf{DS}_{r21}) = -\mathbf{S}_{r12}\times\mathrm{diag}(\pm\omega_i), \quad (\mathbf{BS}_{r11}+\mathbf{A}^T\mathbf{S}_{r21}) = \mathbf{S}_{r22}\times\mathrm{diag}(\pm\omega_i)$$

$$(\mathbf{AS}_{r12}+\mathbf{DS}_{r22}) = \mathbf{S}_{r11}\times\mathrm{diag}(\pm\omega_i), \quad (\mathbf{BS}_{r12}+\mathbf{A}^T\mathbf{S}_{r22}) = -\mathbf{S}_{r21}\times\mathrm{diag}(\pm\omega_i)$$

where the $\pm\omega_i$ takes positive value for positive definite Hamilton function. Substituting into (2.5.1) and using equation (2.4.15') derives

$$\mathrm{diag}(\omega_i)\times\hat{\mathbf{q}} = \mathrm{diag}\left[\sqrt{2Q_i\omega_i}\,\sin[\omega_i(t-P_i)]\right]\times\{\mathbf{1}\}$$

$$\mathrm{diag}(\omega_i)\times\hat{\mathbf{p}} = \mathrm{diag}\left[\sqrt{2Q_i\omega_i}\,\cos[\omega_i(t-P_i)]\right]\times\{\mathbf{1}\}$$

Therefore

$$\dot{\mathbf{Q}} = -\begin{bmatrix} \mathrm{diag}\{\sqrt{2Q_i\omega_i}\,\sin[\omega_i(t-P_i)]\}\times\mathbf{S}_{r12}^T \\ -\mathrm{diag}\{\sqrt{2Q_i\omega_i}\,\cos[\omega_i(t-P_i)]\}\times\mathbf{S}_{r11}^T \end{bmatrix}\times\mathbf{C}$$
$$\times\begin{bmatrix} \mathbf{S}_{r12}\times\mathrm{diag}\{\pm\sqrt{2Q_i\omega_i}\,\sin[\omega_i(t-P_i)]\} \\ -\mathbf{S}_{r11}\times\mathrm{diag}\{\pm\sqrt{2Q_i\omega_i}\,\cos[\omega_i(t-P_i)]\} \end{bmatrix}\times\{\mathbf{1}\} \tag{2.5.2a}$$

$$\dot{\mathbf{P}} = -\begin{bmatrix} \mathrm{diag}[\cos[\omega_i(t-P_i)]/\sqrt{2Q_i\omega_i}\,]\times\mathbf{S}_{r12}^T \\ +\mathrm{diag}[\sin[\omega_i(t-P_i)]/\sqrt{2Q_i\omega_i}\,]\times\mathbf{S}_{r11}^T \end{bmatrix}\times\mathbf{C}$$
$$\times\begin{bmatrix} -\mathbf{S}_{r12}\times\mathrm{diag}\{\pm\sqrt{2Q_i\omega_i}\,\sin[\omega_i(t-P_i)]\} \\ +\mathbf{S}_{r11}\times\mathrm{diag}\{\pm\sqrt{2Q_i\omega_i}\,\cos[\omega_i(t-P_i)]\} \end{bmatrix}\times\{\mathbf{1}\} \tag{2.5.2b}$$

Left multiplying the equation (2.5.2a) with $\{\mathbf{1}\}^T$, the right side of equation becomes a positive definite matrix $\mathbf{C}$ multiplied from left and from right by the vectors of mutually transpose to each other, which is negative definite because of the negative sign. Therefore $\dot{E} = \sum\dot{Q}_i$ is negative, which can be considered as the Lyapunov function (see chapter 6). The system is asymptotically stable. This conclusion is drawn based on that all $a_i = \mathrm{i}$, or the Hamilton function being positive definite. Under such assumption the positive definite damping matrix $\mathbf{C}$ must derive the system being asymptotically stable, a known conclusion. However, the equations (2.5.2a,b) can be used for numerical computation and if there is non-linear factor to be added on, the formulation is not difficult either.

§2.5.2, Case of indefinite Hamilton function

The description is for n-dimensional problem, and a two-degrees of freedom system is used to illustrate. Suppose a 2-D unstable system

$$\ddot{q}_1 + k_1 q_1 = 0\ , \qquad \ddot{q}_2 + k_2 q_2 = 0\ ,\ \ (k_1 < 0, k_2 < 0)$$

after applying the gyroscopic force it derives

$$\ddot{q}_1 + \Gamma\dot{q}_2 + k_1 q = 0,\ \ \ddot{q}_2 - \Gamma\dot{q}_1 + k_2 q = 0$$

or

$$\mathbf{M}\ddot{\mathbf{q}} + \mathbf{G}\dot{\mathbf{q}} + \mathbf{K}\mathbf{q} = \mathbf{0},\ \ \mathbf{M} = \begin{bmatrix} 1 & 0 \\ 0 & 1 \end{bmatrix},\ \ \mathbf{K} = \begin{bmatrix} k_1 & 0 \\ 0 & k_2 \end{bmatrix},\ \ \mathbf{G} = \begin{bmatrix} 0 & \Gamma \\ -\Gamma & 0 \end{bmatrix},$$

$$\mathrm{H}(\mathbf{q},\mathbf{p}) = \mathbf{p}^T\mathbf{D}\mathbf{p}/2 + \mathbf{p}^T\mathbf{A}\mathbf{q} + \mathbf{q}^T\mathbf{B}\mathbf{q}/2 = \mathbf{v}^H(-\mathbf{J}\mathbf{H})\mathbf{v}/2 \tag{2.3.7}$$

where $\mathbf{D} = \mathbf{M}^{-1} = \mathbf{I}$, $\mathbf{A} = \begin{bmatrix} 0 & -\Gamma/2 \\ \Gamma/2 & 0 \end{bmatrix}$, $\mathbf{B} = \begin{bmatrix} k_1 + \Gamma^2/4 & 0 \\ 0 & k_2 + \Gamma^2/4 \end{bmatrix}$.

When $(k_1 < 0, k_2 < 0)$, the Hamilton function is indefinite. The Hamilton matrix and the eigen-problem are

$$\mathbf{H}=\begin{bmatrix}\mathbf{A} & \mathbf{D}\\ -\mathbf{B} & -\mathbf{A}^T\end{bmatrix},\quad \mathbf{H}\psi=\mu\psi$$

the equation for eigenvalues μ is $\det(\mathbf{H}-\mu\mathbf{I})=0$ or in expanded form

$$\mu^4+(\Gamma^2+k_1+k_2)\mu^2+k_1k_2=0$$

If $$k_1k_2>0,\ \Gamma^2+k_1+k_2>0,\ (\Gamma^2+k_1+k_2)-4k_1k_2>0$$

then the roots of μ^2 will be negative, so that μ are purely imaginary numbers, the vibration is stable (but not asymptotically stable). Because $k_1<0, k_2<0$ the first condition is ensured, and if the gyroscopic term Γ is large enough then the system is stable. According to the derivation below (2.4.3), the values of a_1 and a_2 require to check whether the case $a_i=-\mathrm{i}$ appears. Let

$$\bar{\psi}_i^T(-\mathbf{JH})\psi_i=h_i,\qquad i=1,\cdots,n \tag{2.5.3a}$$

and since $-\mathbf{JH}$ is a real symmetric matrix so h_i must be real. Because

$$\mu_{n+i}=-\mathrm{i}\omega_i,\ \psi_{n+i}=a_i\{\bar{\mathbf{q}}_i^T,\bar{\mathbf{p}}_i^T\}^T=a_i\bar{\psi}_i,\quad \bar{a}_ia_i=1 \tag{2.4.3b}$$

so $$\bar{\psi}_{n+i}^T(-\mathbf{JH})\psi_{n+i}=\bar{a}_ia_i\psi_i^T(-\mathbf{JH})\bar{\psi}_i=\bar{a}_ia_i\bar{h}_i=h_i=h_{n+i},\ i=1,\cdots,n \tag{2.5.3b}$$

Since the Hamilton function is assumed **indefinite** and the $2n$ eigenvectors are definitely linearly independent, so there is **at least one** $h_i<0$. Equation (2.5.3a) implies

$$\bar{\psi}_i^T(-\mathbf{JH})\psi_i=-\mu_i\bar{\psi}_i^T\mathbf{J}\psi_i=(-\mathrm{i}\omega_i/a_i)\psi_{n+i}^T\mathbf{J}\psi_i=(\mathrm{i}\omega_i/a_i)=h_i$$

from $h_i<0$ it derives $a_i=-\mathrm{i}$. On the contrary, if all the modes of vibration are $a_i=\mathrm{i}$, for $i=1,\cdots,n$, then the Hamilton function is positive definite, which contradicts to the indefinite Hamilton function assumption. The number of $a_i=-\mathrm{i}$ can be determined from the $\mathbf{LDL}^T$ factorization of matrix $(-\mathbf{JH})$ and checking the number of negative elements in the diagonal matrix $\mathbf{D}$. Therefore after separation of variables, the Hamilton-Jacobi characteristic equation becomes separated too as

$$W=\sum_{i=1}^{n}W_i,\quad \omega_i[(\partial W_i/\partial\hat{q}_i)^2+\hat{q}_i^2]/2=\alpha_i,\quad i=1,\cdots,n \tag{2.5.4}$$

where the selection $\omega_i>0$ is always used, then $\alpha_i>0$. However, the case of indefinite Hamilton function is considered below, and the fundamental gyroscopic system (2.4.2) is a conserved one, because (2.4.2) is a damping free system. The conserved term (energy) for the fundamental system is

$$\begin{aligned}E&=\mathrm{H}(\mathbf{q},\mathbf{p})=-\mathbf{v}^H\mathbf{JHv}/2=\zeta^T\mathbf{S}_\mathrm{r}^T(-\mathbf{JH})\mathbf{S}_\mathrm{r}\zeta/2\\&=\sum_{i=1}^{n}\zeta_{\mathrm{r}i}^T\begin{bmatrix}\pm\omega_i & 0\\ 0 & \pm\omega_i\end{bmatrix}\zeta_{\mathrm{r}i}/2=-\alpha_1+\cdots+\alpha_n\end{aligned} \tag{2.5.5}$$

Because the fundamental system holds an indefinite Hamilton function, when it is diagonalized via a real matrix $\mathbf{S}_\mathrm{r}$, it may not have all positive diagonal value. The sign $\pm$ is thus appearing in the above equation, the selection for signs $\pm$ is that when $a_i = \mathrm{i}$ it takes positive sign, whereas when $a_i = -\mathrm{i}$ it takes negative sign. Under this convention, $\omega_i > 0$ always holds. This sign has been given explicitly in (2.5.5) for the α_i, and without loss of generality, the α_1 has been given the negative sign explicitly. Then the values of all α_i are positive.

Note that, for the original damping system, this conserved function E of fundamental system cannot be used as a Lyapunov function since it is no longer positive definite. Because the negative sign has been explicitly written, all α_i take positive value. Since the fundamental system can completely separate variables, that all α_i, $i = 1,\cdots,n$ are conservative, the vibration is stable but not asymptotically stable. However, if some factor, such as damping, destroys the complete separation of variables, then even E is decreasing because of positive damping, but the system cannot ensure all the α_i decreasing. The increasing of anyone α_i means unstable of the system. The case of indefinite Hamilton function requires solving equation (2.5.4) too. Without loss of generality, $a_1 = -\mathrm{i}$ has been assumed for the fundamental system

$$W_1(\hat{q}_1, \alpha_1) = \sqrt{2\alpha_1/\omega_1} \int \sqrt{1 - \omega_1 \hat{q}_1^2/(2\alpha_1)} \mathrm{d}q_1 + A_1 \tag{2.5.6}$$

According to the same procedure, the action function is

$$S_1(\hat{q}_1, \alpha_1, t) = W_1(\hat{q}_1, \alpha_1) - \alpha_1 t \tag{2.5.7}$$

and then

$$Q_1 = \alpha_1 \tag{2.5.8}$$

$$P_1 = \beta_1 = -\partial S_1/\partial \alpha_1 = -\left(\arcsin \sqrt{\omega_1/(2\alpha_1)}\hat{q}_1\right)\Big/\omega_1 + t \tag{2.5.9}$$

$$\hat{p}_1 = \partial S_1/\partial \hat{q}_1 = \sqrt{(2\alpha_1/\omega_1) - \hat{q}_1^2} \tag{2.5.10}$$

from (2.5.8~10) it gives

$$\hat{q}_1(\alpha_1, \beta_1, t) = \sqrt{2\alpha_1/\omega_1} \sin[\pm \omega_1(t - \beta_1)]$$

$$\hat{p}_1(\alpha_1, \beta_1, t) = \sqrt{2\alpha_1/\omega_1} \cos[\omega_1(t - \beta_1)]$$

The above equations apply to the separable system. However it can be used as a canonical transformation to the original system, which is neither separable nor conservative. The constants α_1, β_1 of the fundamental separable system should be updated as the dual variables Q_1, P_1 of the original system, and the transformation is rewritten as

$$\hat{q}_1(Q_1, P_1, t) = \sqrt{2Q_1/\omega_1} \sin[\pm \omega_1(t - P_1)] \tag{2.5.11a}$$

$$\hat{p}_1(Q_1, P_1, t) = \sqrt{2Q_1/\omega_1} \cos[\omega_1(t - P_1)] \tag{2.5.11b}$$

The derivation is completely similar to that given above. The original system is written in the dual form as

$$\dot{\mathbf{q}} = \mathbf{Aq} + \mathbf{Dp} \tag{2.5.12a}$$

$$\dot{\mathbf{p}} = -\mathbf{Bq} - \mathbf{A}^T\mathbf{p} - \mathbf{C}(\mathbf{Aq} + \mathbf{Dp}) \tag{2.5.12b}$$

The similar transformation derives to (2.4.20'a,b), and for damping only system gives

$$\dot{\mathbf{Q}} = \begin{bmatrix} \mathrm{diag}\{\sqrt{2Q_i\omega_i}\sin[\pm\omega_i(t-P_i)]\}\times\mathbf{S}_{r12}^T \\ -\mathrm{diag}\{\sqrt{2Q_i\omega_i}\cos[\omega_i(t-P_i)]\}\times\mathbf{S}_{r11}^T \end{bmatrix}\times\mathbf{C}\times\begin{bmatrix}(\mathbf{AS}_{r11}+\mathbf{DS}_{r21})\hat{\mathbf{q}} \\ +(\mathbf{AS}_{r12}+\mathbf{DS}_{r22})\hat{\mathbf{p}}\end{bmatrix} \quad (2.5.13\text{'a})$$

$$\dot{\mathbf{P}} = -\begin{bmatrix} \mathrm{diag}\{\cos[\omega_i(t-P_i)]/\sqrt{2Q_i\omega_i}\}\times\mathbf{S}_{r12}^T \\ +\mathrm{diag}\{\sin[\pm\omega_i(t-P_i)]/\sqrt{2Q_i\omega_i}\}\times\mathbf{S}_{r11}^T \end{bmatrix}\cdot\mathbf{C}\cdot\begin{bmatrix}(\mathbf{AS}_{r11}+\mathbf{DS}_{r21})\hat{\mathbf{q}} \\ +(\mathbf{AS}_{r12}+\mathbf{DS}_{r22})\hat{\mathbf{p}}\end{bmatrix} \quad (2.5.13\text{'b})$$

Left multiplying (2.4.6) with $\mathbf{S}_r\mathbf{J}$ and note that $\mathbf{S}_r$ is a symplectic matrix

$$\mathbf{HS}_r = \begin{bmatrix}\mathbf{A} & \mathbf{D} \\ -\mathbf{B} & -\mathbf{A}^T\end{bmatrix}\times\begin{bmatrix}\mathbf{S}_{r11} & \mathbf{S}_{r12} \\ \mathbf{S}_{r21} & \mathbf{S}_{r22}\end{bmatrix} = \mathbf{S}_r\begin{bmatrix}0 & \mathrm{diag}(\pm\omega_i) \\ -\mathrm{diag}(\pm\omega_i) & 0\end{bmatrix}$$

where the sign convention for $\pm\omega_i$ is again that, ***when*** $a_i = \mathrm{i}$ ***it takes*** ω_i, ***whereas when*** $a_i = -\mathrm{i}$ ***it takes*** $-\omega_i$**.** Writing the above equation in block matrix form derives

$$\begin{aligned}&(\mathbf{AS}_{r11}+\mathbf{DS}_{r21}) = -\mathbf{S}_{r12}\cdot\mathrm{diag}(\pm\omega_i),\ (\mathbf{BS}_{r11}+\mathbf{A}^T\mathbf{S}_{r21}) = \mathbf{S}_{r22}\cdot\mathrm{diag}(\pm\omega_i) \\ &(\mathbf{AS}_{r12}+\mathbf{DS}_{r22}) = \mathbf{S}_{r11}\cdot\mathrm{diag}(\pm\omega_i),\ (\mathbf{BS}_{r12}+\mathbf{A}^T\mathbf{S}_{r22}) = -\mathbf{S}_{r21}\cdot\mathrm{diag}(\pm\omega_i)\end{aligned} \quad (2.5.14)$$

Substituting into (2.5.13), afterwards using equations (2.5.11) and (2.4.15'a,b), it derives

$$\mathrm{diag}(\pm\omega_i)\times\hat{\mathbf{q}} = \mathrm{diag}\left(\sqrt{2Q_i\omega_i}\sin\left[\omega_i(t-P_i)\right]\right)\times\{\mathbf{1}\}$$

$$\mathrm{diag}(\pm\omega_i)\times\hat{\mathbf{p}} = \mathrm{diag}\left(\pm\sqrt{2Q_i\omega_i}\cos\left[\omega_i(t-P_i)\right]\right)\times\{\mathbf{1}\}$$

Substituting (2.5.14) into (2.5.13a,b) and using the above equations gives

$$\dot{\mathbf{Q}} = -\begin{bmatrix} \mathrm{diag}\{\sqrt{2Q_i\omega_i}\sin[\pm\omega_i(t-P_i)]\}\times\mathbf{S}_{r12}^T \\ -\mathrm{diag}\{\sqrt{2Q_i\omega_i}\cos[\omega_i(t-P_i)]\}\times\mathbf{S}_{r11}^T \end{bmatrix}\times\mathbf{C} \times\begin{bmatrix}\mathbf{S}_{r12}\times\mathrm{diag}\{\sqrt{2Q_i\omega_i}\sin[\omega_i(t-P_i)]\} \\ -\mathbf{S}_{r11}\times\mathrm{diag}\{\pm\sqrt{2Q_i\omega_i}\cos[\omega_i(t-P_i)]\}\end{bmatrix}\times\{\mathbf{1}\} \quad (2.5.15\text{a})$$

$$\dot{\mathbf{P}} = -\begin{bmatrix} \mathrm{diag}[\cos[\omega_i(t-P_i)]/\sqrt{2Q_i\omega_i}]\times\mathbf{S}_{r12}^T \\ +\mathrm{diag}[\sin[\pm\omega_i(t-P_i)]/\sqrt{2Q_i\omega_i}]\times\mathbf{S}_{r11}^T \end{bmatrix}\times\mathbf{C} \times\begin{bmatrix}\mathbf{S}_{r12}\times\mathrm{diag}\{\sqrt{2Q_i\omega_i}\sin[\omega_i(t-P_i)]\} \\ -\mathbf{S}_{r11}\times\mathrm{diag}\{\pm\sqrt{2Q_i\omega_i}\cos[\omega_i(t-P_i)]\}\end{bmatrix}\times\{\mathbf{1}\} \quad (2.5.15\text{b})$$

The sign selection convention has been repeated several times above which is determined from the canonical transformation but unrelated to the damping factor. The above equation is the extension of equation (2.5.2a,b) for the indefinite Hamilton function case.

For positive definite Hamilton function, all sign $\pm$ selection is positive so the stability is ensured. But for indefinite Hamilton function, such as $a_1 = -\mathrm{i}$, then at right hand side of (2.5.15a) the $\pm$ selection for Q_1 appears negative. In this case $\dot{Q}_1$ may be positive in the vicinity of the original point $Q_i = 0$, which causes instability at the original point. The indefinite Hamilton function determines that there must be a pair of eigen-solutions with $a_i = -\mathrm{i}$, and the negative sign from it decides the instability from the damping factor, which confirms the Chetaev's theorem.

Example 2.3, a simple example is given as

$$\mathbf{M} = \begin{bmatrix} 1 & 0 \\ 0 & 1 \end{bmatrix}, \quad \mathbf{K} = \begin{bmatrix} k_1 & 0 \\ 0 & k_2 \end{bmatrix}, \quad \mathbf{G} = \begin{bmatrix} 0 & \Gamma \\ -\Gamma & 0 \end{bmatrix}, \quad \text{and} \quad \begin{matrix} k_1 = k_2 = -1 \\ \Gamma = 4 \end{matrix},$$

The canonical transformation is required.

Solution: From the given matrices it derives

$$\mathbf{D} = \mathbf{M}^{-1} = \mathbf{I}, \quad \mathbf{A} = \begin{bmatrix} 0 & -2 \\ 2 & 0 \end{bmatrix}, \quad \mathbf{B} = \begin{bmatrix} 3 & 0 \\ 0 & 3 \end{bmatrix}.$$

The Hamilton matrix and the eigen-solutions are

$$\mathbf{H} = \begin{bmatrix} 0 & -2 & 1 & 0 \\ 2 & 0 & 0 & 1 \\ -3 & 0 & 0 & -2 \\ 0 & -3 & 2 & 0 \end{bmatrix},$$

$$\boldsymbol{\psi}_1 = \{1\mathrm{i} \quad 1 \quad \sqrt{3} \quad -\sqrt{3}\mathrm{i}\}^T, \quad \boldsymbol{\psi}_2 = \{1\mathrm{i} \quad 1 \quad -\sqrt{3} \quad \sqrt{3}\mathrm{i}\}^T,$$

$$\mu_1 = \omega_1 \mathrm{i} = (2-\sqrt{3})\mathrm{i} \qquad \mu_2 = \omega_2 \mathrm{i} = (2+\sqrt{3})\mathrm{i}$$

According to equation (2.4.3b), $\mu_{n+i} = -\mathrm{i}\omega_i$, $\boldsymbol{\psi}_{n+i} = a_i\{\overline{\mathbf{q}}_i^T, \overline{\mathbf{p}}_i^T\}^T = a_i\overline{\boldsymbol{\psi}}_i$

$$\boldsymbol{\psi}_3 = a_1\{-1\mathrm{i} \quad 1 \quad \sqrt{3} \quad \sqrt{3}\mathrm{i}\}^T, \quad \boldsymbol{\psi}_4 = a_2\{-1\mathrm{i} \quad 1 \quad -\sqrt{3} \quad -\sqrt{3}\mathrm{i}\}^T,$$

$$\mu_3 = \omega_3 \mathrm{i} = -(2-\sqrt{3})\mathrm{i} \qquad \mu_4 = \omega_4 \mathrm{i} = -(2+\sqrt{3})\mathrm{i}$$

and the symplectic normalizing condition gives

$$\overline{\boldsymbol{\psi}}_3^T \mathbf{J} \boldsymbol{\psi}_1 = a_1 \overline{\boldsymbol{\psi}}_1^T \mathbf{J} \boldsymbol{\psi}_1 = a_1(-\sqrt{3}-\sqrt{3}-\sqrt{3}-\sqrt{3})\mathrm{i} = -4\sqrt{3}\mathrm{i}a_1$$

which ought be -1. This requires $a_1 = -\mathrm{i}$ and the multiplier $3^{-1/4}/2$ is applied to the eigenvectors $\boldsymbol{\psi}_1$ and $\boldsymbol{\psi}_3$ respectively. Again

$$\overline{\boldsymbol{\psi}}_4^T \mathbf{J} \boldsymbol{\psi}_2 = a_2 \overline{\boldsymbol{\psi}}_2^T \mathbf{J} \boldsymbol{\psi}_2 = a_2(\sqrt{3}+\sqrt{3}+\sqrt{3}+\sqrt{3})\mathrm{i} = 4\sqrt{3}\mathrm{i}a_2$$

according to symplectic normalization it should be -1, hence $a_2 = \mathrm{i}$, and also the multiplier $3^{-1/4}/2$ should apply on the eigenvectors $\boldsymbol{\psi}_2$ and $\boldsymbol{\psi}_4$ respectively. The real symplectic matrix $\mathbf{S}_\mathrm{r}$ is composed next, based on (2.4.3c,d) and (2.4.4a)

$$\mathbf{S}_\mathrm{r} = \begin{bmatrix} 0 & 0 & -1 & 1 \\ 1 & 1 & 0 & 0 \\ \sqrt{3} & -\sqrt{3} & 0 & 0 \\ 0 & 0 & \sqrt{3} & \sqrt{3} \end{bmatrix} \Big/ (2\sqrt{3})^{1/2}$$

which is easily verified a symplectic matrix. Block disintegrated as

$$\mathbf{S}_{\mathrm{r}11} = \begin{bmatrix} 0 & 0 \\ 1 & 1 \end{bmatrix} \Big/ (2\sqrt{3})^{1/2}, \qquad \mathbf{S}_{\mathrm{r}12} = \begin{bmatrix} -1 & 1 \\ 0 & 0 \end{bmatrix} \Big/ (2\sqrt{3})^{1/2}$$

$$\mathbf{S}_{\mathrm{r}21} = \begin{bmatrix} \sqrt{3} & -\sqrt{3} \\ 0 & 0 \end{bmatrix} \Big/ (2\sqrt{3})^{1/2}, \qquad \mathbf{S}_{\mathrm{r}22} = \begin{bmatrix} 0 & 0 \\ \sqrt{3} & \sqrt{3} \end{bmatrix} \Big/ (2\sqrt{3})^{1/2}$$

where equation (2.5.14) can be verified by the reader. The above is the separation of variables step for the fundamental system. Further the damping factor is added, the equations (2.5.15a,b) can be derived, the readers are asked to do the calculation themselves. ##

Discussion: The gyroscopic system for rotor dynamics has all eigenvalues purely imaginary, but the corresponding Hamilton function is usually indefinite. The symplectic subspace iteration method can be used to find the eigen-solution composed symplectic subspace, which covers the negative part of the Hamilton function. For rotor dynamics, there are several rotor disks installed on the rotating axis, usually there are only limited number of deformations with negative Hamilton function value. For high speed rotor with angular velocity Ω, the eigenvalues solved from the symplectic eigen-problem (2.3.65), for which the Hamilton function is negative, are nearby $\mu^2 \approx -\Omega^2$. Then the origin shifting technique can be used effectively for the symplectic subspace iterative eigen-solution [22,141]. After solved these eigen-solutions the symplectic subspace, which covers the negative Hamilton function part, the Hamilton function of the complementary symplectic subspace becomes positive definite, then the effective algorithms can be used to find the remaining eigen-solutions. The eigen-equation is (2.3.38) with both matrices positive definite, for which the algorithm given in [51] is quite efficient.

Chapter 3, Probability and stochastic process

The purpose of this chapter is to provide a fundamental knowledge of probability and stochastic process. The interested reader should read such as the books [58~59] for more.

§3.1, Preliminary of probability theory

In nature, quite a number of events appear randomly. An event A is a result of one experiment. The mathematics for such event description is the theory of probability. An experiment is said to be **random**, if the result is not predictable in the ordinary sense before the experiment is carried out. Let $\Pr(A)$ denote the probability of event A to appear. Perform the experiment a large number (say M) of times, and count the number m_A of times that event A occurs. That $\Pr(A)$ can be regarded as m_A/M. If there are totally n different events of experiments $A_i, i=1,2,\cdots,n$, then

$$\sum_{i=1}^{n}\Pr(A_i)=1 \tag{3.1.1}$$

These events $A_i, i=1,2,\cdots,n$ must be mutually exclusive. There are a number of fundamental operations in theory of probability, as listed below.

Union event: At least one of events A or B occurs denoted as $A+B$.

Intersection event: Both of A and B occur, denoted as $A\cdot B$ or AB.

Difference event: A occurs but B does not, denoted as $A-B$.

Inclusion: Occurrence of A implies occurrence of B, denoted as $A\subset B$.

Let A' denotes the complement of event A, that they are mutually exclusive, then $\Pr(A)+\Pr(A')=1$. The probability of intersection event $A\cdot B$ is described as $P(AB)$. If the events A,B,C are mutually independent, their intersection event probability is simply the product of the probability of individual events

$$P(ABC)=P(A)\cdot P(B)\cdot P(C) \tag{3.1.2}$$

The probability of union event $A+B+C$ is $P(A+B+C)$, if the events A,B,C are mutually exclusive then

$$P(A+B+C)=P(A)+P(B)+P(C) \tag{3.1.3}$$

However, if the events A and B are not mutually exclusive then

$$P(A+B)=P(A)+P(B)-P(AB) \tag{3.1.4}$$

If events A and B are mutually exclusive, then $P(AB)=0$, (3.1.4) goes back to (3.1.3).

A very important idea is the conditional probability for interrelated events. The conditional probability of event A to occur under the condition of event B occurs is defined as

$$P(A\,|\,B)=P(AB)/P(B) \tag{3.1.5}$$

provided that $P(B)>0$. If $P(B)=0$ then the conditional probability is undefined. For the special case of events A and B are independent, then according to (3.1.3) the conditional probability simply reduce to the individual probability of event $P(A)$. Because of the events A and B can be permuted each other, so that

$$P(A\mid B)P(B)=P(B\mid A)P(A) \tag{3.1.6}$$

If for the occurrence of event B, one of the mutually exclusive events $A_i, i=1,2,\cdots,n$ must occur, then further consider their conditional probability according to (3.1.6)

$$P(A_i\mid B)=P(B\mid A_i)P(A_i)/P(B),\quad i=1,2,\cdots,n$$

On the other hand $P(B)=\sum_{i=1}^{n}P(B\mid A_i)P(A_i)$, then

$$P(A_i\mid B)=P(B\mid A_i)P(A_i)\Big/\sum_{j=1}^{n}P(B\mid A_j)P(A_j) \tag{3.1.7}$$

it is known as the Bayes theorem.

§3.1.1, Probability distribution function and probability density function

The outcome of events can be discrete (such as through dice) or continuous. The sample spaces of random variables X in the theory of random vibration or state space approach are continuous. Let the ***probability distribution function*** of a continuous random variable X be denoted as

$$F(x)=\Pr(X\le x) \tag{3.1.8}$$

Obviously $X\le x$ is an event. For continuous real random variable X, the bound variable x takes also real continuous value. The probability of random variable X being located within the interval $[x,x+\Delta x)$ can be denoted as

$$\Pr\big(X\in[x,x+\Delta x)\big)=f(x)\Delta x \tag{3.1.9}$$

where $f(x)$ is termed as the ***probabilistic density function*** (p.d.f.) of random variable X. The p.d.f. is never negative and

$$f(x)=\mathrm{d}F(x)/\mathrm{d}x \tag{3.1.10}$$

When the sample space of X is the whole real axis $-\infty<X<\infty$, then according to (3.1.2)

$$F(x)=\int_{-\infty}^{x}f(u)\mathrm{d}u,\qquad F(\infty)=\int_{-\infty}^{\infty}f(u)\mathrm{d}u=1 \tag{3.1.11}$$

Duality variables and state space approach is emphasized in this book. The basic unknowns are composed of the state vector and are multi-dimensional, (n-D). Therefore the joint probability distribution of multi-variables must be considered. The joint probability distribution function for the case of two random variables X,Y is described as

$$F(x,y)=\Pr(X\le x,Y\le y) \tag{3.1.12}$$

Correspondingly, the joint probability density function is

$$f(x,y)=\partial^2F(x,y)/\partial x\partial y \tag{3.1.13}$$

If only the random variable Y is considered, its density function $p(y)$ can be obtained as

$$p(y)=\int_{-\infty}^{\infty} f(x,y)\mathrm{d}x \tag{3.1.14}$$

Under condition of $Y=y$, the probability density function of random variable X is

$$f(x\mid y)=f(x,y)/p(y) \tag{3.1.15}$$

When X,Y are mutually independent random variables, the density function has the form

$$f(x,y)=p(x)\cdot p(y) \tag{3.1.16}$$

For multi-dimensional case, the extension of these equations is straightforward.

§3.1.2, Mathematical expectation, variance and covariance

Probability distribution function or density function is the comprehensive description for the random variable. However in applications, the determining of these functions is difficult, therefore to determine some numerical characteristics of the random variable is the common practice. The most frequently used characteristics are ***mathematical expectation***, ***variance*** and ***covariance*** etc.

For continuous random variable X, the mathematical expectation is defined as

$$\mathrm{E}[X]=\int_{-\infty}^{\infty} xf(x)\mathrm{d}x \tag{3.1.17}$$

That is the integration of possible occurrence x of the random variable X multiplying the probability density function. The mathematical expectation is also termed as ***mean***, ***average***, ***mean-value*** or else ***first order moment etc.*** In general, the mathematical expectation (or average) $\mathrm{E}[g(X)]$ of a function of random variable $g(X)$ is defined as

$$\mathrm{E}[g(X)]=\int_{-\infty}^{\infty} g(x)f(x)\mathrm{d}x \tag{3.1.18}$$

Next, an important numerical characteristic of a random variable X is its ***variance***. The mean square value of X is defined as the mathematical expectation of X^2

$$\mathrm{E}[X^2]=\int_{-\infty}^{\infty} x^2 f(x)\mathrm{d}x \tag{3.1.19}$$

or precisely the second order moment with respect to the original point. The variance of a random variable X is defined as the mean square value with respect to the average value, which reflect the bias with respect to the mathematical expectation, so that it is also termed as ***central second moment***, usually expressed as σ^2

$$\sigma^2=\mathrm{E}[(X-\mathrm{E}(X))^2]=\int_{-\infty}^{\infty}(x-\mathrm{E}(X))^2 f(x)\mathrm{d}x=\mathrm{E}(X^2)-[\mathrm{E}(X)]^2 \tag{3.1.20}$$

The square-root of variance, σ, is a measure of spread in the same unit as X and is termed as ***standard deviation*** of the random variable X. Only for zero-mean random variable, the second moment equals its variance.

The idea of covariance is a measure for different random variables. The definition of covariance of two random variables X,Y is

$$\mathrm{E}[(X-\mathrm{E}(X))\cdot(Y-\mathrm{E}(Y))]=\int_{-\infty}^{\infty}(x-\mathrm{E}(X))(y-\mathrm{E}(Y))f(x,y)\mathrm{d}x\mathrm{d}y \\ =\mathrm{E}(XY)-\mathrm{E}(X)\cdot\mathrm{E}(Y) \tag{3.1,21}$$

i.e. the expectation of product of deviation of the random variables. The ***correlation coefficient*** between two random variables is defined as

$$\rho_{xy}=[\mathrm{E}(XY)-\mathrm{E}(X)\cdot\mathrm{E}(Y)]/(\sigma_x\sigma_y) \tag{3.1.22}$$

The correlation coefficient between random variables X,Y is a proper measure of degree of linear inter-correlation between them. If X,Y are mutually independent then $\rho_{xy}=0$, but the reverse is not true. If Y is a linear function of X, then $\rho_{xy}=\pm 1$.

It is useful to list the following behaviors of numerical characteristics as:

1) The expectation of linear combination of random variables equals the same linear combination of expectations of the random variables. Regardless of the random variables $X_i, i=1,\cdots,n$ being linearly independent or not, it holds $\mathrm{E}[\sum_{i=1}^{n}c_iX_i]=\sum_{i=1}^{n}c_i\mathrm{E}[X_i]$

2) If $X_i, i=1,\cdots,n$ are independent, then $\mathrm{E}[X_1X_2\cdots X_n]=\prod_{i=1}^{n}\mathrm{E}[X_i]$; and the variance of the sum $X=\sum_{i=1}^{n}X_i$ equals the summation of variances of all the random variables X_i, *i.e.* $\sigma_X^2=\sum_{j=1}^{n}\sigma_{X_j}^2$. A constant, c, multiplication cX results the variance multiplying the square of constant, *i.e.* $\sigma_{cX}^2=c^2\sigma_X^2$.

3) As a function of x, $\mathrm{E}[(X-x)^2]$ takes its minimum σ_x^2 when $x=\mathrm{E}[X]$.

Etc.

§3.1.3, Expectation of a random vector and its covariance matrix

A n-dimensional vector $\mathbf{X}$, if all its components X_i are random variables

$$\mathbf{X}=\{X_1 \quad X_2 \quad \cdots \quad X_n\}^T \tag{3.1.23}$$

then $\mathbf{X}$ is a n-dimensional random vector. The mathematical expectation vector, and its variance matrix of a random vector $\mathbf{X}$ can be defined as the mean-value and (co-)variance of these component random variables. The expectation vector (mean-value) is defined as

$$\begin{aligned}\mathrm{E}(\mathbf{X})&=\{\mathrm{E}(X_1) \quad \mathrm{E}(X_2) \quad \cdots \quad \mathrm{E}(X_n)\}^T\\ &=\int_{-\infty}^{\infty}\cdots\int_{-\infty}^{\infty}f(x_1,x_2,\cdots,x_n)\{x_1,x_2,\cdots,x_n\}^T\mathrm{d}x_1\mathrm{d}x_2\cdots\mathrm{d}x_n\\ &=\int_{-\infty}^{\infty}\cdots\int_{-\infty}^{\infty}f(\mathbf{x})\cdot\mathbf{x}\mathrm{d}x_1\mathrm{d}x_2\cdots\mathrm{d}x_n\end{aligned} \tag{3.1.24}$$

Interpreting the probability density function $f(\mathbf{x})$ as a mass density in the n-dimensional space with all the mass equals 1, then $\mathrm{E}(\mathbf{X})$ becomes the position of center of mass. Therefore, it can say that $\mathrm{E}(\mathbf{X})$ is the first order moment of mass distribution.

The variance matrix of a random vector $\mathbf{X}$ is defined as

$$\begin{aligned}\mathbf{P}_{xx} &= \mathrm{E}\left[(\mathbf{X}-\mathrm{E}(\mathbf{X}))\cdot(\mathbf{X}-\mathrm{E}(\mathbf{X}))^{\mathrm{T}}\right] \\ &= \int_{-\infty}^{\infty}\cdots\int_{-\infty}^{\infty}(\mathbf{x}-\mathrm{E}(\mathbf{X}))(\mathbf{x}-\mathrm{E}(\mathbf{X}))^{\mathrm{T}}\cdot f(\mathbf{x})\mathrm{d}x_1\mathrm{d}x_2\cdots\mathrm{d}x_n\end{aligned} \tag{3.1.25}$$

There is a column vector multiplying a row vector under the integration sign, which gives a $n\times n$ matrix. The $\mathbf{P}_{xx}$ can be interpreted as a second moment of distributed mass with respect to the center of mass, often called as ***central second moment***. The i-th diagonal element p_{ii} is the variance of random component X_i, and p_{ij} is the covariance of the components X_i and X_j. It is easily proved, that the variance matrix and the second moment have the relation

$$\mathbf{P}_{xx} = \int_{-\infty}^{\infty}\cdots\int_{-\infty}^{\infty}\mathbf{x}\mathbf{x}^{\mathrm{T}}\cdot f(\mathbf{x})\mathrm{d}x_1\mathrm{d}x_2\cdots\mathrm{d}x_n - [\mathrm{E}(\mathbf{X})][\mathrm{E}(\mathbf{X})]^T \tag{3.1.26}$$

The covariance matrix between a n-dimensional vector $\mathbf{X}$ and a m-dimensional vector $\mathbf{Y}$ is defined as

$$\begin{aligned}\mathbf{P}_{xy} &= \mathrm{E}\left[[\mathbf{X}-\mathrm{E}(\mathbf{X})]\cdot[\mathbf{Y}-\mathrm{E}(\mathbf{Y})]^T\right] \\ &= \int_{-\infty}^{\infty}\cdots\int_{-\infty}^{\infty}[\mathbf{x}-\mathrm{E}(\mathbf{X})][\mathbf{y}-\mathrm{E}(\mathbf{Y})]^T\cdot f(\mathbf{x},\mathbf{y})\mathrm{d}x_1\mathrm{d}x_2\cdots\mathrm{d}x_n\mathrm{d}y_1\mathrm{d}y_2\cdots\mathrm{d}y_m\end{aligned} \tag{3.1.27}$$

where $f(\mathbf{x},\mathbf{y})$ is the joint p.d.f. of $\mathbf{X}$ and $\mathbf{Y}$. Obviously, $\mathbf{P}_{xy}$ is a $n\times m$ matrix, and $\mathbf{P}_{xy} = \mathbf{P}_{yx}^T$. The expectation of a function $g(\mathbf{X})$ of random vector $\mathbf{X}$ is defined as

$$\mathrm{E}[g(\mathbf{X})] = \int_{-\infty}^{\infty}\cdots\int_{-\infty}^{\infty}g(\mathbf{x})f(\mathbf{x})\mathrm{d}x_1\mathrm{d}x_2\cdots\mathrm{d}x_n \tag{3.1.28}$$

All of these are as usual.

§3.1.4, Conditional expectation and covariance of random vector

Giving the joint p.d.f. $f(\mathbf{x},\mathbf{y})$ of random vectors $\mathbf{X}$ and $\mathbf{Y}$, the conditional mean-value (expectation) of $\mathbf{X}$ under the condition $\mathbf{Y}=\mathbf{y}$ is defined as

$$\mathrm{E}(\mathbf{X}\mid\mathbf{Y}=\mathbf{y}) = \mathrm{E}(\mathbf{X}\mid\mathbf{y}) = \int_{-\infty}^{\infty}\cdots\int_{-\infty}^{\infty}\mathbf{x}f(\mathbf{x},\mathbf{y})\mathrm{d}x_1\mathrm{d}x_2\cdots\mathrm{d}x_n \tag{3.1.29}$$

and the corresponding conditional covariance matrix is defined as

$$\begin{aligned}\mathbf{P}_{x|y} &= \mathrm{E}\left[[\mathbf{X}-\mathrm{E}(\mathbf{X}\mid\mathbf{y})]\cdot[\mathbf{X}-\mathrm{E}(\mathbf{X}\mid\mathbf{y})]^T\right] \\ &= \int_{-\infty}^{\infty}\cdots\int_{-\infty}^{\infty}[\mathbf{x}-\mathrm{E}(\mathbf{X}\mid\mathbf{y})][\mathbf{x}-\mathrm{E}(\mathbf{X}\mid\mathbf{y})]^T\cdot f(\mathbf{x},\mathbf{y})\mathrm{d}x_1\mathrm{d}x_2\cdots\mathrm{d}x_n\end{aligned} \tag{3.1.30}$$

The conditional expectation has the behavior

$$\begin{aligned}&\mathrm{E}(\mathbf{A}\cdot\mathbf{X}\mid\mathbf{Y}=\mathbf{y}) = \mathbf{A}\cdot\mathrm{E}(\mathbf{X}\mid\mathbf{Y}=\mathbf{y}), \qquad \text{where } \mathbf{A} \text{ is a given matrix} \\ &\mathrm{E}(\mathbf{X}+\mathbf{Y}\mid\mathbf{Z}=\mathbf{z}) = \mathrm{E}(\mathbf{X}\mid\mathbf{Z}=\mathbf{z}) + \mathrm{E}(\mathbf{Y}\mid\mathbf{Z}=\mathbf{z}) \\ &\mathrm{E}_{\mathbf{y}_1}[\mathrm{E}(\mathbf{X}\mid\mathbf{Y}=\mathbf{y}_1)] = \mathrm{E}(\mathbf{X})\end{aligned} \tag{3.1.31}$$

where $\mathrm{E}_{\mathbf{y}_1}$ means taking the average with respect to $\mathbf{y}_1$ to the conditional ($\mathbf{Y}=\mathbf{y}_1$) expectation of $\mathrm{E}(\mathbf{X}\mid\mathbf{Y}=\mathbf{y}_1)$.

§3.1.5, Characteristic function of random variable

The description above is based on the p.d.f. $f(x)$. The ***characteristic function*** of random variable X is of same importance, and is defined as

$$\phi_X(s) = \mathrm{E}[\exp(\mathrm{i}sx)] = \int_{-\infty}^{\infty} \exp(\mathrm{i}sx) f(x) \mathrm{d}x \tag{3.1.32}$$

obviously, $\phi_X(0) = 1$. It is easily seen that the characteristic function is just the Fourier transform of the p.d.f. Conversely, if characteristic function $\phi_X(s)$ is given, then the p.d.f can be obtained by inverse Fourier transformation.

$$f(x) = (1/2\pi) \int_{-\infty}^{\infty} \exp(-\mathrm{i}sx) \phi_X(s) \mathrm{d}s \tag{3.1.33}$$

Integration by parts shows, that the k-th order moment of random variable can be obtained from the k-th order differential of the characteristic function

$$m_k = \mathrm{E}(X^k) = \mathrm{i}^{-k} \left[\mathrm{d}^k \phi_X(s) / \mathrm{d}s^k \right]_{s=0} \tag{3.1.34}$$

Expanding the characteristic function in power series around $s = 0$ gives

$$\phi_X(s) = 1 + \sum_{k=1}^{\infty} (\mathrm{i}s)^k m_k / k!$$

For a n-dimensional random vector $\mathbf{X} = \{X_1 \quad X_2 \quad \cdots \quad X_n\}^T$, its joint characteristic function can be defined as

$$\phi_{\mathbf{X}}(\mathbf{s}) = \mathrm{E}[\exp(\mathrm{i} \sum_{j=1}^{n} s_j x_j)] = \int_{-\infty}^{\infty} \cdots \int_{-\infty}^{\infty} \exp(\mathrm{i} \sum_{j=1}^{n} s_j x_j) f(\mathbf{x}) \mathrm{d}x_1 \cdots \mathrm{d}x_n \tag{3.1.35}$$

It is a multi-dimensional Fourier transform equation, and correspondingly the p.d.f. $f(\mathbf{x})$ can also be obtained via the inverse Fourier transform.

If all the components $X_i, i = 1, \cdots, n$ of $\mathbf{X}$ are mutually independent, the characteristic function can be expressed as

$$\phi_X(\mathbf{s}) = \prod_{j=1}^{n} \phi_{X_j}(s_j) \tag{3.1.36}$$

i.e. the product of characteristic functions of all the components.

§3.1.6, Normal distribution

This book mainly considers the subjects of vibration problems, linear system control and filter etc. The normal (Gauss) distribution is the main concern. Therefore the later discussion is only limited to the case of Gauss normal distribution. Let us begin with the one-dimensional normal distribution, for which the p.d.f. is

$$f(x) = \frac{1}{\sqrt{2\pi}\sigma} \exp[-\frac{(x-m)^2}{2\sigma^2}] \tag{3.1.37}$$

The shape of function can be seen figure 3.1. The two parameters m and σ, *i.e.* the expectation (average) value and standard deviation determine the function. The area under the curve within the interval $(m-\sigma, m+\sigma)$ equals 0.68, while the area within $(m-2\sigma, m+2\sigma)$ is 0.95. This explains that the probability of the sample value of random variable X departing $\pm 2\sigma$ to its average value m is about

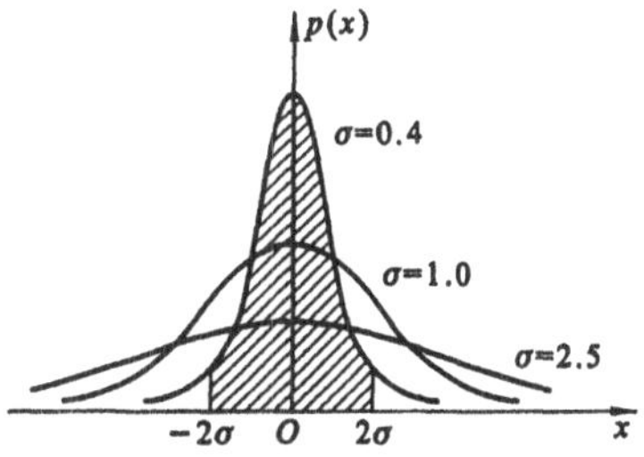

Figure 3.1, *The shape of normal distribution*

0.05.

It can be shown, that the p.d.f. of the summation of several normally distributed random variables is still normal. More important, there is the ***central limit theorem*** in theory of probability. Let $X_1, X_2, \cdots, X_n$ be a series of independent random variables with the same probability density function, then the distribution function of their average value $S_n = (X_1 + X_2 + \cdots + X_n)/n$ tends to normal distribution as $n \to \infty$. More precisely, if n experiment of X are taken from the same p.d.f. with mean value m and variance σ^2, the limiting distribution of $[n^{1/2}(S_n - m)]/\sigma$ as $n \to \infty$ is a normal distribution with zero-mean and variance one. The ***central limit theorem*** has also the version for the case of different p.d.f. functions, [58]. Practical experience verifies also that the average of a large number of independent random variables behaves as its distribution being nearly normal.

Assuming that a n-dimensional random vector $\mathbf{X} = \{X_1 \quad X_2 \quad \cdots \quad X_n\}^T$ has the joint Gauss p.d.f. as

$$f(\mathbf{x}) = \left[(2\pi)^n \cdot \det(\mathbf{P}_{xx})\right]^{-1/2} \exp\left[-(\mathbf{x} - \mathrm{E}(\mathbf{X}))^T \mathbf{P}_{xx}^{-1}(\mathbf{x} - \mathrm{E}(\mathbf{X}))/2\right] \tag{3.1.38}$$

where $\mathbf{P}_{xx}$ is the covariance matrix of $\mathbf{X}$

$$\mathbf{P}_{xx} = \mathrm{E}\left[(\mathbf{X} - \mathrm{E}(\mathbf{X}))(\mathbf{X} - \mathrm{E}(\mathbf{X}))^T\right] \tag{3.1.39}$$

and the corresponding characteristic function be

$$\begin{aligned} \phi_X(\mathbf{s}) &= \mathrm{E}\left[\exp(\mathrm{i}\mathbf{X}^T\mathbf{s})\right] = \int_{-\infty}^{\infty} \cdots \int_{-\infty}^{\infty} \exp(\mathrm{i}\mathbf{x}^T\mathbf{s}) f(\mathbf{x}) \mathrm{d}x_1 \mathrm{d}x_2 \cdots \mathrm{d}x_n \\ &= \exp\left(\mathrm{i}\mathbf{m}_X^T\mathbf{s} - \mathbf{s}^T\mathbf{P}_{xx}\mathbf{s}/2\right) \end{aligned} \tag{3.1.40}$$

where $\mathbf{m}_X = \mathrm{E}(\mathbf{X})$ is the expectation vector of $\mathbf{X}$, similarly $\mathbf{m}_Y = \mathrm{E}(\mathbf{Y})$. Conversely, based on the inverse Fourier transform

$$\begin{aligned} f(\mathbf{x}) &= (2\pi)^{-n} \int_{-\infty}^{\infty} \cdots \int_{-\infty}^{\infty} \exp(-\mathrm{i}\mathbf{s}^T\mathbf{x}) \phi_X(\mathbf{s}) \mathrm{d}s_1 \mathrm{d}s_2 \cdots \mathrm{d}s_n \\ &= \left[(2\pi)^n \cdot \det(\mathbf{P}_{xx})\right]^{-1/2} \exp\left[-[\mathbf{x} - \mathrm{E}(\mathbf{X})]^T \cdot \mathbf{P}_{xx}^{-1} \cdot [\mathbf{x} - \mathrm{E}(\mathbf{X})]/2\right] \end{aligned}$$

The expression (3.1.40) of the joint characteristic function has no matrix inversion, so that it is preferable using the joint characteristic function to give the Gauss distribution, and therefore it can be used even for the covariance matrix $\mathbf{P}_{xx}$ being semi-positive-definite (non-negative definite)..

The joint Gauss distribution for n-D and m-D random vectors $\mathbf{X}$ and $\mathbf{Y}$, respectively, is considered now. Using joint characteristic function gives

$$\phi_{xy}(\mathbf{s}, \mathbf{r}) = \exp\left[\mathrm{i}\begin{Bmatrix}\mathbf{m}_x \\ \mathbf{m}_y\end{Bmatrix}^T \begin{Bmatrix}\mathbf{m}_x \\ \mathbf{m}_y\end{Bmatrix} - \frac{1}{2}\begin{Bmatrix}\mathbf{s} \\ \mathbf{r}\end{Bmatrix}^T \mathbf{P}\begin{Bmatrix}\mathbf{s} \\ \mathbf{r}\end{Bmatrix}\right] \tag{3.1.41}$$

where $\mathbf{s},\mathbf{r}$ are n,m-D deterministic vectors respectively, whereas $\mathbf{P}$ is a $(n+m)\times(n+m)$ symmetric matrix, which can be partitioned as

$$\mathbf{P}=\begin{bmatrix}\mathbf{P}_{xx} & \mathbf{P}_{xy}\\ \mathbf{P}_{yx} & \mathbf{P}_{yy}\end{bmatrix},\quad \begin{matrix}\mathbf{P}_{xx}=\mathrm{E}\left[(\mathbf{X}-\mathbf{m}_x)(\mathbf{X}-\mathbf{m}_x)^T\right]\\ \mathbf{P}_{xy}=\mathrm{E}\left[(\mathbf{X}-\mathbf{m}_x)(\mathbf{Y}-\mathbf{m}_y)^T\right]\end{matrix} \tag{3.1.42}$$

and $$\mathbf{P}_{yy}=\mathrm{E}\left[(\mathbf{Y}-\mathbf{m}_y)(\mathbf{Y}-\mathbf{m}_y)^T\right],\quad \mathbf{P}_{yx}=\mathbf{P}_{xy}^T$$

When $\mathbf{P}$ is positive-definite, the ***partitioned matrix inversion*** gives

$$\mathbf{P}^{-1}=\begin{bmatrix}\mathbf{A} & \mathbf{B}\\ \mathbf{B}^T & \mathbf{C}\end{bmatrix}\begin{matrix}n\\ m\end{matrix},\quad \begin{matrix}\mathbf{A}=(\mathbf{P}_{xx}-\mathbf{P}_{xy}\mathbf{P}_{yy}^{-1}\mathbf{P}_{yx})^{-1}=\mathbf{P}_{xx}^{-1}+\mathbf{P}_{xx}^{-1}\mathbf{P}_{xy}\mathbf{C}\mathbf{P}_{yx}\mathbf{P}_{xx}^{-1}\\ \mathbf{C}=(\mathbf{P}_{yy}-\mathbf{P}_{yx}\mathbf{P}_{xx}^{-1}\mathbf{P}_{xy})^{-1}=\mathbf{P}_{yy}^{-1}+\mathbf{P}_{yy}^{-1}\mathbf{P}_{yx}\mathbf{A}\mathbf{P}_{xy}\mathbf{P}_{yy}^{-1}\end{matrix}$$

$$\mathbf{B}=-\mathbf{A}\mathbf{P}_{xy}\mathbf{P}_{yy}^{-1}=-\mathbf{P}_{xx}^{-1}\mathbf{P}_{xy}\mathbf{C}$$

and the joint Gauss probability density function is given as

$$f(\mathbf{x},\mathbf{y})=\begin{bmatrix}(2\pi)^{n+m}\\ \times\det(\mathbf{P})\end{bmatrix}^{-1/2}\exp\left(-\frac{1}{2}\begin{Bmatrix}\mathbf{x}-\mathbf{m}_x\\ \mathbf{y}-\mathbf{m}_y\end{Bmatrix}^T\begin{bmatrix}\mathbf{A} & \mathbf{B}\\ \mathbf{B}^T & \mathbf{C}\end{bmatrix}\begin{Bmatrix}\mathbf{x}-\mathbf{m}_x\\ \mathbf{y}-\mathbf{m}_y\end{Bmatrix}\right) \tag{3.1.43}$$

If only the p.d.f. of $\mathbf{Y}$ is of concern, using the characteristic function and performing the integration with respect to $\mathbf{x}$ gives

$$f(\mathbf{y})=\left[(2\pi)^m\det(\mathbf{P}_{yy})\right]^{-1/2}\exp\left(-(\mathbf{y}-\mathbf{m}_y)^T\mathbf{P}_{yy}^{-1}(\mathbf{y}-\mathbf{m}_y)/2\right)$$

$$\phi_y(\mathbf{r})=\exp\left(i\mathbf{m}_y^T\mathbf{r}-\mathbf{r}^T\mathbf{P}_{yy}\mathbf{r}/2\right)$$

The conditional Gauss p.d.f $f(\mathbf{x}\mid\mathbf{y})$ is considered now. Using the Bayes theorem of conditional density, after some derivations it gives

$$\begin{aligned}f(\mathbf{x}\mid\mathbf{y})&=f(\mathbf{x},\mathbf{y})/f(\mathbf{y})\\ &=\left[(2\pi)^n\det(\mathbf{Q})\right]^{-1/2}\exp\left(-(\mathbf{x}-\mathbf{m})^T\mathbf{Q}^{-1}(\mathbf{x}-\mathbf{m})/2\right)\end{aligned} \tag{3.1.44}$$

where $\mathbf{m},\mathbf{Q}$ are the conditional average and conditional covariance matrix, respectively, and still keeps Gauss distribution. The average $\mathbf{m}$ depends on the condition $\mathbf{Y}=\mathbf{y}$, but the matrix $\mathbf{Q}$ is independent on the condition $\mathbf{Y}=\mathbf{y}$

$$\mathbf{m}=\mathrm{E}(\mathbf{X}\mid\mathbf{y})=\mathrm{E}(\mathbf{X})+\mathbf{P}_{xy}\mathbf{P}_{yy}^{-1}(\mathbf{y}-\mathrm{E}(\mathbf{Y})) \tag{3.1.45}$$

$$\mathbf{Q}=\mathbf{P}_{X|y}=\mathbf{P}_{xx}-\mathbf{P}_{xy}\mathbf{P}_{yy}^{-1}\mathbf{P}_{yx} \tag{3.1.46}$$

The corresponding characteristic function is

$$\phi_{X|y}(\mathbf{s})=\exp(i\mathbf{m}^T\mathbf{s}-\mathbf{s}^T\mathbf{Q}\mathbf{s}/2) \tag{3.1.47}$$

It is seen from the above, that one feature of the Gauss distribution is that giving only the expectation vector and the covariance matrix, the p.d.f of the random vector has been determined, which brings great convenience for analysis.

§3.1.7, Linear transformation and combination of Gauss random vectors

The action of a Gauss random vector to a **linear system** corresponds to the system input, and the response of the system is considered the system output. The mapping from input to output through a linear system can be considered a linear transformation. The idea of linear transformation involves also linear combination.

It is important to point out that: ***a Gauss random vector input after an arbitrary linear transformation (linear combination) still keeps the output random vector being Gaussian distributed.***

The linear transformation is considered first. Let the n-D (n-dimensional) random Gauss vector $\mathbf{X}$ has the expectation value $\mathrm{E}[\mathbf{X}]$ and covariance matrix $\mathbf{P}_{xx}$. Let $\mathbf{A}$ be a given deterministic $m \times n$ matrix that the linear transformation from random vector $\mathbf{X}$ to the m-D random vector is $\mathbf{Y} = \mathbf{AX}$. It is to prove that $\mathbf{Y}$ is a Gauss distribution random vector, and the expectation of $\mathbf{Y}$ is $\mathrm{E}(\mathbf{Y}) = \mathbf{A}\mathrm{E}(\mathbf{X})$, and its covariance matrix is $\mathbf{P}_{yy} = \mathbf{AP}_{xx}\mathbf{A}^T$.

Proof: The characteristic function of $\mathbf{Y}$ can be derived as

$$\begin{aligned}\phi_y(\mathbf{r}) &= \mathrm{E}\left[\exp\left(\mathrm{i}\mathbf{Y}^T\mathbf{r}\right)\right] = \mathrm{E}\left[\exp\left(\mathrm{i}(\mathbf{AX})^T\mathbf{r}\right)\right] = \mathrm{E}\left[\exp\left(\mathrm{i}\mathbf{X}^T\mathbf{A}^T\mathbf{r}\right)\right] = \phi_x(\mathbf{A}^T\mathbf{r}) \\ &= \exp\left[\mathrm{i}\mathbf{m}_x^T\mathbf{A}^T\mathbf{r} - (\mathbf{A}^T\mathbf{r})^T\mathbf{P}_{xx}\mathbf{A}^T\mathbf{r}/2\right] = \exp\left[\mathrm{i}(\mathbf{Am}_x)^T\mathbf{r} - \mathbf{r}^T(\mathbf{AP}_{xx}\mathbf{A}^T)\mathbf{r}/2\right]\end{aligned}$$

therefore $\mathbf{m}_y = \mathbf{Am}_x, \mathbf{P}_{yy} = \mathbf{AP}_{xx}\mathbf{A}^T$. The characteristic function form determines that the random vector $\mathbf{Y}$ is Gaussian. ##

Next examine the effect of linear combination. Let $\mathbf{X}, \mathbf{Y}$ be the n,m-D Gauss random vectors, respectively, their combined vector $\mathbf{X}_a = \{\mathbf{X}^T, \mathbf{Y}^T\}^T$ is a $(n+m)$-D Gauss random vector. Let the linear combination of $\mathbf{X}, \mathbf{Y}$ be expressed as a p-dimensional random vector $\mathbf{Z} = \mathbf{AX} + \mathbf{BY}$, where $\mathbf{A}, \mathbf{B}$ are the deterministic $p \times n$ and $p \times m$ dimensional given matrices, respectively. It is to prove that $\mathbf{Z}$ is also a Gauss random vector. The proof can be as follows. Evidently, $\mathbf{Z}$ is linearly transformed from $\mathbf{X}_a$ with transformation matrix being $[\mathbf{A}, \mathbf{B}]$. Hence, the above proof determines that $\mathbf{Z}$ is a Gauss random vector, with its mean-value and covariance matrix be

$$\mathbf{m}_z = \mathbf{Am}_x + \mathbf{Bm}_y, \quad \mathbf{P}_{zz} = [\mathbf{A}, \mathbf{B}] \cdot \begin{bmatrix} \mathbf{P}_{xx} & \mathbf{P}_{xy} \\ \mathbf{P}_{yx} & \mathbf{P}_{yy} \end{bmatrix} \cdot \begin{bmatrix} \mathbf{A}^T \\ \mathbf{B}^T \end{bmatrix} \quad \text{##} \qquad (3.1.48)$$

Because, the p.d.f. of a Gauss distribution is determined by its mean-value and covariance matrix, so that if the mutual covariance matrix between two Gauss random vectors $\mathbf{X}, \mathbf{Y}$ is $\mathbf{P}_{xy} = \mathbf{0}$, then the two Gauss random vectors are mutually independent. This statement is valid only for Gauss distribution. For other p.d.f., this statement may be invalid.

§3.1.8, Least square method

Quite often, the probability theory is used to estimate a random object, for which the least square method is the most popular one. Estimating a dynamic object is usually termed as ***filtering***, which is much complicated than static object. Chapter-6 will give detail description for filtering problem. The least square estimation for a static object is given below.

The method of least square was by Gauss. The simplest problem is to estimate the value of one unknown object x. A static object implies that there has no interference from dynamic procedural noise. Expressed with a differential equation

$$\dot{x} = 0 , \qquad x(0) = \hat{x}$$

Suppose the object be measured q times and the results are y_i, $i = 1,2,\cdots,q$. Unavoidably there is interference of measurement noise v_i, $i = 1,2,\cdots,q$. Hence, the measurement equations are

$$y_i = \hat{x} + v_i , \qquad i = 1,2,\cdots,q$$

where the known values are y_i, $i = 1,2,\cdots,q$ and the value $\hat{x}$ is to determine. The real value x of the object cannot be determined, so $\hat{x}$ can only be the optimal estimation.

The criterion for determining $\hat{x}$ is to minimize the ***quadratic index***

$$J = \sum_{j=1}^{q} (y_j - \hat{x})^2 = \min , \quad \text{or} \quad \sum_{j=1}^{q} v_j^2 = \min$$

Carrying out the minimization with respect to $\hat{x}$, *i.e.* $\partial J / \partial \hat{x} = 0$, derives

$$\hat{x} = \sum_{j=1}^{q} y_j \Big/ q$$

Obviously, the index J is the sum of square of the errors (noises), so that it is termed as ***least square method***. Such index corresponds to compute a length in a q-dimensional space, so that it belongs to Euclidean metric. The above index is proposed on the assumption that all the measurements are equally important.

If among the q times of measurement, some measurements use ordinary instrument, however, the others use highly precise instruments. Therefore, the average should emphasize the results measured by the precise instruments, that the weight k_j (credibility) of the measurements from precise instruments must be higher. Therefore the index of least square should be updated as

$$J = \sum_{j=1}^{q} k_j (y_j - \hat{x})^2 = \min , \qquad \hat{x} = \sum k_j y_j \Big/ \sum k_j$$

Larger weight implies much precise or smaller deviation, *i.e.* smaller variance r_j^2. Using variance instead of weight, the index is given as

$$J = \sum_{j=1}^{q} \left[(y_j - \hat{x})^2 \big/ r_j^2 \right] = \min , \quad \textit{i.e.} \quad k_j = r_j^{-2}, \ j = 1,\cdots,q$$

or written in matrix/vector formulation

$$\mathbf{v}^T \mathbf{R}^{-1} \mathbf{v} = \min , \qquad \text{where} \ \ \mathbf{v} = \{v_1, v_2, \cdots, v_q\}^T , \quad \mathbf{R} = \mathrm{diag}[r_1^2, r_2^2 \cdots, r_q^2]$$

Least square method is the simplest form of filtering. The diagonal matrix $\mathbf{R}$ implies that the q times of measurements are independent on each other.

So far the least square method gives a static estimation. It is closely related to static structural mechanics. Let us propose a model in applied mechanics to compare with the least square equations. Suppose there is a point on a one-dimensional axis x, its position $\hat{x}$ is to be estimated. Each measurement y_j can be interpreted as that: there is a spring connecting the point $\hat{x}$ and y_j, the spring stiffness is k_j and the neutral point of the spring is just at y_j, figure 3.2.

After q-times of measurement, which mean number q springs connected to $\hat{x}$, it requires to find the equilibrium point $\hat{x}$. Obviously, $\hat{x}$ is the balance point of all the forces of springs, which can be solved by the ***minimum potential energy variational principle***, *i.e.*

$$U(x)=\sum_{j=1}^{q}k_j(y_j-x)^2\,,\quad U(\hat{x})=\min$$

to find the equilibrium point $\hat{x}$. Potential energy is right away the index J. Let

$$\partial U(x)/\partial x=0\,,\quad \text{it derives}\ \sum_{j=1}^{q}k_j(y_j-\hat{x})=0\,,\ \hat{x}=\sum k_j y_j/\sum k_j \tag{3.1.49}$$

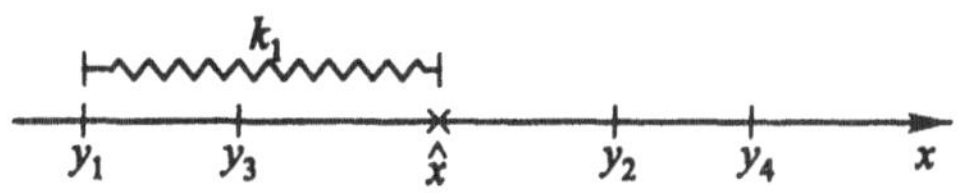

Figure 3.2, *The structural mechanics model for least square method.*

which gives the same result as least square. Therefore weight is just the spring constant k_j, also the inverse of variance r_j^{-2}. Therefore ***least square method corresponds to the minimum potential energy variational principle in applied mechanics***.

Least square method is a sort of estimation, except to find its expectation value $\hat{x}$, the problem of finding the variance of $\hat{x}$ is of concern. Note, $\hat{x}$ is the average of all measurements y_j, however, y_j (before measurement) are random variables in nature, so that its average $\hat{x}$ has also random nature, thus to find its variance is necessary. As mentioned above, ***variance is really the inverse of stiffness, i.e. the flexibility.*** The stiffness of $\hat{x}$ is $\sum k_j$, and the potential energy $U(x)$ is a quadratic function of variables x. The addition of stiffness implies that these springs are parallel. The variance of $\hat{x}$ corresponds to the ***flexibility***, *i.e.*

$$1/\sum k_j=1/\sum r_j^{-2} \tag{3.1.50}$$

The correspondence between variance and the flexibility in structural mechanics is quite useful.

The problem above is for the estimation of a single unknown $\hat{x}$, but with multiple times of measurement. Below, the problem of estimation of multiple, n, unknowns $\mathbf{x}=\{x_1,x_2,\cdots,x_n\}^T$ requires to be investigated. Each measurement is a linear combination of $\mathbf{x}$, totally q-measurements. The measurement equation is

$$\mathbf{y}=\mathbf{C}\mathbf{x}+\mathbf{v} \tag{3.1.51}$$

where $\mathbf{C}$ is a $q\times n$ given matrix, certainly $q>n$ and the rank of matrix $\mathbf{C}$ is n. To find the estimate vector $\hat{\mathbf{x}}$ of $\mathbf{x}$, select the index functional as

$$J=U(\mathbf{x})=\mathbf{v}^T\mathbf{V}^{-1}\mathbf{v}/2=(\mathbf{y}-\mathbf{C}\mathbf{x})^T\mathbf{V}^{-1}(\mathbf{y}-\mathbf{C}\mathbf{x})/2 \tag{3.1.52}$$

where the noise vector $\mathbf{v}$ is zero-mean with variance matrix

$$\mathbf{V} = \mathrm{E}[\mathbf{v}\mathbf{v}^T] \tag{3.1.53}$$

a given symmetric positive definite $q \times q$ matrix. Optimal estimation $\hat{\mathbf{x}}$ should minimize the index functional, *i.e.*

$$\partial J / \partial \mathbf{x} = \mathbf{0}, \qquad \text{which derives} \quad \mathbf{C}^T\mathbf{V}^{-1}\mathbf{C}\hat{\mathbf{x}} - \mathbf{C}^T\mathbf{V}^{-1}\mathbf{y} = \mathbf{0}$$

It solves

$$\hat{\mathbf{x}} = (\mathbf{C}^T\mathbf{V}^{-1}\mathbf{C})^{-1}\mathbf{C}^T\mathbf{V}^{-1}\mathbf{y} \tag{3.1.54}$$

which gives the constraint free multi-dimensional estimation formula. The estimation obtained from variational principle is naturally unbiased. Further, the variance of estimation $\hat{\mathbf{x}}$ is required. Note that the potential energy $U(\mathbf{x})$ functional is quadratic, and the ***coefficient matrix (the stiffness matrix)*** of the quadratic term of vector $\mathbf{x}$ is $\mathbf{C}^T\mathbf{V}^{-1}\mathbf{C}$, and correspondingly the external driven force is $\mathbf{C}^T\mathbf{V}^{-1}\mathbf{y}$.

The above considerations are from structural mechanics. It is necessary to verify directly that, the variance matrix of the estimated vector $\hat{\mathbf{x}}$ is $(\mathbf{C}^T\mathbf{V}^{-1}\mathbf{C})^{-1}$, which is just the inverse of stiffness matrix, *i.e.* the flexibility matrix. The verification is given as follows. From the definition of variance matrix

$$\mathbf{P} = \mathrm{E}[(\mathbf{x} - \hat{\mathbf{x}})(\mathbf{x} - \hat{\mathbf{x}})^T] \tag{3.1.55}$$

According to (3.1.54),

$$\mathbf{x} - \hat{\mathbf{x}} = (\mathbf{C}^T\mathbf{V}^{-1}\mathbf{C})^{-1}\mathbf{C}^T\mathbf{V}^{-1}(\mathbf{C}\mathbf{x} - \mathbf{y}) = -(\mathbf{C}^T\mathbf{V}^{-1}\mathbf{C})^{-1}\mathbf{C}^T\mathbf{V}^{-1}\mathbf{v}$$

so $\quad \mathbf{P} = (\mathbf{C}^T\mathbf{V}^{-1}\mathbf{C})^{-1}\mathbf{C}^T\mathbf{V}^{-1}\left(\mathrm{E}[\mathbf{v}\mathbf{v}^T]\right)\mathbf{V}^{-1}\mathbf{C}(\mathbf{C}^T\mathbf{V}^{-1}\mathbf{C})^{-1} = (\mathbf{C}^T\mathbf{V}^{-1}\mathbf{C})^{-1}$ ##

It is seen again that, the ***variance matrix is the inverse of stiffness matrix***, *i.e.* the ***flexibility matrix***.

The case of $\mathbf{x} = \{x_1, x_2, \cdots, x_n\}^T$ being under constraints is to be considered further. Suppose, there are g linear constraint equations, given as

$$\mathbf{G}\mathbf{x} = \mathbf{w}, \qquad \mathrm{E}\left(\mathbf{w}\mathbf{w}^T\right) = \mathbf{W} \tag{3.1.56}$$

where $\mathbf{G}$ is a $g \times n$ given constraint matrix. Without loss of generality, $\mathbf{G}$ can be considered of full rank and $g < n$. The vector $\mathbf{w}$ is also a zero-mean random vector with its variance $\mathbf{W}$ be a $g \times g$ symmetric positive-definite matrix. After q-times of measurement, the measurement equation is still (3.1.51). The g constraints in combination with the q measurements span the complete n-dimensional space, *i.e.* the rank of the $(g+q) \times n$ matrix

$$\mathbf{C}' = \begin{bmatrix} \mathbf{C} \\ \mathbf{G} \end{bmatrix} \tag{3.1.57}$$

equals n. It is required to estimate the vector $\mathbf{x}$ and also to find the variance matrix $\mathbf{P}$ of the optimal estimate vector $\hat{\mathbf{x}}$.

The solution is as follows. The fundamental part of the index functional can still be as given in (3.1.52). Minimization is necessary, but the constraint noise should also join the index functional, therefore the extended index functional is given as

$$J_A = (\mathbf{Gx})^T \mathbf{W}^{-1}\mathbf{Gx}/2 + (\mathbf{y}-\mathbf{Cx})^T \mathbf{V}^{-1}(\mathbf{y}-\mathbf{Cx})/2 \qquad (3.1.58)$$

It corresponds still to an unconstrained minimization problem of $\mathbf{x}$, which derives

$$\partial J_A / \partial \mathbf{x} = \mathbf{0}, \quad i.e. \quad (\mathbf{C}^T\mathbf{V}^{-1}\mathbf{C} + \mathbf{G}^T\mathbf{W}^{-1}\mathbf{G})\hat{\mathbf{x}} - \mathbf{C}^T\mathbf{V}^{-1}\mathbf{y} = \mathbf{0}$$

where the $n \times n$ matrix $(\mathbf{C}^T\mathbf{V}^{-1}\mathbf{C} + \mathbf{G}^T\mathbf{W}^{-1}\mathbf{G})$ ensures symmetry and positive-definite because the matrix $\mathbf{C}'$ is of full rank. Therefore

$$\hat{\mathbf{x}} = (\mathbf{C}^T\mathbf{V}^{-1}\mathbf{C} + \mathbf{G}^T\mathbf{W}^{-1}\mathbf{G})^{-1}\mathbf{C}^T\mathbf{V}^{-1}\mathbf{y}$$

as before, $\mathbf{C}^T\mathbf{V}^{-1}\mathbf{y}$ is considered the driven external force. Hence

$$\mathbf{P} = (\mathbf{C}^T\mathbf{V}^{-1}\mathbf{C} + \mathbf{G}^T\mathbf{W}^{-1}\mathbf{G})^{-1} \qquad (3.1.59)$$

is again a flexibility matrix. It is suggested to compare with the flexibility matrix after (3.1.55). The matrix $\mathbf{P}$ is certainly symmetric and also ensures non-negative. The idea of flexibility matrix is from structural mechanics, so that one should verify directly the variance of $\hat{\mathbf{x}}$

$$\mathbf{P}_e = \mathrm{E}[(\mathbf{x}-\hat{\mathbf{x}})(\mathbf{x}-\hat{\mathbf{x}})^T]$$

It derives to the same conclusion of $\mathbf{P}_e = \mathbf{P}$. The detail is omitted.

Therefore, one can find the flexibility matrix of the corresponding structural mechanics problem instead of finding the covariance matrix of the system, this analogy method brings much convenience, and will be used in solving the filtering and smoothing problem of linear system, see chapter 6.

§3.2, Preliminary of stochastic process

A process implies that it varies with time. Analysis of stochastic process can be regarded as dynamic probability. Generally speaking, an assembly of random functions of a continuous coordinate is a stochastic process [60,61]. Usually the continuous coordinate is the time t, however in other cases it can also be a space coordinate.

The stochastic processes can be classified into two classes of discrete time and/or continuous time that such classification is from the evaluation time. The classification can also be distinguished from the meaning of function value. Usually the function is continuously evaluated such as force, displacement, current, pressure, temperature etc.; however, some stochastic process must evaluate at discrete time, and is termed as digital signal. In computer processing it appears quite frequently, such as the gray level of graphs usually has 256 levels, money account has a fundamental unit, etc. Here only the continuous-time stochastic process is considered.

Figure 3.3 plots a sample of a stochastic process, which seems no more conclusions can be drawn from it, but only a very complicated sample. If test again, it will give another sampling result. This is an example of a stochastic process. Such problem can only be analyzed by statistical approach.

A stochastic process $X(t)$, can be described with the joint probability distribution of random variables (function values) at different time instances, such as using n time points

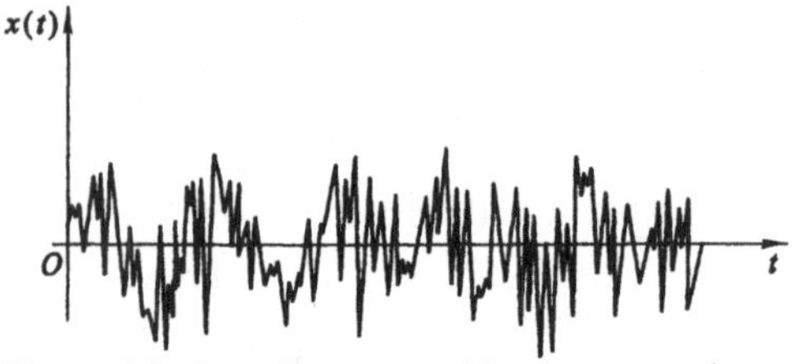

Figure 3.3, *A stochastic variable changes with time*

$$F(x_1,t_1;x_2,t_2;\cdots;x_n,t_n)=\Pr(X(t_1)\le x_1, X(t_2)\le x_2,\cdots,X(t_n)\le x_n) \tag{3.2.1}$$

The number of time points can be $n=1,2,\cdots$. The joint distribution function satisfies two conditions below:

a) The symmetry condition. For arbitrary permutation of

From $(1,2,\cdots,n)$ to $(j_1,j_2,\cdots,j_n)$,

$$F(x_{j_1},t_{j_1};x_{j_2},t_{j_2};\cdots;x_{j_n},t_{j_n})=F(x_1,t_1;x_2,t_2;\cdots;x_n,t_n)$$

b) The consistency condition. For arbitrary $m<n$,

$$F(x_1,t_1;x_2,t_2;\cdots;x_m,t_m;\infty,t_{m+1};\cdots;\infty,t_n)=F(x_1,t_1;x_2,t_2;\cdots;x_m,t_m)$$

The distribution function is exact mathematically, however it is difficult to supply for applications. In applied mechanics and/or linear control theory applications, use is made of the Gauss distributed stochastic processes for a majority of cases. Based on the theorem given in section 3.1.7 that after a linear transformation, the behavior of normal distribution keeps unchanged for the response. So that determining only the mean values and second order moment or variance matrix of the system responses will be enough for probability analysis, which greatly simplifies system analysis.

For a single component Gauss distribution stochastic process $X(t)$, give n time instants $t_1,t_2,\cdots,t_n$, there are n random variables $X(t_1),\cdots,X(t_n)$ and correspondingly a n-dimensional Gauss distribution. Let the mean value and covariance be denoted as

$$\mathrm{E}[X(t_i)]=m(t_i),\quad \mathrm{E}[(X(t_i)-m(t_i))(X(t_j)-m(t_j))]=p(t_i,t_j),\quad i,j=1,\cdots,n$$

Compose the mean values as a vector $\mathbf{m}$ and the covariance as a matrix $\mathbf{P}$, then the Gauss density function is

$$f(x_1,x_2,\cdots;x_n;t_1,t_2;\cdots,t_n)=\left[(2\pi)^n\cdot\det(\mathbf{P})\right]^{-1/2}\exp\left[-(\mathbf{x}-\mathbf{m})^T\mathbf{P}^{-1}(\mathbf{x}-\mathbf{m})/2\right]$$

Using characteristic function

$$\phi_X(\mathbf{s};t_1,t_2,\cdots,t_n)=\exp\left(\mathrm{i}\mathbf{m}^T\mathbf{s}-\mathbf{s}^T\mathbf{P}\,\mathbf{s}/2\right)$$

is more convenient. For a single component stochastic process, it has been so complicated, therefore the selection of two time instants $n=2$ in application is usually adopted.

§3.2.1, Stationary and non-stationary stochastic process

Definition: A stochastic process $X(t)$, if its statistical behavior is independent on the time origin, then $X(t)$ is a stationary stochastic process. On the contrary, if the statistical behavior depends explicitly on the time origin then $X(t)$ is a non-stationary stochastic function.

For a strictly stationary stochastic process the probability distribution should keep unchanged under time coordinate shifting, *i.e.*

$$F(x_1,t_1;x_2,t_2;\cdots;x_n,t_n) = F(x_1,t_1+\tau;x_2,t_2+\tau;\cdots;x_n,t_n+\tau) \tag{3.2.2}$$

where τ is an arbitrary time shift. It implies that the probability distribution depends only on the time difference but not relates to the time origin. Based on equation (3.2.1), the first order probability density function and the expectation can be derived as

$$p(x,t) = p(x,0) = p(x), \qquad \mathrm{E}[x(t)] = m = \mathrm{Const} \tag{3.2.3}$$

For a quadratic correlated function, the probability density function is

$$p(x_1,t_1;x_2,t_2) = p(x_1,x_2,\tau), \quad \tau = t_2 - t_1 \tag{3.2.4}$$

That the quadratic correlated moment $\mu_2(\tau) = \mathrm{E}[X(t)\cdot X(t+\tau)]$ is only a function of time difference τ and the covariance is also a function of τ

$$\mathrm{var}[X(t),X(t+\tau)] = \mathrm{E}[(X(t)-m)(X(t+\tau)-m)] = \mu_2(\tau) - m^2 = r(\tau) \tag{3.2.5}$$

The stationary condition (3.2.2) and (3.2.4) is very strict. The definition of ***wide sense*** stationary stochastic process requires only the mean value be a constant and the covariance function be a function of τ, but disregards the distribution function condition (3.2.2). Wide sense stationary process is also termed as ***weak*** stationary or quadratic stationary process. For Gauss distribution process, its density function is completely determined by its mean value and covariance function. If the covariance is given as (3.2.5) and the mean value m is a constant then the process is also a strictly stationary process. Obviously, we have $r(\tau) = r(-\tau)$.

Further loosen the condition, if we do not require the mean value m being a constant, but only require the covariance function having the form of (3.2.5), then the process is termed as ***covariance stationary***. The disturbance vector $\mathbf{w}$ in LQG optimal control theory is assumed a zero-mean white noise. However, in H_∞ robust control theory, $\mathbf{w}$ is regarded as a covariance stationary white noise, whose mean-value is a non-zero function to be determined. This is the contrast for both the theories.

These definitions can similarly be extended to discrete-time stochastic process.

§3.2.2, Ergodic stationary process

In probability theory, the definition of expectation is the mean value with respect to a large number of samples (results of experiments) termed as ***assembly***. In short, the mean is ***assembly average.*** For stochastic process, it implies that the large number of experiments is at the same time instant, it is quite difficult to do such amount of experiments. But from the assumption of stationary process, a legitimate assumption is that a long time duration sample can be cut off to become a large number of shorter time intervals of same length, then taking average value with respect to these time intervals. In short, use ***time average*** instead of assembly average. Such assumption is based on the consideration, that for a stationary process, the statistical behaviors of any time interval are the same, hence the time intervals cut off from far depart time can be regarded as another sample in the assembly. Such consideration implies an assumption, that for an arbitrarily given state x_g and a given distance ε, a stochastic process will approach to the ε neighbor of the given state x_g, as the time is long enough. This is the ***Ergodic*** assumption.

The definition of ***ergodic assumption*** is: Let $f(X)$ be a given function of a stochastic process $X(t)$, if the average given below is valid with probability 1 (*i.e.* almost sure, a.s.).

Stochastic sequence (discrete time): $\lim_{k\to\infty}\frac{1}{k+1}\sum_{j=0}^{k} f(X_j) = \mathrm{E}[f(X_j)]$ (3.2.6)

Stochastic process: $\lim_{T\to\infty}\int_{-T}^{T} f(X(t))\mathrm{d}t\big/2T = \mathrm{E}[f(X(t))]$ (3.2.7)

Then the stationary stochastic process is ergodic.

An ergodic process must be stationary, but stationary process is unnecessarily ergodic. In applications, the stochastic processes are frequently assumed stationary and also ergodic. According to experience such assumption is almost surely valid. Otherwise, the problem becomes complicated very much, very hard to deal with.

§3.3, Quadratic moment stochastic process (regular process)

Definition: a stochastic process $X(t)$ if

$$\mathrm{E}[(X(t))^2] < \infty \tag{3.3.1}$$

then $X(t)$ is a quadratic moment process, or regular process, [60].

For the solution of stochastic differential equation, the limiting operations, such as continuous, differentiation, integration etc. should be established for stochastic process. Therefore the quadratic moment process is introduced to establish the ***mean square calculus*** for stochastic processes. For deterministic functions or processes the calculus is to a single function. However, for stochastic processes its calculus should be to an assembly.

First, a quadratic moment process exist mean-value $m(t)$ and covariance $v(t,\tau)$ functions

$$m(t) = \mathrm{E}[X(t)], \quad v(t,\tau) = \mathrm{E}[(X(t)-m(t))\times(X(\tau)-m(\tau))]$$

Based on the Schwarz inequality the statement can be proved readily.

Taking limit is the foundation of differentiation and integration. Let us begin with ***mean square limit***. Let $X(t)$ be a quadratic moment stochastic process and t_0 be a given point, for a random variable X_0 such that

$$E(|X(t)-X_0|^2) \to 0, \quad \text{as } t \to t_0 \tag{3.3.2}$$

then $X(t)$ converges to the random variable X_0 as $t \to t_0$ in the sense of ***mean***

square limit

$$X(t) \to X_0(m.s.), \quad \text{as} \quad t \to t_0 \tag{3.3.3}$$

Taking mean value to both sides of the above equation gives

$$\lim_{t \to t_0} \mathrm{E}[X(t)] = \mathrm{E}\left[\lim_{t \to t_0} X(t)\right] = m(t_0) \tag{3.3.4}$$

which implies that for a quadratic moment process taking mean-value operation can permute with the operation of mean square limit.

The theorem below says that the covariance function converges to a finite value is the necessary and sufficient condition of mean square convergence.

The ***mean square convergence*** theorem: Let $X(t)$ be a quadratic moment stochastic process, if its auto-correlation function $r(t,\tau) = E[X(t) \cdot X(\tau)]$ is finite as $t,\tau \to t_0$, then (3.3.3) is valid. On the contrary if (3.3.3) is valid, then $r(t,\tau) = E[X(t) \cdot X(\tau)]$ is finite at $t,\tau \to t_0$, and the limit is $r(t,\tau) \to \mathrm{E}[X_0^2]$.

Proof: Suppose (3.3.3) is valid, it is easy to verify the equality

$$\begin{aligned}&\mathrm{E}[(X(t)-X_0)\cdot(X(\tau)-X_0)] + \mathrm{E}[(X(t)-X_0)\cdot X_0] + \mathrm{E}[(X(\tau)-X_0)\cdot X_0]\\ &= \mathrm{E}[X(t)\cdot X(\tau)] - \mathrm{E}[X_0 \cdot X_0]\end{aligned}$$

According to the Schwarz inequality derives

$$\{\mathrm{E}[(X(t)-X_0)\cdot(X(\tau)-X_0)]\}^2 \le \mathrm{E}[(X(t)-X_0)^2] \times \mathrm{E}[(X(\tau)-X_0)^2] \to 0$$

$$\{\mathrm{E}[(X(t)-X_0)X_0]\}^2 \le \mathrm{E}[(X(t)-X_0)^2] \times \mathrm{E}[X_0^2] \to 0$$

$$\{\mathrm{E}[(X_0(X(\tau)-X_0)]\}^2 \le \mathrm{E}[X_0^2] \times \mathrm{E}[(X(\tau)-X_0)^2] \to 0$$

hence as $t,\tau \to t_0$, $r(t,\tau) = \mathrm{E}[X(t)\cdot X(\tau)] \to E[X_0^2]$.

On the contrary, if $r(t,\tau) \to \gamma < \infty$ as $t,\tau \to t_0$, then

$$\begin{aligned}&\mathrm{E}[(X(t)-X(\tau))^2] = \mathrm{E}[X(t)X(t)] - \mathrm{E}[2X(t)X(\tau)] + \mathrm{E}[X(\tau)X(\tau)]\\ &= r(t,t) - 2r(t,\tau) + r(\tau,\tau) \to \gamma - 2\gamma + \gamma = 0, \quad \text{as } t,\tau \to t_0\end{aligned}$$

The mean square convergence is reached, so that $X(t) \to X_0$ as $t \to t_0$, where X_0 is a random variable, and (3.3.3) is valid. ##

§3.3.1, Continuity and differentiability of a regular stochastic process

The **mean square continuity** can be defined as

$$X(t+h) \to X(t)(m.s.); \quad \text{or} \quad \mathrm{E}[(X(t+h)-X(t))^2] \to 0, \quad \text{as } h \to 0 \tag{3.3.5}$$

Using the mean square convergence theorem derives readily the **mean square continuous theorem:** ***the necessary and sufficient condition for a regular stochastic process*** $X(t)$ ***being continuous at*** t ***is that its auto-correlation function*** $r(t,t+h)$ ***continuous as*** $h \to 0$.

The **mean square differentiation** is defined as

$$[X(t+h)-X(t)]/h \to \dot{X}(t)\,(m.s.), \quad \text{as } h \to 0 \tag{3.3.6}$$

The **mean square differential theorem** reads: ***the necessary and sufficient condition for a regular stochastic process*** $X(t)$ ***to be mean square differentiable***

at t is, its auto-correlation function $r(t,\tau)$ being generalized quadratic differentiable and bounded at (t,t).

Proof: Fixing t, then $Y(h)=[X(t+h)-X(t)]/h$ becomes a stochastic process of h. Let $h \to 0$ then according to the mean square convergence theorem, if the limit of

$$E[Y_h Y_{h'}]=[r(t+h,t+h')-r(t+h,t)-r(t,t+h')+r(t,t)]/hh'$$
$$\to \partial^2 r(t,\tau=t)/\partial t \partial \tau$$

exists as $h,h' \to 0$, then the mean square differential (3.3.6) exists. The limit of the above function is just the generalized quadratic differential of $r(t,\tau)$ at the point (t,t). ##

If the mean square differential of $X(t)$ exists everywhere in the interval (t_0,t_f), then

$$\mathrm{d}m(t)/\mathrm{d}t = \mathrm{dE}[X(t)]/\mathrm{d}t = \mathrm{E}[\dot{X}(t)] \tag{3.3.7}$$

It implies that the operation of taking expectation can permute with the mean square differential operation, if the auto-correlation function $r(t,\tau)$ and the differentials

$$\mathrm{E}[\dot{X}(t)\cdot X(\tau)]=\partial r(t,\tau)/\partial t, \quad \mathrm{E}[X(t)\cdot \dot{X}(\tau)]=\partial r(t,\tau)/\partial \tau$$
$$\mathrm{E}[\dot{X}(t)\cdot \dot{X}(\tau)]=\partial^2 r(t,\tau)/\partial t\, \partial \tau, \quad \text{for } t,\tau \in (t_0,t_f) \tag{3.3.8}$$

exist. Note that the auto-correlation function relates the covariance as

$$r(t,\tau)=v(t,\tau)+m(t)m(\tau)$$

so that the differential relations among them are obvious.

The derivations above have not assumed the stochastic process $X(t)$ being stationary. For wide sense stationary process, the covariance function holds

$$v(t,\tau)=v(t-\tau)=v(\eta), \qquad \eta=t-\tau$$
$$\partial^2 v(t,\tau)/\partial t\, \partial \tau = \partial^2 v(t-\tau)/\partial t\, \partial \tau = -\mathrm{d}^2 v(\eta)/\mathrm{d}\eta^2 \tag{3.3.9}$$

The differential of a wide sense stationary stochastic process is again a wide sense stationary stochastic process. In applications, the wide sense stationary stochastic process appears frequently.

§3.3.2, Mean square integration

The definition of integration of a stochastic process $X(t)$ is similar to the Riemann integration. Let

$$I=\int_{t_0}^{t_f} X(t)\mathrm{d}t \tag{3.3.10}$$

I is also a random variable. It can be proved that for a regular stochastic process $X(t)$ with auto-correlation function $r(t,\tau)$, the necessary and sufficient condition of Riemann integrability of $X(t)$ in the interval (t_0,t_f) is the existence of the double integration

$$\int_{t_0}^{t_f}\int_{t_0}^{t_f} r(t,\tau)\mathrm{d}t\mathrm{d}\tau = \mathrm{E}\left[\int_{t_0}^{t_f} X(t)\mathrm{d}t \int_{t_0}^{t_f} X(\tau)\mathrm{d}\tau\right] = \mathrm{E}[I_t I_\tau]$$

The proof is neglected, [60]. The integration and average operations can be permuted too

$$\mathrm{E}\int_{t_0}^{t_f} X(t)\mathrm{d}t = \int_{t_0}^{t_f} \mathrm{E}X(t)\mathrm{d}t = \int_{t_0}^{t_f} m(t)\mathrm{d}t$$

The usual operations for integration are still valid for stochastic integration, such as

$$\int_{t_0}^{t_f} [aX_1(t) + bX_2(t)]\mathrm{d}t = a\int_{t_0}^{t_f} [X_1(t)]\mathrm{d}t + b\int_{t_0}^{t_f} [X_2(t)]\mathrm{d}t$$

If $\quad Y(t) = \int_a^t X(s)\mathrm{d}s \quad$ then $\quad \dot{Y}(t) = X(t) \quad$ (3.3.11)

However $Y(t)$ is no longer a stationary stochastic process, even if $X(t)$ is stationary.

If a stochastic process $X(t)$ is mean square integrable, and a deterministic function $f(t,s)$ is continuously integrable, then the process

$$Y(t) = \int_a^t f(t,s)X(s)\mathrm{d}s$$

is mean square differentiable, given as

$$\dot{Y}(t) = \int_a^t (\partial f(t,s)/\partial t)\cdot X(s)\mathrm{d}s + f(t,t)X(t) \tag{3.3.12}$$

which is the same as the usual convolution differential. The integration by part formula is also valid

$$\int_a^t f(t,s)\dot{X}(s)\mathrm{d}s = [f(t,s)X(s)]_a^t - \int_a^t [\partial f(t,s)/\partial s]X(s)\mathrm{d}s \tag{3.3.13}$$

The above description is for only one stochastic process $X(t)$. The applications later will be for vector stochastic processes. However, if all its component stochastic processes are regular and mean square continuous, then the vector stochastic process is also mean square continuous. The differentiability and integrability can also be extended similarly.

§3.4, Normal stochastic process

Normal distribution is quadratic integrable, hence normal distributed stochastic processes are regular. Gauss stochastic process is the most frequently used stochastic process in applied mechanics and control theory. Previous equations for single stochastic process can be extended to n-dimensional Gauss stochastic process $\mathbf{X}(t)$. Selecting m time-points $t_1,\cdots,t_m$, if the joint probability distribution of the n-dimensional stochastic vectors $\mathbf{X}(t_1),\cdots,\mathbf{X}(t_m)$ is Gaussian, then $\mathbf{X}(t)$ is a Gauss stochastic vector process. The joint characteristic function is

$$\phi_{\mathbf{X}}(\mathbf{s}_1,\cdots,\mathbf{s}_m) = \exp\left[\mathrm{i}\sum_{j=1}^{m}\mathbf{m}_j^T\mathbf{s}_j - \sum_{i=1}^{m}\sum_{j=1}^{m}\mathbf{s}_i^T\mathbf{P}_{ij}\mathbf{s}_j/2\right], \quad \mathbf{m}_j = \mathrm{E}[\mathbf{X}(t_j)] \tag{3.4.1}$$

where $\mathbf{s}_i, i = 1,\cdots,m$ are all n-dimensional vectors and

$$\mathbf{P}_{ij} = \mathrm{E}[(\mathbf{X}(t_i) - \mathbf{m}_i)(\mathbf{X}(t_j) - \mathbf{m}_j)^T] \tag{3.4.2}$$

Hence the Gauss distribution is completely determined by the m mean-value of n-dimensional vectors $\mathbf{m}_i$ and m^2 covariance matrices $\mathbf{P}_{ij}(i,j = 1,\cdots,m)$ of dimension $n\times n$. Therefore, the Gauss distribution is completely determined. If, for two arbitrary time instants $t, t+\tau$ the ***auto-correlation matrix function***

$$\mathbf{R}(t,t+\tau)=\mathrm{E}\left[\mathbf{X}(t)\mathbf{X}^T(t+\tau)\right]=\mathbf{m}(t)\mathbf{m}^T(t+\tau)+\mathbf{P}(t,t+\tau) \tag{3.4.3}$$

is determined. Here, $\mathbf{P}(t,t+\tau)$ is the ***auto-covariance matrix***.

Both the auto-correlation function and the auto-covariance function of a stationary stochastic process depend only on the time difference τ, $\mathbf{R}(t,t+\tau)=\mathbf{R}(\tau)$ and $\mathbf{P}(t,t+\tau)=\mathbf{P}(\tau)$. Obviously, for stationary process

$$\mathbf{R}(\tau)=\mathbf{R}^T(-\tau) \tag{3.4.4}$$

A ***Gauss stochastic process remains Gaussian under a linear operation***. Such as, a Gauss process after mean square differentiation, mean square integration etc., the results are still Gauss distributed, which is a very useful behavior. There is also the ***central limit theorem***, which implies that the combination of a large number of random factors will derive the result toward Gauss distributed. Hence, the Gauss distributed stochastic process is extensively used in applied mechanics and control theory. In this book, the stochastic processes are almost all Gaussian distributed.

The above equations are derived for real stochastic processes. For complex valued stochastic processes, the corresponding equations are

$$\mathbf{R}(t,t+\tau)=\mathrm{E}\left[\overline{\mathbf{X}}(t)\mathbf{X}^T(t+\tau)\right]=\overline{\mathbf{m}}(t)\mathbf{m}^T(t+\tau)+\mathbf{P}(t,t+\tau) \tag{3.4.3'}$$

$$\mathbf{R}(\tau)=\overline{\mathbf{R}}^T(-\tau)=\mathbf{R}^H(-\tau) \tag{3.4.4'}$$

A bar above denotes taking the complex conjugate value, while ^{H} denotes Hermite transpose.

§3.5, *Markoff process*

The Markoff process is very important for stochastic differential equations. Given arbitrary m time instants $t_1<t_2<\cdots<t_m$, there are n-D vectors $\mathbf{X}_1,\mathbf{X}_2,\cdots,\mathbf{X}_m$ at these time instants. When time reaches t_m, the values $\mathbf{x}_1,\mathbf{x}_2,\cdots,\mathbf{x}_{m-1}$ of the former random vectors $\mathbf{X}_1,\mathbf{X}_2,\cdots,\mathbf{X}_{m-1}$ have been given, and if the random vector $\mathbf{X}_m$ has the conditional distribution as

$$\Pr(\mathbf{X}_m\le\mathbf{x}_m\mid\mathbf{X}_{m-1}=\mathbf{x}_{m-1},\cdots,\mathbf{X}_1=\mathbf{x}_1)=\Pr(\mathbf{X}_m\le\mathbf{x}_m\mid\mathbf{X}_{m-1}=\mathbf{x}_{m-1}) \tag{3.5.1}$$

then such stochastic process is a Markoff process. The characteristic is that probability distribution of the next step depends only on the state of previous step but not relates to all the history, such behavior is termed as Markoff behavior. Expressed with the conditional p.d.f., gives

$$f(\mathbf{x}_m\mid\mathbf{x}_{m-1},\mathbf{x}_{m-2},\cdots,\mathbf{x}_1)=f(\mathbf{x}_m\mid\mathbf{x}_{m-1}) \tag{3.5.2}$$

The above equation is only the one step conditional p.d.f. It can deduce successively

$$f(\mathbf{x}_m,\mathbf{x}_{m-1},\cdots,\mathbf{x}_1)=f(\mathbf{x}_m\mid\mathbf{x}_{m-1})f(\mathbf{x}_{m-1}\mid\mathbf{x}_{m-2})\cdots f(\mathbf{x}_2\mid\mathbf{x}_1)f(\mathbf{x}_1) \tag{3.5.3}$$

which expresses that if the initial p.d.f. $f(\mathbf{x}_0)$ and the transfer p.d.f. $f(\mathbf{x}_k\mid\mathbf{x}_{k-1})$ are given, then the distribution characteristics of the process are completely determined.

A Markoff process can have arbitrary p.d.f. If the distribution function is Gaussian then the process is termed as Gauss-Markoff process. In applications, the majority of Markoff processes are Gaussian, as is in this book.

§3.6, Spectral density of stationary stochastic process

The description with ***auto-correlation function*** gives the time domain characteristics of a stationary stochastic process. In various fields of theory of vibration or system analysis etc., linear system analysis is fundamental. For linear system analysis, the ***frequency domain*** methodology is often preferred. The frequency domain analysis describes how the system response changing with frequency. ***Spectral density*** describes the characteristics of a stationary stochastic process changing with frequency, *i.e.* its frequency domain characteristics. In such description, the external inputs and the system responses are all expressed as the functions of frequency. Hence, ***spectral density*** plays a central role in describing the stationary stochastic processes and the system responses in frequency domain analysis, as that the ***auto-covariance function*** played the role in time domain analysis.

The most frequently used spectral analysis is the transformation between the ***auto-covariance function*** and the ***power spectral density function***, note that these two functions are ***deterministic*** although the process is stochastic. On the other hand, the method of ***direct spectral expansion*** for stochastic process is also of great importance. Both methods will be described below.

§3.6.1, *Wiener-Khintchin relation*

Frequency domain method is closely related to Fourier transformation. Only the absolutely integrable non-periodical function on $(-\infty,\infty)$ can be expressed with the Fourier integration. However, in general a sample function of a stationary stochastic process can hardly be absolutely integrable, hence the stochastic Stieljes integrals must be applied [139], which is termed as ***direct spectral expansion*** (direct spectral analysis) method. Frequency domain analysis of stochastic process can also use the method of ***power spectral density***, which is seen most frequently in practice, and is described first below. The method of power spectral density is equivalent to the direct spectral expansion method. Both methods will be used in random vibration analysis, in chapter 4.

The ***Wiener-Khintchin relation*** relates both the function of power spectral density in frequency domain, and the function of auto-covariance in time domain. That Wiener and Khintchin proposed the following transformation

$$\mathbf{R}_X(\tau)=\int_{-\infty}^{\infty}\mathbf{S}_X(\omega)\exp(\mathrm{i}\omega\tau)\mathrm{d}\omega \tag{3.6.1}$$

$$\mathbf{S}_X(\omega)=\int_{-\infty}^{\infty}\mathbf{R}_X(\tau)\exp(-\mathrm{i}\omega\tau)\mathrm{d}\tau\big/2\pi \tag{3.6.2}$$

independently, hence the relation (3.6.1~2) is nominated with their name. Evidently, $\mathbf{R}_X(\tau)$ and $\mathbf{S}_X(\omega)$ are mutually Fourier and inverse Fourier transform to each other. The auto-covariance function represents the time-domain amplitude statistical information, and the ***spectral density function*** $\mathbf{S}_X(\omega)$ expresses the amplitude statistical information in frequency domain for the corresponding stationary stochastic process. The matrix function $\mathbf{S}_X(\omega)$ terms also as the ***power spectral density*** function. According to (3.4.4'), $\mathbf{S}_X(\omega)$ is a Hermite symmetric matrix, and is also positive definite.

The matrices $\mathbf{S}_X(\omega)$ and $\mathbf{R}_X(\tau)$ are different expressions to a same information source. When system analysis uses the time-domain method, the auto-covariance function will be adopted usually. However, if frequency domain method is applied in system analysis, the spectral density function is preferable. In practical measurement, the instrument responded information is often in the frequency domain. Fast Fourier transform (FFT) was discovered in 1965 by Cooley and Tukey that the frequency domain data $[\mathbf{S}_X(\omega)]$ is very quickly transformed to time domain $[\mathbf{R}_X(\tau)]$. The FFT promoted the instrumentation discipline developed quickly, which demonstrates the effect of high efficiency algorithm.

§3.6.2, *Direct spectral analysis of stationary stochastic process*

The transformed function above is the auto-covariance function $\mathbf{R}_X(\tau)$. The direct spectral analysis for a zero-mean stationary stochastic process $\mathbf{X}(t)$ is also important. Formerly, Wiener developed a generalized harmonic analysis theory, that an arbitrary *deterministic* oscillatory time function $\mathbf{x}(t)$ in $(-\infty,\infty)$ can be expressed as the following Fourier-Stieljes integration

$$\mathbf{x}(t)=\int_{-\infty}^{\infty}\exp(\mathrm{i}\omega t)\mathrm{d}\mathbf{z}(\omega) \tag{3.6.3}$$

where $\mathbf{z}(\omega)$ is a complex valued function uniquely determined by $\mathbf{x}(t)$. When $\mathbf{x}(t)$ decays with $\mathrm{abs}(t)\to\infty$ fast enough, $\mathbf{z}(\omega)$ is differentiable for any ω and the above integration reduces to Fourier integration. However, when $\mathbf{x}(t)$ does not decay and is a non-periodical function, then $\mathbf{z}(\omega)$ is non-differentiable and $|\mathrm{d}\mathbf{z}(\omega)|=O(\sqrt{\mathrm{d}\omega})$, which means $|\mathrm{d}\mathbf{z}(\omega)|$ is far larger than $\mathrm{d}\omega$, therefore the integration in (3.6.3) can only be of Stieljes type. The physical reason behind is that, in the time domain $(-\infty,\infty)$, the energy of non-decay signal is far larger than decay signal. If the process $\mathbf{x}(t)$ is real valued, then $\mathrm{d}\mathbf{z}(\omega)=\mathrm{d}\bar{\mathbf{z}}(-\omega)$.

Now make use of Wiener's result to stationary stochastic process. For a mean square continuous and zero-mean stationary stochastic process $\mathbf{X}(t)$, whose ***spectral expansion***

$$\mathbf{X}(t)=\int_{-\infty}^{\infty}\exp(\mathrm{i}\omega t)\mathrm{d}\mathbf{Z}_X(\omega) \tag{3.6.4}$$

where $\{\mathbf{Z}_X(\omega),-\infty<\omega<\infty\}$ is a complex valued left continuous stochastic process of ω, uniquely determined by $\mathbf{X}(t)$ [139], and $\mathbf{Z}_X(\omega)$ is ***frequency orthogonal increment***, that for arbitrary $\omega_1<\omega_2<\omega_3<\omega_4$, it holds the orthogonal relation

$$\mathrm{E}\left[\left(\bar{\mathbf{Z}}_X(\omega_2)-\bar{\mathbf{Z}}_X(\omega_1)\right)\cdot\left(\mathbf{Z}_X(\omega_4)-\mathbf{Z}_X(\omega_3)\right)^T\right]=\mathbf{0}$$

Expressed in differential form

$$\mathrm{E}\left[\mathrm{d}\overline{\mathbf{Z}}_X(\omega)\mathrm{d}\mathbf{Z}_X^T(\omega_2)\right]=\mathbf{S}_X(\omega)\delta(\omega_2-\omega)\mathrm{d}\omega\mathrm{d}\omega_2 \tag{3.6.5}$$

where a bar above means complex conjugate and $\mathbf{S}_X(\omega)$ is the spectral density of the stationary stochastic process $\mathbf{X}(t)$. Verification is as follows

$$\mathbf{R}(\tau)=\mathrm{E}\left[\overline{\mathbf{X}}(t)\mathbf{X}^T(t+\tau)\right]=\int_{-\infty}^{\infty}\int_{-\infty}^{\infty}\exp[-\mathrm{i}\omega t+\mathrm{i}\omega_2(t+\tau)]\mathrm{E}\left[\mathrm{d}\overline{\mathbf{Z}}_X(\omega)\mathrm{d}\mathbf{Z}_X^T(\omega_2)\right]$$

$$=\int_{-\infty}^{\infty}\exp(\mathrm{i}\omega\tau)\mathbf{S}(\omega)\mathrm{d}\omega$$

which is just the Wiener-Khintchin relation. ##

Therefore the ***power spectral density*** and the ***direct spectral expansion*** (3.6.4) are connected to each other through (3.6.5).

§3.6.3, White noise

In the theory of stochastic process, the classification of stationary stochastic processes is often according to the characteristics of spectral density. Among the others, the ***white noise*** is very frequently used. A white noise is a stationary stochastic process with zero-mean and non-zero constant spectral density $\mathbf{S}_0$. White noise is a model of stochastic excitation to a system, used quite frequently in random vibration theory, in system analysis, signal processing, control theory etc. White noise is only an artificial model for mathematical convenience but there is no exact white noise in real world. However white noise is a good approximation in many cases and mathematically simple, so that it is used extensively as a random excitation model.

Let us begin discussion with discrete time series. A stochastic sequence $\mathbf{X}(k), k=0,1,2\cdots$ with zero-mean

$$\mathbf{m}_k=\mathrm{E}[\mathbf{X}(k)]=\mathbf{0} \tag{3.6.6}$$

and the auto-covariance (auto-correlation) matrix of different time is zero

$$\mathbf{P}_{kj}=\mathrm{E}[\mathbf{X}(k)\mathbf{X}^T(j)]=\mathbf{Q}(k)\delta_{jk}\,,\qquad \delta_{jk}=\begin{cases}1, & k=j\\ 0, & k\neq j\end{cases} \tag{3.6.7}$$

where $\mathbf{Q}(k)$ is a symmetric non-negative (semi-positive-definite) matrix. Such stochastic sequence $\mathbf{X}(k), k=0,1,2\cdots$ is a white noise sequence, and $\mathbf{Q}(k)$ (or $\mathbf{Q}_k$) is the intensity of white noise. For white noise $\mathbf{Q}(k)=\mathbf{Q}_0$ is a constant and is a stationary process, because of (3.6.6), white noise is zero-mean. But $\mathbf{Q}(k)$ can also vary with the time station k, then it is no longer a stationary white noise. Sometimes, non-zero $\mathbf{m}_k$ is also termed as white noise, the drift white noise. Drift white noise is a covariance stationary stochastic process.

The continuous-time white noise, $\mathbf{X}(t)$ can be defined similarly, with zero-mean

$$\mathbf{m}(t)=\mathrm{E}[\mathbf{X}(t)]=\mathbf{0} \tag{3.6.8}$$

and the auto-correlation matrix function or covariance is

$$\mathbf{R}(t,\tau)=\mathrm{E}[\mathbf{X}(t)\mathbf{X}^T(\tau)]=\mathbf{Q}(t)\delta(t-\tau) \tag{3.6.9}$$

where $\mathbf{Q}(t)$ is a symmetric semi-positive-definite matrix, which relates the intensity of the white noise. Written as $\mathbf{Q}(t)$ means the intensity is time t variant, but it is not a stationary white noise process. A stationary stochastic process needs its covariance time-invariant, and white noise requires the covariance being a very narrow band function, which is represented by the δ - function. Stationary process needs $\mathbf{Q}(t)$ being time invariant, a constant function. Because of the δ - function in (3.6.9), even if time difference is very small, white noise is still uncorrelated, which means that white noise has zero inertia. Hence, its time curve jumps arbitrary with no any rule, such as shown before in figure 3.3, even this plot is not really white noise, that white noise does not exist in world.

The auto-covariance function has a corresponding ***spectral density***. Substituting (3.6.9) into (3.6.2) and regarding $\mathbf{Q}(t)$ as a constant matrix $\mathbf{Q}_0$ gives

$$\mathbf{S}(\omega) = \mathbf{Q}_0 / 2\pi = \mathbf{S}_0 \tag{3.6.10}$$

The power spectral density function is a constant, which means that the (dynamic) energy distributed uniformly to all frequencies! It has infinite energy and is impossible. At extremely high frequency, the spectral density cannot keep constant because of inertia. However, white noise is a reasonable mathematical abstraction, which implies extremely wide spectral density function. Corresponding to the consideration above, in equation (3.6.9) the function should not be really a δ -function. Such as band limited white noise

$$S(\omega) = \begin{cases} S_0, & \omega_1 \le \omega \le \omega_2 \\ 0, & \text{for other } \omega \end{cases} \tag{3.6.11}$$

$$R(\tau) = 2S_0(\sin\omega_2\tau - \sin\omega_1\tau)/\tau \tag{3.6.12}$$

The ideal white noise is the limit of $\omega_1 = 0,\ \omega_2 \to \infty$.

The definition for white noise above relates only the auto-covariance function, but has not mentioned the probability distribution function. In applications, normal distribution is usually assumed.

A noise, if it is not white then it must be ***colored***. For the convenience of mathematical treatment, a class of ***rational function spectral density noise*** is introduced, and the spectral density function is expressed as

$$S(\omega) = S_0 \times \frac{\omega^{2n} + a_1\omega^{2n-2} + \cdots + a_n}{\omega^{2m} + b_1\omega^{2m-2} + \cdots + b_m} \tag{3.6.13}$$

Such noise can be generated (driven) by a white noise acting on a linear dynamic system, the system output does have this kind of power spectra. In applications, such kind of ***white noise driven color noise*** is often used as an approximate model of real noise.

§3.6.4, Wiener process

Wiener and Poisson stochastic processes are also very popular in applications. The description should begin with the ***independent incremental process*** (time domain). The definition is:

Give time instants $t_0 < t_1 < \cdots < t_n$ *arbitrarily, such that for a continuous-time stochastic process* $X(t), 0 < t < \infty$ *, if the increments* $X(t_1) - X(t_0), \cdots, X(t_n) - X(t_{n-1})$ *of various time intervals are independent random* *variables, i.e.* $\mathrm{E}\big[[X(t_{i+1})-X(t_i)]\cdot[X(t_{j+1})-X(t_j)]\big]=0$ *when* $i\neq j$ *, then the stochastic process* $X(t)$ *is an independent incremental process.*

If for arbitrary t_2,t_1,h, the random variable increment $X(t_2+h)-X(t_1+h)$ has the same probability distribution as the increment $X(t_2)-X(t_1)$, then the process $X(t)$ is a stationary independent incremental process. It is also a Markoff process. If all these random increments have $\mathrm{E}[X(t_{i+1})-X(t_i)]=0$, zero-mean for all i, then $X(t)$ is termed as an ***orthogonal independent incremental process***, the distribution function of which has the relation

$$f(x_m,x_{m-1},\cdots,x_1)=f(x_m|x_{m-1})f(x_{m-1}|x_{m-2})\cdots f(x_2|x_1)f(x_1)\,.$$

Wiener process is one of the most fundamental stochastic processes. Wiener process provides the mathematical model for such as Brownian motion, thermo-noise in electronic circuits etc. In 1827, Brown discovered that the pollen motion in water is quite irregular, and since then such kind of motion is call the Brownian motion. The statistical interpretation of Brownian motion is one of the successful demonstrations of statistical mechanics. In 1905, Albert Einstein pointed out, that the Brownian motion can be interpreted as the particles induced motion from continuous impacts from the molecules in the surrounding solvent. Let $X(t)$ be the displacement of Brownian motion particle at time t with initial displacement $X(0)=0$. The particle motion is the effect of multiple impacts of surrounding molecules, according to the central limit theorem, the probability of the random variable of each step of displacement $X(t_2)-X(t_1)$ can be regarded as normally distributed with zero-mean. It is conceivable that the increments of motion are certainly independent and also stationary. Then the process is a ***normally distributed stationary independent incremental process***. Such kind of motion is a Wiener process, its definition is:

1) $X(t)$ is a stationary independent incremental process,

2) $X(t)$ is normally distributed,

3) It has zero mean $\mathrm{E}[X(t)]=0$, certainly the mean value of its increments are also zero,

4) Initial displacement is zero, $X(0)=0$.

A stochastic process satisfying the above conditions is a Wiener process.

Although its increments are stationary, the Wiener process itself is not stationary.

Because an arbitrary increment, $X(t_2)-X(t_1)$, is normally distributed, so the distribution function can be determined by its mean-value and variance. $\mathrm{E}[X(t_2)-X(t_1)]=0$ is easily verified. For variance, let us select uniform time increments $t_k=k\eta$, where η is the time step, then the probability distribution function is

$$f_{X(t_k),X(t_1)}(x_k,x_1)=f_{X(t_k)|X(t_1)}(x_k\,|\,x_1)\cdot f_{X(t_1)}(x_1)\,.$$

Denote $f_{X(t_1)}(x_1)\underset{\mathrm{def}}{=}f_\eta(x_1)$ and select $k=2$, according to the first condition

$$f_{X(t_2)|X(t_1)}(x_2\,|\,x_1)=f_{X(t_2)-X(t_1)}(x_2-x_1)=f_\eta(x_2-x_1)$$

and so

$$f_{X(t_2)}(x_2) = f_{2\eta}(x_2) = \int_{-\infty}^{\infty} f_{X(t_2),X(t_1)}(x_2, x_1)\mathrm{d}x_1 = \int_{-\infty}^{\infty} f_\eta(x_2 - x_1)\cdot f_\eta(x_1)\mathrm{d}x_1 \quad (3.6.14)$$

Because the probability distribution is normal and zero-mean, so

$$f_\eta(x) = \frac{1}{\sqrt{2\pi}\sigma(\eta)}\exp\left[-\frac{x^2}{2\sigma^2(\eta)}\right]$$

To determine the function $\sigma^2(\eta)$, according to condition 4, $\sigma^2(0) = 0$, substituting the above equation into (3.6.14) and carrying out the integration with respect to x_1 derives

$$f_{2\eta}(x) = \frac{1}{\sqrt{2\pi}[\sigma(2\eta)]}\exp\left[-\frac{x^2}{2\sigma^2(2\eta)}\right] = \frac{1}{\sqrt{2\pi}[\sqrt{2}\sigma(\eta)]}\exp\left[-\frac{x^2}{4\sigma^2(\eta)}\right]$$

Therefore $\sigma^2(2\eta) = 2\sigma^2(\eta)$, further $\sigma^2(k\eta) = k\sigma^2(\eta)$, where k is an arbitrary positive integer. Because the time step η can be selected arbitrarily, it is derived that

$$\sigma^2(t_2 - t_1) = s^2 \cdot (t_2 - t_1) \quad (3.6.15)$$

where s^2 is a constant. Its physical background is the mean square displacement of a particle for Brownian motion, however for different problem the interpretation should be different. Wiener process is a kind of diffusion, s^2 is proportional to the diffusion constant. Initially $t = 0$, Wiener process is determined zero $X(0) = 0$, which means both mean value and variance are zero. Equation (3.6.15) coincides with this conclusion. For arbitrary time instant t, the condition 3 gives the mean value zero, and from (3.6.15) the mean square deviation is $s\sqrt{t}$.

After the time is discretized as $t_k = k\eta$, the increment $X(t_k) - X(t_{k-1}) = \Delta X_k$ of a Wiener process is also a stochastic process. ΔX_k is zero-mean and normally distributed. Further, because of independent incremental so $\mathrm{E}[\Delta X_i \cdot \Delta X_j] = 0$. From equation (3.6.15) it is seen that

$$\mathrm{E}[\Delta X_i \cdot \Delta X_j] = \delta_{ij} \cdot s^2\eta$$

hence $\Delta X_k, k = 1,2,\cdots$ is a white noise stochastic process. Therefore, Wiener process can be considered as the integration of a stationary white noise.

Chapter 4, Random vibration of structures

Engineering structures are constantly loaded by external forces, among which quite a large part are time-variant and behave randomly. Under the same environment condition, various measurements of time history of dynamic loads are different from time to time. In fact, each time history curve of various measurements is a sample of stochastic processes. The analysis of engineering structure responses under the random environment is naturally a heavily concerned problem for engineers.

The investigation of structural response under random excitations is the subject of theory of random vibration, for which the proper description of the external random excitations is a prerequisite. However, for different structures under various environments the random excitations are quite different from one another. The practical problems such as

1) building under earthquake,
2) roughness of road surface induced vehicle vibration,
3) surface wave action to the ships or sea-platforms,
4) wind-structure interaction for the bridges and/or tall buildings,
5) the action of atmospheric turbulence on aircrafts,
6) combustion and/or jet propulsion induced turbulence excitations, etc.,

are some examples with random nature and are quite different from each other. Hence the first step is to establish the mathematical model for various random excitations. Evidently, all these excitations ought to be described by stochastic processes.

Usually, the excitations are classified as ***stationary*** and ***non-stationary*** stochastic processes. The random excitations are induced from quite a number of uncertainties, usually from independent sources, hence the probability distribution is usually assumed to be normal. A linear system subjected to random excitation with normal distribution, the generated response is still normally distributed. This is a very useful property. According to the ***central limit theorem***, the distribution of the summation of a large number of statistically independent variables behaves approximately normal.

The stochastic processes discussed in the present book are all assumed normally distributed. Note that the normally distributed wide sense stationary stochastic process is a stationary stochastic process.

After the external excitation is determined, the problem of structural dynamic response must be solved, which is an important subject of structural mechanics. Similar to the structural analysis under deterministic external load, the random vibration analysis is also classified as linear and non-linear problems. As is well known, the linear analysis theory of random vibration is well developed, however because of the mathematical difficulty the non-linear analysis is still very difficult both on theory and computation. Hence this chapter concentrates to the description of theory and computer method of random vibration for linear structural systems.

The fundamental frame of linear random vibration theory has been established long time ago. A number of books have been published on random vibration theory, especially on linear theory, see [62~65]. At a first glance, it seems no more problems remained for further investigation. However, the established random vibration theory has not been applied so much in important engineering projects. The reason is the computational complexity of the theory. The use of formerly given theory and method spends tremendous computational expense, which is completely prohibitive to be accepted by the engineers. Therefore the computational complexity is the bottle-neck of applying the linear random vibration theory in the engineering practice. For example, it is commonly recognized in the earthquake engineering community, that using the random vibration method to analyze long span structures is preferable. However, decades of year investigations have not solved this computational problem. Kiureghian and Neuenhofer pointed out that: "*While the random vibration approach is appealing for its statistical nature, it is not yet accepted as a method of analysis by practicing engineers*", in paper [66]. Ernesto and Vanmarcke further pointed out that " *The theoretical framework of a methodology for stochastic-response analysis to random-excitation fields is already available; however, its use by the earthquake-engineering community is viewed as impractical except for simple structures with a small number of degrees of freedom and supports.* " , see [67]. Therefore, to bypass the difficulty of computational complexity, both of them make use of the approximate method—the ***spectral response method***. Even though, the approximate numerical method they use is still very complicated with quite an amount of computational expenses. A warm discussion for the methods and errors etc. was made on the *Transactions of ASCE* [68].

The **frequency domain** analysis is commonly used in linear random vibrations. The problem can be described as, given the power spectral density distribution of the external excitation to find the power spectral density of the structural response. Time domain analysis can also be used, see [62~65,76]. The **time domain** analysis leads usually to the solution of ***Lyapunov differential equation***, for which the precise integration method can also be used to solve the problem highly precisely. Multi-dimensional stochastic differential equation, such as Ornstein-Ulenbeck process [69] also is derived to find the solution of Lyapunov differential equation. In chapter 6, the solution of Lyapunov differential equation will be described in detail by the precise integration method. However in engineering practice, the external random excitations are usually supplied with given power spectral density, so that solving in the **frequency domain** is natural.

The application of random vibration problem relates to a number of engineering disciplines, among which the structural response analysis in earthquake engineering is of great concern. Hence, the structural random response is briefly described below, out of various theories and methods the emphasis is put on the highly efficient ***pseudo-excitation method***, see [70~76], proposed by professor Jia-Hao, Lin. The pseudo excitation method can be used to the analysis of complex structures under random excitations, and the computational efficiency will be $10^2 \sim 10^5$ times faster than the traditional approach. The pseudo excitation method has developed a whole set of algorithm system, which can be applied not only to the solution of stationary random responses, but also to the solution of non-stationary random responses. The pseudo excitation method can also be used to multi-source excitation analysis, even for the response analysis of non-uniformly modulated evolutionary random excitations. The CQC (Complete Quadratic Combination) results can be obtained directly from the pseudo excitation method. The CQC method involves all the covariance between the various vibration modes and is regarded as the exact solution of statistical analysis. The pseudo excitation method can efficiently be applied to the analysis of practical long span structures.

As for non-linear random vibration problems, which are certainly important, however, even the deterministic vibration problems of a few degrees of freedom system are so difficult that from engineers' point of view, the problem is only at the early research stage. Because the superposition principle no longer applies for non-linear system, so the spectral analysis method cannot be used. Even if the input excitation is normally distributed, the responses are no longer distributed normally.

People spend much effort to investigate various methods for the non-linear random vibration of structures, but no method is satisfied, especially for the analysis of multi-degrees of freedom non-linear system. Even the analysis of non-linear system vibration under deterministic external force has been so difficult, so that the difficulty for the non-linear system random vibration is conceivable. Theoretically, the FPK (Fokker-Planck-Kolmogolov) PDE need to be solved, or else to solve a set of non-linear stochastic differential equations using the Ito calculus etc., the interested readers can read the references such as [62,69] etc.

§4.1, Models of random excitation

The differential equation for structural response under random excitation is

$$\mathbf{M}\ddot{\mathbf{x}} + \mathbf{C}\dot{\mathbf{x}} + \mathbf{K}\mathbf{x} = \mathbf{g}(t) \tag{4.1.1}$$

where $\mathbf{x}(t)$ is a n-dimensional displacement vector, $\mathbf{g}(t)$ is the n-dimensional external random excitation vector, $\mathbf{M},\mathbf{C},\mathbf{K}$ are given $n \times n$ time-invariant symmetric non-negative definite matrices of mass, damping and stiffness, respectively. Because the excitation $\mathbf{g}(t)$ is a stochastic process, the induced response vector $\mathbf{x}(t)$ is also a stochastic vector. The random nature is induced from system external excitation, so that the random excitation should be investigated first. Random excitations should be distinguished as stationary and non-stationary, and each class has several random excitation models. The two classes of stationary and non-stationary random excitations are discussed separately.

§4.1.1, Stationary random excitations

The external excitation $\mathbf{g}(t)$ is treated as a stationary stochastic process, which has constant mean-value and the auto-covariance function depends only on the time difference

$$\mathrm{E}[\mathbf{g}(t)] = \mathbf{m}_g, \quad \mathbf{R}_g(t_1,t_2) = \mathbf{R}_g(\tau), \quad \tau = t_2 - t_1 \tag{4.1.2}$$

where $\mathbf{m}_g$ is a n-dimensional vector, and $\mathbf{R}_g$ is a $n \times n$ covariance matrix. In applications, the external force $\mathbf{g}(t)$ is often assumed as covariance stationary stochastic process, *i.e.* the mean value $\mathbf{m}_g(t)$ is also a function of time t. Because the mean value $\mathbf{m}_g(t)$ is a deterministic function, using the superposition principle the influence of a deterministic external force can be considered separately. Then only the zero-mean stationary stochastic process $\mathbf{g}(t)$ should be considered. In this case the auto-correlation matrix function is really the covariance matrix function. The covariance matrix function and the power spectral density function $\mathbf{S}_g(\omega)$ are interrelated with the Wiener-Khintchin relation

$$\mathbf{R}_g(\tau)=\int_{-\infty}^{\infty}\mathbf{S}_g(\omega)\exp(\mathrm{i}\omega\tau)\mathrm{d}\omega \tag{4.1.3}$$

$$\mathbf{S}_g(\omega)=\int_{-\infty}^{\infty}\mathbf{R}_g(\tau)\exp(-\mathrm{i}\omega\tau)\mathrm{d}\tau/2\pi \tag{4.1.4}$$

They constitute a Fourier transformation pair.

The usual random excitation is induced from the combination of quite a number of unpredictable factors. Hence the probability distribution of the excitation is assumed normal. The statistical characteristics of the random excitation can be completely determined by the mean value and the auto-covariance or power spectral density function. The difference between two zero-mean stationary stochastic processes is given by the different auto-covariance or power spectral density function of the two processes.

The above paragraph describes the time dependent variation of excitation, which is similar to the single degree of freedom case. For multi-point excitations, the covariance between various sources should be considered also, called space covariance or multi-point excitation. The different space distribution of excitations is reflected on the different matrices of $\mathbf{S}_g(\omega)$ or $\mathbf{R}_g(\omega)$. For multi-point excitation, the following cases should be considered:

1) Single source in-phase excitations. This is the simplest stationary stochastic excitation;
2) Single source multi-point excitations with different phases at different points; and
3) Multiple sources multi-point excitations with different phases, *i.e.* the combination of arbitrary stationary stochastic excitations.

Single source multi-point excitation means that the multiple points are excited by a single factor. Hence the excitations at the different points interrelate closely to each other, mathematically the ***rank of the power spectral density matrix of these points equals 1***. For single source excitations the phase differences between different points should also be distinguished. In-phase excitation means that all the point excitations have exactly the ***same phase***. For example, in the aseismic analysis of ordinary building structures, the in-phase assumption for excitations is usually adopted and given in the design code. Because the ground size of a building is small that it is only a small fraction of the fundamental ground wave length, so that the in-phase assumption is appropriate.

However, long span structures, such as bridges, pipelines and dams etc., have far apart away ground supports. Earthquake wave speed is finite that the wave front reaches the different ground supports at different instants, which implies that at various ground supports the ***excitation phases are different***, although the excitation source is still single. Such phenomenon appears also in the cases of vehicle running on the rough road surface, sea wave action on ships or platforms, or the case of long span structure on homogeneous ground under earthquake loading, etc.

The general stationary random excitations mean multi-point excitations, for which the power spectral density matrix $\mathbf{S}_g(\omega)$ is general, which means that the random excitations are induced from the combination of various factors. For example, although earthquake is induced from one fault, however which is not a single point, but a local fracture process. And also the wave propagation is not along solely a homogeneous medium, that after multiple reflection or refraction along the way of propagation, when the wave reaches various structural supports, the source is no longer of single factor. Suppose, the ***rank of the power spectral density matrix*** $\mathbf{S}_g(\omega)$ ***is*** r, the excitation phases can also be different. But in general, the matrix $\mathbf{S}_g(\omega)$ is still Hermite symmetric and non-negative definite.

The above description is only from general consideration. More description will be given when the problem of structural response is considered. The power spectral density $\mathbf{S}_g(\omega)$ of external excitation is the system input. The function form of statistical estimation of the input excitations and its quantities should be determined based on a large number of collections of measurements and recordings. To collect these data is a long-term fundamental task, such as earthquake records, wind, sea wave recording spectra, etc.

§4.1.2, Non-stationary random excitations

Generally, the random excitation to engineering structures is usually non-stationary. For quite a number of problems, the simplification as stationary stochastic process gives good approximation to applications. However, there are still problems that the stationary stochastic process simplification cannot model the real situation satisfactorily, such as the structure under wind gust, the maneuver of aircraft, etc. There are many different mathematical descriptions for the non-stationary random excitations. The simple model of uniformly modulated evolutionary random excitation model is quite often applied in structural engineering. ***Uniform modulation*** requires a ***stationary stochastic process*** $\mathbf{g}(t)$ ***as the fundamental excitation model***, for which the power spectral density is denoted as $\mathbf{S}_g(\omega)$. The fundamental excitation model of non-stationary stochastic process can be selected as those stationary stochastic processes described in the last section. Such a fundamental excitation model $\mathbf{g}(t)$ multiplied by a ***deterministic*** scalar function $f(t)$, which is an amplification factor called the excitation ***modulation function,*** gives the ***uniformly modulated evolutionary random excitation model***

$$\mathbf{G}(t) = f(t)\mathbf{g}(t) \tag{4.1.5}$$

The uniformly modulated model is the simplest non-stationary process and is easier to deal with in applications. However, this excitation model is non-stationary only for different time, but the space distribution of excitation is time independent. For such an excitation model, the space distribution can be modeled as for stationary multi-point excitation, *i.e.* the distribution can still be classified as

1) single source in-phase non-stationary excitations;
2) single source multi-phase non-stationary excitations, and
3) multiple sources multi-point non-stationary excitation with different phases.

For the fundamental stationary stochastic process $\mathbf{g}(t)$, given the auto-covariance function $\mathbf{R}_g(\omega)$ or power spectral density distribution $\mathbf{S}_g(\omega)$, the auto-covariance function $\mathbf{R}_G(t_1,t_2)$ or power spectral density $\mathbf{S}_G(\omega,t)$ of the non-stationary process $\mathbf{G}(t)=f(t)\mathbf{g}(t)$ can be extended based on $\mathbf{R}_g(\omega)$ or $\mathbf{S}_g(\omega)$ as

$$\mathbf{R}_G(t_1,t_2)=f(t_1)f(t_2)\mathbf{R}_g(\tau),\quad \tau=t_2-t_1 \tag{4.1.6}$$

$$\mathbf{S}_G(\omega,t)=\left(f(t)\right)^2\mathbf{S}_g(\omega) \tag{4.1.7}$$

Therefore in applications, the first step is to determine the statistical characteristics of the fundamental stationary stochastic process $\mathbf{g}(t)$, *i.e.* $\mathbf{R}_g(\omega)$ or $\mathbf{S}_g(\omega)$, based on large number of records. Thereafter the slowly varying deterministic modulation function $f(t)$ is selected reasonably to model the non-stationary behavior.

To give the deterministic modulation function $f(t)$, a large number of observations are required too. Usually in applications the respective manuals or design codes can be referred. For aseismic structure design, the China design code supplies a modulation curve for the method of spectral response, which is composed of three segments reflecting three stages of random excitations, *i.e.* the initial stage, the stationary stage and the decay stage. This modulation curve can be transformed to be a modulation function $f(t)$ used in equation (4.1.5)

The model of uniformly modulated evolutionary stochastic process is the simplest one among various non-stationary models. The characteristic of uniform modulation is that from begin to end the shape of power spectral density keeps unchanged, only the amplitude is changed with time. For real excitations, the power spectral density may also change with time, *i.e.* the matrix $\mathbf{S}_g(\omega)$ or $\mathbf{R}_g(\omega)$ also change with time. Such as the earthquake excitations, at the beginning stage the excitation is composed of the significant wave number within a wide range, however, the high wave number components are quickly damped out and the excitation becomes dominated by low wave number components. To model such phenomenon, the modulation can be described by the ***non-uniformly modulated evolutionary excitation model*** [77], as given by

$$\mathbf{G}(t)=\int_{-\infty}^{\infty}A(\omega,t)\exp(\mathrm{i}\omega t)\mathrm{d}\mathbf{Z}_G(\omega) \tag{4.1.8}$$

where $A(\omega,t)$ is a deterministic non-uniform modulation function. If using the simplification of $A(\omega,t)=f(t)$, the former uniformly modulate model is resulted. Equation (4.1.8) uses the integration in Stieltjes sense, on which the spectral expansion equation (3.6.4) is based. The factor $\mathrm{d}\mathbf{Z}_G(\omega)$ is the increments in frequency domain, representing a zero-mean complex valued orthogonal incremental stochastic process and its correlation between two frequencies is given as

$$\mathrm{E}[\mathrm{d}\overline{\mathbf{Z}}_G(\omega_1)\mathrm{d}\mathbf{Z}_G^T(\omega_2)]=\mathbf{S}_G(\omega_1)\delta(\omega_2-\omega_1)\mathrm{d}\omega_1\mathrm{d}\omega_2 \tag{4.1.9}$$

where a bar in $\mathrm{d}\overline{\mathbf{Z}}_G$ means taking its complex conjugate value.

Similarly, for the non-uniformly modulated evolutionary excitation model, there are still cases of space distributions as before,

1) single source in-phase excitations;
2) single source multi-phase excitations; and

3) multiple sources multi-point excitations with different phases;

In addition to the non-uniformly modulated evolutionary excitation model (4.1.8), there are a number of other models. However, the model (4.1.8) is used quite often in civil engineering.

LIN's ***pseudo excitation method*** solves the linear response problem for all the above-mentioned random excitations, which contributes not only the computational but also the theoretical progress, see review [78].

§4.2, Response of structures under stationary excitations

Given the model of random excitation, the response analysis is required. The structural response $\mathbf{x}(t)$ is certainly also a stochastic process. The linear governing equation is given as

$$\mathbf{M}\ddot{\mathbf{x}} + \mathbf{C}\dot{\mathbf{x}} + \mathbf{K}\mathbf{x} = \mathbf{g}(t) \tag{4.1.1}$$

for which the superposition principle applies. Also if the input is normally distributed, then so is the response output. Hence it is required to find the mean value and the covariance matrix and then the probability distribution is determined for a normally distributed process.

Let us begin with structural random response analysis of a single degree of freedom system.

§4.2.1, Random response of single degree of freedom system

The dynamic equation of single degree of freedom structural system is

$$m\ddot{x} + c\dot{x} + kx = g(t) \tag{4.2.1}$$

where $g(t)$ and $x(t)$ are both stochastic processes. For single degree of freedom system, they are two functions, not vectors. Using the parametric form

$$\ddot{x} + 2\zeta\omega_1\dot{x} + \omega_1^2 x = g'(t),\ x(0) = 0,\ \dot{x}(0) = 0 \tag{4.2.2}$$

where $\omega_1 = \sqrt{k/m}, \zeta = c/(2\sqrt{km})$ are the natural frequency and damping ratio, respectively, and $g'(t) = g(t)/m$ is still the stochastic external force, for which the mean value and auto-correlation function are denoted as m'_g and $R'_g(\tau)$, respectively. Using Duhamel integration for any sample external force

$$x(t) = \int_0^t h(t-s)g'(s)\mathrm{d}s \tag{4.2.3}$$

where the integration is in the sense of mean square, and the function $h(t-s)$ is a deterministic ***unit impulse response function***, see equation (2.1.15)

$$h(t) = \left(e^{-\varsigma\omega_1 t} \Big/ \sqrt{mk(1-\varsigma^2)} \right) \sin(\sqrt{1-\varsigma^2}\,\omega_1 t)$$

In this solution there is the influence of initial condition (4.2.2), hence although the external force is stationary, the response $x(t)$ is still a non-stationary stochastic process. Taking mathematical expectation gives

$$\begin{aligned} m_x(t) &= \mathrm{E}[x(t)] = \int_0^t h(t-s)\mathrm{E}[g'(s)]\mathrm{d}s = m'_g \int_0^t h(t-s)\mathrm{d}s \\ &= (m'_g/\omega_1^2)\left\{1 - e^{-\varsigma\omega_1 t}\left[(\varsigma\omega_1/\omega_d)\sin\omega_d t + \cos\omega_d t\right]\right\} \end{aligned} \tag{4.2.4}$$

where $\omega_d = \omega_1\sqrt{1-\varsigma^2}$. When $t \to \infty$, because of the damping factor, the response tends to be a stationary stochastic process with the mean value

$$\lim_{t\to\infty} m_x(t) = m'_g/\omega_1^2 \tag{4.2.5}$$

The auto-correlation function can be found from equation (4.2.3) as

$$R_x(t_1,t_2) = \mathrm{E}[x(t_1),x(t_2)] = \int_0^{t_1}\int_0^{t_2} h(t_1-s_1)h(t_2-s_2)R'_g(s_2-s_1)\mathrm{d}s_2\mathrm{d}s_1 \tag{4.2.6}$$

Given the auto-correlation function of external force $R'_g(\tau)$, this integration can be found.

The above analysis is given in the time domain. The integration of an auto-correlation function is not so convenient, and the frequency domain analysis can be used instead. The frequency domain response analysis of a time invariant system under a stationary stochastic process is simple. As $t \to \infty$, the time domain analysis gives stationary stochastic response, then if the external excitation is given in spectral form $g'(t) = g_\omega \exp(\mathrm{i}\omega t)$, then the response is $x(t) = A(\omega)\exp(\mathrm{i}\omega t)$ with no consideration of the initial condition. Substituting spectral expressions into the equation (4.2.2) gives

$$\left[\omega_1^2 + 2\varsigma\omega_1\omega\mathrm{i} - \omega^2\right]A(\omega)\exp(\mathrm{i}\omega t) = g_\omega \exp(\mathrm{i}\omega t).$$

Solving $A(\omega)$ gives

$$x(t) = H(\omega)g_\omega \exp(\mathrm{i}\omega t), \quad H(\omega) = 1/\left[\omega_1^2 + 2\varsigma\omega_1\omega\mathrm{i} - \omega^2\right] \tag{4.2.7}$$

The function $H(\omega)$ is called the frequency response function of the system or simply the transfer function. This is the monochromatic vibration response with frequency ω. The function $H(\omega)$ can be connected with the impulse response function $h(t)$. The function $h(t)$ involves the influence of initial conditions, but as the time evolves very long, the influence of initial conditions tends to be damped out and the response contains only the stationary part. To express in terms of the Duhamel integration of impulse response function

$$\int_0^t h(t-s)\exp(\mathrm{i}\omega s)\mathrm{d}s = \exp(\mathrm{i}\omega t)\int_0^t h(\theta)\exp(-\mathrm{i}\omega\theta)\mathrm{d}\theta$$

when $t \to \infty$ the initial term damps out, and the response reduces to the equation (4.2.7) with no initial condition influence, and it is also for $g_\omega = 1$, therefore

$$H(\omega) = \int_0^\infty h(\theta)\exp(\mathrm{i}\omega\theta)\mathrm{d}\theta \tag{4.2.8}$$

which shows that the system frequency response function $H(\omega)$ is the Fourier transformation of the impulse response function $h(t)$. The frequency response function $H(\omega)$ is integrated as

$$H(\omega,t) = \int_0^t h(\theta)e^{\mathrm{i}\omega\theta}\mathrm{d}\theta = H(\omega)\left\{1 - e^{-(\omega_1\varsigma + \mathrm{i}\omega)t}\left[\cos\omega_d t + \left((\omega_1\varsigma + \mathrm{i}\omega)/\omega_d\right)\sin\omega_d t\right]\right\}$$

where the latter term in the bracket involves the initial condition influence, because $\varsigma > 0$, this term vanishes when $t \to \infty$.

According to the spectral expansion equation (3.6.4) of the stationary stochastic process, that the stochastic process $g'(t)$ is expressed as the composition of various components of multi-chromatic frequencies

$$g'(t)=\int_{-\infty}^{\infty}\exp(\mathrm{i}\omega t)\mathrm{d}Z_g(\omega) \tag{4.2.9}$$

where $Z_g(\omega)$ is an orthogonal incremental stochastic process excitation in the frequency domain and the integration is in Stieltjes sense [139], which corresponding to $g_\omega=\mathrm{d}Z_g(\omega)/\mathrm{d}\omega$. Corresponding to the above spectral expansion for the excitation, the response spectral expansion is given as

$$x(t)=\int_{-\infty}^{\infty}H(\omega,t)\exp(\mathrm{i}\omega t)\mathrm{d}Z_g(\omega) \tag{4.2.10}$$

When the time $t\to\infty$, the influence of initial condition damps out, then

$$x(t)=\int_{-\infty}^{\infty}H(\omega)\exp(\mathrm{i}\omega t)\mathrm{d}Z_g(\omega) \tag{4.2.10'}$$

The spectrum of increment of $Z_g(\omega)$ is given as, [see equation (3.6.5)]

$$\mathrm{E}\left[\mathrm{d}\overline{Z}_g(\omega_1)\mathrm{d}Z_g(\omega_2)\right]=S_g(\omega_1)\delta(\omega_2-\omega_1)\mathrm{d}\omega_1\mathrm{d}\omega_2 \tag{4.2.11}$$

Suppose the mean value of the stochastic excitation $g'(t)$ is $m'_g=0$, then that of the response is also zero-mean $m_x=0$, and the auto-correlation function is expressed as

$$\begin{aligned}R_x(t_1,t_2)&=\mathrm{E}[\overline{x}(t_1),x(t_2)]\\&=\mathrm{E}\left[\int_{-\infty}^{\infty}\overline{H}(\omega_1)\exp(-\mathrm{i}\omega_1t_1)\mathrm{d}\overline{Z}_g(\omega_1)\int_{-\infty}^{\infty}H(\omega_2)\exp(\mathrm{i}\omega_2t_2)\mathrm{d}Z_g(\omega_2)\right]\\&=\int_{-\infty}^{\infty}\int_{-\infty}^{\infty}\overline{H}(\omega_1)H(\omega_2)\exp[\mathrm{i}(\omega_2t_2-\omega_1t_1)]\mathrm{E}[\mathrm{d}\overline{Z}_g(\omega_1)\mathrm{d}Z_g(\omega_2)]\\&=\int_{-\infty}^{\infty}\int_{-\infty}^{\infty}\overline{H}(\omega_1)H(\omega_2)\exp[\mathrm{i}\omega_2(t_2-t_1)-\mathrm{i}(\omega_1-\omega_2)t_1]S_g(\omega_1)\delta(\omega_2-\omega_1)\mathrm{d}\omega_1\mathrm{d}\omega_2\end{aligned}$$

Carrying out the integration gives

$$R_x(t_1,t_2)=\int_{-\infty}^{\infty}\overline{H}(\omega)H(\omega)\exp[\mathrm{i}\omega(t_2-t_1)]S_g(\omega)\mathrm{d}\omega \tag{4.2.12}$$

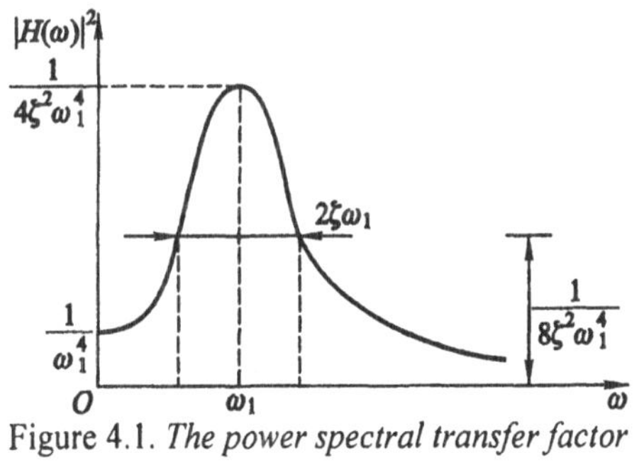

Figure 4.1. *The power spectral transfer factor*

which explains that after a long time the auto-correlation function R_x of the response depends only on the time difference $\tau=t_2-t_1$, and is also an even function of τ. The response spectral density is

$$S_x(\omega)=S_g(\omega)\,|\,H(\omega)\,|^2 \tag{4.2.13}$$

This equation is important for linear stationary random vibration theory. From equation (4.2.7), it gives

$$|\,H(\omega)\,|^2=\left[(\omega^2-\omega_1^2)^2+4\varsigma^2\omega^2\omega_1^2\right]^{-1},\quad \omega_1^2=k/m \tag{4.2.14}$$

After $S_x(\omega)$ is found, the $S_{\dot{x}}(\omega)$, $S_{\ddot{x}}(\omega)$ etc. can also be found as

$$S_{\dot{x}}(\omega)=\omega^2S_x(\omega)=\omega^2\,|\,H(\omega)\,|^2\,S_g(\omega) \tag{4.2.15}$$

$$S_{\ddot{x}}(\omega)=\omega^4S_x(\omega)=\omega^4\,|\,H(\omega)\,|^2\,S_g(\omega),\ \text{etc.} \tag{4.2.16}$$

The power spectral transfer factor $|H(\omega)|^2$ is also very important, and its plot is given in the figure 4.1. The transfer function $H(\omega)$ is the system response under a unit white noise. Usually ς is a small quantity, for example $\varsigma < 0.05$, in such case the power spectral density of the response is a narrow band stochastic process that there is a peak nearby the natural frequency.

§4.2.2, Multi-degrees of freedom system under single source excitation

The random response of linear single degree of freedom system is well developed and is described in a number of books. Similar methodology has been developed for multi-degrees of freedom systems. Theory and equations have been given for a long time, however, when using these equations in practical problems, it faces tremendous computational expense, which causes prohibitive difficulty to practical application and becomes a bottle-neck for using the random vibration theory to engineering.

Prof. J.H. Lin proposed the **pseudo excitation method** to solve this critical problem, which supplies a highly efficient computational algorithm for engineering applications. The proposed algorithm of pseudo excitation method have been applied to a number of important practical engineering structures, including

1) Aseismic analysis of Feng-Man reservoir dam,
2) Random vibration analysis of Hong-Kong Ching-Ma suspension bridge under random wind load and earthquake load,
3) Random vibration analysis of the second Nanking Yangtze river bridge and Dong-Tin lake bridge under earthquake loads,

etc., see [79~82].

FEM model of structure is appropriate for applying the pseudo excitation method. The algorithm of pseudo excitation method is imbedded into the FEM program system JIGFEX/DDJ in china, which has solved a number of random vibration problems for engineering structures, for which the number of degrees of freedom reaches dozens of thousands. In analysis, 300 vibration modes, 1000 wind excitation points, 29 ground excitation points for bridges or 102 ground points for dams are used, etc. The ground wave passage effect and multiple source excitations etc. are taken into consideration.

Hence, the emphasis is put on the pseudo excitation method below in this chapter. Physical insight is emphasized in description, rather than mathematical strictness. The computational effectiveness and efficiency is also stressed.

The meaning of single source implies that, all the excitation points of the structure are excited synchronously. For example, in the case of a usual building structure excited by a stationary stochastic earthquake load, all the ground points are assumed shaking with the same phase angle, called as ***in-phase synchronized***. This is the simplest external excitation. A n-degrees of freedom system under a single source stationary excitation synchronized in-phase, has the dynamic equation

$$\mathbf{M}_n\ddot{\mathbf{y}} + \mathbf{C}_n\dot{\mathbf{y}} + \mathbf{K}_n\mathbf{y} = \mathbf{p}e(t) \tag{4.2.17}$$

where $\mathbf{M}_n$, $\mathbf{K}_n$ and $\mathbf{C}_n$ are $n \times n$ mass, stiffness and damping matrices, respectively, and $\mathbf{p}$, called the partition vector, is a given deterministic n-

dimensional constant vector, and $e(t)$ is a one dimensional normal distributed zero-mean stationary stochastic process, of which the power spectral density $S_e(\omega)$ is given. Because system is linear, hence the response displacement $\mathbf{y}(t)$ is also a zero-mean valued normal distributed stationary stochastic vector to be determined. The vector $\mathbf{y}(t)$ itself cannot be determined but its auto-covariance matrix function (for time domain) $\mathbf{R}_y(\omega)$, or the power spectral density $n \times n$ matrix (frequency domain) $\mathbf{S}_y(\omega)$ is needed.

The FEM models of structures in engineering use more than $n = 10000$ degrees of freedom quite frequently. The analysis for random vibration of structures can also use such FEM model. And usually the vibration modal analysis method is used to pick up the first q-orders of eigen-modes, so as to reduce the degrees of freedom in the analysis, such as $q = 20 \sim 500$ depending on the structure and design requirement. Hence the subspace iteration algorithm is quite frequently used to pick up the first q eigen-solutions $(\omega_j^2, \boldsymbol{\varphi}_j), j = 1, \cdots, q$ out of the n-degrees of freedom eigen-solutions of structure. The eigen-modes $\boldsymbol{\varphi}_j$ are ortho-normalized with respect to the mass matrix $\mathbf{M}_n$. In applications, the generation of mass and stiffness matrices is usually reliable, the estimation of the damping matrix $\mathbf{C}_n$ is not so straightforward, but its value is relatively small and will be described below. Hence the eigen-solutions $(\omega_j^2, \boldsymbol{\varphi}_j), j = 1, \cdots, q$ are usually solved from the eigen-equation

$$(\omega^2 \mathbf{M}_n - \mathbf{K}_n)\boldsymbol{\varphi} = \mathbf{0}$$

where the damping term is removed. For a gyroscopic system, the gyroscopic term should not be neglected, however in dynamic equation (4.2.17), the gyroscopic term has not included.

After solved the first q eigen-modes $\boldsymbol{\varphi}_j$, the $n \times q$ modal matrix $\boldsymbol{\Phi}$

$$\boldsymbol{\Phi} = \begin{bmatrix} \boldsymbol{\varphi}_1 & \boldsymbol{\varphi}_2 & \cdots & \boldsymbol{\varphi}_q \end{bmatrix} \tag{4.2.18}$$

is composed. The displacement vector $\mathbf{y}(t)$ is projected in the q-D subspace, *i.e.*

$$\mathbf{y} = \boldsymbol{\Phi}\mathbf{u} \tag{4.2.19}$$

where $\mathbf{u}$ is a q-D vector to be solved. The eigenvectors $\boldsymbol{\varphi}_1, \boldsymbol{\varphi}_2, \cdots, \boldsymbol{\varphi}_q$ are treated as the basis vectors of the q-D subspace, and the vector $\mathbf{u}$ is the projection vector of $\mathbf{y}$ in the q-D subspace. Substituting the approximation (4.2.19) into the dynamic equation (4.2.17), and left multiplying with the matrix $\boldsymbol{\Phi}^T$, which means the projection of forces on the q-D subspace, gives

$$(\boldsymbol{\Phi}^T \mathbf{M}_n \boldsymbol{\Phi})\ddot{\mathbf{u}} + (\boldsymbol{\Phi}^T \mathbf{C}_n \boldsymbol{\Phi})\dot{\mathbf{u}} + (\boldsymbol{\Phi}^T \mathbf{K}_n \boldsymbol{\Phi})\mathbf{u} = \boldsymbol{\Phi}^T \mathbf{p} e(t)$$

As the columns of matrix $\boldsymbol{\Phi}$ are the normalized eigenvectors with respect to the mass matrix, so that

$$\boldsymbol{\Phi}^T \mathbf{M}_n \boldsymbol{\Phi} = \mathbf{I}_q, \quad \mathbf{K}_q = \boldsymbol{\Phi}^T \mathbf{K}_n \boldsymbol{\Phi} = \mathrm{diag}\left(\omega_1^2, \omega_2^2, \cdots, \omega_q^2\right) \tag{4.2.20}$$

The two reduced matrices of mass and stiffness are diagonalized in the projected subspace. However, there is the damping matrix. It requires to distinct the so-called ***consistent damping matrix*** from general damping matrix. The consistent damping matrix assumes that, the projection matrix $\mathbf{C}_q$ of the original damping matrix $\mathbf{C}_n$ is also diagonalized just as for the mass and stiffness matrices, *i.e.*

$$\mathbf{C}_q = \mathbf{\Phi}^T \mathbf{C}_n \mathbf{\Phi} = \mathrm{diag}\left(2\varsigma_1\omega_1, 2\varsigma_2\omega_2, \cdots, 2\varsigma_q\omega_q\right) \tag{4.2.21}$$

The ***consistent damping matrix*** $\mathbf{C}_n$ is often expressed in the form of (usually called as Rayleigh damping, [8])

$$\mathbf{C}_n = c_m \mathbf{M}_n + c_k \mathbf{K}_n \tag{4.2.22}$$

for which $2\varsigma_j\omega_j = c_m + c_k\omega_j^2, j = 1, \cdots, q$. However, practical structural damping rarely fulfills this assumption. Introducing the projection vector

$$\boldsymbol{\gamma}_q = \mathbf{\Phi}^T \mathbf{p} \tag{4.2.23}$$

then the subspace projected dynamic equation is derived as

$$\ddot{\mathbf{u}} + \mathbf{C}_q \dot{\mathbf{u}} + \mathbf{K}_q \mathbf{u} = \boldsymbol{\gamma}_q e(t) \tag{4.2.24}$$

For consistent damping, the matrix $\mathbf{C}_q$ is also diagonal, so that the resulted equations are separated

$$\ddot{u}_j + 2\varsigma_j\omega_j\dot{u}_j + \omega_j^2 u_j = \gamma_j e(t), \quad j = 1, \cdots q \tag{4.2.25}$$

These equations have been one dimensional, where γ_j is the participation factor of the j-th mode, so that $\boldsymbol{\gamma}_q$ is naturally called the mode participation vector. The right-hand side function $e(t)$, which characterizes the random excitation, is zero-mean, hence the functions $u_j(t)$ are also zero-mean. Therefore the analysis should concentrate on the auto-correlated or power spectral density matrix functions of the responses $u_j(t)$. For consistent damping, equation (4.2.25) has been reduced to number q one dimensional vibration analysis already.

The random response analysis of a time-invariant system under a stationary excitation is easier to solve in the frequency domain. For a given frequency ω, the transfer (frequency response) function is

$$H_j(\omega) = 1/(\omega_j^2 - \omega^2 + \mathrm{i}2\zeta_j\omega_j\omega), \quad j = 1, \cdots q \tag{4.2.26}$$

and the power spectral density of u_j is given by

$$S_{uj}(\omega) = \gamma_j^2 \,|\, H_j(\omega)\,|^2\, S_e(\omega), \quad j = 1, \cdots q \tag{4.2.27}$$

In order to solve the $n \times n$ power spectral density matrix $\mathbf{S}_y(\omega)$ of the response vector $\mathbf{y}(t)$, the easier method of ***direct spectral expansion*** for random excitation $e(t)$ is used

$$e(t) = \int_{-\infty}^{\infty} \exp(\mathrm{i}\omega t)\mathrm{d}Z_e(\omega) \tag{4.2.28}$$

The spectral expansion method implies using the superposition of various monochromatic excitation induced random displacements. The displacement response is computed by

$$\mathbf{y}(t) = \int_{-\infty}^{\infty} \left[\sum_{j=1}^{q} \boldsymbol{\varphi}_j \gamma_j H_j(\omega) \right] \exp(\mathrm{i}\omega t)\mathrm{d}Z_e(\omega) \tag{4.2.29}$$

In the integration, random variable exists only in the orthogonal increments $\mathrm{d}Z_e(\omega)$ in the frequency domain, and all the other factors are deterministic. The correlation between the increments is given as

$$\mathrm{E}[\mathrm{d}\bar{Z}_e(\omega_1)\mathrm{d}Z_e(\omega_2)] = S_e(\omega_1)\delta(\omega_2-\omega_1)\mathrm{d}\omega_1\mathrm{d}\omega_2 \tag{4.2.30}$$

where a bar above represents complex conjugate.

After solved the random response of the displacement vector $\mathbf{y}(t)$, the auto-correlation function is computed as

$$\begin{aligned}\mathbf{R}_y(\tau) = \mathrm{E}[\bar{\mathbf{y}}(t)\mathbf{y}^T(t+\tau)] &= \int_{-\infty}^{\infty}\int_{-\infty}^{\infty}\sum_{i=1}^{q}\sum_{j=1}^{q}[\boldsymbol{\varphi}_j\gamma_j H_j(-\omega_1)\boldsymbol{\varphi}_i^T\gamma_i H_i(\omega_2)] \\ &\qquad \times\exp[-\mathrm{i}\omega_1 t+\mathrm{i}\omega_2(t+\tau)]\cdot\mathrm{E}[\mathrm{d}\bar{Z}_e(\omega_1)\mathrm{d}Z_e(\omega_2)] \\ &= \int_{-\infty}^{\infty}\int_{-\infty}^{\infty}\sum_{i=1}^{q}\sum_{j=1}^{q}[\boldsymbol{\varphi}_j\boldsymbol{\varphi}_i^T H_j(-\omega_1)H_i(\omega_2)\gamma_j\gamma_i] \\ &\qquad \times\exp[\mathrm{i}\omega_2\tau+\mathrm{i}(\omega_2-\omega_1)t)]\cdot S_e(\omega_1)\delta(\omega_2-\omega_1)\mathrm{d}\omega_1\mathrm{d}\omega_2 \\ &= \int_{-\infty}^{\infty}\sum_{i=1}^{q}\sum_{j=1}^{q}\left[\boldsymbol{\varphi}_j\boldsymbol{\varphi}_i^T H_j(-\omega)H_i(\omega)\gamma_j\gamma_i\right]\cdot S_e(\omega)\exp(\mathrm{i}\omega\tau)\mathrm{d}\omega\end{aligned}$$

Therefore according to the transformation relation between auto-correlation function and power spectral density, it derives

$$\mathbf{S}_y(\omega) = \sum_{i=1}^{q}\sum_{j=1}^{q}\left[\boldsymbol{\varphi}_j\boldsymbol{\varphi}_i^T H_j(-\omega)H_i(\omega)\gamma_j\gamma_i\right]\cdot S_e(\omega) \tag{4.2.31}$$

This is the well-known equation for power spectral density matrix of the structural displacement response, the CQC (Complete Quadratic Combination) equation, which includes the cross-correlation of all q participation modes. Therefore equation (4.2.31) is considered exact for $\mathbf{S}_y(\omega)$.

The equation has been expressed clearly. At a first glance, the problem seems completely solved. However, use of the above equation in computation for applications encountered tremendous difficulty from the computational expense, which is estimated now. For statistical estimation, the number of frequency ω points requires at least $p_\omega \geq 100$, it is necessary to compute according to the equation (4.2.31) for all these frequency points. Suppose the degrees of freedom of the system, say $n = 1000 \sim 10000$, the number of multiplication for computing the matrix $\boldsymbol{\varphi}_j\boldsymbol{\varphi}_i^T$ in equation (4.2.31) accounts to $n^2 = 10^6 \sim 10^8$, and to be executed for all the pairs of $(i,j), i \leq j$. The number of participant vibration modes requires $q = 50 \sim 500$. Therefore the number of $p_\omega n^2 q^2/2$ multiplication is accounted. Hence, for simplest single source random excitation analysis, the computational expense is still very large.

For comparing purpose, note that the triangular factorization of the structural global matrix, under the assumption that the average bandwidth is also q, requires the computation of multiplication number estimated as $nq^2/2$. Therefore the computational expense for the power spectral density matrix $\mathbf{S}_y(\omega)$ is roughly np_ω times to the global matrix triangular factorization.

The above estimation of computational expense is only for single source excitation, the practical excitation may be much more complicated. Because of the computational expense, the application of response analysis to structural system under various random excitations faces great computational difficulty. The various problems for stationary and non-stationary structural responses can only be solved practically based on the pseudo-excitation approach.

The computational expense of power spectral density matrix via the traditional method is prohibitively high. Therefore, almost all the classical monographs on random vibration, such as [63,65], recommend a simplified approximation with no exclusion, the SRSS (Square Root of the Sum of Squares) method, for which all the $i \neq j$ terms under the double summation $\sum\sum$ in equation (4.2.31) are completely neglected and the approximate equation

$$\mathbf{S}_y(\omega) \approx S_e(\omega) \cdot \sum_{j=1}^{q} \left[\boldsymbol{\varphi}_j \boldsymbol{\varphi}_j^T H_j(-\omega) H_j(\omega) \gamma_j^2 \right] \tag{4.2.32}$$

is recommended. The computational expense of the SRSS approximation is accounted roughly $1/q$ to the CQC method. However, this approximation can only be used under the prerequisites of small damping and that all participation mode eigenvalues are sparsely spaced. Practically, the first q eigen-frequencies of a three dimensional engineering structure is often clustered. Therefore the applicability of the SRSS approximation to 3-D structures is questionable.

Lin's method of **pseudo excitation** approach is the solution to the prohibitive computational expense of the CQC equation. Presently, the single source stationary stochastic excitation $e(t)$ has the spectral density $S_e(\omega)$, which is a positive function. The pseudo excitation proposes to ***replace the random excitation*** $e(t)$ ***in the equation (4.2.25) with a pseudo deterministic simple harmonic function***

$$e(t) \sim \sqrt{S_e(\omega)} \exp(\mathrm{i}\omega t) \tag{4.2.33}$$

The equation is then updated from a stochastic to a ***deterministic*** forced vibration.

$$\ddot{u}_j + 2\zeta_j \omega_j \dot{u}_j + \omega_j^2 u_j = \gamma_j \sqrt{S_e(\omega)} \exp(\mathrm{i}\omega t) \tag{4.2.34}$$

To solve a 1-D deterministic equation is simply as usual, and is given as

$$u_j = H_j(\omega) \gamma_j \sqrt{S_e(\omega)} \exp(\mathrm{i}\omega t)$$

Therefore the deterministic response of system displacement is

$$\mathbf{y}(t,\omega) = \sum_{j=1}^{q} \boldsymbol{\varphi}_j H_j(\omega) \gamma_j \sqrt{S_e(\omega)} \exp(\mathrm{i}\omega t) \tag{4.2.35}$$

Form which the system spectral density matrix $\mathbf{S}_y(\omega)$ is computed by

$$\mathbf{S}_y(\omega) = \bar{\mathbf{y}} \cdot \mathbf{y}^T \tag{4.2.36}$$

where a bar above represent taking the complex conjugate. Equation (4.2.36) is pseudo excitation approach. Substituting equation (4.2.35) into the above equation derives the double summation form, which returns to the form of equation (4.2.31).

Algebraically, the equation (4.2.36) is identical to equation (4.2.31). Both equations are the CQC result and are statistically the exact solutions. However the

computational expenses are quite different from each other. The meaning of pseudo excitation approach is firstly the high efficiency of computation. Let

$$\mathbf{z}_j(\omega)=\boldsymbol{\varphi}_j H_j(\omega)\gamma_j\sqrt{S_e(\omega)} \tag{4.2.37}$$

the three equations of computation are listed as follows

Usual CQC: $$\mathbf{S}_y(\omega)=\sum_{i=1}^{q}\sum_{j=1}^{q}(\overline{\mathbf{z}}_i\cdot\mathbf{z}_j^T) \tag{4.2.38a}$$

SRSS approximation: $$\mathbf{S}_y(\omega)\approx\sum_{j=1}^{q}(\overline{\mathbf{z}}_j\cdot\mathbf{z}_j^T) \tag{4.2.38b}$$

Pseudo excitation: $$\mathbf{S}_y(\omega)=(\sum_{j=1}^{q}\overline{\mathbf{z}}_j)\cdot(\sum_{i=1}^{q}\mathbf{z}_i^T\cdot) \tag{4.2.38c}$$

The comparison of computational expenses is given below. The main computational expense is the n-dimensional complex conjugate of vector $\mathbf{z}$ multiplying the transpose of $\mathbf{z}$, totally n^2 complex number multiplication, which is a **large computational expense**. The usual CQC method has q^2 times of such vector multiplication; for SRSS approximation it has q times of such multiplication; while for the pseudo excitation method it has only one such multiplication.

The SRSS approximation loses precision, but still requires q times of such multiplication, therefore which achieves nothing with comparison to the pseudo excitation method. The usual CQC method gives exact solution of $\mathbf{S}_y(\omega)$, but the computational expense is q^2 times as that the pseudo excitation method, for which the result is exact too. But why spend q^2 times of computational expense to stay on the traditional CQC approach. The multiplication of $\overline{\mathbf{z}}\cdot\mathbf{z}^T$ is the **main expense** of computations. If $q=100$, the difference of the computational expenses of the two approaches will be 10000 times! Four orders of magnitude! it is too large, at the rate of one minute versus five days.

Visually, for the algebraic identity

$$\begin{aligned}(x_1+x_2+\cdots+x_{100})\times(x_1+x_2+\cdots+x_{100})=\\ x_1^2+x_1x_2+\cdots+x_1x_{100}+\\ x_2x_1+x_2x_2+\cdots+x_2x_{100}+\\ +\cdots\cdots+\\ x_{100}x_1+x_{100}x_2+\cdots+x_{100}x_{100}\end{aligned}$$

where the left and right hand side expressions are equal algebraically, but the computational expense are quite different. The left-hand side expression requires only one multiplication, but the right-hand side requires 10000 times of multiplication. No one would like to execute according to the right-hand side expression. However, comparing the equation (4.2.38c) and (4.2.38a) shows that the latter is just the expanded form, and the pseudo excitation approach is the left-hand side expression, the difference is clear.

The above comparison reveals the essence of difference, which is simply an algebraic identity. However, it is a big problem bothering for decades of years. The method used to solve problem is the simpler the better. The pseudo excitation method does solve the computational problem, however, very efficient and easy to understand, the easier the better, so that it is a very nice method.

The SRSS approximation reduces computational expense, however incurred losing precision of result. But the computational expense of SRSS is still q times as much as the pseudo excitation method, so that it is a both facet lost method, so that it should not be used any longer. Before the pseudo excitation method proposed, using SRSS approximation is understandable, but not now.

§4.2.2.1, Case of inconsistent damping

The derivation above is based on the assumption of ***consistent*** (proportional) ***damping matrix***, see equation (4.2.21) and (4.2.22). The assumption is for the mathematical convenience of separation of variables. However, the applicability of the assumption of consistent damping is restricted. For example, for the installation of dampers, only several devices can be installed therefore the damping matrix is difficult to be modeled as a consistent damping matrix (4.2.22). Hence, the dynamic equation (4.2.24) in the q-dimensional subspace cannot be decomposed and reduced to the form of equation (4.2.25). Then traditional derivation of the CQC method does not work now, that the equation (4.2.32) of double summation equation of CQC is invalid, which raises a theoretical problem to be solved.

Lin's method of pseudo excitation solves the problem easily for the case of ***inconsistent damping*** matrix, because the separation of variables in the q-dimensional subspace is unnecessary. As a matter of fact, the stochastic process factor $e(t)$ at the right-hand side of dynamic equation (4.2.24) can still be substituted with the pseudo excitation of equation (4.2.33)

$$e(t) \sim \sqrt{S_e(\omega)}\exp(\mathrm{i}\omega t)$$

The right-hand side of the above expression has been a deterministic load. Therefore the equation (4.2.24) becomes

$$\ddot{\mathbf{u}} + \mathbf{C}_q\dot{\mathbf{u}} + \mathbf{K}_q\mathbf{u} = \boldsymbol{\gamma}_q\sqrt{S_e(\omega)}\exp(\mathrm{i}\omega t), \quad \boldsymbol{\gamma}_q = \boldsymbol{\Phi}^T\mathbf{p} \tag{4.2.39}$$

where $\mathbf{K}_q = \boldsymbol{\Phi}^T\mathbf{K}_n\boldsymbol{\Phi} = \mathrm{diag}\left(\omega_1^2, \omega_2^2, \cdots, \omega_q^2\right)$. Note that (4.2.39) is a q-D dynamic equation and its stationary solution is available, that the displacement vector is solved in the form

$$\mathbf{u}(t,\omega) = \mathbf{H}_q(\omega)\boldsymbol{\gamma}_q\sqrt{S_e(\omega)}\exp(\mathrm{i}\omega t) \tag{4.2.40}$$

where $\mathbf{H}_q(\omega)$ is a $q \times q$ transfer matrix to be determined. Substituting it into (4.2.39) gives

$$[-\omega^2\mathbf{I}_q - \mathrm{i}\omega\mathbf{C}_q + \mathbf{K}_q]\cdot\mathbf{H}_q(\omega)\boldsymbol{\gamma}_q = \boldsymbol{\gamma}_q$$

and

$$\mathbf{H}_q(\omega) = \left[-\omega^2\mathbf{I}_q - \mathrm{i}\omega\mathbf{C}_q + \mathbf{K}_q\right]^{-1} \tag{4.2.41}$$

Comparing with equation (4.2.26), the sole difference is that the matrix $\mathbf{H}_q(\omega)$ is not a diagonal one. The inversion of a complex valued Hermite matrix is easy. Hence after computed the vector

$$\mathbf{y}(t,\omega) = \boldsymbol{\Phi}\mathbf{H}_q(\omega)\boldsymbol{\gamma}_q\sqrt{S_e(\omega)}\exp(\mathrm{i}\omega t)$$

the power spectral density matrix $\mathbf{S}_y(\omega)$ can still be computed by

$$\mathbf{S}_y(\omega) = \bar{\mathbf{y}} \cdot \mathbf{y}^T \tag{4.2.42}$$

which has the same form with comparison to equation (4.2.36). Therefore

$$\mathbf{S}_y(\omega) = S_e(\omega) \cdot [\mathbf{\Phi H}_q(-\omega)\bar{\boldsymbol{\gamma}}_q] \cdot [\mathbf{\Phi H}_q(\omega)\boldsymbol{\gamma}_q]^T \tag{4.2.43}$$

A n-dimensional complex vector is in the bracket. Evidently the rank of the matrix $\mathbf{S}_y(\omega)$ is 1 and the computational steps are given as:

1) Giving $n \times n$ matrices $\mathbf{M}_n$ and $\mathbf{K}_n$, extract the first q eigen-solutions by means of the subspace iteration method, then compose the $n \times q$ real matrix $\mathbf{\Phi}$ and compute $\boldsymbol{\gamma}_q = \mathbf{\Phi}^T \mathbf{p}$.
2) Matrix inversion of equation (4.2.41), and then compute the $q \times q$ complex matrix $\mathbf{H}_q(\omega)$.
3) Compute the n-dimensional complex valued vector $\mathbf{\Phi H}_q(\omega)\boldsymbol{\gamma}_q$.
4) Use equation (4.2.43) to find $\mathbf{S}_y(\omega)$, which requires n^2 operations of complex multiplication.

Therefore, the pseudo excitation method can readily be used to solve random vibration with inconsistent damping matrix without difficulty, and gives again the exact solution of $\mathbf{S}_y(\omega)$. Usually the damping matrix is often frequency ω dependent, so that the frequency domain solution can fit such a situation.

The solution of random vibration with inconsistent damping matrix demonstrates the versatility of pseudo excitation method, which far extends the traditional CQC method. As a matter of fact, there have been a number of papers published, such as [83], for extending the CQC method to the inconsistent damping case, but these papers can find only approximate solutions with complicated computational method with excessive expense. Clearly, the pseudo excitation method gives exact solution by means of simple but still very efficient algorithm.

§4.2.2.2, Single source multi-point excitation with different phases

Long-span structures are constructed ever increasingly, for which the aseismic considerations are of great importance. Obviously, regardless of the different phases of the excitation points is inappropriate for long-span structures. Treating the different phases of excitations at different ground support points (the ***wave passage effect***) is important in the random vibration analysis of long span structures. The traditional CQC analysis requires separation of variables for the q-D reduced subspace, so that no exact solution can be found for treating wave passage effect. For approximate methods, the computational expense rapidly increases not only with the number q of participation eigen-modes but also with the number of ground supports, that the strict CQC analysis is impractical. A great deal of efforts have been devoted to the extension of the current aseismic response spectrum method in order to consider such kind of multi-phase excitations, see [66,67], however, it requires to introduce a series of approximations, that the verification and validation are difficult.

Using pseudo excitation method to treat the multi-phase problem is straightforward, because the wave passage effect of the ground multi-point excitation is still of single source. The different phases of various earthquake excitations can be determined with the coordinates of the ground supports, wave speed and frequency ω in the spectral expansion, see [84]. The factor of different phases can also be included in the excitation partition vector $\mathbf{p}$ at the right hand side of dynamic equation (4.2.17). Therefore $\mathbf{p}$ is a given complex valued deterministic vector and $e(t)$ is again a zero mean stationary stochastic process, the same as in the former case of in-phase excitation. The case of $\mathbf{p}$ is a complex valued vector has been considered in equation (4.2.43) with (4.2.39). Therefore, by comparing with the in-phase case, only the single source random excitation partition vector $\mathbf{p}$ is updated from a real vector to a complex valued vector, the equations are completely the same, so that the increasing of computational expense is limited. The application of pseudo excitation method is still highly efficient as in the previous cases. The '***wave passage effect***' problem is thus simply solved and the computational result obtained in this way is still precise [85]. The pseudo excitation method not only solves the problem of tremendous computational expense for traditional CQC method but also contributes the theory to problems such as ***inconsistent damping*** and '***wave passage effect***' etc.

§4.2.3, Stationary response of structure to multi-source excitations

The treatment of 'wave passage effect' in last section solves the problem simply and smoothly. For ***multi-source random excitation*** problems the pseudo excitation method can still be applied smoothly to solve the structural response problem.

Multi-sources excitation is nothing more than that the excitation should be expressed with multiple, m, independent sources. Correspondingly, instead of (4.2.17), the dynamic equation should be written as

$$\mathbf{M}_n\ddot{\mathbf{y}} + \mathbf{C}_n\dot{\mathbf{y}} + \mathbf{K}_n\mathbf{y} = \mathbf{P}\mathbf{e}(t) \tag{4.2.44}$$

where $\mathbf{e}(t)$ is a m-dimensional stochastic vector, and $\mathbf{P}$ is a given $n \times m$ matrix. In the single source case there is only one stationary stochastic process of exciting $e(t)$ and the partition is expressed with a n-dimensional vector $\mathbf{p}$. For case of multi-sources excitation, the m components $e_i(t), i = 1,\cdots,m$ of the vector $\mathbf{e}(t)$ are mutually independent zero-mean stationary stochastic processes and the power spectral density matrix is denoted as $\mathbf{S}_e(\omega)$, a $m \times m$ positive definite Hermite matrix. Correspondingly, the partitioning of excitations changes to m groups, which are combined and expressed as a given partitioning matrix $\mathbf{P}$ of $n \times m$ dimension. Because the elements in $\mathbf{P}$ can take complex values so that the 'wave passage effect' is included.

The analysis of random response is again in the projected subspace spanned by the first q eigenvectors of the original $(\mathbf{M}_n, \mathbf{K}_n)$ system. The dynamic equation in the projection space is

$$\ddot{\mathbf{u}} + \mathbf{C}_q\dot{\mathbf{u}} + \mathbf{K}_q\mathbf{u} = \mathbf{\Gamma}_{qm}\mathbf{e}(t) \qquad (4.2.45)$$

Because of $\mathbf{e}(t)$ this equation is still stochastic, where the matrix

$$\mathbf{\Gamma}_{qm} = \mathbf{\Phi}^T\mathbf{P}$$

is $q \times m$ dimensional with $m \le q$ and is given. The stochastic equation (4.2.45) is solved below using the pseudo excitation method.

§4.2.3.1, Spectral expansion

The stationary stochastic process $\mathbf{e}(t)$ can be expressed with the spectral expansion

$$\mathbf{e}(t) = \int_{-\infty}^{\infty} \exp(\mathrm{i}\omega t)\mathrm{d}\mathbf{Z}_e(\omega)$$

where

$$\mathrm{E}\left[\mathrm{d}\mathbf{Z}_e(\omega)\right] = \mathbf{0}\,,\quad \mathrm{E}\left[\mathrm{d}\overline{\mathbf{Z}}_e(\omega)\mathrm{d}\mathbf{Z}_e(\omega_2)^T\right] = \mathbf{S}_e(\omega)\delta(\omega_2 - \omega)\mathrm{d}\omega_2\mathrm{d}\omega$$

which is described in chapter 3. The frequency domain orthogonal incremental process $\mathrm{d}\mathbf{Z}_e(\omega)$, m -dimensioned, is stationary stochastic process $\mathbf{e}(t)$ related, because of the spectral density matrix $\mathbf{S}_e(\omega)$, which is a given deterministic positive definite Hermite matrix for $\mathbf{e}(t)$ and also for $\mathrm{d}\mathbf{Z}_e(\omega)$. Introducing the ***unit orthogonal incremental stochastic process*** $\mathrm{d}\mathbf{Z}(\omega)$ in frequency domain

$$\mathrm{E}\left[\mathrm{d}\mathbf{Z}(\omega)\right] = \mathbf{0}\,,\quad \mathrm{E}\left[\mathrm{d}\overline{\mathbf{Z}}(\omega)\mathrm{d}\mathbf{Z}(\omega_2)^T\right] = \mathbf{I}_m\delta(\omega_2 - \omega)\mathrm{d}\omega_2\mathrm{d}\omega \qquad (4.2.46')$$

to describe the spectral expansion of the stochastic excitation $\mathbf{e}(t)$. The merit of the ***unit orthogonal incremental stochastic process*** is independent on the individual excitation and is unified. Hence

$$\mathbf{e}(t) = \int_{-\infty}^{\infty} \exp(\mathrm{i}\omega t)\mathbf{S}_{\sqrt{e}}(\omega)\mathrm{d}\mathbf{Z}(\omega) \qquad (4.2.46)$$

where $\mathbf{S}_{\sqrt{e}}(\omega)$ is a $m \times m$ matrix. The equation of auto-correlation function of $\mathbf{e}(t)$ is

$$\mathbf{R}_e(\tau) = \mathrm{E}\left[\overline{\mathbf{e}}(\mathrm{t})\mathbf{e}^T(\mathrm{t}+\tau)\right] = \cdots = \int_{-\infty}^{\infty} \exp(\mathrm{i}\omega\tau)\overline{\mathbf{S}}_{\sqrt{e}}(\omega)\mathbf{S}_{\sqrt{e}}^T(\omega)\mathrm{d}\omega \qquad (4.2.47)$$

so that

$$\overline{\mathbf{S}}_{\sqrt{e}}(\omega)\mathbf{S}_{\sqrt{e}}^T(\omega) = \mathbf{S}_e(\omega) \qquad (4.2.48)$$

which shows again that $\mathbf{S}_e(\omega)$ is a $m \times m$ Hermite matrix. The equation (4.2.48) expresses the factorization equation for the matrix $\mathbf{S}_{\sqrt{e}}(\omega)$ based on the given $\mathbf{S}_e(\omega)$.

§4.2.3.2, Response analysis

After the stationary stochastic excitation $\mathbf{e}(t)$ is expanded according to equation (4.2.46), the pseudo excitation is derived and substituted into the equation (4.2.45), which gives

$$\ddot{\mathbf{u}} + \mathbf{C}_q\dot{\mathbf{u}} + \mathbf{K}_q\mathbf{u} = \mathbf{\Gamma}_{qm}\mathbf{S}_{\sqrt{e}}(\omega)\exp(\mathrm{i}\omega t) \qquad (4.2.49)$$

from which the response $\mathbf{u}$, a $q \times m$ matrix, is solved. Then $\mathbf{Y}(\omega) = \mathbf{\Phi u}$, a $n \times m$ matrix function of frequency, is computed from the equation

$$\mathbf{Y}(\omega) = \mathbf{\Phi H}_q(\omega)\mathbf{\Gamma}_{qm}\mathbf{S}_{\sqrt{e}}(\omega) \tag{4.2.50}$$

where the power spectral density matrix of the response is computed by

$$\mathbf{S}_y(\omega) = \overline{\mathbf{Y}} \cdot \mathbf{Y}^T \tag{4.2.51}$$

The equation is quite simple, which is obtained by the same method of ***pseudo excitation method***.

Substituting equation (4.2.50) into (4.2.51), using equation (4.2.48) gives

$$\mathbf{S}_y(\omega) = [\overline{\mathbf{\Phi H}}_q(-\omega)\overline{\mathbf{\Gamma}}_{qm}]\mathbf{S}_e(\omega)[\mathbf{\Phi H}_q(\omega)\mathbf{\Gamma}_{qm}]^T \tag{4.2.52}$$

Hence directly executing the above equation, the factorization (4.2.48) can also be bypassed. However, computing via the factorization (4.2.48) is also efficient.

§4.3, Response under excitation of non-stationary stochastic process

Stationary stochastic process is a simplification of practical excitations. Real excitations are non-stationary. As mentioned above, the traditional CQC computation for structures under stationary excitation has been tremendous, so that the structure under non-stationary excitation will be more difficult. Presently, the uniformly modulated evolutionary stochastic excitation is more acceptable in structural engineering, and is expressed as

$$F(t) = a(t)e(t) \tag{4.3.1}$$

where $e(t)$ is a stationary stochastic excitation, which is regarded simply as a one dimensional zero-mean process with the power spectral density function $S_e(\omega)$ given, and $a(t)$ is a deterministic amplitude modulation function which characterizes the non-stationary behavior of the excitation. The function $a(t)$ should be statistically determined from the recorded data. Usually $a(t)$ is considered a ***slowly varying*** function, *i.e.* the condition $(\mathrm{d}a/\mathrm{d}t)/\omega_l \ll 1$ is fulfilled, where ω_l is the lower bound of significant frequencies.

§4.3.1, Response under uniformly modulated non-stationary excitation

Because $e(t)$ is a zero mean process, so $F(t)$ is also zero mean. The power spectral density can be expressed approximately as

$$S_F(\omega,t) \approx S_e(\omega) \cdot a^2(t) \tag{4.3.2}$$

Traditional CQC analysis for non-stationary excitation is very complicated [86], that the equation derivation and computational expense are both immense. The application of the pseudo excitation method is still quite superior.

For non-stationary uniformly modulated excitation, there are still cases as follows:

1) Single source with in-phase non-stationary excitation;

2) Single source multi-phases non-stationary excitation;
3) Multiple sources multi-point partially coherent non-stationary excitations.

For these excitation cases, the pseudo excitation analysis can be transplanted from the corresponding part with stationary excitations.

The solution of multi-sources excitation problems is selected to give a brief description. The dynamic equation is again (4.2.44), the subspace projected dynamical equation is (4.2.45). The ***spectral expansion*** of the stochastic process is now

$$\mathbf{e}(t) = a(t) \cdot \int_{-\infty}^{\infty} \exp(\mathrm{i}\omega t) \mathbf{S}_{\sqrt{e}}(\omega) \mathrm{d}\mathbf{Z}(\omega) \tag{4.3.3}$$

where the frequency domain unit orthogonal incremental stochastic process $\mathrm{d}\mathbf{Z}(\omega)$, see (4.2.46'), is the same as that described above. Therefore the equation (4.2.49) becomes now

$$\ddot{\mathbf{u}} + \mathbf{C}_q \dot{\mathbf{u}} + \mathbf{K}_q \mathbf{u} = \mathbf{\Gamma}_{qm} \mathbf{S}_{\sqrt{e}}(\omega) a(t) \exp(\mathrm{i}\omega t) \tag{4.3.4}$$

The solution of this equation can be computed by means of the stepwise ***precise integration method*** and the $q \times m$ matrix function $\mathbf{u}(\omega, t)$ is resulted. Afterwards, using $\mathbf{Y} = \mathbf{\Phi u}$ transform back to the original n-dimensional space. The spectral density matrix is also a function of time, *i.e.* it is in the ***time-frequency domain***,

$$\mathbf{S}_y(\omega, t) = \overline{\mathbf{Y}} \mathbf{Y}^T$$

The details can be found from the series of papers [71~76].

§4.3.2, Response under evolutionary modulated non-stationary excitation

The uniformly modulated non-stationary stochastic excitation is the simplest among the non-stationary stochastic processes. Further non-stationary stochastic excitation problems should be considered. The non-uniform amplitude modulation is also considered as ***slowly varying***. For aseismic engineering, at the beginning stage the earthquake excitation is composed of frequencies from low to high components, a wide band stochastic process. However, the high wave-number and also the high frequency components are gradually damped out more quickly with time, and the stochastic excitations are then dominated by low frequency components, *i.e.* the frequency band be biased to the low frequency end. To model such a phenomenon, the non-uniformly modulated ***evolutionary random excitation model*** [77] is used, particularly in earthquake engineering

$$\mathbf{e}(t) = \int_{-\infty}^{\infty} \exp(\mathrm{i}\omega t) A(\omega, t) \mathbf{S}_{\sqrt{e}}(\omega) \mathrm{d}\mathbf{Z}(\omega) \tag{4.3.5}$$

where $A(\omega, t)$ is a given deterministic ***slowly varying*** non-uniform modulation function in the ***frequency-time domain***. Equation (4.3.5) uses the integration in Stieltjes sense, on which the spectral expansion equation (4.2.46') is based. The frequency domain ***unit orthogonal incremental stochastic process*** $\mathrm{d}\mathbf{Z}(\omega)$ is the same as before. The deterministic dynamical equation derived by the pseudo excitation method is

$$\ddot{\mathbf{u}} + \mathbf{C}_q \dot{\mathbf{u}} + \mathbf{K}_q \mathbf{u} = \mathbf{\Gamma}_{qm} \mathbf{S}_{\sqrt{e}}(\omega) A(\omega, t) \exp(\mathrm{i}\omega t) \tag{4.3.6}$$

Its solution can be obtained by the ***precise integration method***. For each specific value of ω in the frequency domain, the integration of equation (4.3.6) need be solved. After solved the $q \times m$ matrix function $\mathbf{u}(\omega,t)$, using $\mathbf{Y} = \mathbf{\Phi u}$ goes back to the original n-dimensional space. The spectral density matrix is also a function of time, that it is in the ***frequency-time domain***,

$$\mathbf{S}_y(\omega,t) = \overline{\mathbf{Y}}\mathbf{Y}^T$$

The details can be found from the series of papers [71~76].

The use of pseudo excitation method to the non-stationary random response problems can be found from a series of papers given by Jia-Hao LIN and others. The theory and computation of non-stationary random vibration of engineering structures is a challenge. Using the traditional approach, it is very difficult to solve practical problems. Lin's pseudo excitation method proposes the present day best approach for solving these problems in engineering. The review of pseudo excitation method can be found from [78,85].

Chapter 5, Elastic system with single continuous coordinate

Theory of elasticity usually treats plane or three dimension deformation problems and is an indispensable foundation of applied mechanics. To solve the problems in the frame of Hamiltonian dual system will be much rational with compare to semi-inverse method in traditional approach. However, elasticity problems have infinite degrees of freedom. For ease of understanding, let us begin with the solution of single continuous coordinate system. Such problems are useful itself, that problems in strength of materials and structural mechanics, such as the Timoshenco's beam theory, are under single continuous coordinate. The semi-analytical approach [7] derived equations naturally falls into single continuous coordinate system. Especially, such system corresponds to the analytical dynamic system, if the single continuous coordinate is regarded as 'time' coordinate. However, the difference is that the boundary condition being initial for analytical dynamics, but being TPBVP presently for structural mechanics.

The importance of single continuous coordinate elastic system analysis is further based on the analogy relationship between structural mechanics and optimal control. When the Kalman filter and LQ (linear quadratic) control problems are discussed in chapter 6, we will find that the Hamilton form dual equations in structural mechanics of the single continuous coordinate system, and the dual equation system in control theory are analogous to each other mathematically. The ***analogy relationship between structural mechanics and optimal control*** is thus established [20~22,87~91]. Especially, if the LQG theory of optimal control analogous to structural mechanics is considered the first stage, then the H_∞ robust control theory corresponds to the eigenvalue problems in structural stability or natural vibration frequency [92~94], such second stage analogy relationship is of fruitful implication. Analogy is quite beneficial to both sides of structural mechanics and optimal control.

So far, in the discussions of analytical dynamics or vibration theory, the degrees of freedom n are limited to be finite. However, the time coordinate is continuous, hence these systems are single continuous coordinated too. The single coordinate is space or time for elasticity or dynamics, respectively. For space coordinate the problem derives to TPBVP of ODE, which corresponds to elliptic type PDEs, however, for single time coordinate, the problem derives to initial value problems which corresponds to evolutionary PDEs. Elastic wave propagation along a strip in frequency domain is also a two point boundary value problem (TPBVP).

For easy understanding, let us begin the discussion with Timoshenco beam theory.

§5.1, Fundamental equations of Timoshenco beam theory

Beam theory with shearing deformation was by S.P. Timoshenco. Only the transverse bending and shearing of plane beam is considered here. Assuming the coordinate z points along the beam axis before deformation, see figure 5.1. The beam transverse cross section is assumed rigid, so that its displacements are denoted with $\tilde{u}(z),\tilde{\psi}(z)$, the line displacement along the transverse coordinate x and rotation of the cross section, respectively. Since only transverse bending is considered, the longitudinal displacement along the axis z is zero. Each beam (infinitesimal) segment along z has bending and shearing deformations, $\tilde{\kappa}=\mathrm{d}\tilde{\psi}/\mathrm{d}z$ and $\tilde{\gamma}=(\mathrm{d}\tilde{u}/\mathrm{d}z-\tilde{\psi})$, respectively. Therefore the strain energy of the beam is

$$\Pi=\int_0^L\left[EJ(\mathrm{d}\tilde{\psi}/\mathrm{d}z)^2/2+kGA(\mathrm{d}\tilde{u}/\mathrm{d}z-\tilde{\psi})^2/2\right]\mathrm{d}z \tag{5.1.1}$$

where EJ is bending stiffness, A is the area of cross section, k is cross section shearing mode coefficient, for rectangular cross section $k\approx 1.2$. The beam internal forces are

$$\tilde{M}=EJ\tilde{\kappa}=EJ\,(\mathrm{d}\tilde{\psi}/\mathrm{d}z),\qquad \tilde{Q}=kGA\tilde{\gamma}=kGA(\mathrm{d}\tilde{u}/\mathrm{d}z-\tilde{\psi}) \tag{5.1.2}$$

The dynamic equations of the beam are

$$\partial\tilde{Q}/\partial z+\tilde{g}=\rho A\,\partial^2\tilde{u}/\partial t^2,\qquad \tilde{Q}-\partial\tilde{M}/\partial z+\tilde{m}=\rho J\,\partial^2\tilde{\psi}/\partial t^2 \tag{5.1.3}$$

where $\tilde{g}$ is the distributed transverse load, $\tilde{m}$ is the distributive moment load, their positive directions coincide with $\tilde{u}$ and $\tilde{\psi}$, respectively. The wavy sign $\sim$ above is to remain identifiers for the frequency domain description.

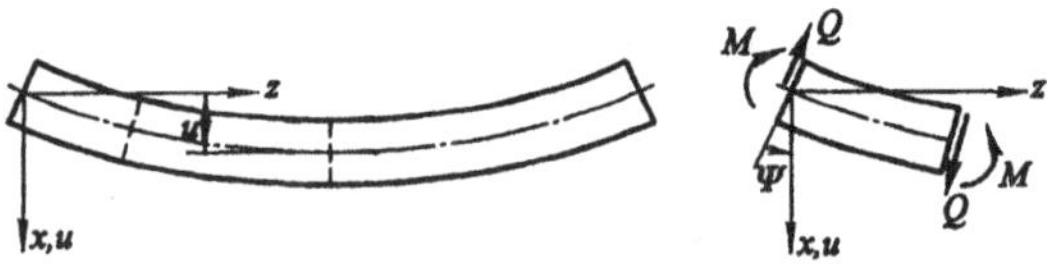

Figure 5.1, *A sketch for Timoshenco beam theory*

Eliminating the internal forces and strains, the dynamic equations expressed in terms of displacements are obtained

$$\begin{aligned}&(\partial/\partial z)\left[kGA\left(\partial\tilde{u}/\partial z-\tilde{\psi}\right)\right]+\tilde{g}=\rho A\,\partial^2\tilde{u}/\partial t^2\\&(\partial/\partial z)\left(EJ\,\partial\tilde{\psi}/\partial z\right)+kGA\left(\partial\tilde{u}/\partial z-\tilde{\psi}\right)+\tilde{m}=\rho J\,\partial^2\tilde{\psi}/\partial t^2\end{aligned} \tag{5.1.4}$$

The initial conditions are

$$\tilde{u}=u_0(z),\quad \tilde{\psi}=\psi_0(z),\qquad \text{when}\quad t=0 \tag{5.1.5}$$

Two end boundary conditions are given at $z=0$ and $z=L$ as

$$\tilde{u}=\text{given}\quad\text{or}\quad\tilde{Q}=\text{given},\qquad\text{and}\qquad\tilde{\psi}=\text{given}\quad\text{or}\quad\tilde{M}=\text{given} \tag{5.1.6}$$

In the analysis of vibration or wave propagation problems, the time coordinate is quite often transformed to the frequency domain. Let

$$\tilde{u}=u(z,\omega)\cdot e^{-i\omega t},\qquad \tilde{\psi}=\psi(z,\omega)\cdot e^{-i\omega t} \tag{5.1.7}$$

where ω is the circular frequency, a parameter now. The dynamic equations become

$$(\mathrm{d}/\mathrm{d}z)\left[kGA(\mathrm{d}u/\mathrm{d}z-\psi)\right]+\rho A\omega^2u+g=0 \tag{5.1.8a}$$

$$(\mathrm{d}/\mathrm{d}z)(EJ\,\mathrm{d}\psi/\mathrm{d}z)+kGA(\mathrm{d}u/\mathrm{d}z-\psi)+\rho J\omega^2\psi+m=0 \tag{5.1.8b}$$

These equations are expressed in displacement method. Introducing the displacement vector

$$\mathbf{q} = \{u, \psi\}^T , \qquad \dot{\mathbf{q}} = \{\dot{u}, \dot{\psi}\}^T \tag{5.1.9}$$

where a dot over the variable $\mathbf{q}$ denotes the derivative with respect to z, $\dot{\mathbf{q}} = \mathrm{d}\mathbf{q}/\mathrm{d}z$. Using matrix/vector formulation will be beneficial to cover the general problem.

§5.2, Potential energy density and mixed energy density

The variational principle corresponding to the dynamic equation (5.1.8) of Timoshenco beam is

$$\delta \int_0^L \left[\left(EJ(\dot{\psi})^2 + kGA(\dot{u} - \psi)^2 - \rho\omega^2 (Au^2 + J\psi^2) \right) / 2 - gu - m\psi \right] \mathrm{d}z = 0 \tag{5.2.1}$$

Because of the inertia term, the potential energy is no longer a minimum but still takes stationary value. To express in terms of vector formulation (5.1.9), the variational formulation becomes

$$\mathrm{L}(\mathbf{q}, \dot{\mathbf{q}}) = \dot{\mathbf{q}}^T \mathbf{K}_{22} \dot{\mathbf{q}}/2 + \dot{\mathbf{q}}^T \mathbf{K}_{21} \mathbf{q} + \mathbf{q}^T \mathbf{K}_{11} \mathbf{q}/2 - \mathbf{g}^T \mathbf{q} \tag{5.2.2}$$

$$\delta \int_0^L \mathrm{L}(\mathbf{q}, \dot{\mathbf{q}}) \mathrm{d}z = 0 \tag{5.2.3}$$

$$\mathbf{K}_{22} = \begin{bmatrix} kGA & 0 \\ 0 & EJ \end{bmatrix}, \mathbf{K}_{21} = \begin{bmatrix} 0 & -kGA \\ 0 & 0 \end{bmatrix}, \mathbf{K}_{11} = -\begin{bmatrix} \rho A\omega^2 & 0 \\ 0 & \rho J\omega^2 - kGA \end{bmatrix} \tag{5.2.4}$$

$$\mathbf{g} = \{g, m\}^T , \qquad \mathbf{K}_{12} = \mathbf{K}_{21}^T$$

For general problems of single continuous coordinate system, the Lagrange function and variational principle can still be written as (5.2.2) and (5.2.3). Equation (5.2.4) is only for Timoshenco beam, and needs to be updated for other problems. The main purpose of this chapter is facing general problems, so that use will only be made of (5.2.2~3) for further discussion. Carrying out the variational calculus, equation (5.2.3) derives to the Lagrange equation

$$(\mathrm{d}/\mathrm{d}z)(\partial \mathrm{L}/\partial \dot{\mathbf{q}}) - \partial \mathrm{L}/\partial \mathbf{q} = \mathbf{0}$$

Comparing to analytical dynamics, the only difference is that the former continuous coordinate time t is changed to be a space continuous coordinate z. The quadratic Lagrange function (5.2.2) derives a linear Euler-Lagrange equation

$$\mathbf{K}_{22} \ddot{\mathbf{q}} + (\mathbf{K}_{21} - \mathbf{K}_{12}) \dot{\mathbf{q}} - \mathbf{K}_{11} \mathbf{q} + \mathbf{g} = \mathbf{0} \tag{5.2.5}$$

The matrix/vector formulation can fit the general problems, but not limited only to Timoshenco beam theory. The solution method given below can work certainly for general case, such that the displacement vector $\mathbf{q}$ is n-dimensional, but not limited two-dimensional as that for Timoshenco beam, which can be used as an example. The semi-analytical method [7] derived equation is n-dimensional.

In the variational equation (5.2.3) and the respective dynamic equation (5.2.5), there is only one kind of variables, *i.e.* the displacement. Introducing dual variables as that in analytical dynamics

$$\mathbf{p} = \partial \mathrm{L}/\partial \dot{\mathbf{q}} , \qquad \text{or} \quad \mathbf{p} = \mathbf{K}_{22} \dot{\mathbf{q}} + \mathbf{K}_{21} \mathbf{q} \tag{5.2.6}$$

which is considered the constitutive rule. Combining with the dynamic equation (5.2.5) compose the dual differential equations as

$$\dot{\mathbf{q}} = -\mathbf{K}_{22}^{-1}\mathbf{K}_{21}\mathbf{q} + \mathbf{K}_{22}^{-1}\mathbf{p} \tag{5.2.7a}$$

$$\dot{\mathbf{p}} = (\mathbf{K}_{11} - \mathbf{K}_{12}\mathbf{K}_{22}^{-1}\mathbf{K}_{21})\mathbf{q} + \mathbf{K}_{12}\mathbf{K}_{22}^{-1}\mathbf{p} - \mathbf{g} \tag{5.2.7b}$$

(5.2.7a) is the compatibility equation, and (5.2.7b) is the dynamic equation. For simplicity, introducing the matrices

$$\mathbf{D} = \mathbf{K}_{22}^{-1}, \quad \mathbf{A} = -\mathbf{K}_{22}^{-1}\mathbf{K}_{21}, \quad \mathbf{B} = \mathbf{K}_{11} - \mathbf{K}_{12}\mathbf{K}_{22}^{-1}\mathbf{K}_{21} \tag{5.2.8}$$

then the dual differential equations become

$$\dot{\mathbf{q}} = \mathbf{A}\mathbf{q} + \mathbf{D}\mathbf{p} + \mathbf{f}_q, \qquad \dot{\mathbf{p}} = \mathbf{B}\mathbf{q} - \mathbf{A}^T\mathbf{p} + \mathbf{f}_p \tag{5.2.9a,b}$$

where $\mathbf{f}_q = \mathbf{0}$, $\mathbf{f}_p = -\mathbf{g}$. Because $\mathbf{K}_{22}$ and $\mathbf{K}_{11}$ are symmetric matrices, $\mathbf{K}_{22}$ positive definite, so $\mathbf{D}$ is positive definite too, and $\mathbf{B}^T = \mathbf{B}$.

Equations (5.2.9a,b) are non-homogeneous, its solution should first solve the corresponding dual homogeneous equation set (the homogeneous compatibility and dynamic equations)

$$\dot{\mathbf{q}} = \mathbf{A}\mathbf{q} + \mathbf{D}\mathbf{p}, \qquad \dot{\mathbf{p}} = \mathbf{B}\mathbf{q} - \mathbf{A}^T\mathbf{p} \tag{5.2.10a,b}$$

Introducing the Hamilton function

$$\mathrm{H}(\mathbf{q},\mathbf{p}) = \mathbf{p}^T\mathbf{D}\mathbf{p}/2 + \mathbf{p}^T\mathbf{A}\mathbf{q} - \mathbf{q}^T\mathbf{B}\mathbf{q}/2 \tag{5.2.11}$$

The corresponding variational principle is

$$S = \int_{z_0}^{z_f}[\mathbf{p}^T\dot{\mathbf{q}} - \mathrm{H}(\mathbf{q},\mathbf{p})]\mathrm{d}t, \quad \delta S = 0 \tag{5.2.12}$$

Performing the variational derivation gives equations (5.2.10a,b), which is a linear Hamilton system. The present dual equations (5.2.10a,b) should also compare with the equations (2.3.9a,b) of gyroscopic system, that they are the same equations except the positive/negative sign before the matrix $\mathbf{B}$. Presently, the continuous coordinate is spatial, so that the boundary conditions should be assigned at the two ends of $z_0 = 0$ and $z_\mathrm{f} = L$. But for gyroscopic system, the continuous coordinate is time, hence the boundary condition should be assigned at the initial point $t = t_0$. The two point boundary value problem has the characteristics of structural mechanics, elasticity etc. Mathematically, their boundary value problem corresponds to the elliptic PDEs. However, the initial boundary condition holds the characteristics of hyperbolic PDEs. To dig further, from analytical dynamics, $S(z_0,\mathbf{q}_0;z_\mathrm{f},\mathbf{q}_\mathrm{f})$ is the action function; however, from structural mechanics, it is the ***deformation energy*** of a given interval (z_0, z_f).

Compose the state vector $\mathbf{v}(z)$ by combining the dual vectors $\mathbf{q},\mathbf{p}$

$$\mathbf{v} = \{\mathbf{q}^T, \mathbf{p}^T\}^T, \qquad \dot{\mathbf{v}} = \mathbf{H}\mathbf{v} \tag{5.2.13}$$

the dual equations are also combined in (5.2.13), where

$$\mathbf{H} = \begin{bmatrix} \mathbf{A} & \mathbf{D} \\ \mathbf{B} & -\mathbf{A}^T \end{bmatrix} \tag{5.2.14}$$

The two end boundary conditions can be determined later for various cases of boundary assignment. The $\mathbf{H}$ is a Hamilton matrix of dimension $2n \times 2n$, with the characteristic of

$$\mathbf{J}=\begin{bmatrix}\mathbf{0} & \mathbf{I}\\ -\mathbf{I} & \mathbf{0}\end{bmatrix},\quad \mathbf{JH}=\begin{bmatrix}\mathbf{B} & -\mathbf{A}^T\\ -\mathbf{A} & -\mathbf{D}\end{bmatrix},\quad (\mathbf{JH})^T=\mathbf{JH}$$

The corresponding Hamilton function $\mathrm{H}(\mathbf{q},\mathbf{p})=-\mathbf{v}^T(\mathbf{JH})\mathbf{v}/2$ is a homogeneous quadratic function of $\mathbf{q},\mathbf{p}$, the characteristic of a conservative linear system.

The Hamilton function $\mathrm{H}(\mathbf{q},\mathbf{p})$ is termed as ***density of mixed energy*** in structural mechanics, and the corresponding Lagrange function $\mathrm{L}(\mathbf{q},\dot{\mathbf{q}})$ is the ***density of potential energy***. Note, formerly in the variational principles of elasticity, the deformation energy is either expressed all in terms of strains ε (and displacements), *i.e.* ***density of strain energy*** $U_0(\varepsilon)$; or else expressed all in terms of stresses, *i.e.* the ***density of complementary energy*** $U_0^*(\sigma)$. Both these expressions are not mixed energy density. The expression of Hamilton function uses partially strains and displacements and partially the dual variables, *i.e.* stresses. The value of Hamilton function is neither density of strain energy nor the density of complementary energy, but the combination of both, hence nominated the term ***density of mixed energy***.

The density of mixed energy can have either positive or negative value, in fact it is indefinite. However, the density of strain energy or density of complementary energy is positive definite usually.

In analytical dynamics, the Lagrange function is composed of (K.E − P.E), where K.E and P.E stand for kinetic and potential energy, respectively. In teaching, it often has the problem why the minus operation, quite a puzzle; however, the corresponding Hamilton function is composed of (K.E + P.E), this rule is easy to understand. Now in structural mechanics, the longitudinal coordinate z is treated as the time coordinate t in analytical dynamics. A parallel mathematical development with the Hamilton system theory in analytical dynamics works too, and the situation is similar. Presently the ***Lagrange function is the density of strain energy***, easy to understand; however, its duality transformed Hamilton function becomes ***density of mixed energy***, not so easy to understand too. The situation is just the reverse of the analytical dynamics case. The cause of such reversion appearing in the dual differential equation is the positive/negative sign difference before the matrix $\mathbf{B}$.

The above discussion is based on linear system. However, we know from analytical dynamics that Hamilton system theory is quite general, never limited only in linear systems. Now the linear structural mechanics and elasticity problems are considered, it certainly reflects the contents of linear Hamilton system. It is emphasized here, that ***Hamilton system theory can also be used for non-linear elasticity***.

So far the general case of n-dimensional displacement $\mathbf{q}$ is considered. The theory is now applied to the problem of Timoshenco beam, for which $n=2$, the dual variables are

$$\mathbf{p}=\mathbf{K}_{22}\dot{\mathbf{q}}+\mathbf{K}_{21}\mathbf{q}=\{kGA(\dot{u}-\psi),\ EJ\dot{\psi}\}^T=\{Q,\ -M\}^T$$

The physical meaning is the generalized internal force, *i.e.* the shearing force Q and bending moment M. The matrices are

$$\mathbf{D}=\begin{bmatrix}(kGA)^{-1} & 0\\ 0 & (EJ)^{-1}\end{bmatrix},\qquad \mathbf{A}=\begin{bmatrix}0 & 1\\ 0 & 0\end{bmatrix},\qquad \mathbf{B}=\begin{bmatrix}-\rho\omega^2 A & 0\\ 0 & -\rho\omega^2 J\end{bmatrix}$$

and the density of mixed energy or the Hamilton function is

$$\mathrm{H}(\mathbf{q},\mathbf{p})=Q\psi-Q^2/(2kGA)-M^2/(2EJ)-\rho\omega^2(Au^2+J\psi^2)/2$$

which is an indefinite quadratic function, with the arguments being the dual unknowns u,ψ,Q,M , but with no derivative with respect to z .

§5.3, Separation of variables, Adjoint symplectic ortho-normality

The solution of the dual equations (5.2.9a,b) can be classified into two kinds of method, namely, 1) direct integration method and 2) method of separation of variables. Direct integration method is popular for initial value problems, however, for ***two point boundary value problem*** (TPBVP) presently, the direct integration may appear serious numerical problem. When dimension n is not large, the precise integration method can be used. Even applying precise integration method it still needs great care. If the TPBVP is solved as an initial value problem, a "brute force" approach, which may appear numerical problem. In order to get highly precise result for TPBVP, the corresponding precise integration approach should turn to solve the ***Riccati differential equation***. Later in section 5.7 the precise integration method for TPBVP will be described in some detail.

Method of separation of variables in combination with the eigen-solution expansion method is also very efficient. One may aware the importance of eigen-solution expansion method by recalling the key-role played by modal analysis in vibration problems. The combination of precise integration and eigen-solution expansion methods will have good result for TPBVP. The method of separation of variables is described first. The homogeneous differential equation is

$$\dot{\mathbf{v}}=\mathbf{H}\mathbf{v} \tag{5.3.1}$$

where $\mathbf{H}$ is a Hamilton matrix. The method of separation of variables intends to find the solution in the form

$$\mathbf{v}(z)=\boldsymbol{\psi}\cdot Z(z) \tag{5.3.2}$$

where $Z(z)$ is a function of z only, but independent on the component number of vector $\mathbf{v}$, and $\boldsymbol{\psi}$ is a $2n$-dimensional vector, independent on z

$$\boldsymbol{\psi}=\{\psi_1,\cdots,\psi_{2n}\}^T \tag{5.3.3}$$

which represents the variation along 'transverse' direction. Substituting (5.3.2) into (5.2.13) derives

$$\mathbf{H}\boldsymbol{\psi}_i=(\dot{Z}/Z)\boldsymbol{\psi}_i\ ,\qquad i=1,2,\cdots,2n$$

In above equation, the left hand side is independent on z, hence the factor $(\dot{Z}/Z)$ must be independent on z too and it is obviously independent on the component number of the vector, hence it must be a constant μ, then $Z(z)=\exp(\mu z)$ and

$$\mathbf{H}\boldsymbol{\psi}=\mu\cdot\boldsymbol{\psi} \tag{5.3.4}$$

which is an eigenvalue problem of the Hamilton matrix.

A $2n \times 2n$ matrix must have $2n$ eigenvalues $\mu_i (i=1,2,\cdots,2n)$. The eigenvalues of a Hamilton matrix has the characteristics as follows. Left multiplying equation (5.3.4) with $\mathbf{J}$ and note that $\mathbf{J}^2 = -\mathbf{I}_{2n}$ gives

$$-\mathbf{JHJ}\cdot\mathbf{J}\psi = \mu\cdot\mathbf{J}\psi \ , \qquad \text{or} \qquad \mathbf{H}^T(\mathbf{J}\psi) = -\mu(\mathbf{J}\psi)$$

which implies that the transpose matrix of $\mathbf{H}$ has the eigen-solution with eigenvalue $-\mu$ and eigenvector $\mathbf{J}\psi$. Because of the transpose matrix and the original matrix must have the same eigenvalue spectra. Hence, the matrix $\mathbf{H}$ must have the eigenvalue $-\mu$. Therefore the $2n$-eigenvalues can be subdivided into two classes:

$$(\alpha)\ \mu_i, \quad \mathrm{Re}(\mu_i) < 0 \ \text{or}\ \mathrm{Re}(\mu_i) = 0 \wedge \mathrm{Im}(\mu_i) > 0\,, \quad (i=1,2,\cdots n) \tag{5.3.5a}$$

$$(\beta)\ \mu_{n+i} = -\mu_i\,, \qquad (i=1,2,\cdots n) \tag{5.3.5b}$$

the eigenvalues in the (α) class are ordered according to $\mathrm{Re}(\mu_i)$, such that the less negative one appears first.

Hamilton matrix $\mathbf{H}$ is asymmetric, hence complex eigenvalues may appear and duplicate eigenvalues should also be considered. Corresponding to the duplicate eigenvalue the ***Jordan normal form*** eigenvector and its subsidiary eigenvectors may appear. In structural mechanics (elasticity), Jordan normal form cannot be avoided. However, the appearance of duplicate eigenvalue of $\mathrm{Re}(\mu) \neq 0$ and the corresponding Jordan form is occasional, such as the simply supported rectangular plate bending. In developing general theory, such occasional situation will not be emphasized, but theoretically it is still interested.

The eigenvalue $\mu = 0$ is a special case, which does not involve in the classification (5.3.5). So the expression (5.3.5) is still not precise enough. In structural static and in theory of elasticity the appearance of $\mu = 0$ is frequent [23] and is a simple case for solution. However, it brings complication on developing theory. Because of $\mu = -\mu = 0$, so that the symplectic adjoint eigenvectors are mixed with the Jordan normal form subsidiary eigen-solutions. The solution methodology for such problem is as follows. The subspace corresponding to the eigenvalue zero of the Hamilton matrix $\mathbf{H}$ should be found first. Then the whole space is subdivided into zero eigenvalue subspace and its complement subspace. The two subspaces must be symplectic orthogonal to each other. The symplectic projection of matrix $\mathbf{H}$ to these two subspaces is in block diagonal form. The submatrix of $\mathbf{H}$ corresponding to the complement subspace has no eigenvalue zero, for which the classification of equation (5.3.5) is appropriate.

The eigen-solutions corresponding to $\mu = 0$ zero in structural mechanics or in elasticity is the most important part in expansion solution, because these eigen-solutions do not decay exponentially. The $\mathrm{Re}(\mu) = 0$ eigen-solutions do not decay exponentially either, which is quite useful in wave propagation problems and corresponding to transmission waves. The $\mathrm{Re}(\mu) \neq 0$ eigen-solutions usually represent local vibrations.

§5.3.1, Adjoint symplectic orthogonality

The requirements from structural vibration and stability and other fields promoted the eigen-problem of a real symmetric matrix undergone deep research. All eigenvalues are real valued and its duplicate eigen-roots have no Jordan form eigenvectors. All eigenvectors are mutually orthogonal to each other, so that they can be ortho-normalized with respect to the mass matrix $\mathbf{M}$. All the eigenvectors span the complete vector space, any vector in this space can be expressed as the linear combination of these eigenvectors, which is call the ***expansion theorem***. See chapter-2.

For Hamiltonian matrix the ***adjoint symplectic ortho-normality*** relationship can be proved instead of the ortho-normality with respect to mass $\mathbf{M}$ for symmetric matrix. For two eigen-solutions of i and j

$$\mathbf{H}\psi_i = \mu_i \psi_i , \qquad \mathbf{H}\psi_j = \mu_j \psi_j$$

Follow the derivation below equation (5.3.4) gives $\mathbf{H}^T\mathbf{J}\psi_i = -\mu_i \mathbf{J}\psi_i$. Left multiplying this equation with ψ_j^T and taking transpose gives (note, a scalar can be transposed arbitrarily)

$$\psi_j^T \mathbf{H}^T \mathbf{J}\psi_i = -\psi_i \mathbf{J}\mathbf{H}\psi_j = -\mu_i \psi_j^T \mathbf{J}\psi_i = \mu_i \psi_i^T \mathbf{J}\psi_j$$

However, left multiplying $\psi_i^T\mathbf{J}$ to the eigen-equation of ψ_j gives $\psi_i^T \mathbf{J}\mathbf{H}\psi_j = \mu_j \psi_i^T \mathbf{J}\psi_j$. Adding together with the previous equation gives

$$(\mu_i + \mu_j)\psi_i^T \mathbf{J}\psi_j = 0 \tag{5.3.6}$$

The ***adjoint symplectic orthogonality*** relationship is derived from this equation as

$$\psi_i^T \mathbf{J}\psi_j = 0 , \qquad \text{when } \mu_i + \mu_j \neq 0 \tag{5.3.7}$$

The case of all eigenvalues are single is considered first. The $2n$ eigenvalues are classified as given in equation (5.3.5). For the eigenvector ψ_i of eigenvalue $\mu_i (i \le n)$, there is only one eigenvector ψ_j $(j = n + i)$ being not symplectic orthogonal to ψ_i, *i.e.* ***symplectic adjoint***, that the other $2n-1$ eigenvectors including ψ_i itself are symplectic orthogonal to ψ_i. It is found that the situation for eigen-solutions parallels to the vibration of gyroscopic systems.

Recall that the orthogonality relationship between two eigenvectors of a symmetric matrix is $\psi_i^T \cdot \psi_j = 0$, which can also be written as $\psi_i^T \mathbf{I}\psi_j = 0$, *i.e.* the ***inner product*** of two vectors equals zero. Comparing with the equation of symplectic orthogonality, the matrix $\mathbf{I}$ is changed as $\mathbf{J}$, which corresponds to the ***metric matrix*** changed from Euclidean to symplectic. Therefore the idea of ***symplectic inner product*** is introduced. The symplectic inner product between two vectors $\mathbf{v}_i$ and $\mathbf{v}_j$ is defined as

$$\mathbf{v}_i^T \mathbf{J}\mathbf{v}_j \equiv -\mathbf{v}_j^T \mathbf{J}\mathbf{v}_i$$

For single eigenvalue case, it should prove that the eigenvectors of adjoint symplectic pair of eigenvalues must not be symplectic orthogonal, *i.e.*

$$\psi_i^T \mathbf{J}\psi_{n+i} = 1 , \qquad i = 1,2,\cdots,n \tag{5.3.8}$$

based on which, because the eigenvector can have an arbitrary constant multiplier, it can always reach ***symplectic normality*** as expressed in (5.3.8). The proof is based on the ***fundamental theorem of linear algebra*** [1] as follows.

Proof: Single root implies that the Jordan form equation $(\mathbf{H}-\mu_i\mathbf{I})\mathbf{v}=\psi_i$ has no solution. Thus the fundamental theorem of linear algebra requires any of the solutions $\mathbf{v}_*$ of its transpose equation $(\mathbf{H}^T-\mu_i\mathbf{I})\mathbf{v}_*=\mathbf{0}$ are not orthogonal to ψ_i, *i.e.* $\mathbf{v}_*^T\psi_i\neq 0$. The transpose equation has eigen-solution $\mathbf{J}\mathbf{v}_*=\psi_{n+i}$, hence $\psi_{n+i}^T\mathbf{J}\psi_i\neq 0$. Selecting appropriate constant factors for the adjoint eigenvectors, symplectic normalization (5.3.8) is reached. ##

The eigenvectors ψ_i and ψ_{n+i} have two constant factors, so that one another condition can be selected as

$$\psi_i^T\psi_i=\psi_{n+i}^T\psi_{n+i}, \qquad i=1,2,\cdots,n \tag{5.3.9}$$

The combination of equations (5.3.7) and (5.3.8) is termed as ***adjoint symplectic ortho-normality*** relationship. Making use of the $2n$ vectors ψ_i as columns composes a $2n\times 2n$ matrix

$$\boldsymbol{\Psi}=[\psi_1,\ \ \psi_2,\ \ \cdots,\ \ \psi_{2n}] \tag{5.3.10}$$

According to the ***adjoint symplectic ortho-normality*** relationship it verifies

$$\boldsymbol{\Psi}^T\mathbf{J}\boldsymbol{\Psi}=\mathbf{J} \tag{5.3.11}$$

A matrix $\boldsymbol{\Psi}$ satisfying the above equation is called a ***symplectic matrix.*** This equation can be used as the definition of symplectic matrix. Such term had been seen in previous chapters. Symplectic matrices have the distinguished behaviors,

1) The product of two symplectic matrices remains symplectic.
2) The inverse of a symplectic matrix is symplectic.
3) The transpose of a symplectic matrix is symplectic.
4) The unit matrix $\mathbf{I}_{2n}$ and $\mathbf{J}$ are symplectic matrices,
5) The determinant of a symplectic matrix equals ± 1.

The matrix $\boldsymbol{\Psi}$ is composed of eigenvectors, evidently the below relation holds

$$\boldsymbol{\Psi}^{-1}\mathbf{H}\boldsymbol{\Psi}=\mathrm{diag}[\mu_1,\mu_2,\cdots,\mu_n;\mu_{n+1},\cdots,\mu_{2n}] \tag{5.3.12}$$

If the determinant of a symplectic matrix equals -1, which is another leaf other than $+1$, then select a column changing its sign derives the matrix determinant to $+1$.

§5.3.2, Expansion theorem

Because the $2n$ eigenvectors are linearly independent, so these eigenvectors span the complete $2n$-dimensional state space. Therefore an arbitrary vector $\mathbf{g}$ in the state space can be linearly combined with these eigenvectors, *i.e.*

$$\mathbf{g}=\sum_{i=1}^{n}(a_i\psi_i+b_i\psi_{n+i}) \tag{5.3.13}$$

where a_i, b_i are coefficients to be determined. Based on the adjoint symplectic ortho-normality relationship gives

$$a_i=-\psi_{n+i}^T\mathbf{J}\mathbf{g},\quad b_i=\psi_i^T\mathbf{J}\mathbf{g} \tag{5.3.14}$$

which is the ***expansion theorem***.

Eigen-problem of a Hamilton matrix, symplectic orthogonality etc. is derived from introducing the dual variables followed with the state space approach, etc. These ideas do not appear in the existing textbooks, such as the series by S.P. Timoshenco. The description given above is from the method of separation of variables etc., not much concern on the physical background. However, the theory is derived from applied mechanics, there must be physical meaning behind. Below in section 5.4.2, it is proved that symplectic orthogonality closely relates to the work reciprocal theorem (Betti) [22,95,96], which gives sound physical background.

§5.4, Multiple eigenvalues and the Jordan normal form

The derivation above is based on the assumption of single eigen-root. However, Hamilton matrix is asymmetric, multiple roots (m-times) and the Jordan normal form may possibly appear. The analysis should extend to such case. A theorem for eigenvalue problem [47,48] of a $n\times n$ matrix $\mathbf{A}$, as follows: ***Given a general*** $n\times n$ ***matrix*** $\mathbf{A}$***, there must be an*** $n\times n$ ***non-singular matrix*** $\mathbf{X}$***, whose elements can be complex numbers, that***

$$\mathbf{AX} = \mathbf{X}\times \mathrm{diag}(\mathbf{J}_1, \ \mathbf{J}_2, \ \cdots, \ \mathbf{J}_t) \tag{5.4.1a}$$

where

$$\mathbf{J}_i = \begin{bmatrix} \lambda_i & 1 & & 0 \\ & \lambda_i & 1 & \\ & & \cdots & 1 \\ 0 & & & \lambda_i \end{bmatrix}, \quad m_i\times m_i \ \textbf{\textit{matrix}} \tag{5.4.1b}$$

and $m_1+m_2+\cdots+m_t=n$. $\mathbf{J}_i$ ***is termed as Jordan block, and there are*** m_i ***eigenvector and subsidiary eigenvectors in*** $\mathbf{X}$ ***correspond to the*** m_i***-multiple eigenvalue*** λ_i. ***The equations are***

$$\begin{aligned} \mathbf{Ax}_{i1} &= \lambda_i\mathbf{x}_{i1} \\ \mathbf{Ax}_{i2} &= \lambda_i\mathbf{x}_{i2} + \mathbf{x}_{i1} \\ &\cdots\cdots \\ \mathbf{Ax}_{im_i} &= \lambda_i\mathbf{x}_{im_i} + \mathbf{x}_{i(m_i-1)} \end{aligned} \tag{5.4.1c}$$

for every Jordan block of $i=1,2,\ldots,t$. This is a theorem for a general $n\times n$ matrix $\mathbf{A}$.

Hamilton matrix has a featured structure, but the arrangement of all the Jordan blocks with equations as given in equation (5.4.1) is **inappropriate**. Let μ_i be a m_i-multiple eigenvalue then $-\mu_i$ is also a m_i-multiple eigenvalue. In order to keep its **Hamilton matrix structure**, the mutually adjoint symplectic dual blocks should be of the forms

$$\mathbf{J}_i = \begin{bmatrix} \mu_i & 1 & & 0 \\ & \mu_i & 1 & \\ & & \cdots & 1 \\ 0 & & & \mu_i \end{bmatrix} \quad -\mathbf{J}_i^T = \begin{bmatrix} -\mu_i & 0 & & 0 \\ -1 & -\mu_i & & \\ & -1 & \cdots & 0 \\ 0 & & -1 & -\mu_i \end{bmatrix} \tag{5.4.2a,b}$$

Hence the original form of Jordan block (5.4.1) should be revised so as to fulfill the

Hamilton matrix form as given in equation (5.4.2a,b). Similarity transformation of a Hamilton matrix $\mathbf{H}$ by means of a symplectic matrix remains Hamiltonian, that

$$\mathbf{H}_p = \mathbf{S}^{-1}\mathbf{H}\mathbf{S}, \qquad \mathbf{S}^T\mathbf{J}\mathbf{S} = \mathbf{J} \tag{5.4.3}$$

since $\mathbf{J}\mathbf{H}_p = \mathbf{J}\mathbf{S}^{-1}\mathbf{H}\mathbf{S} = (\mathbf{J}\mathbf{S}^{-1}\mathbf{J})(\mathbf{J}\mathbf{H}\mathbf{J})\mathbf{J}\mathbf{S} = -\mathbf{S}^T\mathbf{H}^T\mathbf{S}^{-T}\mathbf{J} = (\mathbf{J}\mathbf{S}^{-1}\mathbf{H}\mathbf{S})^T = (\mathbf{J}\mathbf{H}_p)^T$. So $\mathbf{H}_p$ is still a Hamilton matrix.

According to the original composition rule of Jordan form, equation (5.4.1b), the Jordan block of eigenvalue $-\mu_i$ and the corresponding eigen-equations are

$$\begin{bmatrix} -\mu_i & 1 & & 0 \\ & -\mu_i & 1 & \\ & & \cdots & 1 \\ 0 & & & -\mu_i \end{bmatrix}, \qquad \begin{aligned} \mathbf{H}\psi_{n+i} &= -\mu_i\psi_{n+i} \\ \mathbf{H}\psi_{n+i+1} &= -\mu_i\psi_{n+i+1} + \psi_{n+i} \\ &\cdots\cdots \\ \mathbf{H}\psi_{n+i+m_i-1} &= -\mu_i\psi_{n+i+m_i-1} + \psi_{n+i+m_i-2} \end{aligned}$$

which obviously does not coincide with the feature of a Hamilton matrix. To change the Jordan block to the form as given in equation (5.4.2b), the order of (subsidiary)-eigenvectors should be changed, as follows. The subsidiary eigenvectors should change signs alternatively, *i.e.* the vectors $\psi_{n+i+1}, \psi_{n+i+3}, \cdots$ are multiplied by -1 but keep the other vectors unchanged. Thereafter reverse their order. Then the inappropriate Jordan block form can be changed to the form given in (5.4.2b).

When the eigenvector corresponds to a single root, equation (5.3.2) gives the solution of the state vector. When Jordan form appears, its (α) class eigenvalue μ_i corresponding state vector solutions are

$$\begin{aligned} \mathbf{v}_i &= \psi_i \exp(\mu_i z), \\ \mathbf{v}_i^{(1)} &= [\psi_{i+1} + z \cdot \psi_i]\exp(\mu_i z), \\ \mathbf{v}_i^{(2)} &= [\psi_{i+2} + z\psi_{i+1} + (z^2/2!) \cdot \psi_i]\exp(\mu_i z), \end{aligned} \tag{5.4.4}$$

where triple-multiplied eigenvalue μ_i is assumed. Correspondingly, $-\mu_i = \mu_{n+i}$ is also a triple-multiplied eigenvalue. If the original Jordan form solutions are followed, then

$$\begin{aligned} \mathbf{v}_{n+i} &= \psi_{n+i} \exp(-\mu_i z), \\ \mathbf{v}_{n+i}^{(1)} &= [\psi_{n+i+1} + z \cdot \psi_{n+i}]\exp(-\mu_i z), \\ \mathbf{v}_{n+i}^{(2)} &= [\psi_{n+i+2} + z\psi_{n+i+1} + (z^2/2!) \cdot \psi_{n+i}]\exp(-\mu_i z), \end{aligned}$$

where the order of subsidiary eigenvectors has not been changed, hence it cannot coincide with the form of (5.4.2b). Using such (subsidiary)-eigenvectors as columns composed matrix $\mathbf{\Psi}$ cannot ensure a symplectic matrix. In order to keep the adjoint symplectic ortho-normality relationship being valid for the matrix $\mathbf{\Psi}$, the Jordan block must be transformed to behave as in equation (5.4.2b). The necessity of such transformation can be seen from the interpretation of the ***adjoint symplectic ortho-normality relationship*** in applied mechanics, see section 5.4.2.

The above description has not mentioned the zero eigenvalue Jordan normal form. For easily understanding of the zero eigenvalue Jordan form, let us begin with a simple problem.

§5.4.1, *Wave propagation for Timoshenco beam and its extension*

The description above is the general theory for a single continuous coordinate system. For Timoshenco beam theory, its equation is given as

$$\mathbf{H}=\begin{bmatrix} 0 & 1 & 1/kGA & 0 \\ 0 & 0 & 0 & 1/EJ \\ -\rho\omega^2 A & 0 & 0 & 0 \\ 0 & -\rho\omega^2 J & -1 & 0 \end{bmatrix}, \quad \det.(\mathbf{H}-\mu\mathbf{I})=0 \qquad (5.4.5)$$

Expanding the determinant gives

$$\mu^4+\mu^2\rho\omega^2(1/E+1/kG)+\rho^2\omega^4/(EGk)-\rho\omega^2 A/EJ=0$$

Obviously, from this equation, μ and $-\mu$ are simultaneously eigenvalues, which coincides the characteristic of a Hamilton matrix. The equation is quadratic with respect to μ^2 and the condition for two real roots is always valid

$$\rho^2\omega^4(1/E-1/Gk)^2+\rho\omega^2 A/EJ>0$$

hence μ^2 has two real roots. However, there are two cases, let

$$\omega_{cr}^2=kGA/(\rho J) \qquad (5.4.6)$$

Case-1: when $\omega^2>\omega_{cr}^2$, μ^2 has two negative roots and case 2: when $\omega^2<\omega_{cr}^2$, μ^2 has one negative and one positive root.

Case-1: the two roots of μ^2 derive to $\mu_1\mathrm{i},\mu_2\mathrm{i}$ and $-\mu_1\mathrm{i},-\mu_2\mathrm{i}$, therefore the solutions of state vectors are

$$\mathbf{v}_1(z)=\psi_1\exp(\mathrm{i}\mu_1 z), \qquad \mathbf{v}_2(z)=\psi_2\exp(\mathrm{i}\mu_2 z)$$
$$\mathbf{v}_3(z)=\psi_3\exp(-\mathrm{i}\mu_1 z), \quad \mathbf{v}_4(z)=\psi_4\exp(-\mathrm{i}\mu_2 z)$$

which correspond to the solutions

$$\tilde{\mathbf{v}}_1(z,t)=\psi_1\exp[\mathrm{i}(\mu_1 z+\omega t)], \qquad \tilde{\mathbf{v}}_2(z,t)=\psi_2\exp[\mathrm{i}(\mu_2 z+\omega t)]$$
$$\tilde{\mathbf{v}}_3(z,t)=\psi_3\exp[\mathrm{i}(-\mu_1 z+\omega t)], \quad \tilde{\mathbf{v}}_4(z,t)=\psi_4\exp[\mathrm{i}(-\mu_2 z+\omega t)]$$

where $\tilde{\mathbf{v}}=\{\tilde{u},\tilde{\psi},\tilde{Q},-\tilde{M}\}^T$. Obviously, these solutions are two pairs of transmission waves, the phase velocities of wave propagation are ω/μ_1 and ω/μ_2, respectively, each pair traveling toward $+z$ and $-z$ directions, respectively.

Case-2: there are one negative root and one positive root. The negative root of μ^2 gives a pair of transmission waves traveling toward $+z$ and $-z$ directions, respectively, as before; where the positive root of μ^2 gives the two solutions

$$\mathbf{v}_1(z)=\psi_1\exp(-\mu_1 z) \quad \text{and} \quad \mathbf{v}_3(z)=\psi_3\exp(\mu_1 z)$$

Such solutions decay as $\text{abs}(z)\to\infty$, hence the solution resulted behaves the feature of local vibration, and possibly induces resonance, so that it should also be concerned. When the transmission wave mixed with such localized vibration mode, wave induced resonance may appear, and it will be discussed later.

For the special case of $\omega^2=\omega_{cr}^2$, zero roots of μ^2 appears, which gives two zero root of μ. Hence Jordan form appears and the corresponding eigenvector is found from the equation

$$\begin{array}{ccccccccc} 0 & + & \psi & + Q/(kGA) & + & 0 & =0 \\ 0 & + & 0 & + & 0 & -M/EJ & =0 \\ -kGA^2u/J & + & 0 & + & 0 & + & 0 & =0 \\ 0 & - & kGA\psi & - & Q & + & 0 & =0 \end{array}$$

Solving gives $\quad \boldsymbol{\psi}_1^{(0)} = \{u=0,\ \psi=-1,\ Q=kGA,\ -M=0\}^T$ (5.4.7)

Next expanding the first subsidiary eigenvector of Jordan form equation $\mathbf{H}\boldsymbol{\psi}^{(1)} = \boldsymbol{\psi}_1^{(0)}$ gives

$$\begin{array}{ccccccccc} 0 & + & \psi & + Q/(kGA) & + & 0 & = & 0 \\ 0 & + & 0 & + & 0 & -M/EJ & = & -1 \\ -kGA^2u/J & + & 0 & + & 0 & + & 0 & = kGA \\ 0 & - & kGA\psi & - & Q & + & 0 & = & 0 \end{array}$$

Solving gives

$$\boldsymbol{\psi}_1^{(1)} = \{u=-J/A,\ \psi=0,\ Q=0,\ -M=-EJ\}^T \tag{5.4.8}$$

where $\boldsymbol{\psi}^{(1)}$ can linearly superpose the term $a\boldsymbol{\psi}_1^{(0)}$ arbitrarily. The zero eigenvalue is twice duplicated root, so the related eigenvectors have been exhausted.

From the eigenvector $\boldsymbol{\psi}_1^{(0)}$, according to equation (5.3.2), the solution of homogeneous equation (5.2.13) for the state vector is

$$\mathbf{v}_1 = \boldsymbol{\psi}_1^{(0)}, \qquad \tilde{\mathbf{v}}_1(z,t) = \boldsymbol{\psi}_1^{(0)} e^{\mathrm{i}\omega_{cr}t}$$

However, the subsidiary eigenvector is not a solution of the state vector. To find the corresponding solution of state vector, according to equation (5.4.3) the solution of differential equation (5.2.13) for the state vector is composed as

$$\mathbf{v}_2 (= \mathbf{v}_1^{(1)}) = [\boldsymbol{\psi}_1^{(1)} + z\boldsymbol{\psi}_1^{(0)}], \quad \tilde{\mathbf{v}}_2 = \mathbf{v}_2 e^{\mathrm{i}\omega_{cr}t}$$

According to the theory, $\boldsymbol{\psi}_1^{(0)}$ and $\boldsymbol{\psi}_1^{(1)}$ are mutually adjoint symplectic to each other, and can be verified directly.

For $\omega^2 = \omega_{cr}^2$, there is the negative root of $\mu^2 = -A(1+kG/E)/J$, so

$$\mu_{3,4} = \mp \mathrm{i} \cdot \sqrt{(A/J)(1+kG/E)}$$

and the corresponding eigenvectors should satisfy the equations

$$\begin{array}{ccccccccc} -\mu u & + & \psi & + Q/(kGA) & + & 0 & =0 \\ 0 & - & \mu\psi & + & 0 & -M/EJ & =0 \\ -kGA^2u/J & + & 0 & - & \mu Q & + & 0 & =0 \\ 0 & & -kGA\psi & - & Q & + & \mu M & =0 \end{array}$$

Solving the above equations gives

$$\boldsymbol{\psi}_i = \{u=-1/(kGA),\ \psi=1/(\mu_i EJ);\ Q=A/(\mu_i J),\ -M=1\}^T, \quad (i=3,4)$$

These two eigenvectors are mutually symplectic adjoint to each other, as can be verified directly. Except these two symplectic adjoint pairs, the other selection of any two eigenvectors must be symplectic orthogonal to each other, all these can be verified directly, although the conclusion is derived from general theory.

The case of $\omega^2 = \omega_{cr}^2$ gives the Jordan form for $\mu = 0$. Because of $\omega_{cr}^2 \neq 0$, phase velocity of wave propagation (ω/μ) tends to infinity. But the phase velocity implies monochromatic wave, which does not represent energy transfer speed, for which the group velocity of wave should be considered. Looking from the view-point of energy transmission, the group velocity of a wave hump is more meaningful.

A special case of $\omega = 0$ should be considered, for which the eigenvalue equation becomes $\mu^4 = 0$. The zero eigenvalue is 4-tuple, which is a typical case of structural static. The equations for eigenvector and its multiple subsidiary eigenvectors of Jordan normal form can be combined as

$$\begin{array}{lccccc} & \boldsymbol{\Psi}^{(0)} & \boldsymbol{\Psi}^{(1)} & \boldsymbol{\Psi}^{(2)} & \boldsymbol{\Psi}^{(3)} \\ 0+\psi+Q/(kGA)+\quad 0 \quad =0 & 1 & 0 & 0 & 0 \\ 0+0+\quad 0 \quad -M/EJ=0 & 0 & 1 & 0 & EJ/(kGA) \\ 0+0+\quad 0 \quad +\quad 0 \quad =0 & 0 & 0 & 0 & -EJ \\ 0+0-\quad Q \quad +\quad 0 \quad =0 & 0 & 0 & EJ & 0 \end{array} \tag{5.4.9}$$

where the first column of the right hand side is all zero (under the header of $\boldsymbol{\Psi}^{(0)}$), for which solving the equations gives

$$\boldsymbol{\psi}^{(0)} = \{u=1,\ \psi=0;\ Q=0,\ -M=0\}^T$$

where $\boldsymbol{\Psi}^{(0)}$ is the eigenvector. Then $\boldsymbol{\Psi}^{(0)}$ is used as the second column of right hand side (under the header of $\boldsymbol{\psi}^{(1)}$) to solve the first subsidiary eigenvector $\boldsymbol{\psi}^{(1)}$. Solving gives $\boldsymbol{\psi}^{(1)}$. Then $\boldsymbol{\psi}^{(1)}$ is used as the third column vector at the right hand side in equation (5.4.9), (under the header of $\boldsymbol{\psi}^{(2)}$). Because this column vector $\boldsymbol{\psi}^{(1)}$ is symplectic orthogonal to the eigenvector $\boldsymbol{\Psi}^{(0)}$, solving the simultaneous equations gives the second subsidiary eigenvector $\boldsymbol{\psi}^{(2)}$. Then $\boldsymbol{\psi}^{(2)}$ is used as the fourth column vector at the right hand side in equation (5.4.9), (under the header of $\boldsymbol{\psi}^{(3)}$). This column vector is still symplectic orthogonal to the eigenvector $\boldsymbol{\Psi}^{(0)}$, solving the simultaneous equation gives the third subsidiary eigenvector $\boldsymbol{\psi}^{(3)}$. Using $\boldsymbol{\psi}^{(3)}$ as the fifth column at right hand side in (5.4.9). Because this column vector is no longer symplectic orthogonal to the eigenvector $\boldsymbol{\Psi}^{(0)}$, the simultaneous equation has no solution. The Jordan chain ceases here.

Up to here, all the subsidiary Jordan eigenvectors have been found. These $\boldsymbol{\psi}$ vectors are not directly the solutions of the original equation (5.2.13), however the solutions of the original equation can be composed with these vectors as follows:

$$\left.\begin{array}{lr} \mathbf{v}_1 = \boldsymbol{\psi}^{(0)}, & \text{translation} \\ \mathbf{v}_2 = \boldsymbol{\psi}^{(1)} + z\cdot\boldsymbol{\psi}^{(0)}, & \text{rotation} \\ \mathbf{v}_3 = \boldsymbol{\psi}^{(2)} + z\boldsymbol{\psi}^{(1)} + z^2\boldsymbol{\psi}^{(0)}/2, & \text{pure bending} \\ \mathbf{v}_4 = \boldsymbol{\psi}^{(3)} + z\boldsymbol{\psi}^{(2)} + z^2\boldsymbol{\psi}^{(1)}/2 + z^3\boldsymbol{\psi}^{(0)}/3!, & \text{bending with constant shear} \end{array}\right\} \tag{5.4.10}$$

The physical interpretation of these solutions is clear and typical. The relationship between the Jordan subsidiary eigenvectors and the solutions of the original equation is also typical, that in plane elasticity and 3-D elasticity solution of Saint Venant problems, such composition and physical interpretation will appear over and over again [23].

§5.4.2, Physical meaning of symplectic orthogonality—work reciprocity

Adjoint symplectic ortho-normality is a special term of mathematics. Interpreting with ***physical background*** is helpful for understanding. The original equation of structural mechanics is (5.2.13). Using the method of separation of variables, after solved the eigen-solutions $(\mu_i, \boldsymbol{\psi}_i)$ and $(\mu_j, \boldsymbol{\psi}_j)$, the solutions of the original equation are composed of

$$\mathbf{v}_i = \boldsymbol{\psi}_i \exp(\mu_i z), \qquad \mathbf{v}_j = \boldsymbol{\psi}_j \exp(\mu_j z) \tag{5.4.11a,b}$$

Because the original equation (5.2.13) is derived from a conservative system, for which the ***reciprocal theorem of work*** of Betti can certainly be applied.

Taking cross-sections at $z=0$ and $z=z_{\text{b}}$, and the free body $(0, z_{\text{b}})$ is considered. Corresponding to the two solutions of (5.4.11), there are the dual vectors at the cross-sections $\mathbf{q}_{0i}$, $\mathbf{p}_{0i}$ and $\mathbf{q}_{\text{b}i}$, $\mathbf{p}_{\text{b}i}$ for $\mathbf{v}_i$, and also $\mathbf{q}_{0j}$, $\mathbf{p}_{0j}$ and $\mathbf{q}_{\text{b}j}$, $\mathbf{p}_{\text{b}j}$ for $\mathbf{v}_j$, respectively. Using work reciprocity needs to calculate the work done by forces of solution i to the displacements of solution j, and the work done by forces of solution j to the displacements of solution i, respectively. The work done by the cross-section forces $\mathbf{p}_i$ of solution i to the displacements $\mathbf{q}_j$ of solution j is

$$\mathbf{p}_{0i}^T \mathbf{q}_{0j} - \mathbf{p}_{\text{b}i}^T \mathbf{q}_{\text{b}j} = [1-\exp(\mu_i+\mu_j)z]\cdot(\mathbf{p}_{0i}^T \mathbf{q}_{0j})$$

where the negative sign is because of cross-section force $\mathbf{p}_{\text{b}}$ has reverse direction. Such as from (5.4.11) $\mathbf{v}_{0i} = \boldsymbol{\psi}_i$ and $\boldsymbol{\psi}_i = \{\mathbf{q}_{0i}^T, \mathbf{p}_{0i}^T\}^T$. On the other hand, the work done by cross-section forces $\mathbf{p}_j$ of solution j to the displacements $\mathbf{q}_i$ of solution i is expressed as

$$\mathbf{p}_{0j}^T \mathbf{q}_{0i} - \mathbf{p}_{\text{b}j}^T \mathbf{q}_{\text{b}i} = [1-\exp(\mu_i+\mu_j)z](\mathbf{p}_{0j}^T \mathbf{q}_{0i})$$

According to the work reciprocal theorem, they are equal, therefore

$$[1-\exp(\mu_i+\mu_j)z_{\text{b}}](\boldsymbol{\psi}_i^T \mathbf{J} \boldsymbol{\psi}_j) = 0 \tag{5.4.12}$$

From which it concludes that either $(\mu_i+\mu_j)=0$, or $\boldsymbol{\psi}_i$ and $\boldsymbol{\psi}_j$ must be symplectic orthogonal. This proof is from structural mechanics, which explains the interrelation between ***reciprocal theorem and adjoint symplectic orthogonality***, which gives clear physical meaning to symplectic orthogonality.

The derivation above is based on the assumption of single eigen-root. Next, the case of duplicate eigenvalue $\mu_i \neq 0$ of Jordan form is to consider. In this case, the adjoint symplectic orthogonality relationship can also be proved by using the equations (5.4.1~4), with clear physical meaning. The equations for (m_i+1)-tuple Jordan normal form of multiple eigenvalues $\mu_i \neq 0$ are

$$\left.\begin{aligned}
&\mathbf{H}\boldsymbol{\psi}_i^{(0)} = \mu_i \boldsymbol{\psi}_i^{(0)} \\
&\mathbf{H}\boldsymbol{\psi}_i^{(1)} = \mu_i \boldsymbol{\psi}_i^{(1)} + \boldsymbol{\psi}_i^{(0)} \\
&\cdots\cdots \\
&\mathbf{H}\boldsymbol{\psi}_i^{(m_i)} = \mu_i \boldsymbol{\psi}_i^{(m_i)} + \boldsymbol{\psi}_i^{(m_i-1)}
\end{aligned}\right\} \tag{5.4.13}$$

Obviously, the subsidiary eigenvectors $\boldsymbol{\psi}_i^{(1)},\cdots,\boldsymbol{\psi}_i^{(m_i)}$ can arbitrarily superpose on $\boldsymbol{\psi}_i^{(0)}$. According to (5.4.13), the original equation (5.2.13) has solutions

$$\left.\begin{aligned}
&\mathbf{v}_i^{(0)} = \boldsymbol{\psi}_i^{(0)} \exp(\mu_i z) \\
&\mathbf{v}_i^{(1)} = [\boldsymbol{\psi}_i^{(1)} + z\boldsymbol{\psi}_i^{(0)}]\exp(\mu_i z) \\
&\cdots\cdots \\
&\mathbf{v}_i^{(m_i)} = [\boldsymbol{\psi}_i^{(m_i)} + z\boldsymbol{\psi}_i^{(m_i-1)} + \cdots + (z^{m_i}/m_i!)\boldsymbol{\psi}_i^{(0)}]\exp(\mu_i z)
\end{aligned}\right\} \tag{5.4.14}$$

To verify the symplectic orthogonality with other eigen-solutions, assuming there is eigen-solution $(\mu_j, \boldsymbol{\psi}_j)$ with $\mu_i + \mu_j \neq 0$, and the solution of original equation is

$$\mathbf{v}_j = \boldsymbol{\psi}_j \exp(\mu_j z)$$

Using ***work reciprocity theorem*** to the solutions in (5.4.14) successively, for which the method is as before, it can be proved that $\boldsymbol{\psi}_j$ is symplectic orthogonal to all the subsidiary eigenvectors $\boldsymbol{\psi}_i^{(k)} (k = 0,\cdots,m_i)$. The same method proves that all the (subsidiary) eigenvectors $\boldsymbol{\psi}_i^{(k)}$ are mutually symplectic orthogonal, which is under the condition of $\mu_i \neq 0$.

Turn to the symplectic adjoint eigenvectors $(\mu_{n+i} = -\mu_i, \boldsymbol{\psi}_{n+i}^{(0)},\cdots,\boldsymbol{\psi}_{n+i}^{(m_i)})$, for which the multiplicity of eigenvalue must also be $(m_i + 1)$. According to Jordan form equation (5.4.2)

$$\left.\begin{aligned}
&\mathbf{H}\boldsymbol{\psi}_{n+i}^{(0)} = -\mu_i \boldsymbol{\psi}_{n+i}^{(0)} \\
&\mathbf{H}\boldsymbol{\psi}_{n+i}^{(1)} = -\mu_i \boldsymbol{\psi}_{n+i}^{(1)} + \boldsymbol{\psi}_{n+i}^{(0)} \\
&\cdots\cdots \\
&\mathbf{H}\boldsymbol{\psi}_{n+i}^{(m_i)} = -\mu_i \boldsymbol{\psi}_{n+i}^{(m_i)} + \boldsymbol{\psi}_{n+i}^{(m_i-1)}
\end{aligned}\right\} \tag{5.4.15}$$

The solutions of the original equation are

$$\left.\begin{aligned}
&\mathbf{v}_{n+i}^{(0)} = \boldsymbol{\psi}_{n+i}^{(0)} \exp(-\mu_i z) \\
&\mathbf{v}_{n+i}^{(1)} = [\boldsymbol{\psi}_{n+i}^{(1)} + z\boldsymbol{\psi}_{n+i}^{(0)}]\exp(-\mu_i z) \\
&\cdots\cdots \\
&\mathbf{v}_{n+i}^{(m_i)} = [\boldsymbol{\psi}_{n+i}^{(m_i)} + z\boldsymbol{\psi}_{n+i}^{(m_i-1)} + \cdots + (z^{m_i}/m_i!)\boldsymbol{\psi}_{n+i}^{(0)}]\exp(-\mu_i z)
\end{aligned}\right\} \tag{5.4.16}$$

Note that the equation

$$\mathbf{H}\boldsymbol{\psi} = -\mu_i \boldsymbol{\psi} + \boldsymbol{\psi}_{n+i}^{(m_i)}$$

has no solution exist, otherwise the Jordan chain will not cease here. Based on the fundamental theorem of linear algebra, as before, $\boldsymbol{\psi}_i^{(0)}$ must not be symplectic orthogonal to the inhomogeneous vector $\psi_{n+i}^{(m_i)}$, so that they compose a pair of symplectic adjoint vectors. Similarly, $\psi_{n+i}^{(0)}$ and $\psi_i^{(m_i)}$ must also be mutually symplectic adjoint to each other. Below, using work reciprocity proves further.

First, the work reciprocal theorem is applied to $\mathbf{v}_i^{(0)}$ and $\mathbf{v}_{n+i}^{(m_i)}$, and it proves that symplectic orthogonality between $\psi_i^{(0)}$ and any vectors in $\psi_{n+i}^{(j)}, j=0\sim m_i-1$ are all valid. Using the work reciprocity theorem to $\mathbf{v}_i^{(m_i)}$ and $\mathbf{v}_{n+i}^{(0)}$ proves that $\psi_{n+i}^{(0)}$ are symplectic orthogonal to all the vectors $\psi_i^{(j)}, j=0\sim m_i-1$. The symplectic orthogonality relations are symmetric to the adjoint Jordan form series of subsidiary eigenvectors. Note that, if a vector in $\boldsymbol{\Psi}_i$ series is proved symplectic orthogonal to the adjoint series $\boldsymbol{\Psi}_{n+i}$, then the reverse will also be true.

Above, it proves only the symplectic orthogonality of eigenvector $\psi_i^{(0)}$ to the symplectic adjoint Jordan series eigenvectors. Then using the ***work reciprocity theorem*** to $\mathbf{v}_i^{(1)}$ and $\mathbf{v}_{n+i}^{(m_i)}$, based on the proved symplectic orthogonality relations and that $\psi_i^{(1)}$ can superpose an arbitrary $a\psi_i^{(0)}$, it shows that $\psi_i^{(1)}$ is symplectic orthogonal to all the adjoint subsidiary eigenvectors $\psi_{n+i}^{(j)}, j=0,\cdots,m_i-2,m_i$, except $j=m_i-1$, for which $\psi_i^{(1)T}\mathbf{J}\psi_{n+i}^{(m_i-1)}+\psi_i^{(0)T}\mathbf{J}\psi_{n+i}^{(m_i)}=0$, that $\psi_i^{(1)}$ and $\psi_{n+i}^{(m_i-1)}$ are mutually symplectic adjoint pair. However, $\psi_{n+i}^{(m_i-1)}$ should change its sign (multiply -1) to reach symplectic normality. Using the work reciprocal theorem between $\mathbf{v}_i^{(1)}$ and $\mathbf{v}_{n+i}^{(m_i-1)}$ again verifies the symplectic adjoint relation between $\psi_i^{(1)}$ and $\psi_{n+i}^{(m_i-1)}$.

Further, the work reciprocal theorem should be applied between $\mathbf{v}_i^{(2)}$ and $\mathbf{v}_{n+i}^{(m_i)}$. Neglect the details, the result obtained is that $\psi_i^{(2)}$ and $\psi_{n+i}^{(m_i-2)}$ are mutually symplectic adjoint pair, and both $\psi_i^{(2)}$, $\psi_{n+i}^{(m_i-2)}$ are symplectic orthogonal to their other adjoint subsidiary eigenvectors, etc.

After all, the class-(β) subsidiary eigenvectors $\psi_{n+i}^{(j)}$ are adjoint symplectic ortho-normality groups with the class-(α) subsidiary eigenvectors. However, the order should be reversed and also alternatively multiplied with -1, as described in section 5.4. ##

The above proof for Jordan normal form is under the condition of $\mu_i \neq 0$. However, there is the case of zero eigenvalue solutions. Zero eigenvalue is very important as shown in the Timoshenco beam solutions. In general, zero eigenvalue Jordan normal form definitely appears for a class of structural static problems. The (subsidiary) eigenvectors are symplectic orthogonal to other eigenvectors of non-zero eigenvalues. However, the set of eigenvectors with zero eigenvalue itself composes adjoint symplectic ortho-normality group, which is quite different to the set of eigenvectors of non-zero eigenvalue. The set of eigenvectors with zero eigenvalue must have even number of components, *i.e.* m_0 is an odd number. The equations for them are

$$\left.\begin{aligned} &\mathbf{H}\psi^{(0)} = 0 \\ &\mathbf{H}\psi^{(1)} = \psi^{(0)} \\ &\cdots \\ &\mathbf{H}\psi_i^{(m_0)} = \psi^{(m_0-1)} \end{aligned}\right\} \tag{5.4.17}$$

Correspondingly, the solutions of the original equation are

$$\left.\begin{aligned} &\mathbf{v}^{(0)} = \psi^{(0)} \\ &\mathbf{v}^{(1)} = \psi^{(1)} + z\psi^{(0)} \\ &\cdots\cdots \\ &\mathbf{v}^{(m_0)} = \psi_i^{(m_0)} + z\psi^{(m_0-1)} + \cdots + (z^{m_0}/m_0!)\psi^{(0)} \end{aligned}\right\} \tag{5.4.18}$$

It is to explain here, that for a subsidiary eigenvector $\psi^{(j)}$ or higher order, it may substitute

$$\psi^{(j+i)} \text{ with } \psi^{(j+i)} + c_j\psi^{(i)}, \quad i = 0,1,\cdots,m_0 - j \tag{5.4.19}$$

where c_j is arbitrary constants, such that the equation set (5.4.13) is still satisfied.

In order to prove the adjoint sympectic ortho-normality relationship among these Jordan chain subsidiary eigenvectors, the work reciprocal theorem can again be applied, as follows. First, making use of the work reciprocal theorem to the solutions $\mathbf{v}^{(0)}$ and $\mathbf{v}^{(m_0)}$ with two cross-sections at $z = 0$ and $z = z_b$ gives the equation

$$\psi^{(0)T}\mathbf{J}\psi^{(m_0)} = \psi^{(0)T}\mathbf{J}\psi^{(m_0)} + z_b\psi^{(0)T}\mathbf{J}\psi^{(m_0-1)} + \cdots + (z_b^{m_0}/m_0!)\psi^{(0)T}\mathbf{J}\psi^{(0)}$$

the left hand side term cancels the first term at the right hand side. Because z_b can be selected arbitrarily, the other terms must equal to zero individually. So, except vector $\psi^{(m_0)}$, the eigenvector $\psi^{(0)}$ must be symplectic orthogonal to all other subsidiary eigenvectors in this Jordan chain.

The assertion of $\psi^{(0)}$ must be symplectic adjoint to $\psi^{(m_0)}$ can be proved as follows, because after m_0 there has been no Jordan subsidiary eigenvector, so the equation $\mathbf{H}\psi = \psi^{(m_0)}$ has no solution. According to the fundamental theorem of linear algebra, the solution of equation $\mathbf{H}^T\psi = 0$ should not orthogonal to the vector $\psi^{(m_0)}$, *i.e.* $\psi^{(0)^T}\mathbf{J}\psi^{(m_0)}$ must not be zero, *i.e.* symplectic adjoint.

The next step is to prove the symplectic orthogonality of vector $\psi^{(1)}$ with others in the Jordan chain. Selecting solutions $\mathbf{v}^{(1)}$ and $\mathbf{v}^{(m_0-1)}$ using the work reciprocity theorem, because the eigenvector $\psi^{(0)}$ has been proved symplectic orthogonal to all the Jordan chain member until $\mathbf{v}^{(m_0-1)}$, hence the derivation will show that $\psi^{(1)}$ is symplectic orthogonal to all the members of $\psi^{(1)} \sim \psi^{(m_0-2)}$. Making use again the work reciprocal theorem to solutions $\mathbf{v}^{(1)}$ and $\mathbf{v}^{(m_0)}$ verifies

$$\psi^{(0)T}\mathbf{J}\psi^{(m_0)} + \psi^{(1)T}\mathbf{J}\psi^{(m_0-1)} = 0$$

which expresses that $\psi^{(1)}$ and $\psi^{(m_0-1)}$ are mutually symplectic adjoint. The symplectic orthogonal relation for $\psi^{(m_0)}$ and $\psi^{(1)}$ can be reached by selecting the constant c_1 appropriately in the equation (5.4.19).

Thereafter, the symplectic orthogonality between the subsidiary eigenvector $\psi^{(2)}$ and all the vectors in Jordan chain, etc. is to prove. After all, continuously using the work reciprocal theorem in combination with the appropriate selection of the constants in equation (5.4.19), the symplectic adjoint between $\psi^{(i)}$ and $\psi^{(m_0-i)}$, and symplectic orthogonality between $\psi^{(i)}$ and all the other vectors in the Jordan chain can be proved. The steps are similar, and the details are neglected [23]. ##

Zero eigenvalue appears frequently in structural static and also in elasticity, especially the ***Saint Venant problem***. For wave propagation, zero eigenvalue solution appears only in some special case.

§5.5, Expansion solution of the inhomogeneous equation

The eigen-solutions described above are for the homogeneous differential equations (5.2.10). The important application of the eigen-solutions is the ***expansion solution*** using eigenvectors. The inhomogeneous differential equation (5.2.9) can be solved by the precise integration method or by the eigenvector expansion method. Rewrite the inhomogeneous differential equation (5.2.9) as

$$\dot{\mathbf{v}} = \mathbf{H}\mathbf{v} + \mathbf{h} \tag{5.5.1}$$

where the external force vector $\mathbf{h}(z)$ is given. Based on the expansion theorem,

$$\mathbf{v}(z) = \sum_{i=1}^{n} \left[a_i(z)\psi_i + b_i(z)\psi_{n+i} \right] \tag{5.5.2}$$

$$\mathbf{h}(z) = \sum_{i=1}^{n} \left[c_i(z)\psi_i + d_i(z)\psi_{n+i} \right] \tag{5.5.3}$$

because $\mathbf{h}(z)$ is a given function, so that the functions $c_i(z)$ and $d_i(z)$ can be determined by the adjoint symplectic ortho-normality relationship and are known functions. Substituting the above expansions into (5.5.1) and using the ***adjoint symplectic ortho-normality*** relation gives

$$\dot{a}_i = \mu_i a_i + c_i\,, \quad \dot{b}_i = -\mu_i b_i + d_i\,, \qquad i = 1, \cdots, n \tag{5.5.4a,b}$$

These equations are derived under the assumption of single eigen-roots, however, even the Jordan form appears, these equations still valid. These equations have been de-coupled as possible. The solution for these equations has standard methods, the general solution of Duhamel integration

$$\begin{aligned} a_i(z) &= A_i \exp(\mu_i z) + \int_0^z \exp[\mu_i(\zeta - z)]c_i(\zeta)\mathrm{d}\zeta \\ b_i(z) &= B_i \exp[\mu_i(z_f - z)] + \int_z^{z_f} \exp[\mu_i(\zeta - z)]d_i(\zeta)\mathrm{d}\zeta \end{aligned} \tag{5.5.5a,b}$$

can be applied, where A_i, B_i are constants to be determined, such as by means of the two end boundary conditions. The functions $c_i(z)$ and $d_i(z)$ can be determined by the ***adjoint symplectic ortho-normality*** relationship

$$c_i(z) = -\psi_{n+i}^T \mathbf{J}\mathbf{h}(z), \quad d_i(z) = \psi_i^T \mathbf{J}\mathbf{h}(z) \tag{5.5.6a,b}$$

Combining eigenvector expansion and precise integration method, effective methods can be developed for solution.

Note that the non-stationary stochastic excitation of random vibration analysis for large-scale structural analysis system described in last chapter is based on the expansion solution.

§5.6, Two end boundary conditions

The partial differential equations of elasticity are of **elliptic** type, so that the appropriate boundary condition should be assigned along the contour surrounding the domain. For single continuous coordinate system problem, it reduced to the boundary conditions assigned at the two ends. In applied mathematics, it is often called as Two Point Boundary Value Problems or abbreviated as TPBVP, especially in optimal control theory.

The $2n$ first order ODEs has $2n$ integration constants. The two point boundary conditions should supply $2n$ boundary conditions, with each end n conditions. For Timoshenco beam theory, the usual boundary conditions are supplied as

$$\begin{aligned} &\text{free end}: \ Q = 0, M = 0 \\ &\text{simply supported}: \ u = 0, M = 0 \\ &\text{fixed end}: \ u = 0, \psi = 0, \\ &\text{symmetric}: \ Q = 0, \psi = 0. \end{aligned} \tag{5.6.1}$$

two conditions for each end. These are typical end conditions, and there are other forms such as elastic supported etc.

A number of solution methods had been developed. The precise integration method to solve the Riccati differential equation is one among them [41,97~103], which will be given later. The precise integration gives solution up to computer precision, and its importance is also because of the analogy to optimal control theory. The eigenvector expansion method is another effective method. Especially, the eigen-solution expansion method can be applied to the two or three dimensional elasticity problems.

The TPBVP supplies n boundary conditions at each end. A popular method, initial parameter method or shooting method had been developed for a long time and can be described as follows. Except the given n conditions assuming another n parameters at one 'initial' end, *i.e.* n initial parameters, solving the set of $2n$ differential equations and afterwards fix the n initial parameters with the another n end conditions given at the other end. Such initial parameter method is not always effective, that when the real part of the eigenvalues is large, the numerical ill-conditioning may be very serious.

Using eigenvector expansion method to solve the TPBVP should combine the two end boundary conditions and establish $2n$ simultaneous algebraic equations to solve. There are a number of methods to establish the equations. Because of the system is conservative, the matrix of the $2n$ algebraic equation has special feature, *i.e.* the matrix is symmetric, which is the behavior of reciprocity of a conservative system. A variational method is supplied here which can ensure the symmetry of equations.

The boundary condition (5.6.1) is expressed as force equals zero or displacement equals zero, which can also be written as given force or given displacement, respectively, and which can be expressed as boundary conditions, on S_σ or on S_u, in the variational principle, respectively. For two ends $z=0$ and $z=L(=z_{\mathrm{f}})$, the n boundary conditions at each end can be of the mixed type, *i.e.* a part of the conditions being ***given force*** (S_σ) and the other part of conditions being ***given displacement*** (S_u), so that the boundary should be distinguished as $S_{\sigma 0}, S_{u0}$ and $S_{\sigma\mathrm{f}}, S_{u\mathrm{f}}$ for the two ends $z_0=0$ and z_{f}, respectively. Assuming these conditions can be expressed as given the components of dual vectors $\mathbf{q}$ and $\mathbf{p}$, that $S_{\sigma 0}+S_{u0}$ means the n conditions of components

$$\text{At } z=0, \quad [\mathbf{q}=\bar{\mathbf{q}}_0]_{S_{\sigma 0}}; \quad [\mathbf{p}=\bar{\mathbf{p}}_0]_{S_{u0}}$$
$$\text{At } z=z_{\mathrm{f}}, \quad [\mathbf{q}=\bar{\mathbf{q}}_{\mathrm{f}}]_{S_{\sigma\mathrm{f}}}; \quad [\mathbf{p}=\bar{\mathbf{p}}_{\mathrm{f}}]_{S_{u\mathrm{f}}} \tag{5.6.2}$$

where the expressions looked like given all displacements and forces, however, the marks $S_{\sigma\mathrm{f}}$ and $S_{u\mathrm{f}}$ represent that it is for the appropriate components.

A mixed energy variational principle is to propose, such that the dual equations (5.2.9a,b) and also the boundary conditions (5.6.2) are all involved. This variational principle is

$$\delta\left\{\begin{array}{l}\int_0^{z_{\mathrm{f}}}[\mathbf{p}^T\dot{\mathbf{q}}-\mathrm{H}(\mathbf{q},\mathbf{p})]\mathrm{d}z \\ -[\mathbf{p}^T(\mathbf{q}-\bar{\mathbf{q}}_{\mathrm{f}})]_{S_{u\mathrm{f}}}+[\mathbf{p}^T(\mathbf{q}-\bar{\mathbf{q}}_0)]_{S_{u0}}-[\bar{\mathbf{p}}_{\mathrm{f}}^T\mathbf{q}]_{S_{\sigma\mathrm{f}}}+[\bar{\mathbf{p}}_0^T\mathbf{q}]_{S_{\sigma 0}}\end{array}\right\}=0 \tag{5.6.3}$$

Carrying out the variational derivation, that the integration term results the dual equations, so that only the term for two ends are remained as

$$\begin{aligned}&[\delta\mathbf{p}^T(\mathbf{q}-\bar{\mathbf{q}}_0)]_{S_{u0}}-[(\mathbf{p}-\bar{\mathbf{p}}_0)^T\delta\mathbf{q}]_{S_{\sigma 0}} \\ &\quad -[\delta\mathbf{p}^T(\mathbf{q}-\bar{\mathbf{q}}_{\mathrm{f}})]_{S_{u\mathrm{f}}}+[(\mathbf{p}-\bar{\mathbf{p}}_{\mathrm{f}})^T\delta\mathbf{q}]_{S_{\sigma\mathrm{f}}}=0\end{aligned} \tag{5.6.4}$$

Because of the arbitrariness of $\delta\mathbf{p}$ and $\delta\mathbf{q}$ on S_u and on S_σ, respectively, all the boundary conditions of (5.6.2) are obtained. Therefore the variational equation (5.6.4) can be used to substitute the assigned conditions of (5.6.2).

Using the eigenvector expansion method, the differential equations have all been fulfilled, hence the variational equation (5.6.4) is remained to replace the boundary condition (5.6.2). Note that the variational equation is for real vectors however, the eigenvectors may be complex valued. According to that the complex conjugate vector of the complex eigenvector is also an eigenvector, so that their combinations are again real vectors. Let $\boldsymbol{\psi}_i=\{\mathbf{q}_i^T,\mathbf{p}_i^T\}^T$ denote eigenvectors that

$$\mathbf{v}(z)=\sum_{i=1}^{n}\left[A_i\boldsymbol{\psi}_i e^{\mu_i z}+B_i\boldsymbol{\psi}_{n+i}e^{\mu_i(z_{\mathrm{f}}-z)}\right]+\mathbf{v}_h(z)$$

then at both ends

$$\mathbf{v}(0)=\sum_{i=1}^{n}\left[A_i\begin{Bmatrix}\mathbf{q}_i\\ \mathbf{p}_i\end{Bmatrix}+B_i\begin{Bmatrix}\mathbf{q}_{n+i}\\ \mathbf{p}_{n+i}\end{Bmatrix}e^{\mu_i z_f}\right]+\mathbf{v}_h(0)$$

$$\mathbf{v}(z_f)=\sum_{i=1}^{n}\left[A_i e^{\mu_i z_f}\begin{Bmatrix}\mathbf{q}_i\\ \mathbf{p}_i\end{Bmatrix}+B_i\begin{Bmatrix}\mathbf{q}_{n+i}\\ \mathbf{p}_{n+i}\end{Bmatrix}\right]+\mathbf{v}_h(z_f) \tag{5.6.5}$$

where $\mathbf{v}_h$ represents the special solution of inhomogeneous differential equation having no arbitrary constant, and $A_i, B_i (i=1,\cdots,n)$ are arbitrary complex constants to be determined. Note that the solutions used for the variational principle should be real valued, the special solution $\mathbf{v}_h$ can be considered real valued. Hence if μ_i is a complex eigenvalue then its complex conjugate eigen-solution $(\mu_{i+1}=\bar{\mu}_i, \boldsymbol{\psi}_{i+1})$ must be arranged contiguously, so that $A_{i+1}=\bar{A}_i, \boldsymbol{\psi}_{i+1}=\bar{\boldsymbol{\psi}}_i$, where the upper bar $^-$ represents complex conjugate. Therefore the number of arbitrary real constants is still $2n$. The case of $\mathrm{Re}(\mu_i)=0$ needs special attention, that in such case the complex conjugate is really the symplectic adjoint, hence to arrange the complex conjugate pair contiguously is impossible. However, let $B_i=\bar{A}_i$ the real solution can still be achieved and the eigenvectors have the form

$$A_i\exp(\mu_i z-\mathrm{i}\omega t)=A_i\exp[\mathrm{i}(k_i z-\omega t)] \quad \text{and} \quad B_i\exp[\mathrm{i}(-k_i z-\omega t)] \tag{5.6.6}$$

where $\mu_i=\mathrm{i}k_i$, k_i is a real parameter, its physical interpretation is the wave number of transmission wave solution. The solution A_i corresponds to the transmission wave towards $+z$ axis with wave phase velocity $c=\omega/k_i$, and the solution B_i corresponds to the transmission wave towards $-z$.

For the problem of given displacements $\bar{\mathbf{q}}_0, \bar{\mathbf{q}}_f$ at the two ends $z=0$ and $z=z_f$ there is no S_σ boundary, hence the variational principle derives to the boundary conditions in variational form as in equation (5.6.4)

$$[\delta\mathbf{p}^T(\mathbf{q}-\bar{\mathbf{q}}_0)]_{z=0}-[\delta\mathbf{p}^T(\mathbf{q}-\bar{\mathbf{q}}_f)]_{z=z_f}=0 \tag{5.6.7}$$

When solving with eigenvector expansion method, the varying parameters become A_i and B_i. Let the unknown parameters be denoted as vectors

$$\mathbf{a}=\{A_1,\cdots,A_n\}^T,\ \mathbf{b}=\{B_1,\cdots,B_n\}^T \tag{5.6.8a,b}$$

and the functional is a quadratic function of $\mathbf{a}$ and $\mathbf{b}$. The equation corresponding to the variation of $\delta\mathbf{a}$ is derived as

$$\mathbf{Ca}+\mathbf{Db}=\mathbf{h}_1 \tag{5.6.8c}$$

where $\mathbf{C}$ and $\mathbf{D}$ are $n\times n$ matrices, the coefficients are given by

$$c_{11}=\mathbf{p}_1^T\mathbf{q}_1(1-e^{2\mu_1 z_f}),\ c_{12}=\mathbf{p}_1^T\mathbf{q}_2(1-e^{(\mu_1+\mu_2)z_f}),\cdots,\ c_{1n}=\mathbf{p}_1^T\mathbf{q}_n(1-e^{(\mu_1+\mu_n)z_f})$$

......

$$c_{n1}=\mathbf{p}_n^T\mathbf{q}_1(1-e^{(\mu_1+\mu_2)z_f}),\ c_{n2}=\mathbf{p}_n^T\mathbf{q}_2(1-e^{(\mu_n+\mu_2)z_f}),\ \cdots,\ c_{nn}=\mathbf{p}_n^T\mathbf{q}_n(1-e^{2\mu_n z_f})$$

and

$$d_{11}=0,\ d_{12}=\mathbf{p}_1^T\mathbf{q}_{n+2}(e^{\mu_2 z_f}-e^{\mu_1 z_f}),\ \cdots,\ d_{1n}=\mathbf{p}_1^T\mathbf{q}_{2n}(e^{\mu_n z_f}-e^{\mu_1 z_f})$$

......

$$d_{n1}=\mathbf{p}_n^T\mathbf{q}_{n+1}(e^{\mu_1 z_f}-e^{\mu_n z_f}),\ \cdots\cdots,\ d_{nn}=0;$$

On the other hand, the equation derived from $\delta\mathbf{b}$ is

$$\mathbf{Ea}+\mathbf{Fb}=\mathbf{h}_2 \tag{5.6.8d}$$

where $\mathbf{E}$ and $\mathbf{F}$ are also $n\times n$ matrices, the coefficients are given by

$$e_{11}=0,\ e_{12}=\mathbf{p}_{n+1}^T\mathbf{q}_2(e^{\mu_1 z_f}-e^{\mu_2 z_f}),\ \cdots,\ e_{1n}=\mathbf{p}_{n+1}^T\mathbf{q}_n(e^{\mu_1 z_f}-e^{\mu_n z_f}),$$

$$e_{21}=\mathbf{p}_{n+2}^T\mathbf{q}_1(e^{\mu_2 z_f}-e^{\mu_1 z_f}),\ e_{22}=0,\ \cdots,\ e_{2n}=\mathbf{p}_{n+2}^T\mathbf{q}_n(e^{\mu_2 z_f}-e^{\mu_n z_f}),$$

......

$$e_{n1}=\mathbf{p}_{2n}^T\mathbf{q}_1(e^{\mu_n z_f}-e^{\mu_1 z_f}),\ \cdots\cdots,\ e_{nn}=0;$$

and

$$f_{11}=\mathbf{p}_{n+1}^T\mathbf{q}_{n+1}(e^{(\mu_1+\mu_1)z_f}-1),\ \cdots,\ f_{1n}=\mathbf{p}_{n+1}^T\mathbf{q}_{2n}(e^{(\mu_1+\mu_n)z_f}-1),$$

......

$$f_{n1}=\mathbf{p}_{2n}^T\mathbf{q}_{n+1}(e^{(\mu_n+\mu_1)z_f}-1),\ \cdots\cdots,\ f_{2n}=\mathbf{p}_{2n}^T\mathbf{q}_{2n}(e^{2\mu_n z_f}-1);$$

Equation (5.6.8c,d) are the canonical equations for boundary problem when the ends being given displacements. Solving the parameters $A_i, B_i (i=1-n)$ solves the problem.

Several fundamental behaviors should be explained. First the simultaneous equations are symmetric. Examining the matrix $\mathbf{C}$ that its coefficients are symmetric, such as $c_{1n}=c_{n1}$. From the above equations for c_{ij}, the factors in the parenthesis of c_{1n} and c_{n1} are the same, further the equation $\mathbf{p}_1^T\mathbf{q}_n=\mathbf{p}_n^T\mathbf{q}_1$ must be verified, which is just $\boldsymbol{\psi}_1^T\mathbf{J}\boldsymbol{\psi}_n=0$, the verified symplectic orthogonal condition. Hence, $\mathbf{C}$ is a symmetric matrix. Similarly, the matrix $\mathbf{F}$ can be proved symmetric too.

Next the relation $\mathbf{D}^T=\mathbf{E}$ should also be verified, which is again ensured by adjoint symplectic ortho-normality. The pairs of symplectic adjoint eigenvectors appear just at the diagonal elements of matrices $\mathbf{D}$ and $\mathbf{E}$, however, the multiplier in the bracket are all zero. To verify such as $d_{12}=e_{21}$, note that the bracket factors in the above equations for d_{12} and e_{21} are the same, so that it needs only to check that $\mathbf{p}_{n+2}^T\mathbf{q}_1=\mathbf{p}_1^T\mathbf{q}_{n+2}$, which is just the symplectic orthogonality condition $\boldsymbol{\psi}_{n+2}^T\mathbf{J}\boldsymbol{\psi}_1=0$ proved before.

Another problem is complex number arithmetic. The coefficients of complex conjugate eigenvectors are also complex conjugate pair, so there are still two real parameters to be determined. Therefore the complex conjugate pair eigenvectors $\boldsymbol{\psi}_j, \boldsymbol{\psi}_{j+1}$ can be composed of two real vectors. The real pair of vectors (real and imaginary parts of the eigenvector) is still symplectic orthogonal to other eigenvectors, so that the symmetry of the canonical equation keeps unchanged. The verification is as follows: let $\boldsymbol{\psi}_j=\boldsymbol{\psi}_r+\mathrm{i}\boldsymbol{\psi}_i$, then $\boldsymbol{\psi}_{j+1}=\boldsymbol{\psi}_r-\mathrm{i}\boldsymbol{\psi}_i$, where $\boldsymbol{\psi}_r$

and ψ_{i} denote the real and imaginary parts of the eigenvector. The symplectic orthogonality condition between ψ_j and ψ_{j+1} derives $\psi_{\mathrm{r}}^T \mathbf{J} \psi_{\mathrm{i}} = 0$, *i.e.* the symplectic orthogonality behaviour keeps unchanged. The complex conjugate eigenvectors ψ_j and ψ_{j+1} have their symplectic adjoint eigenvectors $\varphi_j = \varphi_{\mathrm{r}} + \mathrm{i}\varphi_{\mathrm{i}}$, and $\varphi_{j+1} = \varphi_{\mathrm{r}} - \mathrm{i}\varphi_{\mathrm{i}}$, note $\psi_{n+j} = \varphi_j$ and $\psi_{n+j+1} = \varphi_{j+1}$. The vectors φ_{r} and φ_{i} are symplectic orthogonal, and using symplectic adjoint relations

$$\psi_j^T \mathbf{J} \varphi_j = 1, \quad \psi_{j+1}^T \mathbf{J} \varphi_{j+1} = 1, \quad \psi_j^T \mathbf{J} \varphi_{j+1} = 0, \quad \psi_{j+1}^T \mathbf{J} \varphi_j = 0$$

The symplectic adjoint relations of ψ_{r} to φ_{r} and ψ_{i} to φ_{i} are verified, all the others are symplectic orthogonal. Hence, using real vectors (they are not eigenvectors now) to do the computation works too.

The above example derives the canonical equations for the case of that the two end boundary conditions are given displacements. According to practical problem on hand, the proposition of boundary condition is different, such as given displacements at the end $z = 0$, but given forces at the end $z = z_{\mathrm{f}}$, (looked like a cantilever beam). Based on the variational equation (5.6.4) for boundary conditions, the canonical equations for various boundary conditions can always be proposed.

Expansion solution with eigenvectors is one of the effective solution methods, for which the key step is to solve all the eigen-solutions of the Hamilton matrix. However, such requirement is not always easy to achieve numerically, only when there is no Jordan normal form appearing then the algorithm (see section 2.3.3.3) is reliable. When the Jordan normal form appears the numerical instability may appear, which may destroy the numerical results. The Jordan normal form itself is numerically unstable, which comes with the theory and causes numerical difficulty. However, there is effective method unaffected from the possible Jordan form, *i.e.* the ***precise integration method***. Even the Jordan form appears, the numerical result computed by the precise integration method still gives numerical results up to the computer precision, which is given in the next section.

§5.7, Interval mixed energy and precise integration method

In the previous sections the methods of separation of variables and eigenvector expansion solution etc. are used to solve the problem, which is termed as the modal analysis method. Although the combination of methods given above is quite effective in the case of no Jordan form appears, however, there are other good methods available. Such as for multiple degrees of freedom vibration system, except the modal analysis method, the time step integration method is used quite frequently. There have been tremendous researches on the method of direct integration algorithms, such as the Runge-Kutta method, the Hobolt method, the Newmark method, the Wilson-θ method, the central difference method, etc. They are all FDM (Finite Difference Method) approaches, but the finite difference approximation is always accompanying with error, and has also numerical problem such as ***stiff, stability*** etc., not so ideal.

Therefore, the ***precise integration method***, which has been given for initial value problem in the introduction, was extended to solve the TPBVP along space coordinate. The precise integration method was first proposed for initial value problems [22,26,27]. In contrast to the usual FDM type algorithms, for which the numerical result is approximate, precise integration uses the ***full computer precision*** in the algorithm, so that the accuracy of numerical results is limited only by the precision of the host computer.

The precise integration method was subsequently extended to cover ***two-point boundary value problems*** (TPBVP) [97~102]. The Riccati differential equation closely relates to TPBVP and so the use of precise integration to solve the Riccati differential equation is applied, for which the numerical solution again achieving full computer precision, which is very attractive. Especially, the precise integration method does not care about the possible Jordan normal form, it always gives highly precise numerical results.

The precise computation has two cruxes, namely

1) ***Using*** 2^N ***algorithm to subdivide the interval length extremely small, such that the approximation error is beyond the computer real number error***
2) ***Keeping track of the incremental part of the related matrices or vectors, rather than their total value,*** such that the round-off error is reduced to the last digit on the host computer

The precision of numerical result obtained from the precise integration method can approach the real value precision of the host computer, very attractive behavior.

§5.7.1, Displacement method analysis

The equation (5.2.2) and (5.2.3) give the Lagrange function and the minimum potential energy variational principle. Since there is the Lagrange function, so that the minimum potential energy variational principle is treated as the Hamilton variational principle in analytical dynamics. The canonical equation form in analytical dynamics is also used. Therefore the longitudinal coordinate z is naturally corresponding to the time coordinate t in analytical dynamics. Let the integration interval and the respective left and right ends in the variational functional (5.2.3) be changed to (z_a, z_b), with $z_a < z_b$. Now (z_a, z_b) is treated as an interval (segment), for which the left and right ends z_a, z_b are also treated as variables. The potential energy density in the interval is the Lagrange function and the physical meaning of the integration

$$U(z_a, z_b, \mathbf{q}_a, \mathbf{q}_b) = \int_{z_a}^{z_b} [\dot{\mathbf{q}}^T \mathbf{K}_{22} \dot{\mathbf{q}} / 2 + \dot{\mathbf{q}}^T \mathbf{K}_{21} \mathbf{q} + \mathbf{q}^T \mathbf{K}_{11} \mathbf{q} / 2 - \mathbf{g}^T \mathbf{q}] \mathrm{d}z \qquad (5.7.1)$$

is the deformation energy of the interval, where $\mathbf{q}$ is a n-dimensional displacement vector. Deformation energy interpretation is from structural mechanics. Mathematically (analytical dynamics), the ***interval deformation energy*** is the ***action function***.

The action function is certainly a function of both ends z_a and z_b of the interval, see section 1.10, and also relates to the end displacements at the boundaries z_a and z_b. As the boundary conditions are assigned as given the two end displacements $\mathbf{q}_a$ and $\mathbf{q}_b$, then equation (5.7.1) gives the ***interval deformation***

energy $U(z_a, z_b, \mathbf{q}_a, \mathbf{q}_b)$. If the interval is treated as a substructure, then the two ends z_a and z_b are the connections to the outside. The combination of two displacement vectors $\mathbf{q}_a$ and $\mathbf{q}_b$ is termed as the external displacements.

If the interval is free from external forces, the external displacement $\mathbf{q}_a, \mathbf{q}_b$ uniquely determines the internal displacements, internal forces etc. of the interval, especially the force vectors $\mathbf{p}_a$ and $\mathbf{p}_b$ at both ends z_a and z_b. Therefore the interval internal displacement $\mathbf{q}$ is expressed as a function of $\mathbf{q}_a$ and $\mathbf{q}_b$, substituting $\mathbf{q}$ into equation (5.7.1), carrying out the integration then the interval deformation energy is obtained, which is a quadratic function of $\mathbf{q}_a, \mathbf{q}_b$ and is written as

$$U(\mathbf{q}_a, \mathbf{q}_b) = \mathbf{q}_a^T \mathbf{K}_{aa} \mathbf{q}_a / 2 + \mathbf{q}_b^T \mathbf{K}_{bb} \mathbf{q}_b / 2 + \mathbf{q}_a^T \mathbf{K}_{ab} \mathbf{q}_b \tag{5.7.2}$$

where $\mathbf{K}_{aa}^T = \mathbf{K}_{aa}$, $\mathbf{K}_{bb}^T = \mathbf{K}_{bb}$. Denote $\mathbf{K}_{ab}^T = \mathbf{K}_{ba}$ for later use. These matrices are all functions of z_a and z_b, but for simplicity, the variables z_a, z_b in function formulation are often omitted. For static problems, no dynamic inertia energy exists, $\mathbf{K}_{aa}$ and $\mathbf{K}_{bb}$ are symmetric and positive definite, and are the end stiffness matrices at z_a and z_b. When dynamic energy is considered, these matrices become dynamic stiffness matrix [46], which may be indefinite but still symmetric.

The above representations have not included the external forces; hence the interval deformation energy is a homogeneous quadratic function. However, if external forces are considered, then the total potential energy will involve the linear terms of $\mathbf{q}_a$ and $\mathbf{q}_b$, and becomes

$$U(\mathbf{q}_a, \mathbf{q}_b) = \left(\mathbf{q}_a^T \mathbf{K}_{aa} \mathbf{q}_a + \mathbf{q}_b^T \mathbf{K}_{bb} \mathbf{q}_b\right)/2 + \mathbf{q}_a^T \mathbf{K}_{ab} \mathbf{q}_b - \mathbf{f}_a^T \mathbf{q}_a - \mathbf{f}_b^T \mathbf{q}_b \tag{5.7.2'}$$

where $\mathbf{f}_a$ and $\mathbf{f}_b$ are induced from the internal distributed external forces within the interval, or from the external forces directly acted at z_a and z_b, respectively. So, $\mathbf{f}_a, \mathbf{f}_b$ are independent on the end (external) displacements $\mathbf{q}_a, \mathbf{q}_b$.

Usually, the whole interval length (z_0, z_f) is subdivided, by the stations $z_0(=0), z_1, z_2, \cdots, z_f$, as a chain of intervals, such as

$$z_0(=0) < z_1 < z_2 < \cdots < z_{k_f - 1} < z_f, \quad z_f = z_{k_f} \tag{5.7.3}$$

totally k_f sub-intervals linked end to end. In principle, the lengths of various sub-intervals can be arbitrarily selected, however, usually uniform length $\eta = z_k - z_{k-1}$ subdivision is preferable, *i.e.* η is independent on k. Treating the interval two end coordinates z_a and z_b as variables makes the interval being an arbitrarily long contiguous interval located inside (z_0, z_f).

Two adjacent intervals can be combined as a longer interval (z_a, z_c), see figure 5.2. After combination, the resulted interval has the two ends at z_a and z_c, and the station z_b becomes internal and so is its displacement vector $\mathbf{q}_b$, which should be eliminated in the combination. The equations for elimination are derived from the minimum potential energy variational principle. Let the deformation energy be expressed as

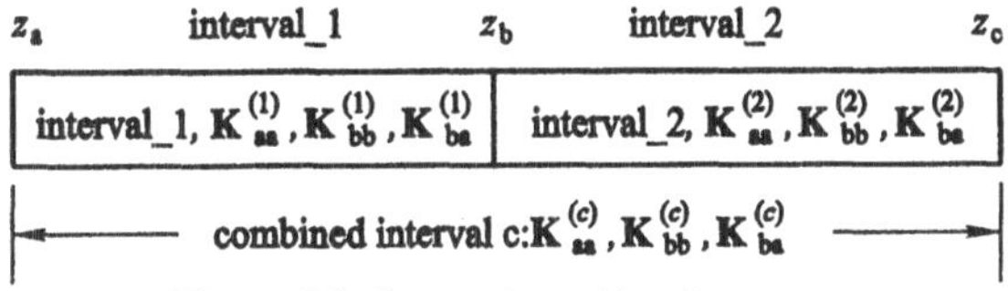

Figure 5.2, *Interval combination*

$$(z_a, z_b): \quad U_1(\mathbf{q}_a, \mathbf{q}_b) = (\mathbf{q}_a^T \mathbf{K}_{aa}^{(1)} \mathbf{q}_a + \mathbf{q}_b^T \mathbf{K}_{bb}^{(1)} \mathbf{q}_b)/2 + \mathbf{q}_a^T \mathbf{K}_{ab}^{(1)} \mathbf{q}_b - \mathbf{f}_a^{(1)^T} \mathbf{q}_a - \mathbf{f}_b^{(1)^T} \mathbf{q}_b$$

$$(z_b, z_c): \quad U_2(\mathbf{q}_b, \mathbf{q}_c) = \left(\mathbf{q}_b^T \mathbf{K}_{aa}^{(2)} \mathbf{q}_b + \mathbf{q}_c^T \mathbf{K}_{bb}^{(2)} \mathbf{q}_c\right)/2 + \mathbf{q}_b^T \mathbf{K}_{ab}^{(2)} \mathbf{q}_c - \mathbf{f}_a^{(2)^T} \mathbf{q}_b - \mathbf{f}_b^{(2)^T} \mathbf{q}_c$$

where the subscripts a and b represent left and right ends, respectively. Even for the interval (z_b, z_c) the same notation is still used. After combination, the potential energy expression has still the same form as

$$(z_a, z_c): \quad U_c(\mathbf{q}_a, \mathbf{q}_c) = (\mathbf{q}_a^T \mathbf{K}_{aa}^{(c)} \mathbf{q}_a + \mathbf{q}_c^T \mathbf{K}_{bb}^{(c)} \mathbf{q}_c)/2 + \mathbf{q}_a^T \mathbf{K}_{ab}^{(c)} \mathbf{q}_c - \mathbf{f}_a^{(c)^T} \mathbf{q}_a - \mathbf{f}_b^{(c)^T} \mathbf{q}_c \tag{5.7.4}$$

where the subscripts a and b for the matrices $\mathbf{K}$ and vectors $\mathbf{f}$ are again considered as the left and right ends, respectively. The superscript (c) represents the combined interval. Using minimum potential energy variational principle derives

$$U_c = \underset{\mathbf{q}_b}{\text{sta.}}(U_1 + U_2), \quad i.e. \;\; \partial(U_1 + U_2)/\partial \mathbf{q}_b = 0$$

where sta. abbreviates taking stationary value. Therefore the equation is given as

$$\mathbf{K}_{ba}^{(1)} \mathbf{q}_a + (\mathbf{K}_{bb}^{(1)} + \mathbf{K}_{aa}^{(2)}) \mathbf{q}_b + \mathbf{K}_{ab}^{(2)} \mathbf{q}_c - \mathbf{f}_a^{(2)} - \mathbf{f}_b^{(1)} = \mathbf{0}$$

Solving $\mathbf{q}_b$ and substituting back into the expression of $(U_1 + U_2)$ gives the deformation energy of the combined interval (z_a, z_c). Comparing with the general form of equation (5.7.4) gives

$$\mathbf{K}_{aa}^{(c)} = \mathbf{K}_{aa}^{(1)} - \mathbf{K}_{ab}^{(1)} (\mathbf{K}_{bb}^{(1)} + \mathbf{K}_{aa}^{(2)})^{-1} \mathbf{K}_{ba}^{(1)} \tag{5.7.5a}$$

$$\mathbf{K}_{bb}^{(c)} = \mathbf{K}_{bb}^{(2)} - \mathbf{K}_{ba}^{(2)} (\mathbf{K}_{bb}^{(1)} + \mathbf{K}_{aa}^{(2)})^{-1} \mathbf{K}_{ab}^{(2)} \tag{5.7.5b}$$

$$\mathbf{K}_{ab}^{(c)} = -\mathbf{K}_{ab}^{(1)} (\mathbf{K}_{bb}^{(1)} + \mathbf{K}_{aa}^{(2)})^{-1} \mathbf{K}_{ab}^{(2)}, \qquad \mathbf{K}_{ba}^{(c)} = \mathbf{K}_{ab}^{(c)^T} \tag{5.7.5c}$$

and

$$\mathbf{f}_a^{(c)} = \mathbf{f}_a^{(1)} - \mathbf{K}_{ab}^{(1)} (\mathbf{K}_{bb}^{(1)} + \mathbf{K}_{aa}^{(2)})^{-1} (\mathbf{f}_a^{(2)} + \mathbf{f}_b^{(1)}) \tag{5.7.6a}$$

$$\mathbf{f}_b^{(c)} = \mathbf{f}_b^{(2)} - \mathbf{K}_{ab}^{(2)} (\mathbf{K}_{bb}^{(1)} + \mathbf{K}_{aa}^{(2)})^{-1} (\mathbf{f}_a^{(2)} + \mathbf{f}_b^{(1)}) \tag{5.7.6b}$$

These are the equations for the elimination and combination of adjacent intervals in the displacement method formulation.

These equations are fundamental in the algorithm of sub-structural analysis program system. This algorithm uses the displacement vector as basic variables, so that it is compatible with the existing FEM program system, for which the displacement method algorithm of sub-structural analysis is the most preferable choice. However, in some special case it has serious problem. Such as for beam element analysis, when the mesh becomes very dense, the numerical ill-conditioning due to round-off operation will destroy the precision of numerical results required.

Because the numerical ill-conditioning problem mentioned above, the stiffness matrix method does not fit the requirement of precise integration method. Later the interval mixed energy will be introduced, which corresponds to the mixed variable system. The numerical behavior will again be analyzed in detail for precise integration method. Here, an explicit reason of ill-conditioning should be pointed out, that in the equations (5.7.5a,b) there is the subtraction for the matrices at the right-hand side. The potential danger of this subtraction is that the trunk part might be cancelled, and the remaining value is only the negligible tail part, therefore the numerical precision is seriously dropped.

§5.7.2, Mixed energy, the dual variables

Displacement method is used in the above analysis, that the two end variables $\mathbf{q}_a$ and $\mathbf{q}_b$ at z_a and z_b, respectively, are all displacement vectors. However, similar to the configuration space in analytical dynamics, the displacement method is not unique. The generalized momentum $\mathbf{p}$ was introduced in the Hamiltonian system in analytical dynamics and it is treated equally important with the generalize displacement $\mathbf{q}$. Combining $\mathbf{q}$ and $\mathbf{p}$ together composes the state vector with dual variables. The dual variable can also be introduced in structural mechanics and combining with the displacement vector gives the dual variables space, and the mixed energy is introduced correspondingly. Here again the case of no external force is considered first, the case of external force existing can be seen later in section 5.9 and 6.5.

According to the expression of deformation energy U, equation (5.7.2), the internal force vector is introduced as the dual vector

$$\mathbf{p}_a = -\partial U/\partial \mathbf{q}_a, \quad \mathbf{p}_b = \partial U/\partial \mathbf{q}_b \tag{5.7.7a,b}$$

Note the negative sign of $\mathbf{p}_a$, which is the requirement of internal force, such that when there is no external force, $(\mathbf{p}_b)_k = (\mathbf{p}_a)_{k+1}$. The negative sign means that the action force and reaction must have reverse direction. Substituting the expression of $U(\mathbf{q}_a, \mathbf{q}_b)$ into the above equations gives

$$\mathbf{p}_a = -\mathbf{K}_{aa}\mathbf{q}_a - \mathbf{K}_{ab}\mathbf{q}_b, \qquad \mathbf{p}_b = \mathbf{K}_{ba}\mathbf{q}_a + \mathbf{K}_{bb}\mathbf{q}_b \tag{5.7.8a,b}$$

Writing as ***complete differential form*** of the interval deformation energy gives

$$\delta U(\mathbf{q}_a, \mathbf{q}_b) = -\mathbf{p}_a^T \delta \mathbf{q}_a + \mathbf{p}_b^T \delta \mathbf{q}_b \tag{5.7.9}$$

This form uses purely displacement as arguments, so that termed as displacement method. Let the ***interval mixed energy*** be defined as

$$V(\mathbf{q}_a, \mathbf{p}_b) = \mathbf{p}_b^T \mathbf{q}_b - U(\mathbf{q}_a, \mathbf{q}_b) \tag{5.7.10}$$

where the vector $\mathbf{q}_b$ at the right hand side is not an independent variable, which should be considered solved from equation (5.7.8b) as

$$\mathbf{q}_b = -\mathbf{K}_{bb}^{-1}(\mathbf{K}_{ba}\mathbf{q}_a - \mathbf{p}_b)$$

Substituting into equation (5.7.10) eliminates $\mathbf{q}_b$, then the mixed energy V of the interval (z_a, z_b) is obtained as a function of $\mathbf{q}_a$ and $\mathbf{p}_b$ only. It is seen that the mixed energy formulation (5.7.10) is obtained from the deformation energy via the ***Legendre transformation*** with respect to the variable $\mathbf{q}_b$. The meaning of mixed energy can be seen from the following partial differentials

$$\partial V/\partial \mathbf{p}_b = \mathbf{q}_b + \mathbf{p}_b^T(\partial \mathbf{q}_b/\partial \mathbf{p}_b) - (\partial U/\partial \mathbf{q}_b)^T(\partial \mathbf{q}_b/\partial \mathbf{p}_b) = \mathbf{q}_b \tag{5.7.11a}$$

where the last equality is from the equation (5.7.7b), and also

$$\partial V/\partial \mathbf{q}_a = \mathbf{p}_b^T(\partial \mathbf{q}_b/\partial \mathbf{q}_a) - \partial U/\partial \mathbf{q}_a - (\partial U/\partial \mathbf{q}_b)^T(\partial \mathbf{q}_b/\partial \mathbf{q}_a) = \mathbf{p}_a \tag{5.7.11b}$$

which uses also the equations (5.7.7a,b). Hence the ***complete differential of mixed energy*** is

$$\delta V(\mathbf{q}_a, \mathbf{p}_b) = \mathbf{q}_b^T \delta \mathbf{p}_b + \mathbf{p}_a^T \delta \mathbf{q}_a \tag{5.7.11}$$

The above equation for complete differential does not limit to that U being a quadratic form. For the special case of a linear system, U is a quadratic form. Substituting $\mathbf{q}_b$ into equation (5.7.10) gives

$$V(\mathbf{q}_a, \mathbf{p}_b) = \mathbf{p}_b^T \mathbf{G} \mathbf{p}_b/2 + \mathbf{p}_b^T \mathbf{F} \mathbf{q}_a - \mathbf{q}_a^T \mathbf{Q} \mathbf{q}_a/2 \tag{5.7.12}$$

where

$$\mathbf{G} = \mathbf{K}_{bb}^{-1}, \quad \mathbf{F} = -\mathbf{K}_{bb}^{-1}\mathbf{K}_{ba}, \quad \mathbf{Q} = \mathbf{K}_{aa} - \mathbf{K}_{ab}\mathbf{K}_{bb}^{-1}\mathbf{K}_{ba} \tag{5.7.13}$$

which gives the expression of mixed energy $V(\mathbf{q}_a, \mathbf{p}_b)$ in the case of no external forces. Substituting the mixed energy expression (5.7.12) into (5.7.11a,b) gives the homogeneous dual equations

$$\mathbf{q}_b = \mathbf{F}\mathbf{q}_a + \mathbf{G}\mathbf{p}_b \tag{5.7.14a}$$

$$\mathbf{p}_a = -\mathbf{Q}\mathbf{q}_a + \mathbf{F}^T\mathbf{p}_b \tag{5.7.14b}$$

The above derivation is mainly from mathematical deduction. However, the physical interpretation for these matrices will be beneficial for understanding.

Substituting $\mathbf{q}_a = \mathbf{0}$ into equation (5.7.14a), *i.e.* under the configuration of end-a clamped, the equation reduces to the case of a force vector $\mathbf{p}_b$ acted on the end-b inducing displacement $\mathbf{q}_b$ at end-b. Hence $\mathbf{G}$ ***is the flexibility matrix at end-*** b ***under the configuration of end-***a ***clamped while end-***b ***free.***

Secondly, substituting $\mathbf{p}_b = \mathbf{0}$ into equation (5.7.14b) gives the force vector $\mathbf{p}_a$ at the end-a under the configuration of given displacement $\mathbf{q}_a$ at end-a with end-b free from force. Hence $\mathbf{Q}$ ***is a stiffness matrix at the end-*** a ***under the configuration of end-*** b ***free and end-*** a ***given displacement.*** The negative sign comes from the force $\mathbf{p}_a$ having reverse direction with respect to the displacement $\mathbf{q}_a$.

Further, let $\mathbf{p}_b = \mathbf{0}$, the equation (5.7.14a) determines that $\mathbf{F}$ ***is the transfer matrix of the displacement*** $\mathbf{q}_a$ ***toward the displacement*** $\mathbf{q}_b$ ***with the end-***b ***being free from force.***

It is seen from the physical interpretation above, that the structural model is always given displacement at the end-a and given force at the end-b. That is a "cantilever beam" type boundary condition. Therefore the mixed energy expression corresponds to the "***cantilever beam***" model.

Go back to section 5.2, there is the mixed energy density (5.2.11) or Hamilton function, which must have close relationship with the mixed energy (5.7.10) of an interval. Also the interrelationship of the dual vectors $\mathbf{q}$ and $\mathbf{p}$ between the two derivations should also be clarified. In section 5.2 the discussion is for the continuous coordinate z, and the ***mixed energy density*** is introduced from deformation energy density by the Legendre transformation. In the present section, the deformation energy of the interval is obtained from integrating the deformation energy density for the interval (z_a, z_b), thereafter introducing the end dual vector (force) by the equation (5.7.7a,b) as well as the ***interval mixed energy***. The interrelationship between the two similar derivations should be clarified.

The displacement vectors $\mathbf{q}_a$ and $\mathbf{q}_b$ should first be considered. From definition, they are the displacement vector $\mathbf{q}(z)$ at the stations z_a and z_b, respectively, *i.e.* $\mathbf{q}(z_a)$ and $\mathbf{q}(z_b)$.

Next, the relationship between two end dual vectors $\mathbf{p}_a$, $\mathbf{p}_b$ and the dual vector $\mathbf{p}(z)$ of continuous coordinate z, see equation (5.2.6), is to clarify. The interval deformation energy can be calculated directly from equation (5.7.8) and (5.7.4) as $U = (\mathbf{p}_b^T\mathbf{q}_b - \mathbf{p}_a^T\mathbf{q}_a)/2$. On the other hand, integrating the deformation energy density gives

$$\begin{aligned} U(z_a, z_b) &= \int_{z_a}^{z_b} [\dot{\mathbf{q}}^T\mathbf{K}_{22}\dot{\mathbf{q}}/2 + \dot{\mathbf{q}}^T\mathbf{K}_{21}\mathbf{q} + \mathbf{q}^T\mathbf{K}_{11}\mathbf{q}/2]\mathrm{d}z \\ &= \int_{z_a}^{z_b} [\mathbf{p}^T\dot{\mathbf{q}}/2 + \mathbf{q}^T\mathbf{K}_{12}\dot{\mathbf{q}}/2 + \mathbf{q}^T\mathbf{K}_{11}\mathbf{q}/2]\mathrm{d}z \\ &= [\mathbf{p}^T \cdot \mathbf{q}/2]_{z_a}^{z_b} + \int_{z_a}^{z_b} \mathbf{q}^T[-\mathbf{K}_{22}\ddot{\mathbf{q}} - (\mathbf{K}_{21} - \mathbf{K}_{12})\dot{\mathbf{q}} + \mathbf{K}_{11}\mathbf{q}]\mathrm{d}z/2 \\ &= [\mathbf{p}^T(z_b)\cdot \mathbf{q}_b - \mathbf{p}^T(z_a)\mathbf{q}_a]/2 \end{aligned}$$

Comparison determines that $\mathbf{p}_a$ and $\mathbf{p}_b$ are again the dual vector $\mathbf{p}(z)$ at the stations z_a and z_b, respectively, *i.e.* $\mathbf{p}(z_a)$ and $\mathbf{p}(z_b)$.

Further, the ***interval mixed energy*** should be related to the ***mixed energy density***. From the definition of mixed energy density equation (5.7.10),

$$\begin{aligned} V(\mathbf{q}_a, \mathbf{p}_b) &= \mathbf{p}_b^T\mathbf{q}_b - \int_{z_a}^{z_b} [\dot{\mathbf{q}}^T\mathbf{K}_{22}\dot{\mathbf{q}}/2 + \dot{\mathbf{q}}^T\mathbf{K}_{21}\mathbf{q} + \mathbf{q}^T\mathbf{K}_{11}\mathbf{q}/2]\mathrm{d}z \\ &= \mathbf{p}_b^T\mathbf{q}_b - \int_{z_a}^{z_b} [\mathbf{p}^T\dot{\mathbf{q}} - \mathrm{H}(\mathbf{q}, \mathbf{p})]\mathrm{d}z \end{aligned} \qquad (5.7.15)$$

where H is the ***mixed energy density*** as given in equation (5.2.11). The derivation here is only for the case of no distribution force, corresponding to homogeneous equations. For the case of external force not equal to zero, the superposition principle can be used, and will be discussed later.

Although the equation (5.7.15) has been given, which expresses that the interval mixed energy is obtained from the integration of Hamilton function. There are matrices $\mathbf{A}, \mathbf{B}$ and $\mathbf{D}$ in H, and there are matrices $\mathbf{F}, \mathbf{G}$ and $\mathbf{Q}$ in the quadratic form (5.7.12) of interval mixed energy, they are certainly interrelated.

The equation (5.7.15) implies, that ***the matrices* $\mathbf{F},\mathbf{G}$ *and* $\mathbf{Q}$ *can be integrated from* $\mathbf{A},\mathbf{B}$ *and* $\mathbf{D}$**. However, to carry out this integration, the ***method of precise integration*** should be used. The computation of $\mathbf{F},\mathbf{G}$ and $\mathbf{Q}$ is unrelated to external forces, hence the precise integration method is introduced for no external force case first, thereafter discuss the external force issue.

§5.7.3, Riccati differential equation and its precise integration

The interval mixed energy is introduced in previous section, in order to compute the interval mixed energy matrices $\mathbf{F}(z_a,z_b)$, $\mathbf{G}(z_a,z_b)$ and $\mathbf{Q}(z_a,z_b)$, the differential equations for these matrices must be derived first, with the matrices $\mathbf{A},\mathbf{B}$ and $\mathbf{D}$ being given. In the derivation below, $\mathbf{A},\mathbf{B}$ and $\mathbf{D}$ can be functions of coordinate z.

Let the interval right end coordinate z_b be increased to $z_b+\Delta z_b$ in the homogeneous dual equation (5.7.14), with Δz_b infinitesimal; with z_a and $\mathbf{q}_a$ unchanged. The right end force $\mathbf{p}_b$ is simultaneously increased to $\mathbf{p}(z_b+\Delta z_b)=\mathbf{p}_b+\dot{\mathbf{p}}_b\Delta z_b$ at $z_b+\Delta z_b$, for which the increment is compatible to the differential equation of the continuous coordinate z. Hence inside the interval the solution of $\mathbf{q}$ and $\mathbf{p}$ have no change (in first order). Certainly $\mathbf{p}_a$ does not change, and the displacement $\mathbf{q}_b$ at z_b has no change either. Only because increment Δz_b at the end-b, the right end displacement becomes $\mathbf{q}_b+\Delta\mathbf{q}_b$ with $\Delta\mathbf{q}_b=\dot{\mathbf{q}}_b\Delta z_b$. So that the differential equations are derived as

$$\mathbf{0}=-(\partial\mathbf{Q}/\partial z_b)\mathbf{q}_a+(\partial\mathbf{F}^T/\partial z_b)\mathbf{p}_b+\mathbf{F}^T\dot{\mathbf{p}}_b$$

$$\dot{\mathbf{q}}_b=(\partial\mathbf{F}/\partial z_b)\mathbf{q}_a+(\partial\mathbf{G}/\partial z_b)\mathbf{p}_b-\mathbf{G}\dot{\mathbf{p}}_b$$

Using differential equation (5.2.10) gives

$$\mathbf{0}=-(\partial\mathbf{Q}/\partial z_b)\mathbf{q}_a+(\partial\mathbf{F}^T/\partial z_b)\mathbf{p}_b+\mathbf{F}^T(\mathbf{B}\mathbf{q}_b-\mathbf{A}^T\mathbf{p}_b)$$

$$\mathbf{A}\mathbf{q}_b+\mathbf{D}\mathbf{p}_b=(\partial\mathbf{F}/\partial z_b)\mathbf{q}_a+(\partial\mathbf{G}/\partial z_b)\mathbf{p}_b+\mathbf{G}(\mathbf{B}\mathbf{q}_b-\mathbf{A}^T\mathbf{p}_b)$$

But the three vectors $\mathbf{q}_a,\mathbf{p}_b$ and $\mathbf{q}_b$ are not completely independent. Eliminating $\mathbf{q}_b$ by substituting with equation (5.7.14a) gives

$$[-\partial\mathbf{Q}/\partial z_b+\mathbf{F}^T\mathbf{B}\mathbf{F}]\mathbf{q}_a+[\partial\mathbf{F}^T/\partial z_b-\mathbf{F}^T\mathbf{A}^T+\mathbf{F}^T\mathbf{B}\mathbf{G}]\mathbf{p}_b=\mathbf{0}$$

$$[\partial\mathbf{F}/\partial z_b-\mathbf{A}\mathbf{F}+\mathbf{G}\mathbf{B}\mathbf{F}]\mathbf{q}_a+[\partial\mathbf{G}/\partial z_b-\mathbf{D}-\mathbf{A}\mathbf{G}-\mathbf{G}\mathbf{A}^T+\mathbf{G}\mathbf{B}\mathbf{G}]\mathbf{p}_b=\mathbf{0}$$

Because the values of $\mathbf{q}_a$ and $\mathbf{p}_b$ correspond to assigning the two end boundary conditions and can be given arbitrarily, hence the three simultaneous PDEs below are resulted

$$\partial\mathbf{F}/\partial z_b=(\mathbf{A}-\mathbf{G}\mathbf{B})\mathbf{F} \tag{5.7.16a}$$

$$\partial\mathbf{G}/\partial z_b=\mathbf{D}+\mathbf{A}\mathbf{G}+\mathbf{G}\mathbf{A}^T-\mathbf{G}\mathbf{B}\mathbf{G} \tag{5.7.16b}$$

$$\partial\mathbf{Q}/\partial z_b=\mathbf{F}^T\mathbf{B}\mathbf{F} \tag{5.7.16c}$$

The fourth equation is only the transposition of equation (5.7.16a). Because the matrices $\mathbf{D}$ and $\mathbf{B}$ are symmetric, thus the matrices $\mathbf{G}$ and $\mathbf{Q}$ keep symmetric too, *i.e.* $\mathbf{G}^T = \mathbf{G},\ \mathbf{Q}^T = \mathbf{Q}$.

The above three differential equations are derived on keeping the left end- z_a unchanged but let z_b increase Δz_b. On the contrary, let z_b keeps unchanged but let z_a increase Δz_a, after similar derivation it gives

$$\partial \mathbf{F}/\partial z_a = -\mathbf{F}(\mathbf{A} - \mathbf{DQ}) \tag{5.7.17a}$$

$$\partial \mathbf{G}/\partial z_a = -\mathbf{FDF}^T \tag{5.7.17b}$$

$$\partial \mathbf{Q}/\partial z_a = -\mathbf{B} - \mathbf{A}^T\mathbf{Q} - \mathbf{QA} + \mathbf{QDQ} \tag{5.7.17c}$$

These are the two sets of three first order simultaneous PDEs for the three matrices $\mathbf{F}(z_a, z_b)$, $\mathbf{G}(z_a, z_b)$ and $\mathbf{Q}(z_a, z_b)$. According to the theory of ODEs, three initial conditions must be supplied. These boundary conditions can be obtained via the limiting value of $(z_b - z_a) \to 0$

$$\mathbf{Q} = \mathbf{0},\ \mathbf{G} = \mathbf{0},\ \mathbf{F} = \mathbf{I}_n, \quad \text{as } z_b \to z_a + 0 \tag{5.7.18}$$

The above derivation does not require that the given matrices $\mathbf{A}, \mathbf{B}$ and $\mathbf{D}$ being coordinate z independent. But when $\mathbf{A}, \mathbf{B}$ and $\mathbf{D}$ are constant matrices, the matrices $\mathbf{F}(z_a, z_b)$, $\mathbf{G}(z_a, z_b)$ and $\mathbf{Q}(z_a, z_b)$ depend only on the interval length

$$\eta = z_b - z_a \tag{5.7.19}$$

and can be written simply as

$$\mathbf{F}(z_a, z_b) = \mathbf{F}(\eta)\ ,\ \partial \mathbf{F}/\partial z_b = \mathrm{d}\mathbf{F}/\mathrm{d}\eta\ ,\ \partial \mathbf{F}/\partial z_a = -\mathrm{d}\mathbf{F}/\mathrm{d}\eta \tag{5.7.20}$$

similarly for $\mathbf{G}, \mathbf{Q}$. In such case, the equations (5.7.16) and (5.7.17) are reduced to

$$\mathrm{d}\mathbf{F}/\mathrm{d}\eta = (\mathbf{A} - \mathbf{GB})\mathbf{F} \tag{5.7.16'a}$$

$$\mathrm{d}\mathbf{G}/\mathrm{d}\eta = \mathbf{D} + \mathbf{AG} + \mathbf{GA}^T - \mathbf{GBG} \tag{5.7.16'b}$$

$$\mathrm{d}\mathbf{Q}/\mathrm{d}\eta = \mathbf{F}^T\mathbf{BF} \tag{5.7.16'c}$$

and

$$\mathrm{d}\mathbf{F}/\mathrm{d}\eta = \mathbf{F}(\mathbf{A} - \mathbf{DQ}) \tag{5.7.17'a}$$

$$\mathrm{d}\mathbf{G}/\mathrm{d}\eta = \mathbf{FDF}^T \tag{5.7.17'b}$$

$$\mathrm{d}\mathbf{Q}/\mathrm{d}\eta = \mathbf{B} + \mathbf{A}^T\mathbf{Q} + \mathbf{QA} - \mathbf{QDQ} \tag{5.7.17'c}$$

and the initial conditions are

$$\mathbf{Q} = \mathbf{0},\ \mathbf{G} = \mathbf{0},\ \mathbf{F} = \mathbf{I}_n, \quad \text{as } \eta \to +0 \tag{5.7.18'}$$

The above two sets of ODEs looked quite different to each other, but they are really the same problem. It is seen that equations (5.7.16'b) and (5.7.17'c) are the Riccati differential equations. The differential equation itself is really non-linear matrix differential equation. If finite difference approximation is used to solve the non-linear set of Riccati differential equation, then there are quite a number of difficulties. However, the Riccati differential equation is derived from a linear system, therefore the method of precise integration can definitely be used for numerical solution, for which the results obtained can reach almost the host computer precision.

From equations (5.7.16) and (5.7.17) it is further seen that when the matrices $\mathbf{A}, \mathbf{B}$ and $\mathbf{D}$ are coordinate z independent, then the interval matrices

$\mathbf{F}(\eta)$, $\mathbf{G}(\eta)$ and $\mathbf{Q}(\eta)$ satisfy the relations

$$\begin{aligned}
&(\mathbf{A}-\mathbf{GB})\mathbf{F}=\mathbf{F}(\mathbf{A}-\mathbf{DQ}) \quad [=\mathrm{d}\mathbf{F}/\mathrm{d}\eta] \\
&\mathbf{FDF}^T=\mathbf{D}+\mathbf{AG}+\mathbf{GA}^T-\mathbf{GBG} \quad [=\mathrm{d}\mathbf{G}/\mathrm{d}\eta] \\
&\mathbf{F}^T\mathbf{BF}=\mathbf{B}+\mathbf{A}^T\mathbf{Q}+\mathbf{QA}-\mathbf{QDQ} \quad [=\mathrm{d}\mathbf{Q}/\mathrm{d}\eta]
\end{aligned} \tag{5.7.17''}$$

§5.7.4, Power series expansion

After derived the set of simultaneous differential equations, how to find its numerical solutions for the non-linear ODEs is a challenging problem. However, when the length η of interval becomes extremely small, the method of Taylor series expansion can be applied smoothly and extremely precisely. One of the cruxes of precise integration method is to subdivide the step size extremely small. For the uniform subdivision of the whole interval (5.7.3), the sub-interval length η is further subdivided into 2^N extremely small interval τ as

$$N=20\ ,\ 2^N=1048576,\quad \tau=\eta/2^N \tag{5.7.19}$$

Then the interval matrices are expanded into power series as

$$\mathbf{F}'(\tau)=\boldsymbol{\varphi}_1\tau+\boldsymbol{\varphi}_2\tau^2+\boldsymbol{\varphi}_3\tau^3+\boldsymbol{\varphi}_4\tau^4+O(\tau^5)\ ,\ \ \mathbf{F}(\tau)=\mathbf{I}+\mathbf{F}'(\tau) \tag{5.7.20a}$$

$$\mathbf{G}(\tau)=\boldsymbol{\gamma}_1\tau+\boldsymbol{\gamma}_2\tau^2+\boldsymbol{\gamma}_3\tau^3+\boldsymbol{\gamma}_4\tau^4+O(\tau^5) \tag{5.7.20b}$$

$$\mathbf{Q}(\tau)=\boldsymbol{\theta}_1\tau+\boldsymbol{\theta}_2\tau^2+\boldsymbol{\theta}_3\tau^3+\boldsymbol{\theta}_4\tau^4+O(\tau^5) \tag{5.7.20c}$$

where $\boldsymbol{\varphi}_i, \boldsymbol{\gamma}_i, \boldsymbol{\theta}_i$, $i=1\sim4$, *i.e.* the coefficient matrices of Taylor expansion, are all $n\times n$ matrices. Substituting equation (5.7.20a) into (5.7.16'b), and note that η in the equations should be changed as τ, comparing the terms of various powers of τ successively gives

$$\begin{aligned}
&\boldsymbol{\gamma}_1=\mathbf{D},\ \ \boldsymbol{\gamma}_2=(\mathbf{A}\boldsymbol{\gamma}_1+\boldsymbol{\gamma}_1\mathbf{A}^T)/2\ ,\ \ \boldsymbol{\gamma}_3=(\mathbf{A}\boldsymbol{\gamma}_2+\boldsymbol{\gamma}_2\mathbf{A}^T-\boldsymbol{\gamma}_1\mathbf{B}\boldsymbol{\gamma}_1)/3, \\
&\boldsymbol{\gamma}_4=(\mathbf{A}\boldsymbol{\gamma}_3+\boldsymbol{\gamma}_3\mathbf{A}^T-\boldsymbol{\gamma}_2\mathbf{B}\boldsymbol{\gamma}_1-\boldsymbol{\gamma}_1\mathbf{B}\boldsymbol{\gamma}_2)/4
\end{aligned} \tag{5.7.21b}$$

Substituting the expansion of $\mathbf{F}(\tau)$ equation (5.7.20a) into the differential equation (5.7.16a'), comparing the various powers of τ gives

$$\begin{aligned}
&\boldsymbol{\varphi}_1=\mathbf{A},\quad \boldsymbol{\varphi}_2=(\mathbf{A}\boldsymbol{\varphi}_1-\boldsymbol{\gamma}_1\mathbf{B})/2,\quad \boldsymbol{\varphi}_3=(\mathbf{A}\boldsymbol{\varphi}_2-\boldsymbol{\gamma}_2\mathbf{B}-\boldsymbol{\gamma}_1\mathbf{B}\boldsymbol{\varphi}_1)/3, \\
&\boldsymbol{\varphi}_4=(\mathbf{A}\boldsymbol{\varphi}_3-\boldsymbol{\gamma}_3\mathbf{B}-\boldsymbol{\gamma}_2\mathbf{B}\boldsymbol{\varphi}_1-\boldsymbol{\gamma}_1\mathbf{B}\boldsymbol{\varphi}_2)/4
\end{aligned} \tag{5.7.21a}$$

Afterwards, for the power series expansion (5.7.20c) of matrix $\mathbf{Q}(\tau)$, based on the differential equation (5.7.16'c) it gives

$$\begin{aligned}
&\boldsymbol{\theta}_1=\mathbf{B},\ \ \boldsymbol{\theta}_2=(\boldsymbol{\varphi}_1^T\mathbf{B}+\mathbf{B}\boldsymbol{\varphi}_1)/2,\ \ \boldsymbol{\theta}_3=(\boldsymbol{\varphi}_2^T\mathbf{B}+\mathbf{B}\boldsymbol{\varphi}_2+\boldsymbol{\varphi}_1^T\mathbf{B}\boldsymbol{\varphi}_1)/3, \\
&\boldsymbol{\theta}_4=(\boldsymbol{\varphi}_3^T\mathbf{B}+\mathbf{B}\boldsymbol{\varphi}_3+\boldsymbol{\varphi}_2^T\mathbf{B}\boldsymbol{\varphi}_1+\boldsymbol{\varphi}_1^T\mathbf{B}\boldsymbol{\varphi}_2)/4
\end{aligned} \tag{5.7.21c}$$

A different set of equations can also be derived using (5.7.17'c), but they give the same numerical results. The merit of power series expansion is that when the step size τ is extremely small, the expansion up to fourth power and the term neglected is $O(\tau^5)$, which has been extremely small quantity. If the matrices of mixed energy of interval length η are of the order of $O(1)$, then because of $\tau\approx\eta\cdot10^{-6}$, the relative error of expansion (5.7.20) is up to the order of $O(\tau^4) \approx \eta^4/10^{24}$. Note that the significant digits for the double precision real number are 16 decimal digits. The neglected term of 10^{-24} has been beyond the double precision for the computer today.

§5.7.5, Interval combination

The interval combination and internal variable elimination are fundamental in sub-structural combination algorithm. Equation (5.7.5) has supplied the fundamental formulae; however it is the expressions of displacement method. The displacement method may induce serious numerical ill-conditioning problem, especially when the mesh subdivision becomes extremely dense as given in equation (5.7.19). The precise integration method takes benefit from using extremely dense meshes to reach high precision. On the basis of usual small interval length η, a refined mesh is introduced by further subdividing the length η into 1048576=2^{20} ***extremely dense intervals***. For such extremely dense mesh size the displacement method cannot be applied properly. The merit of mixed energy representation is that it is ***insensible to the mesh size***, using dense mesh can increase the numerical precision. Therefore ***the mixed energy representation is suitable to the precise integration method***.

The precise integration method needs the equations for interval combination in terms of mixed energy representation. The mixed energy representation of an interval is fundamentally the same algebraically to the deformation energy representation, that the two representations can be transformed to each other. The interval combination equations for mixed energy representation can be derived by means of variational method.

Let the two adjacent intervals be (z_a, z_b) and (z_b, z_c), for which the complete differential forms are, respectively

$$\delta V_1(\mathbf{q}_a, \mathbf{p}_b) = \mathbf{p}_a^T \delta\mathbf{q}_a + \mathbf{q}_b^T \delta\mathbf{p}_b$$

$$\delta V_2(\mathbf{q}_b, \mathbf{p}_c) = \mathbf{p}_b^T \delta\mathbf{q}_b + \mathbf{q}_c^T \delta\mathbf{p}_c$$

The combined interval is (z_a, z_c) and denoted by subscript c, for which the complete differential form is

$$\delta V_c(\mathbf{q}_a, \mathbf{p}_c) = \mathbf{p}_a^T \delta\mathbf{q}_a + \mathbf{q}_c^T \delta\mathbf{p}_c$$

Therefore the mixed energy of the combined interval should be

$$V_c(\mathbf{q}_a, \mathbf{p}_c) = \underset{\mathbf{q}_b, \mathbf{p}_b}{\text{sta.}} [V_1(\mathbf{q}_a, \mathbf{p}_b) + V_2(\mathbf{q}_b, \mathbf{p}_c) - \mathbf{p}_b^T \mathbf{q}_b] \tag{5.7.22}$$

where sta. stands for the vectors $\mathbf{q}_b$ and $\mathbf{p}_b$ taking stationary value. Substituting the expression of V, *i.e.* the equation (5.7.12), into V_c gives

$$\underset{\mathbf{q}_b, \mathbf{p}_b}{\text{sta.}} \begin{bmatrix} \mathbf{p}_b^T \mathbf{F}_1 \mathbf{q}_a + \left(\mathbf{p}_b^T \mathbf{G}_1 \mathbf{p}_b - \mathbf{q}_a^T \mathbf{Q}_1 \mathbf{q}_a\right)/2 - \mathbf{p}_b^T \mathbf{q}_b \\ + \mathbf{p}_c^T \mathbf{F}_2 \mathbf{q}_b + \left(\mathbf{p}_c^T \mathbf{G}_2 \mathbf{p}_c - \mathbf{q}_b^T \mathbf{Q}_2 \mathbf{q}_b\right)/2 \end{bmatrix}$$

from which $\quad -\mathbf{q}_b + \mathbf{G}_1 \mathbf{p}_b + \mathbf{F}_1 \mathbf{q}_a = 0, \quad -\mathbf{p}_b - \mathbf{Q}_2 \mathbf{q}_b + \mathbf{F}_2^T \mathbf{p}_c = 0$

and solves
$$\mathbf{q}_b = (\mathbf{I} + \mathbf{G}_1 \mathbf{Q}_2)^{-1} (\mathbf{F}_1 \mathbf{q}_a + \mathbf{G}_1 \mathbf{F}_2^T \mathbf{p}_c) \tag{5.7.22a}$$

$$\mathbf{p}_b = (\mathbf{I} + \mathbf{Q}_2\mathbf{G}_1)^{-1}(\mathbf{Q}_2\mathbf{F}_1\mathbf{q}_a - \mathbf{F}_2^T\mathbf{p}_c) \tag{5.7.22b}$$

Hence derives the interval combination equations for mixed energy representation

$$\mathbf{F}_c = \mathbf{F}_2(\mathbf{I} + \mathbf{G}_1\mathbf{Q}_2)^{-1}\mathbf{F}_1 \tag{5.7.23a}$$

$$\mathbf{G}_c = \mathbf{G}_2 + \mathbf{F}_2(\mathbf{G}_1^{-1} + \mathbf{Q}_2)^{-1}\mathbf{F}_2^T \tag{5.7.23b}$$

$$\mathbf{Q}_c = \mathbf{Q}_1 + \mathbf{F}_1^T(\mathbf{Q}_2^{-1} + \mathbf{G}_1)^{-1}\mathbf{F}_1 \tag{5.7.23c}$$

It is seen from the set of equations, that the matrices $\mathbf{G}_c$ and $\mathbf{Q}_c$ keep to be symmetric matrices. Note also, that the addition sign appears in the right hand side of the latter two equations. In the case of structural static problem or frequency domain analysis with low ω, $\mathbf{Q}$ and $\mathbf{G}$ are always positive definite matrices, the additional sign ensures the numerical stability of the algorithm.

In displacement method when τ is extremely small, the matrices $\mathbf{K}_{aa}(\tau), \mathbf{K}_{ab}(\tau)$ and $\mathbf{K}_{bb}(\tau)$ in potential energy representation are of the order of $1/\tau^2$. If the computation is carried using matrices $\mathbf{K}$, then initially the matrices are of the order of $1/\tau^2$. With the elimination going on, the interval length will increase to be the usual length η, and the matrices $\mathbf{K}$ will decrease to the order of $O(1)$. It τ is taken extremely small as in the precise integration method then the significant digit will be seriously dropped. However under the mixed energy representation, the initial value of $\mathbf{G}$ and $\mathbf{Q}$ are zero matrices, and when τ is extremely small $\mathbf{Q}(\tau) \approx O(\tau)$ and $\mathbf{G}(\tau) \approx O(\tau)$, hence it does not matter for very small τ. The matrices $\mathbf{Q}$ and $\mathbf{G}$ are positive definite when ω is not large, which implies from the equations (5.7.23b,c) that these matrices are always increasing and can never have the ill-conditioning problem. Even though when ω is not so small, the matrices $\mathbf{Q}$ and $\mathbf{G}$ cannot ensure to be positive definite, but they initially are of the order of $\mathbf{Q}(\tau) \approx O(\tau)$, and will not be dropped as in the potential energy case from the order of $O(\tau^{-2})$.

But there is still the problem of that when τ is very small, the matrix $\mathbf{F}(\tau)$ tends to be a unit matrix. It has been explicitly expressed in the expansion equation (5.7.20a). Using equation (5.7.23a) computing $\mathbf{F}_c$ directly gives again approximately a unit matrix. The unit matrix is a common part, so that the main difference is the part of $O(\tau)$. Hence, the $\mathbf{I}_n$ part should be taken off from the matrix $\mathbf{F}(\tau)$ in numerical computations, that when τ is extremely small the concentration should be put on the matrix $\mathbf{F}'$ in equation (5.7.20a), rather than on $\mathbf{F}$. This step is the second crux of precise integration: ***extract the unit matrix from matrix*** $\mathbf{F}$ ***and concentrate only on its incremental part*** $\mathbf{F}'$. Therefore when the equation (5.7.23) is revised with two intervals of length τ are combined to be an interval of length 2τ, the following equations should be used instead of equations (5.7.23a~c)

$$\mathbf{F}_c' = (\mathbf{F}' - \mathbf{GQ}/2)(\mathbf{I} + \mathbf{GQ})^{-1} + (\mathbf{I} + \mathbf{GQ})^{-1}(\mathbf{F}' - \mathbf{GQ}/2) + \mathbf{F}'(\mathbf{I} + \mathbf{GQ})^{-1}\mathbf{F}' \tag{5.7.24a}$$

$$\mathbf{G}_c = \mathbf{G} + (\mathbf{I} + \mathbf{F}')(\mathbf{G}^{-1} + \mathbf{Q})^{-1}(\mathbf{I} + \mathbf{F}')^T \tag{5.7.24b}$$

$$\mathbf{Q}_c = \mathbf{Q} + (\mathbf{I} + \mathbf{F}')^T(\mathbf{Q}^{-1} + \mathbf{G})^{-1}(\mathbf{I} + \mathbf{F}') \tag{5.7.24c}$$

This set of equations can be used for very small interval length, especially in the 2^N algorithm to compute the fundamental interval matrices of length η.

§5.7.6, Precise integration for the fundamental interval

The Riccati differential equation is non-linear. The purely analytical method can only solve one-dimensional $n = 1$ problem. For multi-dimensional problem only numerical solution can be used. The eigenvector expansion method gives analytical solution in expansion form, as given in the section 5.8, however the eigenvectors need also to be solved numerically.

The coordinate-invariant system is to be solved first, for which the system input are the matrices $\mathbf{A}, \mathbf{B}$ and $\mathbf{D}$ in mixed energy representation, or the matrices $\mathbf{K}_{11}, \mathbf{K}_{22}$ and $\mathbf{K}_{21}$ in deformation energy representation, these input matrices are independent on the coordinate z.

Numerical solution cannot give results at all coordinate z, but can only give at grid points. Uniform grid point scheme is popular, for which the fundamental length η of intervals is reasonable. In precise integration, the selection of length η is quite versatile with comparison to the FEM or FDM.

After selected the fundamental interval length η (step size), which is only for supplying data of output. Similar to the case of precise integration for exponential matrix, the real computation uses the step size as further subdividing η into 2^N extremely dense step size, which is denoted as τ

$$N = 20 \ , \ 2^N = 1048576 \ , \ \tau = \eta/2^N \tag{5.7.19}$$

which has been given a number of times before. For τ step size, power series expansion can reach the double precision real number representation on the host computer. For finding the interval matrices $\mathbf{F}(\eta)$, $\mathbf{G}(\eta)$ and $\mathbf{Q}(\eta)$ of fundamental length η, N times interval combination performs the computation. The algorithm is given in meta-language as:

[Give dimension n and $n \times n$ matrices $\mathbf{K}_{22}, \mathbf{K}_{21}, \mathbf{K}_{11}$]
[Compute $\mathbf{A}, \mathbf{B}, \mathbf{D}$ from (5.2.8)]
[Select step size η, select N, then τ is obtained, see (5.7.19)]
[Using (5.7.20-21), generate $\mathbf{F}'(\tau), \mathbf{G}(\tau), \mathbf{Q}(\tau)$ and keep in $\mathbf{F}'_c, \mathbf{G}_c, \mathbf{Q}_c$]
for $(iter = 0; iter < N; iter++)$ { $[\mathbf{F}' = \mathbf{F}'_c; \mathbf{G} = \mathbf{G}_c; \mathbf{Q} = \mathbf{Q}_c;]$
    [Using (5.7.24') compute $\mathbf{F}'_c, \mathbf{G}_c, \mathbf{Q}_c$]
}
[$\mathbf{F} = \mathbf{I} + \mathbf{F}'_c; \mathbf{G} = \mathbf{G}_c; \mathbf{Q} = \mathbf{Q}_c;$] (5.7.25)

Comment: $\mathbf{F}, \mathbf{G}, \mathbf{Q}$ return the interval mixed energy matrices $\mathbf{F}(\eta), \mathbf{G}(\eta), \mathbf{Q}(\eta)$.

Except the power series expansion (5.7.20), all other derivations are exact. For the power series expansion, 4-terms approximation is used with the precise step size τ, for which the ***error has been beyond the double precision representation of real number of the host computer***. Therefore for TPBVP, the numerical results of precise integration, with the fundamental length η, for Riccati differential equation has reached the host computer precision.

Having found the mixed energy matrices $\mathbf{F}(\eta),\mathbf{G}(\eta),\mathbf{Q}(\eta)$ of fundamental interval η, then arbitrary two point boundary value problem can be solved. For problem with finite length interval $z_{\mathrm{f}}=k_{\mathrm{f}}\eta$, *i.e.* number k_{f} intervals of length η being connected end to end, the simplest solution method is step by step integration $k_{\mathrm{f}}-1$ times to find the mixed energy matrices of the interval $(0,z_{\mathrm{f}})$, for which the interval combination equations have been given in equation (5.7.23). It is to mention here again, that the interval internal force has not been considered yet.

To get the solution, boundary conditions are necessary. Presently, TPBVP is appropriate, since it gives n conditions for both the ends, totally $2n$ conditions.

The simplest boundary condition is, $\mathbf{q}_0=$ given at the $z=0$ end, *i.e.* given $\mathbf{q}_{\mathrm{a}}$ for the interval $(0,z_{\mathrm{f}})$, while $\mathbf{p}_{\mathrm{f}}=$ given at $z=z_{\mathrm{f}}$ end, *i.e.* given $\mathbf{p}_{\mathrm{b}}$ at the right end of whole interval. In such case, the matrices $\mathbf{F},\mathbf{G}$ and $\mathbf{Q}$ in equations (5.9.14a,b) for the whole length interval $(0,z_{\mathrm{f}})$ has been computed. There are $2n$ simultaneous algebraic equations for (5.9.14a,b) with the four n-dimensional vectors of $\mathbf{q}_{\mathrm{a}},\mathbf{p}_{\mathrm{a}}$ and $\mathbf{q}_{\mathrm{b}},\mathbf{p}_{\mathrm{b}}$. If $\mathbf{q}_{\mathrm{a}}$ and $\mathbf{p}_{\mathrm{b}}$ are given, then direct substituting gives $\mathbf{p}_{\mathrm{a}}$ and $\mathbf{q}_{\mathrm{b}}$. After computed two end state vectors, the state vectors of the internal points can be computed by solving with the equations (5.7.22a,b).

The algorithm given above is based on the interval mixed energy, so as to avoid the numerical problem. However, after the mixed energy matrices $\mathbf{F}(\eta),\mathbf{G}(\eta)$ and $\mathbf{Q}(\eta)$ of the fundamental interval η are computed with the precise integration method, transforming to the deformation representation then continuing with the displacement method is also a good choice. Because η is not such a small interval as τ is, so no serious numerical problem will appear.

Equation (5.7.13) gives the transformation from stiffness matrices $\mathbf{K}_{\mathrm{aa}},\mathbf{K}_{\mathrm{bb}}$ and $\mathbf{K}_{\mathrm{ba}}$ of displacement method to the mixed energy matrices $\mathbf{F},\mathbf{G}$ and $\mathbf{Q}$. The reverse transformation equations are

$$\mathbf{K}_{\mathrm{bb}}=\mathbf{G}^{-1},\quad \mathbf{K}_{\mathrm{ba}}=-\mathbf{G}^{-1}\mathbf{F},\quad \mathbf{K}_{\mathrm{aa}}=\mathbf{Q}+\mathbf{F}^{T}\mathbf{G}^{-1}\mathbf{F} \tag{5.7.26}$$

and in equations (5.7.5) and (5.7.6) the interval combination equations have been given there.

The displacement method is easiest to understand and to operate, and FEM uses the displacement method in majority cases. After consummated the precise integration computations for the fundamental interval η, transforming back to the displacement method is also an effective and efficient approach. However continuing the computation in the mixed variable space is also effective and efficient. The description below remains in the mixed variable system.

The boundary condition of elastic support at the end z_f is an interested subject to discuss. Let the elastic support condition be

$$\mathbf{p}_f = -\mathbf{S}_f \mathbf{q}_f \tag{5.7.27}$$

where $\mathbf{S}_f$ is a given $n \times n$ symmetric non-negative definite matrix, usually called the stiffness matrix of an elastic support. The negative sign in equation (5.7.27) is due to that the sign convention of the internal force $\mathbf{p}$ was defined by the negative sign of equation $\mathbf{K}_{aa}$ in equation (5.7.8a) or $\mathbf{Q}$ in equation (5.7.14b).

An elastic support corresponds to a **fictitious** 'interval', for which the mixed energy matrices can be given as $\mathbf{F}_2 = \mathbf{I}$, $\mathbf{G}_2 = \mathbf{0}$ and $\mathbf{Q}_2 = \mathbf{S}_f$. For it is an end elastic support, so that its effect influences the stations $z \le z_f$, *i.e.* no effect will transfer to $z > z_f$, hence $\mathbf{F}_2$ and $\mathbf{G}_2$ can be selected arbitrarily. The stations of $z > z_f$ are meaningless. The fictitious interval is introduced based on physical reasoning, correspondingly the mathematical derivation can be found in section 6.6. Substituting [$\mathbf{F}_1 = \mathbf{F}(z_f - z)$, $\mathbf{Q}_1 = \mathbf{Q}(z_f - z)$, $\mathbf{G}_1 = \mathbf{G}(z_f - z)$] and [$\mathbf{F}_2 = \mathbf{I}$, $\mathbf{G}_2 = \mathbf{0}$, $\mathbf{Q}_2 = \mathbf{S}_f$] into the equation (5.7.23c) gives the interval mixed energy matrices for the interval $(z, z_f]$

$$\mathbf{S}(z) = \mathbf{Q} + \mathbf{F}^T(\mathbf{I} + \mathbf{S}_f\mathbf{G})^{-1}\mathbf{S}_f\mathbf{F} \tag{5.7.28}$$

The semi-closed interval $(z, z_f]$ means that the concentrated elastic support matrix $\mathbf{S}_f$ at the end z_f is included. The matrix $\mathbf{S}(z)$ is the z end stiffness matrix of the semi-closed interval $(z, z_f]$.

Another kind of elastic support problem is an elastic support at the end of $z = 0$, that

$$\mathbf{q}_0 = \mathbf{P}_0 \mathbf{p}_0 \tag{5.7.29}$$

where the given matrix $\mathbf{P}_0$ is again symmetric non-negative definite, *i.e.* the ***flexibility matrix*** of the elastic support. The elastic support is again corresponding to a fictitious 'interval-1', whose interval mixed energy matrices can be selected as [$\mathbf{F}_1 = \mathbf{I}$, $\mathbf{G}_1 = \mathbf{P}_0$ and $\mathbf{Q}_1 = \mathbf{0}$]. End elastic support implies no further transfer is toward $z < 0$, hence it can select arbitrary $\mathbf{F}_1$ and $\mathbf{Q}_1$. Whereas $\mathbf{G}_1$ is the flexibility matrix of the elastic support. The matrices are used as $\mathbf{F}_2 = \mathbf{F}(z)$, $\mathbf{G}_2 = \mathbf{G}(z)$ and $\mathbf{Q}_2 = \mathbf{Q}(z)$ for interval-2, substituting into the equation (5.7.23b) gives

$$\mathbf{P}(z) = \mathbf{G} + \mathbf{F}(\mathbf{I} + \mathbf{P}_0\mathbf{Q})^{-1}\mathbf{P}_0\mathbf{F}^T \tag{5.7.30}$$

where $\mathbf{P}(z)$ is the end elastic flexibility matrix at the z end of the semi-closed interval $[0, z)$, which means that at the $z = 0$ end there is an elastic support with flexibility matrix $\mathbf{P}_0$.

According to the context, the problem discussed is for structural mechanics. However, based on the ***analogy relationship between structural mechanics and optimal control*** to be described later, the computation is extremely important for optimal control theory, they are the fundamental issues both for linear quadratic optimal control and for Kalman-Bucy filtering problems. The matrices $\mathbf{S}(z)$ and $\mathbf{P}(z)$ satisfy the differential equations with the respective initial conditions as

$$\dot{\mathbf{S}} = -\mathbf{B} - \mathbf{A}^T\mathbf{S} - \mathbf{S}\mathbf{A} + \mathbf{S}\mathbf{D}\mathbf{S}, \qquad \mathbf{S}(z_f) = \mathbf{S}_f \tag{5.7.31}$$

$$\dot{\mathbf{P}} = \mathbf{D} + \mathbf{A}\mathbf{P} + \mathbf{P}\mathbf{A}^T - \mathbf{P}\mathbf{B}\mathbf{P}, \qquad \mathbf{P}(0) = \mathbf{P}_0 \tag{5.7.32}$$

Substituting the equation (5.7.28) for $\mathbf{S}(z)$ into the differential equation (5.7.31), making use of the equations (5.7.16) and (5.7.17), the differential equation is verified satisfied; then using the boundary condition (5.7.18) the boundary condition at z_f in (5.7.31) is again verified satisfied. According to the uniqueness theorem of ODE, the solution is unique. Similarly for the matrix $\mathbf{P}(z)$, the differential equation and the initial boundary condition (5.7.32) can also be verified satisfied.

Example 5.1, Let $n = 1$, a 1-D problem can be solved purely analytically. Hence it can be used to check the numerical results of precise integration method. The data is given as

$$\mathbf{A} = 0.8,\ \mathbf{D} = -0.87935,\ \mathbf{B} = 0.64,\ z_f = 0.8,\ \mathbf{S}_f = 0.01.$$

The matrix function $\mathbf{S}(z)$, which is a scalar, is needed to solve from differential equation (5.7.31).

Solution: The differential equation is written as

$$\dot{S} = c + 2aS + bS^2 \tag{5.7.33}$$

This equation can be solved analytically. There are two cases, $\Delta = a^2 - b \times c, \Delta > 0$ or $\Delta < 0$.

1) For $\Delta > 0$, the equation $c + 2aS + bS^2 = 0$ has two real roots p_1 and p_2. The solution of equation (5.7.33) is integrated to be

$$\begin{gathered}(S - p_1)/(S - p_2) = A\exp[b \cdot (p_1 - p_2) \cdot (z - z_f)], \\ A = (S_f - p_1)/(S_f - p_2)\end{gathered} \tag{5.7.34a}$$

   then the solution function $S(z)$ can be computed for various z.

2) For $\Delta < 0$, two complex conjugate roots appear for the quadratic equation. Let $\mu = 2\sqrt{-\Delta}$, then the solution is

$$z - z_f + A = (2/\mu)\mathrm{arctg}[(2/\mu)(a + bS)]$$

   Substituting the boundary condition, when $z = z_f$, $S = S_f$, it determines

$$A = (2/\mu)\mathrm{arctg}[(2/\mu)(a + bS_f)]$$

$$S(z) = (1/b) \times [-a + (\mu/2)\mathrm{tg}(\mu \times (A - z_f + z)/2)] \tag{5.7.34b}$$

Presently $a = -0.8,\ b = -0.87935,\ c = -0.64,\ \Delta = 0.077216 > 0$, *i.e.* case 1.

The numerical results are listed below in table 5.1, where $\mathbf{S}(z)$ is from analytical solution with the value enlarged 1000 times, whereas $\mathbf{S}_p(z)$ is from precise integration.

Table 5.1, the solution of Riccati differential equation.

Z=	0	0.05	0.10	0.15	0.20
S=	1430.93208	1216.51221	1038.756	889.0473	761.2844
S_p=	1430.93206	1216.51220	1038.753	889.0473	761.2844

continue

0.25	0.3	0.35	0.4	0.45	0.5
651.0094	554.8965	470.4127	395.5952	328.8911	269.0920
651.0094	554.8965	470.4127	395.5952	328.8911	269.0920

continue

0.55	0.6	0.65	0.7	0.75	0.8
251.1788	166.3472	121.9273	81.3622	44.1846	0.01
251.1788	166.3472	121.9273	81.3622	44.1846	0.01

The difference is extremely small, which explains that the precise integration result is very precise. Some arithmetic round-off error is unavoidable, so that some errors appearing at the last digit is understandable. ##

Example 5.2, Give $n=4$, let $\eta=2.0$, the fundamental interval mixed energy matrices $\mathbf{F}(\eta),\mathbf{G}(\eta)$ and $\mathbf{Q}(\eta)$ are required, the data is picked up from [25]:p.199

$$\mathbf{A}=\begin{bmatrix} 0 & 0 & 1.0 & 0 \\ 0 & 0 & 0 & 1.0 \\ -2 & 1 & 0 & 0 \\ 0.5 & -0.5 & 0 & 0 \end{bmatrix},\quad \mathbf{B}=\begin{bmatrix} 2.0 & -1.0 & 0 & 0 \\ -1.0 & 1.0 & 0 & 0 \\ 0 & 0 & 1.0 & 0 \\ 0 & 0 & 0 & 2.0 \end{bmatrix},\quad \mathbf{D}=\begin{bmatrix} 0 & 0 & 0 & 0 \\ 0 & 0 & 0 & 0 \\ 0 & 0 & 0 & 0 \\ 0 & 0 & 0 & 0.25 \end{bmatrix}$$

Solution: Problem is clear, precise integration algorithm (5.7.25) gives the results as

$$\mathbf{F}(\eta)=\begin{bmatrix} -.80474 & 0.79784 & 0.29733 & 0.58414 \\ 0.26821 & 0.51292 & 0.34377 & 0.92005 \\ -.26490 & -.03238 & -.79752 & 0.47219 \\ -.12804 & -.23571 & 0.31772 & 0.10195 \end{bmatrix},\quad \mathbf{G}(\eta)=\begin{bmatrix} 0.03646 & 0.08702 & 0.04660 & 0.03031 \\ 0.08702 & 0.26763 & 0.13223 & 0.16037 \\ 0.04660 & 0.13223 & 0.06794 & 0.06407 \\ 0.03031 & 0.06037 & 0.06407 & 0.21841 \end{bmatrix},$$

$$\mathbf{Q}(\eta)=\begin{bmatrix} 3.93011 & -1.91804 & -0.03492 & -0.22521 \\ -1.91804 & 1.89915 & 0.04782 & 0.25011 \\ -0.03492 & 0.04782 & 1.97209 & -0.08174 \\ -0.22521 & 0.25011 & -0.08174 & 2.68508 \end{bmatrix},$$

After calculated the matrices for the fundamental interval η, the other computation is as usual. ##

Section 5.8 below describes the analytical solution based on all the eigen-solutions of the respective Hamiltonian matrix, the numerical results from such analytical solution completely coincides with the numerical results given above.

§5.7.7, Precise integration for asymmetric Riccati equations

So far all the discussed methods are based on the Hamiltonian system theory, for which the related problems belong to a conservative system. However, quite a number of problems are really dissipated, can these methods still be applied? The reply is yes, as shown below in this section.

The Riccati differential equations (5.7.31~32) are called the symmetric Riccati equation, because the solution matrices are always symmetric. These equations are very important for dissipation free wave propagation, optimal control Kalman-Bucy filtering, game theory etc. However, for the problems of wave propagation in dissipative media, radiative transfer, transport theory etc. the dual linear differential equations are given as

$$\dot{\mathbf{q}} = \mathbf{Aq} + \mathbf{Dp}, \quad \dot{\mathbf{p}} = \mathbf{Bq} + \mathbf{Cp} \tag{5.7.35a,b}$$

where $\mathbf{q}$ and $\mathbf{p}$ are, respectively, n- and m-dimensional vectors to be solved. The corresponding multi-dimensional general asymmetric Riccati differential equations are derived as

$$-\dot{\mathbf{S}}(t) = -\mathbf{B} + \mathbf{SA} - \mathbf{CS} + \mathbf{SDS} \tag{5.7.36}$$

where $\mathbf{S}(t)$ is a $m \times n$ dimensional matrix to be solved, $\dot{\mathbf{S}}(t)$ is the time differential of $\mathbf{S}(t)$, the integration domain is $t_0 = 0 \le t \le t_\mathrm{f}$, where t_f is a given finish time instant. Note that the single continuous coordinate is regarded as time in this section, however, one can just regard it as a space coordinate as that in the previous sections. The given time-invariant system matrices $\mathbf{A}, \mathbf{B}, \mathbf{C}$ and $\mathbf{D}$ have dimensions $n \times n$, $m \times n$, $m \times m$ and $n \times m$, respectively. The boundary condition of Riccati differential equation is usually written as

$$\mathbf{S}(t_\mathrm{f}) = \mathbf{S}_\mathrm{f}, \quad \text{when} \quad t = t_\mathrm{f} \tag{5.7.37}$$

such as for linear quadratic optimal control problems. The dual Riccati differential equation is

$$\dot{\mathbf{T}}(t) = -\mathbf{D} - \mathbf{TC} + \mathbf{AT} + \mathbf{TBT} \tag{5.7.38}$$

where $\mathbf{T}(t)$ is a $n \times m$ dimensional matrix function to be determined with the initial condition

$$\mathbf{T}(0) = \mathbf{T}_0 \tag{5.7.39}$$

where $\mathbf{T}_0$ is a given $n \times m$ initial matrix.

For case of infinite time interval, *i.e.* $t_\mathrm{f} \to \infty$, the solutions $\mathbf{S}_\infty$ and $\mathbf{T}_\infty$ tend to be constant matrices and satisfies the ***general algebraic Riccati equations***

$$-\mathbf{B} - \mathbf{CS}_\infty + \mathbf{S}_\infty\mathbf{A} + \mathbf{S}_\infty\mathbf{DS}_\infty = \mathbf{0} \tag{5.7.36'}$$

$$\mathbf{D} - \mathbf{AT}_\infty + \mathbf{T}_\infty\mathbf{C} - \mathbf{T}_\infty\mathbf{BT}_\infty = \mathbf{0} \tag{5.7.38'}$$

However, such a limiting result holds only when the matrices $\mathbf{A}, \mathbf{B}, \mathbf{C}, \mathbf{D}$ correspond to a decay system.

The solution for these Riccati differential equations has two ways, namely the analytical solution based on the eigen-solutions and the precise integration method. The method based on eigen-solutions has a pitfall, that when the Jordan form nearly appears for the eigen-solutions, the numerical results may be questionable because of ill-conditioning problem. However, even in such case the precise integration method still gives the numerical solution up to computer precision but does not suffer from the Jordan form, see [103].

The precise integration for asymmetric Riccati differential equation is introduced in this section, which can follow the parallel way as that for the symmetric Riccati differential equation solutions. In general the physical problems give the $m+n$ boundary conditions in TPBVP form

$$\mathbf{q}(0) = \mathbf{q}_0, \quad \text{when} \quad t = 0; \quad \text{and} \quad \mathbf{p}(t_\mathrm{f}) = \mathbf{p}_\mathrm{f}, \quad \text{when} \quad t = t_\mathrm{f} \tag{5.7.40}$$

where $\mathbf{q}_0, \mathbf{p}_f$ are the given n-, m-dimensional vectors, respectively. The special cases of LQ control or Kalman-Bucy filtering correspond to the conservative systems, respectively, having the restrictions of $m = n$, $\mathbf{C} = -\mathbf{A}^{\mathrm{T}}$ and $\mathbf{B}, \mathbf{D}$ being non-negative definite symmetric matrices.

To find solutions for the general asymmetric Riccati differential equations, we need first to build up the relations connecting the state vectors $\mathbf{q}_a, \mathbf{p}_a$ at $t = t_a$ and $\mathbf{q}_b, \mathbf{p}_b$ at $t = t_b$. If the time interval (t_a, t_b), $t_0 \le t_a < t_b \le t_f$, is considered as a subinterval within the whole integration interval $[0, t_f]$, the relation can be formulated as

$$\mathbf{q}_b = \mathbf{F}\mathbf{q}_a - \mathbf{G}\mathbf{p}_b, \qquad \mathbf{p}_a = \mathbf{Q}\mathbf{q}_a + \mathbf{E}\mathbf{p}_b \tag{5.7.41a,b}$$

where the $n \times n, n \times m, m \times n, m \times m$ dimensioned ***interval matrices*** $\mathbf{F}, \mathbf{G}, \mathbf{Q}, \mathbf{E}$ are to be determined. For time variant linear systems, these matrices depend on both t_a and t_b. The two differential equation sets for $\mathbf{F}, \mathbf{G}, \mathbf{Q}, \mathbf{E}$ can be derived as before, the first set is

$$\begin{aligned} &\partial \mathbf{G} / \partial t_b = \mathbf{A}\mathbf{G} + \mathbf{G}\mathbf{B}\mathbf{G} - \mathbf{D} - \mathbf{G}\mathbf{C}, && \partial \mathbf{Q} / \partial t_b = -\mathbf{E}\mathbf{B}\mathbf{F} \\ &\partial \mathbf{F} / \partial t_b = (\mathbf{G}\mathbf{B} + \mathbf{A})\mathbf{F}, && \partial \mathbf{E} / \partial t_b = \mathbf{E}(\mathbf{B}\mathbf{G} - \mathbf{C}) \end{aligned} \tag{5.7.42}$$

The initial conditions are

$$\mathbf{G} = 0, \quad \mathbf{Q} = 0, \quad \mathbf{E} = \mathbf{I}_m, \quad \mathbf{F} = \mathbf{I}_n \qquad \text{when} \quad t_b = t_a \tag{5.7.43}$$

where $\mathbf{I}_m$ and $\mathbf{I}_n$ are unit matrices with dimensions m and n, respectively. The second set of differential equations for $\mathbf{F}, \mathbf{G}, \mathbf{Q}, \mathbf{E}$ is

$$\begin{aligned} &\partial \mathbf{G} / \partial t_a = \mathbf{F}\mathbf{D}\mathbf{E}, && \partial \mathbf{Q} / \partial t_a = \mathbf{B} - \mathbf{Q}\mathbf{A} + \mathbf{C}\mathbf{Q} - \mathbf{Q}\mathbf{D}\mathbf{Q} \\ &\partial \mathbf{F} / \partial t_a = -\mathbf{F}(\mathbf{A} + \mathbf{D}\mathbf{Q}), && \partial \mathbf{E} / \partial t_a = (\mathbf{C} - \mathbf{Q}\mathbf{D})\mathbf{E} \end{aligned} \tag{5.7.44}$$

The initial condition is similar to (5.7.43) at $t_a = t_b$.

For time invariant system, the matrices $\mathbf{A}, \mathbf{B}, \mathbf{C}, \mathbf{D}$ are independent on t. Hence the matrices $\mathbf{F}, \mathbf{G}, \mathbf{Q}$ and $\mathbf{E}$ will only depend on the length of the interval

$$\eta = t_b - t_a \tag{5.7.45}$$

Therefore the relations

$$\mathbf{Q}(t_a, t_b) = \mathbf{Q}(\eta), \ \partial \mathbf{Q} / \partial t_b = \mathrm{d}\mathbf{Q} / \mathrm{d}\eta = \dot{\mathbf{Q}}, \ \partial \mathbf{Q} / \partial t_a = -\mathrm{d}\mathbf{Q} / \mathrm{d}\eta = -\dot{\mathbf{Q}} \tag{5.7.46}$$

hold for matrix $\mathbf{Q}$, and similarly also hold for $\mathbf{F}, \mathbf{G}, \mathbf{E}$. So that equations (5.7.42) and (5.7.44) can be rewritten as

$$\dot{\mathbf{G}}(\eta) = \mathbf{A}\mathbf{G} + \mathbf{G}\mathbf{B}\mathbf{G} - \mathbf{D} - \mathbf{G}\mathbf{C} \quad = -\mathbf{F}\mathbf{D}\mathbf{E} \tag{5.7.47a}$$

$$\dot{\mathbf{F}}(\eta) = (\mathbf{G}\mathbf{B} + \mathbf{A})\mathbf{F} \quad = \mathbf{F}(\mathbf{A} + \mathbf{D}\mathbf{Q}) \tag{5.7.47b}$$

$$\dot{\mathbf{E}}(\eta) = \mathbf{E}(\mathbf{B}\mathbf{G} - \mathbf{C}) \quad = (\mathbf{Q}\mathbf{D} - \mathbf{C})\mathbf{E} \tag{5.7.47c}$$

$$\dot{\mathbf{Q}}(\eta) = -\mathbf{E}\mathbf{B}\mathbf{F} \quad = -\mathbf{B} + \mathbf{Q}\mathbf{A} - \mathbf{C}\mathbf{Q} + \mathbf{Q}\mathbf{D}\mathbf{Q} \tag{5.7.47d}$$

The dot above now means derivative with respect to η. The two ODE sets in (5.7.47) appear to be quite different from each other, although they all represent the derivatives (forward and backward) of the matrices $\mathbf{F}, \mathbf{G}, \mathbf{Q}$ and $\mathbf{E}$. Based on the interval combination, it can be proved that the latter equal sign in equations (5.7.47) are consistent each other, which means four identities among $\mathbf{F}, \mathbf{G}, \mathbf{Q}$ and

$\mathbf{E}$.

Note that if $\mathbf{Q}$ is treated as a function of t_a, then the latter part of equation (5.7.47d) is the same as differential equation (5.7.36). If an algorithm can be developed to calculate the matrix $\mathbf{Q}$ of (5.7.47d), so that $\mathbf{Q}(\eta)=\mathbf{Q}(t_\mathrm{b}-t_\mathrm{a})$, as a function of t_a, also satisfies the boundary condition (5.7.37), then $\mathbf{Q}$ is the solution $\mathbf{S}$ of (5.7.36) and (5.7.37). The same method also applies to the solutions of equations (5.7.38) and (5.3.39).

Equations (5.7.41a,b) connect the state vectors defined at the two ends t_a and t_b of an interval. Given two contiguous intervals $(t_\mathrm{a},t_\mathrm{b})$ and $(t_\mathrm{b},t_\mathrm{c})$ as shown in Figure 5.2, the combined interior state vectors $\mathbf{q}_\mathrm{b},\mathbf{p}_\mathrm{b}$ defined at t_b can be eliminated, so that the new combined interval is $(t_\mathrm{a},t_\mathrm{c})$, and the corresponding equations similar to equations (5.7.41a,b) connect state vectors defined at the two ends t_a and t_c, respectively.

Equations (5.7.41a,b) for the intervals $(t_\mathrm{a},t_\mathrm{b})$ and $(t_\mathrm{b},t_\mathrm{c})$ are

$$\mathbf{q}_\mathrm{b}=\mathbf{F}_1\mathbf{q}_\mathrm{a}-\mathbf{G}_1\mathbf{p}_\mathrm{b}, \qquad \mathbf{p}_\mathrm{a}=\mathbf{Q}_1\mathbf{q}_\mathrm{a}+\mathbf{E}_1\mathbf{p}_\mathrm{b} \tag{5.7.48a,b}$$

$$\mathbf{q}_\mathrm{c}=\mathbf{F}_2\mathbf{q}_\mathrm{b}-\mathbf{G}_2\mathbf{p}_\mathrm{c}, \qquad \mathbf{p}_\mathrm{b}=\mathbf{Q}_2\mathbf{q}_\mathrm{b}+\mathbf{E}_2\mathbf{p}_\mathrm{c} \tag{5.7.49a,b}$$

respectively. Combining intervals $(t_\mathrm{a},t_\mathrm{b})$ and $(t_\mathrm{b},t_\mathrm{c})$ gives the interval $(t_\mathrm{a},t_\mathrm{c})$, for which

$$\mathbf{q}_\mathrm{c}=\mathbf{F}_\mathrm{c}\mathbf{q}_\mathrm{a}-\mathbf{G}_\mathrm{c}\mathbf{p}_\mathrm{c}, \qquad \mathbf{p}_\mathrm{a}=\mathbf{Q}_\mathrm{c}\mathbf{q}_\mathrm{a}+\mathbf{E}_\mathrm{c}\mathbf{p}_\mathrm{c} \tag{5.7.50a,b}$$

To eliminate the combined interior state vector $\mathbf{q}_\mathrm{b},\mathbf{p}_\mathrm{b}$, solving equations (5.7.48a) and (5.7.49b) gives

$$\mathbf{q}_\mathrm{b}=(\mathbf{I}_n+\mathbf{G}_1\mathbf{Q}_2)^{-1}\mathbf{F}_1\mathbf{q}_\mathrm{a}-(\mathbf{I}_n+\mathbf{G}_1\mathbf{Q}_2)^{-1}\mathbf{G}_1\mathbf{E}_2\mathbf{p}_\mathrm{c} \tag{5.7.51a}$$

$$\mathbf{p}_\mathrm{b}=(\mathbf{I}_m+\mathbf{Q}_2\mathbf{G}_1)^{-1}\mathbf{Q}_2\mathbf{F}_1\mathbf{q}_\mathrm{a}+(\mathbf{I}_m+\mathbf{Q}_2\mathbf{G}_1)^{-1}\mathbf{E}_2\mathbf{p}_\mathrm{c} \tag{5.7.51b}$$

Substituting (5.7.51a,b) into (5.7.49a) and (5.7.48b), respectively, gives equations (5.7.50a,b) with

$$\mathbf{G}_\mathrm{c}=\mathbf{G}_2+\mathbf{F}_2(\mathbf{I}_n+\mathbf{G}_1\mathbf{Q}_2)^{-1}\mathbf{G}_1\mathbf{E}_2 \tag{5.7.52a}$$

$$\mathbf{Q}_\mathrm{c}=\mathbf{Q}_1+\mathbf{E}_1(\mathbf{I}_m+\mathbf{Q}_2\mathbf{G}_1)^{-1}\mathbf{Q}_2\mathbf{F}_1 \tag{5.7.52b}$$

$$\mathbf{F}_\mathrm{c}=\mathbf{F}_2(\mathbf{I}_n+\mathbf{G}_1\mathbf{Q}_2)^{-1}\mathbf{F}_1 \tag{5.7.52c}$$

$$\mathbf{E}_\mathrm{c}=\mathbf{E}_1(\mathbf{I}_m+\mathbf{Q}_2\mathbf{G}_1)^{-1}\mathbf{E}_2 \tag{5.7.52d}$$

which gives the interval matrices combination equations and can be used in stepwise integration.

Using the interval combination equations, both solutions of the Riccati differential equations (5.7.36) and (5.7.38) with boundary conditions (5.7.37) and (5.7.39), respectively, can be obtained based on the interval matrices $\mathbf{F}(\eta),\mathbf{G}(\eta),\mathbf{Q}(\eta)$ and $\mathbf{E}(\eta)$. Note that let $\eta=t_\mathrm{f}-t$, $\mathbf{Q}(\eta)=\mathbf{Q}(t_\mathrm{f}-t)$ as a function of t, satisfies the Riccati differential equation (5.7.36), however, the initial condition (5.7.43) does not coincide with the boundary condition (5.7.37). To satisfy the boundary condition, imagine a virtual interval with interval matrices

$$\mathbf{Q}_*=\mathbf{S}_\mathrm{f},\ \mathbf{G}_*=\mathbf{0},\ \mathbf{E}_*=\mathbf{I}_m,\ \mathbf{F}_*=\mathbf{I}_n$$

This virtual interval being used as the interval 2, to combine (t,t_f) as the interval 1, gives the combined interval matrix $\mathbf{Q}_c$, which is treated as the matrix $\mathbf{S}(t)$. Therefore

$$\mathbf{S}(t)=\mathbf{Q}(\eta)+\mathbf{E}(\eta)(\mathbf{I}_m+\mathbf{S}_f\mathbf{G}(\eta))^{-1}\mathbf{S}_f\mathbf{F}(\eta), \quad \text{with } \eta=t_f-t \tag{5.7.53}$$

To verify that $\mathbf{S}(t)$ satisfies the boundary condition, since $\mathbf{E}\to\mathbf{I}_m$, $\mathbf{F}\to\mathbf{I}_n$, $\mathbf{G}\to\mathbf{0}$ and $\mathbf{Q}\to\mathbf{0}$ as $t\to t_f$, the boundary condition (5.7.37) is easily verified by $\mathbf{S}(t)$ given in equation (5.7.53). To verify that $\mathbf{S}(t)$ satisfies equation (5.7.36), note that $\mathrm{d}\mathbf{X}^{-1}/\mathrm{d}\eta=-\mathbf{X}^{-1}\dot{\mathbf{X}}\mathbf{X}^{-1}$, so

$$\begin{aligned}-\mathrm{d}\mathbf{S}/\mathrm{d}t=\dot{\mathbf{S}}(\eta)&=\dot{\mathbf{Q}}+\dot{\mathbf{E}}(\mathbf{I}_m+\mathbf{S}_f\mathbf{G})^{-1}\mathbf{S}_f\mathbf{F}+\mathbf{E}(\mathbf{I}_m+\mathbf{S}_f\mathbf{G})^{-1}\mathbf{S}_f\dot{\mathbf{F}}\\&\quad-\mathbf{E}(\mathbf{I}_m+\mathbf{S}_f\mathbf{G})^{-1}\mathbf{S}_f\dot{\mathbf{G}}(\mathbf{I}_m+\mathbf{S}_f\mathbf{G})^{-1}\mathbf{S}_f\mathbf{F}\end{aligned}$$

Invoking equations (5.7.47a-d) the following derivation verifies equation (5.7.36)

$$\begin{aligned}-\mathrm{d}\mathbf{S}/\mathrm{d}t&=-\mathbf{B}+\mathbf{QA}-\mathbf{CQ}+\mathbf{QDQ}+(-\mathbf{CE}+\mathbf{QDE})(\mathbf{I}_m+\mathbf{S}_f\mathbf{G})^{-1}\mathbf{S}_f\mathbf{F}\\&\quad+\mathbf{E}(\mathbf{I}_m+\mathbf{S}_f\mathbf{G})^{-1}\mathbf{S}_f(\mathbf{FA}+\mathbf{FDQ})+\mathbf{E}(\mathbf{I}_m+\mathbf{S}_f\mathbf{G})^{-1}\mathbf{S}_f\mathbf{FDE}(\mathbf{I}_m+\mathbf{S}_f\mathbf{G})^{-1}\mathbf{S}_f\mathbf{F}\\&=-\mathbf{B}+\left[\mathbf{Q}+\mathbf{E}(\mathbf{I}_m+\mathbf{S}_f\mathbf{G})^{-1}\mathbf{S}_f\mathbf{F}\right]\mathbf{A}-\mathbf{C}\left[\mathbf{Q}+\mathbf{E}(\mathbf{I}_m+\mathbf{S}_f\mathbf{G})^{-1}\mathbf{S}_f\mathbf{F}\right]\\&\quad+\left[\mathbf{Q}+\mathbf{E}(\mathbf{I}_m+\mathbf{S}_f\mathbf{G})^{-1}\mathbf{S}_f\mathbf{F}\right]\mathbf{D}\left[\mathbf{Q}+\mathbf{E}(\mathbf{I}_m+\mathbf{S}_f\mathbf{G})^{-1}\mathbf{S}_f\mathbf{F}\right]\\&=-\mathbf{B}+\mathbf{SA}-\mathbf{CS}+\mathbf{SDS}\end{aligned}$$

The same composition method gives

$$\mathbf{T}(t)=\mathbf{G}(t)+\mathbf{F}(t)(\mathbf{I}_n+\mathbf{T}_0\mathbf{Q}(t))^{-1}\mathbf{T}_0\mathbf{E}(t) \tag{5.7.54}$$

The verification is similar to the above for $\mathbf{S}(t)$.

It should be noted that if $\mathbf{S}(t)$ converges to $\mathbf{S}_\infty$ when $t_f\to\infty$, the matrix $\mathbf{S}_\infty$ satisfies the general algebraic Riccati equation (5.7.36'). For non-conservative systems, such as the source free transport system and elastic wave propagation with damping, the matrix $\mathbf{S}_\infty$ can be calculated using the interval combination algorithm. The interval combination equations (5.7.52a-d) should be executed recursively, until the matrices $\mathbf{E}$ and $\mathbf{F}$ are nearly zero matrices, the matrices $\mathbf{Q}$ and $\mathbf{G}$ are then $\mathbf{S}_\infty$ and $\mathbf{T}_\infty$, respectively.

So far, the equations (5.7.53) and (5.7.54) are analytical, thus the analytical solutions of the two Riccati differential equations (5.7.36~39) have been reduced to find the solutions of the interval matrices $\mathbf{F}(\eta),\mathbf{G}(\eta),\mathbf{Q}(\eta)$ and $\mathbf{E}(\eta)$. Two approaches can be applied, *i.e.* the precise integration method and the analytical method based on eigen-solutions. The precise integration method is briefly described first.

The sub-structuring technique is extensively used to improve computational efficiency in structural mechanics. Only one substructure needs to be analyzed and be used for all other identical substructures. Equation (5.7.41) corresponds to the sub-structural equation, and (5.7.52) is the elimination equations for the combination of two contiguous sub-structures. Two cruxes must be used in the precise integration, namely

1) Making use the 2^N type algorithm, and
2) Computing only the increment parts of the matrices $\mathbf{F}$ and $\mathbf{E}$ in the 2^N type algorithm.

Let the typical time step length of integration domain be η, subdivide it further uniformly into 2^N subintervals of length τ. If $N=20$ is selected, the length of an extremely small subinterval is then

$$\tau = \eta/2^N = \eta/1048576 \tag{5.7.55}$$

and is extremely small. For such a small time step, Taylor expansion gives extremely high precision, and can be used to solve the non-linear equation set (5.7.47) for the interval τ. Truncating beyond the τ^4 terms gives

$$\mathbf{G}(\tau) \approx \mathbf{g}_1\tau + \mathbf{g}_2\tau^2 + \mathbf{g}_3\tau^3 + \mathbf{g}_4\tau^4 \tag{5.7.56a}$$

$$\mathbf{Q}(\tau) \approx \mathbf{q}_1\tau + \mathbf{q}_2\tau^2 + \mathbf{q}_3\tau^3 + \mathbf{q}_4\tau^4 \tag{5.7.56b}$$

$$\mathbf{F}(\tau) = \mathbf{I}_n + \mathbf{F}', \quad \mathbf{F}' \approx \mathbf{f}_1\tau + \mathbf{f}_2\tau^2 + \mathbf{f}_3\tau^3 + \mathbf{f}_4\tau^4 \tag{5.7.56c}$$

$$\mathbf{E}(\tau) = \mathbf{I}_m + \mathbf{E}', \quad \mathbf{E}' \approx \mathbf{e}_1\tau + \mathbf{e}_2\tau^2 + \mathbf{e}_3\tau^3 + \mathbf{e}_4\tau^4 \tag{5.7.56d}$$

Substituting (5.7.56a) into (5.7.47a) and comparing the coefficients of different powers of τ gives

$$\begin{aligned} &\mathbf{g}_1 = \mathbf{D}, \quad \mathbf{g}_2 = (\mathbf{A}\mathbf{g}_1 - \mathbf{g}_1\mathbf{C})/2, \quad \mathbf{g}_3 = (\mathbf{A}\mathbf{g}_2 - \mathbf{g}_2\mathbf{C} + \mathbf{g}_1\mathbf{B}\mathbf{g}_1)/3, \\ &\mathbf{g}_4 = (\mathbf{A}\mathbf{g}_3 - \mathbf{g}_3\mathbf{C} + \mathbf{g}_2\mathbf{B}\mathbf{g}_1 + \mathbf{g}_1\mathbf{B}\mathbf{g}_2)/4 \end{aligned} \tag{5.7.57}$$

Applying similar procedures to (5.7.47b~d) gives

$$\begin{aligned} &\mathbf{f}_1 = \mathbf{A}, \quad \mathbf{f}_2 = (\mathbf{A}\mathbf{f}_1 - \mathbf{g}_1\mathbf{B})/2, \quad \mathbf{f}_3 = (\mathbf{A}\mathbf{f}_2 - \mathbf{g}_2\mathbf{B} + \mathbf{g}_1\mathbf{B}\mathbf{f}_1)/3, \\ &\mathbf{f}_4 = (\mathbf{A}\mathbf{f}_3 + \mathbf{g}_3\mathbf{B} + \mathbf{g}_2\mathbf{B}\mathbf{f}_1 + \mathbf{g}_1\mathbf{B}\mathbf{f}_2)/4 \end{aligned} \tag{5.7.58}$$

$$\begin{aligned} &\mathbf{e}_1 = -\mathbf{C}, \quad \mathbf{e}_2 = (\mathbf{B}\mathbf{g}_1 - \mathbf{e}_1\mathbf{C})/2, \quad \mathbf{e}_3 = (\mathbf{B}\mathbf{g}_2 - \mathbf{e}_2\mathbf{C} + \mathbf{e}_1\mathbf{B}\mathbf{g}_1)/3, \\ &\mathbf{e}_4 = (\mathbf{B}\mathbf{g}_3 - \mathbf{e}_3\mathbf{C} + \mathbf{e}_2\mathbf{B}\mathbf{g}_1 + \mathbf{e}_1\mathbf{B}\mathbf{g}_2)/4 \end{aligned} \tag{5.7.59}$$

$$\begin{aligned} &\mathbf{q}_1 = -\mathbf{B}, \quad \mathbf{q}_2 = -(\mathbf{B}\mathbf{f}_1 + \mathbf{e}_1\mathbf{B})/2, \quad \mathbf{q}_3 = -(\mathbf{B}\mathbf{f}_2 + \mathbf{e}_2\mathbf{B} + \mathbf{e}_1\mathbf{B}\mathbf{f}_1)/3, \\ &\mathbf{q}_4 = -(\mathbf{B}\mathbf{f}_3 + \mathbf{e}_3\mathbf{B} + \mathbf{e}_2\mathbf{B}\mathbf{f}_1 + \mathbf{e}_1\mathbf{B}\mathbf{f}_2)/4 \end{aligned} \tag{5.7.60}$$

From these interval matrices $\mathbf{F}(\tau), \mathbf{G}(\tau), \mathbf{Q}(\tau)$ and $\mathbf{E}(\tau)$, carrying out $N=20$ interval combination steps recursively, all 1048576 subintervals are combined together. It generates the equation system (5.7.41a,b) connecting state vectors defined at the two ends of the original typical time step of length η. The basic computation of this 2^N type algorithm is the recursive execution of

$$\begin{aligned} &\mathbf{G}_c = \mathbf{G} + \mathbf{F}(\mathbf{I}_n + \mathbf{G}\mathbf{Q})^{-1}\mathbf{G}\mathbf{E}, \quad \mathbf{Q}_c = \mathbf{Q} + \mathbf{E}(\mathbf{I}_m + \mathbf{Q}\mathbf{G})^{-1}\mathbf{Q}\mathbf{F} \\ &\mathbf{F}_c = \mathbf{F}(\mathbf{I}_n + \mathbf{G}\mathbf{Q})^{-1}\mathbf{F}, \quad \mathbf{E}_c = \mathbf{E}(\mathbf{I}_m + \mathbf{Q}\mathbf{G})^{-1}\mathbf{E} \end{aligned} \tag{5.7.61}$$

for $N=20$ times. For each time of interval combination, the calculated matrices $\mathbf{G}_c, \mathbf{Q}_c, \mathbf{F}_c$ and $\mathbf{E}_c$ are substituted into the right-hand side of (5.7.61), to calculate the new matrices of $\mathbf{G}_c, \mathbf{Q}_c, \mathbf{F}_c$ and $\mathbf{E}_c$ for the larger intervals.

Note that all the mathematical formulations before equation (5.7.56a~d) are exact. The sole approximation is truncation errors caused by disregarding the terms of order higher than fourth in equations (5.7.56a~d), such errors cause no effect on the computer because the major truncated terms are of the order of τ^5, *i.e.* of the order of τ^4 times the first term in the expansion. Because $\tau^4 = (\eta/1048576)^4 \approx \eta^4 \times 10^{-24}$, and so is beyond the double precision accuracy of most computers today.

Now the second crux should be emphasized, that direct use of the combination equation (5.7.61) will induce serious round-off errors when the length of the subintervals is very small as τ is. To avoid such kind of ill-conditioning, the matrices $\mathbf{F}$ and $\mathbf{E}$ should be rewritten as

$$\mathbf{F} = \mathbf{I}_n + \mathbf{F}', \quad \mathbf{E} = \mathbf{I}_m + \mathbf{E}', \quad \mathbf{F}_c = \mathbf{I}_n + \mathbf{F}'_c, \quad \mathbf{E}_c = \mathbf{I}_m + \mathbf{E}'_c$$

and replace the equation (5.7.61) as

$$\mathbf{G}_c = \mathbf{G} + (\mathbf{I}_n + \mathbf{F}')(\mathbf{G}^{-1} + \mathbf{Q})^{-1}(\mathbf{I}_n + \mathbf{E}') \tag{5.7.62a}$$

$$\mathbf{Q}_c = \mathbf{Q} + (\mathbf{I}_m + \mathbf{E}')(\mathbf{Q}^{-1} + \mathbf{G})^{-1}(\mathbf{I}_m + \mathbf{F}') \tag{5.7.62b}$$

$$\begin{aligned} \mathbf{F}'_c &= 2\mathbf{F}' + \mathbf{F}'^2 \\ &\quad - (\mathbf{I}_n + \mathbf{F}')[\mathbf{GQ}(\mathbf{I}_n + \mathbf{GQ})^{-1} + (\mathbf{I}_n + \mathbf{GQ})^{-1}\mathbf{GQ}](\mathbf{I}_n + \mathbf{F}')/2 \end{aligned} \tag{5.7.62c}$$

$$\begin{aligned} \mathbf{E}'_c &= 2\mathbf{E}' + \mathbf{E}'^2 \\ &\quad - (\mathbf{I}_m + \mathbf{E}')[\mathbf{QG}(\mathbf{I}_m + \mathbf{QG})^{-1} + (\mathbf{I}_m + \mathbf{QG})^{-1}\mathbf{QG}](\mathbf{I}_m + \mathbf{E}')/2 \end{aligned} \tag{5.7.62d}$$

in the computation. The precise integration algorithm does not suffer from the possible Jordan form.

With appropriate system matrices $\mathbf{A},\mathbf{B},\mathbf{C},\mathbf{D}$, using the interval combination equations (5.7.52a-d) iteratively, until $\mathbf{E}$ and $\mathbf{F}$ are almost null matrices, the matrices $\mathbf{S}_\infty, \mathbf{T}_\infty$ for infinite horizon are also computed as $\mathbf{Q},\mathbf{G}$.

A numerical example is given for demonstration. For verification purpose, after the interval matrices of the fundamental time interval η are computed, the iteration continues until the solution matrices $\mathbf{S}_\infty, \mathbf{T}_\infty$ of the algebraic Riccati equation (ARE) is obtained, which is easier to check by just substituting into the equations (5.7.36') and (5.7.38').

Example 5.3, $n = 4$; $m = 1$; the system matrices are given as

$$\mathbf{A} = \begin{bmatrix} 7.28571 & 9.64286 & -9.75466 & -12.82764 \\ 5.0 & 5.5 & -6.73913 & -9.52717 \\ 8.21429 & 10.60714 & -12.77795 & -15.71584 \\ 0.0 & 0.0 & 0.86957 & -0.17391 \end{bmatrix}, \quad \mathbf{D} = \begin{bmatrix} -6.04037 \\ -4.73913 \\ -7.49224 \\ 0.86957 \end{bmatrix}$$

$$\mathbf{B} = \begin{bmatrix} -8.21429 & 10.60714 & 9.45186 & 14.65606 \end{bmatrix}, \qquad \mathbf{C} = \begin{bmatrix} 4.16615 \end{bmatrix}$$

The interval matrices for a fundamental interval of $\eta = 1.0$ need to compute first and then the two solutions of algebraic Riccati equations are required.

Solution: Selecting $\eta = 1.0$, using precise integration method the interval matrices are computed as

$$\mathbf{F}(1.0) = \begin{bmatrix} 1.08368 & 0.43124 & -0.51467 & -0.01876 \\ -0.15694 & 0.26114 & -0.02745 & -0.01921 \\ -0.02984 & -0.03580 & 0.01193 & -0.18568 \\ 0.25306 & 0.30368 & -0.10122 & 0.25224 \end{bmatrix}, \qquad \mathbf{G}(1.0) = \begin{bmatrix} 0.99996 \\ 0.66667 \\ 1.0 \\ -0.00002 \end{bmatrix},$$

$$\mathbf{Q}(1.0) = \begin{bmatrix} 1.30679 & 1.56817 & -1.52270 & -2.20453 \end{bmatrix}' \qquad \mathbf{E}(1.0) = \begin{bmatrix} 0.00008 \end{bmatrix}$$

Based on these matrices, continuously execute the interval combination algorithm

(5.7.52a~d), which converges quadratically to the solution matrices of the algebraic Riccati equations as

$$\mathbf{S}_{\infty} = [1.30682 \quad 1.56818 \quad -1.52273 \quad -2.20455],$$

$$\mathbf{T}_{\infty}^{T} = [1.00000 \quad 0.66667 \quad 1.00000 \quad 0.00000]$$

Substituting $\mathbf{S}_{\infty}$ into equation (5.7.36'), summation of the two parts $-\mathbf{B}+\mathbf{S}_{\infty}\mathbf{D}\mathbf{S}_{\infty}$ and $-\mathbf{C}\mathbf{S}_{\infty}+\mathbf{S}_{\infty}\mathbf{A}$ dropping down 11 significant digits, which means that the solution $\mathbf{S}_{\infty}$ is very precise. Substituting $\mathbf{T}_{\infty}$ into the equation (5.7.38'), when summing up the two parts $\mathbf{D}-\mathbf{T}_{\infty}\mathbf{B}\mathbf{T}_{\infty}$ and $-\mathbf{A}\mathbf{T}_{\infty}+\mathbf{T}_{\infty}\mathbf{C}$, the residual drops 11 digits also, which means again that the solution is highly precise. ##

The above numerical examples demonstrate that the precise integration method is very precise and effective, and it does not suffer from the possible Jordan normal form and so is highly reliable. However, the eigen-solution based analytical method is still attractive, and is the subject of section 5.8.

§5.8, Eigenvector based solution of Riccati equations

The precise integration method is applied to find the interval matrices and the solutions of two Riccati differential equations. However, the precise integration method is not unique, the method of separation of variables, eigenvector expansion method etc. can also be used to find the interval matrices for a fundamental interval of length η, and so they are still very attractive because of their analytical nature. After that, the same interval combination algorithm can still be applied to do the other computations as given in the last section. As described in the last section, the generalized two Riccati differential equations apply to the general linear systems, so the discussion below is for general case first. However, the Hamilton system based analytical solution has a number of features, it is still worth to have special attentions, and will also be given later.

The eigen-solution method is numerically unstable if Jordan form appears, hence the method described in this section is for the case of Jordan form does not nearly to appear.

In this section, the longitudinal coordinate is denoted as $z \equiv t$.

The dual linear differential equations relate to the two Riccati differential equations, which have been given in dual equations (5.7.35a,b). Introducing the state vector

$$\mathbf{v} = \{\mathbf{q}^T \quad \mathbf{p}^T\}^T \tag{5.8.1}$$

The dual equations can be rewritten as

$$\dot{\mathbf{v}} = \mathbf{H}\mathbf{v}, \quad \mathbf{H} = \begin{bmatrix} \mathbf{A} & \mathbf{D} \\ \mathbf{B} & \mathbf{C} \end{bmatrix} \begin{matrix} n \\ m \end{matrix} \tag{5.8.2}$$

Using the method of separation of variables derives the eigen-problem

$$\mathbf{H}\boldsymbol{\psi} = \mu\boldsymbol{\psi} \tag{5.8.3}$$

But, the eigen-solution will be quite ill-conditioned when the Jordan form is nearly to appear for the eigen-problem (5.8.3), in such case the precise integration method should be used instead, as given in the last section.

To derive the analytical solutions for the general asymmetric Riccati equations, building up the relationship between the state vectors $\mathbf{q}_a, \mathbf{p}_a$ at $t = t_a$ and $\mathbf{q}_b, \mathbf{p}_b$ at $t = t_b$ is necessary, where the time interval (t_a, t_b) is considered a subinterval of the whole integration interval $[0, t_f]$. The relationship has been given as equations (5.7.41a,b).

For time-invariant system, the interval matrices depend only on the length of interval

$$\eta = t_b - t_a \tag{5.8.4}$$

i.e. $\mathbf{F}(\eta), \mathbf{G}(\eta), \mathbf{Q}(\eta)$ and $\mathbf{E}(\eta)$. Finding the analytical solution for these interval matrices should go back to the equation (5.7.41a,b), which can be expressed in combined form

$$\begin{Bmatrix} \mathbf{q}_b \\ \mathbf{p}_a \end{Bmatrix} = \begin{bmatrix} \mathbf{F} & -\mathbf{G} \\ \mathbf{Q} & \mathbf{E} \end{bmatrix} \begin{Bmatrix} \mathbf{q}_a \\ \mathbf{p}_b \end{Bmatrix} \tag{5.8.5}$$

This equation gives the vectors $\mathbf{q}_b$ and $\mathbf{p}_a$ with the two end vectors $\mathbf{q}_a$ and $\mathbf{p}_b$.

Suppose all the eigen-solutions of matrix $\mathbf{H}$ being denoted as

$$\mu_i, \ \psi_i, \quad i = 1, 2, \cdots, (n+m) \tag{5.8.6}$$

If no Jordan normal form nearly to appear, the eigen-matrix composed from all the eigenvectors

$$\boldsymbol{\Psi} = [\psi_1, \psi_2, \cdots, \psi_{n+m}] \tag{5.8.7}$$

is well conditioned, and the numerical result is reliable. The eigen-equation (5.8.3) can be rewritten as

$$\mathbf{H}\boldsymbol{\Psi} = \boldsymbol{\Psi}\mathrm{diag}(\mu_i), \qquad \text{where} \quad \mathrm{diag}(\mu_i) = \mathrm{diag}(\mu_1, \mu_2, \cdots, \mu_{n+m}) \tag{5.8.8}$$

The eigen-matrix $\boldsymbol{\Psi}$ can be partitioned as

$$\boldsymbol{\Psi} = \begin{bmatrix} \boldsymbol{\Psi}_q \\ \boldsymbol{\Psi}_p \end{bmatrix} \begin{matrix} n \\ m \end{matrix} \tag{5.8.9}$$

where $\boldsymbol{\Psi}_q$ and $\boldsymbol{\Psi}_p$ are matrices of dimensions $n \times (n+m)$ and $m \times (n+m)$, respectively. To compose the combined matrix in equation (5.8.5), let

$$\mathbf{D}_\eta = \exp[\mathrm{diag}(\mu_i \eta)] \tag{5.8.10}$$

The combined matrix required is composed as

$$\begin{bmatrix} \mathbf{F}(\eta) & -\mathbf{G}(\eta) \\ \mathbf{Q}(\eta) & \mathbf{E}(\eta) \end{bmatrix} = \begin{bmatrix} \boldsymbol{\Psi}_q \times \mathbf{D}_\eta \\ \boldsymbol{\Psi}_p \end{bmatrix} \times \begin{bmatrix} \boldsymbol{\Psi}_q \\ \boldsymbol{\Psi}_p \times \mathbf{D}_\eta \end{bmatrix}^{-1} \tag{5.8.11}$$

From which, the interval matrices $\mathbf{F}(\eta), \mathbf{G}(\eta), \mathbf{Q}(\eta)$ and $\mathbf{E}(\eta)$ can be picked up. However, the differential equations (5.7.47a~d) and boundary conditions (5.7.43) must be carefully verified satisfied.

The interval matrices composed in equation (5.8.11) are easily verified satisfying the boundary condition (5.7.43). It is clear that when $t_b = t_a$, $\mathbf{D}_\eta = \exp[\mathrm{diag}(0)] = \mathbf{I}_{m+n}$, substituting which into (5.8.11) gives

$$\begin{bmatrix} \mathbf{F}(0) & -\mathbf{G}(0) \\ \mathbf{Q}(0) & \mathbf{E}(0) \end{bmatrix} = \begin{bmatrix} \mathbf{\Psi}_q \\ \mathbf{\Psi}_p \end{bmatrix} \times \begin{bmatrix} \mathbf{\Psi}_q \\ \mathbf{\Psi}_p \end{bmatrix}^{-1} = \mathbf{I}_{m+n}$$

which is the combined form of equation (5.7.43).

To verify the equations (5.7.47a~d), the combined equation should be transformed to its sub-matrix form. Let

$$\begin{bmatrix} \mathbf{\Psi}_q \\ \mathbf{\Psi}_p \times \mathbf{D}_\eta \end{bmatrix}^{-1} \underset{\text{def}}{=} \mathbf{Z}(\eta) \underset{\text{def}}{=} [\mathbf{Z}_\mathrm{a} \quad \mathbf{Z}_\mathrm{b}]$$

where

$$\mathbf{Z}_\mathrm{a}(\eta) = \begin{bmatrix} \mathbf{\Psi}_q \\ \mathbf{\Psi}_p \times \mathbf{D}_\eta \end{bmatrix}^{-1} \times \begin{bmatrix} \mathbf{I}_n \\ \mathbf{0}_{m\times n} \end{bmatrix}, \quad \mathbf{Z}_\mathrm{b}(\eta) = \begin{bmatrix} \mathbf{\Psi}_q \\ \mathbf{\Psi}_p \times \mathbf{D}_\eta \end{bmatrix}^{-1} \times \begin{bmatrix} \mathbf{0}_{n\times m} \\ \mathbf{I}_m \end{bmatrix} \tag{5.8.12a,b}$$

Then the interval matrices are derived as

$$\mathbf{Q}(\eta) = \mathbf{\Psi}_p \times \mathbf{Z}_\mathrm{a}, \qquad \mathbf{F}(\eta) = \mathbf{\Psi}_q \times \mathbf{D}_\eta \times \mathbf{Z}_\mathrm{a} \tag{5.8.13a,b}$$

$$\mathbf{E}(\eta) = \mathbf{\Psi}_p \times \mathbf{Z}_\mathrm{b}, \quad -\mathbf{G}(\eta) = \mathbf{\Psi}_q \times \mathbf{D}_\eta \times \mathbf{Z}_\mathrm{b} \tag{5.8.13c,d}$$

The verification is given below. Equations (5.8.12a,b) can be rewritten as

$$\mathbf{\Psi}_q \times \mathbf{Z}_\mathrm{a} = \mathbf{I}_n, \quad \mathbf{\Psi}_p \times \mathbf{D}_\eta \times \mathbf{Z}_\mathrm{a} = \mathbf{0}_{m\times n} \tag{5.8.14a,b}$$

$$\mathbf{\Psi}_q \times \mathbf{Z}_\mathrm{b} = \mathbf{0}_{n\times m}, \quad \mathbf{\Psi}_p \times \mathbf{D}_\eta \times \mathbf{Z}_\mathrm{b} = \mathbf{I}_m \tag{5.8.14c,d}$$

To verify that the interval matrices (5.8.13a~d) satisfy the differential equations (5.8.47a~d), finding the equations for $\dot{\mathbf{Z}}_\mathrm{a}$ and $\dot{\mathbf{Z}}_\mathrm{b}$ is necessary. To find $\dot{\mathbf{Z}}_\mathrm{b}$, differentiating equations (5.8.14c,d) with respect to η gives

$$\mathbf{\Psi}_q \times \dot{\mathbf{Z}}_\mathrm{b} = \mathbf{0}_{n\times m}, \quad \mathbf{\Psi}_p \times \mathbf{D}_\eta \times \dot{\mathbf{Z}}_\mathrm{b} = -\mathbf{\Psi}_p \times \mathrm{diag}(\mu_i) \times \mathbf{D}_\eta \times \mathbf{Z}_\mathrm{b}$$

Comparing the above equations with the equations (5.8.14c,d) gives

$$\dot{\mathbf{Z}}_\mathrm{b} = -\mathbf{Z}_\mathrm{b} \times \mathbf{\Psi}_p \times \mathrm{diag}(\mu_i) \times \mathbf{D}_\eta \times \mathbf{Z}_\mathrm{b}$$

Expanding equation (5.8.8) gives

$$\mathbf{A\Psi}_q + \mathbf{D\Psi}_p = \mathbf{\Psi}_q \mathrm{diag}(\mu_i), \qquad \mathbf{B\Psi}_q + \mathbf{C\Psi}_p = \mathbf{\Psi}_p \mathrm{diag}(\mu_i).$$

Using equations (5.8.13a~d)

$$\dot{\mathbf{Z}}_\mathrm{b} = -\mathbf{Z}_\mathrm{b}(\mathbf{B\Psi}_q + \mathbf{C\Psi}_p)\mathbf{D}_\eta \mathbf{Z}_\mathrm{b} = \mathbf{Z}_\mathrm{b}(\mathbf{BG} - \mathbf{C}) \tag{5.8.15}$$

The similar procedure for $\dot{\mathbf{Z}}_\mathrm{a}$ derives

$$\dot{\mathbf{Z}}_\mathrm{a} = -\mathbf{Z}_\mathrm{b}\mathbf{BF} \tag{5.8.16}$$

To verify equations (5.7.47a~d), differentiating equation (5.8.13d) with respect to η gives

$$\dot{\mathbf{G}}(\eta) = -\mathbf{\Psi}_q \mathbf{D}_\eta \dot{\mathbf{Z}}_\mathrm{b} - \mathbf{\Psi}_q \mathrm{diag}(\mu_i) \mathbf{D}_\eta \mathbf{Z}_\mathrm{b}$$

Making use of equation (5.8.15) gives

$$\dot{\mathbf{G}}(\eta) = -\mathbf{\Psi}_q \mathbf{D}_\eta \mathbf{Z}_\mathrm{b}(\mathbf{BG} - \mathbf{C}) - (\mathbf{A\Psi}_q + \mathbf{D\Psi}_p)\mathbf{D}_\eta \mathbf{Z}_\mathrm{b}$$

Using equation (5.8.13a~d) derives the equation (5.7.47a) in the form as

$$\dot{\mathbf{G}}(\eta) = \mathbf{G} \cdot (\mathbf{BG} - \mathbf{C}) + \mathbf{AG} - \mathbf{D}$$

Next, differentiating (5.8.13c) derives the equation (5.7.47c), as follows

$$\dot{\mathbf{E}}(\eta) = \mathbf{\Psi}_p \dot{\mathbf{Z}}_\mathrm{b} = \mathbf{\Psi}_p \mathbf{Z}_\mathrm{b}(\mathbf{BG} - \mathbf{C}) = \mathbf{E}(\mathbf{BG} - \mathbf{C})$$

Differentiating equation (5.8.13b) derives the equation (5.7.47b)

$$\dot{\mathbf{F}}(\eta) = \mathbf{\Psi}_q \mathbf{D}_\eta \dot{\mathbf{Z}}_a + \mathbf{\Psi}_q \mathrm{diag}(\mu_i) \mathbf{D}_\eta \mathbf{Z}_a$$
$$= -\mathbf{\Psi}_q \mathbf{D}_\eta \mathbf{Z}_b \mathbf{BF} + (\mathbf{A\Psi}_q + \mathbf{D\Psi}_p) \mathbf{D}_\eta \mathbf{Z}_a = \mathbf{GBF} + \mathbf{AF}$$

Differentiating equation (5.8.13a) derives

$$\dot{\mathbf{Q}}(\eta) = \mathbf{\Psi}_p \dot{\mathbf{Z}}_a = -\mathbf{\Psi}_p \mathbf{Z}_b \mathbf{BF} = -\mathbf{EBF}$$

Up to here, equations (5.7.47a~d) and boundary conditions (5.7.43) are all verified. Based on the uniqueness theorem of ODE [36]-p.50, that (5.8.13a~d) do give the analytical solutions of the interval matrices $\mathbf{F}(\eta), \mathbf{G}(\eta), \mathbf{Q}(\eta)$ and $\mathbf{E}(\eta)$.

However, there is the second set of differential equations in (5.7.47a~d), the verification is given as follows. Equation (5.8.11) can be rewritten as

$$\begin{bmatrix} \mathbf{F}(\eta) & -\mathbf{G}(\eta) \\ \mathbf{Q}(\eta) & \mathbf{E}(\eta) \end{bmatrix} = \begin{bmatrix} \mathbf{\Psi}_q \times \mathbf{D}_\eta \\ \mathbf{\Psi}_p \end{bmatrix} \times \begin{bmatrix} \mathbf{\Psi}_q \\ \mathbf{\Psi}_p \times \mathbf{D}_\eta \end{bmatrix}^{-1} = \begin{bmatrix} \mathbf{\Psi}_q \\ \mathbf{\Psi}_p \times \mathbf{D}_\eta^{-1} \end{bmatrix} \times \begin{bmatrix} \mathbf{\Psi}_q \times \mathbf{D}_\eta^{-1} \\ \mathbf{\Psi}_p \end{bmatrix}^{-1}$$

Let

$$\mathbf{Y}(\eta) \underset{\mathrm{def}}{=} \begin{bmatrix} \mathbf{\Psi}_q \times \mathbf{D}_\eta^{-1} \\ \mathbf{\Psi}_p \end{bmatrix}^{-1} \underset{\mathrm{def}}{=} [\mathbf{Y}_a(\eta) \quad \mathbf{Y}_b(\eta)]$$

where

$$\mathbf{Y}_a(\eta) = \begin{bmatrix} \mathbf{\Psi}_q \times \mathbf{D}_\eta^{-1} \\ \mathbf{\Psi}_p \end{bmatrix}^{-1} \times \begin{bmatrix} \mathbf{I}_n \\ \mathbf{0}_{m\times n} \end{bmatrix}, \quad \mathbf{Y}_b(\eta) = \begin{bmatrix} \mathbf{\Psi}_q \times \mathbf{D}_\eta^{-1} \\ \mathbf{\Psi}_p \end{bmatrix}^{-1} \times \begin{bmatrix} \mathbf{0}_{n\times m} \\ \mathbf{I}_m \end{bmatrix} \tag{5.8.17a,b}$$

Afterward the second form of interval matrices are derived as

$$\mathbf{Q}(\eta) = \mathbf{\Psi}_p \times \mathbf{D}_\eta^{-1} \times \mathbf{Y}_a, \qquad \mathbf{F}(\eta) = \mathbf{\Psi}_q \times \mathbf{Y}_a \tag{5.8.18a,b}$$

$$\mathbf{E}(\eta) = \mathbf{\Psi}_p \times \mathbf{D}_\eta^{-1} \times \mathbf{Y}_b, \quad -\mathbf{G}(\eta) = \mathbf{\Psi}_q \times \mathbf{Y}_b \tag{5.8.18c,d}$$

The equations (5,8,17a,b) can be rewritten as

$$\mathbf{\Psi}_q \times \mathbf{D}_\eta^{-1} \times \mathbf{Y}_a = \mathbf{I}_n, \quad \mathbf{\Psi}_p \times \mathbf{Y}_a = \mathbf{0}_{m\times n} \tag{5.8.19a,b}$$

$$\mathbf{\Psi}_q \times \mathbf{D}_\eta^{-1} \times \mathbf{Y}_b = \mathbf{0}_{n\times m}, \quad \mathbf{\Psi}_p \times \mathbf{Y}_b = \mathbf{I}_m \tag{5.8.19c,d}$$

To verify the second form of interval matrix equation (5.7.47a~d), similar procedure is applied. The first step is to derive $\dot{\mathbf{Y}}_a$ and $\dot{\mathbf{Y}}_b$. Differentiate equation (5.8.19a,b)

$$\mathbf{\Psi}_q \mathbf{D}_\eta^{-1} \dot{\mathbf{Y}}_a = \mathbf{\Psi}_q \mathrm{diag}(\mu_i) \mathbf{D}_\eta^{-1} \mathbf{Y}_a, \quad \mathbf{\Psi}_p \dot{\mathbf{Y}}_a = \mathbf{0}$$

Comparison with (5.8.19a,b) determines

$$\dot{\mathbf{Y}}_a = \mathbf{Y}_a \mathbf{\Psi}_q \mathrm{diag}(\mu_i) \mathbf{D}_\eta^{-1} \mathbf{Y}_b = \mathbf{Y}_a (\mathbf{A\Psi}_q + \mathbf{B\Psi}_p) \mathbf{D}_\eta^{-1} \mathbf{Y}_b = \mathbf{Y}_a (\mathbf{A} + \mathbf{DQ})$$

The same method derives

$$\dot{\mathbf{Y}}_b = \mathbf{Y}_a \mathbf{DE}$$

Then the differential equations (5.7.47a~d) can be verified as follows

$$\dot{\mathbf{Q}}(\eta) = \mathbf{\Psi}_p \mathbf{D}_\eta^{-1} \dot{\mathbf{Y}}_a - \mathbf{\Psi}_p \mathrm{diag}(\mu_i) \mathbf{D}_\eta^{-1} \mathbf{Y}_a$$
$$= \mathbf{\Psi}_p \mathbf{D}_\eta^{-1} \mathbf{Y}_a (\mathbf{A} + \mathbf{DQ}) - (\mathbf{B\Psi}_q + \mathbf{C\Psi}_p) \mathbf{D}_\eta^{-1} \mathbf{Y}_a = \mathbf{Q}(\mathbf{A} + \mathbf{DQ}) - \mathbf{B} - \mathbf{CQ}$$

$$\dot{\mathbf{G}}(\eta) = -\mathbf{\Psi}_q \dot{\mathbf{Y}}_b = -\mathbf{\Psi}_q \mathbf{Y}_a \mathbf{DE} = -\mathbf{FDE}$$

$$\dot{\mathbf{E}}(\eta)=\mathbf{\Psi}_p\mathbf{D}_\eta^{-1}\dot{\mathbf{Y}}_b-\mathbf{\Psi}_p\mathrm{diag}(\mu_i)\mathbf{D}_\eta^{-1}\mathbf{Y}_b$$
$$=\mathbf{\Psi}_p\mathbf{D}_\eta^{-1}\mathbf{Y}_a\mathbf{DE}-(\mathbf{B\Psi}_q+\mathbf{C\Psi}_p)\mathbf{D}_\eta^{-1}\mathbf{Y}_a=(\mathbf{QD}-\mathbf{C})\mathbf{E}$$
$$\dot{\mathbf{F}}(\eta)=\mathbf{\Psi}_p\dot{\mathbf{Y}}_a=\mathbf{\Psi}_p\mathbf{Y}_a(\mathbf{A}+\mathbf{DQ})=\mathbf{F}(\mathbf{A}+\mathbf{DQ})$$

Therefore, all the differential equations for the interval matrices are verified. Based on the uniqueness theorem of ODE, the interval matrices composed are really the analytical solution required.

The eigen-solutions based analytical solutions are proved above, the numerical computation algorithm and example are necessary to demonstrate its effectiveness and are given below.

The analytical solution is based on all the eigen-solutions, which should be solved numerically. Note that the eigen-solutions may have complex conjugate pairs. However, the original equations are all real valued, so do the solutions of the Riccati equations. Therefore, a real valued algorithm should be found. The matrix $\mathbf{H}$ is real, so the complex eigenvectors appear in complex conjugate pairs. The usual algorithm, such as the HQR2 given in [42]-II.15, supplies the real and imaginary parts of the complex conjugate eigenvectors as contiguous two real vectors. Such form of "eigenvector matrix", denoted as $\mathbf{\Psi}_r$, corresponds to the complex eigen-matrix $\mathbf{\Psi}$ multiplied by the matrix $\mathbf{C}_d$ from right hand side, *i.e.*

$$\mathbf{\Psi}_r=\mathbf{\Psi}\times\mathbf{C}_d,\quad \mathbf{C}_d=\mathrm{diag}[1,\cdots,\mathbf{C}_c,\cdots,\mathbf{C}_c,\cdots],\quad \mathbf{C}_c=\begin{bmatrix}0.5 & -0.5i\\ 0.5 & 0.5i\end{bmatrix} \tag{5.8.20}$$

The same treatment applies also to the matrix $\mathbf{\Psi D}_\eta$. Note that equation (5.8.11) can be rewritten as

$$\begin{bmatrix}\mathbf{F}(\eta) & -\mathbf{G}(\eta)\\ \mathbf{Q}(\eta) & \mathbf{E}(\eta)\end{bmatrix}=\begin{bmatrix}\mathbf{\Psi}_q\times\mathbf{D}_\eta\\ \mathbf{\Psi}_p\end{bmatrix}\mathbf{C}_d\times\mathbf{C}_d^{-1}\begin{bmatrix}\mathbf{\Psi}_q\\ \mathbf{\Psi}_p\times\mathbf{D}_\eta\end{bmatrix}^{-1}$$
$$=\begin{bmatrix}\mathbf{\Psi}_q\times\mathbf{D}_\eta\times\mathbf{C}_d\\ \mathbf{\Psi}_p\times\mathbf{C}_d\end{bmatrix}\times\begin{bmatrix}\mathbf{\Psi}_q\times\mathbf{C}_d\\ \mathbf{\Psi}_p\times\mathbf{D}_\eta\times\mathbf{C}_d\end{bmatrix}^{-1}$$

where the sub-matrices have been all real valued.

The algorithm for analytical solution is simply given in the form

[Input n,m and $\mathbf{A},\mathbf{B},\mathbf{C},\mathbf{D}$; input η, the step size, and the finish time $t_f=k_f\eta$]

[Compose the matrix $\mathbf{H}$ according to equation (5.8.2)]

[Solve the eigen-matrix $\mathbf{\Psi}_r$ and the corresponding eigenvalues μ_i, $i=1,\cdots,n+m$]

[Compute $\mathbf{\Psi}\times\mathbf{D}_\eta\times\mathbf{C}_d$, and pickup the sub-matrices $\mathbf{\Psi}_{rq}$, $\mathbf{\Psi}_{rp}$, etc.]

[According to equation (5.8.11) compute the interval matrices $\mathbf{F}(\eta),\mathbf{G}(\eta),\mathbf{Q}(\eta)$ and $\mathbf{E}(\eta)$]

[Compute the interval matrices at all the grid stations $t_k=k\eta, k=1,\cdots k_f$, based on (5.7.52a~d)]

[Input the boundary matrices $\mathbf{S}_f,\mathbf{T}_0$, and compute all the matrices $\mathbf{S}(t_k),\mathbf{T}(t_k)$]

With appropriate system matrices $\mathbf{A},\mathbf{B},\mathbf{C},\mathbf{D}$, using the interval combination equations (5.7.52a~d), the matrices $\mathbf{S}_\infty,\mathbf{T}_\infty$ for infinite horizon can also be computed.

Example 5.4, $n=4$; $m=1$; the system matrices are the same as those given in example 5.3. The solutions of two asymmetric Riccati differential equations are required.

Solution: According to equation (5.8.2) compose the matrix $\mathbf{H}$, finding the eigen-solutions. The composed matrix $\mathbf{H}$ is not far apart to having Jordan normal form. The eigenvalues are all real and computed as

9.999993 −1.999990 −0.500003 −0.499997 −3.000003

and the eigen-matrix is

$$\mathbf{\Psi}=\begin{bmatrix} 0.53882 & -0.57276 & -0.65465 & 30920.193 & 0.24649 \\ 0.35921 & 0.0 & 0.65465 & -30919.819 & 0.36973 \\ 0.53882 & -1.43188 & 0.32732 & -15459.899 & 0.73947 \\ 0.0 & 0.0 & 0.0 & 0.24945 & -0.49298 \\ 0.53882 & -1.43188 & 0.32732 & 15459.806 & 0.86272 \end{bmatrix}$$

It is seen that the third and fourth columns are almost parallel vectors and their eigenvalues are −0.500003 and −0.499997 respectively, which means the Jordan form is nearly to appear. In such case, the computer precise integration should be executed. Selecting $\eta=1.0$, the interval matrices are computed as

$$\mathbf{F}(1.0)=\begin{bmatrix} 1.08368 & 0.43124 & -0.51467 & -0.01876 \\ -0.15694 & 0.26114 & -0.02745 & -0.01921 \\ -0.02984 & -0.03580 & 0.01193 & -0.18568 \\ 0.25306 & 0.30368 & -0.10122 & 0.25224 \end{bmatrix},\quad \mathbf{G}(1.0)=\begin{bmatrix} 0.99996 \\ 0.66667 \\ 1.0 \\ -0.00002 \end{bmatrix}$$

$$\mathbf{Q}(1.0)=\begin{bmatrix} 1.30679 & 1.56817 & -1.52270 & -2.20453 \end{bmatrix}' \qquad \mathbf{E}(1.0)=\begin{bmatrix} 0.00008 \end{bmatrix}$$

However, the eigen-matrix based analytical solution can still be computed, suprisingly, the same numerical result is obtained for this example, but the results may be dangerous for other problems.

Based on these matrices, the algebraic Riccati equations are solved as

$$\mathbf{S}_\infty=\begin{bmatrix} 1.30682 & 1.56818 & -1.52273 & -2.20455 \end{bmatrix}$$

$$\mathbf{T}_\infty^{\mathrm{T}}=\begin{bmatrix} 1.00000 & 0.66667 & 1.00000 & 0.00000 \end{bmatrix}$$

which are the same as obtained by the precise integration method. Substituting into the algebraic Riccati equations (5.7.36') and (5.7.38'), the checking results is quite satisfactory as described in the last section. ##

Example 5.5: $n=7$; $m=2$; the system matrices are given as

$$\mathbf{A}=\begin{bmatrix} -5.2 & 0.5 & 0.0 & 0.0 & 2.0 & -.5 & 1.6 \\ 0.0 & -1.5 & 1.6 & 0.0 & 0.0 & -.1 & 0.8 \\ 0.0 & 0.0. & -2.0 & 0.8 & 0.0 & -.2 & 0.5 \\ 0.0 & 0.0 & 0.0 & -2. & 0.5 & -.1 & 0.2 \\ 0.0 & 0.0 & 0.0 & 0.0 & -1.0 & -.8 & 0.2 \\ 0.4 & 0.6 & 0.8 & 0.0 & -.2 & -8.0 & 0.5 \\ -.4 & -.2 & 1.8 & 3.0 & -.2 & 0.8 & -5.0 \end{bmatrix},\quad \mathbf{D}=\begin{bmatrix} -10.0 & 0.0 \\ 0.0 & 0.1 \\ 0.0 & -2.0 \\ 0.0 & 0.0 \\ 0.0 & -0.4 \\ -6.0 & -5.0 \\ -7.0 & 0.0 \end{bmatrix}$$

$$\mathbf{B}=\begin{bmatrix}-10.0 & 0.0 & 0.0 & 0.0 & 0.0 & 0.0 & 0.0\\ 0.0 & -1.0 & 0.0 & 0.0 & 0.0 & 0.0 & 0.0\end{bmatrix},\quad \mathbf{C}=\begin{bmatrix}7.0 & 1.0\\ 2.0 & 6.0\end{bmatrix}$$

The eigenvalues are computed as

$$\begin{matrix}13.01995 & 5.85708 & -10.30548 & -7.04055 & -1.62845 & -2.00943 & -0.92412\\ & & & \pm 0.62496i & \pm 0.66731i & & \end{matrix}$$

Selecting $\eta = 0.1$, the interval matrices are computed as

$$\mathbf{F}(0.1)=\begin{bmatrix}0.42309 & 0.02864 & 0.00998 & 0.01440 & 0.12160 & -.02249 & 0.08041\\ -.00736 & 0.85940 & 0.13959 & 0.01477 & -.00122 & -.00525 & 0.06049\\ -.00158 & -.00722 & 0.82086 & 0.07113 & 0.00148 & -.01137 & 0.03539\\ -.00134 & -.00036 & 0.00103 & 0.82100 & 0.04286 & -.00704 & 0.01428\\ 0.00278 & -.00245 & -.00108 & 0.00217 & 0.90576 & -.05168 & 0.01389\\ -.06395 & -.02393 & 0.05366 & 0.00622 & -.02356 & 0.45231 & 0.02118\\ -.13705 & -.01869 & 0.12723 & 0.21648 & -.02951 & 0.04449 & 0.59808\end{bmatrix}$$

$$\mathbf{Q}(0.1)=\begin{bmatrix}0.49131 & -.00733 & -.00167 & 0.00279 & 0.04157 & -.00876 & 0.02983\\ -.03782 & 0.06923 & 0.00470 & -.00009 & -.00529 & -.00084 & 0.00142\end{bmatrix}$$

$$\mathbf{G}^T(0.1)=\begin{bmatrix}0.50729 & 0.01025 & -.00514 & 0.00166 & -.00779 & 0.26138 & 0.34621\\ -.02258 & 0.00111 & 0.13540 & -.00079 & 0.01888 & 0.26339 & 0.00564\end{bmatrix}$$

$$\mathbf{E}(0.1)=\begin{bmatrix}0.36315 & -.04256\\ -.08643 & 0.55299\end{bmatrix}$$

Based on these matrices, continuously execute the interval combination algorithm (5.7.52a~d), which converges quickly to the solution matrices of the algebraic Riccati equations as

$$\mathbf{S}_\infty=\begin{bmatrix}0.55650 & 0.00949 & 0.00492 & 0.01077 & 0.08270 & -.01489 & 0.05143\\ -.06545 & 0.12707 & 0.02357 & -.00162 & -.03427 & 0.00413 & -.00586\end{bmatrix}$$

$$\mathbf{T}_\infty^T=\begin{bmatrix}0.58396 & 0.01167 & -.02365 & 0.00325 & -.01290 & 0.27318 & 0.38479\\ -.03739 & 0.03747 & 0.24420 & -.00250 & 0.01833 & 0.35975 & 0.04374\end{bmatrix}$$

Substituting $\mathbf{S}_\infty$ into equation (5.7.36'), the summation of the two parts $-\mathbf{B}+\mathbf{S}_\infty\mathbf{D}\mathbf{S}_\infty$ and $-\mathbf{C}\mathbf{S}_\infty+\mathbf{S}_\infty\mathbf{A}$ dropping down 12 significant digits means that the solution $\mathbf{S}_\infty$ is very precise. Substituting $\mathbf{T}_\infty$ into the equation (5.7.38'), when summing up the two parts $\mathbf{D}-\mathbf{T}_\infty\mathbf{B}\mathbf{T}_\infty$ and $-\mathbf{A}\mathbf{T}_\infty+\mathbf{T}_\infty\mathbf{C}$, the residual drops 12 significant digits, which means again that the solution is highly precise.

The precise integration method gives the same numerical result with the eigen-solutions based analytical method. Both the methods check to each other. ##

§5.8.1, Analytical solution applied to the symmetric Riccati equations

The method of analytical solution based on the eigen-solutions applies also to the symmetric Riccati equations for conservative system, which is a special case to the asymmetric Riccati equations. However, the eigen-solutions for a conservative system have special features, for which the matrix $\mathbf{H}$ is Hamiltonian as $n = m$, $\mathbf{C} = -\mathbf{A}^{\mathrm{T}}$ and $\mathbf{B}, \mathbf{D}$ being non-negative symmetric matrices.

For a conservative system the $2n \times 2n$-dimensioned eigen-matrix $\boldsymbol{\Psi}$ defined in (5.8.7) is a symplectic matrix, $\boldsymbol{\Psi}^T\mathbf{J}\boldsymbol{\Psi} = \mathbf{J}$, and can be block decomposed into four $n \times n$ sub-matrices as

$$\boldsymbol{\Psi} = \begin{bmatrix} \mathbf{Q}_\alpha & \mathbf{Q}_\beta \\ \mathbf{P}_\alpha & \mathbf{P}_\beta \end{bmatrix} \begin{matrix} n \\ n \end{matrix} \tag{5.8.22}$$

Assuming that the eigenvalue μ has no Jordan normal form, then

$$\mathbf{H}\boldsymbol{\Psi} = \boldsymbol{\Psi} \begin{bmatrix} \text{diag}(\mu_i) & \mathbf{0} \\ \mathbf{0} & -\text{diag}(\mu_i) \end{bmatrix}, \quad \text{or} \quad \mathbf{H} = \boldsymbol{\Psi} \begin{bmatrix} \text{diag}(\mu_i) & \mathbf{0} \\ \mathbf{0} & -\text{diag}(\mu_i) \end{bmatrix} \boldsymbol{\Psi}^{-1}$$

and then

$$\mathbf{A} = \mathbf{Q}_\alpha \text{diag}(\mu_i) \mathbf{P}_\beta^T + \mathbf{Q}_\beta \text{diag}(\mu_i) \mathbf{P}_\alpha^T \tag{5.8.23a}$$

$$\mathbf{D} = -\mathbf{Q}_\alpha \text{diag}(\mu_i) \mathbf{Q}_\beta^T - \mathbf{Q}_\beta \text{diag}(\mu_i) \mathbf{Q}_\alpha^T \tag{5.8.23b}$$

$$\mathbf{B} = \mathbf{P}_\alpha \text{diag}(\mu_i) \mathbf{P}_\beta^T + \mathbf{P}_\beta \text{diag}(\mu_i) \mathbf{P}_\alpha^T \tag{5.8.23c}$$

also

$$\mathbf{Q}_\alpha^T \mathbf{B} \mathbf{Q}_\beta - \mathbf{Q}_\alpha^T \mathbf{A}^T \mathbf{P}_\beta - \mathbf{P}_\alpha^T \mathbf{A} \mathbf{Q}_\beta - \mathbf{P}_\alpha^T \mathbf{D} \mathbf{P}_\beta = -\text{diag}(\mu_i) \tag{5.8.24a}$$

$$\mathbf{Q}_\alpha^T \mathbf{B} \mathbf{Q}_\alpha - \mathbf{Q}_\alpha^T \mathbf{A}^T \mathbf{P}_\alpha - \mathbf{P}_\alpha^T \mathbf{A} \mathbf{Q}_\alpha - \mathbf{P}_\alpha^T \mathbf{D} \mathbf{P}_\alpha = \mathbf{0} \tag{5.8.24b}$$

$$\mathbf{Q}_\beta^T \mathbf{B} \mathbf{Q}_\beta - \mathbf{Q}_\beta^T \mathbf{A}^T \mathbf{P}_\beta - \mathbf{P}_\beta^T \mathbf{A} \mathbf{Q}_\beta - \mathbf{P}_\beta^T \mathbf{D} \mathbf{P}_\beta = \mathbf{0} \tag{5.8.24c}$$

In a fundamental interval $(0, \eta)$, the solution of homogeneous equation (5.2.13) is

$$\mathbf{v}(z) = \sum_{i=1}^{n} [\boldsymbol{\psi}_i a_i \exp(\mu_i z) + \boldsymbol{\psi}_{n+i} b_i \exp(\mu_i (\eta - z))] \tag{5.8.25a}$$

or given in vector form

$$\mathbf{v}(z) = \begin{bmatrix} \mathbf{Q}_\alpha \\ \mathbf{P}_\alpha \end{bmatrix} \exp[\text{diag}(\mu_i z)] \cdot \mathbf{a} + \begin{bmatrix} \mathbf{Q}_\beta \\ \mathbf{P}_\beta \end{bmatrix} \exp(\text{diag}[\mu_i(\eta - z)]) \cdot \mathbf{b} \tag{5.8.25b}$$

the coefficient vectors $\mathbf{a}, \mathbf{b}$ should be determined by the end boundary conditions. According to the physical interpretation of matrices $\mathbf{Q}(\eta), \mathbf{F}(\eta)$ gives the boundary conditions of the TPBVP as, at $z = 0$ end, the displacement $\mathbf{v}(0)$ is the n columns of a unit matrix $\mathbf{I}_n$ in turn; while at $z = \eta$ it is free from force. Then the vector $\mathbf{a}$ and $\mathbf{b}$ solutions in turn, respectively, compose two $n \times n$ matrices $\mathbf{A}_1$ and $\mathbf{B}_1$, respectively. The equations for solving matrices $\mathbf{A}_1$ and $\mathbf{B}_1$ are

$$\mathbf{Q}_\alpha \mathbf{A}_1 + \mathbf{Q}_\beta \mathbf{D}_{c\eta} \mathbf{B}_1 = \mathbf{I}_n, \quad [z = 0, \text{ unit displacement}] \tag{5.8.26a}$$

$$\mathbf{P}_\alpha \mathbf{D}_{c\eta} \mathbf{A}_1 + \mathbf{P}_\beta \mathbf{B}_1 = \mathbf{0}, \quad [z = \eta, \text{ free}] \tag{5.8.26b}$$

where introduced the matrix

$$\mathbf{D}_{c\eta} = \exp[\text{diag}(\mu_i \eta)]$$

From the solved two $n \times n$ matrices $\mathbf{A}_1, \mathbf{B}_1$, the $\mathbf{Q}(\eta)$ is computed as the force at $z = 0$ end

$$\mathbf{Q}(\eta) = -(\mathbf{P}_\alpha \mathbf{A}_1 + \mathbf{P}_\beta \mathbf{D}_{c\eta} \mathbf{B}_1) \tag{5.8.26c}$$

and $\mathbf{F}(\eta)$ is the displacement at $z = \eta$ as

$$\mathbf{F}(\eta) = \mathbf{Q}_\alpha \mathbf{D}_{c\eta} \mathbf{A}_1 + \mathbf{Q}_\beta \mathbf{B}_1 \tag{5.8.26d}$$

On the other hand, based on the physical interpretation of $\mathbf{G}(\eta)$ and $\mathbf{F}^T(\eta)$, there is no displacement at the end $z=0$, but there is unit force matrix at the end $z=\eta$. Correspondingly, the vectors **a** and **b** in equation (5.8.25b) have also n each, their combinations are two $n \times n$ matrices $\mathbf{A}_2, \mathbf{B}_2$. Therefore at the two ends $z=0$ and $z=\eta$ the equations establish as

$$\mathbf{Q}_\alpha \mathbf{A}_2 + \mathbf{Q}_\beta \mathbf{D}_{c\eta} \mathbf{B}_2 = \mathbf{0}, \qquad [z=0\text{, fixed}] \tag{5.8.27a}$$

$$\mathbf{P}_\alpha \mathbf{D}_{c\eta} \mathbf{A}_2 + \mathbf{P}_\beta \mathbf{B}_2 = \mathbf{I}, \quad [z=\eta\text{, unit force}] \tag{5.8.27b}$$

From which, solves the two $n \times n$ matrices $\mathbf{A}_2$ and $\mathbf{B}_2$. Thereafter compute

$$\mathbf{G}(\eta) = \mathbf{Q}_\alpha \mathbf{D}_{c\eta} \mathbf{A}_2 + \mathbf{Q}_\beta \mathbf{B}_2 \tag{5.8.27c}$$

$$\mathbf{F}^T(\eta) = \mathbf{P}_\alpha \mathbf{A}_2 + \mathbf{P}_\beta \mathbf{D}_{c\eta} \mathbf{B}_2 \tag{5.8.27d}$$

All the equations above explains that if the eigen-matrix $\mathbf{\Psi}$ has been found, solving the interval matrices $\mathbf{F}(\eta), \mathbf{G}(\eta), \mathbf{Q}(\eta)$ needs only algebraic operations, and no iteration is necessary. Although these equations are established by physical reasoning, however, the proof is also available, especially the asymmetric matrix case has been shown valid.

Therefore to find all the eigen-solutions is necessary, that its numerical solution becomes a critical issue. For conservative system, the eigen-solution of a Hamilton matrix is described below.

§5.8.2, Algorithm for the eigen-solutions of a Hamilton matrix

The analytical solution of Riccati differential equations explains the importance of eigen-solutions of a Hamilton matrix **H**

$$\mathbf{H}\psi = \mu\psi \tag{5.3.4}$$

Application needs the numerical result of the eigen-matrix $\mathbf{\Psi}$, which is composed of all the eigenvectors $\psi_i, (i=1,\cdots,n; n+1,\cdots,2n)$ of $\mathbf{H}$. In section 5.3 it had shown that $\mathbf{\Psi}$ is a symplectic matrix.

A Hamilton matrix has its special structure, hence the eigenvalue μ can be classified as given by equation (5.3.5a,b), into α and β two groups, and the computation for the eigenvalue problem of a Hamilton matrix can be derived to solve an eigen-problem of a $n \times n$ matrix [55]. In solving the vibration problem of a gyroscopic system, the eigenvalue and eigenvector computation had been described in detail, where the key step of which is to derive a symplectic eigen-problem of a skew-symmetric matrix, see section 2.3.3. The same strategy is also used here, deriving the eigen-problem (5.3.4) of the Hamilton matrix **H** to become a symplectic eigen-problem of a skew-symmetric matrix.

For solving the gyroscopic system, the description begins from the given matrices $\mathbf{M}, \mathbf{G}$ and $\mathbf{K}$, *i.e.* the mass, gyroscopic and stiffness matrices. Presently, the description can begin from the matrices $\mathbf{K}_{22}, \mathbf{K}_{21}$ and $\mathbf{K}_{11}$ of the displacement method, but directly from Hamilton matrix is preferred

$$\mathbf{H} = \begin{bmatrix} \mathbf{A} & \mathbf{D} \\ \mathbf{B} & -\mathbf{A}^T \end{bmatrix}, \quad \mathbf{D} = \mathbf{D}^T, \quad \mathbf{B} = \mathbf{B}^T \tag{5.2.14}$$

Left multiplying equation (5.3.4) with the matrix **J** gives the symplectic eigen-problem of a symmetric matrix

$$\mathbf{B}_H \psi = \mu \mathbf{J} \psi , \quad \text{where } \mathbf{B}_H = \mathbf{JH}, \ \mathbf{B}_H^T = \mathbf{B}_H$$

then the symplectic eigen-problem of the skew-symmetric matrix is simply derived as

$$\mathbf{A}_H \psi = \mu^2 \mathbf{J} \psi , \quad \text{where } \mathbf{A}_H = -\mathbf{B}_H \mathbf{J} \mathbf{B}_H \tag{5.8.28}$$

Obviously, we have $\mathbf{A}_H^T = -\mathbf{A}_H$. Hence reordering the eigenvector $\mathbf{\Psi}$ as

$$\psi' = \{q_1, p_1;\ q_2, p_2;\ \cdots;\ q_n, p_n;\}^T \tag{5.8.29}$$

Certainly, the matrices should also be reordered correspondingly and the canonical form of a skew-symmetric symplectic eigen-problem, as described in section 2.3.3, is obtained, where vector $\mathbf{w}_b$ in equation (2.3.61) is the vector ψ' here.

Therefore, the eigenvalue μ^2 of (5.8.28) and the respective two eigenvectors can be regarded as found. The two eigen-solutions of equation (5.3.4), (μ_i, ψ_i) and $(\mu_{n+i} = -\mu_i, \psi_{n+i})$, are certainly corresponding to the duplicated eigen-solutions of equation (5.8.28)

$$(\mu_i^2, \psi_i), \quad \text{and} \quad (\mu_i^2, \psi_{n+i}) \tag{5.8.30}$$

but the reverse is not true. Because any linear combination of the symplectic adjoint eigenvectors ψ_i and ψ_{n+i} is also an eigenvector of equation (5.8.28), but the duplicate eigenvectors, denoted as ψ_{ia} and ψ_{ib}, are not necessarily the eigenvector of the original eigen-equation (5.3.4). The two dimensional subspace of the duplicate eigenvectors ψ_{ia} and ψ_{ib} of equation (5.8.28) is spanned by ψ_i and ψ_{n+i}. To find the two eigenvecrtors ψ_i and ψ_{n+i} of the Hamilton matrix $\mathbf{H}$, the symplectic projection matrix of $\mathbf{H}$ to the two dimensional subspace is required. The problem reduced to be a 2-D symplectic eigenvalue problem and is easy to solve. Hence the problem now is how to calculate the 2-D symplectic projection matrix. From equation

$$\mathbf{\Psi}^{-1} \mathbf{H} \mathbf{\Psi} = \text{diag}[\text{diag}(\mu), -\text{diag}(\mu)] \tag{5.3.12}$$

since $\mathbf{\Psi}$ is a symplectic matrix, it is derived to be

$$\mathbf{\Psi}^T \mathbf{J} \mathbf{H} \mathbf{\Psi} = \begin{bmatrix} 0 & -\text{diag}(\mu) \\ -\text{diag}(\mu) & 0 \end{bmatrix} \tag{5.8.31}$$

Note that the skew-symmetric symplectic eigen-problem (5.8.28) corresponds to sort the eigenvectors in α and β classes. Sorting $\mathbf{\Psi}$ in the order of equation (5.8.29) gives the vector $\mathbf{\Psi}'$ and the right hand side of equation (5.8.31) becomes the form of

$$-\text{diag}\left(\begin{bmatrix} 0 & \mu_1 \\ \mu_1 & 0 \end{bmatrix}, \begin{bmatrix} 0 & \mu_2 \\ \mu_2 & 0 \end{bmatrix}, \cdots, \begin{bmatrix} 0 & \mu_n \\ \mu_n & 0 \end{bmatrix} \right) \tag{5.8.32}$$

The reordering corresponds to a permutation, such as for $n = 4$, the permutation matrix is

$$\mathbf{P}_e = \begin{bmatrix} 1 & 0 & 0 & 0 & 0 & 0 & 0 & 0 \\ 0 & 0 & 0 & 0 & 1 & 0 & 0 & 0 \\ 0 & 1 & 0 & 0 & 0 & 0 & 0 & 0 \\ 0 & 0 & 0 & 0 & 0 & 1 & 0 & 0 \\ 0 & 0 & 1 & 0 & 0 & 0 & 0 & 0 \\ 0 & 0 & 0 & 0 & 0 & 0 & 1 & 0 \\ 0 & 0 & 0 & 1 & 0 & 0 & 0 & 0 \\ 0 & 0 & 0 & 0 & 0 & 0 & 0 & 1 \end{bmatrix}, \quad \begin{aligned} &\mathbf{P}_e^{-1} = \mathbf{P}_e^T \\ &\psi' = \mathbf{P}_e \psi \\ &\psi = \mathbf{P}_e^T \psi' \end{aligned}$$

Because only the two symplectic adjoint eigenvectors can be linearly composed to be the eigenvectors of equation (5.8.28), hence in equation (5.8.32) only these diagonal sub-block-matrices can be changed to be 2×2 symmetric matrices. Evidently, the left-hand-side of equation (5.8.31) gives the fundamental rule of adjoint symplectic subspace projection. The duplicate adjoint symplectic eigenvectors ψ_{ia} and ψ_{ib} of equation (5.8.28) corresponding to the eigenvalue μ_i^2 can be used as the basis to compose the two dimensional subspace

$$\mathbf{\Psi}_2 = [\psi_{ia}, \psi_{ib}], \quad \text{which gives } \mathbf{\Psi}_2^T \mathbf{J} \mathbf{\Psi}_2 = \mathbf{J}_1 = \begin{bmatrix} 0 & 1 \\ -1 & 0 \end{bmatrix} \tag{5.8.33}$$

From which the two dimensional subspace is composed, and the symplectic projection is

$$\mathbf{\Psi}_2^T \mathbf{J} \mathbf{H} \mathbf{\Psi}_2 \underset{\text{def}}{=} \mathbf{J}_1 \mathbf{H}_1 \tag{5.8.34}$$

Obviously, the left-hand-side is a symmetric matrix, for the matrix $\mathbf{JH}$ is symmetric, and the right hand side is also a 2×2 symmetric matrix, so that $\mathbf{H}_1$ is a Hamilton matrix corresponding to $n=1$ and $\mathbf{J}_1$ is a 2×2 matrix. Such kind of subspace projection, keeping the structure of a Hamilton matrix unchanged, is called a ***symplectic conservative*** subspace. For the eigenvalue μ_i^2 correspondent two-dimensional subspace, the eigenvectors of the original matrix $\mathbf{H}$ must be linearly composed of the vectors ψ_{ia} and ψ_{ib}, as

$$\psi = c_1 \psi_{ia} + c_2 \psi_{ib} = \mathbf{\Psi}_2 \mathbf{c}, \quad \mathbf{c} = \{c_1, c_2\}^T \tag{5.8.35}$$

The eigen-equation is

$$\mathbf{H}\mathbf{\Psi}_2 \mathbf{c} = \mu \mathbf{\Psi}_2 \mathbf{c}$$

Multiplying $\mathbf{\Psi}_2^T \mathbf{J}$ from the left gives

$$\mathbf{J}_1 \mathbf{H}_1 \mathbf{c} = \mu \mathbf{J}_1 \mathbf{c} \quad \Rightarrow \quad \mathbf{H}_1 \mathbf{c} = \mu \mathbf{c} \tag{5.8.36}$$

which is an eigen-problem of $n=1$ projection Hamilton matrix $\mathbf{H}_1$, the solution is easily solved as

$$\mu = \mu_i, \quad \mathbf{c}_1 = \{c_{11}, c_{21}\}^T; \qquad \text{and} \qquad \mu = -\mu_i, \quad \mathbf{c}_2 = \{c_{12}, c_{22}\}^T$$

After solved the two eigenvectors in the subspace, substituting into equation (5.8.35) the two symplectic adjoint eigenvectors of the original matrix $\mathbf{H}$ are obtained.

Therefore, the algorithm for the eigen-solutions of a Hamiltonian matrix is given as

[Given $\mathbf{H}$, transforming to skew-symmetric symplectic eigen-problem, see (5.8.28)]

[Solve all the duplicate eigen-solutions $(\mu_i^2, \psi_{ia}, \psi_{ib}), i = 1, \cdots, n$ of eigen-problem

(5.8.28)]

Comment: the key step, its algorithm had been given in section 2.3.3

for ($i=1$; $i \le n$; $i++$) {

[Compose $\mathbf{\Psi}_2$, then compute the project Hamilton matrix $\mathbf{H}_1$, find its eigen-solution]

[According to (5.8.35), compute the eigenvectors ψ_i, ψ_{n+i} of the original matrix $\mathbf{H}$]

} Comment: all the eigen-solutions of $\mathbf{H}$ are obtained, and $\mathbf{\Psi}$ is composed (5.8.37)

§5.8.3, *Transform to real value computation*

The solution of the Riccati equation should be a real valued matrix, and the differential equations (5.7.16) and (5.7.17) are both real equations. However, the analytical solution needs eigen-matrix and the eigen-solutions of a Hamilton matrix may appear complex roots and complex eigenvectors. Computing directly based on the equations (5.8.26) and (5.8.27), the complex algebraic equations may appear. Avoiding such complex arithmetic brings convenient for implementation.

For the case of purely imaginary eigenvalues, *i.e.* $\mathrm{Re}(\mu) \neq 0$, the real arithmetic can also be attained. According to the classification of equation (5.3.5a,b), both the complex conjugate pair eigenvalues locate in the same class and can be sorted contiguously, the respective eigenvectors are also a complex conjugate pair. The two complex conjugate eigenvectors can be transformed as the two linear combinations of their real and imaginary vector parts. For real eigenvector no transform is needed and correspondingly an element 1 is appeared at the diagonal of the transformation matrix. However, complex conjugate eigenvector pair need the transformation matrix a 2×2 diagonal block

$$\mathbf{C}_c = \begin{bmatrix} 1/2 & -\mathrm{i}/2 \\ 1/2 & \mathrm{i}/2 \end{bmatrix}, \quad \mathbf{C}_d = \mathrm{diag}(1,\cdots,\mathbf{C}_c,\cdots,\mathbf{C}_c\cdots) \tag{5.8.38a}$$

where $\mathbf{C}_d$ is the $n\times n$ transformation matrix composed of 1 and/or $\mathbf{C}_c$ as the diagonal blocks. Its inverse matrix is

$$\mathbf{C}_c^{-1} = \begin{bmatrix} 1 & 1 \\ \mathrm{i} & -\mathrm{i} \end{bmatrix}, \quad \mathbf{C}_d^{-1} = \mathrm{diag}(1,\cdots,\mathbf{C}_c^{-1},\cdots,\mathbf{C}_c^{-1}\cdots) \tag{5.8.38b}$$

In equations (5.8.26) and (5.8.27), the matrices $\mathbf{Q}_\alpha, \mathbf{Q}_\beta, \mathbf{P}_\alpha, \mathbf{P}_\beta$ are all complex valued, whose corresponding real valued matrices can be expressed as

$$\mathbf{Q}_{\alpha r} = \mathbf{Q}_\alpha \mathbf{C}_d, \quad \mathbf{Q}_{\beta r} = \mathbf{Q}_\beta \mathbf{C}_d, \quad \mathbf{P}_{\alpha r} = \mathbf{P}_\alpha \mathbf{C}_d, \quad \mathbf{P}_{\beta r} = \mathbf{P}_\beta \mathbf{C}_d \tag{5.8.39a}$$

$$\mathbf{Q}_\alpha = \mathbf{Q}_{\alpha r} \mathbf{C}_d^{-1}, \quad \mathbf{Q}_\beta = \mathbf{Q}_{\beta r} \mathbf{C}_d^{-1}, \quad \mathbf{P}_\alpha = \mathbf{P}_{\alpha r} \mathbf{C}_d^{-1}, \quad \mathbf{P}_\beta = \mathbf{P}_{\beta r} \mathbf{C}_d^{-1} \tag{5.8.39b}$$

where the subscript r is used to distinguish with the complex valued matrices. Transformation matrix $\mathbf{C}_d$ not only applies to the matrices $\mathbf{Q}_\alpha$ etc., but also when multiplies the matrix $\mathbf{D}_\eta$, transforms it to be a real matrix. Therefore the equations (5.8.26a,b) become

$$\mathbf{Q}_{\alpha r}\mathbf{C}_d^{-1}\mathbf{A}_1 + \mathbf{Q}_{\beta r}\mathbf{C}_d^{-1}\mathbf{D}_\eta\mathbf{C}_d\mathbf{C}_d^{-1}\mathbf{B}_1 = \mathbf{I}_n$$

$$\mathbf{P}_{\alpha r}\mathbf{C}_d^{-1}\mathbf{D}_\eta\mathbf{C}_d\mathbf{C}_d^{-1}\mathbf{A}_1 + \mathbf{P}_{\beta r}\mathbf{Q}_\alpha\mathbf{C}_d^{-1}\mathbf{B}_1 = \mathbf{0}$$

Denoting $e^{\mu_i\eta} = a_\eta + ib_\eta$, then the complex sub-matrix of $\mathbf{D}_\eta$ is $\text{diag}(a_\eta + ib_\eta, a_\eta - ib_\eta)$. Let μ_1 be a negative real number, then

$$\mathbf{C}_d^{-1} \cdot \text{diag}\left[e^{\mu_i\eta}\right] \cdot \mathbf{C}_d = \text{diag}\left(e^{\mu_1\eta}, \cdots, \begin{bmatrix} a_\eta & b_\eta \\ -b_\eta & a_\eta \end{bmatrix}_i, \cdots, \begin{bmatrix} a_\eta & b_\eta \\ -b_\eta & a_\eta \end{bmatrix}_k, \cdots \right) = \mathbf{D}_\eta$$

becomes a real matrix. Hence the equations become

$$\begin{aligned} &\mathbf{Q}_{\alpha r}\mathbf{A}_{r1} + \mathbf{Q}_{\beta r}\mathbf{D}_\eta\mathbf{B}_{r1} = \mathbf{I}_n, \quad && \mathbf{A}_{r1} = \mathbf{C}_d^{-1}\mathbf{A}_1 \\ &\mathbf{P}_{\alpha r}\mathbf{D}_\eta\mathbf{A}_{r1} + \mathbf{P}_{\beta r}\mathbf{B}_{r1} = \mathbf{0}, && \mathbf{B}_{r1} = \mathbf{C}_d^{-1}\mathbf{B}_1 \end{aligned} \tag{5.8.40a,b}$$

where the coefficient matrices have been real matrices, hence the solution matrices $\mathbf{A}_{r1}, \mathbf{B}_{r1}$ are real also. Afterwards the computation can be

$$\mathbf{Q}(\eta) = -(\mathbf{P}_{\alpha r}\mathbf{A}_{r1} + \mathbf{P}_{\beta r}\mathbf{D}_\eta\mathbf{B}_{r1}) \tag{5.8.40c}$$

$$\mathbf{F}(\eta) = \mathbf{Q}_{\alpha r}\mathbf{D}_\eta\mathbf{A}_{r1} + \mathbf{Q}_{\beta r}\mathbf{B}_{r1} \tag{5.8.40d}$$

Similarly

$$\mathbf{Q}_{\alpha r}\mathbf{A}_{r2} + \mathbf{Q}_{\beta r}\mathbf{D}_\eta\mathbf{B}_{r2} = \mathbf{0} \tag{5.8.41a}$$

$$\mathbf{P}_{\alpha r}\mathbf{D}_\eta\mathbf{A}_{r2} + \mathbf{P}_{\beta r}\mathbf{B}_{r2} = \mathbf{I}_n \tag{5.8.41b}$$

The matrices $\mathbf{A}_{r2}, \mathbf{B}_{r2}$ solved from which are also real matrices. Afterwards compute as

$$\mathbf{G}(\eta) = \mathbf{Q}_{\alpha r}\mathbf{D}_\eta\mathbf{A}_{r2} + \mathbf{Q}_{\beta r}\mathbf{B}_{r2} \tag{5.8.41c}$$

$$\mathbf{F}^T(\eta) = \mathbf{P}_{\alpha r}\mathbf{A}_{r2} + \mathbf{P}_{\beta r}\mathbf{D}_\eta\mathbf{B}_{r2} \tag{5.8.41d}$$

Up to here, all the computations for the matrices and equations have been transformed to real arithmetic. But the above transformation is under the limitations of that the Hamilton matrix $\mathbf{H}$ has no purely imaginary root. In the optimal control or Kalman-Bucy filtering problems, the controllability and observability are their fundamental requirements, such requirements ensure the interval index integration being positive definite, see section 6.7. From the point of view of structural mechanics, it means the interval deformation energy being positive definite, which corresponds to the fundamental property of material stability in structural mechanics. These properties ensure that the Hamilton matrix $\mathbf{H}$ has no purely imaginary eigen-roots. Therefore the above real arithmetic analytical solution based on eigen-solutions for the Riccati differential equations can really be applied.

Example 5.6: Given $n = 4$, and the sub-matrices of the Hamilton matrix $\mathbf{H}$ are given as

$$\mathbf{A} = \begin{bmatrix} 0 & 0 & 1.0 & 0 \\ 0 & 0 & 0 & 1.0 \\ -2.0 & 1.0 & 0 & 0 \\ 0.5 & -0.5 & 0 & 0 \end{bmatrix}, \quad \mathbf{B} = \begin{bmatrix} 2.0 & -1.0 & 0 & 0 \\ -1.0 & 1.0 & 0 & 0 \\ 0 & 0 & 1.0 & 0 \\ 0 & 0 & 0 & 2.0 \end{bmatrix},$$

$$\mathbf{D} = \begin{bmatrix} 0 & 0 & 0 & 0 \\ 0 & 0 & 0 & 0 \\ 0 & 0 & 0 & 0 \\ 0 & 0 & 0 & 0.25 \end{bmatrix}, \qquad \mathbf{H} = \begin{bmatrix} \mathbf{A} & \mathbf{D} \\ \mathbf{B} & -\mathbf{A}^T \end{bmatrix}$$

Select $\eta = 2.0$ for the fundamental interval length, the matrices $\mathbf{Q}(\eta), \mathbf{G}(\eta), \mathbf{F}(\eta)$

are required numerically.

Solution: This problem can be solved by the precise integration method, but it can also be solved by means of eigen-solutions analytically. First the eigen-solutions of the Hamilton matrix are solved as

$$\mu_\alpha = \begin{matrix} -0.428314 & -0.428314 & -0.165241 & -0.165241 \\ +0.440528i & -0.440528i & +1.505554i & -1.505554i \end{matrix}$$

$$\mathbf{Q}_{\alpha r} = \begin{bmatrix} -.11372 & -.02428 & 0.01863 & 0.00583 \\ -.23540 & -.00538 & -.00156 & -.01067 \\ 0.05940 & -.03970 & -.01186 & 0.02709 \\ 0.10319 & -.10140 & 0.01632 & -.00058 \end{bmatrix}, \quad \mathbf{Q}_{\beta r} = \begin{bmatrix} -.34227 & 0.26643 & 0.53020 & -.59221 \\ -.58037 & 0.65920 & -.42158 & -.12204 \\ -.02923 & 0.26490 & -.80399 & -.89610 \\ 0.04182 & 0.53801 & -.25340 & 0.61455 \end{bmatrix},$$

$$\mathbf{P}_{\alpha r} = \begin{bmatrix} 0.07480 & 0.05441 & -.23084 & -.14777 \\ 0.27624 & 0.07206 & 0.12360 & 0.08142 \\ -.09235 & 0.12475 & 0.09899 & -.15631 \\ -.24148 & 0.39335 & -.04763 & 0.06565 \end{bmatrix}, \quad \mathbf{P}_{\beta r} = \begin{bmatrix} -.33491 & 0.08297 & 9.36760 & -6.06387 \\ -.86881 & 0.61575 & -5.10090 & 3.20835 \\ 0.13451 & 0.56311 & -4.12429 & -6.30339 \\ 0.54349 & 1.63361 & 1.62991 & 2.87255 \end{bmatrix}$$

From the eigenvalues, the matrix $\mathbf{D}_\eta$ is computed as

$$\mathbf{D}_\eta = \mathrm{diag}\left(\begin{bmatrix} 0.270183 & 0.327534 \\ -0.327534 & 0.270183 \end{bmatrix}, \begin{bmatrix} -0.712468 & 0.093497 \\ -0.093497 & -0.712468 \end{bmatrix} \right)$$

From which the matrices $\mathbf{A}_{r2}, \mathbf{B}_{r2}$ are computed by equations (5.8.41a,b), and the matrices $\mathbf{A}_{r1}, \mathbf{B}_{r1}$ are solved using equations (5.8.40a,b). Thereafter, the matrices $\mathbf{Q}(\eta), \mathbf{G}(\eta), \mathbf{F}(\eta)$ are computed. The numerical results obtained in this way are completely in agreement with those obtained with the precise integration method, ten more decimal digit coincidences are reached. ##

Using entirely different two methods, the precise integration method and the analytical solution method based on eigen-solutions, both methods give the same numerical results, which mean that both are highly precise algorithms. Numerical data of the three matrices $\mathbf{Q}(\eta), \mathbf{G}(\eta), \mathbf{F}(\eta)$ have been listed in section 5.7.6.

In case of no purely imaginary eigen-root, the real arithmetic algorithm given above works quite well. However, which still has a limitation of no Jordan normal form appears and must be bore in mind. On the other hand, the precise integration method does not care about the possible Jordan normal form appearing.

§5.8.4, Transformation for purely imaginary eigenvalues

The analytical solution equation (5.8.5) and (5.8.6) do not exclude the case of purely imaginary roots presented. In problems of structural static and/or linear quadratic optimal control or Kalman-Bucy filtering, the purely imaginary eigenvalue never appears. However for the problems of elastic wave propagation, eletro-magnetic wave-guide etc., purely imaginary eigenvalues are critical for applications, because it represents transmission waves, that energy or signal propagation to far away place is based on such transmission waves. The real arithmetic computation for the solutions of Riccati equations and the fundamental interval mixed energy matrices is described in the present section.

For purely imaginary eigen-root the complex conjugate equals to its negative value, in other words, its symplectic adjoint value coincides with its complex conjugate. Therefore the complex conjugate of a class α eigen-root belongs to the class β. Hence when put the real part of its eigenvector in the sub-matrices $\mathbf{Q}_{\alpha r}, \mathbf{P}_{\alpha r}$, its imaginary part of eigenvector must be in the sub-matrices $\mathbf{Q}_{\beta r}, \mathbf{P}_{\beta r}$ with the same column number. Denoting $\mathbf{B}_1' = \mathrm{diag}(e^{\mu\eta})\mathbf{B}_1 = \mathbf{D}_{c\eta}\mathbf{B}_1$ in equations (5.8.26a,b) for the columns corresponding to purely imaginary roots, hence

$$\mathbf{Q}_\alpha \mathbf{A}_1 + \mathbf{Q}_\beta \mathbf{B}_1' = \mathbf{I}, \qquad \mathbf{P}_\alpha \mathbf{D}_{c\eta} \mathbf{A}_1 + \mathbf{P}_\beta \mathbf{D}_{c\eta}^{-1} \mathbf{B}_1' = \mathbf{0}$$

which are still in complex numerical form. It should be noted that the columns corresponding to the purely imaginary eigenvalues, in matrices $\mathbf{Q}_\alpha$ and $\mathbf{Q}_\beta$, and in matrices $\mathbf{P}_\alpha \mathbf{D}_{c\eta}$ and $\mathbf{P}_\beta \mathbf{D}_{c\eta}^{-1}$, respectively, must be complex conjugate to each other. Therefore the corresponding rows in matrices $\mathbf{A}_1$ and $\mathbf{B}_1'$ must also be complex conjugate to each other, and are denoted as

$$\mathbf{A}_1 = (\mathbf{A}'_{pr} - i\mathbf{B}'_{pr})/2 \ , \qquad \mathbf{B}_1' = (\mathbf{A}'_{pr} + i\mathbf{B}'_{pr})/2$$

where the negative sign and dividing 2 are only for convenience. The matrices $\mathbf{A}'_{pr}$ and $\mathbf{B}'_{pr}$ are both real valued. For corresponding columns in complex valued matrices, denote

$$\mathbf{Q}_\alpha = \mathbf{Q}_{\alpha r} + i\mathbf{Q}_{\beta r}, \ \mathbf{Q}_\beta = \mathbf{Q}_{\alpha r} - i\mathbf{Q}_{\beta r}; \quad \mathbf{P}_{\alpha r} = \mathrm{Re}(\mathbf{P}_\alpha), \ \mathbf{P}_{\beta r} = \mathrm{Im}(\mathbf{P}_\alpha)$$

Therefore the equations to be solved become

$$\mathbf{Q}_{\alpha r}\mathbf{A}'_{pr} + \mathbf{Q}_{\beta r}\mathbf{B}'_{pr} = \mathbf{I} \tag{5.8.42a}$$

$$\mathbf{P}_{\alpha r\eta}\mathbf{A}'_{pr} + \mathbf{P}_{\beta r\eta}\mathbf{B}'_{pr} = \mathbf{0} \tag{5.8.42b}$$

where $\qquad \mathbf{P}_{\alpha r\eta} = \mathrm{Re}(\mathbf{P}_\alpha \mathbf{D}_{c\eta}), \ \ \mathbf{P}_{\beta r\eta} = \mathrm{Im}(\mathbf{P}_\alpha \mathbf{D}_{c\eta})$

The problem of that equations (5.8.40a,b) cannot involve purely imaginary eigenvalues is then compensated. After solved the matrices $\mathbf{A}'_{pr}$ and $\mathbf{B}'_{pr}$, (which are solved simultaneously with matrices $\mathbf{A}_{r1}$ and $\mathbf{B}_{r1}$), the interval matrices are computed as

$$\mathbf{Q}(\eta) = -\mathbf{P}_{\alpha r}\mathbf{A}'_{pr} - \mathbf{P}_{\beta r}\mathbf{B}'_{pr} \tag{5.8.42c}$$

$$\mathbf{F}(\eta) = \mathbf{Q}_{\alpha r\eta}\mathbf{A}'_{pr} + \mathbf{Q}_{\beta r\eta}\mathbf{B}'_{pr} \tag{5.8.42d}$$

where $\ \mathbf{Q}_{\alpha r\eta} = \mathrm{Re}(\mathbf{Q}_\alpha \mathbf{D}_{c\eta})$, $\mathbf{Q}_{\beta r\eta} = \mathrm{Im}(\mathbf{Q}_\alpha \mathbf{D}_{c\eta})$, $\mathrm{diag}(e^{\mu\eta}) = \mathbf{D}_{c\eta}$ (5.8.42e)

The computation for the matrices $\mathbf{G}(\eta)$ and $\mathbf{F}^T(\eta)$ are similar, that only the equations are listed below. The equations for solution are

$$\mathbf{Q}_{\alpha r}\mathbf{A}_{pr}^{(2)} + \mathbf{Q}_{\beta r}\mathbf{B}_{pr}^{(2)} = \mathbf{0} \tag{5.8.43a}$$

$$\mathbf{P}_{\alpha r\eta}\mathbf{A}_{pr}^{(2)} + \mathbf{P}_{\beta r\eta}\mathbf{B}_{pr}^{(2)} = \mathbf{I} \tag{5.8.43b}$$

where the superscript $^{(2)}$ is only a notation. After solving the matrices $\mathbf{A}_{pr}^{(2)}$ and $\mathbf{B}_{pr}^{(2)}$

$$\mathbf{G}(\eta) = \mathbf{Q}_{\alpha r\eta}\mathbf{A}_{pr}^{(2)} + \mathbf{Q}_{\beta r\eta}\mathbf{B}_{pr}^{(2)} \tag{5.8.43c}$$

$$\mathbf{F}^T(\eta) = \mathbf{P}_{\alpha r}\mathbf{A}_{pr}^{(2)} + \mathbf{P}_{\beta r}\mathbf{B}_{pr}^{(2)} \tag{5.8.43d}$$

Certainly these equations are only for purely imaginary eigenvalues. As for non-zero real part eigenvalues, the equations have been given in equation (5.8.40) and (5.8.41) already.

Example 5.7: Dimension $n=5$, the sub-matrices of the Hamilton matrix $\mathbf{H}$ are given as

$$\mathbf{A}=\begin{bmatrix} -0.2 & 0.5 & 0 & 0 & 0 \\ 0 & -0.5 & 1.6 & 0 & 0 \\ 0 & 0 & -0.2 & 0.8 & 0 \\ 0 & 0 & 0 & -0.25 & 7.5 \\ 0 & 0 & 0 & 0 & -0.1 \end{bmatrix}, \quad \begin{array}{l} \mathbf{B}=\mathrm{diag}(1.0,\ 0.0,\ 0.0,\ -1.0,\ 0.0), \\ \mathbf{D}=\mathrm{diag}(0,\ 0,\ 0,\ 0,\ 0.09). \end{array}$$

to compute the eigen-solution based interval matrices.

Solution: Solving the eigen-solutions as

$$\mathrm{Re}(\mu)= \quad 0 \quad -.7878 \quad -.7878 \quad 0 \quad -1.5259$$

$$\mathrm{Im}(\mu)= \quad 0.8150 \quad 0.4050 \quad -.4050 \quad 1.4747 \quad 0$$

$$\mathbf{Q}_{\alpha r}=\begin{bmatrix} 0.5788 & 0.3275 & 0.1216 & -.1566 & -.1295 \\ 0.1379 & -.4835 & 0.1223 & -.2413 & 0.3435 \\ -.4492 & 0.0560 & -.1444 & 0.3280 & -.2203 \\ -.4916 & 0.0320 & 0.1345 & 0.7441 & 0.3651 \\ 0.0232 & -.0096 & -.0079 & -.0764 & -.0621 \end{bmatrix}, \quad \mathbf{Q}_{\beta r}=\begin{bmatrix} 0.0574 & 0.1974 & -.3247 & 0.0606 & -.0995 \\ 0.9664 & 0.1270 & -.8015 & -.4376 & -.3435 \\ 0.3723 & -.1007 & -.6773 & -.3592 & -.4350 \\ -.3646 & -.4672 & -.7853 & 0.5148 & -.9384 \\ -.0656 & -.1071 & -.0834 & 0.1635 & -.2222 \end{bmatrix}$$

$$\mathbf{P}_{\alpha r}=\begin{bmatrix} -.0979 & -.2406 & -.2217 & 0.0549 & 0.0751 \\ 0.2790 & -.0604 & -.1051 & -.0244 & 0.0185 \\ 0.5550 & -.0240 & -.1800 & -.0320 & 0.0172 \\ 0.0232 & 0.0138 & 0.0038 & -.2473 & 0.2133 \\ 0.6196 & 0.1087 & 0.0175 & -2.764 & 0.9839 \end{bmatrix}, \quad \mathbf{P}_{\beta r}=\begin{bmatrix} -.6861 & 0.4858 & -.2176 & 0.0988 & -.0751 \\ -.2312 & -.4617 & -.2716 & 0.0268 & 0.0366 \\ 0.4116 & 0.5067 & 1.0884 & -.0222 & -.0441 \\ -.0655 & 0.1497 & -.0460 & 0.5292 & 0.7632 \\ 0.1375 & -1.432 & -.3413 & -1.071 & -4.014 \end{bmatrix}$$

Selecting $\eta=2.0$, computation gives

$$\mathbf{Q}(\eta)=\begin{bmatrix} 1.3766 & 0.5006 & 0.5167 & 0.1725 & 0.2274 \\ 0.5006 & 0.2466 & 0.2942 & 0.1042 & 0.1387 \\ 0.5167 & 0.2942 & 0.3877 & 0.1437 & 0.1920 \\ 0.1725 & 0.1042 & 0.1437 & -.4420 & 0.7797 \\ 0.2274 & 0.1387 & 0.1920 & 0.7797 & 21.743 \end{bmatrix}, \quad \mathbf{G}(\eta)=\begin{bmatrix} -.0089 & -.0652 & -.1211 & -.2850 & -.0412 \\ -.0652 & -.3748 & -.5986 & -1.252 & -.1683 \\ -.1211 & -.5986 & -.8341 & -1.491 & -.1718 \\ -.2850 & -1.252 & -1.491 & -1.966 & -.1052 \\ -.0412 & -.1683 & -.1718 & -.1052 & 0.0410 \end{bmatrix}$$

$$\mathbf{F}(\eta)=\begin{bmatrix} 0.6715 & 0.5049 & 0.8881 & 0.3750 & 0.4652 \\ 0.0140 & 0.3773 & 1.6274 & 0.8197 & 0.4965 \\ 0.0305 & 0.0201 & 0.7005 & 0.2297 & -1.3161 \\ 0.0787 & 0.0514 & 0.0762 & -.8628 & -5.9511 \\ 0.0119 & 0.0077 & 0.0114 & -.1766 & -1.0065 \end{bmatrix} \qquad \#\#$$

The numerical results of fundamental interval mixed energy matrices can be used for various purposes. The precise integration method is also used to compute these matrices, the numerical results check with those obtained by the eigenvectors based analytical approach, still have ten more decimal digit coincidences, which expresses again that both numerical methods are highly precise.

Such class of structural mechanic problem has analogy relationship to the linear quadratic optimal control problems, see [20~22,87~91]. The ***positive definiteness of deformation energy*** in structural mechanics corresponds to the ***controllability and observability*** of state space linear control system, see section 6.7. On the other hand, the ***positive definiteness of deformation energy*** implies that structural system can simulate the ***least square method*** in probability application, because it is shown in chapter 3 that the variance matrix in least square method corresponds to the flexibility matrix in structural analysis. The analysis of the structural system with single continuous coordinate can be further investigated, such as the sub-structural combination can be used to realize the stepwise integration for strip domain structural analysis. Such kind methodology can also be used in filtering problem.

The purely imaginary eigenvalue of a Hamilton matrix has some special feature, it closely relates to energy transmission process, for which the complex eigenvectors are necessary. The problems of power flow and wave scattering etc. will be described in later sections.

§5.9, Stepwise integration by means of sub-structural combination

The main step of precise integration method for TPBVP given in the previous sections, is the computation of interval mixed energy matrices $\mathbf{F}(\eta),\mathbf{G}(\eta)$ and $\mathbf{Q}(\eta)$ for a typical fundamental interval with length η. A typical interval is a substructure, so that the sub-structural combination algorithm given by equations (5.7.5a~c) in the displacement formulation or by equations (5.7.23a~c) in mixed variable system can be used for stepwise integration problems. Using sub-structural combination method corresponds to the discretized structural analysis problem. The problem can be proposed as: to find the n-dimensional displacement vector $\mathbf{q}(z)$, with the boundary condition of an elastic support at the left end of $z=0$, for which the deformation energy is expressed by

$$\Pi_0(\mathbf{q}_0)=(\mathbf{q}_0-\hat{\mathbf{q}}_0)^T\mathbf{P}_0^{-1}(\mathbf{q}_0-\hat{\mathbf{q}}_0)/2 \tag{5.9.1}$$

where $\hat{\mathbf{q}}_0$ is a given displacement, the neutral point, and $\mathbf{P}_0^{-1}$ is the stiffness matrix of the elastic support, and therefore $\mathbf{P}_0$ is the flexibility matrix of the elastic support. The force vector

$$\mathbf{p}_0=\partial\Pi_0/\partial\mathbf{q}_0=\mathbf{P}_0^{-1}(\mathbf{q}_0-\hat{\mathbf{q}}_0),\qquad \text{then } \mathbf{q}_0=\mathbf{P}_0\mathbf{p}_0+\hat{\mathbf{q}}_0$$

is introduced. To analyze the structure by the sub-structural combination method, a typical sub-structure with length η is computed first, then the stepwise integration is initiated from the elastic support at $z=0$, with each step a typical substructure (with length η) is attached at the right-end for totally k_f times. Therefore the nodal points are naturally generated as

$$z_0=0,\ z_1=\eta,\cdots\cdots,\ z_k=k\eta,\cdots\cdots,\ z_f=k_f\eta$$

For each typical substructure of length η, the elastic behavior is described by the mixed energy matrices of $\mathbf{F}(\eta),\mathbf{G}(\eta)$ and $\mathbf{Q}(\eta)$. These interval mixed energy matrices represent the elastic behaviors only, but cannot represent if there were distributed external forces acting upon within the typical substructure, such as (z_{k-1},z_k), denoted as $\#k$. The elastic behavior can be unified for all substructures but the distributed external forces may be different for different substructures. The distributed forces can also be represented by external forces $\mathbf{f}_{ak}$, $\mathbf{f}_{bk}$ at the two ends of interval $\#k$, which are interval dependent, since distributed forces are interval dependent. The computation of $\mathbf{f}_{ak}$ and $\mathbf{f}_{bk}$ can also use sub-structural combination algorithm in precise integration method. $\mathbf{f}_{ak}$ is a n- dimensional force vector acting on the substructure at the left-end station z_{k-1}, whereas $\mathbf{f}_{bk}$ is also n- dimensional but acting on the substructure at the right-end station z_k.

Using energy method to solve this problem, the simplest is the minimum potential energy variational principle. Note that

$$U_k(\mathbf{q}_a,\mathbf{q}_b)=U_k^{(1)}(\mathbf{q}_a,\mathbf{q}_b)+U_k^{(2)}(\mathbf{q}_a,\mathbf{q}_b)$$

where $U_k^{(1)}(\mathbf{q}_a,\mathbf{q}_b)=\mathbf{q}_a^T\mathbf{K}_{aa}\mathbf{q}_a/2+\mathbf{q}_b^T\mathbf{K}_{bb}\mathbf{q}_b/2+\mathbf{q}_b^T\mathbf{K}_{ba}\mathbf{q}_a$ is the deformation energy of substructure $\#k$, $U_k^{(2)}(\mathbf{q}_a,\mathbf{q}_b)=-\mathbf{f}_{ak}^T\mathbf{q}_a-\mathbf{f}_{bk}^T\mathbf{q}_b$ is the external force potential energy of $\#k$ and $U_k(\mathbf{q}_a,\mathbf{q}_b)$ is the total potential energy of $\#k$

$$U_k(\mathbf{q}_a,\mathbf{q}_b)=\mathbf{q}_a^T\mathbf{K}_{aa}\mathbf{q}_a/2+\mathbf{q}_b^T\mathbf{K}_{bb}\mathbf{q}_b/2+\mathbf{q}_b^T\mathbf{K}_{ba}\mathbf{q}_a-\mathbf{f}_{ak}^T\mathbf{q}_a-\mathbf{f}_{bk}^T\mathbf{q}_b \tag{5.9.2}$$

where the subscripts $_{a,b}$ denote, respectively, the left- and right-end, *i.e.* $k-1$ and k stations, respectively. Therefore the minimum potential energy of the whole structure is

$$\min_{\mathbf{q}}\left[\sum_{k=1}^{k_f}U_k(\mathbf{q}_{k-1},\mathbf{q}_k)+(\mathbf{q}_0-\hat{\mathbf{q}}_0)^T\mathbf{P}_0^{-1}(\mathbf{q}_0-\hat{\mathbf{q}}_0)/2\right] \tag{5.9.3}$$

Carrying out the variational operation derives the equilibrium equation

$$\mathbf{K}_{ba}\mathbf{q}_{k-1}+(\mathbf{K}_{bb}+\mathbf{K}_{aa})\mathbf{q}_k+\mathbf{K}_{ab}\mathbf{q}_{k+1}=\mathbf{f}_{b,k-1}+\mathbf{f}_{a,k} \tag{5.9.4}$$

This equation is derived by the displacement method, hence it looks nothing related to the mixed energy matrices $\mathbf{F}(\eta),\mathbf{G}(\eta)$ and $\mathbf{Q}(\eta)$. To see the relation with mixed energy matrices, note that

$$V_k(\mathbf{q}_a,\mathbf{p}_b)=\mathbf{p}_b^T\mathbf{G}\mathbf{p}_b/2+\mathbf{p}_b^T\mathbf{F}\mathbf{q}_a-\mathbf{q}_a^T\mathbf{Q}\mathbf{q}_a+\mathbf{p}_b^T\mathbf{r}_{bk}+\mathbf{q}_a^T\mathbf{r}_{ak} \tag{5.9.5}$$

and there is a transformation between mixed energy and potential energy. The constant term is useless and can be disregarded. The interrelation between them is the Legendre transformation

$$\mathbf{p}_b=\partial U/\partial\mathbf{q}_b=\mathbf{K}_{ba}\mathbf{q}_a+\mathbf{K}_{bb}\mathbf{q}_b-\mathbf{f}_b,\quad \mathbf{q}_b=\mathbf{F}\mathbf{q}_a+\mathbf{G}\mathbf{p}_b+\mathbf{G}\mathbf{f}_b \tag{5.9.6}$$

where $\mathbf{F},\mathbf{Q},\mathbf{G}$ are the same as in (5.7.13), but the derivation before has no external forces. The mixed energy is obtained as

$$V(\mathbf{q}_a,\mathbf{p}_b)=\mathbf{p}_b^T\mathbf{q}_b-U(\mathbf{q}_a,\mathbf{q}_b)$$

where the vector $\mathbf{q}_b$ should be substituted with equation (5.9.6), and the subscript $_k$ is taken off. After some algebraic derivation gives

$$\mathbf{r}_a(=\mathbf{r}_{k-1})=\mathbf{f}_a+\mathbf{F}\mathbf{f}_b,\quad \mathbf{r}_b(=\mathbf{r}_k)=\mathbf{G}\mathbf{f}_b \tag{5.9.7}$$

which explains that the non-homogeneous terms $\mathbf{f}_a$ and $\mathbf{f}_b$ can be mutually transformed with the mixed energy non-homogeneous terms $\mathbf{r}_a$ and $\mathbf{r}_b$.

The relation between mixed energy and potential energy is given by

$$U_k(\mathbf{q}_{k-1},\mathbf{q}_k)=\max_{\mathbf{p}_k}[\mathbf{p}_k^T\mathbf{q}_k-V_k(\mathbf{q}_{k-1},\mathbf{p}_k)] \tag{5.9.8}$$

Substituting the above representation into minimum potential energy principle (5.9.3) gives

$$\min_{\mathbf{q}}\max_{\mathbf{p}}\left[\sum_{k=1}^{k_f}\left[\mathbf{p}_k^T\mathbf{q}_k - V_k(\mathbf{q}_{k-1},\mathbf{p}_k)\right]+(\mathbf{q}_0-\hat{\mathbf{q}}_0)^T\mathbf{P}_0^{-1}(\mathbf{q}_0-\hat{\mathbf{q}}_0)/2\right] \tag{5.9.9}$$

Carrying out the variational operation derives the dual equations

$$\mathbf{q}_{k+1} = \mathbf{F}\mathbf{q}_k + \mathbf{G}\mathbf{p}_{k+1} + \mathbf{r}_{bk}, \quad \mathbf{p}_k = -\mathbf{Q}\mathbf{q}_k + \mathbf{F}^T\mathbf{p}_{k+1} + \mathbf{r}_{ak} \tag{5.9.10}$$

From the above dual equations and the variational principle (5.9.9), it is seen that they are in the same form with comparison to the Kalman filtering problem, see section 6.5. It gives the ***analogy relationship between optimal filtering and structural mechanics.***

The successive combination algorithm of typical substructures is described as follows. The elastic support at the initial point z_0 can be regarded as an initial substructure. Its total potential energy is expressed by equation (5.9.1). The successive composition corresponds to mathematical induction. Suppose that the $(k-1)$-th step combination of typical substructure $\#(k-1)$ has been consummated, which gives a substructure $[z_0, z_{k-1})$ extending from the z_0 elastic support up to the station z_{k-1}, for which the right end displacement vector $\mathbf{q}_{k-1}$ is still a variable (external displacement) to be determined. The total potential energy of substructure $[z_0, z_{k-1})$ can be expressed by the right end displacement $\mathbf{q}_{k-1}$ as

$$\Pi_{k-1}(\mathbf{q}_{k-1}) = (\mathbf{q}_{k-1}-\hat{\mathbf{q}}_{k-1})^T\mathbf{P}_{k-1}^{-1}(\mathbf{q}_{k-1}-\hat{\mathbf{q}}_{k-1})/2 \tag{5.9.1'}$$

Note here, that $\Pi_{k-1}(\mathbf{q}_{k-1})$ denote the potential energy of the chain $[z_0, z_{k-1})$. Obviously, the right-end equilibrium (neutral) displacement vector is $\hat{\mathbf{q}}_{k-1}$, and the flexibility matrix is $\mathbf{P}_{k-1}$, that all other displacements at the inside stations including z_0 had been eliminated as the internal variables of the substructure $[z_0, z_{k-1})$. It is a substructure chain, from z_0 to z_{k-1}.

The next stepwise integration is to combine the typical substructure $\#k$ to the combined sub-structural chain $[z_0, z_{k-1})$, that the stepwise integration process is recursive in nature, and $U_k(\mathbf{q}_{k-1}, \mathbf{q}_k)$ is the potential energy of substructure $\#k$. After this step of combination, the sub-structural chain extends its right end to the station z_k. The extended substructure right end equilibrium displacement $\hat{\mathbf{q}}_k$ and its flexibility matrix $\mathbf{P}_k$ are to be determined. In this step, the (eliminated) displacement vector $\bar{\mathbf{q}}_{k-1}$ at station z_{k-1} and the flexibility matrix $\mathbf{P}_{k-1|k}$ at z_{k-1} are also of concern. However, the station $\#(k-1)$ has been an internal station now, therefore different notations are used. To explain clearer, the flexibility matrix $\mathbf{P}_{k-1|k}$ corresponds to that the sub-structural chain has extended the right end to station z_k, so that the $\mathbf{P}_{k-1|k}$ is the flexibility at an internal station z_{k-1}. The displacement vector $\bar{\mathbf{q}}_{k-1}$ is also written as $\bar{\mathbf{q}}_{k-1|k}$, to emphasize that the sub-structural chain has extended to the station z_k, but the displacement vector is evaluated at the internal station z_{k-1}.

To carry out the one step computation, the minimum potential energy principle is used. The total potential energy $U_k(\mathbf{q}_a, \mathbf{q}_b)$ of the typical substructure $\#k$ is given in equation (5.9.2), where the subscripts a, b represent, respectively, the left and right end, *i.e.* $k-1, k$ stations. Therefore the minimum potential energy variational principle of the sub-structural chain extended up to the station $\#k$ is given as

$$\min_{\mathbf{q}_{k-1}, \mathbf{q}_k} [\Pi_{k-1}(\mathbf{q}_{k-1}) + U_k(\mathbf{q}_{k-1}, \mathbf{q}_k)] \tag{5.9.11}$$

However, the precise integration algorithm supplies mixed energy matrices, it is much convenient using the mixed energy $V_k(\mathbf{q}_a, \mathbf{p}_b)$ as given in (5.9.5) in the computation. The transformation between mixed energy and potential energy is the Legendre transformation (5.9.6) and their relation has been given in equation (5.9.8). Substituting this expression into the potential energy variational principle (5.9.11) gives

$$\min_{\mathbf{q}_{k-1}, \mathbf{q}_k} \max_{\mathbf{p}_k} [\Pi_{k-1}(\mathbf{q}_{k-1}) + \mathbf{p}_k^T \mathbf{q}_k - V_k(\mathbf{q}_{k-1}, \mathbf{p}_k)] \tag{5.9.12}$$

Carrying out the variational operation gives

$$\delta \mathbf{q}_{k-1}: \quad \mathbf{q}_{k-1} = (\mathbf{I} + \mathbf{P}_{k-1}\mathbf{Q})^{-1}\hat{\mathbf{q}}_{k-1} + (\mathbf{P}_{k-1}^{-1} + \mathbf{Q})^{-1}(\mathbf{r}_{k-1} + \mathbf{F}^T \mathbf{p}_k)$$

$$\delta \mathbf{p}_k: \quad \mathbf{q}_k = \mathbf{G}\mathbf{p}_k + \mathbf{F}\mathbf{q}_{k-1} + \mathbf{r}_k$$

$$\delta \mathbf{q}_k: \quad \mathbf{p}_k = \mathbf{0}$$

Eliminating $\mathbf{q}_{k-1}$ gives

$$\mathbf{q}_k = \mathbf{P}_k \mathbf{p}_k + \hat{\mathbf{q}}_k \tag{5.9.13}$$

$$\mathbf{P}_k = \mathbf{G} + \mathbf{F}(\mathbf{P}_{k-1}^{-1} + \mathbf{Q})^{-1}\mathbf{F}^T \tag{5.9.14}$$

$$\hat{\mathbf{q}}_k = \mathbf{F}(\mathbf{I} + \mathbf{P}_{k-1}\mathbf{Q})^{-1}\hat{\mathbf{q}}_{k-1} + \mathbf{F}(\mathbf{P}_{k-1}^{-1} + \mathbf{Q})^{-1}\mathbf{r}_{k-1} + \mathbf{r}_k \tag{5.9.15}$$

Therefore, the right-end equilibrium displacement vector $\hat{\mathbf{q}}_k$ of the sub-structural chain and the respective flexibility matrix $\mathbf{P}_k$ are found. Next step is from the station z_k forward by combining the typical substructure $\#(k+1)$ recursively.

Certainly, instead of the equation (5.9.1') there has been the potential energy

$$\Pi_k(\mathbf{q}_k) = (\mathbf{q}_k - \hat{\mathbf{q}}_k)^T \mathbf{P}_k^{-1}(\mathbf{q}_k - \hat{\mathbf{q}}_k)/2 \tag{5.9.1''}$$

which is derived from equation (5.9.12), keeping the variable $\mathbf{q}_k$ not eliminated and taking variational calculation. The potential energy expressed as a function of right-end displacement is obtained as

$$\Pi_k(\mathbf{q}_k) = \min_{\mathbf{q}_{k-1}} \max_{\mathbf{p}_k} [\Pi_{k-1}(\mathbf{q}_{k-1}) + \mathbf{p}_k^T \mathbf{q}_k - V_k(\mathbf{q}_{k-1}, \mathbf{p}_k)]$$

Therefore the situation of recursive derivation becomes clearer.

Equations (5.9.14) and (5.9.15) give the formulae for computing $\mathbf{P}_k$ and $\hat{\mathbf{q}}_k$, they locate at the front of the combined sub-structural chain. The displacement vector and flexibility matrix of the internal point z_{k-1} can also be given as

$$\bar{\mathbf{q}}_{k-1} = (\mathbf{I} + \mathbf{P}_{k-1}\mathbf{Q})^{-1}\hat{\mathbf{q}}_{k-1} + \mathbf{P}_{k-1|k}\mathbf{r}_{k-1}, \quad \mathbf{P}_{k-1|k} = (\mathbf{P}_{k-1}^{-1} + \mathbf{Q})^{-1} \tag{5.9.16}$$

The substructure front has been at the station z_k, so that the station z_{k-1} is one typical substructure lagged behind. These equations apply only for one typical substructure behind, however, because of the length η can be selected arbitrarily, so the equations are still quite general, see section 6.6.

The meaning of these equations does not limit to only structural analysis. When Kalman filter is considered, it will be seen that the analysis is similar. Hence the discrete-time Kalman filter problem corresponds to successive substructure combination. The continuous-time Kalman-Bucy filtering will be shown corresponding to the single continuous coordinate structural analysis problem. The problem considered in this chapter is, therefore, quite analogy theory oriented.

In optimal control theory, there is linear quadratic (LQ) control problem, which is also analogous to the sub-structural chain analysis [20~22,87~88]. Suppose that an elastic support with stiffness matrix $\mathbf{S}_f$ locates at the right end station k_f. Therefore the whole potential energy is

$$\min_{\mathbf{q}}\left[\sum_{k=1}^{k_f} U_k(\mathbf{q}_{k-1},\mathbf{q}_k)+\mathbf{q}_f^T\mathbf{S}_f\mathbf{q}_f/2\right] \tag{5.9.17}$$

Using the mixed energy variational principle similarly derives

$$\min_{\mathbf{q}}\max_{\mathbf{p}}\left[\sum_{k=1}^{k_f}\left[\mathbf{p}_k^T\mathbf{q}_k - V_k(\mathbf{q}_{k-1},\mathbf{p}_k)\right]+\mathbf{q}_f^T\mathbf{S}_f\mathbf{q}_f/2\right] \tag{5.9.18}$$

Performing the variational operation derives the dual equations

$$\begin{aligned}\mathbf{q}_{k+1} &= \mathbf{F}\mathbf{q}_k + \mathbf{G}\mathbf{p}_{k+1} + \mathbf{r}_{bk} \\ \mathbf{p}_k &= -\mathbf{Q}\mathbf{q}_k + \mathbf{F}^T\mathbf{p}_{k+1} + \mathbf{r}_{ak}\end{aligned}, \qquad k = 0,1,\cdots,k_f-1 \tag{5.9.19}$$

and boundary conditions

$$\mathbf{p}_f = -\mathbf{S}_f\mathbf{q}_f + \mathbf{r}_{bf}, \quad \mathbf{p}_0 = \mathbf{0}$$

The LQ control correspondent structural analysis problem is a special class with zero external forces $\mathbf{r}_{ak} = \mathbf{r}_{bk} = \mathbf{0}$, and the analogy relationship will be considered later in dealing with the LQ control theory.

§5.10, Influence function of single continuous coordinate system

Previous sections described the algorithms for the solution of Riccati differential equations and sub-structural combination algorithm, which provides the basis for the computation of ***influence function***. The influence function is quite useful in determining the most unfavorable moving load in bridge engineering design, which can be performed by the dynamic programming method [104]. Especially the influence function is quite useful to the inverse problem and parameter identification etc., therefore, the precise integration computation of influence function is important. Based on the analogy relationship between structural mechanics and optimal control, the influence function is quite useful to system identification and decentralized control analysis etc.

Let the domain of problem be $z \in [z_0 = 0, z_f]$, and the two end boundary conditions are

$$\mathbf{q}_0 = \mathbf{P}_0\mathbf{p}_0 + \hat{\mathbf{q}}_0, \text{ when } z = z_0; \qquad \text{and} \quad \mathbf{p}_f = -\mathbf{S}_f\mathbf{q}_f, \text{ when } z = z_f \tag{5.10.1}$$

Traditionally, the influence function is described by the displacement method. For duality system, it turns to describe on the basis of n-dimensional dual vectors $\mathbf{q}$ and $\mathbf{p}$, for which the dual differential equations are given before

$$\begin{aligned}\dot{\mathbf{q}} &= \mathbf{A}\mathbf{q} + \mathbf{D}\mathbf{p} + \mathbf{f}_q(z) \\ \dot{\mathbf{p}} &= \mathbf{B}\mathbf{q} - \mathbf{A}^T\mathbf{p} + \mathbf{f}_p(z)\end{aligned} \tag{5.2.9a,b}$$

where $\mathbf{f}_q$ and $\mathbf{f}_p$ are the given distributed external force. Note, the two differential equations are interpreted as compatibility and equilibrium, hence the interpretations of the forces $\mathbf{f}_q$ and $\mathbf{f}_p$ are also different (even the units).

The traditional idea for an influence function is described as follows. The response displacement $\mathbf{q}(z)$ (or other quantity) at a fixed location z is of concern, which is induced from a unit concentrated force applied at the location ζ. As ζ varying in the whole interval $0 < \zeta < z_f$, the displacement $\mathbf{q}_z$ varies too and is the function of the force location, denoted as $\mathbf{q}_z(\zeta)$, which is the influence function and can also be denoted as $\mathbf{q}(z,\zeta)$. For duality system, both the unit forces and the response dual vectors are twice as much with comparison to the traditional influence function.

Since it is to find the influence function, so there is unit force acting at the point ζ, and the response dual vectors $\mathbf{q}(z,\zeta)$ and $\mathbf{p}(z,\zeta)$ are needed, *i.e.* the following dual differential equations need to be solved

$$\begin{aligned}\dot{\mathbf{q}}_a &= \mathbf{A}\mathbf{q}_a + \mathbf{D}\mathbf{p}_a + \mathbf{k}_a\delta(z-\zeta) \\ \dot{\mathbf{p}}_a &= \mathbf{B}\mathbf{q}_a - \mathbf{A}^T\mathbf{p}_a\end{aligned} \tag{5.10.2a,b}$$

$$\begin{aligned}\dot{\mathbf{q}}_b &= \mathbf{A}\mathbf{q}_b + \mathbf{D}\mathbf{p}_b \\ \dot{\mathbf{p}}_b &= \mathbf{B}\mathbf{q}_b - \mathbf{A}^T\mathbf{p}_b + \mathbf{k}_b\delta(z-\zeta)\end{aligned} \tag{5.10.3a,b}$$

where the subscript a and b are used to distinct the two cases of the external forces, never confusing with the interval left and right ends. For duality system, the unit force vector turns out to be two unit external vectors denoted as $\mathbf{k}_a$ and $\mathbf{k}_b$ corresponding to the two equations, respectively, where the physical meanings are that $\mathbf{k}_b$ is the external force, and $\mathbf{k}_a$ is the deformation incompatibility. Both $\mathbf{k}_a$ and $\mathbf{k}_b$ should be taken as the n unit vectors of columns of unit matrix $\mathbf{I}_n$ in turn, respectively, for the two equations. Therefore the two sets of dual equations (5.10.2a,b) and (5.10.3a,b) have $2n$ solutions of $\mathbf{q}$ and $\mathbf{p}$. The traditional influence function theory in structural mechanics is based on one kind of variables, *i.e.* the displacement method, however, presently the influence function is for duality system, so that the unit forces are for two kinds of concentrated forces and the functions are also of dual variables. To solve both the dual equations (5.10.2a,b) and (5.10.3a,b), the precise integration can be applied effectively.

Because of the external force acting at the point ζ, the whole interval (segment) is subdivided into two intervals $z_0 \le z < \zeta$ and $\zeta < z \le z_f$ with no external force, and can be solved separately. As in section 5.7, using precise integration method should subdivide the whole interval into uniform intervals of fundamental interval length η. Both the influence function's response point z and the external force acting point ζ are located on the grid points.

To solve the homogeneous dual differential equations for both the intervals $z_0 \le z < \zeta$ and $\zeta < z \le z_f$, the solutions of the respective Riccati differential equations can be used. The precise integration method is selected for the solutions, because it does not care about the possible Jordan normal form and is always reliable. According to that given in section 5.7.6, the mixed energy matrices $\mathbf{F}(\eta), \mathbf{G}(\eta), \mathbf{Q}(\eta)$ of the fundamental interval η can be regarded as computed.

For the interval $z_0 \le z < \zeta$, let the displacement vector be

$$\mathbf{q}_a(z) = \mathbf{P}(z) \cdot \mathbf{p}_a(z) + \hat{\mathbf{q}}_a(z), \qquad z_0 \le z < \zeta \tag{5.10.4}$$

Substituting into equation (5.10.2a,b) derives the forward integration of Riccati ODE (filter type) as

$$\dot{\mathbf{P}}(z) = \mathbf{D} + \mathbf{AP} + \mathbf{PA}^T - \mathbf{PBP}, \quad \mathbf{P}(0) = \mathbf{P}_0 \tag{5.10.5}$$

and the 'cantilever' displacement equation

$$\dot{\hat{\mathbf{q}}}_a(z) = [\mathbf{A} - \mathbf{P}(z)\mathbf{B}]\hat{\mathbf{q}}_a, \qquad \hat{\mathbf{q}}_a(0) = \mathbf{0} \tag{5.10.6}$$

Based on the mixed energy matrices $\mathbf{F}(\eta), \mathbf{G}(\eta), \mathbf{Q}(\eta)$ of fundamental interval η, the Riccati equation (5.10.5) can be solved by the interval combination equation (5.7.23), including the initial condition.

The solution $\hat{\mathbf{q}}_a(z)$ should be found from (5.10.6), which is a homogeneous differential equation with null initial condition, so that the solution in $z_0 \le z < \zeta$ is $\hat{\mathbf{q}}_a(z) = \mathbf{0}$, and from the uniqueness theorem of ODE, it is the unique solution. Therefore

$$\mathbf{q}_a(z) = \mathbf{P}(z) \cdot \mathbf{p}_a(z), \qquad \text{in } z_0 \le z < \zeta \tag{5.10.4'}$$

especially, $\mathbf{q}_a(\zeta - 0) = \mathbf{P}(\zeta)\mathbf{p}_a(\zeta - 0)$. Further, from equation (5.10.2a) it can be derived that $\mathbf{q}_a(\zeta + 0) - \mathbf{q}_a(\zeta - 0) = \mathbf{k}_a$. Next, the solution in the interval $\zeta < z \le z_f$ is selected as

$$\mathbf{p}_a(z) = -\mathbf{S}(z) \cdot \mathbf{q}_a(z), \qquad \zeta < z \le z_f \tag{5.10.7}$$

Substituting into equation (5.10.2a,b) gives the backward integration (control type) Riccati ODE as

$$\dot{\mathbf{S}}(z) = -\mathbf{B} - \mathbf{A}^T\mathbf{S} - \mathbf{SA} + \mathbf{SDS}, \qquad \mathbf{S}(z_f) = \mathbf{S}_f \tag{5.10.8}$$

After solving the matrix $\mathbf{S}(z)$, the differential equation for the displacement vector $\mathbf{q}_a(z)$ is

$$\dot{\mathbf{q}}_a(z) = [\mathbf{A} - \mathbf{DS}(z)]\mathbf{q}_a, \quad \zeta < z \le z_f \tag{5.10.9}$$

with initial condition:

$$\mathbf{q}_a(\zeta) = \mathbf{P}(\zeta) \cdot \mathbf{p}_a(\zeta) + \mathbf{k}_a, \qquad \text{as} \quad z = \zeta$$

The boundary conditions uniquely determine the solution matrices $\mathbf{S}(z)$ and $\mathbf{P}(z)$, and the Riccati differential equation (5.10.8) can be solved by precise integration, see section 5.7. The differential equation (5.10.9) for displacement vector $\mathbf{q}_a(z)$ can also be precisely integrated, however the initial condition at $z = \zeta$ must be given by $\mathbf{p}_a(\zeta)$. Based on the differential equation (5.10.2b), $\mathbf{p}_a(\zeta)$ is continuous at $z=\zeta$. After $\mathbf{p}_a(\zeta)$ is solved, the vector $\mathbf{p}_a(z)$ in interval $\zeta<z\le z_f$ is determined in principle. To determine $\mathbf{p}_a(\zeta)$, the variational method can be used as follows.

The variational principle of dual equations (5.2.9a,b) is

$$J=\int_{z_0}^{z_f}[\mathbf{p}^T\dot{\mathbf{q}}-H(\mathbf{q},\mathbf{p})+\mathbf{p}^T\mathbf{f}_q-\mathbf{q}^T\mathbf{f}_p]\mathrm{d}z-\mathbf{q}_f^T\mathbf{S}_f\mathbf{q}_f/2-\mathbf{q}_0^T\mathbf{P}_0^{-1}\mathbf{q}_0/2$$

$$H(\mathbf{q},\mathbf{p})=\mathbf{p}^T\mathbf{A}\mathbf{q}-\mathbf{q}^T\mathbf{B}\mathbf{q}/2+\mathbf{p}^T\mathbf{D}\mathbf{p}/2,\qquad \delta J=0$$

For equations (5.10.2a,b), it reduces to

$$J_a=\int_{z_0}^{z_f}[\mathbf{p}_a^T\dot{\mathbf{q}}_a-H(\mathbf{q}_a,\mathbf{p}_a)]\mathrm{d}z+\mathbf{p}_a^T(\zeta)\mathbf{k}_a-(\mathbf{q}_{af}^T\mathbf{S}_f\mathbf{q}_{af}+\mathbf{q}_{a0}^T\mathbf{P}_0^{-1}\mathbf{q}_{a0})/2$$
$$H(\mathbf{q}_a,\mathbf{p}_a)=\mathbf{p}_a^T\mathbf{A}\mathbf{q}_a-\mathbf{q}_a^T\mathbf{B}\mathbf{q}_a/2+\mathbf{p}_a^T\mathbf{D}\mathbf{p}_a/2,\qquad \delta J_a=0 \tag{5.10.10}$$

where the subscript a stands for the solution of equations (5.10.2). The integration of both intervals are computed using the solutions of Riccati equations as

$$\int_{z_0}^{\zeta}[\mathbf{p}_a^T\dot{\mathbf{q}}_a-H(\mathbf{q}_a,\mathbf{p}_a)]\mathrm{d}z-\mathbf{q}_{a0}^T\mathbf{P}_0^{-1}\mathbf{q}_{a0}/2=[\mathbf{p}_a^T\mathbf{q}_a/2]_{z=\zeta}=[\mathbf{p}_a^T\mathbf{P}\mathbf{p}_a^T/2]_{z=\zeta}$$

$$\int_{\zeta}^{z_f}[\mathbf{p}_a^T\dot{\mathbf{q}}_a-H(\mathbf{q}_a,\mathbf{p}_a)]\mathrm{d}z-\mathbf{q}_{af}^T\mathbf{S}_f\mathbf{q}_{af}/2=[\mathbf{q}_a^T\mathbf{S}\mathbf{q}_a/2]_{z=\zeta}=[\mathbf{p}_a^T\mathbf{S}^{-1}\mathbf{p}_a/2]_{z=\zeta}$$

Because $\mathbf{p}_a(z)$ is continuous at $z=\zeta$ so that the variational principle (5.10.10) becomes

$$\delta[\mathbf{p}_a^T\mathbf{k}_a+\mathbf{p}_a^T\mathbf{P}\mathbf{p}_a/2+\mathbf{p}_a^T\mathbf{S}^{-1}\mathbf{p}_a/2]_{z=\zeta}=0$$

Therefore

$$\mathbf{p}_a(\zeta)=-\left[\mathbf{P}(\zeta)+\mathbf{S}^{-1}(\zeta)\right]^{-1}\mathbf{k}_a \tag{5.10.11}$$

After solving $\mathbf{p}_a(\zeta)$, the dual vector functions $\mathbf{q}_a(z,\zeta)$ and $\mathbf{p}_a(z,\zeta)$ are determined in principle in the two intervals of $z_0\le z<\zeta$ and $\zeta<z\le z_f$. Their numerical solutions will be given later in section 5.10.2. When $\mathbf{k}_a$ is taken as the n column vectors of unit matrix $\mathbf{I}_n$ in turn, the n vectors of $\mathbf{p}_a(\zeta)$ compose a $n\times n$ matrix as

$$\mathbf{P}_a(\zeta,\zeta)=-\left[\mathbf{P}(\zeta)+\mathbf{S}^{-1}(\zeta)\right]^{-1} \tag{5.10.12}$$

where the former ζ in the matrix function $\mathbf{P}_a(\zeta,\zeta)$ means that the value is taken at $z=\zeta$, and the latter ζ means the point of unit force $\mathbf{k}_a\delta(z-\zeta)$. Note that $\mathbf{P}_a$ and $\mathbf{P}$ have different meanings. From equation (5.10.2a) gives $\mathbf{q}_a(\zeta+0)-\mathbf{q}_a(\zeta-0)=\mathbf{k}_a$, more precisely $\mathbf{q}_a(z,\zeta)$ is discontinuous at $z=\zeta$, the discontinuity is given as $\mathbf{q}_a(\zeta+0,\zeta)=\mathbf{k}_a+\mathbf{q}_a(\zeta-0,\zeta)$. It gives the initial condition for the integration of $\mathbf{q}_a(z,\zeta)$ in interval $\zeta<z\le z_f$. When $\mathbf{k}_a$ is selected from the n columns of a unit matrix $\mathbf{I}_n$ in turn, the solved n sets of vectors $\mathbf{q}_a(z,\zeta)$ and $\mathbf{p}_a(z,\zeta)$ compose the matrices $\mathbf{Q}_a(z,\zeta)$ and $\mathbf{P}_a(z,\zeta)$. The above procedure explains the solution steps for dual equations (5.10.2a,b) and the corresponding numerical algorithm will be given later.

The methodology for solving the dual equations (5.2.11a,b) is parallel to the above, so that only the difference is explained. The subscript should be changed from a to b for the variational principle

$$J_b = \int_{z_0}^{z_f} [\mathbf{p}_b^T \dot{\mathbf{q}}_b - H(\mathbf{q}_b, \mathbf{p}_b)] \mathrm{d}z - \mathbf{q}_b^T(\zeta)\mathbf{k}_b - \mathbf{q}_{bf}^T \mathbf{S}_f \mathbf{q}_{bf}/2 - \mathbf{q}_{b0}^T \mathbf{P}_0^{-1} \mathbf{q}_{b0}/2$$

$$H(\mathbf{q}_b, \mathbf{p}_b) = \mathbf{p}_b^T \mathbf{A} \mathbf{q}_b - \mathbf{q}_b^T \mathbf{B} \mathbf{q}_b/2 + \mathbf{p}_b^T \mathbf{D} \mathbf{p}_b/2, \qquad \delta J_b = 0$$

According to equation (5.10.3a), $\mathbf{q}_b$ is continuous at $z = \zeta$. From the solutions of Riccati equations

$$\int_{z_0}^{\zeta} [\mathbf{p}_b^T \dot{\mathbf{q}}_b - H(\mathbf{q}_b, \mathbf{p}_b)] \mathrm{d}z - \mathbf{q}_{b0}^T \mathbf{P}_0^{-1} \mathbf{q}_{b0}/2 = [\mathbf{q}_b^T \mathbf{P}^{-1} \mathbf{q}_b/2]_{z=\zeta}$$

$$\int_{\zeta}^{z_f} [\mathbf{p}_b^T \dot{\mathbf{q}}_b - H(\mathbf{q}_b, \mathbf{p}_b)] \mathrm{d}z - \mathbf{q}_{bf}^T \mathbf{S}_f \mathbf{q}_{bf}/2 = [\mathbf{q}_b^T \mathbf{S} \mathbf{q}_b/2]_{z=\zeta}$$

substituting into the variational principle gives

$$\delta[-\mathbf{q}_b^T \cdot \mathbf{k}_b - \mathbf{q}_b^T \mathbf{P}^{-1}(\zeta)\mathbf{q}_b/2 - \mathbf{q}_b^T \mathbf{S}(\zeta)\mathbf{q}_b/2] = 0$$

Therefore

$$\mathbf{q}_b(\zeta, \zeta) = -\left[\mathbf{P}^{-1}(\zeta) + \mathbf{S}(\zeta)\right]^{-1} \cdot \mathbf{k}_b \tag{5.10.13}$$

Solving $\mathbf{q}_b(\zeta, \zeta)$ means that the initial conditions at $z = \zeta$ have been given for the two intervals $z_0 \le z < \zeta$ and $\zeta < z \le z_f$, and then the dual vectors $\mathbf{q}_b(z, \zeta)$ and $\mathbf{p}_b(z, \zeta)$ can be solved for both intervals. Integrating equation (5.10.3a) gives $\mathbf{p}_b(\zeta + 0, \zeta) - \mathbf{p}_b(\zeta - 0, \zeta) = \mathbf{k}_b$. When $\mathbf{k}_b$ turns out to be the n column vectors of a unit matrix $\mathbf{I}_n$, and the n solved vectors $\mathbf{q}_b(\zeta, \zeta)$ compose a $n \times n$ matrix as

$$\mathbf{Q}_b(\zeta, \zeta) = -[\mathbf{P}^{-1}(\zeta) + \mathbf{S}(\zeta)]^{-1} \tag{5.10.14}$$

This equation gives response displacements at the point ζ of unit force, which can be used as initial condition to solve dual vectors $\mathbf{q}_b(z, \zeta)$ and $\mathbf{p}_b(z, \zeta)$ in the two contiguous intervals, totally n groups. Thereafter, two $n \times n$ matrices $\mathbf{Q}_b(z, \zeta)$ and $\mathbf{P}_b(z, \zeta)$ are composed for both the intervals $z_0 \le z < \zeta$ and $\zeta < z \le z_f$.

According to symmetry and positive definiteness of the Riccati matrices (solutions of Riccati equations) $\mathbf{S}(\zeta)$ and $\mathbf{P}(\zeta)$, the symmetry and negative definiteness of the matrices $\mathbf{Q}_b(\zeta, \zeta)$ and $\mathbf{P}_a(\zeta, \zeta)$ readily follows. The matrix-functions $\mathbf{Q}_b(z, \zeta)$ and $\mathbf{P}_a(z, \zeta)$ are functions of single continuous coordinate z and are continuous at $z = \zeta$, but $\mathbf{Q}_a(z, \zeta)$ and $\mathbf{P}_b(z, \zeta)$ are discontinuous. Hence, one must distinguish

$$\mathbf{Q}_a(\zeta + 0, \zeta) = -\mathbf{S}^{-1}(\zeta)\mathbf{P}_a(\zeta, \zeta) = [\mathbf{I} + \mathbf{P}\mathbf{S}]_{\zeta}^{-1} \tag{5.10.15a}$$

and

$$\mathbf{Q}_a(\zeta - 0, \zeta) = \mathbf{P}(\zeta) \cdot \mathbf{P}_a(\zeta, \zeta) = -[\mathbf{I} + \mathbf{S}^{-1}\mathbf{P}^{-1}]_{\zeta}^{-1} \tag{5.10.15b}$$

Equation $\mathbf{Q}_a(\zeta + 0, \zeta) - \mathbf{Q}_a(\zeta - 0, \zeta) = \mathbf{I}_n$ is readily verified based on the identity $(\mathbf{I} + \mathbf{Y})^{-1} + \mathbf{Y}(\mathbf{I} + \mathbf{Y})^{-1} \equiv \mathbf{I}$, where $\mathbf{Y} = \mathbf{P}\mathbf{S}$ is selected. Similarly,

$$\begin{aligned} \mathbf{P}_b(\zeta - 0, \zeta) &= \mathbf{P}^{-1}\mathbf{Q}_b(\zeta, \zeta) = -[\mathbf{I} + \mathbf{S}\mathbf{P}]_{\zeta}^{-1} \\ \mathbf{P}_b(\zeta + 0, \zeta) &= [\mathbf{I} + \mathbf{P}^{-1}\mathbf{S}^{-1}]_{\zeta}^{-1} \end{aligned} \tag{5.10.16}$$

The difference between them is also $\mathbf{I}_n$ as anticipated.

Having found the matrices $\mathbf{Q}_a(\zeta-0,\zeta)$, $\mathbf{Q}_a(\zeta+0,\zeta)$, $\mathbf{Q}_b(\zeta,\zeta)$, $\mathbf{P}_a(\zeta,\zeta)$ as well as $\mathbf{P}_b(\zeta-0,\zeta)$ and $\mathbf{P}_b(\zeta+0,\zeta)$, the influence functions $\mathbf{Q}_a(z,\zeta)$, $\mathbf{P}_a(z,\zeta)$; $\mathbf{Q}_b(z,\zeta)$ and $\mathbf{P}_b(z,\zeta)$ at arbitrary coordinate z are to be found next. The precise integration method is applied also for this purpose. Based on these $n\times n$ matrix functions, an $2n\times 2n$ ***impulse influence matrix function***

$$\mathbf{\Phi}(z,\zeta) \underset{\text{def}}{=} \begin{bmatrix} \mathbf{Q}_a & \mathbf{Q}_b \\ \mathbf{P}_a & \mathbf{P}_b \end{bmatrix} \tag{5.10.17}$$

is composed. The differential equation for $\mathbf{\Phi}(z,\zeta)$ should be clarified. Matrices $\mathbf{Q}_a(z,\zeta)$, $\mathbf{P}_a(z,\zeta)$ are the influence of first equation set (5.10.2a,b) under unit concentrated force $\mathbf{I}_n\delta(z-\zeta)$; and $\mathbf{Q}_b(z,\zeta)$, $\mathbf{P}_b(z,\zeta)$ are the influence of second equation set (5.10.3a,b) under the action of $\mathbf{I}_n\delta(z-\zeta)$. The combined differential equation is

$$\dot{\mathbf{\Phi}}(z,\zeta) = \mathbf{H}(z)\mathbf{\Phi}(z,\zeta) + \mathbf{I}_{2n}\delta(z-\zeta), \quad \mathbf{H}(z) = \begin{bmatrix} \mathbf{A}(z) & \mathbf{D}(z) \\ \mathbf{B}(z) & -\mathbf{A}^T(z) \end{bmatrix} \tag{5.10.18}$$

where the system matrix is written as $\mathbf{H}(z)$, which implies that the impulse influence matrix function is defined for variable coefficient system.

§5.10.1, Reciprocal theorems of the impulse influence matrix functions

The sub-matrices of impulse influence matrix function hold three reciprocal theorems. There must be two sets of forces for reciprocal theorems. Suppose the first set of forces acts at the point $z=\tau$, and the second set of forces acts at the point $z=\zeta$. The differential equations for the sub-matrices, as functions of coordinate z, are

$$\dot{\mathbf{Q}}_a(z,\tau) = \mathbf{A}(z)\mathbf{Q}_a(z,\tau) + \mathbf{D}(z)\mathbf{P}_a(z,\tau) + \mathbf{I}_n\delta(z-\tau) \tag{5.10.19a}$$

$$\dot{\mathbf{P}}_a(z,\tau) = \mathbf{B}(z)\mathbf{Q}_a(z,\tau) - \mathbf{A}^T(z)\mathbf{P}_a(z,\tau) \tag{5.10.19b}$$

$$\dot{\mathbf{Q}}_b(z,\tau) = \mathbf{A}(z)\mathbf{Q}_b(z,\tau) + \mathbf{D}(z)\mathbf{P}_b(z,\tau) \tag{5.10.19c}$$

$$\dot{\mathbf{P}}_b(z,\tau) = \mathbf{B}(z)\mathbf{Q}_b(z,\tau) - \mathbf{A}^T(z)\mathbf{P}_b(z,\tau) + \mathbf{I}_n\delta(z-\tau) \tag{5.10.19d}$$

and

$$\dot{\mathbf{Q}}_a(z,\zeta) = \mathbf{A}(z)\mathbf{Q}_a(z,\zeta) + \mathbf{D}(z)\mathbf{P}_a(z,\zeta) + \mathbf{I}_n\delta(z-\zeta) \tag{5.10.20a}$$

$$\dot{\mathbf{P}}_a(z,\zeta) = \mathbf{B}(z)\mathbf{Q}_a(z,\zeta) - \mathbf{A}^T(z)\mathbf{P}_a(z,\zeta) \tag{5.10.20b}$$

$$\dot{\mathbf{Q}}_b(z,\zeta) = \mathbf{A}(z)\mathbf{Q}_b(z,\zeta) + \mathbf{D}(z)\mathbf{P}_b(z,\zeta) \tag{5.10.20c}$$

$$\dot{\mathbf{P}}_b(z,\zeta) = \mathbf{B}(z)\mathbf{Q}_b(z,\zeta) - \mathbf{A}^T(z)\mathbf{P}_b(z,\zeta) + \mathbf{I}_n\delta(z-\zeta) \tag{5.10.20d}$$

The boundary conditions are

$$\begin{aligned} &\mathbf{P}_a(0,\tau) = \mathbf{P}_0^{-1}\mathbf{Q}_a(0,\tau), \quad \mathbf{P}_b(z_f,\tau) = -\mathbf{S}_f\mathbf{Q}_b(z_f,\tau) \\ &\mathbf{P}_a(0,\zeta) = \mathbf{P}_0^{-1}\mathbf{Q}_a(0,\zeta), \quad \mathbf{P}_b(z_f,\zeta) = -\mathbf{S}_f\mathbf{Q}_b(z_f,\zeta) \end{aligned} \tag{5.10.21}$$

Writing the matrices as $\mathbf{A}(z)$ etc. explicitly means that the theorems are valid also for variant coefficient system.

Carrying out the integration

$$\int_{z_0}^{z_f} [\mathbf{P}_a^T(z,\zeta)\times(19\text{a}) - \mathbf{P}_a^T(z,\zeta)\times(19\text{b}) + (20\text{b})^T \times \mathbf{Q}_a(z,\tau) - (20\text{a})^T \times \mathbf{P}_a(z,\tau)]\mathrm{d}z$$

gives [comment: where (19a) represents (5.10.19a), the same below]

$$\mathbf{P}_a^T(\tau,\zeta) - \mathbf{P}_a(\zeta,z) = \int_{z_0}^{z_f}\left[\mathrm{d}\left(\mathbf{P}_a^T(z,\zeta)\mathbf{Q}_a(z,\tau)\right) - \mathrm{d}\left(\mathbf{Q}_a^T(z,\zeta)\mathbf{P}_a(z,\tau)\right)\right] = \mathbf{0}$$

where the last equality is based on the boundary condition. The above equation gives the ***first reciprocal theorem***. Equation (5.10.12) proposing a symmetric matrix is a special case of the theorem at $\zeta = \tau$.

Next, the following integration is to compute

$$\int_{z_0}^{z_f}[\mathbf{P}_b^T(z,\zeta)\times(19\text{c}) - \mathbf{Q}_b^T(z,\zeta)\times(19\text{d}) + (20\text{d})^T \times \mathbf{Q}_b(z,\tau) - (20\text{c})^T \times \mathbf{P}_b(z,\tau)]\mathrm{d}z$$

Using boundary conditions gives

$$\mathbf{Q}_b(\zeta,\tau) - \mathbf{Q}_b^T(\tau,\zeta) = \int_{z_0}^{z_f}\left[\mathrm{d}\left(\mathbf{P}_b^T(z,\zeta)\mathbf{Q}_b(z,\tau)\right) - \mathrm{d}\left(\mathbf{Q}_b^T(z,\zeta)\mathbf{P}_b(z,\tau)\right)\right] = \mathbf{0}$$

which is the ***second reciprocal theorem.*** Equation (5.10.14) proposing a symmetric matrix is a special case at $\zeta = \tau$ of the theorem.

Computing the integration

$$\int_{z_0}^{z_f}[\mathbf{P}_b^T(z,\zeta)\times(19\text{a}) - \mathbf{Q}_b^T(z,\zeta)\times(19\text{b}) + (20\text{d})^T \times \mathbf{Q}_a(z,\tau) - (20\text{c})^T \times \mathbf{P}_a(z,\tau)]\mathrm{d}z$$

then using the boundary conditions, it derives

$$\mathbf{P}_b^T(\zeta,\tau) + \mathbf{Q}_a(\tau,\zeta) = \int_{z_0}^{z_f}\left[\mathrm{d}\left(\mathbf{P}_b^T(z,\zeta)\mathbf{Q}_a(z,\tau)\right) - \mathrm{d}\left(\mathbf{Q}_b^T(z,\zeta)\mathbf{P}_a(z,\tau)\right)\right] = \mathbf{0}$$

which is the ***third reciprocal theorem***. Integrating

$$\int_0^{t_f}[\mathbf{P}_b^T(z,\zeta)\times(20\text{a}) - \mathbf{Q}_b^T(z,\tau)\times(20\text{b}) + (19\text{d})^T \times \mathbf{Q}_a(z,\zeta) - (19\text{c})^T \times \mathbf{P}_a(z,\zeta)]\mathrm{d}z$$

gives $\mathbf{P}_b^T(\tau,\zeta) + \mathbf{Q}_a(\zeta,\tau) = \mathbf{0}$, the third reciprocal theorem. All these can concisely be expressed as

$$\mathbf{P}_a^T(\tau,\zeta) = \mathbf{P}_a(\zeta,\tau) \tag{5.10.22}$$

$$\mathbf{Q}_b(\zeta,\tau) = \mathbf{Q}_b^T(\tau,\zeta) \tag{5.10.23}$$

$$\mathbf{Q}_a^T(\tau,\zeta) = -\mathbf{P}_b(\zeta,\tau) \tag{5.10.24}$$

To express in terms of the ***impulse influence matrix function***, let

$$\boldsymbol{\Phi}(z,\tau) \underset{\text{def}}{=} \begin{bmatrix} \mathbf{Q}_a(z,\tau) & \mathbf{Q}_b(z,\tau) \\ \mathbf{P}_a(z,\tau) & \mathbf{P}_b(z,\tau) \end{bmatrix} \left(= \begin{bmatrix} \boldsymbol{\Phi}_{11}(z,\tau) & \boldsymbol{\Phi}_{12}(z,\tau) \\ \boldsymbol{\Phi}_{21}(z,\tau) & \boldsymbol{\Phi}_{22}(z,\tau) \end{bmatrix} \right) \tag{5.10.25}$$

It is easy to verify that

$$\mathbf{J}\boldsymbol{\Phi}^T(z,\tau)\mathbf{J} = \begin{bmatrix} \mathbf{Q}_a(\tau,z) & \mathbf{Q}_b(\tau,z) \\ \mathbf{P}_a(\tau,z) & \mathbf{P}_b(\tau,z) \end{bmatrix} = \boldsymbol{\Phi}(\tau,z),$$

or given as

$$\begin{aligned} [\mathbf{J}\boldsymbol{\Phi}(z,\tau)]^T &= \mathbf{J}\boldsymbol{\Phi}(\tau,z) \\ [\boldsymbol{\Phi}(z,\tau)\mathbf{J}]^T &= \boldsymbol{\Phi}(\tau,z)\mathbf{J} \end{aligned} \tag{5.10.26}$$

The matrix $\mathbf{J}\boldsymbol{\Phi}(z,\tau)$ will appear frequently and the equation (5.10.26) expresses its ***reciprocal symmetry***. The general theorem proved above is very important however, numerical result is necessary for applications. The precise integration numerical computation for a coordinate invariant system is given in the next section. The ***impulse influence matrix function***, as described in previous section, satisfies the differential equation (5.10.18).

The impulse influence function can be used to solve the non-homogeneous dual differential equations (5.2.9a,b). Let the initial value $\hat{\mathbf{q}}_0$ be substituted by the initial impulse $\mathbf{f}_{q0} = \hat{\mathbf{q}}_0\delta(z - z_0)$, $\mathbf{f}_{p0} = \mathbf{0}$ and be superimposed on $\mathbf{f}_q$, $\mathbf{f}_p$, then the solution of equations (5.2.9a,b) can be expressed as

$$\begin{Bmatrix} \mathbf{q}(z) \\ \mathbf{p}(z) \end{Bmatrix} = \int_{z_0}^{z_f} \mathbf{\Phi}(z,\tau) \begin{Bmatrix} \mathbf{f}_q(\tau) \\ \mathbf{f}_p(\tau) \end{Bmatrix} d\tau \tag{5.10.27}$$

The solution of non-homogeneous dual differential equation is very useful for sub-system analysis and system identification, etc.

The positive definiteness of the diagonal matrices $\mathbf{Q}_b(z,z)$ and/or $\mathbf{P}_a(z,z)$ are clearly seen from the equations (5.10.14) and (5.10.12) based on the positive definiteness of the solution matrices of the Riccati differential equations $\mathbf{S}(z)$ and $\mathbf{P}(z)$. The positive definiteness of the matrices

$$\mathbf{K}_q(z,\zeta) \underset{\text{def}}{=} \begin{bmatrix} \mathbf{Q}_b(z,z) & \mathbf{Q}_b(z,\zeta) \\ \mathbf{Q}_b(\zeta,z) & \mathbf{Q}_b(\zeta,\zeta) \end{bmatrix} \tag{5.1028a}$$

$$\mathbf{K}_p(z,\zeta) \underset{\text{def}}{=} \begin{bmatrix} \mathbf{P}_a(z,z) & \mathbf{P}_a(z,\zeta) \\ \mathbf{P}_a(\zeta,z) & \mathbf{P}_a(\zeta,\zeta) \end{bmatrix} \tag{5.10.28b}$$

should be proved further, where $z \neq \zeta$. Based on the reciprocal symmetry theorem the symmetry of the two matrices is readily verified. To prove positive definiteness of the matrix $\mathbf{K}_q(z,\zeta)$, the positive definiteness of the strain energy is required,

$$E = \int_{z_0}^{z_f} [-\mathbf{p}^T\dot{\mathbf{q}} + \mathbf{p}^T\mathbf{A}\mathbf{q} - \mathbf{q}^T\mathbf{B}\mathbf{q}/2 + \mathbf{p}^T\mathbf{D}\mathbf{p}/2]\mathrm{d}z + \mathbf{q}_f^T\mathbf{S}_f\mathbf{q}_f/2 + \mathbf{q}_0^T\mathbf{P}_0^{-1}\mathbf{q}_0/2 > 0$$

which is always valid for structural mechanics. In optimal control problem, the combination of ***controllability and observability*** ensures the positive definiteness of the "strain energy", see section 6.7.

Denoting z and ζ as ζ_1 and ζ_2, respectively, and selecting the component numbers j_1 and j_2, j_3, compose the dual differential equations (5.10.3a,b) as

$$\begin{aligned} &\dot{\mathbf{q}} = \mathbf{A}\mathbf{q} + \mathbf{D}\mathbf{p} \\ &\dot{\mathbf{p}} = \mathbf{B}\mathbf{q} - \mathbf{A}^T\mathbf{p} + \mathbf{k}_1\delta(z-\zeta_1) + \mathbf{k}_2\delta(z-\zeta_2) \\ &\mathbf{k}_1 = \{0,\cdots,\underset{j_1}{a_1},\cdots,0\}^T,\ \mathbf{k}_2 = \{0,\cdots,\underset{j_2}{a_2},\cdots,\underset{j_3}{a_3},\cdots,0\}^T \end{aligned} \tag{5.10.3'}$$

where the arbitrary load factors a_1 and a_2, a_3 are located at j_1 and j_2, j_3, respectively. According to (5.10.27) the solution can be given as

$$\mathbf{q}(z) = \mathbf{Q}_b(z,\zeta_1)\cdot\mathbf{k}_1 + \mathbf{Q}_b(z,\zeta_2)\cdot\mathbf{k}_2$$

$$\mathbf{p}(z) = \mathbf{P}_b(z,\zeta_1)\cdot\mathbf{k}_1 + \mathbf{P}_b(z,\zeta_2)\cdot\mathbf{k}_2$$

Substituting this solution into the integration of positive definite strain energy E. Based on the dual equations (5.10.3a,b') and boundary conditions, the strain energy E is computed as

$$2E = \int_{z_0}^{z_f} \mathbf{q}^T[\mathbf{k}_1\delta(z-\zeta_1)+\mathbf{k}_2\delta(z-\zeta_2)]\mathrm{d}z = \left[\mathbf{q}^T(\zeta_1)\mathbf{k}_1 + \mathbf{q}^T(\zeta_2)\mathbf{k}_2\right]$$

$$= \mathbf{k}_1^T \cdot \mathbf{Q}_b(\zeta_1,\zeta_1)\mathbf{k}_1 + \mathbf{k}_2^T\mathbf{Q}_b(\zeta_2,\zeta_1)\mathbf{k}_1 + \mathbf{k}_1^T\mathbf{Q}_b(\zeta_1,\zeta_2)\mathbf{k}_2 + \mathbf{k}_2^T\mathbf{Q}_b(\zeta_2,\zeta_2)\mathbf{k}_2 > 0$$

which is a positive quadratic function of arbitrary parameters a_1 and a_2, a_3. It implies positive definiteness of the matrix

$$\begin{bmatrix} \mathbf{Q}_{b\mathrm{j}_1\mathrm{j}_1}(\zeta_1,\zeta_1) & \mathbf{Q}_{b\mathrm{j}_1\mathrm{j}_2}(\zeta_1,\zeta_2) & \mathbf{Q}_{b\mathrm{j}_1\mathrm{j}_3}(\zeta_1,\zeta_2) \\ \mathbf{Q}_{b\mathrm{j}_2\mathrm{j}_1}(\zeta_2,\zeta_1) & \mathbf{Q}_{b\mathrm{j}_2\mathrm{j}_2}(\zeta_2,\zeta_2) & \mathbf{Q}_{b\mathrm{j}_2\mathrm{j}_3}(\zeta_2,\zeta_2) \\ \mathbf{Q}_{b\mathrm{j}_3\mathrm{j}_1}(\zeta_2,\zeta_1) & \mathbf{Q}_{b\mathrm{j}_3\mathrm{j}_2}(\zeta_2,\zeta_2) & \mathbf{Q}_{b\mathrm{j}_3\mathrm{j}_3}(\zeta_2,\zeta_2) \end{bmatrix}$$

This behavior is valid for arbitrary component numbers $\mathrm{j}_1, \mathrm{j}_2, \mathrm{j}_3 \le n$ and arbitrary parameters a_1 and a_2, a_3. Although the positive definiteness is proved for 3-D subspace, but the method applies to arbitrary dimensions, so that the positive definiteness of the matrix $\mathbf{K}_q(z,\zeta)$ is proved.

The positive definiteness of the matrix $\mathbf{K}_p(z,\zeta)$ can also be similarly proved.

§5.10.2, Precise integration of impulse influence matrix function

The equations (5.10.12) and (5.10.14) presented two sub-matrices at the force acting point ζ, but the ***impulse influence matrix function*** $\mathbf{\Phi}(z,\zeta)$ needs to compute all the sub-matrices $\mathbf{Q}_a(z,\zeta)$, $\mathbf{Q}_b(z,\zeta)$, $\mathbf{P}_a(z,\zeta)$ and $\mathbf{P}_b(z,\zeta)$ at any coordinate z. Two cases of $z<\zeta$ and $z>\zeta$ should be computed separately. The $n \times n$ system matrices $\mathbf{A}, \mathbf{B}$ and $\mathbf{D}$ are assumed coordinate-invariant. First the differential equation of the submatrix $\mathbf{P}_a(z,\zeta)$ is derived as follows, substituting the matrix form of equation (5.10.4')

$$\mathbf{Q}_a(z,\zeta) = \mathbf{P}(z)\cdot\mathbf{P}_a(z,\zeta)$$

into the equation (5.10.19b), then taking transpose gives

$$\dot{\mathbf{P}}_a^T(z,\zeta) = \mathbf{P}_a^T(z,\zeta)\cdot[\mathbf{P}(z)\mathbf{B}-\mathbf{A}]$$

Based on the inverse matrix derivative rule $\mathrm{d}(\mathbf{X}^{-1})/\mathrm{d}t = -\mathbf{X}^{-1}\dot{\mathbf{X}}\mathbf{X}^{-1}$ gives

$$\dot{\mathbf{L}}_a(z,\zeta) = [\mathbf{A}-\mathbf{P}(z)\mathbf{B}]\mathbf{L}_a(z,\zeta), \qquad \text{where } \mathbf{L}_a = \mathbf{P}_a^{-T} \tag{5.10.29}$$

Equation (5.10.12) gives $\mathbf{L}_a(\zeta,\zeta) = \mathbf{P}_a^{-T}(\zeta,\zeta) = -[\mathbf{P}(\zeta)+\mathbf{S}^{-1}(\zeta)]$, *i.e.* the initial condition. Equation (5.10.19) is a ***variant coefficient*** ODE. Usually a variant coefficient differential equation is difficult to solve analytically. However, the equation (5.10.29) is derived from a ***coordinate-invariant*** system, solving Riccati differential equation by means of the precise integration method as given in section 5.7, the $n \times n$ matrices $\mathbf{F}(\eta), \mathbf{G}(\eta), \mathbf{Q}(\eta)$ are computed simultaneously, where in the interval $z<\zeta$ the matrix $\mathbf{F}_\mathrm{f}(\eta)$ satisfies the differential equation (5.7.16). To express the interval $z<\zeta$ of filter type, the subscript $_\mathrm{f}$ is used

$$\dot{\mathbf{F}}_\mathrm{f}(z) = [\mathbf{A}-\mathbf{P}(z)\mathbf{B}]\mathbf{F}_\mathrm{f}(z), \qquad \text{initial condition: } \mathbf{F}_\mathrm{f}(0) = \mathbf{I}_n \tag{5.10.29'}$$

This differential equation is just (5.10.29). Therefore

$$\mathbf{L}_a(z,\zeta)=\mathbf{P}_a^{-T}(z,\zeta)=-\mathbf{F}_f(z)\mathbf{F}_f^{-1}(\zeta)[\mathbf{P}(\zeta)+\mathbf{S}^{-1}(\zeta)]$$

And then

$$\begin{aligned}\mathbf{P}_a(z,\zeta)&=-\mathbf{F}_f^{-T}(z)\mathbf{F}_f^{T}(\zeta)[\mathbf{P}(\zeta)+\mathbf{S}^{-1}(\zeta)]^{-1}\\ \mathbf{Q}_a(z,\zeta)&=-\mathbf{P}(z)\mathbf{F}_f^{-T}(z)\mathbf{F}_f^{T}(\zeta)[\mathbf{P}(\zeta)+\mathbf{S}^{-1}(\zeta)]^{-1}\end{aligned},\quad z<\zeta \tag{5.10.30}$$

Next step is to solve $\mathbf{Q}_b(z,\zeta)=-\mathbf{P}(z)\cdot\mathbf{P}_b(z,\zeta)$ for $z<\zeta$. Differential equation is still (5.10.29). According to equation (5.10.14), the initial condition is $\mathbf{L}_b(\zeta-0,\zeta)=\mathbf{P}_b^{-T}(\zeta-0,\zeta)=[\mathbf{I}+\mathbf{P}(\zeta)\mathbf{S}(\zeta)]$. Therefore it solves

$$\begin{aligned}\mathbf{P}_b(z,\zeta)&=-\mathbf{F}_f^{-T}(z)\mathbf{F}_f^{T}(\zeta)[\mathbf{I}+\mathbf{S}(\zeta)\mathbf{P}(\zeta)]^{-1}\\ \mathbf{Q}_b(z,\zeta)&=-\mathbf{P}(z)\mathbf{F}_f^{-T}(z)\mathbf{F}_f^{T}(\zeta)[\mathbf{I}+\mathbf{S}(\zeta)\mathbf{P}(\zeta)]^{-1}\end{aligned},\quad z<\zeta \tag{5.10.31}$$

It needs to find the solutions in the interval $z>\zeta$. When solving the Riccati matrix $\mathbf{S}(z)$ using precise integration method in section 5.7, the matrix $\mathbf{F}_c(z)$ is obtained simultaneously, for which the differential equation is

$$\mathrm{d}\mathbf{F}_c(z)/\mathrm{d}z=-\mathbf{F}_c(z)\times[\mathbf{A}-\mathbf{DS}(z)]\ ,\quad \mathbf{F}_c(0)=\mathbf{I}_n$$

from which, it derives the differential equation

$$\mathrm{d}(\mathbf{F}_c^{-1})/\mathrm{d}z=[\mathbf{A}-\mathbf{DS}(z)]\mathbf{F}_c^{-1},\quad \mathbf{F}_c^{-1}(0)=\mathbf{I}_n \tag{5.10.32}$$

On the other hand, writing the equation (5.10.9) in matrix form gives

$$\dot{\mathbf{Q}}_a(z,\zeta)=[\mathbf{A}-\mathbf{DS}(z)]\mathbf{Q}_a(z,\zeta) \tag{5.10.9'}$$

and $\mathbf{Q}_a(\zeta+0,\zeta)=-\mathbf{S}^{-1}(\zeta)\mathbf{P}_a(\zeta,\zeta)=[\mathbf{I}+\mathbf{PS}]_\zeta^{-1}$ gives the initial condition (5.10.15a). So

$$\begin{aligned}\mathbf{Q}_a(z,\zeta)&=\mathbf{F}_c^{-1}(z)\mathbf{F}_c(\zeta)[\mathbf{I}+\mathbf{P}(\zeta)\mathbf{S}(\zeta)]^{-1}\\ \mathbf{P}_a(z,\zeta)&=-\mathbf{S}(z)\mathbf{F}_c^{-1}(z)\mathbf{F}_c(\zeta)[\mathbf{I}+\mathbf{P}(\zeta)\mathbf{S}(\zeta)]^{-1}\end{aligned},\quad z>\zeta \tag{5.10.33}$$

Similarly

$$\begin{aligned}\mathbf{Q}_b(z,\zeta)&=-\mathbf{F}_c^{-1}(z)\mathbf{F}_c(\zeta)[\mathbf{P}^{-1}(\zeta)+\mathbf{S}(\zeta)]^{-1}\\ \mathbf{P}_b(z,\zeta)&=\mathbf{S}(z)\mathbf{F}_c^{-1}(z)\mathbf{F}_c(\zeta)[\mathbf{P}^{-1}(\zeta)+\mathbf{S}(\zeta)]^{-1}\end{aligned},\quad z>\zeta \tag{5.10.34}$$

Up to here, the four sub-matrices of the impulse influence matrix function have all been solved by the precise integration method.

The three reciprocal theorems should be verified. However, only the check of consistency of the three reciprocal theorems is verified below. Substituting $\mathbf{P}_a$ into the first reciprocal theorem, equation (5.10.22) and, noting the variable substitution rule gives

$$-\mathbf{F}_f^{-T}(\zeta)\mathbf{F}_f^{T}(\tau)[\mathbf{P}(\tau)+\mathbf{S}^{-1}(\tau)]^{-1}=-[\mathbf{I}+\mathbf{P}(\zeta)\mathbf{S}(\zeta)]^{-T}\mathbf{F}_c^{T}(\zeta)\mathbf{F}_c^{-T}(\tau)\mathbf{S}(\tau)$$

On the other hand, the second reciprocal theorem gives

$$-\mathbf{P}(\zeta)\mathbf{F}_f^{-T}(\zeta)\mathbf{F}_f^{T}(\tau)[\mathbf{I}+\mathbf{S}(\tau)\mathbf{P}(\tau)]^{-1}=-[\mathbf{P}^{-1}(\zeta)+\mathbf{S}(\zeta)]^{-1}\mathbf{F}_c^{T}(\zeta)\mathbf{F}_c^{-T}(\tau)$$

To check that the above two equations are consistent to each other, left multiplying $\mathbf{P}(\zeta)$ and right multiplying $\mathbf{S}^{-1}(\tau)$ the equation of first reciprocal theorem derives the equation of the second reciprocal theorem. The third reciprocal theorem gives

$$-[\mathbf{P}(\zeta)+\mathbf{S}^{-1}(\zeta)]^{-1}\mathbf{F}_f(\zeta)\mathbf{F}_f^{-1}(\tau)\mathbf{P}(\tau)=-\mathbf{S}(\zeta)\mathbf{F}_c^{-1}(\zeta)\mathbf{F}_c(\tau)[\mathbf{P}^{-1}(\tau)+\mathbf{S}(\tau)]^{-1}$$

This equality should also be consistent with the previous two. Taking transpose, left multiplying $\mathbf{P}^{-1}(\tau)$ and interchanging τ and ζ, the first reciprocal theorem equation is obtained. Then the consistency is checked for the three reciprocal theorems. ##

The above explicit computational equations are only for coordinate-invariant systems. These reciprocal equations imply the mutual relation between the matrix functions $\mathbf{F}_c(z)$ and $\mathbf{F}_f(z)$. Evidently, the computation of matrix functions $\mathbf{P}(z)$, $\mathbf{S}(z)$, $\mathbf{F}_c(z)$ and $\mathbf{F}_f(z)$ is the key step. The precise integration method proposed in section 5.7 solves the computational problem. But directly using these equations (5.10.30~31) and (5.10.33~34) in computation has still a pitfall that the inverse matrices $\mathbf{F}_c^{-1}(z)$ and $\mathbf{F}_f^{-1}(z)$ appear in the equations. When z becomes large, both $\mathbf{F}_c(z)$ and $\mathbf{F}_f(z)$ tend to be null matrices and the inversion will cause numerical ill-conditioning in computation. To bypass such problem, the strategy of stepwise go forward can be used. The whole interval (z_0, z_f) has been subdivided into N_{int} fundamental intervals with length $\eta = (z_f - z_0)/N_{\text{int}}$. Note that the forms $\mathbf{F}_f(\zeta)\mathbf{F}_f^{-1}(z)$ and $\mathbf{F}_c^{-1}(\zeta)\mathbf{F}_c(z)$, ($\zeta > z$) always appear simultaneously, hence the matrices $\mathbf{F}_c^{-1}(z+\eta)\mathbf{F}_c(z)$ and $\mathbf{F}_f(z+\eta)\mathbf{F}_f^{-1}(z)$ can be computed and stored at the station z. The two matrices are interrelated to each other, that using the first reciprocal theorem derives

$$\begin{aligned}&[\mathbf{P}(z+\eta)\mathbf{S}(z+\eta)+\mathbf{I}]^{-1}\mathbf{F}_f(z+\eta)\mathbf{F}_f^{-1}(z)\\&\quad=\mathbf{F}_c^{-1}(z+\eta)\mathbf{F}_c(z)[\mathbf{I}+\mathbf{P}(z)\mathbf{S}(z)]^{-1}\end{aligned} \tag{5.10.35}$$

The precise integration of the sub-matrices of mixed energy needs to compute the sub-matrices $\mathbf{Q}(\eta), \mathbf{G}(\eta), \mathbf{F}(\eta)$ for a single fundamental interval η and the matrices $\mathbf{S}$ and $\mathbf{P}$ of the Riccati differential equations. Using the interval combination (5.7.23a) gives

$$\mathbf{F}_c(z) = \mathbf{F}_c(z+\eta)[\mathbf{I}+\mathbf{G}(\eta)\mathbf{S}(z+\eta)]^{-1}\mathbf{F}(\eta)$$

which derives

$$\mathbf{F}_c^{-1}(z+\eta)\mathbf{F}_c(z) \underset{\text{def}}{=} \mathbf{F}_S(z+\eta, z) = [\mathbf{I}+\mathbf{G}(\eta)\mathbf{S}(z+\eta)]^{-1}\mathbf{F}(\eta)\,, \quad \zeta < z \tag{5.10.36a}$$

Similarly

$$\mathbf{F}_f(z+\eta)\mathbf{F}_f^{-1}(z) \underset{\text{def}}{=} \mathbf{F}_P(z+\eta, z) = \mathbf{F}(\eta)[\mathbf{I}_n+\mathbf{P}(z)\mathbf{Q}(\eta)]^{-1}\,, \quad \zeta < z \tag{5.10.36b}$$

All of these equations explain that the individual matrices $\mathbf{F}_f$ and $\mathbf{F}_c$ are unnecessary to compute, only after the mixed energy submatrices $\mathbf{Q}(\eta), \mathbf{G}(\eta), \mathbf{F}(\eta)$ of a single step size η and the Riccati matrices $\mathbf{P}(z), \mathbf{S}(z)$ are computed, then the computation of (5.10.36a,b) is to perform. Such algorithm can be free from the unnecessary numerical ill-conditioning mentioned above.

§5.10.3, Numerical example of impulse influence matrix function

Example 5.8: Suppose $n=4$, the system matrices $\mathbf{A}, \mathbf{B}, \mathbf{D}$ are given as

$$\mathbf{A}=\begin{bmatrix}0&0&1&0\\0&0&0&1\\-2&1&0&0\\0.5&-.5&0&0\end{bmatrix},\quad \mathbf{B}=\begin{bmatrix}2&-1&0&0\\-1&1&0&0\\0&0&1&0\\0&0&0&2\end{bmatrix},\quad \begin{matrix}\mathbf{D}=\mathrm{diag}[0,0,0,0.25]\\ \mathbf{S}_\mathrm{f}=0.8\times\mathbf{I}_4\\ \mathbf{P}_0=0.01\times\mathbf{I}_4\end{matrix},\quad \begin{matrix}z_0=0\\ z_\mathrm{f}=4.8\end{matrix}$$

$N_\mathrm{int}=8$ uniform fundamental intervals are selected with totally 9 stations, step size $\eta=0.6$.

Solution: According to these data, using the precise integration method compute the Riccati matrices $\mathbf{P}(z),\mathbf{S}(z)$, and the combined matrices $\mathbf{F}_c^{-1}(z+\eta)\mathbf{F}_c(z)$ and $\mathbf{F}_\mathrm{f}(z+\eta)\mathbf{F}_\mathrm{f}^{-1}(z)$. Then according to the formulas derived above, the ***impulse influence matrix functions*** $\mathbf{\Phi}(z,\zeta)$ at all stations are computed next. Both the force acting point ζ and response points z have all nine stations. There are too much data, which cannot be listed exhaustively. It is checked that the reciprocal symmetric property fulfills perfectly. Part of the numerical result is listed below.

For $\zeta=1.2, z=1.8$:

$$\mathbf{Q}_a=\begin{bmatrix}0.07883&0.30581&0.58639&-.04619\\0.49933&0.37676&0.04520&0.18014\\-1.10129&0.39596&-.09067&0.10277\\0.13921&-.30450&0.24182&0.11922\end{bmatrix},\quad \mathbf{Q}_b=\begin{bmatrix}0.01227&0.05573&0.03492&0.04555\\0.01513&0.07396&0.02920&0.08715\\-.00357&0.03037&0.00575&0.05004\\-.00056&-.00720&0.00145&-.00019\end{bmatrix}$$

$$\mathbf{P}_a=\begin{bmatrix}-.66924&0.65917&2.95551&-.84519\\0.82821&0.17219&-1.3120&0.80800\\-3.0575&1.26814&-.44357&0.40347\\0.95667&-.97658&0.62629&0.62808\end{bmatrix},\quad \mathbf{P}_b=\begin{bmatrix}0.02636&0.09271&0.10637&-.00904\\0.00797&0.06985&-.00718&0.15208\\-.00721&0.10816&0.01988&0.17599\\0.00490&0.00815&0.01241&0.05105\end{bmatrix}$$

and for $\zeta=1.8, z=1.2$:

$$\mathbf{Q}_a=\begin{bmatrix}-.02636&-.00797&0.00721&-.00490\\-.09271&-.06985&-.10816&-.00815\\-.10637&0.00718&-.01988&-.01241\\0.00904&-.15208&-.17599&-.05105\end{bmatrix},\quad \mathbf{Q}_b=\begin{bmatrix}0.01227&0.01513&-.00357&-.00056\\0.05573&0.07396&0.03037&-.00720\\0.03492&0.02920&0.00575&0.00145\\0.04555&0.08715&0.05004&-.00019\end{bmatrix}$$

$$\mathbf{P}_a=\begin{bmatrix}-.66924&0.82821&-3.0575&0.95667\\0.65917&0.17219&1.26814&-.97658\\2.95551&-1.3120&-.44357&0.62629\\-.84519&0.80800&0.40347&0.62808\end{bmatrix},\quad \mathbf{P}_b=\begin{bmatrix}-.07883&-.49933&1.10129&-.13921\\-.30581&-.37676&-.39596&0.30450\\-.58639&-.04520&0.09067&-.24182\\0.04619&-.18014&-.10277&-.11922\end{bmatrix}$$

The reciprocal symmetry is checked perfectly for the two groups of data. Because these matrices are computed by precise integration method, hence the numerical results almost reach computer precision.

If $N_\mathrm{int}=16$ is selected with step size $\eta=0.3$, the submatrices at the same grid points give almost the same numerical results, that the precise integration method gives certainly very precise numerical results. ##

§5.10.4, Application

Equation (5.2.9a,b) proposes the problem of structure response under external forces $\mathbf{f}_q(z)$ and $\mathbf{f}_p(z)$, which is solved with the equation (5.10.27). However, the structural analysis needs also to solve the eigen-problems as given by dual equations

$$\dot{\mathbf{q}} = \mathbf{A}\mathbf{q} + \mathbf{D}\mathbf{p}$$
$$\dot{\mathbf{p}} = \mathbf{B}\mathbf{q} - \mathbf{A}^T\mathbf{p} - \omega^2\mathbf{M}\mathbf{q} \tag{5.10.37a,b}$$

where $\mathbf{M}$ is a non-negative symmetric $n \times n$ mass matrix and ω^2 is the eigenvalue to be determined, whose physical meaning is the natural frequency. On the other hand, the other form of structural eigenvalue problem derives to

$$\dot{\mathbf{q}} = \mathbf{A}\mathbf{q} + \mathbf{D}\mathbf{p} - \omega^2\mathbf{F}_e\mathbf{p}$$
$$\dot{\mathbf{p}} = \mathbf{B}\mathbf{q} - \mathbf{A}^T\mathbf{p} \tag{5.10.38a,b}$$

where $\mathbf{F}_e$ is a non-negative symmetric $n \times n$ geometric matrix and ω^2 is the eigenvalue to be determined, whose physical meaning is the loading factor. These eigenvalue problems can be reduced to an integral equation by the method of ***impulse influence matrix*** $\mathbf{\Phi}(z,\zeta)$. Comparing equations (5.2.9a,b) with (5.10.37a,b) finds the substitution of $\mathbf{f}_q(z) \sim \mathbf{0}$ and $\mathbf{f}_p(z) \sim \omega^2\mathbf{M}\mathbf{q}$, and then the equation (5.10.27) is reduced to

$$\mathbf{q}(z) = -\omega^2 \int_{z_0}^{z_f} \mathbf{\Phi}_{12}(z,\tau)\mathbf{M}\mathbf{q}(\tau)\mathrm{d}\tau \tag{5.10.392}$$

$$\mathbf{p}(z) = -\omega^2 \int_{z_0}^{z_f} \mathbf{\Phi}_{22}(z,\tau)\mathbf{M}\mathbf{q}(\tau)\mathrm{d}\tau \tag{5.10.39b}$$

Using equation (5.10.25), the former one gives the integral equation

$$\mathbf{q}(z) = -\omega^2 \int_{z_0}^{z_f} \mathbf{Q}_b(z,\tau)\mathbf{M}\mathbf{q}(\tau)\mathrm{d}\tau \tag{5.10.39'}$$

Because of the second reciprocal theorem (5.10.23), it can be reduced to an eigen-problem of a symmetric kernel integral equation, the mathematical theory of which was given perfectly, see [1]. After solving the eigen-pair $\omega^2, \mathbf{q}(z)$, the dual vector $\mathbf{p}(z)$ computation is simply a quadrature of (5.10.39b).

Similarly for (5.10.38a,b), it derives the equations

$$\mathbf{p}(z) = -\omega^2 \int_{z_0}^{z_f} \mathbf{P}_a(z,\tau)\mathbf{F}_e\mathbf{p}(\tau)\mathrm{d}\tau, \qquad \mathbf{q}(z) = -\omega^2 \int_{z_0}^{z_f} \mathbf{Q}_a(z,\tau)\mathbf{F}_e\mathbf{p}(\tau)\mathrm{d}\tau \tag{5.10.40a,b}$$

where the former one can be reduced to an eigen-problem of integral equation with symmetric kernel, which is due to the first reciprocal theorem, and the latter equation is simply the integration computation of $\mathbf{q}(z)$ after the eigen-problem is solved.

Numerical solution should first reduce the integral equation to be of symmetric kernel. The matrix $\mathbf{M}$ is factorized as $\mathbf{M} = \mathbf{L}\mathbf{L}^T$ first, then the equation (5.10.39') is derived as

$$\mathbf{L}^T\mathbf{q}(z) = -\omega^2 \int_{z_0}^{z_f} [\mathbf{L}^T\mathbf{Q}_b(z,\tau)\mathbf{L}] \cdot [\mathbf{L}^T\mathbf{q}(\tau)]\mathrm{d}\tau \tag{5.10.41}$$

which has been an integration equation eigenvalue problem with symmetric kernel $\mathbf{L}^T\mathbf{Q}_b(z,\tau)\mathbf{L}$ and the unknown vector $\mathbf{L}^T\mathbf{q}(z)$. After solving the eigen-solution pair ω^2, $\mathbf{L}^T\mathbf{q}(z)$, the eigenvector of the equation (5.10.39') can be found simply from the equation

$$\mathbf{q}(z) = -\omega^2 \int_{z_0}^{z_f} \mathbf{Q}_b(z,\tau)\mathbf{L}[\mathbf{L}^T\mathbf{q}(\tau)]\mathrm{d}\tau .$$

The solution of integral equation with symmetric kernel can use the numerical integration method. For eigenvalue problem it reduces to an eigenvalue problem with symmetric matrix and typical solution algorithm can be invoked, see [42,43].

The conclusion is that the impulse influence matrix $\boldsymbol{\Phi}(z,\zeta)$ can be precisely computed at the grid points, the numerical result approaches the exact solution, which supplies good basis for further developments. Based on the analogy between structural mechanics and optimal control, the impulse influence matrix $\boldsymbol{\Phi}(z,\zeta)$ is also quite useful in optimal control problems.

§5.11, Power flow

Elastic wave propagation is a complex problem, for which there are a number of monographs and papers [105~113]. The elastic wave-guide is an important part, that its domain is a strip or a prism. Discretizing along the cross-section derives a single continuous coordinate problem.

The displacement method is frequently used in the traditional analysis of elastic wave-guides. From Timoshenco beam analysis, it is seen that the analysis of wave-guide problems can also use the state space approach. Therefore, the damping free wave propagation becomes a problem of Hamilton system theory, and then the method of separation of variables, eigen-problem of a Hamilton matrix, adjoint symplectic ortho-normality and expansion theorem etc. can be used for solving the elastic wave-guide problem. Here, the problem is assumed transversely discretized along the cross-section, and wave-guide problem is reduced to be a multi-variable system along a single continuous coordinate [114~118]. For the problems of theory of elasticity, the variables on the cross-section are also continuous functions, which is in the infinite dimensional symplectic space. In such case, the transverse discretization method (semi-analytical method) can be used, which reduce the problem to finite dual variables transversely, then the method of separation of variables, symplectic eigen-solution expansion, etc. can be used to find the numerical solution.

Assuming, a wave-guide with n-displacements along the cross-section (for Timoshence beam theory, $n = 2$) is analyzed by means of the frequency domain method. Given frequency ω, that both the displacement vector $\mathbf{q}$ and internal force vector $\mathbf{p}$ hold a factor of $e^{-\mathrm{i}\omega t}$ and the displacement, velocity and internal force are expressed as

$$\mathrm{Re}\{\mathbf{q}e^{-\mathrm{i}\omega t}\},\ \mathrm{Re}\{-\mathrm{i}\omega\mathbf{q}e^{-\mathrm{i}\omega t}\},\ \mathrm{Re}\{\mathbf{p}e^{-\mathrm{i}\omega t}\} \tag{5.11.1}$$

Therefore the ***time average*** power flow across the station z can be computed as

$$W_z = -(\omega/2\pi)\int_0^{2\pi/\omega} \mathrm{Re}\{\mathbf{p}^T e^{-\mathrm{i}\omega t}\}\cdot\mathrm{Re}\{-\mathrm{i}\omega\mathbf{q}e^{-\mathrm{i}\omega t}\}\mathrm{d}t \tag{5.11.2}$$

where $\mathbf{p} = \mathbf{p}(z)$ etc. The above equation is general. Using the expansion theorem

$$\mathbf{q} = \mathbf{Q}_\alpha\tilde{\mathbf{a}}(z) + \mathbf{Q}_\beta\tilde{\mathbf{b}}(z), \qquad \mathbf{p} = \mathbf{P}_\alpha\tilde{\mathbf{a}}(z) + \mathbf{P}_\beta\tilde{\mathbf{b}}(z)$$

where $\tilde{\mathbf{a}}(z)$ and $\tilde{\mathbf{b}}(z)$ are n-dimensional vectors to be solved. According to the solution obtained from expansion method

$$\tilde{\mathbf{a}}(z)=\mathbf{D}_{c\eta}\mathbf{a}\,,\quad \tilde{\mathbf{b}}(z)=\mathbf{D}_{c\eta}^{-1}\mathbf{b}$$

then
$$\mathbf{q}=\mathbf{Q}_\alpha\mathbf{D}_{c\eta}\mathbf{a}+\mathbf{Q}_\beta\mathbf{D}_{c\eta}^{-1}\mathbf{b},\quad \mathbf{p}=\mathbf{P}_\alpha\mathbf{D}_{c\eta}\mathbf{a}+\mathbf{P}_\beta\mathbf{D}_{c\eta}^{-1}\mathbf{b} \tag{5.11.3}$$

where $\mathbf{a}$ and $\mathbf{b}$ are vectors to be determined, and

$$\mathbf{D}_{c\eta}=\mathrm{diag}(e^{\mu_i z})=\mathrm{diag}[e^{\mu_1 z},e^{\mu_2 z},\cdots,e^{\mu_n z}]$$

where $\eta=z$ for the subscript. Substituting equation (5.11.3) into W_z gives

$$\begin{aligned}W_z&=(\frac{\mathrm{i}\omega^2}{8\pi})\int_0^{2\pi/\omega}\{(\mathbf{a}^T\mathbf{D}_{c\eta}\mathbf{P}_\alpha^T+\mathbf{b}^T\mathbf{D}_{c\eta}^{-1}\mathbf{P}_\beta^T)e^{-\mathrm{i}\omega t}+(\bar{\mathbf{a}}^T\bar{\mathbf{D}}_{c\eta}\bar{\mathbf{P}}_\alpha^T+\bar{\mathbf{b}}^T\bar{\mathbf{D}}_{c\eta}^{-1}\bar{\mathbf{P}}_\beta^T)e^{-\mathrm{i}\omega t}\}\times\\&\qquad\{(\mathbf{Q}_\alpha\mathbf{D}_{c\eta}\mathbf{a}+\mathbf{Q}_\beta\mathbf{D}_{c\eta}^{-1}\mathbf{b})e^{-\mathrm{i}\omega t}-(\bar{\mathbf{Q}}_\alpha\bar{\mathbf{D}}_{c\eta}\bar{\mathbf{a}}+\bar{\mathbf{Q}}_\beta\bar{\mathbf{D}}_{c\eta}^{-1}\bar{\mathbf{b}})e^{-\mathrm{i}\omega t}\}\mathrm{d}t\\&=(\mathrm{i}\omega/4)\{\bar{\mathbf{a}}^T\bar{\mathbf{D}}_{c\eta}(\bar{\mathbf{P}}_\alpha^T\mathbf{Q}_\alpha-\bar{\mathbf{Q}}_\alpha^T\mathbf{P}_\alpha)\mathbf{D}_{c\eta}\mathbf{a}+\bar{\mathbf{b}}^T\bar{\mathbf{D}}_{c\eta}^{-1}(\bar{\mathbf{P}}_\beta^T\mathbf{Q}_\alpha-\bar{\mathbf{Q}}_\beta^T\mathbf{P}_\alpha)\mathbf{D}_{c\eta}\mathbf{a}\\&\qquad\bar{\mathbf{a}}^T\bar{\mathbf{D}}_{c\eta}(\bar{\mathbf{P}}_\alpha^T\mathbf{Q}_\beta-\bar{\mathbf{Q}}_\alpha^T\mathbf{P}_\beta)\mathbf{D}_{c\eta}^{-1}\mathbf{b}+\bar{\mathbf{b}}^T\bar{\mathbf{D}}_{c\eta}^{-1}(\bar{\mathbf{P}}_\beta^T\mathbf{Q}_\beta-\bar{\mathbf{Q}}_\beta^T\mathbf{P}_\beta)\mathbf{D}_{c\eta}^{-1}\mathbf{b}\}\\&=(\omega/4)\begin{bmatrix}\bar{\mathbf{a}}^T\bar{\mathbf{D}}_{c\eta}\mathbf{A}_\alpha\mathbf{D}_{c\eta}\mathbf{a}-\bar{\mathbf{b}}^T\bar{\mathbf{D}}_{c\eta}^{-1}\mathbf{A}_\beta\mathbf{D}_{c\eta}^{-1}\mathbf{b}\\+\bar{\mathbf{a}}^T\bar{\mathbf{D}}_{c\eta}\mathbf{A}_\gamma\mathbf{D}_{c\eta}^{-1}\mathbf{b}+\bar{\mathbf{b}}^T\bar{\mathbf{D}}_{c\eta}^{-1}\mathbf{A}_\gamma^T\mathbf{D}_{c\eta}\mathbf{a}\end{bmatrix}\end{aligned} \tag{5.11.4}$$

where

$$\begin{aligned}&\mathbf{A}_\alpha=\mathrm{i}(\bar{\mathbf{P}}_\alpha^T\mathbf{Q}_\alpha-\bar{\mathbf{Q}}_\alpha^T\mathbf{P}_\alpha),\ \mathbf{A}_\beta=-\mathrm{i}(\bar{\mathbf{P}}_\beta^T\mathbf{Q}_\beta-\bar{\mathbf{Q}}_\beta^T\mathbf{P}_\beta)\\&\mathbf{A}_\gamma=\mathrm{i}(\bar{\mathbf{P}}_\alpha^T\mathbf{Q}_\beta-\bar{\mathbf{Q}}_\alpha^T\mathbf{P}_\beta)\end{aligned} \tag{5.11.5a~c}$$

These matrices are independent on z, and they represent energy transmission or the characteristics of power flow. Selecting arbitrarily the constant vectors $\mathbf{a}$ and $\mathbf{b}$, equation (5.11.4) can be used to compute the power flow of arbitrary wave along the wave-guide.

The above analysis applies to constant cross-section wave-guide of single continuous coordinate z. A related problem of which is the wave propagation along periodical structure. For a given frequency ω, precisely integrating the length η of a period of wave-guide obtains the respective mixed energy matrices $\mathbf{F}(\eta),\mathbf{G}(\eta),\mathbf{Q}(\eta)$. The continuous wave-guide can be considered as a periodical wave-guide with the mixed energy matrices of the fundamental period substructure being given. Hence the periodical structural wave-guide analysis can be given similarly. However, the periodical structure wave propagation has some features, and will be considered later in section 5.14.

§5.11.1, Algebraic Riccati equation (ARE)

The discussion of power flow can begin with the analysis of semi-infinite strip domain $0\le z<\infty$. The solution selected must be decayed when $z\to\infty$, *i.e.* the class-α eigen-solutions. If there is no ***purely imaginary eigenvalue***, then all the eigen-solutions decay. It is easily recognized that such solutions will not carry energy to infinity. Only the purely imaginary class-α eigenvalue $\mu=ik,k>0$, the corresponding solution has a factor

$$\exp[\mathrm{i}(kz-\omega t)]$$

which implies that the solutions propagate along the axis z with wave speed

$c = \omega/k$ toward $+\infty$ and these solutions transmit energy.

When there is no ***purely imaginary eigenvalue***, then only decayed solutions exist at $z \to \infty$. Taking $\mathbf{p}_b \to 0$ in the interval expression (5.7.14) gives

$$\mathbf{p}_a = -\mathbf{Q}_\infty \mathbf{q}_a \tag{5.11.6a}$$

where the subscript-a represents left-hand end of an interval, which can be an arbitrary coordinate z, such as $z = 0$. Matrix $\mathbf{Q}_\infty$ is the left end stiffness matrix of the semi-infinitely long wave-guide, a dynamic stiffness matrix. Substituting $\mathbf{p} = -\mathbf{Q}_\infty \mathbf{q}$ into the dual differential equations (5.2.10a,b) derives the equation to be satisfied by $\mathbf{Q}_\infty$

$$\mathbf{B} + \mathbf{Q}_\infty \mathbf{A} + \mathbf{A}^T \mathbf{Q}_\infty - \mathbf{Q}_\infty \mathbf{D} \mathbf{Q}_\infty = \mathbf{0} \tag{5.11.7a}$$

the ARE. The above derivation is for wave-guide theory. This equation is very important for optimal control theory. The controllability and observability in optimal control theory ensure that the strain energy of arbitrary interval is positive definite, hence there is no purely imaginary eigen-root, which derives also the ARE.

From the symplectic orthogonality condition, the matrix $\mathbf{Q}_\alpha^T \mathbf{P}_\alpha$ is symmetric, so is $\mathbf{P}_\alpha \mathbf{Q}_\alpha^{-1}$. In fact

$$\mathbf{Q}_\infty = -\mathbf{P}_\alpha \mathbf{Q}_\alpha^{-1} \tag{5.11.8a}$$

Substituting all the columns of matrix $\mathbf{P}_\alpha$ in turn, as the vector $\mathbf{p}_a$, all the respective columns of $\mathbf{Q}_\alpha$ as vectors $\mathbf{q}_a$ and $\mathbf{p}_b = 0$ into equation (5.7.14b) gives the above equation. This equation is a special case of equation (5.8.5). The matrices $\mathbf{P}_\alpha$ and $\mathbf{Q}_\alpha$ are complex valued and are inconvenient for computation. In case of no purely imaginary eigen-root, using (5.8.19) and (5.8.18) gives

$$\mathbf{Q}_\infty = -\mathbf{P}_\alpha \mathbf{Q}_\alpha^{-1} = -\mathbf{P}_{\alpha r} \mathbf{C}_d^{-1} (\mathbf{Q}_{\alpha r} \mathbf{C}_d^{-1})^{-1} = -\mathbf{P}_{\alpha r} \mathbf{Q}_{\alpha r}^{-1} \tag{5.11.8'a}$$

and the computation becomes real arithmetic. When there is purely imaginary eigen-root, the $n \times n$ matrix $\mathbf{C}_d$ is not able to bring the matrices real. Such case corresponds to wave propagation and will be considered below.

The case of semi-infinite interval $-\infty < z \le 0$ is considered now. For which, only the class-β eigen-solutions should be selected. In case of no purely imaginary eigen-root, only decay solutions exist for $z \to -\infty$. In equation (5.7.14a) $\mathbf{q}_a \to 0$ for the left end-a, then it gives

$$\mathbf{q}_b = \mathbf{G}_\infty \mathbf{p}_b \tag{5.11.6b}$$

where the subscript-b means right-hand end. $\mathbf{G}_\infty$ is the flexibility matrix of semi-infinite strip wave-guide. From the dual differential equations (5.2.10a,b) derives the equation for $\mathbf{G}_\infty$ as

$$\mathbf{D} + \mathbf{A} \mathbf{G}_\infty + \mathbf{G}_\infty \mathbf{A}^T - \mathbf{G}_\infty \mathbf{B} \mathbf{G}_\infty = \mathbf{0} \tag{5.11.7b}$$

the ARE, which is very important in Kalman-Bucy filter. The controllability and observability conditions (chapter 6) ensure that there is no purely imaginary eigenvalue. However, the wave propagation solution requires ***purely imaginary eigenvalue***.

Having found all the eigen-solutions, the ARE solution can be computed as

$$\mathbf{G}_\infty = \mathbf{Q}_\beta \mathbf{P}_\beta^{-1} \tag{5.11.8b}$$

According to the adjoint symplectic ortho-normality condition, $\mathbf{G}_\infty$ is easily verified a symmetric matrix. Equation (5.11.8b) is obtained by substituting the columns of $\mathbf{P}_\beta$ in turn as the vector $\mathbf{p}_b$ and the corresponding columns of $\mathbf{Q}_\beta$ as $\mathbf{q}_b$ into equation (5.7.14a). The decay condition at $z \to -\infty$ determines $\mathbf{q}_a = \mathbf{0}$. This equation is a special case of equation (5.8.6). However, the matrices $\mathbf{Q}_\beta$ and $\mathbf{P}_\beta$ in the above equation are complex valued, which is inconvenient for computations. In case of no purely imaginary eigen-root, the equations (5.8.18~19) can be used to derive the equation

$$\mathbf{G}_\infty = \mathbf{Q}_\beta \mathbf{P}_\beta^{-1} = \mathbf{Q}_{\beta r} \mathbf{C}_d^{-1} (\mathbf{P}_{\beta r} \mathbf{C}_d^{-1})^{-1} = \mathbf{Q}_{\beta r} \mathbf{P}_{\beta r}^{-1} \tag{5.11.8'b}$$

which is computed in real arithmetic.

§5.11.2, Transmission waves

The last section points out that in case of no purely imaginary eigen-root, the two AREs (algebraic Riccati equations) can be solved within real arithmetic. Although the complex conjugate eigenvalues may appear and correspondingly the complex conjugate eigenvectors located both in class-α or in class-β, hence the solution matrices of the AREs must be symmetric and can be computed with equations (5.11.8'a,b).

When there are purely imaginary eigenvalues, the two complex conjugate eigenvalues locate separately in class-α and in class-β, hence the matrix $\mathbf{C}_d$ in (5.8.18) cannot be involved in the transformation within class-α nor within class-β. When using the class-α eigen-solutions only the transmission wave toward $+\infty$ exists but the transmission waves coming from $+\infty$ does not exist because the class-β eigen-solutions are excluded. Such sort of solutions coincides with the energy radiation condition.

The solution satisfying the radiation condition must be strictly distinguished from the solution of perfectly reflected at $z \to +\infty$. The radiation condition means that energy can only flow to $z \to +\infty$ with no return, it implies energy dissipation. However, perfect reflection at $z \to +\infty$ means that the radiated energy will be coming back from $+\infty$ with no energy loss. Such solution can also be interpreted that a wave coming from $+\infty$ which means energy coming but the system has no dissipation so that the energy is radiated. The energy balance solution is composed of not only the class-α purely imaginary eigen-solution but also the class-β purely imaginary eigen-solution.

The precise integration solutions described before solve the differential equations (5.7.16) and (5.7.17) with the two point boundary conditions, for which the condition at $z = z_f$ is $\mathbf{p}_f = \mathbf{0}$. This boundary condition corresponds to energy reflected perfectly. Let $\mathbf{p}_f = \mathbf{0}$ at some point of z_f very large, then the integrated interval matrices $\mathbf{F}, \mathbf{G}, \mathbf{Q}$ are all real valued. The analytical solution given in the previous section is also determined from such two end boundary conditions hence the obtained interval matrices are all real valued.

Below, consider the class-α solution only. Let the matrix $\mathbf{Q}_\infty$ be defined by (5.11.8a). When there is purely imaginary eigen-root, then although $\mathbf{Q}_\infty$ satisfies the ARE (5.11.7a) but it is still complex valued. From which the power flow of pure class-α solution can be computed as

$$\mathbf{A}_\alpha = \mathrm{i}(\overline{\mathbf{P}}_\alpha^T \mathbf{Q}_\alpha - \overline{\mathbf{Q}}_\alpha^T \mathbf{P}_\alpha) = -\mathrm{i}\overline{\mathbf{Q}}_\alpha^T (\overline{\mathbf{Q}}_\infty - \mathbf{Q}_\infty)\mathbf{Q}_\alpha$$
$$W_z^{(\alpha)} = (\omega/4)\bar{\mathbf{a}}^T \overline{\mathbf{D}}_{c\eta} \mathbf{A}_\alpha \mathbf{D}_{c\eta}\mathbf{a}, \qquad \eta = z \tag{5.11.9a}$$

where $\mathbf{Q}_\infty$ is still a symmetric (but not Hermitian symmetric) matrix, complex valued. The symmetry is ensured from symplectic orthogonality.

For two given coordinates z_a and $z_\mathrm{b} > z_\mathrm{a}$, the power flow balance condition requires $W_{z\mathrm{a}} = W_{z\mathrm{b}}$. To verify, from equation (5.11.2) derives

$$W_{z\mathrm{a}} = (\mathrm{i}\omega/4)\{\overline{\mathbf{p}}_\mathrm{a}^T \mathbf{q}_\mathrm{a} - \mathbf{p}_\mathrm{a}^T \overline{\mathbf{q}}_\mathrm{a}\} \quad \text{and} \quad W_{z\mathrm{b}} = (\mathrm{i}\omega/4)\{\overline{\mathbf{p}}_\mathrm{b}^T \mathbf{q}_\mathrm{b} - \mathbf{p}_\mathrm{b}^T \overline{\mathbf{q}}_\mathrm{b}\}$$

Using equations (5.7.14a,b) and the symmetry of the matrices $\mathbf{Q}$ and $\mathbf{G}$ gives

$$W_{z\mathrm{a}} = (\mathrm{i}\omega/4)\{\overline{\mathbf{p}}_\mathrm{b}^T \overline{\mathbf{F}} \mathbf{q}_\mathrm{a} - \mathbf{p}_\mathrm{b}^T \mathbf{F} \overline{\mathbf{q}}_\mathrm{a}\} \quad \text{and} \quad W_{z\mathrm{b}} = (\mathrm{i}\omega/4)\{\overline{\mathbf{p}}_\mathrm{b}^T \mathbf{F} \mathbf{q}_\mathrm{a} - \mathbf{p}_\mathrm{b}^T \overline{\mathbf{F}} \overline{\mathbf{q}}_\mathrm{a}\}$$

Because the matrix $\mathbf{F}$ is also real valued, so $W_{z\mathrm{a}} = W_{z\mathrm{b}}$ follows, which gives the constant power flow theorem along the z axis.

The class-β solutions is considered next. According to equation (5.11.8b), the matrix $\mathbf{G}_\infty$ is obtained. If the purely imaginary eigenvalue exists, then the matrix $\mathbf{G}_\infty$ is complex valued and is different with that obtained from the two end boundary conditions. The matrix $\mathbf{G}_\infty$ satisfies the ARE (5.11.7b), and then the power flow of class-β solution can be derived as

$$\mathbf{A}_\beta = -\mathrm{i}(\overline{\mathbf{P}}_\beta^T \mathbf{Q}_\beta - \overline{\mathbf{Q}}_\beta^T \mathbf{P}_\beta) = \mathrm{i}\overline{\mathbf{Q}}_\beta^T (\overline{\mathbf{G}}_\infty - \mathbf{G}_\infty)\mathbf{Q}_\beta$$
$$W_z^{(\beta)} = -(\omega/4)\bar{\mathbf{b}}^T \overline{\mathbf{D}}_{c\eta}^{-1} e^{-\bar{\mu} z} \mathbf{A}_\beta \mathbf{D}_{c\eta}^{-1}\mathbf{b} \tag{5.11.9b}$$

§5.11.3, On power flow orthogonality

Examining equation (5.11.9a), because $W_z^{(\alpha)}$ is constant all along the coordinate z, hence

$$\overline{\mathbf{D}}_{c\eta} \mathbf{A}_\alpha \mathbf{D}_{c\eta} e^{\mu z} = \mathbf{A}_\alpha$$

icking up the (i,j) element from the above matrix equation gives $\mathbf{A}_{\alpha ij}(1 - e^{\bar{\mu}_i z} e^{\mu_j z}) = 0$. For non-diagonal element $i \neq j$, the factor of the parenthesis is not zero, hence $A_{\alpha ij} = 0$ when $i \neq j$. Even for the diagonal element $\mathbf{A}_{\alpha ii}$, if μ_i is not a purely imaginary eigen-root, still $\mathbf{A}_{\alpha ii} = 0$. Only for purely imaginary eigen-root, $A_{\alpha ii}$ can be non-zero. Hence the power flow toward right-hand direction ($z \to \infty$) is

$$W_z^{(\alpha)} = (\omega/4)\sum_{i=1}^{n} \mathbf{A}_{\alpha ii} \left| a_i^2 \right| \tag{5.11.10a}$$

The conclusion can be drawn as that the decay. eigen-solution has no power flow and there is no energy flow between two different eigen-solutions either. The power flow exists only for purely imaginary eigenvalue solution. It is the orthogonality relationship of power flow.

For class- β eigen-solutions, the analysis is similar.

There remains the problem of mutual power flow between toward-left and toward-right eigen-solutions. For which, the conclusion of that power flow is a constant along the z axis can still be applied. From the equation (5.11.5c) and (5.11.8) give

$$\begin{aligned} \mathbf{A}_\gamma &= i\overline{\mathbf{Q}}_\alpha^T(\overline{\mathbf{Q}}_\infty + \mathbf{G}_\infty^{-1})\mathbf{Q}_\beta \\ \overline{\mathbf{a}}^T\mathbf{A}_\gamma\mathbf{b} + \mathbf{a}^T\overline{\mathbf{A}}_\gamma\overline{\mathbf{b}} &= \overline{\mathbf{a}}^T\overline{\mathbf{D}}_{c\eta}\mathbf{A}_\gamma\mathbf{D}_{c\eta}^{-1}\mathbf{b} + \mathbf{a}^T\mathbf{D}_{c\eta}\overline{\mathbf{A}}_\gamma\overline{\mathbf{D}}_{c\eta}^{-1}\overline{\mathbf{b}} \end{aligned} \tag{5.11.9c}$$

The above equations are valid for arbitrary vectors $\mathbf{a}$ and $\mathbf{b}$. Two cases are selected as

Case 1: $a_i = 1, b_j = i$. Case 2: $a_i = 1, b_j = 1$, $(i, j \le n)$

Substituting into the above equalities gives $A_{\gamma ij}\{\exp[(\overline{\mu}_i - \mu_j)z] - 1\} = 0$. Hence the conclusion can be drawn that the mutual power flow is zero except $\overline{\mu}_i = \mu_j (i, j \le n)$. It is again a power flow orthogonality theorem. It determines that the purely imaginary eigenvalue solutions (pass-band solution) have no mutual power flow. Only for symplectic adjoint eigen-solutions, the class- α and class- β eigen-solutions can have mutual power flow, but such mutual action can be applied only for finite length interval.

Power flow orthogonality is not enough to describe all behaviors, that there must be power flow positive definiteness. Certainly the power flow positive definiteness can be only for purely imaginary eigenvalue solutions, *i.e.* the pass-band eigen-solutions. For class- α transmission waves, $\mu_\alpha = \mathrm{i}k, k > 0$, the factor $\exp[\mathrm{i}(kz - \omega t)]$ implies transmission toward right, *i.e.* the power flow toward right should be positive. It is expressed by the element $A_{\alpha ii}$ in equation (5.11.10a), *i.e.* the i-th diagonal element in the matrix $\mathbf{A}_\alpha$ of equation (5.11.9a), being positive. On the other hand, the class- β transmission wave of $\mu = -ik$ is toward left, which implies the power flow toward right should be negative. Hence, for class- β wave

$$W_z^{(\beta)} = -(\omega/4)\sum_{i=1}^{n} A_{\beta ii}\left|b_i^2\right| \tag{5.11.10b}$$

and the i-th diagonal element $A_{\beta ii}$ of matrix (5.11.9b) must be positive, which gives the positive definiteness of power flow.

The above discussion is based on the physical interpretation. Mathematical proof is necessary which will be given in the next section. All the previous discussions on power flow are based on eigenvector expansion method and the wave-guide is considered invariant along the coordinate z. Practically, the strip domain must have ends, and internal to the wave-guide there may be non-uniform junctions. At these non-uniform parts of the wave-guide, the wave scatters. The wave scattering analysis is very important in applications.

§5.12, Wave scattering analysis

The eigenvector expansion method is quite effective for the analysis of wave scattering, which appears at the ends of wave-guide or at the junctions.

The problem of an elastic body connected by two semi-infinite elastic wave-guides is to analyze, see figure 5.3.

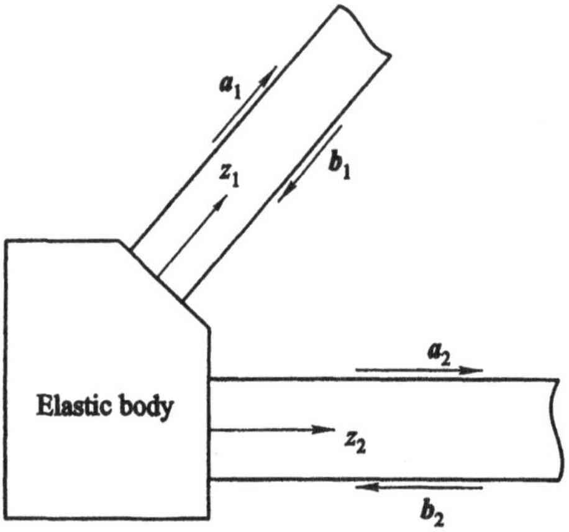

Figure 5.3, *An elastic body attached with two wave-guides*

A problem with multiple semi-infinite wave-guides can similarly be analyzed. The two wave-guides attached can have different properties. Let the dimensions be n_1 and n_2, respectively, and have found the $n_1 \times n_1$ sub-matrices $\mathbf{Q}_{\alpha 1}, \mathbf{Q}_{\beta 1}, \mathbf{P}_{\alpha 1}, \mathbf{P}_{\beta 1}$ of the eigen-matrix $\mathbf{\Psi}_1$ and the $n_2 \times n_2$ sub-matrices $\mathbf{Q}_{\alpha 2}, \mathbf{Q}_{\beta 2}, \mathbf{P}_{\alpha 2}, \mathbf{P}_{\beta 2}$ of eigen-matrix $\mathbf{\Psi}_2$, respectively, for two wave-guides. The elastic body is represented with a $(n_1 + n_2) \times (n_1 + n_2)$ dimensioned external dynamic stiffness matrix $\mathbf{R}(\omega)$, which should be Hermite symmetric. Given in block matrix form

$$\mathbf{R}(\omega) = \begin{bmatrix} \mathbf{R}_{11} & \mathbf{R}_{12} \\ \mathbf{R}_{21} & \mathbf{R}_{22} \end{bmatrix} \begin{matrix} n_1 \\ n_2 \end{matrix} \tag{5.12.1}$$

$$\mathbf{R}_{11}^H = \mathbf{R}_{11}, \ \mathbf{R}_{22}^H = \mathbf{R}_{22}, \qquad \mathbf{R}_{21}^H = \mathbf{R}_{12} \tag{5.12.2}$$

where the superscript ^{H} represents Hermite transpose, *i.e.* transposition of its complex conjugate, $\mathbf{M}^H = \overline{\mathbf{M}}^T$.

The two wave-guides have incoming waves of the class-β eigen-solutions denoted as $\mathbf{b}_1$ and $\mathbf{b}_2$ of the two wave-guides, respectively. Only the components corresponding to the ***purely imaginary eigenvalue*** can be non-zero. Because of the scattering of elastic body, the various class-α eigen-solutions will be generated and expressed by the vectors $\mathbf{a}_1$ and $\mathbf{a}_2$, respectively, where the component corresponding to the ***purely imaginary eigenvalue*** radiates power to infinity. To find the vectors $\mathbf{a}_1$ and $\mathbf{a}_2$, a set of algebraic equations must be established. After solving the vectors $\mathbf{a}_1$ and $\mathbf{a}_2$, the energy balance equation should be verified, *i.e.* the power flow poured in by the incoming wave vectors $\mathbf{b}_1$ and $\mathbf{b}_2$ must equal to the power flow radiated by the scattering vectors $\mathbf{a}_1$ and $\mathbf{a}_2$.

Selecting longitudinal coordinates z_1 and z_2 for the two wave-guides, respectively, such that the origin $z_1 = 0$ and $z_2 = 0$ locate at the attached points to the elastic body. According to the expansion theorem, the solutions can be expressed as

$$\mathbf{q}_1 = \mathbf{Q}_{1\alpha} \exp[\mathrm{diag}(\mu_{1i} z_1)]\mathbf{a}_1 + \mathbf{Q}_{1\beta} \exp[\mathrm{diag}(-\mu_{1i} z_1)]\mathbf{b}_1 \tag{5.12.3a}$$

$$\mathbf{p}_1 = \mathbf{P}_{1\alpha} \exp[\mathrm{diag}(\mu_{1i} z_1)]\mathbf{a}_1 + \mathbf{P}_{1\beta} \exp[\mathrm{diag}(-\mu_{1i} z_1)]\mathbf{b}_1 \tag{5.12.3b}$$

$$\mathbf{q}_2 = \mathbf{Q}_{2\alpha} \exp[\mathrm{diag}(\mu_{2i} z_2)]\mathbf{a}_2 + \mathbf{Q}_{2\beta} \exp[\mathrm{diag}(-\mu_{2i} z_2)]\mathbf{b}_2 \tag{5.12.4a}$$

$$\mathbf{p}_2 = \mathbf{P}_{2\alpha} \exp[\mathrm{diag}(\mu_{2i} z_2)]\mathbf{a}_2 + \mathbf{P}_{2\beta} \exp[\mathrm{diag}(-\mu_{2i} z_2)]\mathbf{b}_2 \tag{5.12.4b}$$

and at the two attached points $z_1 = 0$ and $z_2 = 0$ the equations of the elastic body are

$$\mathbf{p}_1 = -(\mathbf{R}_{11}\mathbf{q}_1 + \mathbf{R}_{12}\mathbf{q}_2), \quad \mathbf{p}_2 = -(\mathbf{R}_{21}\mathbf{q}_1 + \mathbf{R}_{22}\mathbf{q}_2) \tag{5.12.5a,b}$$

Substituting equations (5.12.3~4) into the above equations and distinguishing the scattering and incoming terms at the two sides of equations gives

$$(\mathbf{R}_{11}\mathbf{Q}_{1\alpha} + \mathbf{P}_{1\alpha})\mathbf{a}_1 + \mathbf{R}_{12}\mathbf{Q}_{2\alpha}\mathbf{a}_2 = -(\mathbf{R}_{11}\mathbf{Q}_{1\beta} + \mathbf{P}_{1\beta})\mathbf{b}_1 - \mathbf{R}_{12}\mathbf{Q}_{2\beta}\mathbf{b}_2 \tag{5.12.6a}$$

$$\mathbf{R}_{21}\mathbf{Q}_{1\alpha}\mathbf{a}_1 + (\mathbf{R}_{22}\mathbf{Q}_{2\alpha} + \mathbf{P}_{2\alpha})\mathbf{a}_2 = -\mathbf{R}_{21}\mathbf{Q}_{1\beta}\mathbf{b}_1 - (\mathbf{R}_{22}\mathbf{Q}_{2\beta} + \mathbf{P}_{2\beta})\mathbf{b}_2 \tag{5.12.6b}$$

where the incoming wave vectors $\mathbf{b}_1, \mathbf{b}_2$ are given, so that the scattered waves are

$$\begin{Bmatrix} \mathbf{a}_1 \\ \mathbf{a}_2 \end{Bmatrix} = -\begin{bmatrix} \mathbf{R}_{11}\mathbf{Q}_{1\alpha} + \mathbf{P}_{1\alpha} & \mathbf{R}_{12}\mathbf{Q}_{2\alpha} \\ \mathbf{R}_{21}\mathbf{Q}_{1\alpha} & \mathbf{R}_{22}\mathbf{Q}_{2\alpha} + \mathbf{P}_{2\alpha} \end{bmatrix}^{-1} \times \begin{bmatrix} \mathbf{R}_{11}\mathbf{Q}_{1\beta} + \mathbf{P}_{1\beta} & \mathbf{R}_{12}\mathbf{Q}_{2\beta} \\ \mathbf{R}_{21}\mathbf{Q}_{1\beta}\mathbf{b}_1 & \mathbf{R}_{22}\mathbf{Q}_{2\beta} + \mathbf{P}_{2\beta} \end{bmatrix} \begin{Bmatrix} \mathbf{b}_1 \\ \mathbf{b}_2 \end{Bmatrix} \tag{5.12.7}$$

Having solved the vectors $\mathbf{a}_1$ and $\mathbf{a}_2$ of scattering waves from the above equation, the energy balance must be checked. Based on the proved power orthogonality theorem, the incoming waves $\mathbf{b}_1$ and $\mathbf{b}_2$ have only transmission components, the mutual power between incoming and reflected waves equal zero. The power flow can only be carried by purely imaginary eigen-root solutions. Left multiplying equation (5.12.6a) with $\mathrm{i}(\mathbf{a}_1^H\mathbf{Q}_{1\alpha}^H + \mathbf{b}_1^H\mathbf{Q}_{1\beta}^H)$ and picking out the real part, using equation (5.11.5) gives

$$\mathbf{a}^H\mathbf{A}_\alpha\mathbf{a} = -\mathrm{i}\mathbf{a}^H(\mathbf{P}_\alpha^H\mathbf{Q}_\alpha - \mathbf{Q}_\alpha^H\mathbf{P}_\alpha)\mathbf{a} = -2\mathbf{a}^H\mathrm{Re}(\mathrm{i}\mathbf{Q}_\alpha^H\mathbf{P}_\alpha)\mathbf{a}$$

$$\mathbf{b}^H\mathbf{A}_\beta\mathbf{b} = \mathrm{i}\mathbf{b}^H(\mathbf{P}_\beta^H\mathbf{Q}_\beta - \mathbf{Q}_\beta^H\mathbf{P}_\beta)\mathbf{b} = 2\mathbf{b}^H\mathrm{Re}(\mathrm{i}\mathbf{Q}_\beta^H\mathbf{P}_\beta)\mathbf{b}$$

$$\mathrm{Re}[\mathrm{i}\mathbf{a}^H\mathbf{Q}_\alpha^H\mathbf{P}_\beta\mathbf{b} + \mathrm{i}\mathbf{b}^H\mathbf{Q}_\beta^H\mathbf{P}_\alpha\mathbf{a}] = \mathrm{Re}[-\mathrm{i}\mathbf{a}^H(\mathbf{P}_\alpha^H\mathbf{Q}_\beta - \mathbf{Q}_\alpha^H\mathbf{P}_\beta)\mathbf{b}] = \mathrm{Re}[\mathbf{a}^H\mathbf{A}_\gamma\mathbf{b}] = 0$$

where the last equality is because the vector $\mathbf{b}$ involves only the components corresponding to the purely imaginary eigenvalue. Based on that $\mathbf{R}_{11}$ is Hermite symmetry, so that

$$\mathrm{Re}[\mathrm{i}(\mathbf{a}_1^H\mathbf{Q}_{1\alpha}^H+\mathbf{b}_1^H\mathbf{Q}_{1\beta}^H)\mathbf{R}_{11}(\mathbf{Q}_{1\alpha}\mathbf{a}_1+\mathbf{Q}_{1\beta}\mathbf{b}_1)]=0$$

etc. Further, based on these equalities derives the equation

$$-\mathbf{a}_1^H\mathbf{A}_{1\alpha}\mathbf{a}_1/2+\mathbf{b}_1^H\mathbf{A}_{1\beta}\mathbf{b}_1/2+\mathrm{Re}[\mathrm{i}(\mathbf{a}_1^H\mathbf{Q}_{1\alpha}^H+\mathbf{b}_1^H\mathbf{Q}_{1\beta}^H)\mathbf{R}_{12}(\mathbf{Q}_{2\alpha}\mathbf{a}_2+\mathbf{Q}_{2\beta}\mathbf{b}_2)]=0$$

Left multiplying equation (5.12.6b) with $\mathrm{i}(\mathbf{a}_2^H\mathbf{Q}_{2\alpha}^H+\mathbf{b}_2^H\mathbf{Q}_{2\beta}^H)$ and pick out the real part gives

$$-\mathbf{a}_2^H\mathbf{A}_{2\alpha}\mathbf{a}_2/2+\mathbf{b}_2^H\mathbf{A}_{2\beta}\mathbf{b}_2/2+\mathrm{Re}[\mathrm{i}(\mathbf{a}_2^H\mathbf{Q}_{2\alpha}^H+\mathbf{b}_2^H\mathbf{Q}_{2\beta}^H)\mathbf{R}_{21}(\mathbf{Q}_{1\alpha}\mathbf{a}_1+\mathbf{Q}_{1\beta}\mathbf{b}_1)]=0$$

Because $\mathbf{R}_{12}$ and $\mathbf{R}_{21}$ are mutually Hermite transpose matrices, so that the Re-part terms in the above two equations add to zero. Adding the two above equations together gives

$$\mathbf{a}_1^H\mathbf{A}_{1\alpha}\mathbf{a}_1/2+\mathbf{a}_2^H\mathbf{A}_{2\alpha}\mathbf{a}_2/2=\mathbf{b}_1^H\mathbf{A}_{1\beta}\mathbf{b}_1/2+\mathbf{b}_2^H\mathbf{A}_{2\beta}\mathbf{b}_2/2$$

which explains that the total power flow of the incoming waves equals to the total power flow of the radiative waves, which coincides with the energy conservation principle.

For an elastic body connected with multiple wave-guides, the method of analysis is the same. In fact, the two wave-guides can be combined together forming a large wave-guide with $n=n_1+n_2$ and $\mathbf{Q}_\alpha=\mathrm{diag}(\mathbf{Q}_{1\alpha},\mathbf{Q}_{2\alpha})$, etc. So that the problem is reduced to an elastic body connected with one wave-guide of dimension $n=n_1+n_2$.

§5.13, Wave induced resonance

Vibration is closely related with wave propagation problems. Resonance is extremely important phenomenon in vibration theory, so that similar phenomenon should be considered in wave propagation. The resonance cavity in acoustic wave-guide or in electro-magnetic wave-guide theory belongs to the problem of same class.

It is seen from the above scattering analysis above, although there may be multiple wave-guides connected to an elastic body, it can still be treated as a combined large wave-guide. Hence the model used here is one semi-infinite connected wave-guide, whose one end connects an elastic body. The equation (5.12.7) becomes

$$\mathbf{a}=-(\mathbf{R}\mathbf{Q}_\alpha+\mathbf{P}_\alpha)^{-1}(\mathbf{R}\mathbf{Q}_\beta+\mathbf{P}_\beta)\mathbf{b} \tag{5.13.1}$$

Since there is matrix inversion, the case of singular matrix must be considered. Because $\mathbf{R}$ is a Hermite matrix and $\mathbf{Q}_\infty=-\mathbf{P}_\alpha\mathbf{Q}_\alpha^{-1}$ is a complex valued symmetric matrix, since there is ***purely imaginary eigenvalue***. A complex valued symmetric matrix is never a Hermite symmetric matrix hence the matrix $\mathbf{R}\mathbf{Q}_\alpha+\mathbf{P}_\alpha$ may not be a null matrix, but its determinant may be zero and then ***resonance*** appears. In such resonance case, however, it does not mean that all the components of the scattering vector $\mathbf{a}$ tend to infinity. In fact, there is the energy flow balance condition

$$\mathbf{a}^H \mathbf{A}_\alpha \mathbf{a}/2 = \mathbf{b}^H \mathbf{A}_\beta \mathbf{b}/2$$

hence, the reflected purely imaginary eigenvalue wave is still finite but the **local vibration components** may be very large.

The ***resonance condition*** is

$$\det(\mathbf{R} - \mathbf{Q}_\infty) \approx 0$$

It is a complex matrix, hence corresponds to two real valued equations, which is different to the usual multiple degrees of freedom vibration system, for which the determinant is zero. Since, there exists the term radiating energy to infinity, which means damping. Hence, the amplitude can be very large for wave induced resonance, but it may be still finite.

A numerical example is given below to show that wave induced resonance really exists.

Example 5.9, Suppose $n = 4$, which corresponds to a two degrees of freedom system that $n = 4$ is the state space model. Suppose at some frequency ω, the system matrices are

$$\mathbf{A} = \begin{bmatrix} 0 & 0 & 1 & 0 \\ 0 & 0 & 0 & 0.3 \\ -2 & 0 & 0 & 0 \\ 1 & 1.5 & 0 & 0 \end{bmatrix}, \quad \mathbf{B} = \text{diag}[2 \quad 1 \quad 0.1 \quad -.1], \quad \mathbf{D} = [4.1 \quad 1 \quad 0.1 \quad 0.01]$$

The eigenvalues of the corresponding Hamiton matrix are computed as (class-α)

$$0.339315\text{i}, \quad -.461340, \quad \begin{matrix} -1.492474 \\ +0.349041\text{i} \end{matrix} \quad \begin{matrix} -1.492474 \\ -0.349041\text{i} \end{matrix};$$

and the matrices are computed as

$$\mathbf{Q}_\infty = \begin{bmatrix} 0.7224 & 0.0525 & -.0393 & -.0074 \\ 0.0525 & 0.4428 & 0.0699 & 0.3015 \\ -.0393 & 0.0699 & 0.4172 & -.0010 \\ -.0074 & 0.3015 & -.0010 & -.0220 \end{bmatrix} + \text{i} \begin{bmatrix} -.0009 & -.0170 & 0.0043 & 0.0051 \\ -.0170 & -.3333 & 0.0843 & 0.1002 \\ 0.0043 & 0.0843 & -.0213 & -.0253 \\ 0.0051 & 0.1002 & -.0253 & -.0301 \end{bmatrix}$$

$$\mathbf{A}_\alpha = \text{diag}[2 \quad 0 \quad 0 \quad 0]; \quad \mathbf{A}_\beta = [2 \quad 0 \quad 0 \quad 0].$$

If the stiffness matrix of the elastic body is

$$\mathbf{R}_0 = \begin{bmatrix} 0.722 & 0.0525 & -.0393 & -.0074 \\ 0.0525 & 0.443 & 0.0699 & 0.301 \\ -.0393 & 0.0699 & 0.417 & -.001 \\ -.0074 & 0.301 & -.001 & -.022 \end{bmatrix}, \quad \mathbf{R} = \mathbf{R}_0$$

then according to the equation (5.13.1), when $b_1 = 1,\ b_2 = b_3 = b_4 = 0$ it solves

$$a_1 = \begin{matrix} 0.26718516 \\ -0.96364521\text{i} \end{matrix} \quad a_2 = \begin{matrix} 4.5885 \\ +6.0338\text{i} \end{matrix} \quad a_3 = \begin{matrix} 119.2092 \\ -6.0873\text{i} \end{matrix} \quad a_4 = \begin{matrix} 25.9849 \\ -116.5018\text{i} \end{matrix}$$

To check the balance of power flow, the input power is $\mathbf{b}^T \mathbf{A}_\beta \mathbf{b}/2 = 1$; the reflected (radiated) power can be computed by the wave component a_1, which gives $\mathbf{a}^T \mathbf{A}_\alpha \mathbf{a}/2 = a_1^H a_1 = 0.999999998$ coincidence is very good. However, the numerical result reveals that the other scattered wave components are quite large, that resonance appears. For this example, the resonance extent is very small, if selects $\mathbf{R} = 0.99\mathbf{R}_0$, that the deviation is small, but when $b_1 = 1,\ b_2 = b_3 = b_4 = 0$ the numerical result solves as

$$a_1 = 0.3102 - .9507i \quad a_2 = -.0702 - .0968i \quad a_3 = 8.1850 + 0.8366i \quad a_4 = -1.7438 + 8.0408i$$

The power flow balance checks still OK, but the amplitude is greatly reduced. ##

Resonance may have devastating consequences; however, it can also be applied for some purpose. Wave induced resonance should be carefully monitored. Here, only a simple description is made for attention.

§5.14, Wave propagation along periodical structures

In aeronautical, mechanical and electrical engineering, analysis of wave propagation along a periodical structure shows that the eigenvalues exhibit energy band behavior similar to that in solid state physics [119]. Such that a wave with frequency ω, which is in a pass-band, can propagate along the periodical structure, whereas otherwise ω is in a stop-band and the wave decays to zero over long distance. The energy analysis used to find ***pass- and stop-bands*** is also important in other practical disciplines. In particular, periodical electro-magnetic wave-guides, e.g. gratings, behave analogously.

The basic unit in analysis of a periodical wave-guide is its fundamental period, *i.e.* its shortest repeating length. For a given frequency ω, its longitudinal ***wave number*** analysis determines that when an eigenvalue is located on the ***unit circle*** for a periodical structure, in which case the wave can propagate to infinity without decay. Hence this ω is said to be in the ***pass-band***, whereas when ω is in the ***stop-band,*** the wave number is not on the unit circle and so the wave decays for long distance propagation along a periodical structure. Therefore pass- or stop-band analysis is very important, because it can be used to filter electro-magnetic (optical) signals. Note that the energy band analysis of periodical structures and the analysis of electro-magnetic wave-guides are analogous and so a method used in either field can be transformed to the other. The key analysis step for a periodical structure is to ***compute the dynamic stiffness matrix of its typical periodical sub-structure***. (Whereas for a periodical electro-magnetic wave-guide it is to generate the respective 'stiffness matrix' related to the two end electric field vectors of a fundamental period)

A frequency ω must either lie in a pass-band or in a stop-band. The pass-band solution can be derived from a Rayleigh quotient analysis. All the eigen-solutions of the Rayleigh quotient laying in a given frequency range can be found by using the W-W (Wittrick-Williams) algorithm and its extension, as described in Chapter-2, which were developed in the context of the vibration and wave propagation problems of structural mechanics [40,41,114~117]. The W-W algorithm is applied here in the relevant structural mechanics context.

§5.14.1, Dynamic stiffness matrix of a fundamental substructure

The formulation is given in the frequency domain ω. A periodical structure is really a ***sub-structural chain*** [118], which is composed of ***fundamental periodical sub-structures*** (fundamental segment) linked together end-to-end. The left-hand and right-hand ends of a sub-structure are indicated by subscripts a and b respectively.

Let the dynamic stiffness matrix corresponding to the two end displacement vectors of a ***fundamental intervalt*** for a given frequency ω be

$$\mathbf{K}(\omega)=\mathbf{K}=\begin{bmatrix}\mathbf{K}_{\mathrm{aa}} & \mathbf{K}_{\mathrm{ab}}\\ \mathbf{K}_{\mathrm{ba}} & \mathbf{K}_{\mathrm{bb}}\end{bmatrix},\quad \mathbf{K}_{\mathrm{aa}}=\mathbf{K}_{\mathrm{aa}}^{T},\ \mathbf{K}_{\mathrm{bb}}=\mathbf{K}_{\mathrm{bb}}^{T},\ \mathbf{K}_{\mathrm{ba}}=\mathbf{K}_{\mathrm{ab}}^{T} \tag{5.14.1}$$

Here $\mathbf{K}$ is a symmetric matrix, the positive definiteness of which is not assured; $\mathbf{K}_{\mathrm{aa}}$, $\mathbf{K}_{\mathrm{bb}}$ and $\mathbf{K}_{\mathrm{ab}}$ are ($n\times n$) submatrices and, the fundamental period is the shortest repeating length of the periodical structure. According to the dynamic stiffness matrix theory [46], knowing only the dynamic stiffness corresponding to the external displacements of the substructure is insufficient on its own to express the eigenvalue behavior, because the internal eigenvalue count $J(\omega_{\#})$ is also needed [40]. The dynamic stiffness matrix is not the only available representation of the dynamic behavior of a substructure, because the mixed energy representation is also available [114~117] and is very important for the precise integration of the mixed energy matrices of the substructure.

§5.14.2, Energy band and eigen-solutions of a symplectic matrix

After the $2n\times 2n$ dynamic stiffness matrix $\mathbf{K}(\omega)$ of the fundamental substructure has been generated, the energy band analysis for the periodical structure or sub-structural chain [118] can be computed. The dynamic deformation energy $U_{\mathrm{e}}(\mathbf{q}_{\mathrm{a}},\mathbf{q}_{\mathrm{b}};\omega)$ of a fundamental sub-structure can be expressed in terms of the ***dynamic stiffness matrix*** $\mathbf{K}(\omega)$ as

$$U_{\mathrm{e}}(\mathbf{q}_{\mathrm{a}},\mathbf{q}_{\mathrm{b}};\omega)=\begin{Bmatrix}\mathbf{q}_{\mathrm{a}}\\ \mathbf{q}_{\mathrm{b}}\end{Bmatrix}^{T}\mathbf{K}(\omega)\begin{Bmatrix}\mathbf{q}_{\mathrm{a}}\\ \mathbf{q}_{\mathrm{b}}\end{Bmatrix}/2 \tag{5.14.2}$$

where $\mathbf{q}_{\mathrm{a}}$ and $\mathbf{q}_{\mathrm{b}}$ are its two end displacement vectors and ω is only a parameter, which is sometimes not ever stated explicitly, as in equation (5.14.3) below which is obtained by applying the variational principle to the chain of m substructures

$$U(\mathbf{q}_0,\mathbf{q}_n)=\sum_{i=1}^{i=m}U_i(\mathbf{q}_{i-1},\mathbf{q}_i),\quad \delta\Pi\,|_{\mathbf{q}_i,i=1,\cdots,m-1}=0 \tag{5.14.3}$$

Introducing the dual vector

$$\mathbf{p}_{\mathrm{a}}=-\partial U_{\mathrm{e}}/\partial\mathbf{q}_{\mathrm{a}}=-(\mathbf{K}_{\mathrm{aa}}\mathbf{q}_{\mathrm{a}}+\mathbf{K}_{\mathrm{ab}}\mathbf{q}_{\mathrm{b}}) \tag{5.14.4a}$$

$$\mathbf{p}_{\mathrm{b}}=\partial U_{\mathrm{e}}/\partial\mathbf{q}_{\mathrm{b}}=(\mathbf{K}_{\mathrm{ba}}\mathbf{q}_{\mathrm{a}}+\mathbf{K}_{\mathrm{bb}}\mathbf{q}_{\mathrm{b}}) \tag{5.14.4b}$$

The **equilibrium equation** of the i-th station is derived from equation (5.14.3) as

$$\partial U_i(\mathbf{q}_{i-1},\mathbf{q}_i)/\partial\mathbf{q}_i+\partial U_{i+1}(\mathbf{q}_i,\mathbf{q}_{i+1})/\partial\mathbf{q}_i=\mathbf{p}_{\mathrm{b},i}-\mathbf{p}_{\mathrm{a},i+1}=\mathbf{0}$$

Note that equation (5.14.4) expresses the dual vectors $\mathbf{p}_a, \mathbf{p}_b$ by means of the original displacement vectors $\mathbf{q}_a$ and $\mathbf{q}_b$ in the displacement method. However, by introducing the ***state vector***

$$\mathbf{v}_j \underset{\text{def}}{=} \begin{Bmatrix} \mathbf{q}_j \\ \mathbf{p}_j \end{Bmatrix}$$

the equilibrium equation can be transformed into the following state vector transfer form, in which the right-hand end vectors $\mathbf{q}_b$ and $\mathbf{p}_b$ are expressed in terms of the left-hand end vectors $\mathbf{q}_a$ and $\mathbf{p}_a$. Hence from equation (5.14.4)

$$\mathbf{v}_b = \mathbf{S}\mathbf{v}_a \tag{5.14.5}$$

i.e. $\mathbf{v}_{j+1} = \mathbf{S}\mathbf{v}_j$ for a ***fundamental substructure*** (fundamental interval)

$$\mathbf{S} = \begin{bmatrix} \mathbf{S}_{11} & \mathbf{S}_{12} \\ \mathbf{S}_{21} & \mathbf{S}_{22} \end{bmatrix}, \quad \left. \begin{array}{ll} \mathbf{S}_{11} = -\mathbf{K}_{ab}^{-1}\mathbf{K}_{aa}, & \mathbf{S}_{12} = -\mathbf{K}_{ab}^{-1} \\ \mathbf{S}_{21} = \mathbf{K}_{ba} - \mathbf{K}_{bb}\mathbf{K}_{ab}^{-1}\mathbf{K}_{aa}, & \mathbf{S}_{22} = -\mathbf{K}_{bb}\mathbf{K}_{ab}^{-1} \end{array} \right\} \tag{5.14.6}$$

where $\mathbf{S}$ is the transfer matrix, which is ***symplectic***, *i.e.* it satisfies

$$\mathbf{S}^T\mathbf{J}\mathbf{S} = \mathbf{J}, \quad \text{where } \mathbf{J} = \begin{bmatrix} \mathbf{0} & \mathbf{I}_n \\ -\mathbf{I}_n & \mathbf{0} \end{bmatrix} \tag{5.14.7}$$

Using the method of separation of variables to solve the transfer matrix equation (5.14.5) yields the eigenvalue problem

$$\mathbf{S}\boldsymbol{\psi} = \mu\boldsymbol{\psi} \tag{5.14.8}$$

After the eigen-pair $(\mu, \boldsymbol{\psi})$ has been found, the solution of the original transfer equation (5.14.5) is $\mathbf{v}_i = \boldsymbol{\psi}\mu^i$, where i denotes the i-th junction (station) of the sub-structural chain. Now the symplectic eigenvalue problem has the following characteristics.

If μ is an eigenvalue then so also is μ^{-1}, as follows. Left multiplying equation (5.14.8) by $\mathbf{S}^T\mathbf{J}$ and using equation (5.14.7) gives $\mathbf{S}^T(\mathbf{J}\boldsymbol{\psi}) = \mu^{-1}(\mathbf{J}\boldsymbol{\psi})$, which shows that μ^{-1} is an eigenvalue of $\mathbf{S}^T$ for which the eigenvector is $\mathbf{J}\boldsymbol{\psi}$. However, any eigenvalue of $\mathbf{S}^T$ is also an eigenvalue of $\mathbf{S}$. Hence the $2n$ eigenvalues of $\mathbf{S}$ can be subdivided into the two classes

$$\alpha)\quad \mu_j,\ \operatorname{abs}(\mu_j) < 1 \text{ or } \operatorname{abs}(\mu_j) = 1 \wedge \operatorname{Im}(\mu_j) > 0,\quad j = 1,\cdots,n \tag{5.14.9a}$$

$$\beta)\quad \mu_{n+j} = \mu_j^{-1},\quad j = 1,\cdots,n \tag{5.14.9b}$$

where μ_j and μ_{n+j} are mutually ***symplectic adjoint*** eigenvalues.

The following derivation applies for any two eigen-solutions, denoted by j and k, of the symplectic matrix $\mathbf{S}$:

$$\begin{array}{ll} \mathbf{S}\boldsymbol{\psi}_j = \mu_j\boldsymbol{\psi}_j, & \mathbf{S}\boldsymbol{\psi}_k = \mu_k\boldsymbol{\psi}_k \\ \mathbf{S}^T(\mathbf{J}\boldsymbol{\psi}_j) = \mu_j^{-1}(\mathbf{J}\boldsymbol{\psi}_j), & \boldsymbol{\psi}_k^T \cdot \mathbf{S}^T = \mu_k\boldsymbol{\psi}_k^T \\ \boldsymbol{\psi}_k^T \cdot (\mathbf{S}^T\mathbf{J}) \cdot \boldsymbol{\psi}_j = \mu_j^{-1}\boldsymbol{\psi}_k^T \cdot \mathbf{J} \cdot \boldsymbol{\psi}_j, & \boldsymbol{\psi}_k^T \cdot (\mathbf{S}^T\mathbf{J}) \cdot \boldsymbol{\psi}_j = \mu_k\boldsymbol{\psi}_k^T \cdot \mathbf{J} \cdot \boldsymbol{\psi}_j \end{array}$$

Hence, subtraction of the equations in the final line gives

$$(\mu_k - \mu_j^{-1})\psi_k^T \cdot \mathbf{J} \cdot \psi_j = 0$$

Therefore either the two eigen-solutions j and $k = n + j$ are mutually ***symplectic adjoint*** to each other, so that two constant factors can be selected to achieve ***adjoint symplectic normalization***, or the eigen-solutions j and k are ***symplectic orthogonal.*** In other words

$$\psi_j^T \cdot \mathbf{J} \cdot \psi_k = \delta_{j,k-n} \tag{5.14.10}$$

which is called ***adjoint symplectic ortho-normality***. Composing a $2n \times 2n$ matrix $\mathbf{\Psi}$ by using all the eigenvectors gives

$$\mathbf{\Psi} = [\psi_1 \quad \cdots \quad \psi_n; \quad \psi_{n+1} \quad \cdots \quad \psi_{2n}] \tag{5.14.11}$$

and then using the adjoint symplectic ortho-normality relationship yields the matrix identity $\mathbf{\Psi}^T \mathbf{J} \mathbf{\Psi} = \mathbf{J}$, so that $\mathbf{\Psi}$ is called a ***symplectic*** matrix.

When the eigenvalue is located on the unit circle $\mu = e^{i\theta}$, the corresponding solution of the original equation is $\mathbf{v}_j = \psi \mu^j$. Therefore the vector $\mathbf{v}_j$ does not decay because $\text{abs}(\mu^j) = \text{abs}(e^{i(j\theta)}) = 1$ and so the solution gives a transmission wave.

For the simplest case of $n = 1$, the eigen-equation is $\mu^2 - (S_{11} + S_{22})\mu + 1 = 0$. Then for

$$(S_{11} + S_{22})^2 < 4 \quad \text{or} \quad \text{abs}[(K_{11} + K_{22}) / K_{12}] < 2 \tag{5.14.12}$$

a transmission wave appears, i.e. ω is in the pass-band. Changing the above inequality to equality and solving with respect to ω gives the boundary between the pass- and stop-bands.

The general case is multi-dimensional (n-dimensional) and so solving the symplectic eigenvalue problem (5.14.8), requires prior solution of a skew-symmetric symplectic eigen-problem, as follows. Left multiplying equation (5.14.8) by $\mathbf{S}^T\mathbf{J}$ to give $\mathbf{S}^T(\mathbf{J}\psi) = \mu^{-1}(\mathbf{J}\psi)$ and then using equation (5.14.6) gives $\mathbf{S}^{-1}\psi = \mu^{-1}\psi$. Combining this with equation (5.14.8) gives

$$\mathbf{A}\psi = (\mu + \mu^{-1})\mathbf{J}\psi, \quad \mathbf{A} = \mathbf{J}(\mathbf{S} + \mathbf{S}^{-1}), \quad \mathbf{A}^T = -\mathbf{A} \tag{5.14.13}$$

where the skew-symmetric nature of $\mathbf{A}$ is readily verified from equation (5.14.7). Solving to find the pairs of eigen-solutions of equation (5.14.13) makes solution for the eigen-solutions of the original equation (5.14.8) easy. Moreover an algorithm given in section 2.3.3.3 is available for solving this ***symplectic eigen-problem for a skew-symmetric matrix.***

§5.14.3, Pass-band analysis for periodical wave-guides

Because the electro-magnetic wave-guide analysis also generates a stiffness matrix, an electro-magnetic stiffness matrix, the above analysis can be used in such wave-guide problems. The pass-band wave number of a periodical wave-guide is $\mu = \exp(i\theta)$. Here the phase angle θ is the only parameter and is real, so that the eigen-problem can be solved inversely, by finding the frequency ω for a given wave number μ. For a class-α eigenvalue μ (see equation (5.14.9a)) this gives $0 \le \theta \le \pi$. In structural vibration the frequency eigenvalue problem is usually to find ω with the boundary conditions given. However, for wave propagation problems, the boundary condition becomes a given wave number and the requirement is to find the frequency ω. The boundaries between the pass- and stop-bands are of great concern and are given by solving to find ω for $\mu = 1$ and $\mu = -1$.

In practical applications, all the eigen-frequencies ω in a selected frequency range $0 < \omega^2 < \omega_\#^2$ are required and the W-W algorithm for counting the number of eigenvalues for a dynamic stiffness matrix is indeed for this purpose.

Substituting $\mu = 1$ into the eigen-equation $\mathbf{S}\boldsymbol{\psi} = \mu\boldsymbol{\psi}$ gives the eigen-equation for ω as

$$\mathbf{K}_{qq}(\omega)\mathbf{q} = 0, \quad \mathbf{K}_{qq}(\omega) = \mathbf{K}_{aa} + \mathbf{K}_{ab} + \mathbf{K}_{ba} + \mathbf{K}_{bb} \tag{5.14.14}$$

where all the sub-matrices are functions of ω. The other boundary between pass- and stop-bands is obtained by substituting $\mu = -1$, which gives

$$(\mathbf{K}_{aa} + \mathbf{K}_{bb} - \mathbf{K}_{ab} - \mathbf{K}_{ba})\mathbf{q} = 0 \tag{5.14.15}$$

Note that both of equations (5.14.14) and (5.14.15) are eigenvalue problems of $n \times n$ symmetric matrices, and so the W-W eigenvalue counting algorithm is applicable. Note also that $\mathbf{K}_{aa}(\omega)$, $\mathbf{K}_{bb}(\omega)$ and $\mathbf{K}_{ba}(\omega)$ are all submatrices of the dynamic stiffness matrix $\mathbf{K}(\omega)$ of the interval (z_a, z_b).

Equations (5.14.14) and (5.14.15) apply only for the ***boundaries*** $\theta = 0$ and π ***of the pass-band,*** respectively. However for an arbitrary phase angle $0 \le \theta \le \pi$, the equation $\mathbf{S}\boldsymbol{\psi} = \mu\boldsymbol{\psi}$ yields the eigen-equation for ω as

$$\mathbf{K}_{qq}(\omega)\mathbf{q} = 0, \quad \mathbf{K}_{qq}(\omega) = \mathbf{K}_{ab}\mu + (\mathbf{K}_{aa} + \mathbf{K}_{bb}) + \mu^{-1}\mathbf{K}_{ba} \tag{5.14.16}$$

where μ is a given parameter and all the sub-matrices are functions of ω. Because the parameter $\mu = \exp(\mathrm{i}\theta)$ is a complex number, the dynamic stiffness matrix $\mathbf{K}_{qq}(\omega)$ is no longer real and symmetric but is instead Hermitian, as follows

$$\mathbf{K}_{aa}^H = \mathbf{K}_{aa}, \quad \mathbf{K}_{bb}^H = \mathbf{K}_{bb}, \quad \mathbf{K}_{ab}^H = \mathbf{K}_{ba} \tag{5.14.17}$$

$$\begin{aligned} \mathbf{K}_{qq}^H &= \mathbf{K}_{ab}^H \exp(-\mathrm{i}\theta) + (\mathbf{K}_{aa}^H + \mathbf{K}_{bb}^H) + \mathbf{K}_{ba}^H \exp(\mathrm{i}\theta) \\ &= \mathbf{K}_{ba} \exp(-\mathrm{i}\theta) + (\mathbf{K}_{aa} + \mathbf{K}_{bb}) + \mathbf{K}_{ab} \exp(\mathrm{i}\theta) = \mathbf{K}_{qq} \end{aligned} \tag{5.14.18}$$

Therefore equation (5.14.16) is an eigen-problem for a Hermite matrix, which is a problem for which the W-W algorithm still applies. The equation looked simple, however, its precise computation and the execution of the W-W algorithm need further explanation.

The description for the displacement vector (electric field vectors) $\mathbf{q}_a$ and $\mathbf{q}_b$ at the two ends a and b of the fundamental period must coincide with each other, but unless $\theta = 0$ or π the values of $\mathbf{q}_a$ and $\mathbf{q}_b$ are different due to phase changes. The origin of the fundamental period should be chosen such that $\mathbf{q}_a$ (and also $\mathbf{q}_b$) has the least dimension possible. In applying W-W algorithm, the count of the number of eigenvalues internal to the fundamental period, $J(\omega_{\#})$, is necessary, where $\omega_{\#}$ is a given upper bound of a frequency range for which the lower bound is zero. Hence there are number $J_{\rm i}(\omega_{\#})$ eigenvalues in the range $0 \le \omega^2 \le \omega_{\#}^2$ when the two ends of the fundamental period are regarded as ***clamped***, *i.e.* $\mathbf{q}_{\rm a} = \mathbf{q}_{\rm b} = \mathbf{0}$.

§5.14.4, Dynamic stiffness of a fundamental period and eigenvalue count

The eigen-equation of the fundamental interval $(z_{\rm a}, z_{\rm b})$ described above is based on the external dynamic stiffness matrix $\mathbf{K}(\omega)$ of the fundamental interval of the wave-guide. However, $\mathbf{K}(\omega_{\#})$ represents only the external behavior so that, as for the dynamic problem there is internal behavior of the fundamental period and so the internal eigenvalue count $J_{\rm i}(\omega_{\#})$ is required. Hence the boundaries between pass- and stop-bands in the frequency ω domain are still the eigen-problems of a symmetric matrix given by equations (5.14.14) and (5.14.15). Therefore the matrix $\mathbf{K}(\omega)$ should be computed precisely, by methods which depend on the type of wave-guide, etc. Below, a fundamental period of plane electro-magnetic wave-guide is considered with its frequency ω given.

Let the fundamental period be composed of several constant cross-section intervals (segments). Then combining two adjacent intervals 1 and 2 gives a longer interval c, see Figure.5.4, with the equations for the combination being

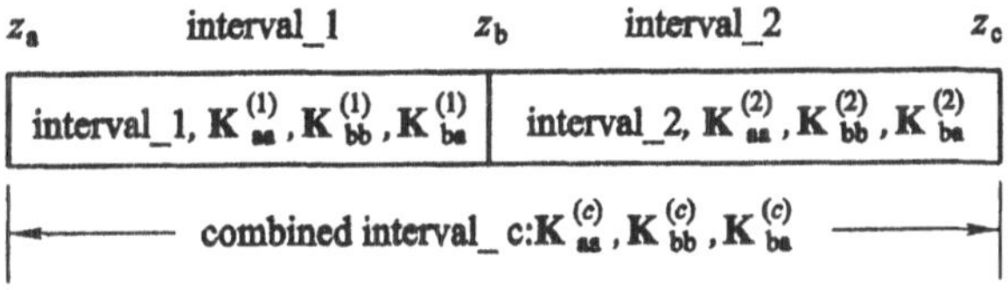

Figure 5.4, Interval combination

$$\mathbf{K}^{(c)}_{\rm aa} = \mathbf{K}^{(2)}_{\rm aa} - \mathbf{K}^{(2)}_{\rm ab}(\mathbf{K}^{(1)}_{\rm bb} + \mathbf{K}^{(2)}_{\rm aa})^{-1}\mathbf{K}^{(2)}_{\rm ba} \tag{5.14.19a}$$

$$\mathbf{K}^{(c)}_{\rm bb} = \mathbf{K}^{(2)}_{\rm bb} - \mathbf{K}^{(2)}_{\rm ba}(\mathbf{K}^{(1)}_{\rm bb} + \mathbf{K}^{(2)}_{\rm aa})^{-1}\mathbf{K}^{(2)}_{\rm ab} \tag{5.14.19b}$$

$$\mathbf{K}^{(c)}_{\rm ab} = -\mathbf{K}^{(1)}_{\rm ab}(\mathbf{K}^{(1)}_{\rm bb} + \mathbf{K}^{(2)}_{\rm aa})^{-1}\mathbf{K}^{(2)}_{\rm ab} \ , \quad \mathbf{K}^{(c)}_{\rm ba} = \mathbf{K}^{(c)T}_{\rm ab} \tag{5.14.19c}$$

where the superscripts denote intervals 1,2 and c. Note that although the two ends of interval 1 are marked as a and b, it is not necessarily a fundamental interval. The above equation applies only for the combination of external $n \times n$ stiffness matrices. However, for dynamic stiffness matrix analysis (and hence also for the present analogous electro-magnetic stiffness matrix analysis), the internal eigenvalue count should also be combined, and the recurrence equation for this eigenvalue count is the W-W algorithm

$$J_c(\omega_\#) = J_1(\omega_\#) + J_2(\omega_\#) + s\{\mathbf{R}\} \tag{5.14.20}$$

where $\omega_\#$ is a given frequency bound and $\mathbf{R}(\omega_\#) = \mathbf{R} = (\mathbf{K}_{bb}^{(1)} + \mathbf{K}_{aa}^{(2)})$ is the internal stiffness matrix when combining the two intervals. $J_1(\omega_\#)$ and $J_2(\omega_\#)$ are the internal eigenvalue counts of intervals 1 and 2 respectively, and $s\{\cdots\}$ represents the eigenvalue count operator such that $s\{\mathbf{R}\}$ is the number of negative entries of the diagonal matrix $\mathbf{D}$ obtained by the factorization $\mathbf{R} = \mathbf{L}^T\mathbf{D}\mathbf{L}$, see [115]. Equation (5.14.20) enables all the eigenvalues in the given range $0 \le \omega^2 < \omega_\#^2$ to be found with certainty by many alternative methods, of which the simplest is the bisection method.

When the fundamental period has more than two intervals it can be assembled by applying the above equations recursively to add one interval out a time.

The problem has now been reduced to performing computations for each constant cross section interval. The interval is a continuum and so has an infinite number of degrees of freedom in its cross-section. However, in practical computation these are usually approximated by a finite number of degrees of freedom, n, via a discretisation procedure, *i.e.* the semi-analytical approach is used. This discretisation has two alternative forms. In the first, which is called the spectral method and is the one adopted here, the field is assumed to be the linear combination of n independent basis functions, which represent the transverse cross-section distribution. Thus the n multipliers of these basis functions are the unknown functions of n which need to be found. A set of n simultaneous differential equations derived from the variational principle are then solved by a numerical method, e.g. the precise integration method. The second discretisation procedure is identical to the first except that at the first step FEM is used for the cross-section, to give the precise form [102] of the finite strip method.

For ease of presentation, a simple problem is now introduced as an example. As mentioned, the electro-magnetic wave-guide problem is analogous to the sub-structural analysis, so that the example is such a wave-guide.

Example 5.10, Consider a plane electro-magnetic wave-guide shown in figure 5.5, where z is a longitudinal coordinate along which the wave propagates. Suppose that the core film is a vacuum surrounded by perfect conductors with $\mu_0\varepsilon_0 = 1/c^2$, $c = 2.998\times10^8$ m/sec (=the velocity of light in a vacuum), $\varepsilon_1 = \varepsilon_2 = \varepsilon_0$, $x_2 = 1.25\times10^{-6}$ m, $x_1 = 0.8x_2$ and $l_1 = l_2 = 1\times10^{-7}$ m. The problem is to find the eigenvalue ω/c for any given wave phase angle for which $0 \le \theta \le \pi$.

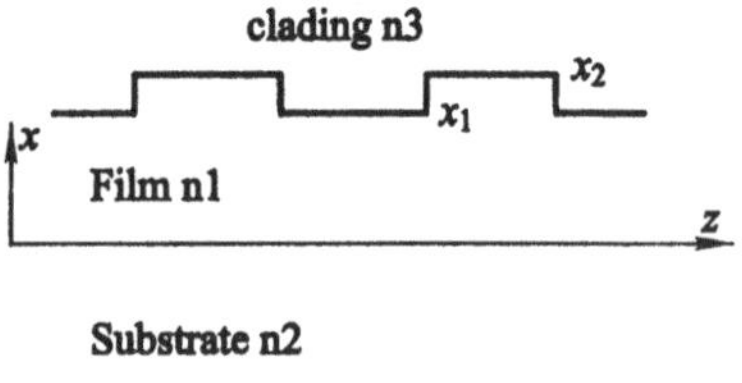

Figure 5.5, *A plane wave-guide*

Solution: For this problem, the fundamental substructure is composed of two uniform intervals between which the cross-section changes abruptly. Let the fundamental period of the wave-guide be $-l_1 < z < l_2$, with intervals $-l_1 < z < 0$ and $0 < z < l_2$ which are of thickness x_1 and x_2, respectively, and are represented by m and n basis functions, respectively. Now the abrupt change of cross-section $x_1 < x_2$ at $z = 0$, results the following conditions there

$$e_y(x,+0) = \begin{cases} e_y(x,-0), & x < x_1 \\ 0, & x_1 \le x < x_2 \end{cases}, \quad h_x(x,+0) = \begin{cases} h_x(x,-0), & x < x_1 \\ 0, & x_1 \le x < x_2 \end{cases}, \quad \text{at } z = 0$$

where h_x is the component of magnetic field vector. The variational method can be used to process these junction conditions by transforming the interval $0 < z < l_2$ to $-0 < z < l_2 + 0$, *i.e.* the thickness of the two external cross-sections changes to x_1 and the number of terms in the expansion changes to m. For brevity the derivation of this transformation matrix is omitted, but the resulting matrix $\mathbf{T}$ at the stations $z = 0$ and $z = l_2$ needed for treating the cross-section change gives

$$\begin{aligned} q_{ai}^{(2)} &= \sum_{j=1}^{m} T_{ij} \cdot q_{aj}^{(1)}, \qquad i = 1, \cdots, n, \qquad \text{or} \quad \mathbf{q}_a^{(2)} = \mathbf{T}\mathbf{q}_a^{(1)} \\ T_{ij} &= \sin[(j - ix_1/x_2)\pi]/[\pi(jx_2/x_1 - i)] \\ &\quad - \sin[(j + ix_1/x_2)\pi]/[\pi(jx_2/x_1 + i)] \end{aligned} \tag{5.14.21}$$

where the vectors $\mathbf{q}_a^{(2)}$ and $\mathbf{q}_a^{(1)}$ are of orders n and m, respectively, so that $\mathbf{T}$ is $n \times m$. A similar derivation for the right-hand end gives $\mathbf{q}_b^{(2)} = \mathbf{T}\mathbf{q}_b^{(1)}$. Here subscripts a and b represent the left- and right-hand ends, and the superscripts $^{(1)}$ and $^{(2)}$ denote series expansion for thickness x_1 or x_2, respectively. Therefore the ***electro-magnetic interval stiffness matrix*** $\mathbf{K}_2^{(2)}$, which is for the end vectors $\mathbf{q}_a^{(2)}$ and $\mathbf{q}_b^{(2)}$ of a wave-guide thickness x_2, should be transformed by using the equations

$$\begin{Bmatrix} \mathbf{q}_a^{(2)} \\ \mathbf{q}_b^{(2)} \end{Bmatrix} = \begin{bmatrix} \mathbf{T} & \mathbf{0} \\ \mathbf{0} & \mathbf{T} \end{bmatrix} \begin{Bmatrix} \mathbf{q}_a^{(1)} \\ \mathbf{q}_b^{(1)} \end{Bmatrix}, \qquad \mathbf{K}_2^{(1)} = \begin{bmatrix} \mathbf{T}^T & \mathbf{0} \\ \mathbf{0} & \mathbf{T}^T \end{bmatrix} \mathbf{K}_2^{(2)} \begin{bmatrix} \mathbf{T} & \mathbf{0} \\ \mathbf{0} & \mathbf{T} \end{bmatrix} \tag{5.14.22}$$

for the two end electric field (displacement) vectors $\mathbf{q}_a^{(1)}$ and $\mathbf{q}_b^{(1)}$ with thickness x_1. The matrix $\mathbf{K}_2^{(2)}$ is a $2n \times 2n$ matrix composed of four diagonal $n \times n$ matrices, whereas the transformation makes $\mathbf{K}_2^{(1)}$ into a fully populated $2m \times 2m$ matrix. (Note, that for a general cross-section wave-guide the semi-analytical method discretizing the cross-section results in a $\mathbf{K}_2^{(2)}$ which is fully populated) ** to be continued.

After transformation, the stiffness matrix $\mathbf{K}_2^{(1)}$ represents the behavior of interval-2 but relates to the contracted cross-section. However, the fundamental period interval is composed of both the interval-1 and interval -2. Therefore the electro-magnetic stiffness matrix $\mathbf{K}_1^{(1)}$ is also needed by the interval combination algorithm (see equations (5.14.19) and (5.14.20)) which is applied to give the

combined electro-magnetic stiffness matrix $\mathbf{K}_c$ for the fundamental period of wave-guide and its associated J_c. Thus the computation of $\mathbf{K}_c$ at given ω is a fundamental step in the analysis of a periodical wave-guide and corresponds to the ***dynamic stiffness matrix*** in structural analysis of the fundamental substructure, which is also computed for given ω.

Because the ***stiffness matrix*** is fundamental to FEM, after introducing potential energy and its stiffness matrix the analysis of electro-magnetic wave-guides can use the same methodology as in structural mechanics. Thus computation of the eigenvalue count for a fundamental period of a wave-guide has three principal steps. The first two are to compute the eigenvalues counts internal to each ***uniform*** interval by using the precise integration method and for the combination of intervals within the fundamental period. Then the final step gives equation (5.14.23) as follows. For a pair θ, $\omega_\#$ of given phase angle and given frequency, the eigenvalue count of the pass-band wave-guide J_p can be computed from the equation

$$J_p = J_f + s\{\mathbf{K}_{qq}\} \tag{5.14.23}$$

where $s\{\mathbf{K}_{qq}\}$ is the eigenvalue count of the Hermitian matrix given by equation (5.14.18) and J_f is the value of the eigenvalue count for the fundamental period yielded by the first two steps.

§5.14.5, Computation of the pass-band eigenvalues

As shown in section 5.14.4, the boundaries of the pass-band are given by equations (5.14.14) and (5.14.15), for which the phase angles are $\theta = 0, \pi$ respectively. For other values of θ and a given frequency ω, the electro-magnetic stiffness matrix $\mathbf{K}_c(\omega)$ of the fundamental period of a wave-guide can be computed as described in the previous sections. Note that $\mathbf{K}_c(\omega)$ is a Hermitian matrix, *i.e.* $\mathbf{K}_c(\omega) = \mathbf{K}_c^H(\omega)$, where superscript ^{H} denotes Hermitian transposition. Because the frequency domain is multiplied by $\exp(-\mathrm{i}\omega t)$, equation (5.14.2) for the energy should be slightly modified, by replacing the superscript T as H to give

$$U_e(\mathbf{q}_a, \mathbf{q}_b; \omega) = \begin{Bmatrix} \mathbf{q}_a \\ \mathbf{q}_b \end{Bmatrix}^H \mathbf{K}(\omega) \begin{Bmatrix} \mathbf{q}_a \\ \mathbf{q}_b \end{Bmatrix} / 2 \tag{5.14.2'}$$

To find all the pass-band eigenvalues of a periodical wave-guide, the electro-magnetic stiffness matrix $\mathbf{K}_c(\omega)$ of the fundamental period should be used as the matrix $\mathbf{K}(\omega)$ in equations (5.14.14~16). Hence the matrix $\mathbf{K}_{qq}(\omega)$ in equation (5.14.16) is Hermitian, so that the W-W algorithm still applies. For a given $\omega_\#$ the eigenvalue count of the Hermitian matrix $\mathbf{K}_{qq}(\omega_\#)$ can be computed by the triangular factorization

$$\mathbf{K}_{qq}(\omega_\#) = \mathbf{LDL}^H, \quad s\{\mathbf{K}_{qq}\} = \text{number of negative entries in } \mathbf{D} \tag{5.14.24}$$

where $\mathbf{D}$ is a real diagonal matrix. Therefore all the eigenvalues can be found, e.g. by the bisection method.

** Continuing computation of the previous numerical example 5.10 gives the results in Table 5.2

Table 5.2. *The value of* $10^{-6}\cdot\omega/c$ *with given* θ *solved with* $n=24, m=24$ *are listed for the example 5.10.*

$\theta=0°$	15°	30°	45°	60°	75°	90°
3.117	4.849	4.883	4.994	6.063	7.207	8.393
6.234	6.368	6.755	7.354	8.118	9.005	9.980
9.349	9.439	9.704	10.131	10.698	11.386	12.172
12.462	12.530	12.731	13.059	13.504	14.055	14.699
15.572	15.626	15.788	16.053	16.417	16.873	17.413
18.676	18.721	18.856	19.079	19.386	19.773	20.234
21.771	21.810	21.925	22.116	22.380	22.714	23.111
24.849	24.882	24.982	25.147	25.373	25.109	23.828

Continue

105°	120°	135°	150°	165°	180°
9.603	10.825	12.045	13.241	14.337	14.928
11.017	12.097	13.201	14.303	15.326	15.886
13.036	13.962	14.930	15.914	16.846	17.101
15.423	16.213	17.055	17.925	17.696	17.365
18.028	18.709	19.442	18.805	18.485	17.914
20.764	21.269	20.024	19.553	18.768	19.190
22.547	21.360	20.730	20.212	19.735	19.246
23.171	21.949	21.826	20.749	20.968	20.843

Periodical electro-magnetic wave-guides have important applications and determining their characteristics involves an eigen-problem, which can be solved by using symplectic mathematics and the methodology applied in structural mechanics. Thus the electro-magnetic stiffness matrix (or impedence) of a fundamental period of a periodical wave-guide has been introduced and then symplectic mathematics has been applied as in the analysis of sub-structural chains. The energy band analysis for the periodical wave-guide is then carried out on the same basis as in structural analysis. ##

In the above text, the application of duality methodology is shown for various topics, however, only a beginning exploration has been made. The application of this methodology can be far extended to other fields, further work is anticipated.

Chapter 6, Linear optimal control, theory and computation

Control systems are found throughout in nature and in technology. Automatic control theory was initiated from applied mechanics. Classical control system design was mostly developed during the first sixty years of the twentieth century, and has been wide used in various engineering disciplines. For which, the mathematical basis is the theory of ODE, the Laplace transformation, the stability theory, the transfer functions, etc., and is characterized by the use of root locus, frequency response method, and Nyquist contour etc. for analysis and design. The classical controller design methodology is iterative, and is effective for single-input, single-output linear time-invariant system analysis and design. The disadvantage of classical control theory is often fail to get intuitive insight for high order systems and for multi-input, multi-output (MIMO) systems, and fail to describe the system internal behavior.

Under the computer impact, the linear quadratic Gaussian (LQG) optimal control theory was developed during the fifties and sixties and refined in the seventies of the twentieth century, which means the transition from classical control theory to the modern control theory characterized by the state space approach. This design method has been successfully used in a wide range of applications in aerospace and many other areas.

The modern control theory is not a simple extension along the way of classical control theory, but changes the methodology which updates the theoretical basis, that the fundamental variables changes from single to state variables etc. The state space description involves the system ***internal variables,*** so that it is no longer only the input-output relation description as that in classical control theory. The state space description can directly discuss the system performance in finite time domain, and the controllability and observability of the system mean deeper understanding of system structure.

All real systems are non-linear, however the mathematical analysis for non-linear system is usually very difficult. Linear system is a good approximation to real system and is the fundamental part of control theory.

Control system design requires a mathematical model of the plant, and there are many forms of the models, including differential equation models, transfer function models and state space models etc. The state space model is a set of first-order linear differential equations in matrix/vector form. The state space model is ideal for computer aided design, since computers typically work well with matrices and vectors. In addition, many powerful results from linear algebra can be applied when using state space models. The state space model is particularly convenient for control system analysis and design, because powerful mathematical method, computer aided analysis and design software are available. Furthermore, the same basic equations can be used to describe low- and high-order systems, as well as single-input single-output (SISO) and multi-input multi-output (MIMO) systems, etc.

Control theory updated the theoretical basis according to its own requirement, and at a first glance, the control theory has developed far apart from the applied mechanics on theoretical basis. However, it is found that the mathematics of modern control theory has a one-one correspondence relationship with some kind of problems in structural mechanics, which establishes the **analogy relationship between structural mechanics and optimal control**. The analogy relationship means that the theories and algorithms developed in both sides can be mutually transplanted to each other, which is quite beneficial to both sides of optimal control and applied mechanics.

The methods of precise integration, the algorithm for eigen-solutions of Hamilton matrix, adjoint symplectic ortho-normality and the precise integration solutions of the Riccati differential equations etc. described in the previous chapters can be used in the modern control theory. The interdisciplinary research shortens the distance between both the disciplines, which is quite beneficial to the teaching and research in both sides.

§6.1, State space of linear system

A dynamical system is usually described by a set of ordinary differential or difference equations. System theory or control theory is usually developed based on a dynamical system. When the set of differential or difference governing equations is linear, then the system is linear. Although the real system has non-linear factors, however, the motion around the vicinity of the **nominal** orbit, *i.e.* the perturbation, can be described by the linear system theory. The linear system theory is convenient for its mathematical treatment, so that the linear system theory is very important in control theory and engineering.

§6.1.1, Input-output description and state space description

The mathematical description of a control system can be classified as two fundamental kinds:
1) External description, or input-output description; and
2) State space description; and is sketched in figure 6.1.

$$\begin{array}{ccc}
\textbf{input:} & \boxed{\textbf{state vector:}} & \textbf{output:} \\
u = \{u_1, \cdots, u_m\}^T \longrightarrow & x = \{x_1, \cdots, x_n\}^T & \longrightarrow \quad y = \{y_1, \cdots, y_q\}^T
\end{array}$$

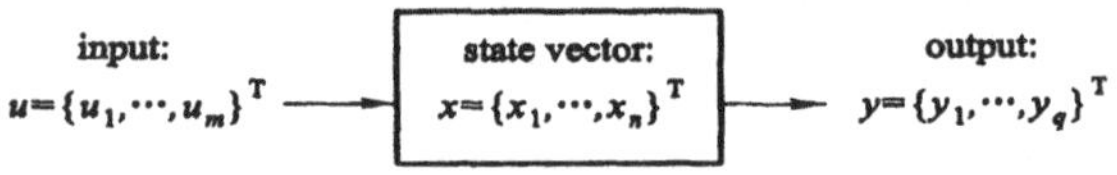

Figure 6.1, *The state space method*

The vectors $\mathbf{u} = \{u_1, u_2, ..., u_m\}^T$ and $\mathbf{y} = \{y_1, y_2, ..., y_q\}^T$ are the input and output of the system, respectively, and are considered the **external variables** of the system. However, the system dynamics cannot be described completely with these variables, that the system dynamics is described by the **state variables** denoted as $x_1, x_2, ..., x_n$, or combined as the **state vector**

$$\mathbf{x} = \{x_1, x_2, \cdots, x_n\}^T \tag{6.1.1}$$

which is a function of time t. Part of the state vector is under the measurement all the time and the result of measurement is regarded as the output vector. The state variables completely present the behavior of the system, but some part of the state (internal variables) is not involved in the output variables.

The traditional control theory treats the system as a 'black box', and put the attention to the variation of output variables $\mathbf{y}$ with the changing of input variables $\mathbf{u}$. Although the system is originally a multiple degrees of freedom one, but for mathematical convenience the methodology is to eliminate as many variables as possible, in order to simplify the basic equation to be solved as a single-input single-output high order ODE. Recall that in engineering mechanics, the traditional methodology is also of such classical consideration. For a time-invariant linear system with only one input variable u and one output variable y, the external mathematical description is a linear ODE with constant coefficients

$$\begin{aligned} &y^{(n)} + a_{n-1}y^{(n-1)} + \dots + a_1 y^{(1)} + a_0 y = \\ &\qquad b_{n-1}u^{(n-1)} + b_{n-2}u^{(n-2)} + \dots + b_1 u^{(1)} + b_0 u \end{aligned} \tag{6.1.2}$$

where the coefficients a_i, b_i are all real constants, and $y^{(i)} = \mathrm{d}^i y / \mathrm{d}t^i$. Assuming the initial condition of the input and output variables u, y are all zero, then taking Laplace transformation to the equation (6.1.2) gives the system frequency domain description as

$$\tilde{y}(s) = G(s) \times \tilde{u}(s) \tag{6.1.3}$$

$$\mathrm{L}(y) = \tilde{y}(s) = \int_0^\infty e^{-st} y(t) dt, \qquad \mathrm{L}^{-1}(\tilde{y}) = y(t) = \frac{1}{2\pi i} \int_{\sigma - i\infty}^{\sigma + i\infty} e^{ts} \tilde{y}(s) ds \tag{6.1.4}$$

where $\tilde{u}(s), \tilde{y}(s)$ are the Laplace transformation functions of the input and output $u(t), y(t)$, and

$$G(s) = \frac{b_{n-1}s^{n-1} + \dots + b_1 s + b_0}{s^n + a_{n-1}s^{n-1} + \dots + a_1 s + a_0} = \frac{B(s)}{A(s)} \tag{6.1.5}$$

is the ***transfer function*** of the system.

The classical control theory emphasizes the analysis of system input versus output and their transfer function, and the main concern is the ***system stability***. However, the relation of input versus output is incomplete for system description, because which cannot describe the whole picture internal to the black box. On the other hand, the state vector describes also the internal state of system, so that gives the complete dynamical characteristics of the system. It is an important progress to describe the system with the ***state space method***.

State space is not a new idea, that in classical dynamics the state variables had long been proposed. Hamilton dynamics practically laid the system on state variable description. The state of a dynamical system is defined as a minimum set of variables $x_1(t), x_2(t) \dots, x_n(t)$ which completely characterizes the system behavior in time domain. Equation (6.1) uses $\mathbf{x}(t)$ to denote the state vector. According to the mathematical theory of ordinary differential equations see [36], under the condition of no uncertain external disturbance, give the initial state $\mathbf{x}_0$ at $t = t_0$ then the dynamic equation uniquely determines the system state evolution $\mathbf{x}(t)$. Simply, if the input is given certainly, then the motion is also determined.

A dynamical system is described by a set of ODEs. However, the coefficients of the set of ODEs are not so certain, especially the system is always under external random disturbances all the time, hence only based on the given initial state $\mathbf{x}_0$ at the time t_0, the motion estimation of long time performance under random disturbances is not enough. Therefore a control system requires monitoring its state continuously, that the necessity of the measurement vector $\mathbf{y}$ output is laid on the estimation of the current state $\mathbf{x}(t)$. The ideal measurement is to get the current state $\mathbf{x}(t)$ completely, but it is very difficult or even impossible. Therefore, the measured vector $\mathbf{y}$ is only q-dimensional, $q \le n$. The mathematical model of state space description is given as the dynamic equation and initial condition

$$\dot{\mathbf{x}}(t) = \mathbf{f}(\mathbf{x}, \mathbf{u}, t), \qquad \mathbf{x}(0) = \mathbf{x}_0 \tag{6.1.6}$$

with the measurement output

$$\mathbf{y} = \mathbf{g}(\mathbf{x}, \mathbf{u}, t) \tag{6.1.7}$$

where $\mathbf{f}$ and $\mathbf{g}$ are vector functions. The above representation is general but somewhat vague, because there are random disturbances etc., and the determination of functions $\mathbf{f}$ and $\mathbf{g}$ is not so easy. Both the functions $\mathbf{f}$ and $\mathbf{g}$ can be non-linear. Generally speaking, real systems are all non-linear, however, the solution of non-linear system is very difficult mathematically. Fortunately, for quite a number of practical systems the differential equation can be treated as a linear system approximately. If only the motion of system nearby the nominal motion $\mathbf{x}_*(t)$, $\mathbf{u}_*(t)$ is considered, then the solution can be written as

$$\begin{aligned} &\mathbf{x}(t) = \mathbf{x}_*(t) + \xi(t), \quad \mathbf{u}(t) = \mathbf{u}_*(t) + \eta(t) \\ &\mathbf{y}(t) = \mathbf{y}_*(t) + \zeta(t), \quad \mathbf{y}_*(t) = \mathbf{g}(\mathbf{x}_*, \mathbf{u}_*, t) \end{aligned} \tag{6.1.8}$$

The functions $\mathbf{f}$ and $\mathbf{g}$ in equations (6.1.6) and (6.1.7) can be expanded in Taylor series at the vicinity of $\mathbf{x}_*$ and $\mathbf{u}_*$ with the higher power of ξ and η be neglected (linear approximation)

$$\begin{aligned} &\mathbf{f}(\mathbf{x}, \mathbf{u}, t) \simeq \mathbf{f}(\mathbf{x}_*, \mathbf{u}_*, t) + (\partial \mathbf{f} / \partial \mathbf{x})_*^T \xi(t) + (\partial \mathbf{f} / \partial \mathbf{u})_*^T \eta(t) \\ &\mathbf{g}(\mathbf{x}, \mathbf{u}, t) \simeq \mathbf{g}(\mathbf{x}_*, \mathbf{u}_*, t) + (\partial \mathbf{g} / \partial \mathbf{x})_*^T \xi(t) + (\partial \mathbf{g} / \partial \mathbf{u})_*^T \eta(t) \end{aligned} \tag{6.1.9}$$

In the above equation the derivative of a vector function with respect to a vector variable follows again the rule (1.7.20)

$$\frac{\partial \mathbf{f}}{\partial \mathbf{x}} \equiv \begin{bmatrix} \dfrac{\partial f_1}{\partial x_1} & \cdots & \dfrac{\partial f_n}{\partial x_1} \\ \vdots & \ddots & \vdots \\ \dfrac{\partial f_1}{\partial x_n} & \cdots & \dfrac{\partial f_n}{\partial x_n} \end{bmatrix} = \mathbf{A}(t), \qquad \begin{aligned} \mathbf{C}(t) &= (\partial \mathbf{g}/\partial \mathbf{x})_*^T \\ \mathbf{B}_u(t) &= (\partial \mathbf{f}/\partial \mathbf{u})_*^T \\ \mathbf{D}_u(t) &= (\partial \mathbf{g}/\partial \mathbf{u})_*^T \end{aligned} \tag{6.1.10}$$

Therefore, the dynamic and measurement equations (6.1.6) and (6.1.7) are approximately given as

$$\begin{aligned} \dot{\xi} &= \mathbf{A}(t)\xi + \mathbf{B}_u \eta + [\mathbf{f}(\mathbf{x}_*, \mathbf{u}_*, t) - \dot{\mathbf{x}}_*] \\ \zeta &= \mathbf{C}(t)\xi + \mathbf{D}_u(t)\eta \end{aligned}$$

which are linear equations of ξ, η and ζ. Because linear system is easier to be analyzed mathematically, hence usually the linear approximation of systems is often used. The computation of non-linear system is usually solved based on its approximate linear system solution by iterative method. Usually, the state variables are preferable to be written as $\mathbf{x}(t)$, $\mathbf{u}(t)$ and $\mathbf{y}(t)$ instead of ξ, η and ζ, so that the equations are written as

$$\dot{\mathbf{x}}(t) = \mathbf{A}(t)\mathbf{x} + \mathbf{B}_u(t)\mathbf{u} + \mathbf{w} \tag{6.1.11}$$

$$\mathbf{y} = \mathbf{C}(t)\mathbf{x} + \mathbf{D}_u(t)\mathbf{u} \tag{6.1.12}$$

where $\mathbf{w}(t) = \mathbf{f}(\mathbf{x}_*, \mathbf{u}_*, t) - \dot{\mathbf{x}}_*$ is the non-homogeneous term. The corresponding homogeneous equations are

$$\dot{\mathbf{x}}(t) = \mathbf{A}(t)\mathbf{x} + \mathbf{B}_u(t)\mathbf{u} \tag{6.1.11'}$$

$$\mathbf{y} = \mathbf{C}(t)\mathbf{x} + \mathbf{D}_u(t)\mathbf{u} \tag{6.1.12'}$$

with initial condition $\mathbf{x}(0) = \mathbf{x}_0$, a given vector. The state, control-input and measurement vectors are denoted as $\mathbf{x} = \{x_1, x_2, \ldots, x_n\}^T$, $\mathbf{u} = \{u_1, u_2, \ldots, u_m\}^T$, $\mathbf{y} = \{y_1, y_2, \ldots, y_q\}^T$, respectively. The matrices $\mathbf{A}, \mathbf{B}_u, \mathbf{C}$ and $\mathbf{D}_u$ are given with dimensions $n \times n$, $n \times m$, $q \times n$ and $q \times m$, respectively. For time-variant system these matrices are functions of time t.

For linearized approximate system the matrices $\mathbf{A}(t), \mathbf{B}_u(t), \mathbf{C}(t), \mathbf{D}(t)$ are still dependent on time t. It gives a time variant linear system of differential equations. Comparatively, the time variant linear system has been easier than non-linear system mathematically, but the behavior of a time variant system varies from time to time, and it is not so easy on computations either. Time-invariant (steady state) system, for which the matrices $\mathbf{A}, \mathbf{B}_u, \mathbf{C}, \mathbf{D}$ are invariant with time, is much easier on theory and also on computation. Using steady state system instead of time variant system approximately, some useful results can still be obtained. Hence the time invariant system is investigated most often, and it will be discussed carefully below.

§6.1.1.1, Continuous-time and discrete-time systems

For some kind of problems, the state variables are naturally given at discrete-time, such as many economic problems, the ecology problems etc. The discrete-time model can also be derived from continuous-time problem. The state space linear discrete-time description are given by the equations

$$\mathbf{x}(k+1) = \mathbf{F}(k)\mathbf{x}(k) + \mathbf{B}(k)\mathbf{u}(k) \tag{6.1.13}$$

$$\mathbf{y}(k) = \mathbf{C}(k)\mathbf{x}(k) + \mathbf{D}(k)\mathbf{u}(k) \tag{6.1.14}$$

which give a time-variant system. When the matrices $\mathbf{F}, \mathbf{B}, \mathbf{C}, \mathbf{D}$ are independent on k, the system is time-invariant, where k corresponds to the time coordinate.

Computer simulation is frequently used to evaluate control system performance. Computer simulation can be used to solve outputs, when the system is of high order or is subjected to complicated inputs that are not easily amenable to analytic solutions. In addition, the effects of time-variations, delays, and non-linearities can be evaluated using simulation. Computer simulation is also typically used to verify analytic results prior to hardware implementation.

Computer simulation requires that the continuous-time plant and the controller models be approximated by discrete-time systems, *i.e.* by finite difference equations.

Digital controllers also utilize difference equation models when operating. One method of generating a digital controller is to design a continuous-time controller and then approximate this controller in discrete-time. The generation of discrete-time approximation from a continuous-time system is therefore of fundamental importance in simulation and control system design.

A number of methods are used to form a discrete-time approximation of a continuous-time system, such as Euler's method, the zero-order holding approximation and so on [132]. Let the time step be denoted as η. The Euler's method approximate the time derivative as

$$(\mathbf{x}(k\eta+\eta)-\mathbf{x}(k\eta))/\eta \approx \dot{\mathbf{x}}(k\eta) = \mathbf{A}\mathbf{x}(k\eta)+\mathbf{B}\mathbf{u}(k\eta)$$

and then a difference equation for the state is simply obtained as

$$\mathbf{x}(k\eta+\eta) = \mathbf{x}(k\eta) + \eta[\mathbf{A}\mathbf{x}(k\eta)+\mathbf{B}\mathbf{u}(k\eta)]$$

The zero-order holding approximation is obtained for time-invariant system. The state difference equations is obtained by solving the equation (6.1.11') while assuming the input is constant over the time step, *i.e.* $\mathbf{u}(t)=\mathbf{u}(k\eta)$ and $\mathbf{w}(t)=\mathbf{w}(k\eta)$. The state at time of one step further $(k+1)\eta$ solves

$$\mathbf{x}(k\eta+\eta) = \mathbf{F}\cdot\mathbf{x}(k\eta) + \int_0^{\eta}\exp[\mathbf{A}(\eta-\tau)]\cdot[\mathbf{B}\mathbf{u}(k\eta)+\mathbf{w}(k\eta)]\mathrm{d}\tau$$

where $\mathbf{F}=\exp[\mathbf{A}\eta]$, which can be computed by the ***precise integration method.*** Based on the above equation, the discrete-time dynamic equation can be easily computed as

$$\mathbf{x}(k\eta+\eta) = \mathbf{F}\cdot\mathbf{x}(k\eta)+\mathbf{G}\mathbf{u}(k\eta)+(\mathbf{F}-\mathbf{I})\mathbf{w}(k\eta), \quad \mathbf{G}=(\mathbf{F}-\mathbf{I})\cdot\mathbf{B} \tag{6.1.15}$$

$$\mathbf{y}(k\eta) = \mathbf{C}\mathbf{x}(k\eta)+\mathbf{D}\mathbf{u}(k\eta) \tag{6.1.16}$$

System analysis can also be classified into deterministic system and/or stochastic system. The deterministic system means, that not only the system parameters are deterministic but also the input variables (including the control and disturbances) are also changing deterministically. For a stochastic system, the input variables (control and disturbances) are considered a stochastic processes, even more that the system parameters or the structural characteristics may also involve some random variables. The characteristic of a stochastic system is that there are no deterministic state responses, that only the statistical parameters such as mean values and variance matrices can be available. It corresponds to the random vibration problems or to random structural analysis.

The state space approach is emphasized in the following description, but the classical approach is also described at some places. The continuous-time and time-invariant system is of much concern. Based on the ***analogy between structural mechanics and optimal control***, the theory and algorithms in structural mechanics are introduced into the optimal control theory so as to get some useful ideas and progresses.

Example 6.1, A single degree of freedom mass-spring vibration system is shown in figure 6.2. The force F and the damping dashpot velocity v are inputs, the displacement x of the mass m is output. The formulation of the state space equation is required.

Solution: The forces acting on the mass are: the inertial force $m\ddot{x}$, the damping force $c(\dot{x}-v)$, elastic force of the spring kx and external force $-F$. Hence the dynamical equation is

$$m\ddot{x}+c(\dot{x}-v)+kx-F=0$$

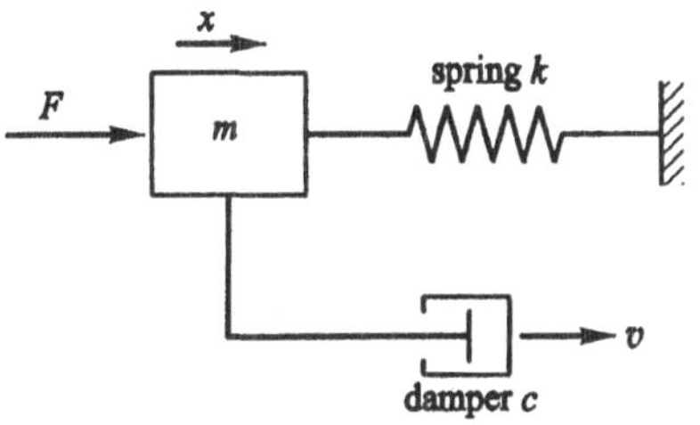

Figure 6.2, *A mass-spring vibration system*

The state variables are selected as $x_1=x,\ x_2=\dot{x}$, therefore the dimension of the state and input vectors are $n=2$ and $m_u=2$, respectively

$$\mathbf{x}=\{x,\dot{x}\}^T,\quad \mathbf{u}=\{F,v\}^T$$

The state space dynamic and measurement equations are given as

$$\dot{x}_1=x_2,\quad \dot{x}_2=-kx_1/m-cx_2/m+(cv+F)/m$$

$$y=x_1$$

From the state space dynamic equation and the measurement equation finds that

$$\mathbf{A}=\begin{bmatrix}0 & 1\\ -k/m & -c/m\end{bmatrix},\quad \mathbf{B}=\begin{bmatrix}0 & 0\\ 1/m & c/m\end{bmatrix},\quad \mathbf{C}=\begin{bmatrix}1 & 0\end{bmatrix}$$

These equations can be modeled with the block diagram of state variables, Figure 6.3. The block diagram of state variables can ease the use of an analog computer. There are only integrators, additive and multiplier operators and connection lines. The output variables can be picked up from the block diagram according to the output equation. The operator $1/s$ corresponds to integration, and parameter s is the variable of Laplace transformation. ##

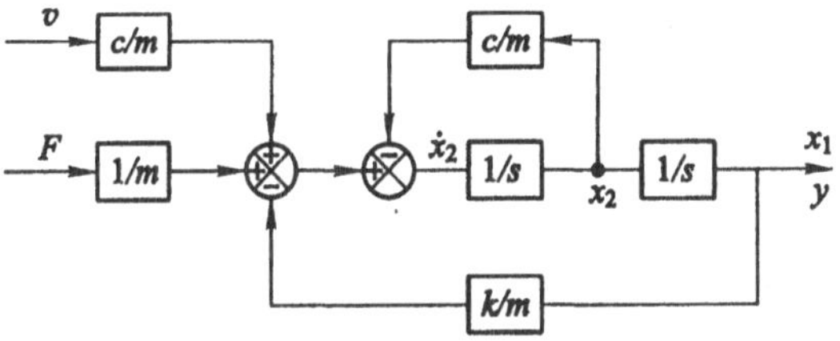

Figure 6.3, *The state variable block diagram of example 1.*

§6.1.2, State space description of single-input single-output system

The differential equation of single-input single-output linear steady continuous-time system is

$$\begin{aligned} y^{(n)}+a_{n-1}y^{(n-1)}+\cdots+a_1y^{(1)}+a_0y= \\ b_{n-1}u^{(n-1)}+b_{n-2}u^{(n-2)}+\cdots+b_1u^{(1)}+b_0u \end{aligned} \tag{6.1.2}$$

Its corresponding ***state space realization*** is required such that the input-output relation is kept unchanged. The selection of state space can be different, for which the dynamic equations can also be different, therefore there are a number of methods of state space realization. The normal form of state space realization is required. For equation (6.1.2), the transfer function $\tilde{G}(s)$ has given in equation (6.1.5). The observable normal form and the control normal form of state space descriptions are given below, respectively.

1) Transform to observable normal form

Because in equation (6.1.2) there are differentials of the input u so that the state variables are selected as

$$x_n = y$$
$$x_i = \dot{x}_{i+1} + a_i y - b_i u, \qquad i = n-1, n-2, \cdots, 1$$

Differentiate the above equation and eliminates the variables, and finally using equation (6.1.2) gives

$$\begin{aligned} \dot{x}_1 &= \ddot{x}_2 + a_1\dot{y} - b_1\dot{u} \\ &= x_3^{(3)} + a_1\dot{y} - b_1\dot{u} + a_2 y^{(2)} - b_2 u^{(2)} \\ &= \cdots\cdots = -a_0 x_n + b_0 u \end{aligned} \tag{6.1.17}$$

Therefore the following equations are obtained

Dynamic equations:
$$\left.\begin{aligned} \dot{x}_1 &= -a_0 x_n + b_0 u \\ \dot{x}_2 &= x_1 - a_0 x_n + b_1 u \\ &\cdots \\ \dot{x}_n &= x_{n-1} - a_{n-1}x_n + b_{n-1}u \end{aligned}\right\} \tag{6.1.18}$$

The measurement equation: $\quad y = x_n \qquad (6.1.19)$

Or in matrix/vector form

$$\dot{\mathbf{x}} = \mathbf{Ax} + \mathbf{Bu}, \qquad \mathbf{y} = \mathbf{Cx}$$

$$\mathbf{A} = \begin{bmatrix} 0 & 0 & \dots & 0 & -a_0 \\ 1 & & & & -a_1 \\ 0 & 1 & & & -a_2 \\ \vdots & & \ddots & 0 & \vdots \\ 0 & & 0 & 1 & -a_{n-1} \end{bmatrix}, \quad \mathbf{B} = \begin{bmatrix} b_0 \\ b_1 \\ b_2 \\ \vdots \\ b_{n-1} \end{bmatrix}, \quad \begin{array}{c} \mathbf{C} = [0, \ \cdots, \ 0, \ 1] \\ \mathbf{D}_u = \mathbf{0} \end{array} \tag{6.1.20}$$

2) Transform to control normal form

Introducing an intermediate variables $\tilde{z}(s)$ rewrite the equation (6.1.3) as

$$\tilde{y}(s)/\tilde{B}(s) = \tilde{u}(s)/\tilde{A}(s) = \tilde{z}(s)$$

which correspond to the equation of inverse Laplace transform

$$z^{(n)} + a_{n-1}z^{(n-1)} + \cdots + a_1 z^{(1)} + a_0 z = u$$
$$b_{n-1}z^{(n-1)} + \cdots + b_1 z^{(1)} + b_0 z = y$$

where the function $\tilde{B}(s)$ and $\tilde{A}(s)$ has been given in equation (6.1.5). Introducing the state variables as

$$x_1 = z, \quad x_2 = z^{(1)}, \quad x_3 = z^{(2)}, \ \ldots, \quad x_n = z^{(n-1)}$$

then the dynamic equations are

$$\begin{aligned}\dot{x}_1 &= x_2 \\ \dot{x}_2 &= x_3 \\ &\cdots \\ \dot{x}_n &= -a_0x_1 - a_1x_2 - \cdots - a_{n-1}x_n + u\end{aligned} \tag{6.1.21}$$

and the output equation is given as

$$y = b_0x_1 + b_1x_2 + \cdots + b_{n-1}x_n \tag{6.1.22}$$

The corresponding matrices of state space realization are

$$\mathbf{A} = \begin{bmatrix} 0 & 1 & & & \\ & 0 & 1 & & \\ & & \cdots & \cdots & \\ & & & 0 & 1 \\ -a_0 & -a_1 & \cdots & & -a_{n-1} \end{bmatrix}, \quad \mathbf{B}_u = \begin{bmatrix} 0 \\ 0 \\ \vdots \\ 0 \\ 1 \end{bmatrix}, \quad \begin{matrix} \mathbf{C} = [b_0 \quad b_1 \quad \cdots \quad b_{n-1}] \\ \mathbf{D}_u = \mathbf{0} \end{matrix} \tag{6.1.23}$$

This form of state space realization is called the control normal form.

§6.1.3, Integration of linear time-invariant systems

Integration of dynamic equation (6.1.11) is certainly of concern, which is a linear non-homogeneous differential equation. Its homogeneous equation should be solved first

$$\dot{\mathbf{x}}(t) = \mathbf{A}\mathbf{x}, \qquad \mathbf{x}(0) = \mathbf{x}_0 \tag{6.1.24}$$

This is a problem of fundamental importance. The solution of (6.1.24) can be written as

$$\mathbf{x}(t) = \exp(\mathbf{A}t) \times \mathbf{x}_0 = \mathbf{\Phi}(t) \times \mathbf{x}_0 \tag{6.1.25}$$

where the power series expansion of the exponential matrix is the same as exponential function

$$\exp(\mathbf{A}t) = \mathbf{I}_n + \mathbf{A}t + (\mathbf{A}t)^2/2! + \cdots + (\mathbf{A}t)^k/k! + \cdots = \mathbf{\Phi}(t) \tag{6.1.26}$$

which is termed as the ***state transition matrix.***

The behaviors of matrix exponentiation function are given as:

1) $\lim_{t\to 0} \exp(\mathbf{A}t) = \mathbf{I}_n$
2) $\exp[\mathbf{A}(t_1 + t_2)] = \exp(\mathbf{A}t_1) \times \exp(\mathbf{A}t_2)$, further $[\exp(\mathbf{A}t)]^m = \exp(\mathbf{A}t \cdot m)$.
3) $[\exp(\mathbf{A}t)]^{-1} = \exp(-\mathbf{A}t)$ gives the inverse matrix for exponential matrix.
4) Generally speaking, the exponential matrix does not apply the commute rule of multiplication, *i.e.*

$$\exp(\mathbf{A}t) \times \exp(\mathbf{B}t) \neq \exp(\mathbf{B}t) \times \exp(\mathbf{A}t), \quad \text{when} \quad \mathbf{AB} \neq \mathbf{BA}$$

But $\exp(\mathbf{A}t) \times \exp(\mathbf{B}t) = \exp(\mathbf{B}t) \times \exp(\mathbf{A}t) = \exp[(\mathbf{A} + \mathbf{B})t]$, when $\mathbf{AB} = \mathbf{BA}$

i.e. when the matrices $\mathbf{A}, \mathbf{B}$ are commute for multiplication, the corresponding exponential matrices can also commute for multiplication.

5) The differential equation for exponential matrix function is

$$(\mathrm{d}/\mathrm{d}t)\exp(\mathbf{A}t) = \mathbf{A} \times \exp(\mathbf{A}t)$$

The exponential matrix is the impulse response of differential equation, when input is only the control vector $\mathbf{u}$, then the response of state vector is computed by

$$\mathbf{x}(t) = \exp(\mathbf{A}t)\mathbf{x}_0 + \int_0^t \exp[\mathbf{A}(t-\tau)]\mathbf{B}_u\mathbf{u}(\tau)\mathrm{d}\tau \tag{6.1.27}$$

In control theory the control input term can be selected arbitrarily. The appropriate selection of the input vector $\mathbf{u}$ can control the performance of state vector $\mathbf{x}(t)$. Let the initial time be t_0, the above equation becomes

$$\mathbf{x}(t;t_0,\mathbf{x}_0,\mathbf{u}) = \mathbf{\Phi}(t-t_0)\mathbf{x}_0 + \int_{t_0}^t \mathbf{\Phi}(t-\tau)\mathbf{B}_u\mathbf{u}(\tau)\mathrm{d}\tau \tag{6.1.27'}$$

The computation of exponential matrix is very important, **19 dubious ways** are given in the review paper [29], which means the problem was not solved satisfactorily at that time. Afterwards, the book [30] investigated again, that further investigation is needed. ***Precise integration method*** gives the algorithm, which computes the exponential of $n \times n$ matrix up to computer precision. The precise integration method uses the ***combination of two cruxes***, that only one of the cruxes cannot solve the problem perfectly. The precise integration method for initial value problem has been given in the introduction. The numerical results are given at the grids (time stations) of uniform time step η

$$t_0,\ t_0+\eta,\ t_0+2\eta,\cdots\cdots \tag{6.1.28}$$

Note that the exponential matrix function can also be computed by the eigenvector expansion method, which is similar to the modal expansion method in structural vibration theory. The problem requires to distinguish if Jordan normal form appears or not for the $n \times n$ matrix $\mathbf{A}$. When the Jordan form does not appear, then the matrix $\mathbf{A}$ has n linearly independent eigenvectors

$$\mathbf{A}\boldsymbol{\varphi}_i = \mu_i\boldsymbol{\varphi}_i, \qquad i = 1,2,\cdots,n \tag{6.1.29}$$

A matrix is composed using all the eigenvectors $\boldsymbol{\varphi}_i$ as columns

$$\mathbf{\Phi}_a = [\boldsymbol{\varphi}_1,\boldsymbol{\varphi}_2,\cdots,\boldsymbol{\varphi}_n] \tag{6.1.30}$$

$$\mathbf{A}\mathbf{\Phi}_a = \mathbf{\Phi}_a\mathrm{diag}(\mu_1,\mu_2,\cdots,\mu_n) = \mathbf{\Phi}_a\mathrm{diag}(\mu_i),\quad \mathbf{A} = \mathbf{\Phi}_a\mathrm{diag}(\mu_i)\mathbf{\Phi}_a^{-1} \tag{6.1.29'}$$

Let

$$\mathbf{\Phi}(t) = \mathbf{\Phi}_a\mathrm{diag}[\exp(\mu_i t)]\mathbf{\Phi}_a^{-1} \tag{6.1.31}$$

then obviously, $\mathbf{\Phi}(0) = \mathbf{I}_n$; and

$$\dot{\mathbf{\Phi}}(t) = \mathbf{\Phi}_a\mathrm{diag}[\mu_i\exp(\mu_i t)]\mathbf{\Phi}_a^{-1} = \mathbf{\Phi}_a\mathrm{diag}(\mu_i)\mathrm{diag}[\exp(\mu_i t)]\mathbf{\Phi}_a^{-1}$$
$$= \mathbf{A}\mathbf{\Phi}_a\mathrm{diag}[\exp(\mu_i t)]\mathbf{\Phi}_a^{-1} = \mathbf{A}\mathbf{\Phi}$$

The differential equation and initial condition are all coincided with the exponential matrix function, so

$$\exp(\mathbf{A}t) = \mathbf{\Phi}(t) \tag{6.1.26}$$

Therefore the equation (6.1.31) gives the analytical equation for state transition matrix, where a key step is to find all the eigen-solutions in order to compose the matrix $\mathbf{\Phi}_a$, and $\mu_i,(i=1,2,\cdots,n)$.

Jordan normal form does can appear for the matrix $\mathbf{A}$. Jordan normal form is pretty mathematically, but is numerically unstable computationally. When a very small error appears in the numerical computation, then the duplicate eigenvalue will split into two very close eigenvalues and the two corresponding eigenvectors become almost two parallel vectors with very large value, and then the matrix $\mathbf{\Phi}_a$ behaves very much ill-conditioned. The numerical result is then **dubious**.

In many cases, the matrix $\mathbf{A}$ does not appear the Jordan normal form, and then the eigen-matrix computing equation (6.1.31) is attractive. However, the precise integration method does not care if there is possibly the Jordan normal form appearing, which always gives almost exact numerical result.

After the state transition matrix is computed, the state response computation problem of equation (6.1.27) follows, and then the output is obtained as

$$\mathbf{y}(t) = \mathbf{C\Phi}(t - t_0)\mathbf{x}_0 + \int_{t_0}^{t} \mathbf{C\Phi}(t - \tau)\mathbf{B}_u \mathbf{u}(t)\mathrm{d}\tau + \mathbf{D}_u \mathbf{u}(t) \tag{6.1.32}$$

The input-output relation is of great concern for control theory, for which the ***input-output impulse response function*** $\mathbf{G}(t-\tau)$ is used for the m input $\mathbf{u}(t)$ induced q output $\mathbf{y}(t)$

$$\mathbf{G}(t-\tau) = \begin{bmatrix} g_{11}(t-\tau) & \cdots & g_{1m}(t-\tau) \\ g_{21}(t-\tau) & \cdots & g_{2m}(t-\tau) \\ \vdots & & \vdots \\ g_{q1}(t-\tau) & \cdots & g_{qm}(t-\tau) \end{bmatrix} \tag{6.1.33a}$$

$$\mathbf{y}(t) = \int_{t_0}^{t} \mathbf{G}(t-\tau)\mathbf{u}(\tau)d\tau \tag{6.1.33}$$

Comparing to equation (6.1.32), the $q \times m$ dimensional ***input-output impulse response matrix*** is calculated as

$$\mathbf{G}(t-\tau) = \mathbf{C\Phi}(t-\tau)\mathbf{B}_u + \mathbf{D}_u \delta(t-\tau) \tag{6.1.34}$$

The eigen-equation of the matrix $\mathbf{A}$ is given as

$$\det(\mathbf{A} - \mu\mathbf{I}_n) = 0$$

Expanding it as a n-th order polynomial equation gives

$$\mu^n + \alpha_{n-1}\mu^{n-1} + \cdots + \alpha_1\mu + \alpha_0 = 0$$

The Cayley-Hamilton theorem determines that if the matrix $\mathbf{A}$ substitutes the variable μ in the above polynomial equation, the obtained matrix must be zero. This is

$$\mathbf{A}^n + \alpha_{n-1}\mathbf{A}^{n-1} + \cdots + \alpha_1\mathbf{A} + \alpha_0\mathbf{I}_n = \mathbf{0} \tag{6.1.35}$$

This equation explains that $\mathbf{A}^n$ can be expressed with the linear combination of $\mathbf{A}^{n-1},\cdots,\mathbf{A},\mathbf{I}_n$, and then $\mathbf{A}^{n+1}$ and all the higher powers of $\mathbf{A}$ can also be expressed as linear combinations of $\mathbf{A}^{n-1},\cdots,\mathbf{I}_n$. The proof of Cayley-Hamilton theorem can be found in references [47,48]. When Jordan normal form does not appear, $\mathbf{A}^m = \mathbf{\Phi}_a \mathrm{diag}(\mu_i^m)\mathbf{\Phi}_a^T$ and the proof is simply given as

$$\begin{aligned} &\mathbf{A}^n + \alpha_{n-1}\mathbf{A}^{n-1} + \cdots + \alpha_1\mathbf{A} + \alpha_0\mathbf{I}_n = \\ &\qquad \mathbf{\Phi}_a \mathrm{diag}\left(\mu_i^n + \alpha_{n-1}\mu_i^{n-1} + \cdots + \alpha_1\mu_i + \alpha_0\right)\mathbf{\Phi}_a^{-1} = \mathbf{0} \end{aligned}$$

The above computations are given in the time domain. The frequency domain analysis is quite often used in the control theory, especially in input-output representation of classical control theory.

§6.1.4, Frequency domain analysis, transfer function

The frequency domain representation of state transition matrix and the transfer matrix function are of great concern, that the stability analysis is usually based on these functions. The state transition matrix comes from equation (6.1.24). Taking Laplace transform

$$(s\mathbf{I}_n - \mathbf{A})\tilde{\mathbf{x}}(s) = \mathbf{x}_0, \qquad \tilde{\mathbf{x}}(s) = (s\mathbf{I}_n - \mathbf{A})^{-1}\mathbf{x}_0$$

Obviously

$$\tilde{\mathbf{\Phi}}(s) = (s\mathbf{I}_n - \mathbf{A})^{-1}$$

which is the frequency domain representation of the state transition matrix.

Obviously, the input-output transfer matrix function in frequency domain is

$$\tilde{\mathbf{G}}(s) = \mathbf{C}(s\mathbf{I}_n - \mathbf{A})^{-1}\mathbf{B}_u + \mathbf{D}_u \tag{6.1.36}$$

Taking Laplace transform to the dynamic and measurement equations gives

$$(s\mathbf{I}_n - \mathbf{A})\tilde{\mathbf{x}} - \mathbf{B}_u\tilde{\mathbf{u}} = \mathbf{x}_0, \qquad \tilde{\mathbf{y}} = \mathbf{C}\tilde{\mathbf{x}} + \mathbf{D}_u\tilde{\mathbf{u}}$$

The vector $\mathbf{x}_0$ is unrelated hence taken as zero, then eliminating $\tilde{\mathbf{x}}$ gives equation (6.1.36). From equation (6.1.34) taking Laplace transformation gives equation (6.1.36) too. Obviously, the transfer matrix function is a rational fraction.

§6.1.5, Controllability and observability of a linear system

In designing control systems, it is important to know whether any control law is effective, optimal or not. Note that the outside world controls the system performance only through the input vector $\mathbf{u}$. This issue is not because of the structure of control law but the structure of dynamical system. The definition of ***controllability*** is:

Give an arbitrary state vector $\mathbf{x}_0$ at time t_0, if and only if it is possible by means of the selection of input vector $\mathbf{u}(t)$ to transfer the system state vector to $\mathbf{x}(t_1) = 0$ at a finite time $t_1 > t_0$, then the system is called controllable.

Controllability is the property regarded to the couple of matrices $(\mathbf{A}, \mathbf{B}_u)$ in the dynamic equation (6.1.15).

In designing control system, it is important to monitor the state of system. Although the state vector can describe the internal variables of system, but only the measurement vector $\mathbf{y}$ can be accessed from the outside world to estimate the state of system. The observability is defined in terms of the ability to estimate the state. This issue is also a problem of system structure. The definition of ***observability*** is:

A system is said to be observable if and only if its state $\mathbf{x}(t_0)$ at any time t_0, can be determined from the knowledge of the measurement and input vectors $\mathbf{y}(t)$ and $\mathbf{u}(t)$, respectively, in a finite period of time $t_0 \le t \le t_1$.

Observability is the property regarded to the couple of matrices $(\mathbf{A}, \mathbf{C})$ in the system equations.

The ***observability*** is used to investigate, if the system internal variables (state vector) can be reflected by the measurement $\mathbf{y}$; and the ***controllability*** is used to investigate, if the state vector can be operated with the input vector $\mathbf{u}$.

The Cayley-Hamilton theorem for a $n \times n$ matrix $\mathbf{A}$ can be used to prove the ***controllability and observability*** of a system. Based on this theorem, the power matrix $\mathbf{A}^n$ (and higher powers of $\mathbf{A}$) can be expressed with a linear combination of the matrices $\mathbf{I}, \mathbf{A}, \mathbf{A}^2, \cdots, \mathbf{A}^{n-1}$.

§6.1.5.1, Controllability of a steady system

For a time-invariant system, the solution of dynamic equation is

$$\mathbf{x}(t) = \mathbf{\Phi}(t_1 - t_0)\mathbf{x}_0 + \int_{t_0}^{t_1} \mathbf{\Phi}(t_1 - t)\mathbf{B}_u \mathbf{u}(t)\mathrm{d}t = \mathbf{0}$$

where $\mathbf{\Phi}$ is the exponential matrix function (6.1.26). According to the requirement of controllability, the above equation can be expressed as

$$\mathbf{x}_0 = -\int_0^t e^{-\mathbf{A}(t-t_0)}\mathbf{B}_u \mathbf{u}(t)\mathrm{d}t$$

Expanding the exponential matrix as a power series and using the Cayley-Hamilton theorem gives

$$e^{-\mathbf{A}(t-t_0)} = \sum_{i=0}^{n-1} \alpha_i (t - t_0)\mathbf{A}^i$$

where α_i is a determined function of $t - t_0$. Substituting into the previous equation gives

$$\mathbf{x}_0 = -\sum_{i=0}^{n-1} \mathbf{A}^i \mathbf{B}_u \mathbf{u}_i, \quad \mathbf{u}_i = \int_{t_0}^{t_1} \alpha_i (t - t_0)\mathbf{u}(t)\mathrm{d}t$$

Writing the above equation in matrix form gives

$$\mathbf{x}_0 = -[\mathbf{B}_u \quad \mathbf{A}\mathbf{B}_u \quad \mathbf{A}^2\mathbf{B}_u \ldots. \quad \mathbf{A}^{n-1}\mathbf{B}_u] \times [\mathbf{u}_0^T \quad \mathbf{u}_1^T \quad \ldots.. \quad \mathbf{u}_{n-1}^T]^T$$

Let the former $n \times mn$ matrix be denoted as

$$\mathbf{Q}_c = [\mathbf{B}_u, \quad \mathbf{A}\mathbf{B}_u, \quad \ldots. \quad , \mathbf{A}^{n-1}\mathbf{B}_u] \tag{6.1.37}$$

According to the condition that $\mathbf{x}_0$ is an arbitrary initial state vector, hence, $\mathbf{Q}_c$ must be of full rank, *i.e.*

$$\mathrm{rank}(\mathbf{Q}_c) = n \tag{6.1.37a}$$

is the necessary condition of controllability of the system. It is also a sufficient condition, because of $\mathbf{u}(t)$ can be selected arbitrarily, the vectors $\mathbf{u}_0, \mathbf{u}_1, \cdots, \mathbf{u}_{n-1}$ can take arbitrary value. Hence if $\mathbf{Q}_c$ is of full rank, then $\mathbf{x}_0$ can have arbitrary value.

The necessary and sufficient condition of controllability can also be expressed by the symmetric and positive definite condition of the Gram matrix

$$\mathbf{W}_c(t) = \int_0^t \exp(\mathbf{A}\tau)\mathbf{B}_u \mathbf{B}_u{}^T \exp(\mathbf{A}^T\tau)\mathrm{d}\tau > 0 \tag{6.1.37b}$$

In section 6.7, we will come back to this matrix.

§6.1.5.2, Observability of a steady system

With no loss of generality, let $\mathbf{u} \equiv 0$ in examining the system observability. Because $\mathbf{u}$ is given, the effect of which can be substituted by a correction of $\mathbf{y}$.

To uniquely determine the arbitrary initial state vector $\mathbf{x}_0$, the measurement $\mathbf{y}(t)$ in the time period $(0,t)$ is available. According to equation (6.1.32), $(t_0 = 0)$

$$\mathbf{y}(t) - \mathbf{D}_u\mathbf{u} - \mathbf{C}\int_0^t e^{\mathbf{A}(t-\tau)}\mathbf{B}_u\mathbf{u}(\tau)\mathrm{d}\tau = \mathbf{C}e^{\mathbf{A}t}\mathbf{x}_0$$

Let $\mathbf{u} = \mathbf{0}$ and using the Cayley-Hamilton theorem gives

$$\mathbf{y}(t) = \mathbf{C}\sum_{k=0}^{n-1}\alpha_k(t)\mathbf{A}^k\mathbf{x}_0 = \begin{bmatrix}\alpha_0(t)\mathbf{I}_q & \alpha_1\mathbf{I}_q & \cdots & \alpha_{n-1}\mathbf{I}_q\end{bmatrix} \times \begin{bmatrix}\mathbf{C} \\ \mathbf{CA} \\ \vdots \\ \mathbf{CA}^{n-1}\end{bmatrix}\mathbf{x}_0$$

where $\mathbf{I}_q$ is a q -dimensional unit matrix. The former matrix $[\alpha_0(t)\mathbf{I}_q,\cdots,\alpha_{n-1}(t)\mathbf{I}_q]$ is of full rank. The $nq\times n$ matrix

$$\mathbf{Q}_o = \begin{bmatrix}\mathbf{C} \\ \mathbf{CA} \\ \vdots \\ \mathbf{CA}^{n-1}\end{bmatrix} \tag{6.1.38}$$

is called the observability matrix. If this $nq\times n$ matrix is not of full rank, then there is a subspace of the initial vector $\mathbf{x}_0$, that any vector in this subspace will not influence the measurement $\mathbf{y}(t)$, *i.e.* not measurable, which cannot fulfill the measurement condition. Hence

$$\mathrm{rank}(\mathbf{Q}_o) = n \tag{6.1.38a}$$

is a necessary condition of observability. It is also a sufficient condition. The proof is neglected.

The necessary and sufficient condition of observability can also use the positive definiteness condition of the observability Gram matrix

$$W_o(t) = \int_0^t \exp(\mathbf{A}^T\tau)\mathbf{C}^T\mathbf{C}\exp(\mathbf{A}\tau)\mathrm{d}\tau \tag{6.1.38b}$$

at any time of $t > 0$. The derivation of equations (6.1.37b) and (6.1.38b) of Gram matrices will be given in section 6.7. The controllability and observability conditions given above are used only for steady (time-invariant) system, but there are time-variant systems or even non-linear systems, for which the controllability and observability will be touched also in section 6.7.

§6.1.6, Linear transformation

Suppose the original system is expressed by

$$\dot{\mathbf{x}} = \mathbf{A}\mathbf{x} + \mathbf{B}_u\mathbf{u} \tag{6.1.15}$$

$$\mathbf{y} = \mathbf{C}\mathbf{x} + \mathbf{D}_u\mathbf{u} \tag{6.1.16}$$

Because the selection of state vector $\mathbf{x}$ is arbitrary in mathematical modeling, so that the state vector can be selected as one another vector denoted as $\bar{\mathbf{x}}$, which is obtained from the original state vector $\mathbf{x}$ by a non-singular linear transformation $\mathbf{P}$, that

$$\mathbf{x} = \mathbf{P}\bar{\mathbf{x}} \tag{6.1.39}$$

Substituting the above equation into (6.1.15) and then left multiplying $\mathbf{P}^{-1}$ gives

$$\overline{\mathbf{A}} = \mathbf{P}^{-1}\mathbf{A}\mathbf{P}, \qquad \overline{\mathbf{B}}_u = \mathbf{P}^{-1}\mathbf{B}_u, \qquad \overline{\mathbf{C}} = \mathbf{C}\mathbf{P} \qquad \overline{\mathbf{D}}_u = \mathbf{D}_u \tag{6.1.40}$$

So that the system equations are transformed as

$$\dot{\overline{\mathbf{x}}} = \overline{\mathbf{A}}\overline{\mathbf{x}} + \overline{\mathbf{B}}_u\mathbf{u}\,, \qquad \mathbf{y} = \overline{\mathbf{C}}\overline{\mathbf{x}} + \overline{\mathbf{D}}_u\mathbf{u}$$

The bar sign above, such as $\overline{\mathbf{x}}$ etc., are just for distinguishing the original quantities, but not the complex conjugate. Because the difference between matrices $\overline{\mathbf{A}}$ and $\mathbf{A}$ is only a similarity transformation, hence

1) The original and transformed systems have the same eigenvalue, and the difference between the corresponding eigenvectors is also a transformation of (6.1.39).
2) After linear transformation, the state transition matrix $\mathbf{\Phi}(t)$ also undergoes a similarity transformation

$$\overline{\mathbf{\Phi}}(t) = \exp(\mathbf{P}^{-1}\mathbf{A}\mathbf{P}t) = \mathbf{P}^{-1}\exp(\mathbf{A}t)\mathbf{P} = \mathbf{P}^{-1}\mathbf{\Phi}(t)\mathbf{P}$$

3) After linear transformation the system transfer matrix does not change, which is verified as

$$\begin{aligned}\overline{\widetilde{\mathbf{G}}} &= \overline{\mathbf{C}}(s\mathbf{I} - \overline{\mathbf{A}})^{-1}\overline{\mathbf{B}}_u + \mathbf{D}_u = \mathbf{C}\mathbf{P}(s\mathbf{I} - \mathbf{P}^{-1}\mathbf{A}\mathbf{P})^{-1}\mathbf{P}^{-1}\mathbf{B}_u + \mathbf{D}_u \\ &= \mathbf{C}\mathbf{P}[\mathbf{P}^{-1}(s\mathbf{I} - \mathbf{A})\mathbf{P}]^{-1}\mathbf{P}^{-1}\mathbf{B}_u + \mathbf{D}_u = \mathbf{C}(s\mathbf{I} - \mathbf{A})\mathbf{B}_u + \mathbf{D}_u = \widetilde{\mathbf{G}}\end{aligned}$$

4) Linear transformation does not change the controllability of the original system, because

$$\overline{\mathbf{Q}}_c = \left[\overline{\mathbf{B}}_u, \overline{\mathbf{A}}\overline{\mathbf{B}}_u, \cdots, \overline{\mathbf{A}}^{n-1}\overline{\mathbf{B}}_u\right] = \mathbf{P}^{-1}\left[\mathbf{B}_u, \mathbf{A}\mathbf{B}_u, \cdots, \mathbf{A}^{n-1}\mathbf{B}_u\right] = \mathbf{P}^{-1}\mathbf{Q}_c$$

and $\mathbf{P}$ is of full rank, therefore the rank of $\overline{\mathbf{Q}}_c$ is the same as $\mathbf{Q}_c$.
5) Linear transformation does not change the observability of the original system.

Linear transformation is very useful for a series of applications, such as expansion solution by eigen-solutions, structural factorization to the control normal form or observable normal form etc.

§6.1.7, Realization of a transfer function in state space

Two kinds of transfer functions are given in previous paragraphs, namely the single-input single-output transfer function in (6.1.5) and the multi-input multi-output transfer function (matrix) in (6.1.36). Both of them are given in the frequency domain.

These transfer functions are derived from the time domain equations to the frequency domain. On the contrary, for a given rational fractional transfer function $\widetilde{\mathbf{G}}(s)$, the system matrices $\mathbf{A}, \mathbf{B}_u, \mathbf{C}, \mathbf{D}_u$ can be found correspondingly. The system $(\mathbf{A}, \mathbf{B}_u, \mathbf{C}, \mathbf{D}_u)$ obtained this way is called a realization of the transfer function $\widetilde{\mathbf{G}}(s)$ in the state space model.

For the case of single-input single-output transfer function, the function should be reduced to be a real rational fractional function

$$\widetilde{\mathbf{G}}(s) = b_n + B(s)/A(s)\,, \qquad \begin{aligned} B(s) &= b_{n-1}s^{n-1} + \cdots + b_1 s + b_0 \\ A(s) &= s^n + a_{n-1}s^{n-1} + \cdots + a_1 s + a_0 \end{aligned}$$

where the polynomials $B(s)$ and $A(s)$ have no common factor. In such case, the state space realization $(\mathbf{A},\mathbf{b}_u,\mathbf{c},\mathbf{D}_u)$ can have a number of expressions including the control normal form and the observable normal form. These forms are obtained by the equation (6.1.20) for observable normal form, and by the equation (6.1.23) for control normal form, with the matrix be given as $\mathbf{D}_u = [b_n]$ in both cases.

For the case of multi-input multi-output, corresponding to a given rational fractional transfer function $\tilde{\mathbf{G}}(s)$, the system matrices $\mathbf{A},\mathbf{B}_u,\mathbf{C},\mathbf{D}_u$ can also be found for the realization in the state space. The realization can be simply written as $(\mathbf{A},\mathbf{B}_u,\mathbf{C},\mathbf{D}_u)$. Realization of a transfer matrix function has the fundamental characteristics as follows:

1) Realization is not unique, that for a given $\tilde{\mathbf{G}}(s)$, there are a number of correspondent realizations with different dimensions. Even for the same dimension, the realization is also not unique.
2) Among all the realizations of $\tilde{\mathbf{G}}(s)$, there is a class of minimum dimensional realizations called the ***minimal realization***. The minimal realization is a simplest external equivalence of system with the input-output frequency domain characteristic of $\tilde{\mathbf{G}}(s)$.
3) Among all the realizations corresponding to the transfer matrix function $\tilde{\mathbf{G}}(s)$, there is no necessary algebraic equivalence relation, but for minimal realizations there are mutual algebraic equivalent relation.
4) If the real system is controllable and observable, and if the minimal realization of the transfer matrix $\tilde{\mathbf{G}}(s)$ is also controllable and observable, then this minimal realization reflects the structure of the real system.
5) If the given transfer function $\tilde{\mathbf{G}}(s)$ is a strictly real fractional function, then the realization has the form $(\mathbf{A},\mathbf{B}_u,\mathbf{C})$, *i.e.* $\mathbf{D}_u = \mathbf{0}$. If $\tilde{\mathbf{G}}(s)$ is a real but not strictly real, then the realization form is $(\mathbf{A},\mathbf{B}_u,\mathbf{C},\mathbf{D}_u)$, and $\lim_{s\to\infty}\tilde{\mathbf{G}}(s) = \mathbf{D}_u$.

There is a theorem for minimal realization: ***Suppose*** $\tilde{\mathbf{G}}(s)$ ***be a strictly real fractional function, then the necessary and sufficient condition for*** $(\mathbf{A},\mathbf{B}_u,\mathbf{C})$ ***being a minimal realization is that*** $(\mathbf{A},\mathbf{B}_u)$ ***be controllable and*** $(\mathbf{A},\mathbf{C})$ ***be observable.*** The proof is neglected, see [16~18].

§6.1.8, Duality principle for controllability and observability

For two time-invariant linear systems Γ_1 and Γ_2, the dynamic and measurement equations are

$$\Gamma_1: \quad \dot{\mathbf{x}} = \mathbf{A}\mathbf{x} + \mathbf{B}_u\mathbf{u}, \qquad \mathbf{y} = \mathbf{C}\mathbf{x}$$

$$\Gamma_2: \quad \mathbf{x}_2 = -\mathbf{A}^T\mathbf{x}_2 - \mathbf{C}^T\mathbf{v}, \quad \mathbf{y}_2 = \mathbf{B}_u^T\mathbf{x}_2$$

respectively, where $\mathbf{x}$ and $\mathbf{x}_2$ are all n-dimensional state space vectors; $\mathbf{u}$ and $\mathbf{y}_2$ are both m-dimensional vectors, and $\mathbf{y},\mathbf{v}$ are q-dimensional vectors. It is to show that these two systems are mutually dual to each other. The state vector impulse response matrix (transition matrix) of system Γ_1 is

$$\mathbf{\Phi}(t,t_0)=\exp[\mathbf{A}(t-t_0)]$$

and the state vector impulse response matrix of system Γ_2 is

$$\mathbf{\Phi}_2(t,t_0)=\exp[-\mathbf{A}^T(t-t_0)]=\left[\exp[\mathbf{A}\cdot(t_0-t)]\right]^T=\mathbf{\Phi}^T(t_0,t)$$

The controllability of one system corresponds to the observability of its dual system, and the observability corresponds to the controllability of its dual system. The verification is that the controllability of Γ_2 is the full rank of the matrix

$$\mathbf{Q}_{c2}=\left[\mathbf{C}^T,-\mathbf{A}^T\mathbf{C}^T,\mathbf{A}^{2T},\cdots,(-\mathbf{A}^T)^{n-1}\mathbf{C}^T\right]=\begin{bmatrix}\mathbf{C}\\ \mathbf{CA}\\ \mathbf{CA}^2\\ \vdots\\ \mathbf{CA}^{n-1}\end{bmatrix}\cdot[\,\mathbf{I}_q,-\mathbf{I}_q,\cdots,(-1)^{n-1}\mathbf{I}_q\,]$$

Therefore $\mathrm{rank}(\mathbf{Q}_{c2})=\mathrm{rank}(\mathbf{Q}_o)$, where $\mathbf{Q}_o$ is the observability matrix of the system Γ_1. On the other hand, the observability matrix of Γ_2 is

$$\mathbf{Q}_{o2}=\begin{bmatrix}\mathbf{B}_u^T\\ -\mathbf{B}_u^T\mathbf{A}^T\\ \vdots\\ \mathbf{B}_u^T(-\mathbf{A}^T)^{n-1}\end{bmatrix}^T=[\mathbf{B}_u,\mathbf{AB}_u,\cdots,\mathbf{A}^{n-1}\mathbf{B}_u]\times\mathrm{diag}[\mathbf{I}_m,-\mathbf{I}_m,\cdots,(-1)^{n-1}\mathbf{I}_m]$$

Therefore $\mathrm{rank}(\mathbf{Q}_{o2})=\mathrm{rank}(\mathbf{Q}_c)$, where $\mathbf{Q}_c$ is the controllability matrix of system Γ_1.

Simple examples are given to demonstrate the controllability and observability.

Example 6.2, Suppose $n=2$, $q=1,m=1$, $\mathbf{A}=\begin{bmatrix}-1 & 0\\ 0 & 2\end{bmatrix}$, $\mathbf{B}_u=\begin{bmatrix}1\\ 0\end{bmatrix}$, $\mathbf{C}=[1\ \ 1]$, analyze the controllability and observability of the system.

Solution: Computing gives $\mathbf{Q}_c=[\mathbf{B}_u,\mathbf{AB}_u]=\begin{bmatrix}1 & -1\\ 0 & 0\end{bmatrix}$, because $\mathrm{rank}(\mathbf{Q}_c)=1<n$, so the system is uncontrollable. However, computing gives that $\mathbf{Q}_o=\begin{bmatrix}\mathbf{C}\\ \mathbf{CA}\end{bmatrix}=\begin{bmatrix}1 & 1\\ 1 & 2\end{bmatrix}$, because $\mathrm{rank}(\mathbf{Q}_o)=2=n$, so the system is observable.

If $\mathbf{B}_u=[1\ \ 0.001]^T$, then $\mathbf{Q}_c=\begin{bmatrix}1 & -1\\ 0.001 & 0.002\end{bmatrix}$ is computed, it gives $\mathrm{rank}(\mathbf{Q}_c)=2$, which means that the system is controllable. However, the system is nearly uncontrollable. ##

As a matter of fact, the cases of uncontrollable or unobservable are occasional.

If the data changes a small amount for an uncontrollable or unobservable system, then the system turns back to be controllable and observable. But for such nearly uncontrollable or unobservable system, the performance is quite ill-conditioned. ##

Example 6.3, Suppose $n=4;\ q=1;\ m=1;$ the system matrices are given as

$$\mathbf{A}=\begin{bmatrix}0 & 0 & 1 & 0\\ 0 & 0 & 0 & 1\\ -2 & 1 & 0 & 0\\ 0.5 & -0.5 & 0 & 0\end{bmatrix},\quad \mathbf{B}_u=\begin{bmatrix}0\\0\\0\\0.5\end{bmatrix},\quad \mathbf{C}=[1\ \ 0\ \ 0\ \ 0]$$

the controllability and observability are required to analyze.

Solution: It is computed that

$$\mathbf{Q}_c=\begin{bmatrix}0 & 0 & 1 & 0\\ 0 & 0.5 & 0 & -0.25\\ 0 & 0 & 0.5 & 0\\ 0.5 & 0 & -0.25 & 0\end{bmatrix},\quad \operatorname{rank}(\mathbf{Q}_c)=4(=n)$$

So, the system is controllable. Then computes

$$\mathbf{Q}_o=\begin{bmatrix}1 & 0 & 0 & 0\\ 0 & 0 & 1 & 0\\ -2 & 1 & 0 & 0\\ 0 & 0 & -2 & 1\end{bmatrix},\quad \operatorname{rank}(\mathbf{Q}_o)=4(=n)$$

the system is also observable. ##

§6.1.9, Discrete-time control

For a continuous-time control system, because of that the actuator can act only at the discrete-time instants, or because of that the data sampling analysis needs discrete-time, facing the problem of transforming from the continuous-time to discrete-time control system. The transformation should have the assumption:

1) The sampling period η is a constant, *i.e.* equal sampling period. The time required for data sampling should be far less than η. Denote as

$$\mathbf{y}_k=\mathbf{y}(k\eta),\qquad t_0=0$$

2) The control vector input is kept constant in the time interval η. That is

$$\mathbf{u}(t)=\mathbf{u}(k\eta)=\mathbf{u}_k,\qquad \text{when}\quad k\eta\le t<(k+1)\eta$$

Certainly, the step size η should be selected small, so that the Shannon sampling theorem is satisfied.

Sampling theorem: Suppose the highest frequency contained in a continuous-time signal $x(t)$ is $f_{\max}=B$ Hz. Then, if $x(t)$ is sampled periodically at a rate $f_s=1/T_s>2B$, the signal can be exactly reproduced from the sample values $x(k), k=-\infty,\cdots,\infty,$ using the interpolation rule

$$x(t)=\sum_{k=-\infty}^{\infty}x(kT_s)h(t-kT_s),\qquad h(t)=\sin(\pi t/T_s)/(\pi t/T_s)$$

see [120],p.12. The lower bound on the sampling frequency, denoted as $f_s^*=2B$, is called the Nyquist sample rate or simply Nyquist rate. In applications, it must be

$f_s > f_s^*$, but not equal. The frequency $f_n = f_s/2$ is generally referred to as the Nyquist frequency.

§6.2, Theory of stability

Stability is a fundamental characteristic of a system. Stability analysis of system dynamics is one of the most important parts in the theory of control and system analysis. The system dynamics has two kinds of stability definitions, *i.e.*

1) The **system external stability** defined with the input-output relation, and
2) The **system internal stability** defined by the state dynamics under zero input condition.

These are described as follows.

1) The external stability: Corresponding to a bounded input $\mathbf{u}(t)$, *i.e.*

$$\|\mathbf{u}(t)\| \le k_1 < \infty \qquad t_0 \le t < \infty \tag{6.2.1}$$

if the produced output $\mathbf{y}(t)$ is also bounded, *i.e.*

$$\|\mathbf{y}(t)\| \le k_2 < \infty, \qquad t_0 \le t < \infty \tag{6.2.2}$$

then the system is external stable, *i.e.* Bounded-Input Bounded-Output stable, or simply BIBO stable, where $\|*\|$ represents the norm of a vector.

For a linear steady dynamical system with zero initial condition, let $\mathbf{G}(t)$ denote its impulse response matrix function, or expressed in the frequency domain with the transfer matrix function $\widetilde{\mathbf{G}}(s)$, then the necessary and sufficient condition of system BIBO stability is

$$\int_0^\infty |g_{ij}(t)| \mathrm{d}t \le k < \infty \qquad i = 1,2,\cdots,q; j = 1,2,\cdots,m \tag{6.2.3}$$

where $g_{ij}(t)$ is the element of $\mathbf{G}(t)$, or in the frequency domain that the real part of the poles of every element $\widetilde{g}_{ij}(s)$ of the transfer matrix function $\widetilde{\mathbf{G}}(s)$ be negative. That is all the poles located at the left half plane of the complex plane.

2) The internal stability: For a linear steady system

$$\dot{\mathbf{x}} = \mathbf{A}\mathbf{x} + \mathbf{B}_u\mathbf{u}, \qquad \mathbf{x}(0) = \text{given} \tag{6.2.4}$$

$$\mathbf{y} = \mathbf{C}\mathbf{x} + \mathbf{D}_u\mathbf{u} \tag{6.2.5}$$

No input means $\mathbf{u}(t) = \mathbf{0}$. When the initial state is arbitrary, the state response behaves that

$$\mathbf{x} = \mathbf{\Phi}(t,t_0)\mathbf{x}_0 \to \mathbf{0}, \qquad \text{when} \quad t \to \infty \tag{6.2.6}$$

then the system is called ***internally stable***, and also internally asymptotically stable. As is well known that, ***the necessary and sufficient conditions of the asymptotical stability for the differential equation*** $\dot{\mathbf{x}}(t) = \mathbf{A}\mathbf{x}$ ***is that the real part of all the eigenvalues of the matrix*** $\mathbf{A}$ ***being negative.*** This is a fundamental theorem for stability.

If the linear steady system given by (6.2.4) and (6.2.5) is internally stable, then it is BIBO table.

The BIBO stability cannot ensure internal stability of the system. However, if a linear steady system is controllable and observable, then the internal stability and external stability are equivalent.

§6.2.1, Stability analysis under Lyapunov meaning

Stability of motion usually treats the system with no external input. The case of a non-linear system can be described by

$$\dot{\mathbf{x}} = \mathbf{f}(\mathbf{x},t), \qquad \mathbf{x}(t_0) = \mathbf{x}_0, \qquad t \geq t_0 \tag{6.2.7}$$

where $\mathbf{x}$ and $\mathbf{f}$ denote the n-dimensional state vector and vector function, respectively. For general non-linear system the stability analysis is very complicated. For a linear time-dependent system, let the equation be

$$\dot{\mathbf{x}} = \mathbf{A}(t)\mathbf{x}, \qquad \mathbf{x}(t_0) = \mathbf{x}_0, \qquad t \geq t_0 \tag{6.2.8}$$

The superposition method can be applied for linear equations, which brings convenience for the analysis. For periodical function $\mathbf{A}(t)$, there is the Floquet method, see chapter 2, which derives the system to be a discrete-time system and is easier for stability analysis.

For a general system, only simple stability description can be made. Suppose the equation (6.2.7) has a solution

$$\mathbf{x}(t) = \mathbf{\Phi}(t; \mathbf{x}_0, t_0), \qquad t \geq t_0 \tag{6.2.9}$$

If an equilibrium point $\mathbf{x} = \mathbf{x}_e$ is a solution, then the analysis of stability in the neighborhood of this equilibrium point is necessary. If the motion (6.2.9) is periodical, *i.e.* a limit cycle, then the stability analysis is necessary, that if the state is slightly depart to this limit cycle at some time instant, can the following motion go back to the limit cycle?

The perturbation analysis nearby the limit cycle derives to the linear differential equations with periodical coefficient matrix $\mathbf{A}(t)$, which lead the problem to be solved by the Floquet method. Checking the eigenvalues of the Floquet matrix determines the stability of the limit cycle.

For a general non-linear dynamical system, the solution usually enters a chaotic motion for arbitrary initial condition. Quite often the state vector moves at a nearly periodic orbit, however, although the motion is governed by a deterministic differential equation but the solution behaves quasi-random. Much of what is known about this topic today has been obtained by numerical simulation studies. This fact makes it difficult to be precise about the exact nature of the trajectory. Indeed, one cannot even make sure, from a finite-length numerical simulation, that a solution is not periodic. From a large number of simulations, it is highly plausible that the trajectory seems almost periodic. However, it comes from simulation experience but not from mathematical proof. This situation brings tremendous difficulty to stability analysis for such chaotic motion, because even a nominal motion cannot be determined clearly, that the analytical solution is usually hopeless, only numerical solution can be invoked. Also, the chaotic motion is very sensitive to a very small departure of the initial value, the so-called **butterfly effect**. The conclusion is that there is a long way to go for non-linear system chaos analysis.

A simpler problem is the stability analysis at an equilibrium point $\mathbf{x}_e = \mathbf{0}$. Assuming the state is slightly departed to the equilibrium point, it requires to consider the subsequent motion if it tends to return the equilibrium point automatically (asymptotically stable), or at least limited in a small neighborhood of the equilibrium point (stable but not asymptotically).

Stability in the sense of Lyapunov can be expressed as: Give a small quantity $\varepsilon > 0$ arbitrarily, there exists a neighbor area $\delta(\varepsilon, t_0) > 0$, such that for any initial point $\mathbf{x}_0$, within the δ neighbor of the equilibrium point, the induced subsequent motion satisfies the condition

$$\|\boldsymbol{\Phi}(t; \mathbf{x}_0, t_0) - \mathbf{x}_e\| < \varepsilon, \qquad \text{when} \quad \|\mathbf{x}_0 - \mathbf{x}_e\| < \delta(\varepsilon, t_0), \quad t > t_0 \tag{6.2.10}$$

then the system is stable at the equilibrium point.

The asymptotic stability is much useful for engineering applications, *i.e.*

$$\lim_{t\to\infty} \boldsymbol{\Phi}(t; \mathbf{x}_0, t_0) = \mathbf{x}_e, \quad \text{when} \quad \|\mathbf{x}_0 - \mathbf{x}_e\| < \delta(\varepsilon, t_0), \quad t > t_0 \tag{6.2.11}$$

This definition requires that the initial point $\mathbf{x}_0$ is located nearby the equilibrium point. Global asymptotic stability is defined as

$$\lim_{t\to\infty} \boldsymbol{\Phi}(t; \mathbf{x}_0, t_0) = \mathbf{x}_e \tag{6.2.12}$$

with no limitation of the initial point $\mathbf{x}_0$. These are only general view and definitions, as how to fulfill these stability conditions is a problem to be investigated.

§6.2.2, Lyapunov method of stability analysis

Let us begin with the stability analysis of an autonomous system. Consider the dynamic equation

$$\dot{\mathbf{x}} = \mathbf{f}(\mathbf{x}), \qquad \mathbf{x}(t_0) = \mathbf{x}_0, \qquad t \geq t_0 \tag{6.2.13}$$

The ***Lyapunov second method***, or the direct method, is used frequently. The direct method uses an auxiliary function, called as ***Lyapunov function***, denoted as $V(\mathbf{x})$, which is a scalar function of state with continuous partial derivatives. The time-derivative of this function along the trajectory is

$$\dot{V}(\mathbf{x}) \underset{\text{def}}{=} \frac{\mathrm{d}V(\mathbf{x})}{\mathrm{d}t} = \sum_{j=1}^{n} \frac{\partial V}{\partial x_j} \dot{x}_j = \left(\frac{\partial V}{\partial \mathbf{x}}\right)^T \mathbf{f}(\mathbf{x}) \tag{6.2.14}$$

and is also a continuous function of the state vector $\mathbf{x}$. Further, both V and $\dot{V}$ are required to be defined in an open domain Ω including the equilibrium point $\mathbf{x}_e$, and satisfying the conditions:

1) $V(\mathbf{x}) > V(\mathbf{x}_e)$, for $\mathbf{x} \neq \mathbf{x}_e$ in the domain Ω. Without loss of generality, let $V(\mathbf{x}_e) = 0$ and $\mathbf{x}_e = \mathbf{0}$. There exist two continuous and non-decreasing scalar functions $\alpha(\|\mathbf{x}\|)$ and $\beta(\|\mathbf{x}\|)$, with $\alpha(0) = 0$ and $\beta(0) = 0$, such that

$$\alpha(\|\mathbf{x}\|) \leq V(\mathbf{x}) \leq \beta(\|\mathbf{x}\|)$$

2) If Ω is unbounded, then as $|\mathbf{x}| \to \infty$, $V(\mathbf{x}) \to \infty$

3) $\dot{V}(\mathbf{x}) < 0$, for $\mathbf{0} \neq \mathbf{x} \in \Omega$.

If such a Lyapunov scalar function $V(\mathbf{x})$ can be found, then this system is asymptotically stable. See [121],chap.5.

proof: Suppose initially the state vector $\mathbf{x}_0 \in \Omega$ is departed to the equilibrium point $\mathbf{x}_e$, and the subsequent trajectory of motion be denoted as $\mathbf{x} = \boldsymbol{\varphi}(t; \mathbf{x}_0, t_0)$. Because the function $\beta(\|\mathbf{x}\|)$ is continuous and non-decreasing, so that for any real value $\varepsilon > 0$ there is $\delta(\varepsilon) > 0$ such that $\beta(\delta) \le \alpha(\varepsilon)$. Based on that $\dot{V}(\mathbf{x}) < 0$,

$$V(\boldsymbol{\varphi}(t; \mathbf{x}_0, t_0)) - V(\mathbf{x}_0) = \int_{t_0}^{t} \dot{V}(\boldsymbol{\varphi}(\tau; \mathbf{x}_0, t_0)) \mathrm{d}\tau < 0$$

It tends to a non-negative limit as $t \to \infty$, when along any trajectory stay in Ω. Note that a trajectory can never across itself, so that its limiting point $\mathbf{x}_l$ is an equilibrium point. If $\mathbf{x}_l$ is other than the equilibrium point $\mathbf{x}_e$, then the condition 3, $\dot{V}(\mathbf{x}_l) < 0$, implies $\dot{\mathbf{x}}_l$ is not zero, which means that $\mathbf{x}_l$ is not an equilibrium point, a contradiction. Then $\mathbf{x}_l$ must coincide with $\mathbf{x}_e$, so that the motion tends to the point $\mathbf{x}_e$ asymptotically. ##

The idea of the Lyapunov function $V(\mathbf{x})$ of the direct method (his second method) is somewhat close to the consideration of system energy. Usually the damping effect decreases the system energy. Such consideration is based on the physical reasoning and is quite helpful. However, generally $V(\mathbf{x})$ is not necessarily the energy function, such as for problem with indefinite Hamilton function. There is no general method found today to select such a Lyapunov function for all problems.

For a linear time invariant system

$$\dot{\mathbf{x}} = \mathbf{A}\mathbf{x}, \qquad \mathbf{x}(t_0) = \mathbf{x}_0, \qquad t \ge t_0 \tag{6.2.15}$$

which is obtained by linear expansion method nearby the equilibrium point $\mathbf{x}_e = \mathbf{0}$. The Lyapunov first method is to solve all eigenvalues $\lambda_i, (i = 1, \cdots, n)$ of the matrix $\mathbf{A}$, *i.e.* the roots of equation $\det(s\mathbf{I} - \mathbf{A}) = 0$. Stability requires that all the roots locate in the left half plane of s, *i.e.*

$$\mathrm{Re}(\lambda_i) < 0, \qquad i = 1, \cdots, n \tag{6.2.16}$$

which ensures that $\exp(\mathbf{A}t) \to \mathbf{0}$, as $t \to \infty$.

Using the Lyapunov second method to equation (6.2.15), the Lyapunov function is to be constructed. For linear steady system the Lyapunov function can be selected as a quadratic function of $\mathbf{x}$, *i.e.*

$$V(\mathbf{x}) = \mathbf{x}^T \mathbf{P} \mathbf{x} \tag{6.2.17}$$

where $\mathbf{P}$ is a $n \times n$ symmetric positive definite matrix to be determined. Differentiating gives

$$\dot{V}(\mathbf{x}) = \dot{\mathbf{x}}^T \mathbf{P} \mathbf{x} + \mathbf{x}^T \mathbf{P} \dot{\mathbf{x}} = \mathbf{x}^T (\mathbf{A}^T \mathbf{P} + \mathbf{P}\mathbf{A}) \mathbf{x} = -\mathbf{x}^T \mathbf{Q} \mathbf{x} \tag{6.2.18}$$

where

$$\mathbf{A}^T \mathbf{P} + \mathbf{P}\mathbf{A} = -\mathbf{Q} \tag{6.2.19}$$

which is called the ***algebraic Lyapunov equation***, a linear simultaneous equation for the elements of matrix $\mathbf{P}$.

Arbitrarily select a ***positive definite symmetric matrix*** $\mathbf{Q}$, the matrix $\mathbf{P}$ is obtained by solving the algebraic Riccati equation,. For n-dimensional problem, there are $n \times (n+1)/2$ unknown elements of $\mathbf{P}$ to be solved. If the solved matrix $\mathbf{P}$ is positive definite, then the Lyapunov function is found, and based on the theorem the system is stable.

If the condition (6.2.16) is satisfied then

$$\mathbf{P} = \int_0^\infty \exp(\mathbf{A}^T t)\mathbf{Q}\exp(\mathbf{A}t)\mathrm{d}t \tag{6.2.20}$$

As a matter of fact, the ***Lyapunov differential equation***

$$\dot{\mathbf{P}}(t) = \mathbf{A}^T\mathbf{P}(t) + \mathbf{P}(t)\mathbf{A} + \mathbf{Q}, \quad \mathbf{P}(0) = \mathbf{0} \tag{6.2.21}$$

is solved by

$$\mathbf{P}(t) = \int_0^t \exp[\mathbf{A}^T(t-\tau)]\mathbf{Q}\exp[\mathbf{A}(t-\tau)]\mathrm{d}\tau \tag{6.2.22}$$

Directly substituting into the differential equation (6.2.21) verifies the conclusion. Note that the differential equation can be solved by the ***precise integration method.*** In section 6.4 when the problem of prediction is considered, the Lyapunov differential equation appears again. The solution requires computing the integration (6.2.22), where $\mathbf{Q}$ can be a function of time τ. The precise integration method is given there in some detail.

Linear time-invariant system gives the simplest differential equation. For non-linear differential equation, there is no general method to find the Lyapunov function, and more investigations can be found in such as [121,122]. Both the stability of time-variant differential equations of state vector in LQ control and of Kalman-Bucy filter can be proved by the Lyapunov second method, and will be given in section 6.7.

As described above, the selection of Lyapunov function V requires further investigation.

§6.3, Prediction, filtering and smoothing

In science and technology, estimation is always required. Estimation can be subdivided into two classes, *i.e.* state estimation and parameter estimation. The state estimation is for a given system to estimate the state under noise disturbance, where the state $\mathbf{x}(t)$ is a vector of stochastic process. The parameter estimation means to identify the parameters for the system itself. Parameter estimation is also often used for curve fitting, which must be distinguished to the system parameter identification. The least square criterion is used most frequently in estimations. For optimal control problems, the state estimation is one of the main concerns to be investigated.

State estimation here is mainly for dynamical system, the static estimation is only a special case of dynamical system. Suppose the system dynamic and measurement equations are, respectively,

$$\dot{\mathbf{x}}(t) = \mathbf{A}(t)\mathbf{x} + \mathbf{B}_w(t)\mathbf{w}(t) + \mathbf{B}_u(t)\mathbf{u}(t), \tag{6.3.1}$$

$$\mathbf{y}(t) = \mathbf{C}_y(t)\mathbf{x}(t) + \mathbf{v}(t) \tag{6.3.2}$$

where $\mathbf{x}(t), \mathbf{y}(t), \mathbf{u}(t)$ are n, q, m -dimensional vectors of state, measurement and control input, respectively. The vectors $\mathbf{v}(t)$ and $\mathbf{w}(t)$ are the q, l -dimensional measurement and dynamic noises, respectively. The given system matrices

$\mathbf{A}, \mathbf{B}_w, \mathbf{B}_u, \mathbf{C}_y$ have appropriate dimensions. Let the variable t denote the present time. Estimation means to estimate the state vector $\mathbf{x}(t)$ based on the measured data $\mathbf{y}(\tau), \tau \le t$. The continuous-time system usually estimates the state using Kalman-Bucy filtering. The estimation of discrete-time system is called as Kalman filtering. Filtering means the estimation of state of the system at the present time.

The Linear Quadratic Gaussian (LQG) optimal control theory requires to find the feedback control vector $\mathbf{u}(t)$ based on the state vector $\mathbf{x}(t)$, however, the state vector has not been completely measured directly. Hence, the filtering estimated vector $\hat{\mathbf{x}}(t)$ is used in place of the state vector $\mathbf{x}(t)$. Therefore, Kalman filter is used as an integral component in the LQG optimal control theory. The Kalman filter has been utilized is a wide range of applications, not only for LQG control but also as a tool of signal processing.

Usually, the disturbances $\mathbf{w}$ and $\mathbf{v}$ are assumed to be zero mean white noises, independent on each other, of which the statistical parameters are given as

$$\begin{aligned} \mathrm{E}[\mathbf{w}(t)] &= \mathbf{0} \\ \mathrm{E}[\mathbf{v}(t)] &= \mathbf{0} \end{aligned}, \quad \begin{aligned} \mathrm{E}[\mathbf{w}(t)\mathbf{w}^T(\tau)] &= \mathbf{W}(t)\delta(t-\tau) \\ \mathrm{E}[\mathbf{v}(t)\mathbf{v}^T(\tau)] &= \mathbf{V}(t)\delta(t-\tau) \\ \mathrm{E}[\mathbf{w}(t)\mathbf{v}^T(\tau)] &= \mathbf{0} \end{aligned} \tag{6.3.3}$$

where $\delta(t-\tau)$ denotes the Dirac-function and $\mathbf{W}, \mathbf{V}$ are symmetric and positive definite covariance matrices of the white noises, respectively.

The initial state is also a stochastic vector, of which the mathematical expectation and covariance matrices are given as

$$\mathrm{E}[\mathbf{x}(0)] = \hat{\mathbf{x}}_0, \quad \mathrm{E}[(\mathbf{x}(0)-\hat{\mathbf{x}}_0)(\mathbf{x}(0)-\hat{\mathbf{x}}_0)^T] = \mathbf{P}_0 \tag{6.3.4}$$

and also assuming that the noises are independent on the initial state, that

$$\mathrm{E}[(\mathbf{x}(0)-\hat{\mathbf{x}}_0)\mathbf{w}^T] = \mathbf{0}$$

Suppose the measurement vector is $\mathbf{y}(\tau), 0 \le \tau \le t_1$ and t is the present time. There are three classes of estimation problems, namely 1) $t > t_1$; 2) $t = t_1$; 3) $t < t_1$. Based on the dynamic equation, the initial condition and the measurement vector, the problem is to find the optimal estimation $\hat{\mathbf{x}}(t)$ of the state vector $\mathbf{x}(t)$ and also the covariance matrix $\mathbf{P}(t)$ of the error

$$\mathrm{E}[\mathbf{x}(t)] = \hat{\mathbf{x}}(t), \quad \mathrm{E}[(\mathbf{x}(t)-\hat{\mathbf{x}}(t))(\mathbf{x}(t)-\hat{\mathbf{x}}(t))^T] = \mathbf{P}(t) \tag{6.3.5}$$

The descriptions for the three classes of measurement are:

1) For the class of $t_1 < t$, the problem is ***prediction***. In the time interval $(0, t_1)$ the estimation is supported by measurement, however, the time t under investigation is ahead of t_1 that in the interval (t_1, t) there is no measurement available. The motion in interval (t_1, t) can only be solved by the dynamic equation only. Firstly, the filter analysis is carried out until t_1, after solving the mean value $\hat{\mathbf{x}}(t_1)$ and the variance matrix $\mathbf{P}(t_1)$. The next step is, based

on the estimation $\hat{\mathbf{x}}(t_1)$ and $\mathbf{P}(t_1)$ at t_1 as the initial conditions, to do the measurement free analysis for time interval (t_1, t), *i.e.* the ***prediction.***

2) The measurement interval reaches the present time, *i.e.* $t_1 = t$, and the estimation of $\hat{\mathbf{x}}(t)$ is required. It is a popular statement, called as ***Filtering***. Real time response needs filtering analysis.
3) For class of $t < t_1$, the problem is called as ***Smoothing***. Smoothing using longer time period measured data to estimate the state at time instant t. For example, bringing the data recorded in situ testing, afterwards analyze in the laboratory, which has been the off-line computation.

For prediction, the available data is less than the filter, so that the precision of prediction estimation is lower than filter. For smoothing the available data is more than filter, hence the precision of smoothing estimation is higher than filter. That is to say that the variance of prediction is larger than filter and the variance of smoothing is smaller than filter.

The majority of real systems are non-linear, in this sense it is best to do the prediction, filtering and smoothing directly for non-linear systems. However, the general method for the analysis of a non-linear system is very difficult, even impossible today. In applications, the first step is to expand the solution nearby the **nominal** state, so that the perturbation is governed by a set of linear equations. Quite often, the perturbation method gives a good approximation. Mathematically, a linear system is far easier than a non-linear one, that a number of methods can be applied. Hence, the linear system estimation is considered below. Special attention is put on the algorithms. Generally speaking, the solution of a non-linear system can be found on the basis of the solution of approximate linear perturbation by iteration.

The modern control theory closely depends on the filtering analysis. It is seen from figure 6.4, that the whole control time interval, from the initial time t_0 until the finish time t_f, is subdivided by the present time instant t, into past $(0, t)$ and future (t, t_f) two time intervals.

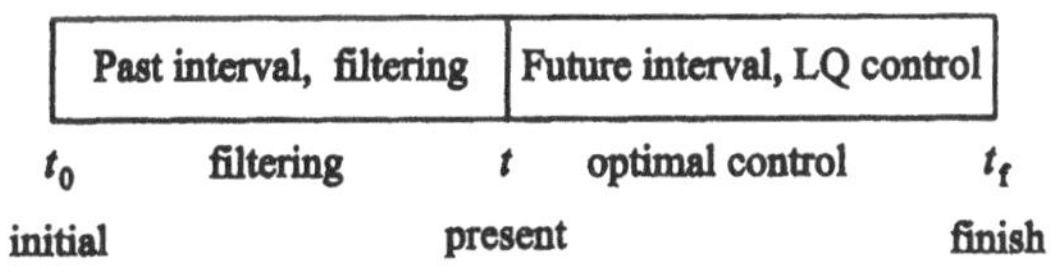

Figure 6.4, *Past and futute intervals subdivided at* t

For **past time interval**, the control vector $\mathbf{u}$ had been selected and performed and has become the history, that it must be treated as a given vector. The analysis for the past time interval should be based on the knowledge of the system model and the measured data to determine the state at the present instant. Because of dynamic and measurement noises of system, the real state at present instant cannot be found exactly, only the optimal estimation $\hat{\mathbf{x}}(t)$ and the variance matrix $\mathbf{P}(t)$ can be solved. That is to say, ***the analysis for past time interval is filtering***, which will be described below in some detail. The problem stated in such way is the direct problem. As the identification of system model itself, *i.e.* the system parameter identification, is a further requirement and is classified as the inverse problem. The related problem is adaptive control.

For **future time interval,** the required analysis is ***Linear Quadratic (LQ) optimal control*** with the initial condition being the filtered state vector $\hat{\mathbf{x}}(t)$ at the present time instant t, in combination with the variance matrix $\mathbf{P}(t)$.

At the **present instant** t, the ***feedback control vector*** $\mathbf{u}(t)$ ***is required.*** Certainly this is based on the results of filtering analysis for the past time interval and control analysis for the future time interval. The feedback control vector $\mathbf{u}(t)$ must be supplied at the real time. Hence the computation of $\mathbf{u}(t)$ must be subdivided into two parts, the ***off-line*** computation and the ***on-line*** computation. The computations unrelated to the measured date $\mathbf{y}$ can be computed beforehand, *i.e.* ***off-line***, and keep the results in the databank, while the computations relate to the measurements, *i.e.* ***on-line***, can only be executed at the ***real time***. Because, the control vector $\mathbf{u}(t)$ demands to be supplied at the real time, *i.e.* as quickly as possible, so that the ***on-line computation should be reduced to the minimum***. The next three sections are devoted to the three classes of estimation problems.

§6.4, Prediction and its computation

Prediction, a very attractive subject, decision making must be based on the result of prediction. The present section predicts the state vector for a system governed by linear equations. Two simple examples are given first.

Example 6.4, A device with mass M is installed on the top of a column, and the ground is excited because of earthquake motion. The mass vibration equation is governed by

$$M\ddot{x}+C\dot{x}+Kx=f(t), \qquad x(0)=\dot{x}(0)=0 \tag{6.4.1}$$

The excitation force $f(t)$ induced by the ground motion is random and treated as a stochastic process so that only the statistical parameters are available. Hence the response $x(t)$ is also a stochastic process to be solved, for which the solution is also the statistical parameters. This problem is simply random vibration of a one-dimensional problem. ##

Example 6.5, The financial analysis needs to determine the short-term interest of the money market. Quite a number of events continuously come from time to time, such as the jobless rate given by the society, index of price market, the average personal income, selling rate of estate and retailing index etc. These factors influence the short-term interest rate fluctuating, and the effects of these events can be combined together and modeled by a Gaussian white noise. However, the short-term interest rate $r(t)$ may not vary far depart to the mean value θ and has a trend to go back the mean value θ. Hence the stochastic process $r(t)$ is considered satisfying the stochastic differential equation

$$\dot{r}(t) = K \cdot (\theta - r) + \sigma X(t) \ , \qquad r(0) = \text{given} \tag{6.4.2}$$

where $X(t)$ is a unit white noise and σ is its intensity, K represents the recovering proportional parameter of the interest rate. The parameters σ and K can be constants and the stochastic process $r(t)$ is to be solved. This is the simplest model for interest rate, called the Vasicek model [61]. ##

There are quite a number of prediction examples, for which the differential equation to be solved involves stochastic process terms. The two examples given above have only stochastic terms but the system parameters have no random factor, hence simple. This section discusses the problems with only stochastic input term and only mean square calculus is required.

For the problems of stochastic control and stochastic system prediction etc. the mathematical models are always reduced to solve stochastic differential equations. Let $x(t)$ denotes the solution, because of the stochastic input, so that $x(t)$ is also a stochastic process. Strictly speaking, a random variable needs the distribution function, but this requirement is difficult to fulfill. Even to supply a distribution function for the initial state x_0 is not so easy. Therefore in system analysis, the detail distribution function is not demanded, but requires only the statistical characteristic parameters of the solution $x(t)$. The ***central limit theorem*** in probability theory determines that the combination of a large number of random factors tends to be a Gauss distribution and the Gauss distribution determination needs only the mean value and the variance. For a stochastic process, the mean value function $\hat{x}(t)$ and the auto-correlation function $R(t,t_1)$ are needed, and for $R(t,t_1)$, the most important is the variance function $R(t) \equiv R(t,t)$. Therefore except declared specially, the stochastic process is always regarded as Gauss distributed, and the solution becomes to find the mean value and auto-correlation function or variance function.

Although the process is stochastic but the mean value $\hat{x}(t)$ and the mean square $R(t)$ functions (or variance) are deterministic. Hence it is found below that the solution of a stochastic differential equation is transformed to solve a set of ordinary differential equations. Therefore, the combination of analytical and numerical integration methods can be applied effectively.

§6.4.1, Mathematical model for prediction

Prediction applies to a system without measurement, corresponding to open loop. Using state space method description, the dynamic equation and output function can be written as

$$\dot{\mathbf{x}} = \mathbf{f}(\mathbf{x},\mathbf{u},t) + \mathbf{B}_w(\mathbf{x},t)\mathbf{w}(t) + \mathbf{B}_u(\mathbf{x},t)\mathbf{u}(t) \tag{6.4.3}$$

$$\mathbf{z} = \mathbf{g}(\mathbf{x},\mathbf{u},t) \tag{6.4.4}$$

where $\mathbf{x}$, $\mathbf{z}$ and $\mathbf{u}$ are the n-, q- and m-dimensional state, output and deterministic control input vectors, respectively. And $\mathbf{f}$ is a n-dimensional given function, $\mathbf{w}$ is a l-dimensional dynamic noise. The matrices $\mathbf{B}_w$ and $\mathbf{B}_u$ are $n \times l$, $n \times m$-dimensioned matrix functions, respectively, and $\mathbf{g}$ is a p-dimensional given vector function.

The output vector $\mathbf{z}$ is not measurement vector $\mathbf{y}$, that they should be strictly distinguished. The measurement vector $\mathbf{y}$ can be used to check the state vector, however, the output vector $\mathbf{z}$ is obtained from the state vector $\mathbf{x}$ to be used for other purpose, which is not the measured data and cannot be used for checking.

The formulation of the above equations is non-linear. The general solution for non-linear stochastic differential equations is very difficult presently, so that usually a linear set of differential equations is solved.

$$\dot{\mathbf{x}}(t) = \mathbf{A}(t)\mathbf{x}(t) + \mathbf{B}_u(t)\mathbf{u}(t) + \mathbf{B}_w(t)\mathbf{w}(t) \tag{6.4.5}$$

$$\mathbf{z}(t) = \mathbf{C}(t)\mathbf{x}(t) + \mathbf{D}(t)\mathbf{u}(t) \tag{6.4.6}$$

where $\mathbf{A}$ is the $n \times n$ plant matrix, $\mathbf{B}_u$ is $n \times m$ control input matrix, $\mathbf{B}_w$ is $n \times l$ disturbance input matrix, $\mathbf{C}$ is a $q \times n$ output matrix and $\mathbf{D}$ is a $q \times m$ control output matrix usually be $\mathbf{0}$. For a linear time-invariant (steady) system the above system matrices are constants.

The above equations are continuous-time mathematical model for linear systems. For discrete-time problems a linear system equations are expressed as

$$\mathbf{x}(k+1) = \mathbf{A}_k\mathbf{x}(k) + \mathbf{B}_{uk}\mathbf{u}(k) + \mathbf{B}_{wk}\mathbf{w}(k) \tag{6.4.7}$$

$$\mathbf{z}(k) = \mathbf{C}_k\mathbf{x}(k) + \mathbf{D}_k\mathbf{u}(k) \tag{6.4.8}$$

For a time-invariant system, $\mathbf{A}_k, \mathbf{B}_{uk}, \mathbf{B}_{wk}, \mathbf{C}_k$ etc. are independent on k. Because there is the dynamic noise vector $\mathbf{w}$, hence the state vector $\mathbf{x}$ and output vector $\mathbf{z}$ are stochastic processes too. For linear systems, the assumption of, $\mathbf{w}$ being Gaussian distributed, makes that $\mathbf{x}$ and $\mathbf{z}$ are Gaussian distributed too. Hence both vector stochastic processes $\mathbf{x}$ and $\mathbf{z}$ can be expressed by the mean value functions $\hat{\mathbf{x}}(t)$, $\hat{\mathbf{z}}(t)$ and their corresponding variance matrices.

Both differential and difference equations require initial conditions, which are given as

$$\hat{\mathbf{x}}(0) = \hat{\mathbf{x}}_0 ; \qquad \mathbf{G}(0) = \mathbf{G}_0 \tag{6.4.9}$$

where $\hat{\mathbf{x}}(t)$ is the mean value of n-dimensional state vector, and $\mathbf{G}(t)$ is the $n \times n$ variance matrix, symmetric and non-negative definite, all deterministic terms.

Writing the equation (6.4.5) in differential form

$$\mathrm{d}\mathbf{x} = (\mathbf{B}_u\mathbf{u} + \mathbf{A}\mathbf{x})\mathrm{d}t + \mathbf{B}_w \mathrm{d}\mathbf{w}_e$$

called as Ornstein-Uhlenbeck process [61,69], which is quite useful in a number of disciplines, where $\mathbf{w}_e$ is a Wiener process. The time differential of a Wiener process is a white noise, so that it corresponds to the equation (6.4.5) with $\mathbf{w}$ being a white noise there.

§6.4.2, Prediction of one dimensional system

The equation (6.4.5) is a linear stochastic differential equation. The dynamic noise $\mathbf{w}$ is assumed Gaussian distributed white noise, so that the response is also Gaussian. Gauss distribution is determined by the mean value and the variance, solving which gives the solution of linear equations of prediction. Therefore the stochastic differential equation is transformed to differential equations for the unknown deterministic functions $\hat{\mathbf{x}}(t)$ and $\mathbf{G}(t)$.

The simplest case of one-dimensional problem is considered first. The dynamic equation is

$$\dot{x}(t) = \alpha(\theta - x) + \sigma w \tag{6.4.10}$$

and the prediction of state function $x(t), 0 < t < T$ is required, where θ is the given equilibrium point with the initial conditions

$$\hat{x}(0) = \hat{x}_0, \quad G(0) = G_0, \qquad \text{when } t = 0 \tag{6.4.11}$$

and w is a white noise with

$$\mathrm{E}(w) = 0, \qquad \mathrm{var}[w(t), w(\tau)] = R_w(t)\delta(t - \tau) \tag{6.4.12}$$

where R_w is the same function of W in equation (6.3.3).

The basic requirement of prediction is to solve $\hat{x}(t)$ first. According to the ***least square*** principle, the index of disturbance (analogous to the potential energy in mechanics) should be minimized

$$J = \int_0^T [w(t)R_w^{-1}(t)w(t)/2]\mathrm{d}t + G_0^{-1}(x_0 - \hat{x}_0)^2/2, \qquad \min J \tag{6.4.13}$$

under the conditions that the dynamic equation and initial conditions should be satisfied beforehand, which means that the minimization of J is conditional. Using the Lagrange multiplier method, introducing the dual function $\lambda(t)$ of the dynamic equation gives

$$\begin{gathered} J_A = \int_0^T [\lambda\dot{x} - \lambda\sigma w + \lambda\alpha x - \lambda\alpha\theta + R_w^{-1}w^2/2]\mathrm{d}t + G_0^{-1}(x_0 - \hat{x}_0)^2/2, \\ \delta J_A = 0 \end{gathered} \tag{6.4.14}$$

where J_A is the extended index. The variations of the functions w, x and λ are independent to each other. Minimization with respect to w first gives

$$w = \sigma R_w \lambda \tag{6.4.15}$$

Substituting back into equation (6.4.14) gives

$$J_A = \int_0^T [\lambda\dot{x} + \lambda\alpha x - \lambda\alpha\theta - \sigma^2 R_w\lambda^2/2]\mathrm{d}t + G_0^{-1}(x_0 - \hat{x}_0)^2/2, \quad \delta J_A = 0 \tag{6.4.16}$$

There are two kinds of variables x, λ in this variational principle, and x_0 is also a variable. These two functions x and λ are dual to each other, and are all functions of stochastic process. Carrying out the variational derivation and using integration by parts gives

$$\begin{aligned} \delta J_A = \int_0^T [\delta\lambda(\dot{x} - \alpha(\theta - x) - \sigma^2 R_w\lambda) + \delta x(\alpha\lambda - \dot{\lambda})]\mathrm{d}t \\ + \lambda(T)\delta x(T) + [-\lambda(0) + G_0^{-1}(x_0 - \hat{x}_0)]\delta x_0 = 0 \end{aligned}$$

From which, the dual equations are derived as

$$\begin{aligned} \dot{x} &= -\alpha x + \sigma^2 R_w \lambda + \alpha\theta \\ \dot{\lambda} &= \alpha\lambda \end{aligned} \tag{6.4.17}$$

with the initial condition

$$x_0 = \hat{x}_0 + G_0\lambda(0), \quad \text{when } t = 0 \tag{6.4.18}$$

The problem is how to solve the dual differential equations. Because $x(t)$ is a stochastic differential equation with Gaussian distribution, so that it can be expressed as the sum of a deterministic mean-value function denoted as $\hat{x}(t)$, and a zero-mean Gaussian stochastic process. The function λ is also a zero-mean Gaussian stochastic process. From the initial condition (6.4.18) for x, the form of solution has been seen. Let

$$x(t) = \hat{x}(t) + G(t)\lambda(t) \tag{6.4.19}$$

Substituting into (6.4.17), because of $\dot{x} = \dot{\hat{x}} + \dot{G}\lambda + G\dot{\lambda} = \dot{\hat{x}} + \dot{G}\lambda + G\alpha\lambda$, hence

$$\dot{\hat{x}} + \dot{G}\lambda = -\alpha\hat{x} + \alpha\theta - 2G\alpha\lambda + \sigma^2 R_w \lambda$$

The above equation is composed of two kinds of terms, namely the deterministic term

$$\dot{\hat{x}} = -\alpha\hat{x} + \alpha\theta, \qquad \hat{x}(0) = \hat{x}_0 \tag{6.4.20}$$

and the stochastic terms about λ. Eliminating λ gives

$$\dot{G} = -2\alpha G + \sigma^2 R_w, \qquad G(0) = G_0 = \text{given} \tag{6.4.21}$$

which is also a deterministic equation. Solving the mean value of x from equation (6.4.20) gives

$$\hat{x}(t) = \theta + (\hat{x}_0 - \theta)e^{-\alpha t} \tag{6.4.22}$$

As $t \to \infty$, the mean value tends to θ. From equation (6.4.21) solves

$$G(t) = \sigma^2 R_w/(2\alpha) + \left(G_0 - \sigma^2 R_w/(2\alpha)\right)\cdot e^{-2\alpha t} \tag{6.4.23}$$

As $t \to \infty$, the function tends to a constant $\sigma^2 R_w/(2\alpha)$. From equation (6.4.19) gives

$$\lambda(t) = G^{-1}(t)[x(t) - \hat{x}(t)] \tag{6.4.24}$$

A special case should be noticed, *i.e.* the case of $\alpha = 0$, then it gives $\hat{x} = \hat{x}_0$. Taking the limit $\alpha \to 0$ on (6.4.23) gives $G(t) = G_0 + \sigma^2 R_w t$. When $\hat{x}_0 = 0$, $G_0 = 0$ and $\sigma^2 R_w = 1$, it gives $G(t) = t$. Such stochastic process is just a Wiener process, which represents Brownian motion Gaussian distributed.

The solution obtained for the stochastic differential equation (6.4.10) can be found from such as books [61,69], where the solution is obtained via the Ito calculus, a complicated derivation. Here, only the usual means are applied, simple and easy to understand.

§6.4.3, Prediction of multi-degrees of freedom system

Prediction means that there is no measured data to verify the estimation. The equations are (6.4.5) and (6.4.6).

The solution method for multi-degrees of freedom system is almost the same as single-degree of freedom. The state is a stochastic process vector $\mathbf{x}(t)$, which should satisfy the dynamic equation (6.4.5) and the initial condition (6.4.9) beforehand. Under these conditions, the index of disturbance energy

$$J = \int_0^T [\mathbf{w}^T \mathbf{W}^{-1}\mathbf{w}/2]\mathrm{d}t + (\mathbf{x}_0 - \hat{\mathbf{x}}_0)^T \mathbf{G}^{-1}{}_0(\mathbf{x}_0 - \hat{\mathbf{x}}_0)/2, \quad \min J = 0 \tag{6.4.25}$$

is minimized. Introducing the Lagrange multiplier vector $\boldsymbol{\lambda}$ for the dynamic equation constraint gives

$$\begin{aligned} J_A = \int_0^T [\boldsymbol{\lambda}^T(\dot{\mathbf{x}} - \mathbf{A}\mathbf{x} - \mathbf{B}_u\mathbf{u} - \mathbf{B}_w\mathbf{w}) + \mathbf{w}^T\mathbf{W}^{-1}\mathbf{w}/2]\mathrm{d}t \\ + (\mathbf{x}_0 - \hat{\mathbf{x}}_0)^T \mathbf{G}_0^{-1}(\mathbf{x}_0 - \hat{\mathbf{x}}_0)/2, \qquad \delta J_A = 0 \end{aligned} \tag{6.4.26}$$

where the independently varied functions are $\mathbf{x}, \boldsymbol{\lambda}$ and $\mathbf{w}$, and the vector $\mathbf{u}$ is only a deterministic input. The variation of disturbance vector $\mathbf{w}$ can be performed first, which gives

$$\mathbf{w} = \mathbf{W}\mathbf{B}_w^T\boldsymbol{\lambda} \tag{6.4.27}$$

Substituting $\mathbf{w}$ back into J_A obtains

$$J_A = \int_0^T [\boldsymbol{\lambda}^T(\dot{\mathbf{x}} - \mathbf{A}\mathbf{x} - \mathbf{B}_u\mathbf{u}) - \boldsymbol{\lambda}^T\mathbf{B}_w\mathbf{W}\mathbf{B}_w^T\boldsymbol{\lambda}/2]\mathrm{d}t + (\mathbf{x}_0 - \hat{\mathbf{x}}_0)^T\mathbf{G}_0^{-1}(\mathbf{x}_0 - \hat{\mathbf{x}}_0)/2 \tag{6.4.28}$$

which has two kinds of independently varied functions $\mathbf{x}$ and $\boldsymbol{\lambda}$. Taking $\delta J_A = 0$ derives the dual differential equations

$$\begin{aligned} \dot{\mathbf{x}} &= \mathbf{A}\mathbf{x} + \mathbf{B}_w\mathbf{W}\mathbf{B}_w^T\boldsymbol{\lambda} + \mathbf{B}_u\mathbf{u} \\ \dot{\boldsymbol{\lambda}} &= -\mathbf{A}^T\boldsymbol{\lambda} \end{aligned} \tag{6.4.29}$$

with the initial condition

$$\mathbf{x}_0 = \hat{\mathbf{x}}_0 + \mathbf{G}_0\boldsymbol{\lambda}_0, \quad \text{when } t = 0 \tag{6.4.30}$$

where $\mathbf{x}$ and $\boldsymbol{\lambda}$ are Gauss distributed vector functions of stochastic processes, and $\hat{\mathbf{x}}_0, \mathbf{G}_0$ are the deterministic initial mean vector and variance matrix, respectively. The solution $\mathbf{x}(t)$ is substituted as

$$\mathbf{x}(t) = \hat{\mathbf{x}}(t) + \mathbf{G}(t)\boldsymbol{\lambda}(t) \tag{6.4.31}$$

Differentiation gives $\dot{\mathbf{x}} = \dot{\hat{\mathbf{x}}} + \dot{\mathbf{G}}\boldsymbol{\lambda} + \mathbf{G}\dot{\boldsymbol{\lambda}} = \dot{\hat{\mathbf{x}}} + \dot{\mathbf{G}}\boldsymbol{\lambda} - \mathbf{G}\mathbf{A}^T\boldsymbol{\lambda}$. Then substituting into (6.3.29) gives

$$\dot{\hat{\mathbf{x}}} + \dot{\mathbf{G}}\boldsymbol{\lambda} - \mathbf{G}\mathbf{A}^T\boldsymbol{\lambda} = \mathbf{A}\hat{\mathbf{x}} + \mathbf{A}\mathbf{G}\boldsymbol{\lambda} + \mathbf{B}_w\mathbf{W}\mathbf{B}_w^T\boldsymbol{\lambda} + \mathbf{B}_u\mathbf{u} \tag{6.4.32}$$

The above equation has the deterministic terms and zero mean stochastic terms $\boldsymbol{\lambda}$ together, so that making distinction derives two equations as

$$\begin{aligned} \dot{\hat{\mathbf{x}}} &= \mathbf{A}\hat{\mathbf{x}} + \mathbf{B}_u\mathbf{u}, & \hat{\mathbf{x}}(0) &= \hat{\mathbf{x}}_0 \\ \dot{\mathbf{G}} &= \mathbf{G}\mathbf{A}^T + \mathbf{A}\mathbf{G} + \mathbf{B}_w\mathbf{W}\mathbf{B}_w^T, & \mathbf{G}(0) &= \mathbf{G}_0 \end{aligned} \tag{6.4.33,34}$$

Equation (6.4.33) is a deterministic vector ODE for the mean value function $\hat{\mathbf{x}}$, which can be integrated for a given input vector $\mathbf{u}$. Equation (6.4.34) is also a linear deterministic ODE for a $n \times n$ symmetric matrix function $\mathbf{G}$, which holds $n \times (n+1)/2$ component functions. Both equations (6.4.33,34) are sets of ODEs, the existence of solution is definite, but for applications, numerical solutions are necessary. For which, the precise integration method is described in next section. Equation (6.4.33) is the usual vector ODE, and the equation (6.4.34) is called as Lyapunov ODE. Note that from the Ornstein-Ulenbeck process, the set of equations (6.4.33) and (6.4.34) is also derived.

Here again, the Ito calculus is bypassed for the solution of the stochastic differential equation. The derivation here is simple and straightforward.

§6.4.4, Precise time integration

The state space analysis of linear dynamical system requires solving the equation

$$\dot{\mathbf{x}}(t) = \mathbf{A}(t)\mathbf{x}(t) + \mathbf{r}(t) \tag{6.4.35}$$

where $\mathbf{x}(t)$ is a n-dimensional vector to be solved, $\mathbf{A}$ is a given $n \times n$ matrix, and $\mathbf{r}(t)$ is a given non-homogeneous input vector. The initial condition is

$$\mathbf{x}(0) = \mathbf{x}_0 = \text{given}, \qquad \text{when } t = t_0 = 0 \tag{6.4.36}$$

From theory of ODE, the homogeneous equation should be solved first

$$\dot{\mathbf{x}} = \mathbf{A}\mathbf{x} \tag{6.4.37}$$

If the matrix $\mathbf{A}$ is time-variant, the solution can also be expressed by the ***state transition matrix*** $\mathbf{\Phi}(t, t_0)$, or unit impulse response matrix, which satisfies

$$\dot{\mathbf{\Phi}}(t,t_0) = \mathbf{A}(t)\mathbf{\Phi}(t,t_0), \qquad \mathbf{\Phi}(t_0,t_0) = \mathbf{I}_n \tag{6.4.38}$$

Based on the superposition principle, the solution can be expressed by the Duhamel integration

$$\mathbf{x}(t) = \mathbf{\Phi}(t,0)\mathbf{x}_0 + \int_0^t \mathbf{\Phi}(t,\tau)\mathbf{r}(\tau)\mathrm{d}\tau \tag{6.4.39}$$

The theory is elegant, but the problem is numerical computation now. For a general time-variant matrix $\mathbf{A}(t)$, the computation of the matrix $\mathbf{\Phi}(t,\tau)$ is not so easy. When $\mathbf{A}$ is time-invariant

$$\mathbf{\Phi}(t,t_0) = \mathbf{\Phi}(t - t_0) = \exp[\mathbf{A}(t - t_0)] \tag{6.4.40}$$

the numerical computation of the exponential matrix is as follows. The expansion

$$\exp(\mathbf{A}t) = \mathbf{\Phi}(t) = \mathbf{I} + \mathbf{A}t + (\mathbf{A}t)^2/2 + \cdots + (\mathbf{A}t)^k/k! + \cdots \tag{6.4.41}$$

is an exponential function. Note that the matrix multiplication is in general non-commutative, *i.e.* $\mathbf{AB} \neq \mathbf{BA}$, so that generally speaking $\exp(\mathbf{A}) \cdot \exp(\mathbf{B}) \neq \exp(\mathbf{A} + \mathbf{B})$. Only when the matrices are commutative for multiplication then

$$\exp(\mathbf{A}) \cdot \exp(\mathbf{B}) = \exp(\mathbf{A} + \mathbf{B}), \qquad \text{when } \mathbf{AB} = \mathbf{BA} \tag{6.4.42}$$

It is easy to verify that the unit impulse response matrix $\mathbf{\Phi}(t,t_0)$ has the characteristics

$$\mathbf{\Phi}(t,t_0) = \mathbf{\Phi}(t,t_1) \times \mathbf{\Phi}(t_1,t_0) \tag{6.4.43}$$

and when the system is time-invariant, it reduces to

$$\mathbf{\Phi}(t) = \mathbf{\Phi}(t - \tau)\mathbf{\Phi}(\tau) \tag{6.4.43'}$$

These have been described in section 0.1 already.

Numerical integration needs to have a time step η, that the uniform step-size time instants are given as $t_0 = 0, t_1 = \eta, \cdots, t_k = k\eta, \cdots$. For the homogeneous equation of linear time invariant system (6.4.37)

$$\mathbf{x}(\eta) = \mathbf{x}_1 = \mathbf{T}\mathbf{x}_0, \qquad \mathbf{T} = \exp(\mathbf{A}\eta) \tag{6.4.44}$$

and the recursive integrations are $\mathbf{x}_2 = \mathbf{T} \cdot \mathbf{x}_1, \cdots, \mathbf{x}_{k+1} = \mathbf{T} \cdot \mathbf{x}_k, \cdots\cdots$, which is only matrix-vector multiplication. The problem reduces to compute the matrix $\mathbf{T}$ as given in equation (6.4.44), and this key step has been solved in section 0.1 by the precise integration method.

Matrix exponentiation is widely applied, and is one of the most frequently computed matrix functions. The paper [29] reviewed 19 **dubious** algorithms and later, the book [30] means further investigation is needed. It should be mentioned that the eigenvector expansion method, in the case of no duplicate eigenvalues, is still effective and is described below.

The eigen-solution for the matrix $\mathbf{A}$ is given as

$$\mathbf{AY} = \mathbf{Y}\mathrm{diag}(\mu_1, \cdots, \mu_n)$$

where $\mathbf{Y}$ is the eigenvector composed matrix and μ_i is the respective eigenvalues, and $\mathrm{diag}(\mu_i)$ represents a diagonal matrix. [Note $\mathbf{A}$ is not necessarily symmetric matrix, so μ_i can possibly have duplicate eigenvalues of the Jordan normal form]. Hence it is derived as

$$\exp(\mathbf{A}) = \mathbf{Y}\exp[\mathrm{diag}(\mu_i)]\mathbf{Y}^{-1} = \mathbf{Y}\mathrm{diag}[\exp(\mu_i t)]\mathbf{Y}^{-1}$$

Evidently, the above equation is the analytical solution of exponential matrix, however, which is established based on all the eigen-solutions of matrix $\mathbf{A}$. The difficulty comes from the possible Jordan normal form. In such case, the eigen-solution of matrix $\mathbf{A}$ is numerically unstable.

However, the precise integration method gives the numerical solution approaching computer precision, even in the case of Jordan form appearing, the numerical result is always stable and has the same precision.

§6.4.4.1, Precise integration of inhomogeneous equations

After the exponential matrix $\mathbf{T} = \exp(\mathbf{A}\eta)$ is computed precisely, the time step integration of dynamic equation (6.4.35) can be executed. The time-invariant system is discussed first. If there is no input ($\mathbf{r} = \mathbf{0}$), then the integration is only a series of matrix-vector multiplication and the computation is precise. However, when there is the input vector $\mathbf{r}$, the expression of $\mathbf{r}$ is needed. But it is not always available precisely within the time-step. Then various approximations are needed.

The special solution of inhomogeneous equation can be expressed as the convolution in (6.4.39)

$$\mathbf{x}_r(t) = \int_0^t \mathbf{\Phi}(t-\tau)\mathbf{r}(\tau)\mathrm{d}\tau \tag{6.4.45}$$

which is exact, but the matrix $\mathbf{\Phi}$ is computed only at the grid points. Suppose that the time integration has been reached $t_k = k\eta$ that

$$\mathbf{x}_k = \mathbf{\Phi}(t_k)\cdot\mathbf{x}_0 + \int_0^{t_k} \mathbf{\Phi}(t_k-\tau)\mathbf{r}(\tau)\mathrm{d}\tau \tag{6.4.45'}$$

Next step is to find the value at t_{k+1}. The integration can be transformed initiated from t_k

$$\mathbf{x}_{k+1} = \mathbf{T}\cdot\mathbf{x}_k + \int_{t_k}^{t_{k+1}} \mathbf{\Phi}(t_k-t)\mathbf{r}(t)\mathrm{d}t \tag{6.4.46}$$

where remains the integration of one time step, and the expression of the force vector $\mathbf{r}$ is needed. If the expression of $\mathbf{r}$ is not available, but only the values at the grid points t_k and t_{k+1} are available, then the simplest way is to assume that within the time step $\mathbf{r}$ varies linearly

$$\dot{\mathbf{x}} = \mathbf{A}\mathbf{x} + \mathbf{r}_0 + \mathbf{r}_1 \cdot (t - t_k), \qquad \text{and} \qquad \text{when } t = t_k, \quad \mathbf{x} = \mathbf{x}_k$$

where $\mathbf{r}_0, \mathbf{r}_1$ are obtained from the two end values $\mathbf{r}(t_k)$, $\mathbf{r}(t_{k+1})$ of interval $t \in [t_k, t_{k+1}]$. Integration gives

$$\mathbf{x} = \mathbf{\Phi}(t - t_k) \cdot \left[\mathbf{x}_k + \mathbf{A}^{-1}(\mathbf{r}_0 + \mathbf{A}^{-1}\mathbf{r}_1)\right] - \mathbf{A}^{-1}\left[\mathbf{r}_0 + \mathbf{A}^{-1}\mathbf{r}_1 + (t - t_k)\mathbf{r}_1\right]$$

Substituting $t = t_{k+1}$ gives

$$\mathbf{x}_{k+1} = \mathbf{T}\left[\mathbf{x}_k + \mathbf{A}^{-1}(\mathbf{r}_0 + \mathbf{A}^{-1}\mathbf{r}_1)\right] - \mathbf{A}^{-1}\left[\mathbf{r}_0 + \mathbf{A}^{-1}\mathbf{r}_1 + \mathbf{r}_1 \cdot \eta\right] \tag{6.4.47}$$

This is the time step integration formula for the force term being linear within the time step.

Example 6.6, Solve the two degrees of freedom of elastic vibration

$$\mathbf{M}\ddot{\mathbf{v}} + \mathbf{C}\dot{\mathbf{v}} + \mathbf{K}\mathbf{v} = \mathbf{f}, \quad \mathbf{v}(0) = 0, \quad \dot{\mathbf{v}}(0) = 0,$$

$$\mathbf{M} = \begin{bmatrix} 2 & 0 \\ 0 & 2 \end{bmatrix}, \quad \mathbf{C} = 0, \quad \mathbf{K} = \begin{bmatrix} 6 & -2 \\ -2 & 4 \end{bmatrix}$$

Solution: Introduce the dual vector $\mathbf{p} = \mathbf{M}\mathbf{v}$, and compose the state vector $\mathbf{x} = \{\mathbf{v}^T, \mathbf{p}^T\}^T$,

$$\mathbf{A} = \begin{bmatrix} 0 & 0 & 0.5 & 0 \\ 0 & 0 & 0 & 1 \\ -6 & 2 & 0 & 0 \\ 2 & -4 & 0 & 0 \end{bmatrix}, \quad \mathbf{r} = \begin{Bmatrix} 0 \\ 0 \\ 0 \\ 10 \end{Bmatrix}$$

Step size $\eta = 0.28$ is selected, and the input force is $\mathbf{r}_0 = \mathbf{r}, \mathbf{r}_1 = \mathbf{0}$. Numerical results are given as

k=1	2	3	4	5	6	7	8	9	10
v1=0	0.003	0.176	0.486	0.996	1.657	2.338	2.861	3.052	2.806
v2=0	0.382	1.412	2.781	4.094	5.291	4.986	4.227	3.457	2.806

continue

11	12
2.131	1.157
2.484	2.489

For this example, the precision reaches more than 12 decimal digits (not listed here). This example is picked from reference [8] chapter 8, where the numerical results by various FDM style numerical integration methods are plotted, but all of them are clearly depart to the correct result. ##

The linear interpolation within the time step for the input vector $\mathbf{r}$ is a rough approximation. For some problems, the input vectors are usually exponential functions, trigonometric functions polynomials or their products etc. For such kinds of inputs, the time step integration can also be found analytically, see [31]. Using these solutions, the numerical results based on these equations will have high precision.

1) The input of trigonometric function

$$\mathbf{r}(t) = \mathbf{r}_1 \sin(\omega t) + \mathbf{r}_2 \cos(\omega t) \tag{6.4.48}$$

where $\mathbf{r}_1$ and $\mathbf{r}_2$ are constant vectors, ω is the parameter of excitation frequency. The solution is

$$\mathbf{x}_r(t) = \mathbf{a}\sin(\omega t) + \mathbf{b}\cos(\omega t) \tag{6.4.49}$$

$$\mathbf{a} = \left(\omega\mathbf{I} + \mathbf{A}^2/\omega\right)^{-1}\left(\mathbf{r}_2 - \mathbf{A}\mathbf{r}_1/\omega\right),\ \ \mathbf{b} = \left(\omega\mathbf{I} + \mathbf{A}^2/\omega\right)^{-1}\left(-\mathbf{r}_1 - \mathbf{A}\mathbf{r}_2/\omega\right)$$

and the precise integration solution, the HPD-S(Sinusoidal) scheme, is

$$\mathbf{x}_{k+1} = \mathbf{T}[\mathbf{x}_k - \mathbf{a}\sin(\omega t_k) - \mathbf{b}\cos(\omega t_k)] + \mathbf{a}\sin(\omega t_{k+1}) + \mathbf{b}\cos(\omega t_{k+1}) \tag{6.4.50}$$

where $\eta = t_{k+1} - t_k$. When the load varies in the time step η being exactly sinusoidal, then the equation (6.4.50) gives the exact result. It should be pointed out, that for a damping free vibration system, when ω exactly equals one of the eigen-frequencies the matrix inversion in equation (6.4.49) may not be possible. However, for a vibration system the damping factor always exists, therefore the equation (6.4.49) is enough.

A purely sinusoidal input corresponds to a constant modulation.

2) The polynomial modulated input:

$$\mathbf{r}(t) = (\mathbf{r}_0 + \mathbf{r}_1 t + \mathbf{r}_2 t^2)(\alpha\sin\omega t + \beta\cos\omega t) \tag{6.4.51}$$

The special solution is

$$\left.\begin{array}{l}
\mathbf{x}_r(t) = (\mathbf{a}_0 + \mathbf{a}_1 t + \mathbf{a}_2 t^2)\sin\omega t + (\mathbf{b}_0 + \mathbf{b}_1 t + \mathbf{b}_2 t^2)\cos\omega t \\
\left.\begin{array}{l}\mathbf{a}_i = \left(\mathbf{A}^2 + \omega^2\mathbf{I}\right)^{-1}\left(-\mathbf{A}\mathbf{p}_{ia} + \omega\mathbf{p}_{ib}\right) \\
\mathbf{b}_i = \left(\mathbf{A}^2 + \omega^2\mathbf{I}\right)^{-1}\left(-\omega\mathbf{p}_{ia} - \mathbf{H}\mathbf{p}_{ib}\right)\end{array}\right. \quad i = 2,1,0 \\
\mathbf{p}_{2a} = \alpha\,\mathbf{r}_2,\ \ \mathbf{p}_{2b} = \beta\,\mathbf{r}_2 \\
\mathbf{p}_{1a} = \alpha\,\mathbf{r}_1 - 2\mathbf{a}_2,\ \ \mathbf{p}_{1b} = \beta\,\mathbf{r}_1 - 2\mathbf{b}_2 \\
\mathbf{p}_{0a} = \alpha\,\mathbf{r}_0 - \mathbf{a}_1,\ \ \mathbf{p}_{0b} = \beta\,\mathbf{r}_0 - \mathbf{b}_1
\end{array}\right\} \tag{6.4.52}$$

where the parameters $\alpha, \beta, \mathbf{r}_0, \mathbf{r}_1, \mathbf{r}_2; \mathbf{a}, \mathbf{a}_0, \mathbf{a}_1, \mathbf{a}_2; \mathbf{b}, \mathbf{b}_0, \mathbf{b}_1, \mathbf{b}_2$ are all constants.

3) The modulation with exponential function:

$$\mathbf{r}(t) = e^{\alpha t}(\mathbf{r}_1\sin\omega t + \mathbf{r}_2\cos\omega t) \tag{6.4.53}$$

The special solution is found as

$$\left.\begin{array}{l}
\mathbf{x} = e^{\alpha t}(\mathbf{a}\sin\omega t + \mathbf{b}\cos\omega t) \\
\mathbf{a} = \left[(\alpha\mathbf{I} - \mathbf{A})^2 + \omega^2\mathbf{I}\right]^{-1}\left[(\alpha\,\mathbf{I} - \mathbf{A})\mathbf{r}_1 + \omega\,\mathbf{r}_2\right] \\
\mathbf{b} = \left[(\alpha\mathbf{I} - \mathbf{A})^2 + \omega^2\mathbf{I}\right]^{-1}\left[(\alpha\,\mathbf{I} - \mathbf{A})\mathbf{r}_2 - \omega\,\mathbf{r}_1\right]
\end{array}\right\} \tag{6.4.54}$$

where the parameters $\alpha, \mathbf{r}_1, \mathbf{r}_2 \mathbf{a}, \mathbf{b}$ are constants.

Because, the characteristic of input function is known, the precision of numerical integration is greatly improved, and with high efficiency.

§6.4.5, Precise integration of the Lyapunov differential equation

The precise integration method given in the previous section is appropriate for the mean value differential equation (6.4.33). However there is the ODE (6.4.34) for the variance, *i.e.* the Lyapunov ODE. Rewriting the equation as

$$\dot{\mathbf{G}}(t) = \mathbf{G}(t)\mathbf{A}^T + \mathbf{A}\mathbf{G}(t) + \mathbf{D}(t), \quad \mathbf{G}(0) = \mathbf{G}_0 \tag{6.4.55}$$

where $\mathbf{G}(t)$ is the $n \times n$ matrix to be solved, $\mathbf{A}$ is a given time-invariant matrix,

$\mathbf{G}_0$ is a $n \times n$ symmetric non-negative matrix, and $\mathbf{D}(t)$ is a given symmetric non-negative disturbance matrix.

When the real part of all the eigenvalues of the matrix $\mathbf{A}$ are negative, *i.e.* when $\mathbf{A}$ represents an asymptotically stable system, then as $t \to \infty$ the solution of Lyapunov ODE $\mathbf{G}(t) \to \mathbf{G}_\infty$. The iteration of precise integration can also solve the matrix $\mathbf{G}_\infty$, which satisfies the equation

$$\mathbf{G}_\infty \mathbf{A}^T + \mathbf{A}\mathbf{G}_\infty + \mathbf{D}_0 = \mathbf{0} \tag{6.4.56}$$

called the ***algebraic Lyapunov equation***, where $\mathbf{D}_0$ is a constant symmetric non-negative disturbance matrix. The numerical results obtained by the precise integration method usually have 10 more significant digits.

Lyapunov ODE is linear so that the superposition principle applies. Because that $\mathbf{G}$ is a symmetric matrix, which has $n_2 = n(n+1)/2$ unknown variable function, the number of initial conditions is n_2 too. The homogeneous ODE is to be solved first

$$\dot{\mathbf{G}} = \mathbf{A}\mathbf{G} + \mathbf{G}\mathbf{A}^T, \quad \mathbf{G}(0) = \mathbf{G}_0 \tag{6.4.55a}$$

The solution is

$$\mathbf{G}(t) = \mathbf{\Phi}(t)\mathbf{G}_0\mathbf{\Phi}^T(t) \tag{6.4.57}$$

where $\dot{\mathbf{\Phi}}(t) = \mathbf{A}\mathbf{\Phi}(t), \mathbf{\Phi}(0) = \mathbf{I}$ is the unit impulse response matrix. The verification is as follows, substituting directly the $\mathbf{G}$ in equation (6.4.57) into equation (6.4.55a) gives

$$\dot{\mathbf{G}} = \dot{\mathbf{\Phi}}\mathbf{G}_0\mathbf{\Phi}^T + \mathbf{\Phi}\mathbf{G}_0\dot{\mathbf{\Phi}}^T = (\mathbf{A}\mathbf{\Phi})\mathbf{G}_0\mathbf{\Phi}^T + \mathbf{\Phi}\mathbf{G}_0(\mathbf{A}\mathbf{\Phi})^T = \mathbf{A}\mathbf{G} + \mathbf{G}\mathbf{A}^T$$

That the differential equation is verified, and the initial condition verification is straightforward as

$$\mathbf{G}(0) = \mathbf{\Phi}(0)\mathbf{G}_0\mathbf{\Phi}^{\mathrm{T}}(0) = \mathbf{G}_0$$

According to the uniqueness theorem of ODE [36], (6.4.57) really gives the unique solution of ODE and initial condition (6.4.55a). The matrix $\mathbf{G}_0$ has n_2 independent parameters, so that the solution matrix (6.4.57) supplies all the basis (unit response) solutions of the homogeneous Lyapunov ODE.

Based on the basis solutions of the homogeneous equation, the solution of the non-homogeneous equation (6.4.55) can be obtained by the Duhamel integration as

$$\mathbf{G}(t) = \mathbf{\Phi}(t)\mathbf{G}_0\mathbf{\Phi}^{\mathrm{T}}(t) + \int_0^t \sum_{i=1}^{n}\sum_{j=1}^{n} D_{ij}(s)G_{ij}(t-s)\mathrm{d}s \tag{6.4.58}$$

where D_{ij} is i-th row, j-th column element of the matrix $\mathbf{D}$, and G_{ij} is the elements of basis solution of the homogeneous Lyapunov ODE, which is obtained from equation (6.4.57) with the initial matrix $\mathbf{G}_0$ be composed of $G_{0ij} = 1$ and the other elements be 0. Using the Duhamel integration it is proved that $\mathbf{G}(t)$ can be written in the closed form

$$\mathbf{G}(t) = \mathbf{\Phi}(t)\mathbf{G}_0\mathbf{\Phi}^T(t) + \int_0^t \mathbf{\Phi}(t-s)\mathbf{D}(s)\mathbf{\Phi}^T(t-s)\mathrm{d}s \tag{6.4.58'}$$

Although this closed form solution is simply obtained, but numerical result is still required. Based on the precise integration result of $\mathbf{\Phi}(t)$, the precise integration of Lyapunov ODE can also be proposed, as in the following sections.

The initial value response solution is (6.4.57), of which the precise numerical solution $\mathbf{G}(t)$ can be obtained based on the computed matrix $\mathbf{\Phi}(t)$ by only matrix multiplication. The solution of inhomogeneous equation is required now. The fundamental solution of equation (6.4.55) is for the case of $\mathbf{D}(t)=\mathbf{D}_0=$ constant matrix.

§6.4.5.1, *Precise integration of the algebraic Lyapunov equation*

The equation (6.4.56) is a set of linear algebraic equation with number $n_2=n(n+1)/2$ unknowns. The meaning of the solution of algebraic Lyapunov equation is the limit of solution of the corresponding Lyapunov differential equation as $t\to\infty$. Existence of the limit depends on the distribution of eigenvalues $\mu_i, i\le n$ of the matrix $\mathbf{A}$. When all eigenvalues $\mu_i, i\le n$ of matrix $\mathbf{A}$ are located in the left half of the μ plane, *i.e.* $\mathrm{Re}(\mu_i)<0$, then the limit exists, because $\mathbf{\Phi}(t)\to\mathbf{0}$ exponentially when $t\to\infty$. Hence the integration term in equation (6.4.58) converges definitely and the contribution of the initial term tends to zero.

In computation, the integration below should be carried out first

$$\mathbf{G}(t)=\int_0^t \mathbf{\Phi}(t-s)\mathbf{D}_0\mathbf{\Phi}^T(t-s)\mathrm{d}s \tag{6.4.59}$$

After this integration is performed, the result of point t ***is considered the initial point of the next integration step with the initial condition of*** $\mathbf{G}_0=\mathbf{G}(t)$. Therefore

$$\begin{aligned}\mathbf{G}(2t)&=\mathbf{\Phi}(t)\mathbf{G}_0\mathbf{\Phi}^{\mathrm{T}}(t)+\int_0^t \mathbf{\Phi}(t-s)\mathbf{D}_0\mathbf{\Phi}^{\mathrm{T}}(t-s)\mathrm{d}s\\&=\mathbf{\Phi}(t)\mathbf{G}(t)\mathbf{\Phi}^{\mathrm{T}}(t)+\mathbf{G}(t)\end{aligned} \tag{6.4.60}$$

The interpretation of the above equation is that the initial point is t with initial value $\mathbf{G}(t)$, then integrating forward further a step length of t. Therefore the integration interval has become $(0,t)+(t,2t)=(0,2t)$. Note that one crux of precise integration method is the 2^N algorithm, which ***computes the*** $\mathbf{\Phi}(2t)$ ***based on the computation of*** $\mathbf{\Phi}(t)$. Now, the effect of equation (6.4.60) is, ***compute the functions*** $\mathbf{G}(2t)$ ***and*** $\mathbf{\Phi}(2t)$ ***based on the computed matrices*** $\mathbf{\Phi}(t)$ ***and*** $\mathbf{G}(t)$. Therefore the 2^N algorithm for Lyapunov differential equation can be described as, based on the computed functions $\mathbf{\Phi}(2^m t),\mathbf{G}(2^m t)$, further compute $\mathbf{\Phi}(2^{m+1}t),\mathbf{G}(2^{m+1}t)$ for $m=1,2,\cdots$, etc. see [123,124]. The length of time interval is doubled each iteration, which is the merit of 2^N algorithm. The computation of $\mathbf{\Phi}$ has been described in section 6.4.4 in detail, now the 2^N algorithm for Lyapunov ODE should compute the functions $\mathbf{\Phi}$ and $\mathbf{G}$ together.

To use the 2^N algorithm an initial time interval $t=\tau$ solution of (6.4.60) is needed, where τ is very small. Initially $\mathbf{G}_0=\mathbf{0}$, when $t=0$. The Taylor series expansion taking up to the fourth order τ^4 is enough, that

$$\mathbf{\Phi}(\tau)=\mathbf{I}+\mathbf{T}_a,\quad \mathbf{T}_a\approx\mathbf{A}\tau+(\mathbf{A}\tau)^2\times[\mathbf{I}+(\mathbf{A}\tau)/3+(\mathbf{A}\tau)^2/12]/2 \tag{6.4.41'}$$

Substituting into the equation and integrating term by term gives

$$\begin{aligned}\mathbf{G}(\tau)\approx\int_0^\tau\mathbf{\Phi}(s)\mathbf{D}_0\mathbf{\Phi}^{\mathrm{T}}(s)\mathrm{d}s&=\mathbf{D}_0\tau+(\mathbf{A}\mathbf{D}_0+\mathbf{D}_0\mathbf{A}^{\mathrm{T}})\tau^2/2\\&+(\mathbf{A}^2\mathbf{D}_0+2\mathbf{A}\mathbf{D}_0\mathbf{A}^{\mathrm{T}}+\mathbf{D}_0\mathbf{A}^{2\mathrm{T}})\tau^3/6\\&+\tau^4(\mathbf{A}^3\mathbf{D}_0+3\mathbf{A}^2\mathbf{D}_0\mathbf{A}^{\mathrm{T}}+3\mathbf{A}\mathbf{D}_0\mathbf{A}^{2\mathrm{T}}+\mathbf{D}_0\mathbf{A}^{3\mathrm{T}})/24\end{aligned} \tag{6.4.61}$$

The neglected term has been of the order of $O(\tau^5)$.

When τ is very small, the expression (6.4.41') is very precise; but as $t\to\infty$, $\mathbf{\Phi}(t)\to\mathbf{0}$, which means $\mathbf{T}_a\to-\mathbf{I}$. Therefore as $t\to\infty$ the addition of $\mathbf{I}+\mathbf{T}_a$ may cause significant ill-conditioning numerical problem and lose precision. To solve this problem, select a moderately large time size η, then let $\tau=\eta/2^N$ where $N=20$ and compute the matrix $\mathbf{T}_a(\eta)$ and $\mathbf{G}(\eta)$. Afterward, the addition $\mathbf{T}(\eta)=\mathbf{I}+\mathbf{T}_a(\eta)$ is performed, because $\mathbf{T}_a(\eta)$ is no longer very small and the addition has been no significant error. Therefore the algorithm is as follow:

[Give $\mathbf{A},\mathbf{D}_0$; select η and let $\tau=\eta/2^N, N=20$; select the error tolerance ε]
[From (6.4.41), (6.4.61) compute $\mathbf{T}_a(\tau),\mathbf{G}(\tau)$, initiate matrices of step size τ]
for $(i=0;i<N;i++)$ { $\mathbf{G}=\mathbf{G}+(\mathbf{I}+\mathbf{T}_a)\times\mathbf{G}\times(\mathbf{I}+\mathbf{T}_a)$;
$\mathbf{T}_a=2\mathbf{T}_a+\mathbf{T}_a*\mathbf{T}_a$; }
$\mathbf{T}=\mathbf{I}+\mathbf{T}_a$;
Do { $\mathbf{G}=\mathbf{G}+\mathbf{T}\times\mathbf{G}\times\mathbf{T}^T$; $\mathbf{T}=\mathbf{T}\times\mathbf{T}$;} while ($\|\mathbf{T}\|>\varepsilon$); (6.4.62)

Where $\|\mathbf{T}\|$ represents the norm of matrix $\mathbf{T}$, such as the maximum of absolute value of the elements of $\mathbf{T}$, the error tolerance ε can be selected as 10^{-8}, say. When the iteration converges, the matrix $\mathbf{G}$ is the solution of the algebraic Lyapunov equation. Asymptotic convergence reaches when all the eigenvalues of the matrix $\mathbf{A}$ have negative real parts.

Example 6.7, Give the matrices

$$\mathbf{A}=\begin{bmatrix}-0.25&1.0&0\\0&-0.25&1.0\\0&0&-0.25\end{bmatrix};\qquad \mathbf{D}_0=\begin{bmatrix}10.0&1.0&5.0\\1.0&7.0&4.0\\5.0&4.0&9.0\end{bmatrix};$$

the solution of Lyapunov differential equation is needed.

Solution: Select the time step size $\eta=0.50$, the corresponding matrices are computed numerically

$$\mathbf{T}=\mathbf{\Phi}(\eta)=\begin{bmatrix}0.88250&0.44125&0.11031\\0&0.88250&0.44215\\0&0&0.88250\end{bmatrix}$$

$$\mathbf{G}(\eta)=\begin{bmatrix} 5.11362 & 1.97947 & 2.79160 \\ 1.97947 & 4.25601 & 2.72356 \\ 2.79160 & 2.72356 & 3.98159 \end{bmatrix}$$

Although the matrix $\mathbf{A}$ appears duplicated eigen-root of Jordan normal form, the precision of matrix $\mathbf{T}$ still has more than ten decimal digits. The matrix $\mathbf{G}(\eta)$ is the value of transient process at $t = 0.5$. Continue the iteration gives

$$\mathbf{G}_\infty=\begin{bmatrix} 2332.00000 & 578.00000 & 98.00000 \\ 578.00000 & 190.00000 & 44.00000 \\ 98.00000 & 44.00000 & 18.00000 \end{bmatrix}$$

The verification can be, computing $\mathbf{AG}_\infty + \mathbf{G}_\infty \mathbf{A}^T$ and comparing with $-\mathbf{D}_0$, the coincidence of numerical digits reach ten more decimal digits. For this example, it can be verified by hand. ##

Other numerical examples are neglected for saving space. The precise integration method can also be used to compute the transient process of arbitrary interval $[0,t_\mathrm{f}]$. The convergence of iteration for $\mathbf{G}_\infty$ requires the asymptotic stability of matrix $\mathbf{A}$, but the finite time transient process does not require it, however t_f should not be very large.

§6.4.5.2, Integration of asymmetric Lyapunov equation

In application, there is further the asymmetric Lyapunov equation, *i.e.* the matrices $\mathbf{D}$ and $\mathbf{G}(t)$ are of dimension $n\times m$, satisfying the linear differential equation

$$\dot{\mathbf{G}}(t)=\mathbf{AG}+\mathbf{GB}^T+\mathbf{D}, \qquad \mathbf{G}(0)=\mathbf{G}_0 \tag{6.4.63}$$

where $\mathbf{A}$ and $\mathbf{B}$ are, respectively, $n\times n$ and $m\times m$ matrices. To solve the equation (6.4.63) for time-invariant matrices $\mathbf{A}$ and $\mathbf{B}$, the two impulse response matrix function should be solved first

$$\dot{\mathbf{\Phi}}_a=\mathbf{A\Phi}_a, \qquad \mathbf{\Phi}_a(0)=\mathbf{I}_n \tag{6.4.64}$$

$$\dot{\mathbf{\Phi}}_b=\mathbf{B\Phi}_b, \qquad \mathbf{\Phi}_b(0)=\mathbf{I}_m \tag{6.4.65}$$

Obviously $\mathbf{\Phi}_a(t)=\exp(\mathbf{A}t)$ and $\mathbf{\Phi}_b(t)=\exp(\mathbf{B}t)$. For a given time step η, the precise integration can be used to compute the matrices $\mathbf{\Phi}_a(\eta)$ and $\mathbf{\Phi}_b(\eta)$ as described before, then based on inspection, the solution of homogeneous equation is proposed as

$$\mathbf{G}(t)=\mathbf{\Phi}_a(t)\mathbf{G}_0\mathbf{\Phi}_b^T(t) \tag{6.4.66}$$

which is directly verified as

$$\begin{aligned}\dot{\mathbf{G}} &= \dot{\mathbf{\Phi}}_a\mathbf{G}_0\mathbf{\Phi}_b^T+\mathbf{\Phi}_a\mathbf{G}_0\dot{\mathbf{\Phi}}_b^T \\ &= \mathbf{A\Phi}_a\mathbf{G}_0\mathbf{\Phi}_b^T+\mathbf{\Phi}_a\mathbf{G}_0(\mathbf{B\Phi}_b)^T=\mathbf{AG}+\mathbf{GB}^T\end{aligned}, \qquad \mathbf{G}(0)=\mathbf{G}_0\,.$$

According to the same method as for the solution of symmetric Lyapunov equation (6.4.66), the solution of the non-homogeneous equation (6.4.63) is given as

$$\mathbf{G}(t)=\mathbf{\Phi}_a(t)\mathbf{G}_0\mathbf{\Phi}_b^T(t)+\int_0^t\mathbf{\Phi}_a(t-s)\mathbf{D}_0\mathbf{\Phi}_b^T(t-s)\mathrm{d}s \tag{6.4.67}$$

Let
$$\mathbf{G}_d(t)=\int_0^t \mathbf{\Phi}_a(t-s)\mathbf{D}(s)\mathbf{\Phi}_b^T(t-s)\mathrm{d}s \tag{6.4.68}$$
Direct verification gives
$$\dot{\mathbf{G}}_d(t)=\mathbf{\Phi}_a(0)\mathbf{G}(t)\mathbf{\Phi}_b^T(0)+\int_0^t\begin{bmatrix}\dot{\mathbf{\Phi}}_a(t-s)\mathbf{D}(s)\mathbf{\Phi}_b^T(t-s)\\ +\mathbf{\Phi}_a(t-s)\mathbf{D}(s)\dot{\mathbf{\Phi}}_b^T(t-s)\end{bmatrix}\mathrm{d}s$$
$$=\mathbf{D}(t)+\mathbf{A}\mathbf{G}_d+\mathbf{G}_d\mathbf{B}^T$$
hence $\mathbf{G}_d(t)$ is the solution of the non-homogeneous equation (6.4.63) with the initial value $\mathbf{G}_0=0$.

The solution matrix $\mathbf{G}_{d\infty}$ of the algebraic asymmetric Lyapunov equation is also of concern
$$\mathbf{A}\mathbf{G}_{d\infty}+\mathbf{G}_{d\infty}\mathbf{B}^T+\mathbf{D}_0=\mathbf{0} \tag{6.4.69}$$
where $\mathbf{D}_0$ is a given $n\times m$ matrix. When all the eigenvalues of matrices $\mathbf{A}$ and $\mathbf{B}$ have negative real parts, the convergence of iteration is ensured, where initially the matrix is selected as $\mathbf{G}_0=\mathbf{0}$.

The equation (6.4.68) has given the matrix $\mathbf{G}_d$ where the matrix $\mathbf{D}(s)=\mathbf{D}_0=\text{const}$. The same approach, as in solving the symmetric Lyapunov equation, is applied. To derive the additional theorem for $\mathbf{G}_d(t)$, treating the time t as initial time and using the solution (6.4.67) gives
$$\mathbf{G}_d(2t)=\mathbf{G}_d(t)+\mathbf{\Phi}_a(t)\mathbf{G}_d(t)\mathbf{\Phi}_b^T(t) \tag{6.4.70}$$
Deriving similarly as below the equation (6.4.60), the matrices $\mathbf{\Phi}_a,\mathbf{\Phi}_b$ and $\mathbf{G}_d$ are iteratively computed simultaneously, computing these matrices at the time instants $t,2t,2^2t,\cdots,2^kt,\cdots$, until converge.

The initial time interval τ is selected extremely small, using Taylor series expansion
$$\mathbf{\Phi}_a(\tau)=\mathbf{I}+\mathbf{T}_a,\ \ \mathbf{T}_a\approx\mathbf{A}\tau+(\mathbf{A}\tau)^2\cdot[\mathbf{I}_n+(\mathbf{A}\tau)/3+(\mathbf{A}\tau)^2/12]/2$$
$$\mathbf{\Phi}_b(\tau)=\mathbf{I}+\mathbf{T}_b,\ \ \mathbf{T}_b\approx\mathbf{B}\tau+(\mathbf{B}\tau)^2\cdot[\mathbf{I}_m+(\mathbf{B}\tau)/3+(\mathbf{B}\tau)^2/12]/2$$
and
$$\begin{aligned}\mathbf{G}_d(\tau)\approx{}&\mathbf{D}_0\tau+\tau^2(\mathbf{A}\mathbf{D}_0+\mathbf{D}_0\mathbf{B}^T)/2+\tau^3(\mathbf{A}^2\mathbf{D}_0+2\mathbf{A}\mathbf{D}_0\mathbf{B}^T+\mathbf{D}_0\mathbf{B}^{2T})/6\\&+\tau^4(\mathbf{A}^3\mathbf{D}_0+3\mathbf{A}^2\mathbf{D}_0\mathbf{B}^T+3\mathbf{A}\mathbf{D}_0\mathbf{B}^{2T}+\mathbf{D}_0\mathbf{B}^{3T})/24\end{aligned} \tag{6.4.71}$$
where the first truncated term has been of the order of $O(\tau^5)$.

When τ is extremely small, the expansion (6.4.71) is very precise, but as $t\to\infty$ the asymptotic stability of matrices $\mathbf{A}$ and $\mathbf{B}$ gives $\mathbf{\Phi}_a(t)\to\mathbf{0}$ and $\mathbf{\Phi}_b(t)\to\mathbf{0}$, which means $\mathbf{T}_a\to-\mathbf{I}_n$ and $\mathbf{T}_b\to-\mathbf{I}_m$, hence in the computation of $\mathbf{\Phi}_a(t),\mathbf{\Phi}_b(t)$ the ill-conditioning problem still exists as $t\to\infty$. The method to avoid such problem is at an appropriate time t_0 to transform from $\mathbf{T}_a,\mathbf{T}_b$ to $\mathbf{\Phi}_a,\mathbf{\Phi}_b$ when $t\geq t_0$. Select $\tau=t_0/2^N, N=20, 2^N=1048576$, and the truncated first term of equation (6.4.71) is of the order of $\tau^5/120$. Comparing to the first term gives a factor of order $\tau^4/120 = t_0^{\ 4}\cdot(1048576)^{-4}/120 \approx 10^{-26}\cdot t_0^{\ 4}$. Note the computer real number double precision is decimal digits of $O(10^{-16})$, so that the truncation error has been beyond the real number precision on the computer. The transformation at t_0 is also used in the symmetric matrix case, where $t_0 = \eta$ is selected.

Algorithm of iterative solution of asymmetric Lyapunov equation is given as

[Give $n, m, \mathbf{A}, \mathbf{B}, \mathbf{D}_0$, select t_0, let $\tau = t_0/2^N$, Select tolerated error ε]
[From (6.4.71) compute $\mathbf{T}_a(\tau), \mathbf{T}_b(\tau), \mathbf{G}_d(\tau)$]
for $(iter = 0;\ iter < N;\ iter{+}{+})$ { $\mathbf{G}_d = \mathbf{G}_d + (\mathbf{I}_n + \mathbf{T}_a)\mathbf{G}_d(\mathbf{I}_m + \mathbf{T}_b)^T$;
$\mathbf{T}_a = 2\mathbf{T}_a + \mathbf{T}_a * \mathbf{T}_a$; $\mathbf{T}_b = 2\mathbf{T}_b + \mathbf{T}_b * \mathbf{T}_b$; }
[$\mathbf{\Phi}_a = \mathbf{I}_n + \mathbf{T}_a$; $\mathbf{\Phi}_b = \mathbf{I}_m + \mathbf{T}_b$;]
Do {$\mathbf{G}_d = \mathbf{G}_d + \mathbf{\Phi}_a \times \mathbf{G}_d \times \mathbf{\Phi}_b^T$; $\mathbf{\Phi}_a = \mathbf{\Phi}_a \times \mathbf{\Phi}_a$; $\mathbf{\Phi}_b = \mathbf{\Phi}_b \times \mathbf{\Phi}_b$;
} while $((\|\mathbf{\Phi}_a\| > \varepsilon) \vee (\|\mathbf{\Phi}_b\| > \varepsilon))$;
Comment: Convergence of the iteration gives $\mathbf{G}_d = \mathbf{G}_{d\infty}$, the solution of asymmetric algebraic Lyapunov equation. (6.4.72)

The algebraic Lyapunov equation is linear. The above algorithm can be used for the transient process of Lyapunov differential equation, which is quite interested.

Occasionally the solution of the homogeneous Riccati differential equation is required

$$\dot{\mathbf{P}} = \mathbf{B}^T\mathbf{P} + \mathbf{P}\mathbf{A} + \mathbf{P}\mathbf{D}\mathbf{P}, \quad \mathbf{P}(0) = \mathbf{P}_0 \tag{6.4.73}$$

where $\mathbf{A}, \mathbf{B}, \mathbf{D}, \mathbf{P}$ are $n \times n$ matrices, the unknown function is $\mathbf{P}(t)$. The transformation is as follows, denoting $\mathbf{P}^{-1} = \mathbf{G}$, then $\mathbf{G}\mathbf{P} = \mathbf{I}$ and $\dot{\mathbf{G}}\mathbf{P} + \mathbf{G}\dot{\mathbf{P}} = \mathbf{0}$, so $\dot{\mathbf{G}} = -\mathbf{P}^{-1}\dot{\mathbf{P}}\mathbf{P}^{-1}$. Substituting into (6.4.73) gives

$$\dot{\mathbf{G}} = -\mathbf{A}\mathbf{G} - \mathbf{G}\mathbf{B}^T - \mathbf{D}, \quad \mathbf{G}(0) = \mathbf{P}_0^{-1} \tag{6.4.73'}$$

which is the Lyapunov differential equation and solution method is the same.

Example 6.8, Give the matrices $\mathbf{A}, \mathbf{B}, \mathbf{D}_0, (n = 3, m = 4)$ as

$$\mathbf{A} = \begin{bmatrix} -.25 & 1.0 & 0 \\ 0 & -.25 & 1.0 \\ 0 & 0 & -.25 \end{bmatrix}, \quad \mathbf{B} = \begin{bmatrix} -4 & 2 & 1 & 1 \\ 0 & -3 & 2 & 1 \\ 1 & -4 & -9 & -1 \\ 0.5 & 1 & 0 & -2 \end{bmatrix}, \quad \mathbf{D}_0 = \begin{bmatrix} 10.0 & 2.0 & 1.0 & 1.0 \\ 2.0 & 5.0 & 2.0 & 1.0 \\ 1.0 & 2.0 & 9.0 & -1.0 \end{bmatrix}$$

The solution of algebraic Lyapunov equation is required.
Solution: selecting $t_0 = 0.4, N = 20$, and the precise integration gives

$$\mathbf{\Phi}_a = \begin{bmatrix} 0.90484 & 0.36193 & 0.07239 \\ 0 & 0.90484 & 0.36193 \\ 0 & 0 & 0.90484 \end{bmatrix}, \quad \mathbf{\Phi}_b = \begin{bmatrix} 0.22915 & 0.16557 & 0.07389 & 0.15075 \\ 0.03624 & 0.23865 & 0.08184 & 0.11327 \\ 0.01946 & -.15539 & -.01068 & -.09548 \\ 0.06932 & 0.15715 & 0.03783 & 0.49126 \end{bmatrix}$$

$\mathbf{G}_d =$

$$\begin{bmatrix} 2.30703 & 0.75230 & 0.00405 & 0.66121 \\ 0.86257 & 1.29359 & -.13090 & 0.52891 \\ 0.50351 & 0.73135 & 0.75064 & -.10115 \end{bmatrix},$$

$\mathbf{G}_{d\infty} =$

$$\begin{bmatrix} 4.2397 & 1.8118 & -.61134 & 3.0382 \\ 1.9680 & 2.0729 & -.60909 & 1.9043 \\ 0.9231 & 1.0508 & 0.59374 & 0.2277 \end{bmatrix}$$

Using the converged matrix $\mathbf{G}_{d\infty}$ of iteration check the value of $\mathbf{AG}_{d\infty} + \mathbf{G}_{d\infty}\mathbf{B}^T$ with $\mathbf{D}_0$, ten more digits coincidence is reached. ##

§6.4.5.3, Solution with modulated input

The precise integration above is applied to the case of constant inhomogeneous term of input matrix $\mathbf{D}(t) = \mathbf{D}_0$. It corresponds to a constant intensity of white noise disturbance suddenly applied at $t = 0$. In practical applications such as in earthquake engineering, the intensity changes with time. Such problem is of practical importance.

If the matrix $\mathbf{D}(t)$ in equation (6.4,63) is a power function, such as

$$\mathbf{D}(t) = \mathbf{D}_1 \cdot t, \quad [\mathbf{G}(0) = \mathbf{0}] \tag{6.4.74}$$

the precise integration is given as follows. The other cases can be extended similarly. The transient analysis is considered. Give the uniformly allocated time instants with duration η

$$t_0 = 0, \ t_1 = \eta, \ \cdots, \ t_k = k\eta, \ \cdots \tag{6.4.75}$$

To compute the matrices $\mathbf{G}(t_k)$ successively (subscript d is neglected). According to (6.4.67)

$$\mathbf{G}^{(1)}(t) = \int_0^t \mathbf{\Phi}_a(t-s)\mathbf{D}_1 s \mathbf{\Phi}_b^T(t-s)\mathrm{d}s \tag{6.4.76a}$$

where the superscript $^{(1)}$ represents the integration of power 1. Define the matrix

$$\mathbf{G}^{(0)}(t) = \int_0^t \mathbf{\Phi}_a(t-s)\mathbf{D}_1 \mathbf{\Phi}_b^T(t-s)\mathrm{d}s \tag{6.4.76}$$

The matrices $\mathbf{G}^{(0)}$ and $\mathbf{G}^{(1)}$ at $t = 0, \eta, 2\eta, \cdots$ can be computed step by step successively. Based on the computed matrices $\mathbf{G}^{(0)}(\eta), \mathbf{G}^{(1)}(\eta)$ and $\mathbf{G}^{(0)}(t_k), \mathbf{G}^{(1)}(t_k)$, the matrices $\mathbf{G}^{(0)}(t_k+\eta), \mathbf{G}^{(1)}(t_k+\eta)$ are to be computed. Obviously, $\mathbf{G}^{(1)}(t_{k+1}) \equiv \mathbf{G}^{(1)}(t_k+\eta)$, and also $\mathbf{\Phi}_a(t+\eta) = \mathbf{\Phi}_a(t)\mathbf{\Phi}_a(\eta)$ and the same for $\mathbf{\Phi}_b$, so that the derivation can be

$$\begin{aligned} \mathbf{G}^{(0)}(t_{k+1}) &= \int_0^{t_{k+1}} \mathbf{\Phi}_a(t_{k+1}-s)\mathbf{D}_1\mathbf{\Phi}_b^T(t_{k+1}-s)\mathrm{d}s \\ &= \mathbf{\Phi}_a(\eta)\mathbf{G}^{(0)}(t_k)\mathbf{\Phi}_b^T(\eta) + \int_{t_k}^{t_{k+1}} \mathbf{\Phi}_a(\eta+t_k-s)\mathbf{D}_1\mathbf{\Phi}_b^T(\eta+t_k-s)[-\mathrm{d}(t_k-s)] \\ &= \mathbf{\Phi}_a(\eta)\mathbf{G}^{(0)}(t_k)\mathbf{\Phi}_b^T(\eta) + \int_0^{\eta} \mathbf{\Phi}_a(\eta-s')\mathbf{D}_1\mathbf{\Phi}_b^T(\eta-s')\mathrm{d}s' \\ &= \mathbf{\Phi}_a(\eta)\mathbf{G}^{(0)}(t_k)\mathbf{\Phi}_b^T(\eta) + \mathbf{G}^{(0)}(\eta) \end{aligned} \tag{6.4.77a}$$

Similarly

$$\mathbf{G}^{(0)}(2t) = \mathbf{\Phi}_a(t)\mathbf{G}^{(0)}(t)\mathbf{\Phi}_b^T(t) + \mathbf{G}^{(0)}(t) \tag{6.4.77b}$$

The computation for $\mathbf{G}^{(1)}$ can be

$$\mathbf{G}^{(1)}(t_{k+1})=\int_0^{t_k+\eta}\mathbf{\Phi}_a(t_{k+1}-s)\mathbf{D}_1 s\mathbf{\Phi}_b^T(t_{k+1}-s)\mathrm{d}s=\mathbf{\Phi}_a(\eta)\mathbf{G}^{(1)}(t_k)\mathbf{\Phi}_b^T(\eta)$$
$$+\int_{t_k}^{t_k+\eta}\mathbf{\Phi}_a(\eta+t_k-s)\mathbf{D}_1 s\mathbf{\Phi}_b^T(\eta+t_k-s)\mathrm{d}(s-t_k) \tag{6.4.78a}$$
$$=\mathbf{\Phi}_a(\eta)\mathbf{G}^{(1)}(t_k)\mathbf{\Phi}_b^T(\eta)+\int_0^{\eta}\mathbf{\Phi}_a(\eta-s')\mathbf{D}_1\cdot(s'+t_k)\mathbf{\Phi}_b^T(\eta-s')\mathrm{d}s'$$
$$=\mathbf{\Phi}_a(\eta)\mathbf{G}^{(1)}(t_k)\mathbf{\Phi}_b^T(\eta)+t_k\cdot\mathbf{G}^{(0)}(\eta)+\mathbf{G}^{(1)}(\eta)$$

Similar derivation gives

$$\mathbf{G}^{(1)}(2t)=\mathbf{\Phi}_a(t)\mathbf{G}^{(1)}(t)\mathbf{\Phi}_b^T(t)+t\mathbf{G}^{(0)}(t)+\mathbf{G}^{(1)}(t) \tag{6.4.78b}$$

The above equations are exact, in which the equations (6.4.77b) and (6.4.78b) can be used to the 2^N type time integration, and the equations (6.4.77a) and (6.4.78a) can be used for recursive time integration with constant step size η. For the 2^N type integration, an initial time interval is necessary, which can select a very small time interval τ, $\tau=\eta/2^N$ with $N=20$. For the extremely small step size τ the Taylor series expansion can be used, the approximation of $\mathbf{\Phi}_a(\tau),\mathbf{\Phi}_b(\tau)$ have been given in equation (6.4.71), and also

$$\mathbf{G}^{(0)}(\tau)\approx\mathbf{D}_1\tau+\tau^2(\mathbf{A}\mathbf{D}_1+\mathbf{D}_1\mathbf{B}^T)/2+\tau^3(\mathbf{A}^2\mathbf{D}_1+2\mathbf{A}\mathbf{D}_1\mathbf{B}^T+\mathbf{D}_1\mathbf{B}^{2T})/6$$
$$+\tau^4(\mathbf{A}^3\mathbf{D}_1+3\mathbf{A}^2\mathbf{D}_1\mathbf{B}^T+3\mathbf{A}\mathbf{D}_1\mathbf{B}^{2T}+\mathbf{D}_1\mathbf{B}^{3T})/24 \tag{6.4.79}$$

$$\mathbf{G}^{(1)}(\tau)\approx\mathbf{D}_1\tau^2/2+(\mathbf{A}\mathbf{D}_1+\mathbf{D}_1\mathbf{B}^T)\tau^3/6$$
$$+(\mathbf{A}^2\mathbf{D}_1+2\mathbf{A}\mathbf{D}_1\mathbf{B}^T+\mathbf{D}_1\mathbf{B}^{2T})\tau^4/24 \tag{6.4.80}$$

where the truncated parts have been of the order of $O(\tau^5)$, which exceeds the double precision of real word of the current computer.

Therefore the algorithm for transient process is given as:

[Give the step size η, and matrices $\mathbf{A},\mathbf{B},\mathbf{D}_1$; select $N=20;\tau=\eta/2^N$;]

[Using (6.4.71) compute $\mathbf{T}_a,\mathbf{T}_b$. Use (6.4.79), (6.4.80) for computing $\mathbf{G}_0=\mathbf{G}^{(0)}(\tau);\mathbf{G}_1=\mathbf{G}^{(1)}(\tau);\ tt=\tau$;]

for $(iter=0;\ iter<N;\ iter++)$ { Comment: 2^N algorithm.

$\mathbf{G}_1=\mathbf{G}_1+(\mathbf{I}_n+\mathbf{T}_a)\mathbf{G}_1(\mathbf{I}_m+\mathbf{T}_b)^T+tt*\mathbf{G}_0$; Comment: equation (6.4.78b).

$\mathbf{G}_0=\mathbf{G}_0+(\mathbf{I}_n+\mathbf{T}_a)\mathbf{G}_0(\mathbf{I}_m+\mathbf{T}_b)^T$; Comment: (6.4.77b).

$\mathbf{T}_a=2\times\mathbf{T}_a+\mathbf{T}_a\times\mathbf{T}_a$; $\mathbf{T}_b=2\times\mathbf{T}_b+\mathbf{T}_b\times\mathbf{T}_b$; $tt=2\times tt$; }

$\mathbf{G}_1^{(1)}=\mathbf{G}_1$; $\mathbf{G}_1^{(0)}=\mathbf{G}_0$; $tt=\eta$; $\mathbf{\Phi}_a=\mathbf{I}_n+\mathbf{T}_a$; $\mathbf{\Phi}_b=\mathbf{I}_m+\mathbf{T}_b$; Comment: initiation.

for $(k=1;\ k\le k_{\max};\ k++)$ {Comment: time step integration.

$\mathbf{G}_{k+1}^{(1)}=\mathbf{\Phi}_a\mathbf{G}_k^{(1)}\mathbf{\Phi}_b^T+\mathbf{G}_1+tt\times\mathbf{G}_0$; Comment: equation (6.4.78a)

$\mathbf{G}_{k+1}^{(0)}=\mathbf{\Phi}_a\mathbf{G}_k^{(0)}\mathbf{\Phi}_b^T+\mathbf{G}_0$; Comment: equation (6.4.77a)

$tt=tt+\eta$; Comment: step forward

} (6.4.81)

The stepwise integration is consummated.

Discussion: the external disturbance (6.4.74) is the simplest one. The method used for the above derived integration equation (6.4.78) takes the benefit of the identity $t = t_k + (t - t_k)$. For the disturbance of $\mathbf{D} = \mathbf{D}_2 t^2$, then the identity is $t^2 = t_k^{\ 2} + 2t_k(t - t_k) + (t - t_k)^2$, based on which the similar method can be developed for the precise integration equations. Then it can be said that for any polynomial of $\mathbf{D}(t)$ the precise integration equations can be found.

For the disturbance of exponential function $\mathbf{D}(t) = \mathbf{D}_e \exp(\mu t)$, based on the identity $e^{\mu t} = e^{\mu t_k} \cdot e^{\mu(t - t_k)}$ and integration by parts the equation for precise integration can also be developed.

For the disturbance of sinusoidal functions the integration by parts and the additional theorem

$$\sin(\omega t) = \sin(\omega t_k)\cos[\omega(t - t_k)] + \cos(\omega t_k)\sin[\omega(t - t_k)],$$

$$\cos(\omega t) = \cos(\omega t_k)\cos[\omega(t - t_k)] - \sin(\omega t_k)\sin[\omega(t - t_k)],$$

can be used to derive the precise integration equations for the Lyapunov differential equation.

For saving space, the detail derivation and numerical example are neglected. The application of the Lyapunov differential equation solution can be found for the prediction of random vibration problems, see [124].

§6.4.6, Disturbance of colored noise

In the previous sections for multi-degrees of system, the dynamic noise expressed by the equation (6.4.25) is assumed a white noise. When the practical input is somewhat correlated such as for a narrow band noise, the noise should be considered colored. In applications, the narrow band colored noise is often modeled as the response of a linear system excited by a white noise. Under this assumption the system response analysis can use the method of ***extended state space method***, so that the problem is still reduced to be an analysis of a linear system excited by a white noise.

The mathematical model of that, the dynamic noise is colored and the measurement noise is white, is considered. In the prediction problem, there is no measurement feedback analysis, hence

$$\dot{\mathbf{x}}(t) = \mathbf{A}(t)\mathbf{x}(t) + \mathbf{B}(t)\mathbf{w}(t) \tag{6.4.82}$$

where $\mathbf{w}(t)$ is a m-dimensional colored noise, which is assumed modeled by the linear system

$$\dot{\mathbf{w}}(t) = \mathbf{F}\mathbf{w}(t) + \mathbf{H}\mathbf{r}(t) \tag{6.4.83}$$

where $\mathbf{r}(t)$ is a given intensity white noise, l-dimensional, and $\mathbf{F}$, $\mathbf{H}$ are given matrices with appropriate dimensions.

A time-invariant system is considered for numerical computations. The two equations of (6.4.82) and (6.4.83) can be combined as

$$\dot{\mathbf{x}}_* = \mathbf{A}_*\mathbf{x}_* + \mathbf{B}_*\mathbf{r} \tag{6.4.84}$$

where

$$\mathbf{x}_* = \begin{Bmatrix} \mathbf{x} \\ \mathbf{w} \end{Bmatrix}, \quad \mathbf{A}_* = \begin{bmatrix} \mathbf{A} & \mathbf{B} \\ \mathbf{0} & \mathbf{F} \end{bmatrix} \begin{matrix} n \\ m \end{matrix}, \quad \mathbf{B}_* = \begin{bmatrix} \mathbf{0} \\ \mathbf{H} \end{bmatrix} \tag{6.4.85}$$

which is the extended mathematical model, $\mathbf{x}_*$ is the ***extended state vector*** and $\mathbf{A}_*$ is the extended plant matrix. The white noise $\mathbf{r}(t)$ is independent on the extended vector $\mathbf{x}_*$. In such case, the prediction computation is considered below.

As in the case of colored measurement noise, which relates only to filter problem, the respective analysis will be considered in the section 6.5.

§6.4.6.1, Some comments for precise integration of the extended system

Precise integration needs to compute the exponential matrix

$$\mathbf{T}_* = \exp(\mathbf{A}_* \eta) \tag{6.4.86}$$

where η is the given integration step size. If the computational expense is disregarded then the precise integration for a full matrix $\mathbf{A}_*$ can also give satisfactory results, however there is many multiplication of zero. To avoid these useless operations, the upper block diagonal form of matrix $\mathbf{A}_*$ can be applied, called simply as R-blocked form. The product and sum of two R-blocked matrices are again R-blocked matrix, that

$$\begin{bmatrix} \mathbf{A}_1 & \mathbf{A}_2 \\ \mathbf{0} & \mathbf{A}_3 \end{bmatrix} \times \begin{bmatrix} \mathbf{A}_4 & \mathbf{A}_5 \\ \mathbf{0} & \mathbf{A}_6 \end{bmatrix} = \begin{bmatrix} \mathbf{A}_1\mathbf{A}_4 & (\mathbf{A}_1\mathbf{A}_5 + \mathbf{A}_2\mathbf{A}_6) \\ \mathbf{0} & \mathbf{A}_3\mathbf{A}_6 \end{bmatrix} \tag{6.4.87}$$

where the product of diagonal sub-matrices are unrelated to the off-diagonal block. All of these characteristics can be used in the computations.

For precise integration algorithm, the initialization equation (6.4.48), and the operations of matrix $\mathbf{A}_*$ are all addition and multiplication, hence the resulted matrix $\mathbf{T}_a$ is also R-blocked. Then in the N times of iterations, there is all addition and multiplication, hence the preservation of R-blocked is kept unchanged, the matrix $\mathbf{T}_*$ is again R-blocked. The definition of an exponential matrix uses only addition and multiplication, keeps R-blocked naturally

$$\mathbf{T}_a = \begin{bmatrix} \mathbf{T}_{an} & \mathbf{T}_{nm} \\ \mathbf{0} & \mathbf{T}_{am} \end{bmatrix} \tag{6.4.88}$$

Therefore the algorithm (6.4.71) is updated as

for $(iter = 0;\ iter < N;\ iter{+}{+})\{\ \mathbf{T}_{nm} = \mathbf{T}_{an} \cdot \mathbf{T}_{nm} + \mathbf{T}_{nm} \cdot \mathbf{T}_{am};$

$\mathbf{T}_{an} = 2\mathbf{T}_{an} + \mathbf{T}_{an} \cdot \mathbf{T}_{an};\ \mathbf{T}_{am} = 2\mathbf{T}_{am} + \mathbf{T}_{am} \cdot \mathbf{T}_{am};\ \}$

[Composing $\mathbf{T}_a$; then $\mathbf{T}_* = \mathbf{I}_{n+m} + \mathbf{T}_a$;] (6.4.89)

To find the response of non-homogeneous term, inverse matrix is needed. The inverse matrix of a R-blocked matrix is still a R-blocked matrix, which is given by

$$\begin{bmatrix} \mathbf{A}_1 & \mathbf{A}_2 \\ \mathbf{0} & \mathbf{A}_3 \end{bmatrix}^{-1} = \begin{bmatrix} \mathbf{A}_1^{-1} & -\mathbf{A}_1^{-1}\mathbf{A}_2\mathbf{A}_3^{-1} \\ \mathbf{0} & \mathbf{A}_3^{-1} \end{bmatrix} \tag{6.4.90}$$

Therefore the non-homogeneous term computation in section 6.4.4.2 can be performed in the R-blocked matrix form.

The R-blocked matrix $\mathbf{T}_*$ can be utilized in the precise integration of the Lyapunov matrix. The computation of $\mathbf{T}_* \times \mathbf{G} \times \mathbf{T}_*^T$ in equation (6.4.62) can also be in blocked form

$$\begin{bmatrix} \mathbf{T}_n & \mathbf{T}_{nm} \\ \mathbf{0} & \mathbf{T}_m \end{bmatrix} \begin{bmatrix} \mathbf{G}_n & \mathbf{G}_{nm} \\ \mathbf{G}_{nm}^T & \mathbf{G}_m \end{bmatrix} \begin{bmatrix} \mathbf{T}_n^T & \mathbf{0} \\ \mathbf{T}_{nm}^T & \mathbf{T}_m^T \end{bmatrix} = \begin{bmatrix} \mathbf{G}'_n & \mathbf{G}'_{nm} \\ \mathbf{G}'^T_{nm} & \mathbf{G}'_m \end{bmatrix}$$

$$\mathbf{G}'_n = \mathbf{T}_n\mathbf{G}_n\mathbf{T}_n^T + \mathbf{T}_{nm}\mathbf{G}_{nm}^T\mathbf{T}_n^T + \mathbf{T}_n\mathbf{G}_{nm}\mathbf{T}_{nm}^T + \mathbf{T}_{nm}\mathbf{G}_m\mathbf{T}_{nm}^T;$$

$$\mathbf{G}'_{nm} = \mathbf{T}_n\mathbf{G}_{nm}\mathbf{T}_m^T + \mathbf{T}_{nm}\mathbf{G}_m\mathbf{T}_m^T, \qquad \mathbf{G}'_m = \mathbf{T}_m\mathbf{G}_m\mathbf{T}_m^T$$

Especially, when the matrix $\mathbf{T}_*$ is R-blocked, then all the block sub-matrices of the matrix $\mathbf{G}$ can be integrated mutually independently and the integration of the upper-right block is just the asymmetric Lyapunov differential equation.

§6.5, Kalman filtering

In control systems, feedback is extremely important; for monitoring the performance of a system, the state vector is required. The performance of system is under disturbances that the system practical situation cannot be determined purely by prediction, so that measurement is necessary. However, measurement is not able to measure all the state variables, that the measured data can only be a q-dimensional subspace vector $\mathbf{y}$ out of the n-dimensional state space vector $\mathbf{x}$. Also these measured data are not exact that they are under measurement disturbances too. Therefore, based on the measured data $\mathbf{y}$ to estimate the stochastic state vector $\mathbf{x}$ is necessary.

In 1960, R.E. Kalman proposed the recurrence algorithm for linear optimal filter. Kalman filter does not require the computer to store all the data measured previously, that according to the measured data $\mathbf{y}_k$ at the present time step t_k and the estimated state $\mathbf{x}_{k-1}$ of the previous time step t_{k-1}, the present time state $\mathbf{x}_k$ is estimated recurrently. So that the computational expense and memory requirement are greatly reduced and is easier for real time processing. Also the Kalman filter can be used to estimate the non-stationary stochastic process. Based on these merits, the Kalman filter is widely used in control theory, in aerospace technology and in other fields.

Kalman filter was first proposed for discrete-time system, and the continuous-time version of Kalman-Bucy filter was quickly developed. In fact, the same kind of proposition can be used to ***prediction, filtering and smoothing*** three kinds of estimations. Prediction means that the measured data $\mathbf{y}$ is known only before the instant t_1, but want to estimate the state at the instant $t_2 > t_1$, that in the duration $t_1 \le t < t_2$ the estimation is prediction. Filter means the measurement is available up to the time t_1, the state at same time t_1 is required to estimate, which does not object the ***causality***. Smoothing is often used to the problem of that the data is recorded in situ and brought back to laboratory for processing, which means that the data available is all the time interval of $0 \le t < t_2$ and the state estimation at $t_1 < t_2$ is required, see [125].

The whole set of theory and methodology was proposed and developed under the computer impact. Hence computation composes an indispensable part in theory and methodology of system analysis. Although a number of books have been published, however, the computation methods are still far from enough.

Based on the analogy theory between structural mechanics and optimal control, combining the methods of analytical solution and precise integration, the theory and computational methods are systematically reorganized and part of the contents appears first time in this book.

§6.5.1, Model of linear estimation

Kalman model is based on the state space approach. The theory and computation are mainly for linear systems, and the external disturbances are considered as Gaussian distributed stochastic processes. According to theory of probability, under the Gaussian distributed stochastic disturbances, a linear system excited state variables are still Gaussian distributed. This conclusion proposes great convenience for development, that it needs only to find the mean value and the variance of the state vector, and then the distribution of state vector is obtained.

There are two versions of dynamic equation and measurement, namely the ***discrete-time*** and ***continuous-time*** versions of system, although the two versions are closely interrelated. The discrete-time version is described first.

§6.5.1.1, Model of discrete-time system

The dynamic, measurement and output equations of discrete-time system are given as

$$\mathbf{x}_{k+1} = \mathbf{F}_k \mathbf{x}_k + \mathbf{B}_k \mathbf{w}_k + \mathbf{B}_{uk} \mathbf{u}_k \tag{6.5.1}$$

$$\mathbf{y}_k = \mathbf{C}_k \mathbf{x}_k + \mathbf{v}_k \tag{6.5.2}$$

$$\mathbf{z}_k = \mathbf{C}_{zk} \mathbf{x}_k \tag{6.5.3}$$

where $k = 0,1,2,\cdots$ denote the time steps, $\mathbf{x}_k$, $\mathbf{u}_k$, $\mathbf{y}_k$ and $\mathbf{z}_k$ are the n-, m-, q- and p-dimensional state, deterministic control input, measurement, and output vectors, respectively. The disturbance noises are denoted as $\mathbf{w}_k$, $\mathbf{v}_k$, the l- and q-dimensional dynamic and measurement vectors, respectively. Because of the dynamic and measurement noises $\mathbf{w}_k$ and $\mathbf{v}_k$, so that $\mathbf{x}_k$, $\mathbf{y}_k$ and $\mathbf{z}_k$ are all stochastic processes. Matrix $\mathbf{F}_k$ is the $n \times n$ plant matrix, $\mathbf{B}_{uk}$ is the $n \times m$ input matrix, $\mathbf{B}_k$ is the $n \times l$ matrix, $\mathbf{C}_k$ is $q \times n$ measurement matrix, and $\mathbf{C}_{zk}$ is $p \times n$ output matrix, that all these matrices are deterministic. The initial conditions are

$$\hat{\mathbf{x}}_0 = \text{given}, \qquad \mathbf{P}_0 = \text{given} \tag{6.5.4}$$

i.e. the initial mean value of state and variance matrix are given. The estimation of present state vector $\mathbf{x}_k$ is based on the measured data of vectors $\mathbf{y}_k, \mathbf{y}_{k-1}, \cdots, \mathbf{y}_0$, afterward the output vector $\mathbf{z}_k$ is also estimated. The estimation of state vector $\mathbf{x}_k$ and therefore the output vector $\mathbf{z}_k$ at the present time step k is called as filtering. For the same instant k, the distinction of ***pre-verification*** (before the verification of $\mathbf{y}_k$) and ***post-verification*** should be made. The pre- and post-verification of mean states and variances are denoted by $\hat{\mathbf{x}}_k$, $\mathbf{P}_k$ and by $\hat{\mathbf{x}}'_k$, $\mathbf{P}'_k$, respectively. Note that these descriptions are the filtering.

As described in the last section, the prediction is considered that the measurement time j is lagged behind the present time k. It can be considered that the state estimation at the time step j is filtering (post verification), which is regarded as the initial condition (0-th time step) for the later $(k-j)$ time steps of prediction. The algorithm of prediction had been given in the previous section. Therefore, the analysis for the former time interval $[0, j]$ is filtering and smoothing, and for the later time interval (j, k) is prediction.

The algorithm for smoothing of the estimation time step k uses the measured data after the time step k, which objects to the causality. The estimation algorithms for both filtering and smoothing can use similar method in structural mechanics. The analogy principle will be given as that the ***mean value vector corresponds to the displacement vector in structural mechanics*** and the ***variance matrix corresponds to the flexibility matrix in structural mechanics***. This analogy principle has been seen in chapter 3 for least square method.

§6.5.1.2, Model of continuous-time system

The fundamental differential equations for continuous-time system are

$$\text{Dynamic:} \quad \dot{\mathbf{x}}(t) = \mathbf{A}(t)\mathbf{x}(t) + \mathbf{B}_w(t)\mathbf{w}(t) + \mathbf{B}_u(t)\mathbf{u}(t) \tag{6.5.5}$$

$$\text{Output:} \quad \mathbf{z}(t) = \mathbf{C}_z(t)\mathbf{x}(t) + \qquad \mathbf{D}(t)\mathbf{u}(t) \tag{6.5.6}$$

$$\text{Measurement:} \quad \mathbf{y}(t) = \mathbf{C}_y(t)\mathbf{x}(t) + \quad \mathbf{v}(t) \tag{6.5.7}$$

The meanings of vectors coincide to that given in discrete-time system.

Because it is the linear system estimation, hence the Gaussian distributed stochastic noise input induced response is also Gaussian distributed stochastic process. The initial conditions are the given mean value and variance

$$\hat{\mathbf{x}}(0) = \hat{\mathbf{x}}_0 = \text{given}, \quad \mathbf{P}(0) = \mathbf{P}_0 = \text{given} \tag{6.5.8}$$

Below, the filtering problem is concentrated.

§6.5.2, Kalman filtering analysis for discrete-time linear system

Estimation of the output vector $\mathbf{z}_k$ entirely depends on the estimation of state vector $\mathbf{x}_k$. Taking mathematical expectation to equation (6.5.3) gives

$$\hat{\mathbf{z}}_k = \mathbf{C}_{zk}\hat{\mathbf{x}}_k \tag{6.5.9}$$

Subtracting with equation (6.5.3) gives $\mathbf{z}_k - \hat{\mathbf{z}}_k = \mathbf{C}_{zk}(\mathbf{x}_k - \hat{\mathbf{x}}_k)$, and the variance is

$$\begin{aligned}\mathbf{P}_{zk} &= \mathrm{E}[(\mathbf{z}_k - \hat{\mathbf{z}}_k)(\mathbf{z}_k - \hat{\mathbf{z}}_k)^T] = \mathrm{E}[\mathbf{C}_{zk}(\mathbf{x}_k - \hat{\mathbf{x}}_k)(\mathbf{x}_k - \hat{\mathbf{x}}_k)^T \mathbf{C}_{zk}{}^T] \\ &= \mathbf{C}_{zk}\mathbf{P}_k\mathbf{C}_{zk}{}^T\end{aligned} \tag{6.5.10}$$

for which, finding the estimation $\hat{\mathbf{x}}_k$ and $\mathbf{P}_k$ of state $\mathbf{x}_k$ have been sufficient. Hence the emphasis should be put on the equations (6.5.1) and (6.5.2).

The simplest case is that the disturbances $\mathbf{w}_k$ and $\mathbf{v}_k$ are mutually independent white noises

$$\begin{aligned}&\mathrm{E}(\mathbf{w}_k) = \mathbf{0}, \quad \mathrm{var}[\mathbf{w}_k, \mathbf{w}_j] = \mathbf{W}_k\delta_{kj}, \quad \mathrm{covar}[\mathbf{w}_k, \mathbf{v}_j] = \mathbf{0} \\ &\mathrm{E}(\mathbf{v}_k) = \mathbf{0}, \quad \mathrm{var}[\mathbf{v}_k, \mathbf{v}_j] = \mathbf{V}_k\delta_{kj}\end{aligned} \tag{6.5.11}$$

Suppose that the filter analysis reaches the time step k_t, *i.e.* the related time steps are $k = 0,1,\cdots,k_t$; and the measured data are $\mathbf{y}_0, \mathbf{y}_1, \cdots, \mathbf{y}_{t-1}$ and the filtering vector $\hat{\mathbf{x}}_t$ is to determine (pre-verification). The available measurement data in the estimation of $\mathbf{x}_t$ are $\mathbf{y}_k, k < k_t$, so that the causality is satisfied. The initial estimation (6.5.8) of $\mathbf{x}(0)$ is considered uncorrelated to the later disturbance vectors $\mathbf{w}_k, \mathbf{v}_k$.

The principle for finding the estimation is still to minimize the energy index of disturbances. For $\hat{\mathbf{x}}_t$ filtering (estimation), the problem is to find the state vector subject to the dynamic and measurement equations, so as to minimize the energy of disturbances J_t (or the error index)

$$J_t = \sum_{k=0}^{k_t}[\mathbf{w}_k^T\mathbf{W}_k^{-1}\mathbf{w}_k + \mathbf{v}_k^T\mathbf{V}_k^{-1}\mathbf{v}_k^T]/2 + (\mathbf{x}_0 - \hat{\mathbf{x}}_0)^T\mathbf{P}_0^{-1}(\mathbf{x}_0 - \hat{\mathbf{x}}_0)/2, \ \min_{\mathbf{x}} J_t \tag{6.5.12}$$

where the subscript k_t is written as t for ease of notation. It is a ***conditional minimization*** problem, that the conditions are dynamic and measurement equations. This problem is a conditional ***least square***. The formulation reaches the time station (step) k_t. As time k_t flowing forward, each step proposes a least square problem, so that there are a series of least square problems. After $\hat{\mathbf{x}}_t$, the pre-verification filtered state vector, is obtained, the vector $\mathbf{y}_t$ is measured at the present time k_t, therefore the post-verification $\hat{\mathbf{x}}'_t$ of k_t should also be filtered (estimated), and then follows with the pre-verification filtering at the time station $k_t + 1$. Note that the measurement equation has no finite difference factor, so that the disturbance $\mathbf{v}_k$ can be eliminated first, that

$$J_t = \frac{1}{2}\left[\sum_{k=0}^{k_t}[\mathbf{w}_k{}^T\mathbf{W}_k^{-1}\mathbf{w}_k + (\mathbf{y}_k - \mathbf{C}_k\mathbf{x}_k)^T\mathbf{V}_k^{-1}(\mathbf{y}_k - \mathbf{C}_k\mathbf{x}_k)] + (\mathbf{x}_0 - \hat{\mathbf{x}}_0)^T\mathbf{P}_0^{-1}(\mathbf{x}_0 - \hat{\mathbf{x}}_0)\right]$$

Introducing the dual vectors $\boldsymbol{\lambda}_k, k = 1,2,\cdots$ (Lagrange multipliers) corresponding to the constraint of dynamic equation gives the extended index J_{et}

$$J_{et} = \sum_{k=0}^{k_t}[\boldsymbol{\lambda}_{k+1}^T(\mathbf{x}_{k+1} - \mathbf{F}_k\mathbf{x}_k - \mathbf{B}_k\mathbf{w}_k - \mathbf{B}_{uk}\mathbf{u}_k) + (\mathbf{y}_k - \mathbf{C}_k\mathbf{x}_k)^T\mathbf{V}_k^{-1}(\mathbf{y}_k - \mathbf{C}_k\mathbf{x}_k)/2$$
$$+ \mathbf{w}_k{}^T\mathbf{W}_k^{-1}\mathbf{w}_k/2] + (\mathbf{x}_0 - \hat{\mathbf{x}}_0)^T\mathbf{P}_0^{-1}(\mathbf{x}_0 - \hat{\mathbf{x}}_0)/2, \qquad \delta J_{et} = 0 \tag{6.5.13}$$

where the independently varied vectors are $\mathbf{x}, \boldsymbol{\lambda}, \mathbf{w}$, three kinds of variables. Because the vector $\mathbf{w}_k$ appears in equation (6.5.13) with no finite difference factor, (*i.e.* only $\mathbf{w}_k$, neither $\mathbf{w}_{k+1}$ nor $\mathbf{w}_{k-1}$, is in the functional), so that minimizing with respect to $\mathbf{w}_k$ is performed first, which gives

$$\mathbf{w}_k = \mathbf{W}_k\mathbf{B}_k^T\boldsymbol{\lambda}_{k+1} \tag{6.5.14'}$$

Substituting into equation (6.5,13) to eliminate $\mathbf{w}_k$ from J_{et} gives

$$J_{et} = \sum_{k=0}^{k_t}[\boldsymbol{\lambda}_{k+1}^T(\mathbf{x}_{k+1} - \mathbf{F}_k\mathbf{x}_k - \mathbf{B}_{uk}\mathbf{u}_k) - \boldsymbol{\lambda}_{k+1}^T(\mathbf{B}_k\mathbf{W}_k\mathbf{B}_k^T)\boldsymbol{\lambda}_{k+1}/2 - \mathbf{y}_k{}^T\mathbf{V}_k^{-1}\mathbf{C}_k\mathbf{x}_k$$
$$+ \mathbf{x}_k^T(\mathbf{C}_k{}^T\mathbf{V}_k^{-1}\mathbf{C}_k)\mathbf{x}_k/2] + (\mathbf{x}_0 - \hat{\mathbf{x}}_0)^T\mathbf{P}_0^{-1}(\mathbf{x}_0 - \hat{\mathbf{x}}_0)/2, \qquad \delta J_{et} = 0 \tag{6.5.14}$$

which has been a unconditional variational principle with two kinds of independently varied vectors $\mathbf{x}, \boldsymbol{\lambda}$. From the variational principle, it derives

$$[\delta\boldsymbol{\lambda}_{k+1}^T]: \quad \mathbf{x}_{k+1} = \mathbf{F}_k\mathbf{x}_k + (\mathbf{B}_k\mathbf{W}_k\mathbf{B}_k{}^T)\boldsymbol{\lambda}_{k+1} + \mathbf{B}_{uk}\mathbf{u}_k, \quad k = 0, \cdots, k_t \tag{6.5.15a}$$

$$[\delta\mathbf{x}_k^T]: \quad \boldsymbol{\lambda}_k = -\mathbf{C}_k{}^T\mathbf{V}_k^{-1}\mathbf{C}_k\mathbf{x}_k + \mathbf{F}_k{}^T\boldsymbol{\lambda}_{k+1} + \mathbf{C}_k{}^T\mathbf{V}_k^{-1}\mathbf{y}_k, \quad k = 1, \cdots, k_t \tag{6.5.15b}$$

$$[\delta\mathbf{x}_0^T]: \quad (\mathbf{P}_0^{-1} + \mathbf{C}_0^T\mathbf{V}_0^{-1}\mathbf{C}_0)(\mathbf{x}_0 - \hat{\mathbf{x}}_0) = \mathbf{C}_0^T\mathbf{V}_0^{-1}(\mathbf{y}_0 - \mathbf{C}_0\hat{\mathbf{x}}_0) + \mathbf{F}_0^T\boldsymbol{\lambda}_1 \tag{6.5.16}$$

$$[\delta\mathbf{x}_{t+1}^T]: \quad \hat{\boldsymbol{\lambda}}_{t+1} = \mathbf{0} \tag{6.5.17}$$

Note, in the solution of equation system (6.5.14-17), only $\mathbf{x}_t$ gives the filter value at station k_t, as for other stations $k < k_t$ the vectors $\mathbf{x}_k$ solved are smoothing values, because the solution depends on the measurements later than the k station. To concentrate the analysis from k_t to k_{t+1}, assuming first that the pre-verification $\hat{\mathbf{x}}_t$ has been found at the station k_t, *i.e.*

$$\mathbf{x}_t = \hat{\mathbf{x}}_t + \mathbf{P}_t\,\boldsymbol{\lambda}_t \tag{6.5.18}$$

The pre-verification filtering $\hat{\mathbf{x}}_{t+1}$ at the station k_{t+1} is required, which means to analyze one time step forward. One time step forward involves two successive sub-steps, *i.e.* the post-verification at station k_t and then the pre-verification filtering (prediction) at station k_{t+1}. The one time step forward needs to solve the dual finite difference equations (6.5.15a,b), where $k = k_t$ and $\mathbf{x}_t$ is expressed as that given in equation (6.5.18) before the measurement $\mathbf{y}_t$. Substituting (6.5.18) into (6.5.15a,b) gives

$$\begin{aligned}\mathbf{x}_{k+1} &= \mathbf{F}_k\mathbf{P}_k\boldsymbol{\lambda}_k + \mathbf{B}_k\mathbf{W}_k\mathbf{B}_k^T\boldsymbol{\lambda}_{k+1} + \mathbf{F}_k\hat{\mathbf{x}}_k + \mathbf{B}_{uk}\mathbf{u}_k \\ \boldsymbol{\lambda}_k &= -\mathbf{C}_k^T\mathbf{V}_k^{-1}\mathbf{C}_k\mathbf{P}_k\boldsymbol{\lambda}_k + \mathbf{F}_k^T\boldsymbol{\lambda}_{k+1} + \mathbf{C}_k^T\mathbf{V}_k^{-1}(\mathbf{y}_k - \mathbf{C}_k\hat{\mathbf{x}}_k)\end{aligned}, \quad k = k_t \tag{6.5.19}$$

and combined as

$$\mathbf{P}_k\boldsymbol{\lambda}_k = \mathbf{P}'_k\mathbf{F}_k^T\boldsymbol{\lambda}_{k+1} + \mathbf{P}'_k\mathbf{C}_k^T\mathbf{V}_k^{-1}(\mathbf{y}_k - \mathbf{C}_k\hat{\mathbf{x}}_k) \tag{6.5.20}$$

where $\mathbf{P}'_k$ is given in (6.5.25). Eliminating $\mathbf{P}_k\boldsymbol{\lambda}_k$ from (6.5.19) gives

$$\mathbf{x}_{t+1} = \hat{\mathbf{x}}_{t+1} + \mathbf{P}_{t+1}\boldsymbol{\lambda}_{t+1} \tag{6.5.21}$$

$$\hat{\mathbf{x}}_{t+1} = \mathbf{F}_t \cdot [\hat{\mathbf{x}}_t + \mathbf{K}_t(\mathbf{y}_t - \mathbf{C}_t\hat{\mathbf{x}}_t)] + \mathbf{B}_{u,t}\mathbf{u}_t = \mathbf{F}_t\hat{\mathbf{x}}'_t + \mathbf{B}_{u,t}\mathbf{u}_t \tag{6.5.22}$$

$$\hat{\mathbf{x}}'_t = \hat{\mathbf{x}}_t + \mathbf{K}_t(\mathbf{y}_t - \mathbf{C}_t\hat{\mathbf{x}}_t) \tag{6.5.23}$$

$$\mathbf{K}_t = \mathbf{P}'_t\mathbf{C}_t^T\mathbf{V}_t^{-1} \tag{6.5.24}$$

$$\mathbf{P}'_t = (\mathbf{P}_t^{-1} + \mathbf{C}_t^T\mathbf{V}_t^{-1}\mathbf{C}_t)^{-1} \tag{6.5.25}$$

$$\mathbf{P}_{t+1} = \mathbf{F}_t\mathbf{P}'_t\mathbf{F}_t^T + \mathbf{B}_t\mathbf{W}_t\mathbf{B}_t^T \tag{6.5.26}$$

These equations supply a recurrence situation, that assuming at the time step k_t the pre-verification filter vector $\hat{\mathbf{x}}_t$ and the variance matrix $\mathbf{P}_t$ have been found, which corresponds to the initial condition (6.5.4). Therefore from equation (6.5.25) the matrix $\mathbf{P}'_t$ is computed. Then, from the equation (6.5.24) the ***gain matrix*** $\mathbf{K}_t$ is computed. Further, from equation (6.5.23) computes the vector $\hat{\mathbf{x}}'_t$. The filter mean value $\hat{\mathbf{x}}'_t$ and the variance matrix $\mathbf{P}'_t$ of $\mathbf{x}_t$ give the results of first sub-step, *i.e.* the ***post-verification*** at the station k_t. Continuing, from equation (6.5.22) computes $\hat{\mathbf{x}}_{t+1}$, and then from equation (6.5.26) computes $\mathbf{P}_{t+1}$. The computation of one time step forward from k_t to k_{t+1} station is consummated, (which corresponds to one time step prediction). Note that the equations (6.5.18) and (6.5.21) are the same form, but for different time steps of k_t and k_{t+1}, respectively. Therefore these equations are the recursive computation for one time step forward. Hence, from the initial condition (6.5.8') of $k_t = 0$, the $k_t = 1$ filter is deduced; forward further, from $k_t = 1$ deduce $k_t = 2, \cdots$, and so on. The mathematical induction algorithm is realized.

Let the equations (6.5.15a,b) and (6.5.16) be satisfied, only remain $\boldsymbol{\lambda}_{t+1}$ as variable, the index

$$J_{et} = (\mathbf{x}_{t+1} - \hat{\mathbf{x}}_{t+1})^T\mathbf{P}_{t+1}^{-1}(\mathbf{x}_{t+1} - \hat{\mathbf{x}}_{t+1})/2 + \text{Const}$$

is obtained. In fact, the above expression can also be derived from

$$J_{et} = \boldsymbol{\lambda}_{t+1}^T[\mathbf{x}_{t+1} - \mathbf{F}_t\mathbf{x}_t - \mathbf{B}_{ut}\mathbf{u}_t - \mathbf{B}_t\mathbf{W}_t\mathbf{B}_t^T\boldsymbol{\lambda}_{t+1}/2] + \mathbf{x}_t^T\mathbf{C}_t^T\mathbf{V}_t^{-1}\mathbf{C}_t\mathbf{x}_t/2$$
$$+ (\mathbf{x}_t - \hat{\mathbf{x}}_t)^T\mathbf{P}_t^{-1}(\mathbf{x}_t - \hat{\mathbf{x}}_t)/2 - \mathbf{x}_t^T\mathbf{C}_t^T\mathbf{V}_t^{-1}\mathbf{y}_t, \qquad \max_{\boldsymbol{\lambda}_{t+1}}[\min_{\mathbf{x}_t} J_{et}] \tag{6.5.14''}$$

To find the pre-verification filter vector at $t+1$, the max in the above equation is applied; however, the computation of J_{et} needs only a minimization inside the bracket. The mean value of filter vector $\hat{\mathbf{x}}_{t+1}$ reaches at equation (6.5.17), which can also be verified by the equation (6.5.21). For the sake of simplicity, the front station k_t is written as k again.

The derivation here for discrete-time Kalman filter equations uses the variational method, which is quite simple. The equations obtained are the same as usual, for which the comparison is listed below.

Usual	$\hat{\mathbf{x}}_k(k \mid k-1)$	$\hat{\mathbf{x}}_k(k \mid k)$	$\mathbf{P}(k \mid k-1)$	$\mathbf{P}(k \mid k)$
Here	$\hat{\mathbf{x}}_k$	$\hat{\mathbf{x}}'_k$	$\mathbf{P}_k$	$\mathbf{P}'_k$

For linear system with Gaussian distributed stochastic process the estimation from the variational principle is naturally ***unbiased***.

In the recurrence equations (6.5.22)~(6.5.26), the variance matrices $\mathbf{P}_k$, $\mathbf{P}'_k$ and the gain matrix $\mathbf{K}_k$ are independent on the measurement $\mathbf{y}$, therefore these matrices can be computed and stored beforehand (off-line). In real time applications, the computations are only for the equation (6.5.22) and (6.5.23).

In the whole derivation, the deterministic input $\mathbf{u}_k$ appears only as an additional term in filter dynamic equation (6.5.22) of mean value, but has no other effect. Hence the input $\mathbf{u}_k$ is neglected below.

The comparison between the equations of Kalman filter and the equations in structural mechanics is interested, which means the analogy relation between the two fields. Comparing both the variational principles of (6.5.14) and (5.9.9), listed as

$$V_k(\mathbf{q}_a,\mathbf{p}_b) = \mathbf{p}_b^T\mathbf{G}\mathbf{p}_b/2 + \mathbf{p}_b^T\mathbf{F}\mathbf{q}_a - \mathbf{q}_a^T\mathbf{Q}\mathbf{q}_a + \mathbf{p}_b^T\mathbf{r}_{bk} + \mathbf{q}_a^T\mathbf{r}_{ak} \tag{5.9.5}$$

$$\min_{\mathbf{q}}\max_{\mathbf{p}}\left[\sum_{k=1}^{k_f}\left[\mathbf{p}_k^T\mathbf{q}_k - V_k(\mathbf{q}_{k-1},\mathbf{p}_k)\right] + (\mathbf{q}_0-\hat{\mathbf{q}}_0)^T\mathbf{P}_0^{-1}(\mathbf{q}_0-\hat{\mathbf{q}}_0)/2\right] \tag{5.9.9}$$

The comparison is given as

Structural mechanics	**Kalman filter**
Displacement and internal force $\mathbf{q},\mathbf{p}$	Dual vectors $\mathbf{x},\boldsymbol{\lambda}$
Right end k of interval ·$[0,k)$ and $k+1$	k_t of time interval $[0,k_t)$, and k_{t+1}
The matrices $\mathbf{F},\mathbf{G},\mathbf{Q}$	$\mathbf{F},\mathbf{C}^T\mathbf{V}^{-1}\mathbf{C},\mathbf{B}\mathbf{W}\mathbf{B}^T$
Equivalent external forces $\mathbf{r}_{bk},\mathbf{r}_{ak}$	$\mathbf{B}_u\mathbf{u}_k,\mathbf{C}^T\mathbf{V}^{-1}\mathbf{y}_k$
Mixed energy $V_k(\mathbf{q}_a,\mathbf{p}_b)$	Mixed energy $V_k(\mathbf{x}_a,\boldsymbol{\lambda}_b)$
Dual equations (5.9.10)	Dual equations (6.5.15a,b)
Potential energy $\Pi_k(\mathbf{q}_k)$ of interval $[0,k)$	Index J_{e,k_t} of time interval $[0,k_t)$

And so on.

Some important ideas must be explained at this point. In the equation set (6.5.15a,b) and (6.5.16~17), the filter is for the $\hat{\mathbf{x}}_{t+1}$ of time station $t+1$. It corresponds to the boundary condition of $\hat{\boldsymbol{\lambda}}_{t+1}=\mathbf{0}$. For the station t the value $\boldsymbol{\lambda}_t$ is not zero, because the corresponding estimation is post-verification $\hat{\mathbf{x}}'_t$. For the earlier time stations $k<t$ the value $\boldsymbol{\lambda}_k$ is not a null vector, that the smoothing state vector $\bar{\mathbf{x}}_k$ (or denoted as $\bar{\mathbf{x}}(k,t+1)$) is obtained. It should be emphasized that the equation set (6.5.15~17) gives a number of smoothing solutions with the estimation time step goes forward. However, only the filtering solution is of concern here.

§6.5.2.1, Correlated dynamic and measurement noises

When considering the colored noise, the case of correlated dynamic and measurement noises $\mathbf{w}_k$ and $\mathbf{v}_k$ is needed, that the equation (6.5.11) should be changed as

$$\mathrm{E}(\mathbf{w}_k)=\mathbf{0},\quad \mathrm{E}(\mathbf{v}_k)=\mathbf{0},$$

$$\mathrm{var}[\mathbf{w}_i,\mathbf{w}_j]=\mathbf{W}_i\delta_{ij},\quad \mathrm{var}[\mathbf{v}_i,\mathbf{v}_j]=\mathbf{V}_i\delta_{ij},\quad \mathrm{covar}[\mathbf{w}_i,\mathbf{v}_j]=\mathbf{S}_i\delta_{ij} \tag{6.5.27}$$

and the index is given as

$$J=\sum_{k=0}^{k_f}\frac{1}{2}\begin{Bmatrix}\mathbf{w}_k\\ \mathbf{y}_k-\mathbf{C}_k\mathbf{x}_k\end{Bmatrix}^T\begin{bmatrix}\mathbf{W}_k & \mathbf{S}_k\\ \mathbf{S}_k^T & \mathbf{V}_k\end{bmatrix}^{-1}\begin{Bmatrix}\mathbf{w}_k\\ \mathbf{y}_k-\mathbf{C}_k\mathbf{x}_k\end{Bmatrix}+\frac{1}{2}(\mathbf{x}_0-\hat{\mathbf{x}}_0)^T\mathbf{P}_0^{-1}(\mathbf{x}_0-\hat{\mathbf{x}}_0)$$

$$\min J \tag{6.5.28}$$

Because the energy of noise must be positive, so that the matrix in the above equation is positive definite and the inverse matrix exists. The block matrix inverse equation is given as

$$\begin{bmatrix}\mathbf{Q} & \mathbf{S}\\ \mathbf{S}^T & \mathbf{R}\end{bmatrix}^{-1}=\begin{bmatrix}(\mathbf{Q}-\mathbf{S}\mathbf{R}^{-1}\mathbf{S}^T)^{-1} & -\mathbf{Q}^{-1}\mathbf{S}(\mathbf{R}-\mathbf{S}^T\mathbf{Q}^{-1}\mathbf{S})^{-1}\\ -\mathbf{R}^{-1}\mathbf{S}^T(\mathbf{Q}-\mathbf{S}\mathbf{R}^{-1}\mathbf{S}^T)^{-1} & (\mathbf{R}-\mathbf{S}^T\mathbf{Q}^{-1}\mathbf{S})^{-1}\end{bmatrix} \tag{6.5.29}$$

from which derives the ***matrix inversion lemma***, which is useful later

$$(\mathbf{R}-\mathbf{S}^T\mathbf{Q}^{-1}\mathbf{S})^{-1}=\mathbf{R}^{-1}+\mathbf{R}^{-1}\mathbf{S}^T(\mathbf{Q}-\mathbf{S}\mathbf{R}^{-1}\mathbf{S}^T)^{-1}\mathbf{S}\mathbf{R}^{-1} \tag{6.5.30}$$

The minimization of the index (6.5.28) is conditional. Introducing the Lagrange multiplier $\boldsymbol{\lambda}_k$ to release the constraint gives the extended index

$$J_e=\sum_{k=0}^{k_f}\begin{bmatrix}\boldsymbol{\lambda}_{k+1}^T(\mathbf{x}_{k+1}-\mathbf{F}_k\mathbf{x}_k-\mathbf{B}_k\mathbf{w}_k)+\mathbf{w}_k^T(\mathbf{W}_k-\mathbf{S}_k\mathbf{V}_k^{-1}\mathbf{S}_k^T)^{-1}\mathbf{w}_k/2\\ +(\mathbf{y}_k-\mathbf{C}_k\mathbf{x}_k)^T(\mathbf{V}_k-\mathbf{S}_k^T\mathbf{W}_k^{-1}\mathbf{S}_k)^{-1}(\mathbf{y}_k-\mathbf{C}_k\mathbf{x}_k)/2\\ -\mathbf{w}_k^T\mathbf{W}_k^{-1}\mathbf{S}_k(\mathbf{V}_k-\mathbf{S}_k^T\mathbf{W}_k^{-1}\mathbf{S}_k)^{-1}\cdot(\mathbf{y}_k-\mathbf{C}_k\mathbf{x}_k)\end{bmatrix} \tag{6.5.31}$$

$$+(\mathbf{x}_0-\hat{\mathbf{x}}_0)^T\mathbf{P}_0^{-1}(\mathbf{x}_0-\hat{\mathbf{x}}_0)/2,\qquad \delta J_e=0$$

The variation of the above functional is unconditional that the three kinds of variables $\mathbf{x},\boldsymbol{\lambda}$ and $\mathbf{w}$ are independent. Minimizing with respect to vector $\mathbf{w}$ gives

$$\mathbf{w}_k=(\mathbf{W}_k-\mathbf{S}_k\mathbf{V}_k^{-1}\mathbf{S}_k^T)\mathbf{B}_k^T\boldsymbol{\lambda}_{k+1}+\mathbf{S}_k\mathbf{V}_k^{-1}(\mathbf{y}_k-\mathbf{C}_k\mathbf{x}_k) \tag{6.5.32}$$

where the identity $(\mathbf{Q}-\mathbf{S}\mathbf{R}^{-1}\mathbf{S}^T)\mathbf{Q}^{-1}\mathbf{S}(\mathbf{R}-\mathbf{S}^T\mathbf{Q}^{-1}\mathbf{S})^{-1}\equiv\mathbf{S}\mathbf{R}^{-1}$ is used in the derivation. Substituting $\mathbf{w}_k$ back into the functional J_e, after some matrix algebraic derivation, gives

$$J_e=\sum_{k=0}^{k_f}\begin{bmatrix}\boldsymbol{\lambda}_{k+1}^T(\mathbf{x}_{k+1}-\mathbf{F}_k\mathbf{x}_k)-\boldsymbol{\lambda}_{k+1}^T\mathbf{B}_k(\mathbf{W}_k-\mathbf{S}_k\mathbf{V}_k^{-1}\mathbf{S}_k^T)\mathbf{B}_k^T\boldsymbol{\lambda}_{k+1}/2\\ -\boldsymbol{\lambda}_{k+1}^T\mathbf{B}_k\mathbf{S}_k\mathbf{V}_k^{-1}(\mathbf{y}_k-\mathbf{C}_k\mathbf{x}_k)+(\mathbf{y}_k-\mathbf{C}_k\mathbf{x}_k)^T\mathbf{V}_k^{-1}(\mathbf{y}_k-\mathbf{C}_k\mathbf{x}_k)/2\end{bmatrix} \tag{6.5.31'}$$

$$+(\mathbf{x}_0-\hat{\mathbf{x}}_0)^T\mathbf{P}_0^{-1}(\mathbf{x}_0-\hat{\mathbf{x}}_0)/2,\qquad \delta J_e=0$$

There are two kinds of independent variables $\mathbf{x},\boldsymbol{\lambda}$ in the functional. Variational derivation derives the dual equations as

$$\mathbf{x}_{k+1}=(\mathbf{F}_k-\mathbf{J}_k\mathbf{C}_k)\mathbf{x}_k+\mathbf{B}_k(\mathbf{W}_k-\mathbf{S}_k\mathbf{V}_k^{-1}\mathbf{S}_k^T)\mathbf{B}_k^T\boldsymbol{\lambda}_{k+1}+\mathbf{J}_k\mathbf{y}_k,\quad k=0,\cdots \tag{6.5.33a}$$

$$\boldsymbol{\lambda}_k=-\mathbf{C}_k^T\mathbf{R}_k^{-1}\mathbf{C}_k\mathbf{x}_k+(\mathbf{F}_k-\mathbf{J}_k\mathbf{C}_k)^T\boldsymbol{\lambda}_{k+1}+\mathbf{C}_k^T\mathbf{V}_k^{-1}\mathbf{y}_k,\quad k=1,\cdots \tag{6.5.33b}$$

$$(\mathbf{P}_0^{-1}+\mathbf{C}_0^T\mathbf{V}_0^{-1}\mathbf{C}_0)(\mathbf{x}_0-\hat{\mathbf{x}}_0)=\mathbf{C}_0^T\mathbf{V}_0^{-1}(\mathbf{y}_0-\mathbf{C}_0\mathbf{x}_0)+(\mathbf{F}_0-\mathbf{J}_0\mathbf{C}_0)^T\boldsymbol{\lambda}_1 \qquad (6.5.34)$$

$$\mathbf{J}_k=\mathbf{B}_{k+1}\mathbf{S}_k\mathbf{V}_k^{-1}, \quad (k=0,1,\cdots) \qquad (6.5.35)$$

Obviously, this set of equations has the same structure with equations (6.5.15) and (6.5.16), that only the additional terms $\mathbf{S}_k$ and $\mathbf{J}_k$ reflect the effect of correlation. Solving it, the expression (6.5.21) and (6.5.18) still have the same form. Substituting into equation (6.5.33) gives

$$\mathbf{x}_{k+1}=\mathbf{F}_{*k}\mathbf{P}_k\boldsymbol{\lambda}_k+\mathbf{B}_k\mathbf{W}_{*k}\mathbf{B}_k^T\boldsymbol{\lambda}_{k+1}+\mathbf{F}_{*k}\hat{\mathbf{x}}_k+\mathbf{J}_k\mathbf{y}_k, \quad \mathbf{F}_{*k}=\mathbf{F}_k-\mathbf{J}_k\mathbf{C}_k$$

$$\boldsymbol{\lambda}_k=-\mathbf{C}_k^T\mathbf{V}_k^{-1}\mathbf{C}_k\mathbf{P}_k\boldsymbol{\lambda}_k+\mathbf{W}_{*k}^T\boldsymbol{\lambda}_{k+1}+\mathbf{C}_k^T\mathbf{V}_k^{-1}(\mathbf{y}_k-\mathbf{C}_k\hat{\mathbf{x}}_k),\ \mathbf{W}_{*k}=\mathbf{W}_k-\mathbf{S}_k\mathbf{V}_k^{-1}\mathbf{S}_k^T$$

Eliminating $\boldsymbol{\lambda}_k$ and note that the subscript k is just k_t, gives

$$\mathbf{x}_{k+1}=\mathbf{P}_{k+1}\boldsymbol{\lambda}_{k+1}+\hat{\mathbf{x}}_{k+1} \qquad (6.5.36)$$

$$\hat{\mathbf{x}}_{k+1}=\mathbf{F}_{*k}\cdot[\hat{\mathbf{x}}_k+\mathbf{K}_k(\mathbf{y}_k-\mathbf{C}_k\hat{\mathbf{x}}_k)]+\mathbf{J}_k\mathbf{y}_k=\mathbf{F}_k\hat{\mathbf{x}}'_k+\mathbf{J}_k(\mathbf{y}_k-\mathbf{C}_k\hat{\mathbf{x}}'_k) \qquad (6.5.37)$$

$$\hat{\mathbf{x}}'_k=\hat{\mathbf{x}}_k+\mathbf{K}_k(\mathbf{y}_k-\mathbf{C}_k\hat{\mathbf{x}}_k) \qquad (6.5.38)$$

$$\mathbf{K}_k=\mathbf{P}'_k\mathbf{C}_k^T\mathbf{V}_k^{-1} \qquad (6.5.39)$$

$$\mathbf{P}'_k=(\mathbf{P}_k^{-1}+\mathbf{C}_k^T\mathbf{V}_k^{-1}\mathbf{C}_k)^{-1} \qquad (6.5.40)$$

$$\mathbf{P}_{k+1}=\mathbf{F}_{*k}\mathbf{P}'_k\mathbf{F}_{*k}^T+\mathbf{B}_k\mathbf{W}_{*k}\mathbf{B}_k^T \qquad (6.5.41)$$

The derivation of these equations explains that for the case of $\mathbf{w}_k, \mathbf{v}_k$ being correlated, the computation still has the same form, only the equations having more terms. These equations are useful for the case of that the white noise assumption is substituted by colored noise.

Filtering treats the mean vector $\hat{\mathbf{x}}_0$ and variance matrix $\mathbf{P}_0$ just as the given initial values. The correction of estimation of $\mathbf{x}_0$ and the corresponding variance $\mathbf{P}_0$ based on the measurements later is disregarded, that such correction is considered in the smoothing analysis.

§6.5.3, Continuous-time Kalman-Bucy filtering analysis

The basic equations of continuous-time Kalman filter can be obtained as a limiting process of discrete-time Kalman filter by taking the time interval duration $\eta \to 0$. Similar to the derivation of discrete-time case, the variational method is used also to derive the basic equations of continuous-time filtering, see also [126]. As seen in the discrete-time case, that the effect of deterministic input $\mathbf{u}$ is only a given term in the differential equation for mean state vector $\hat{\mathbf{x}}(t)$.

$$\dot{\mathbf{x}}=\mathbf{A}\mathbf{x}+\mathbf{B}_w\mathbf{w}+\mathbf{B}_u\mathbf{u} \qquad (6.5.5')$$

$$\mathbf{y}=\mathbf{C}\mathbf{x}+\mathbf{v} \qquad (6.5.7')$$

The basic assumption is made for the stochastic disturbances $\mathbf{w}$ and $\mathbf{v}$ that

$$\mathrm{E}(\mathbf{w}(t))=\mathbf{0}, \quad \mathrm{var}[\mathbf{w}(t),\mathbf{w}(\tau)]=\mathbf{W}(t)\delta(t-\tau); \qquad (6.5.42a)$$

$$\mathrm{E}(\mathbf{v}(t))=\mathbf{0}, \quad \mathrm{var}[\mathbf{v}(t),\mathbf{v}(\tau)]=\mathbf{V}(t)\delta(t-\tau); \qquad (6.5.42b)$$

$$\mathrm{covar}[\mathbf{w}(t),\mathbf{v}(\tau)]=\mathbf{0}; \qquad (6.5.42c)$$

where both $\mathbf{W}(t)$ and $\mathbf{V}(t)$ are symmetric positive definite matrices. The initial condition is for $\mathbf{x}(0)$, which is a Gaussian distributed random vector, and is independent on $\mathbf{w}$ and $\mathbf{v}$, *i.e.* $\text{covar}[\mathbf{x}_0, \mathbf{v}] = \mathbf{0}$ and $\text{covar}[\mathbf{x}_0, \mathbf{w}] = \mathbf{0}$. The initial conditions are given as

$$E[\mathbf{x}(0)] = \hat{\mathbf{x}}_0, \quad \text{var}[\mathbf{x}_0, \mathbf{x}_0] = \mathbf{P}_0, \quad \text{known} \tag{6.5.43}$$

or

$$\mathbf{x}_0 = \mathbf{x}(0) = \hat{\mathbf{x}}_0 + \mathbf{P}_0 \boldsymbol{\lambda}_0 \tag{6.5.43a}$$

where $\boldsymbol{\lambda}_0$ is a zero mean Gaussian distributed vector for filtering, that from $E[(\mathbf{x}_0 - \hat{\mathbf{x}}_0)(\mathbf{x}_0 - \hat{\mathbf{x}}_0)^T] = \mathbf{P}_0$, it derives $E[\boldsymbol{\lambda}_0 \boldsymbol{\lambda}_0^T] = \mathbf{P}_0^{-1}$, note, it is only for filtering. As time is going on, it becomes a smoothing solution at time $t = 0$.

Derivation of filter equation should be based on the measured vector $\mathbf{y}$ to estimate $\mathbf{x}$. The principle is that the quadratic functional index of noises J is minimized

$$J = \int_0^t [\mathbf{w}^T \mathbf{W}^{-1} \mathbf{w} + \mathbf{v}^T \mathbf{V}^{-1} \mathbf{v}] \mathrm{d}\tau/2 + (\mathbf{x}_0 - \hat{\mathbf{x}}_0)^T \mathbf{P}_0^{-1} (\mathbf{x}_0 - \hat{\mathbf{x}}_0)/2, \ \min_{\mathbf{x}} J \tag{6.5.44}$$

Substituting with equation (6.5.7') gives

$$J = \int_0^t [\mathbf{w}^T \mathbf{W}^{-1} \mathbf{w} + (\mathbf{y} - \mathbf{C}\mathbf{x})^T \mathbf{V}^{-1} (\mathbf{y} - \mathbf{C}\mathbf{x})] \mathrm{d}\tau/2 + (\mathbf{x}_0 - \hat{\mathbf{x}}_0)^T \mathbf{P}_0^{-1} (\mathbf{x}_0 - \hat{\mathbf{x}}_0)/2$$

$$\min_{\mathbf{x}} J$$

where the varying functions are $\mathbf{w}$ and $\mathbf{x}$, but there is the constraint condition of dynamic equation (6.5.5'), so that it is a **conditional minimization** problem. Introducing the Lagrange multiplier vector function $\boldsymbol{\lambda}(t)$ derives to the unconditional variational principle of the extended index J_{At}

$$\begin{aligned} J_{At} = \int_0^t & [\boldsymbol{\lambda}^T (\dot{\mathbf{x}} - \mathbf{A}\mathbf{x} - \mathbf{B}_w \mathbf{w} - \mathbf{B}_u \mathbf{u}) + (\mathbf{y} - \mathbf{C}\mathbf{x})^T \mathbf{V}^{-1} (\mathbf{y} - \mathbf{C}\mathbf{x})/2 \\ & + \mathbf{w}^T \mathbf{W}^{-1} \mathbf{w}/2] \mathrm{d}\tau + (\mathbf{x}_0 - \hat{\mathbf{x}}_0)^T \mathbf{P}_0^{-1} (\mathbf{x}_0 - \hat{\mathbf{x}}_0)/2, \qquad \delta J_{At} = 0 \end{aligned} \tag{6.5.45}$$

where the independently varying functions are $\mathbf{x}, \boldsymbol{\lambda}$ and $\mathbf{w}$. The function $\mathbf{w}$ is unrelated to the time derivative, so that the minimization is carried out first

$$\mathbf{w} = \mathbf{W} \mathbf{B}_w^T \boldsymbol{\lambda} \tag{6.5.46}$$

Substituting into equation (6.5.45) eliminates $\mathbf{w}$, which derives the variational principle with two kinds of independent variables $\mathbf{x}$ and $\boldsymbol{\lambda}$ as

$$\begin{aligned} J_{At} = \int_0^t & [\boldsymbol{\lambda}^T (\dot{\mathbf{x}} - \mathbf{A}\mathbf{x} - \mathbf{B}_u \mathbf{u}) + (\mathbf{y} - \mathbf{C}\mathbf{x})^T \mathbf{V}^{-1} (\mathbf{y} - \mathbf{C}\mathbf{x})/2 \\ & - \boldsymbol{\lambda}^T (\mathbf{B}_w \mathbf{W} \mathbf{B}_w^T) \boldsymbol{\lambda}/2] \mathrm{d}\tau + (\mathbf{x}_0 - \hat{\mathbf{x}}_0)^T \mathbf{P}_0^{-1} (\mathbf{x}_0 - \hat{\mathbf{x}}_0)/2, \quad \delta J_{At} = 0 \end{aligned} \tag{6.5.45'}$$

where the input vector $\mathbf{u}$ and the measurement $\mathbf{y}$ are given vectors, so no variations for them. Performing the variational derivation gives the dual differential equations

$$\dot{\mathbf{x}} = \mathbf{A}\mathbf{x} + \mathbf{B}_w \mathbf{W} \mathbf{B}_w^T \boldsymbol{\lambda} + \mathbf{B}_u \mathbf{u} \tag{6.5.47a}$$

$$\dot{\boldsymbol{\lambda}} = \mathbf{C}^T \mathbf{V}^{-1} \mathbf{C}\mathbf{x} - \mathbf{A}^T \boldsymbol{\lambda} - \mathbf{C}^T \mathbf{V}^{-1} \mathbf{y} \tag{6.5.47b}$$

The initial conditions have been given in equation (6.5.43a), where $\hat{\mathbf{x}}_0$ is given. The variance matrix $\mathbf{P}_0$ has been explicitly written in the variational principle. It is important to introduce the time interval, the filtering interval is $[0, t)$, where at

$t_0 = 0$ the initial condition is non-zero, so that the close end of the interval is used, and the end t progresses continuously. The filtering problem, satisfying the causality condition, gives the initial state to estimate the state $\mathbf{x}(t)$ at the end t. For the dual differential equations (6.5.47a,b), the initial condition appears as both the mean state $\hat{\mathbf{x}}_0$ and the variance matrix $\mathbf{P}_0$ given, whereas the natural boundary condition at the end t is derived from the variational principle as $\hat{\boldsymbol{\lambda}}(t) = 0$.

Comparing the dual equations of structural mechanics in chapter 5

$$\dot{\mathbf{q}} = \mathbf{A}\mathbf{q} + \mathbf{D}\mathbf{p} + \mathbf{f}_q, \qquad \dot{\mathbf{p}} = \mathbf{B}\mathbf{q} - \mathbf{A}^T\mathbf{p} + \mathbf{f}_p \tag{5.2.9a,b}$$

the analogy relation between structural mechanics and Kalman-Bucy filter is found as

Structural mechanics	**Kalman-Bucy filter**
Displacement, internal force $\mathbf{q},\mathbf{p}$	Dual vectors $\mathbf{x},\boldsymbol{\lambda}$
Space interval $[z_0, z)$	Time interval $[t_0, t)$
$\mathbf{A}, \mathbf{B}, \mathbf{D}$	$\mathbf{A},\ \mathbf{C}^T\mathbf{V}^{-1}\mathbf{C},\ \mathbf{B}_w\mathbf{W}\mathbf{B}_w^T$
Equivalent external forces $\mathbf{f}_q, \mathbf{f}_p$	$\mathbf{B}_u\mathbf{u}, -\mathbf{C}^T\mathbf{V}^{-1}\mathbf{y}$
Dual equations (5.2.9a,b)	Dual equations (6.5.47)
The action function S of interval $[z_0, z)$	The index J_{At} of time-interval $[t_0, t)$

and so on.

The solution of the dual equations (6.5.47) can be as follows: A Gaussian stochastic process can be expressed as the sum of a mean value function $\hat{\mathbf{x}}(t)$ and a zero-mean Gaussian stochastic process, with the variance matrix $\mathbf{P}(t)$ to be determined, *i.e.*

$$\mathbf{x} = \hat{\mathbf{x}}(t) + \mathbf{P}(t)\boldsymbol{\lambda}(t) \tag{6.5.48}$$

This form of filter solution is similar to the discrete-time case of equation (6.5.18). Substituting into the dual equations (6.5.47a,b) the derivation follows as

$$\dot{\mathbf{x}} = \dot{\hat{\mathbf{x}}} + \dot{\mathbf{P}}\boldsymbol{\lambda} + \mathbf{P}\dot{\boldsymbol{\lambda}} = \mathbf{A}\hat{\mathbf{x}} + \mathbf{A}\mathbf{P}\boldsymbol{\lambda} + \mathbf{B}_w\mathbf{W}\mathbf{B}_w^T\boldsymbol{\lambda} + \mathbf{B}_u\mathbf{u}$$

$$\mathbf{P}\dot{\boldsymbol{\lambda}} = \mathbf{P}\mathbf{C}^T\mathbf{V}^{-1}\mathbf{C}\hat{\mathbf{x}} + \mathbf{P}\mathbf{C}^T\mathbf{V}^{-1}\mathbf{C}\mathbf{P}\boldsymbol{\lambda} - \mathbf{P}\mathbf{A}^T\boldsymbol{\lambda} - \mathbf{P}\mathbf{C}^T\mathbf{V}^{-1}\mathbf{y}$$

Eliminating $\mathbf{P}\dot{\boldsymbol{\lambda}}$ from the two equations gives

$$\dot{\hat{\mathbf{x}}} + \dot{\mathbf{P}}\boldsymbol{\lambda} = \mathbf{A}\hat{\mathbf{x}} + (\mathbf{A}\mathbf{P} + \mathbf{P}\mathbf{A}^T + \mathbf{B}_w\mathbf{W}\mathbf{B}_w^T - \mathbf{P}\mathbf{C}^T\mathbf{V}^{-1}\mathbf{C}\mathbf{P})\boldsymbol{\lambda}$$
$$- \mathbf{P}\mathbf{C}^T\mathbf{V}^{-1}\mathbf{C}\hat{\mathbf{x}} + \mathbf{P}\mathbf{C}^T\mathbf{V}^{-1}\mathbf{y} + \mathbf{B}_u\mathbf{u}$$

In the above equation, there are the deterministic terms with no factor of $\boldsymbol{\lambda}$, and the stochastic terms with the factor $\boldsymbol{\lambda}$. The distinction of them gives the two equations

$$\dot{\hat{\mathbf{x}}}(t) = \mathbf{A}\hat{\mathbf{x}} + \mathbf{P}\mathbf{C}^T\mathbf{V}^{-1}(\mathbf{y} - \mathbf{C}\hat{\mathbf{x}}) + \mathbf{B}_u\mathbf{u}, \qquad \hat{\mathbf{x}}(0) = \hat{\mathbf{x}}_0 \tag{6.5.49}$$

$$\dot{\mathbf{P}}(t) = \mathbf{B}_w\mathbf{W}\mathbf{B}_w^T + \mathbf{A}\mathbf{P} + \mathbf{P}\mathbf{A}^T - \mathbf{P}\mathbf{C}^T\mathbf{V}^{-1}\mathbf{C}\mathbf{P}, \qquad \mathbf{P}(0) = \mathbf{P}_0 \tag{6.5.50}$$

where (6.5.50) is called the ***matrix Riccati differential equation,*** see also equation (5.7.32) in section §5.7.6. The symmetric matrix function $\mathbf{P}(t)$ to be solved is of dimension $n \times n$. When the system is controllable and observable $\mathbf{P}(t)$ is positive definite, see section 6.7. The linear differential equation (6.5.49) is for solving the mean state value, *i.e.* the filter vector $\hat{\mathbf{x}}(t)$, and is called as ***filter differential equation,*** which can also be written as

$$\dot{\hat{\mathbf{x}}}(t) = \mathbf{A}\hat{\mathbf{x}} + \mathbf{K}(\mathbf{y} - \mathbf{C}\hat{\mathbf{x}}) + \mathbf{B}_u\mathbf{u}\,, \quad \mathbf{K}(t) = \mathbf{P}(t)\mathbf{C}^T\mathbf{V}^{-1} \tag{6.5.51}$$

where the matrix $\mathbf{K}$ is called as the ***gain matrix***. The above derivation applies also to linear time-variant system.

The numerical solution of Riccati differential equation is very important for applications. Comparing to prediction, whose equation for variance matrix is Lyapunov differential equation. Filtering holds the measurement, so that the additional term $\mathbf{PC}^T\mathbf{V}^{-1}\mathbf{CP}$ appears in the equation, which is a quadratic function of $\mathbf{P}(t)$ and then the Riccati differential equation for variance matrix is non-linear. However, it is derived from the linear dual equations of a linear system. Based on the analogy between structural mechanics and optimal control, the solution variance matrix of the Riccati differential equation corresponds to the ***flexibility matrix at the end*** t ***of interval*** $[0,t)$. As was seen in Chapter 5, for time-invariant system the flexibility matrix can be solved by the precise integration method. The flexibility matrix can also be found by the analytical method based on all the eigen-solutions of the dual equations, as is given in section 5.8. For filtering, the analytical method is similar, and will be considered later. Not only the Riccati differential equation can be solved analytically, but also the filter equation can find the analytical solution, which is quite useful and is not known before.

The filter solution considers always the front time station $\tau = t$. The length of interval $[0,t)$ is ever increasing with the time t. The corresponding smoothing solution can be denoted as $\bar{\mathbf{x}}(\tau,t)$, with $\tau < t$, which will be described in section 6.6.

Riccati differential equation can also be applied to infinite horizon problem, *i.e.* $t_f \to \infty$. The solution has a transient stage near by $t = 0$, afterwards when $t \to \infty$, $\mathbf{P}(t) \to \mathbf{P}_\infty$. The limit matrix $\mathbf{P}_\infty$ satisfies the algebraic Riccati equation

$$\mathbf{B}_w\mathbf{W}\mathbf{B}_w^T + \mathbf{A}\mathbf{P}_\infty + \mathbf{P}_\infty\mathbf{A}^T - \mathbf{P}_\infty\mathbf{C}^T\mathbf{V}^{-1}\mathbf{C}\mathbf{P}_\infty = \mathbf{0} \tag{6.5.52}$$

Disregarding the transient stage of process, the gain matrix $\mathbf{K}_\infty = \mathbf{P}_\infty\mathbf{C}^T\mathbf{V}^{-1}$ in the equation (6.5.49a) is time-invariant. Then in the filter equation (6.5.49a), the coefficient matrix $(\mathbf{A} - \mathbf{K}_\infty\mathbf{C})$ is also time-invariant, so that the precise integration method is simply applied in solving the filter equation. With the time step size η is given, the transition matrix

$$\mathbf{\Phi}_\infty = \exp[(\mathbf{A} - \mathbf{K}_\infty\mathbf{C})\eta] \tag{6.5.53}$$

can be computed beforehand, so that the real time computation needs only the matrix-vector multiplication. The matrix $\mathbf{\Phi}_\infty$ is simply only for infinite horizon time-invariant system, however for finite horizon time-invariant system, although the coefficient matrix $(\mathbf{A}-\mathbf{KC})$ is time-variant, the precise integration method still applies. Certainly, the derivation is not so simple as infinite horizon case, see section 6.5.7 later.

Note that the algebraic Riccati equation (6.5.52) is quadratic to the unknown matrix $\mathbf{P}_\infty$, hence the solution is not unique. However, the required $\mathbf{P}_\infty$ is a ***symmetric positive definite matrix***, a very important condition, under which the solution is unique. Analytical solution of Riccati differential equation will be given in later sections based on all eigen-solutions of the corresponding Hamilton matrix, the matrix $\mathbf{P}_\infty$ corresponding to the class (β) eigen-solutions, see equation (6.5.117). The solution of eigen-problem of a Hamilton matrix can be found in section 5.3.

Inspection of equations (6.5.49) and (6.5.50) determines that the matrix $\mathbf{P}(t)$ is unrelated to the measurement $\mathbf{y}$, and that the state mean value $\hat{\mathbf{x}}(t)$ depends on the $\mathbf{y}$ linearly. Hence the integration of $\hat{\mathbf{x}}$ with given $\mathbf{y}$ can apply the superposition principle, which is very important for precise integration of $\hat{\mathbf{x}}(t)$.

§6.5.3.1, Correlated dynamic and measurement noises

For considering the input of colored noise, the analysis of noise correlation between $\mathbf{w}$ and $\mathbf{v}$ is needed. The equation (6.5.42c) should be updated as

$$\mathrm{covar}[\mathbf{w}(t),\mathbf{v}(\tau)] = \mathbf{S}(t)\delta(t-\tau) \tag{6.5.42'c}$$

and the index J should be minimized

$$J=\int_0^t \frac{1}{2}\begin{Bmatrix}\mathbf{w}\\ \mathbf{v}\end{Bmatrix}^T \mathbf{W}_e^{-1}\begin{Bmatrix}\mathbf{w}\\ \mathbf{v}\end{Bmatrix}\mathrm{d}\tau+(\mathbf{x}_0-\hat{\mathbf{x}}_0)^T\mathbf{P}_0^{-1}(\mathbf{x}_0-\hat{\mathbf{x}}_0)/2,\quad \mathbf{W}_e=\begin{bmatrix}\mathbf{W} & \mathbf{S}\\ \mathbf{S}^T & \mathbf{V}\end{bmatrix} \tag{6.5.54}$$

$$\min_{\mathbf{x}} J$$

which is still a conditional minimization, that the dynamic and measurement equations must be satisfied beforehand. The energy of noise must be positive definite, so that $\mathbf{W}_e$ is symmetric and positive definite. To compose the quadratic index, using the blocked inverse matrix equation (6.5.29), and then introducing the Lagrange multiplier, it gives

$$J_e=\int_0^t\begin{bmatrix}\boldsymbol{\lambda}^T(\dot{\mathbf{x}}-\mathbf{A}\mathbf{x}-\mathbf{B}_w\mathbf{w}-\mathbf{B}_u\mathbf{u})+\mathbf{w}^T(\mathbf{W}-\mathbf{S}\mathbf{V}^{-1}\mathbf{S}^T)^{-1}\mathbf{w}/2\\ +(\mathbf{y}-\mathbf{C}\mathbf{x})^T(\mathbf{V}-\mathbf{S}^T\mathbf{W}^{-1}\mathbf{S})^{-1}(\mathbf{y}-\mathbf{C}\mathbf{x})/2\\ -\mathbf{w}^T\mathbf{W}^{-1}\mathbf{S}(\mathbf{V}-\mathbf{S}^T\mathbf{W}^{-1}\mathbf{S})^{-1}(\mathbf{y}-\mathbf{C}\mathbf{x})\end{bmatrix}\mathrm{d}\tau \tag{6.5.55}$$

$$+(\mathbf{x}_0-\hat{\mathbf{x}}_0)^T\mathbf{P}_0^{-1}(\mathbf{x}_0-\hat{\mathbf{x}}_0)/2,\qquad \delta J_e=0$$

which has been an unconditional variational principle with three kinds of variables $\mathbf{x}$, $\boldsymbol{\lambda}$ and $\mathbf{w}$. Minimizing with respect to $\mathbf{w}$ first and using the identity $(\mathbf{W}-\mathbf{S}\mathbf{V}^{-1}\mathbf{S}^T)\mathbf{W}^{-1}\mathbf{S}(\mathbf{V}-\mathbf{S}^T\mathbf{W}^{-1}\mathbf{S})^{-1}\equiv\mathbf{S}\mathbf{V}^{-1}$, it is derived as

$$\mathbf{w}=(\mathbf{W}-\mathbf{S}\mathbf{V}^{-1}\mathbf{S}^T)\mathbf{B}_w^T\boldsymbol{\lambda}+\mathbf{S}\mathbf{V}^{-1}(\mathbf{y}-\mathbf{C}\mathbf{x}) \tag{6.5.56}$$

substituting it back into equation (6.5.55) derives

$$J_e = \int_0^{t_f}[\boldsymbol{\lambda}^T(\dot{\mathbf{x}} - \mathbf{A}\mathbf{x} - \mathbf{B}_u\mathbf{u}) + (\mathbf{y} - \mathbf{C}\mathbf{x})^T\mathbf{V}^{-1}(\mathbf{y} - \mathbf{C}\mathbf{x})/2 - \boldsymbol{\lambda}^T\mathbf{B}_w\mathbf{S}\mathbf{V}^{-1}(\mathbf{y} - \mathbf{C}\mathbf{x}) \\ - \boldsymbol{\lambda}^T\mathbf{B}_w(\mathbf{W} - \mathbf{S}\mathbf{V}^{-1}\mathbf{S}^T)\mathbf{B}_w^T\boldsymbol{\lambda}/2]\mathrm{d}\tau + (\mathbf{x}_0 - \hat{\mathbf{x}}_0)^T\mathbf{P}_0^{-1}(\mathbf{x}_0 - \hat{\mathbf{x}}_0)/2, \quad \delta J_e = 0 \tag{6.5.55'}$$

where the matrix inversion lemma (6.5.30) is used in the derivation. Equation (6.5.55') is a variational principle with two kinds of variables $\mathbf{x}, \boldsymbol{\lambda}$. The dual equations derived from which are

$$\dot{\mathbf{x}} = (\mathbf{A} - \mathbf{J}\mathbf{C})\mathbf{x} + \mathbf{B}_w(\mathbf{W} - \mathbf{S}\mathbf{V}^{-1}\mathbf{S}^T)\mathbf{B}_w^T\boldsymbol{\lambda} + \mathbf{B}_u\mathbf{u} + \mathbf{J}\mathbf{y} \tag{6.5.57a}$$

$$\dot{\boldsymbol{\lambda}} = \mathbf{C}^T\mathbf{V}^{-1}\mathbf{C}\mathbf{x} - (\mathbf{A} - \mathbf{J}\mathbf{C})^T\boldsymbol{\lambda} - \mathbf{C}^T\mathbf{V}^{-1}\mathbf{y}, \quad [\mathbf{J} = \mathbf{B}_w\mathbf{S}\mathbf{V}^{-1}] \tag{6.5.57b}$$

From the positive definiteness of matrix $\mathbf{W}_e$, the matrix $(\mathbf{W} - \mathbf{S}\mathbf{V}^{-1}\mathbf{S}^T)$ is also positive definite. Because the stochastic process $\mathbf{x}(t)$ is Gaussian, so that its solution form is still of (6.5.48), then the derivation is as follows

$$\dot{\mathbf{x}} = \dot{\hat{\mathbf{x}}} + \dot{\mathbf{P}}\boldsymbol{\lambda} + \mathbf{P}\dot{\boldsymbol{\lambda}} = (\mathbf{A} - \mathbf{J}\mathbf{C})\hat{\mathbf{x}} + (\mathbf{A} - \mathbf{J}\mathbf{C})\mathbf{P}\boldsymbol{\lambda} + \mathbf{B}_w(\mathbf{W} - \mathbf{S}\mathbf{V}^{-1}\mathbf{S}^T)\mathbf{B}_w^T\boldsymbol{\lambda} + \mathbf{B}_u\mathbf{u} + \mathbf{J}\mathbf{y}$$

$$\mathbf{P}\dot{\boldsymbol{\lambda}} = \mathbf{P}\mathbf{C}^T\mathbf{V}^{-1}\mathbf{C}\hat{\mathbf{x}} + \mathbf{P}\mathbf{C}^T\mathbf{V}^{-1}\mathbf{C}\mathbf{P}\boldsymbol{\lambda} - \mathbf{P}(\mathbf{A} - \mathbf{J}\mathbf{C})\boldsymbol{\lambda} - \mathbf{P}\mathbf{C}^T\mathbf{V}^{-1}\mathbf{y}$$

Eliminating $\mathbf{P}\dot{\boldsymbol{\lambda}}$ gives

$$\dot{\hat{\mathbf{x}}} + \dot{\mathbf{P}}\boldsymbol{\lambda} = (\mathbf{A} - \mathbf{J}\mathbf{C} - \mathbf{P}\mathbf{C}^T\mathbf{V}^{-1}\mathbf{C})\hat{\mathbf{x}} + \mathbf{P}\mathbf{C}^T\mathbf{V}^{-1}\mathbf{y} + \mathbf{B}_u\mathbf{u} + \mathbf{J}\mathbf{y} \\ + \mathbf{B}_w(\mathbf{W} - \mathbf{S}\mathbf{V}^{-1}\mathbf{S}^T)\mathbf{B}_w^T\boldsymbol{\lambda} + (\mathbf{A} - \mathbf{J}\mathbf{C})\mathbf{P}\boldsymbol{\lambda} + \mathbf{P}(\mathbf{A} - \mathbf{J}\mathbf{C})^T\boldsymbol{\lambda} - \mathbf{P}\mathbf{C}^T\mathbf{V}^{-1}\mathbf{C}\mathbf{P}\boldsymbol{\lambda}$$

This differential equation is composed of deterministic and stochastic parts of a Gaussian process. Hence, differential equations for deterministic mean value $\hat{\mathbf{x}}(t)$ and for variance $\mathbf{P}(t)$ matrix of the stochastic terms are derived separately as

$$\dot{\hat{\mathbf{x}}} = \mathbf{A}\hat{\mathbf{x}} + \mathbf{B}_u\mathbf{u} + \mathbf{K}(\mathbf{y} - \mathbf{C}\hat{\mathbf{x}}), \tag{6.5.58}$$

where

$$\mathbf{K} = \mathbf{P}\mathbf{C}^T\mathbf{V}^{-1} + \mathbf{J} \tag{6.5.59}$$

is the ***gain matrix***, and

$$\dot{\mathbf{P}} = \mathbf{B}_w(\mathbf{W} - \mathbf{S}\mathbf{V}^{-1}\mathbf{S}^T)\mathbf{B}_w^T + (\mathbf{A} - \mathbf{J}\mathbf{C})\mathbf{P} + \mathbf{P}(\mathbf{A} - \mathbf{J}\mathbf{C})^T - \mathbf{P}\mathbf{C}^T\mathbf{V}^{-1}\mathbf{C}\mathbf{P} \tag{6.5.60}$$

with the initial conditions

$$\hat{\mathbf{x}}(0) = \hat{\mathbf{x}}_0, \qquad \mathbf{P}(0) = \mathbf{P}_0 \tag{6.5.43'}$$

The above proposes the fundamental differential equations for the case of $\mathbf{w}$ and $\mathbf{v}$ being correlated. The matrix $\mathbf{J} = \mathbf{B}_w\mathbf{S}\mathbf{V}^{-1}$ represents the correlation term, that the matrices $\mathbf{A}$ and $\mathbf{W}$ in Riccati equation are changed as $(\mathbf{A} - \mathbf{J}\mathbf{C})$ and $(\mathbf{W} - \mathbf{S}\mathbf{V}^{-1}\mathbf{S}^T)$, respectively. Note that, the above derivation is also valid for time-variant system.

For time-invariant infinite horizon filter, the initial condition is unrelated. The Riccati differential equation reduces to be the algebraic Riccati equation, which is obtained by letting $\dot{\mathbf{P}} = \mathbf{0}$ in equation (6.5.60). Making use of the precise integration method, its solution can be obtained as the limit of solution of Riccati differential equation as $t_f \to \infty$, *i.e.* $\mathbf{P}_\infty = \mathbf{P}(t \to \infty)$. Infinite horizon problem is a long-term effect, that the initial value $\mathbf{P}_0$ induced transient effect decays. The respective gain matrix $\mathbf{K}_\infty$ is also time-invariant, and the precise integration method applies too. Precise integration method gives the stable numerical result closed to the computer precision, which is important for applications.

§6.5.3.2, Continuous-time Kalman-Bucy filter under colored noises

Disturbances are from dynamic and measurement noises $\mathbf{w}(t)$ and $\mathbf{v}(t)$, two sources respectively. In the analysis above, the assumption for correlation functions are (6.5.42a~c) and/or (6.5.42'c), the white noise has the factor $\delta(t-\tau)$. However white noise is an approximation, the real noise is always colored.

In the analysis of prediction, section 6.4.6 gives the method for colored noise $\mathbf{w}$, which is driven by a white noise from another linear system, and an extended state space method is applied. For the present case, if the colored noise $\mathbf{w}$ is driven by the same method but $\mathbf{v}$ is assumed still white, then the extended state space method in 6.4.6 can also be transplanted here. Hence in this section, only the case of $\mathbf{w}(t)$ being white and the measurement noise $\mathbf{v}(t)$ being colored is considered. The model of colored $\mathbf{v}(t)$ is still assumed driven by a white noise $\mathbf{r}(t)$ via another linear system. The system mathematical model is given as

$$\dot{\mathbf{x}} = \mathbf{A}\mathbf{x} + \mathbf{B}\mathbf{w} \tag{6.5.61a}$$

$$\mathbf{y} = \mathbf{C}\mathbf{x} + \mathbf{v} \tag{6.5.61b}$$

$$\dot{\mathbf{v}} = \mathbf{F}\mathbf{v} + \mathbf{H}\mathbf{r} \tag{6.5.61c}$$

where $\mathbf{u} = \mathbf{0}$ is assumed, which has limited influence to the analysis, and $\mathbf{r}$ is a white noise to drive the measurement noise $\mathbf{v}$, and $\mathbf{r}$ is not correlated with $\mathbf{w}$.

The set (6.5.61a~c) are linear differential equations, differentiating equation (6.5.61b) gives

$$\dot{\mathbf{v}} = \dot{\mathbf{y}} - \mathbf{C}\dot{\mathbf{x}} - \dot{\mathbf{C}}\mathbf{x} = \dot{\mathbf{y}} - (\mathbf{C}\mathbf{A} + \dot{\mathbf{C}})\mathbf{x} - \mathbf{C}\mathbf{B}\mathbf{w}$$

Eliminating $\dot{\mathbf{v}}$ with (6.5.61c), then eliminating $\dot{\mathbf{x}}$ with (6.5.61a) gives

$$\dot{\mathbf{y}} = \mathbf{F}\mathbf{v} + (\mathbf{C}\mathbf{A} + \dot{\mathbf{C}})\mathbf{x} + \mathbf{C}\mathbf{B}\mathbf{w} + \mathbf{H}\mathbf{r}$$

Using equation (6.5.61b) again, eliminating $\mathbf{v}$ gives

$$\mathbf{z} = \mathbf{C}_z\mathbf{x} + \mathbf{v}_z, \quad \text{where } \mathbf{z} = \dot{\mathbf{y}} - \mathbf{F}\mathbf{y} \tag{6.5.62a}$$

$$\mathbf{C}_z = \mathbf{C}\mathbf{A} + \dot{\mathbf{C}} - \mathbf{F}\mathbf{C} \tag{6.5.62b}$$

$$\mathbf{v}_z = \mathbf{C}\mathbf{B}\mathbf{w} + \mathbf{H}\mathbf{r} \tag{6.5.62c}$$

Therefore (6.5.62a) becomes a new measurement equation, where the vector $\mathbf{v}_z$ is the corresponding measurement white noise with mean value zero and variance matrix

$$\mathrm{var}(\mathbf{v}_z(t), \mathbf{v}_z(\tau)) = \mathbf{V}_z\delta(t-\tau), \quad \mathbf{V}_z = \mathbf{C}\mathbf{B}\mathbf{W}\mathbf{B}^T\mathbf{C}^T + \mathbf{H}\mathbf{V}_r\mathbf{H}^T \tag{6.5.63a}$$

where $\mathbf{V}_r$ is the intensity matrix of the driving white noise $\mathbf{r}(t)$. The $\mathbf{v}_z$ and $\mathbf{w}$ are correlated white noises and the covariance matrix is

$$\mathrm{covar}[\mathbf{w}(t), \mathbf{v}_z(t)] = \mathbf{S}_z\delta(t-\tau), \quad \mathbf{S}_z = \mathbf{W}\mathbf{B}^T\mathbf{C}^T \tag{6.5.63b}$$

Thus, the equations (6.5.61a) and (6.5.62a) compose the system dynamic equation and the measurement equation, respectively. However, the measurement noise $\mathbf{v}_z$ and the dynamic noise $\mathbf{w}$ are correlated, so that the method given in previous section 6.5.3.1 can be applied. The same derivation gives the filter and Riccati differential equations and the gain matrix etc. as

$$\dot{\hat{\mathbf{x}}} = \mathbf{A}\hat{\mathbf{x}} + \mathbf{K}[\dot{\mathbf{y}} - \mathbf{F}\mathbf{y} - (\mathbf{C}\mathbf{A} + \dot{\mathbf{C}} - \mathbf{F}\mathbf{C})\hat{\mathbf{x}}] \tag{6.5.64}$$

$$\mathbf{K} = [\mathbf{P} \cdot (\mathbf{CA} + \dot{\mathbf{C}} - \mathbf{FC})^T + \mathbf{BWB}^T\mathbf{C}^T] \cdot (\mathbf{CBWB}^T\mathbf{C}^T + \mathbf{HVH}^T)^{-1} \tag{6.5.65}$$

$$\mathbf{J} = \mathbf{BWB}^T\mathbf{C}(\mathbf{CBWB}^T\mathbf{C}^T + \mathbf{HVH}^T)^{-1} \tag{6.5.66}$$

$$\dot{\mathbf{P}} = \mathbf{B}(\mathbf{W} - \mathbf{S}_z\mathbf{V}_z^{-1}\mathbf{S}_z^T)\mathbf{B}^T + (\mathbf{A} - \mathbf{JC}_z)\mathbf{P} + \mathbf{P}(\mathbf{A} - \mathbf{JC}_z)^T + \mathbf{PC}_z^T\mathbf{V}_z^{-1}\mathbf{C}_z\mathbf{P} \tag{6.5.67}$$

The above derivation gives the fundamental equations for some cases of filter problems. Because of the Gaussian distribution, the mean value vector $\hat{\mathbf{x}}(t)$ and the variance matrix $\mathbf{P}(t)$ are always derived to solve the corresponding differential equations. The differential equation for the mean vector $\hat{\mathbf{x}}$ is linear, and the inhomogeneous term is the linear combination of the measurement $\mathbf{y}(\tau)$, hence $\mathbf{y}$ looked a driven 'external force'. The linear property comes from that the system is originally linear. For the variance matrix $\mathbf{P}(t)$, the equations (6.5.50), (6.5.60) and (6.5.67) etc. are always the Riccati differential equation, which is independent on the measurement vector $\mathbf{y}$. This explains that it is the system characteristics. If the strength of white noise is very large, *i.e.* $\mathbf{V} \to \infty$, which implies that the measurement result being useless. Then the quadratic term in the Riccati differential tends a null matrix, and then the Riccati differential equation is reduced to be a Lyapunov equation, so that the problem reduced to be only a prediction, as anticipated.

Riccati differential equation itself is non-linear, but it is derived from a linear system. Looking from applied mechanics side, the solution matrix corresponds to the ***end flexibility matrix***. Finding the precise integration solution matrix of the Riccati differential equation is a meaningful problem. Certainly, if all the eigen-solutions of the corresponding Hamilton matrix are solved then the solution can be solved analytically. The situation is entirely similar to that given in chapter 5, and the solution is determined from the analogy relationship between structural mechanics and optimal control. Below, the precise integration method is described first, and then the analytical solution method is given afterwards.

§6.5.4, Interval mixed energy

The fundamental differential equations for Kalman-Bucy filter are derived above, that the equation (6.5.49) for the mean vector $\hat{\mathbf{x}}$ and the equation (6.5.50) for the variance matrix are required to solve. However, the traditional finite difference approximation has the problem of error prone even for time invariant linear differential equation and is not so favorable. Using precise integration method, the numerical result can reach the full computer precision, so that the solution of Riccati differential equation (6.5.50) should also use the precise integration method. Precise integration method needs to introduce the ***interval mixed energy***, which is also useful in dealing with the computation of smoothing problems in section 6.6. Note that the interval mixed energy has been heavily used in Chapter 5.

Time step integration should have a time step size denoted as η, and then the grid points are

$$t_0 = 0, \quad t_1 = \eta, \quad \cdots, \quad t_k = k\eta, \quad \cdots \tag{6.5.68}$$

Precise integration no longer uses the finite difference approximation for this time interval η. Instead, the interval mixed energy for a given time interval (t_a, t_b) is introduced as follows

$$V(\mathbf{x}_a, \boldsymbol{\lambda}_b) = \boldsymbol{\lambda}_b^T \mathbf{x}_b - \int_{t_a}^{t_b} [\boldsymbol{\lambda}^T \dot{\mathbf{x}} - \mathrm{H}(\mathbf{x}, \boldsymbol{\lambda}) - \mathbf{x}^T \mathbf{C}^T \mathbf{V}^{-1} \mathbf{y} - \boldsymbol{\lambda}^T \mathbf{B}_u \mathbf{u}] \mathrm{d}t \tag{6.5.69}$$

$$\mathrm{H}(\mathbf{x}, \boldsymbol{\lambda}) = \boldsymbol{\lambda}^T \mathbf{A}\mathbf{x} + \boldsymbol{\lambda}^T \mathbf{B}\mathbf{W}\mathbf{B}^T \boldsymbol{\lambda}/2 - \mathbf{x}^T \mathbf{C}^T \mathbf{V}^{-1} \mathbf{C}\mathbf{x}/2 \tag{6.5.70}$$

where the matrix $\mathbf{B}_w$ is simply written as $\mathbf{B}$. Equation (6.5.69) defined mixed energy V is a function of state vector $\mathbf{x}_a$ at time t_a and the dual vector $\boldsymbol{\lambda}_b$ at time t_b, where $t_0 \le t_a < t_b \le t_f$. Measurement $\mathbf{y}$ is a given vector function. In the interval (t_a, t_b), the vectors $\mathbf{x}$ and $\boldsymbol{\lambda}$ should make the functional V be a stationary value, *i.e.*

$$\delta V(\mathbf{x}_a, \boldsymbol{\lambda}_b) = (\delta\boldsymbol{\lambda}_b)^T \cdot \mathbf{x}_b + \boldsymbol{\lambda}_b^T \cdot \delta\mathbf{x}_b - \int_{t_a}^{t_b} [(\delta\boldsymbol{\lambda})^T \cdot (\dot{\mathbf{x}} - \mathbf{A}\mathbf{x} - \mathbf{B}\mathbf{W}\mathbf{B}^T\boldsymbol{\lambda} - \mathbf{B}_u \mathbf{u})$$

$$+ \delta\mathbf{x}^T (-\dot{\boldsymbol{\lambda}} + \mathbf{C}^T \mathbf{V}^{-1} \mathbf{C}\mathbf{x} - \mathbf{A}^T \boldsymbol{\lambda} - \mathbf{C}^T \mathbf{V}^{-1} \mathbf{y})] \mathrm{d}t - [\boldsymbol{\lambda}^T \cdot \delta\mathbf{x}]_{t_a}^{t_b}$$

where the variations $\delta\boldsymbol{\lambda}$ and $\delta\mathbf{x}$ are arbitrarily selected in the time interval, which gives the dual differential equations

$$\dot{\mathbf{x}} = \mathbf{A}\mathbf{x} + \mathbf{B}\mathbf{W}\mathbf{B}^T\boldsymbol{\lambda} + \mathbf{B}_u \mathbf{u} \tag{6.5.47'a}$$

$$\dot{\boldsymbol{\lambda}} = \mathbf{C}^T \mathbf{V}^{-1} \mathbf{C}\mathbf{x} - \mathbf{A}^T \boldsymbol{\lambda} - \mathbf{C}^T \mathbf{V}^{-1} \mathbf{y} \tag{6.5.47'b}$$

where $\mathbf{B} = \mathbf{B}_w$. Therefore

$$\delta V(\mathbf{x}_a, \boldsymbol{\lambda}_b) = \mathbf{x}_b^T \cdot \delta\boldsymbol{\lambda}_b + \boldsymbol{\lambda}_a^T \cdot \delta\mathbf{x}_a \equiv (\partial V/\partial\boldsymbol{\lambda}_b)^T \delta\boldsymbol{\lambda}_b + (\partial V/\partial\mathbf{x}_a)^T \cdot \delta\mathbf{x}_a \tag{6.5.71}$$

which derives

$$\mathbf{x}_b = \partial V/\partial\boldsymbol{\lambda}_b, \qquad \boldsymbol{\lambda}_a = \partial V/\partial\mathbf{x}_a \tag{6.5.71a}$$

From the definition of interval mixed energy (6.5.69), it is seen that the mixed energy $V(\mathbf{x}_a, \boldsymbol{\lambda}_b)$ is a quadratic function of the arguments $\mathbf{x}_a$ and $\boldsymbol{\lambda}_b$, and the linear terms are induced of the measurement vector $\mathbf{y}$. The general form of a quadratic function is

$$V(\mathbf{x}_a, \boldsymbol{\lambda}_b) = \boldsymbol{\lambda}_b^T \mathbf{F}\mathbf{x}_a + \boldsymbol{\lambda}_b^T \mathbf{G}\boldsymbol{\lambda}_b/2 - \mathbf{x}_a^T \mathbf{Q}\mathbf{x}_a/2 + \boldsymbol{\lambda}_b^T \mathbf{r}_x + \mathbf{x}_a^T \mathbf{r}_\lambda \tag{6.5.72}$$

where $\mathbf{Q}, \mathbf{F}, \mathbf{G}$ are $n \times n$ matrices, $\mathbf{Q}^T = \mathbf{Q}$, $\mathbf{G}^T = \mathbf{G}$. These three matrices determine the quadratic terms, and $\mathbf{r}_x, \mathbf{r}_\lambda$ are n-dimensional vectors, which determine the linear term. The matrices $\mathbf{Q}, \mathbf{F}, \mathbf{G}$ relate only to the system matrices $\mathbf{A}$, $\mathbf{C}^T \mathbf{V}^{-1} \mathbf{C}$ and $\mathbf{B}\mathbf{W}\mathbf{B}^T$, and $\mathbf{r}_x, \mathbf{r}_\lambda$ linearly relate to the measurement vector $\mathbf{y}$ and the control vector $\mathbf{u}$. Substituting (6.5.72) into (6.5.71a) derives the interval dual equations

$$\mathbf{x}_b = \mathbf{F}\mathbf{x}_a + \mathbf{G}\boldsymbol{\lambda}_b + \mathbf{r}_x \tag{6.5.73a}$$

$$\boldsymbol{\lambda}_a = -\mathbf{Q}\mathbf{x}_a + \mathbf{F}^T \boldsymbol{\lambda}_b + \mathbf{r}_\lambda \tag{6.5.73b}$$

where $\mathbf{Q}, \mathbf{F}, \mathbf{G}$ and $\mathbf{r}_x, \mathbf{r}_\lambda$ are functions of t_a and t_b, such as $\mathbf{Q} = \mathbf{Q}(t_a, t_b)$ etc. The boundary conditions are derived as

$$\mathbf{G} \to \mathbf{0}, \quad \mathbf{Q} \to \mathbf{0}, \quad \mathbf{F} \to \mathbf{I}_n, \quad \mathbf{r}_x \to \mathbf{0}, \quad \mathbf{r}_\lambda \to \mathbf{0}, \quad \text{when } t_b \to t_a \tag{6.5.74}$$

Above is mathematical derivation, however, physical interpretation is beneficial.

Vector $\mathbf{x}_a$ is the state (displacement) at the end-t_a and $\boldsymbol{\lambda}_b$ is the 'force' vector at the end-t_b. Vector $\mathbf{r}_x$ is $\mathbf{y}$ induced state (displacement) at the end-t_b under the conditions of $\mathbf{x}_a = \mathbf{0}$ and $\boldsymbol{\lambda}_b = \mathbf{0}$; and the vector $\mathbf{r}_\lambda$ is $\mathbf{y}$ induced force at end-t_a under also the boundary conditions of $\mathbf{x}_a = \mathbf{0}, \boldsymbol{\lambda}_b = \mathbf{0}$. The $\mathbf{F}$ is a ***transfer matrix***, under the conditions of $\mathbf{y} = \mathbf{0}$ and $\boldsymbol{\lambda}_b = \mathbf{0}$, whose columns give the state vector $\mathbf{x}_b$ at the end-t_b induced from the boundary condition of the state $\mathbf{x}_a$ being given as the columns of $\mathbf{I}_n$ at end-t_a. The $\mathbf{G}$ is a ***flexibility matrix*** at end-t_b that $\mathbf{x}_b = \mathbf{G}\boldsymbol{\lambda}_b$ is interpreted as the force $\boldsymbol{\lambda}_b$ induced state vector $\mathbf{x}_b$. The $\mathbf{Q}$ is a ***stiffness matrix*** at end-t_a that $\boldsymbol{\lambda}_a = -\mathbf{Q}\mathbf{x}_a$ is interpreted as the displacement $\mathbf{x}_a$ induced force vector $\boldsymbol{\lambda}_a$. Physical interpretation is easier for structural mechanics arguments.

§6.5.4.1, Interval combination

Algebraically, the operation of interval mixed energy is ***interval combination***. Two contiguous time intervals (t_a, t_b) and (t_b, t_c), marked with 1 and 2 respectively, can be combined together becoming a longer time interval (t_a, t_c). The corresponding interval matrices $\mathbf{Q}, \mathbf{F}, \mathbf{G}$ can be marked with subscripts $1, 2, c$ respectively. See figure 6.5

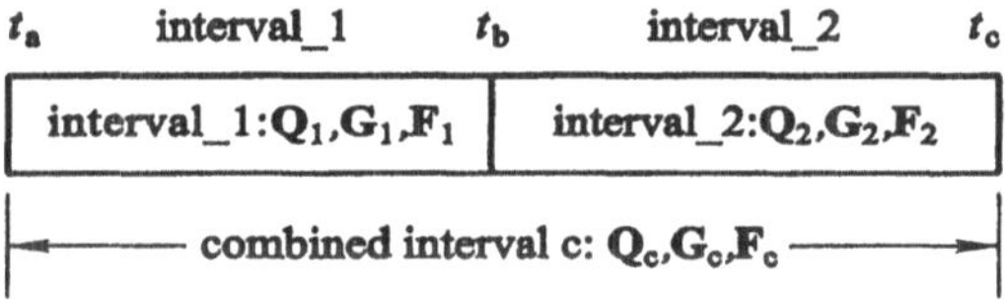

Figure 6.5, *Interval combination*

The mixed energy V_c of the combined time interval (t_a, t_c) is composed of time intervals 1 and 2

$$V_c(\mathbf{x}_a, \boldsymbol{\lambda}_c) = \min_{\boldsymbol{\lambda}_b} \max_{\mathbf{x}_b} [V_1(\mathbf{x}_a, \boldsymbol{\lambda}_b) + V_2(\mathbf{x}_b, \boldsymbol{\lambda}_c) - \boldsymbol{\lambda}_b^T \mathbf{x}_b] \tag{6.5.75}$$

which is obtained from elimination of $\mathbf{x}_b$ and $\boldsymbol{\lambda}_b$, *i.e.* $\min_{\boldsymbol{\lambda}_b} \max_{\mathbf{x}_b}$. Using (6.5.73) gives

$$\boldsymbol{\lambda}_a = -\mathbf{Q}_1 \mathbf{x}_a + \mathbf{F}_1^T \boldsymbol{\lambda}_b \mathbf{r}_{\lambda 1}, \quad \mathbf{x}_b = \mathbf{F}_1 \mathbf{x}_a + \mathbf{G}_1 \boldsymbol{\lambda}_b + \mathbf{r}_{x1}$$

$$\boldsymbol{\lambda}_b = -\mathbf{Q}_2 \mathbf{x}_b + \mathbf{F}_2^T \boldsymbol{\lambda}_c \mathbf{r}_{\lambda 2}, \quad \mathbf{x}_c = \mathbf{F}_2 \mathbf{x}_b + \mathbf{G}_2 \boldsymbol{\lambda}_c + \mathbf{r}_{x2}$$

where the simultaneous equations of $\mathbf{x}_b =$ and $\boldsymbol{\lambda}_b =$ can be solved as

$$\mathbf{x}_b = (\mathbf{I}_n + \mathbf{G}_1 \mathbf{Q}_2)^{-1} (\mathbf{F}_1 \mathbf{x}_a + \mathbf{G}_1 \mathbf{F}_2 \boldsymbol{\lambda}_c + \mathbf{r}_{x1} + \mathbf{G}_1 \mathbf{r}_{\lambda 2}) \tag{6.5.76a}$$

$$\boldsymbol{\lambda}_b = (\mathbf{I}_n + \mathbf{Q}_2 \mathbf{G}_1)^{-1} (-\mathbf{Q}_2 \mathbf{F}_1 \mathbf{x}_a + \mathbf{F}_2^T \boldsymbol{\lambda}_c - \mathbf{Q}_2 \mathbf{r}_{x1} + \mathbf{r}_{\lambda 2}) \tag{6.5.76b}$$

Substituting back into equation (6.5.75), or substituting into the equations for $\boldsymbol{\lambda}_a$ and $\mathbf{x}_c$ gives

$$\mathbf{Q}_c = \mathbf{Q}_1 + \mathbf{F}_1^T(\mathbf{Q}_2^{-1} + \mathbf{G}_1)^{-1}\mathbf{F}_1 \tag{6.5.77a}$$

$$\mathbf{G}_c = \mathbf{G}_2 + \mathbf{F}_2(\mathbf{G}_1^{-1} + \mathbf{Q}_2)^{-1}\mathbf{F}_2^T \tag{6.5.77b}$$

$$\mathbf{F}_c = \mathbf{F}_2(\mathbf{I}_n + \mathbf{G}_1\mathbf{Q}_2)^{-1}\mathbf{F}_1 \tag{6.5.77c}$$

$$\mathbf{r}_{\lambda c} = \mathbf{r}_{\lambda 1} + \mathbf{F}_1^T(\mathbf{I}_n + \mathbf{Q}_2\mathbf{G}_1)^{-1}(\mathbf{r}_{\lambda 2} - \mathbf{E}_2\mathbf{r}_{x1}) \tag{6.5.78a}$$

$$\mathbf{r}_{xc} = \mathbf{r}_{x2} + \mathbf{F}_2(\mathbf{I}_n + \mathbf{G}_1\mathbf{Q}_2)^{-1}(\mathbf{r}_{x1} + \mathbf{G}_1\mathbf{r}_{\lambda 2}) \tag{6.5.78b}$$

The combined interval matrices and vectors $\mathbf{r}$ of the mixed energy V_c are obtained from the composition of intervals 1 and 2. The equation (6.5.77) and (6.5.78) are the interval combination and elimination equations, see also [127], and the equations (6.5.76a,b) are called as ***back substitution*** of the internal dual vectors, which is quite useful in solving the smoothing problem.

It is noted that the combined interval matrices $\mathbf{Q}_c, \mathbf{G}_c, \mathbf{F}_c$ are obtained only from the composition sub-interval matrices $\mathbf{Q}_1, \mathbf{G}_1, \mathbf{F}_1$ and $\mathbf{Q}_2, \mathbf{G}_2, \mathbf{F}_2$ but unrelated to the measurement $\mathbf{y}$.

The interval combination is a kind of operation, which is associative. Suppose there are three contiguous intervals, as shown in figure 6.6, to be combined as one interval c. This can be performed with two processes of successive interval combinations. The first is carrying out the combination of intervals 1 and 2 to obtain the interval a, thereafter the interval a is combined with interval 3 to obtain the interval c. The second process is to execute the interval combination of intervals 2 and 3 to obtain the interval b, thereafter the intervals 1 and b are combined to obtain the interval c. The difference between the two combination processes is the order of combination. Examining the operations in the interval combination, there is only matrix multiplication, inversion and addition. According to that the matrix multiplication satisfying the associative rule $(\mathbf{AB})\mathbf{C} = \mathbf{A}(\mathbf{BC})$, it is determined that the interval combination satisfies the ***associative*** rule too.

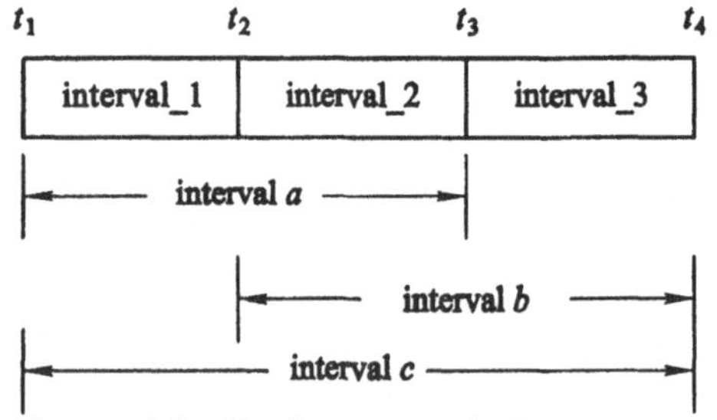

Figure 6.6, *Combination of three contiguous intervals*

Direct verification of the associative rule of interval combination can be found from [22]. Because of the importance of interval combination, this operation is denoted with the sign '$\backsim$' as

$$(t_a, t_b) \backsim (t_b, t_c) = (t_a, t_c) \tag{6.5.79}$$

which implies that the respective interval matrices and vectors is combined with the equations (6.5.77) and (6.5.78). Algebraically, the operation $\backsim$ is regarded as multiplication, and the associative rule of interval combination is the associative rule of multiplication operation $\backsim$

$$[(t_1,t_2) \backsim (t_2,t_3)] \backsim (t_3,t_4) = (t_1,t_2) \backsim [(t_2,t_3) \backsim (t_3,t_4)]$$
$$= (t_1,t_2) \backsim (t_2,t_3) \backsim (t_3,t_4) \tag{6.5.79a}$$

It should be pointed out, that the interval combination and precise integration method given below is entirely similar to that given in chapter 5 for structural mechanics, which is the consequence of the ***analogy relationship between structural mechanics and optimal control***.

§6.5.4.2, Differential equations for the interval matrices and vectors

The interval mixed energy matrices $\mathbf{Q}(t_a,t_b),\mathbf{G}(t_a,t_b),\mathbf{F}(t_a,t_b)$ and interval vectors $\mathbf{r}_x(t_a,t_b),\mathbf{r}_\lambda(t_a,t_b)$, should satisfy the corresponding differential equations, respectively, and are derived as follows. If the state vector $\mathbf{x}_a,\boldsymbol{\lambda}_a$ is fixed at the end-t_a, then the state vector $\mathbf{x}_b,\boldsymbol{\lambda}_b$ at the other end-t_b is determined too. Differentiating the equations (6.5.73a,b) with respect to t_b gives

$$\frac{\partial \mathbf{x}_b}{\partial t_b} = \frac{\partial \mathbf{F}}{\partial t_b}\mathbf{x}_a + \frac{\partial \mathbf{G}}{\partial t_b}\boldsymbol{\lambda}_b + \mathbf{G}\frac{\partial \boldsymbol{\lambda}_b}{\partial t_b} + \frac{\partial \mathbf{r}_x}{\partial t_b},$$

$$\mathbf{0} = -\frac{\partial \mathbf{Q}}{\partial t_b}\mathbf{x}_a + \frac{\partial \mathbf{F}^T}{\partial t_b}\boldsymbol{\lambda}_b + \mathbf{F}^T\frac{\partial \boldsymbol{\lambda}_b}{\partial t_b} + \frac{\partial \mathbf{r}_\lambda}{\partial t_b}$$

Using the equation (6.5.47) with $t = t_b$ substituted, gives

$$\partial\boldsymbol{\lambda}_b/\partial t_b = \mathbf{C}^T\mathbf{V}^{-1}\mathbf{C}\mathbf{x}_b - \mathbf{A}^T\boldsymbol{\lambda}_b - \mathbf{C}^T\mathbf{V}^{-1}\mathbf{y}_b, \quad \partial\mathbf{x}_b/\partial t_b = \mathbf{A}\mathbf{x}_b + \mathbf{B}\mathbf{W}\mathbf{B}^T\boldsymbol{\lambda}_b + \mathbf{B}_u\mathbf{u}_b$$

Substituting into the above equations gives

$$(\partial\mathbf{F}/\partial t_b)\mathbf{x}_b + (\partial\mathbf{G}/\partial t_b - \mathbf{G}\mathbf{A}^T - \mathbf{B}\mathbf{W}\mathbf{B}^T)\boldsymbol{\lambda}_b - (\mathbf{A} - \mathbf{G}\mathbf{C}^T\mathbf{V}^{-1}\mathbf{C})\mathbf{x}_b$$
$$+ \partial\mathbf{r}_x/\partial t_b - \mathbf{G}\mathbf{C}^T\mathbf{V}^{-1}\mathbf{y}_b - \mathbf{B}_u\mathbf{u}_b = \mathbf{0}$$

$$\mathbf{F}^T\mathbf{C}^T\mathbf{V}^{-1}\mathbf{C}\mathbf{x}_b - (\partial\mathbf{Q}/\partial t_b)\mathbf{x}_a + (\partial\mathbf{F}^T/\partial t_b - \mathbf{F}^T\mathbf{A}^T)\boldsymbol{\lambda}_b$$
$$+ \partial\mathbf{r}_\lambda/\partial t_b - \mathbf{F}^T\mathbf{C}^T\mathbf{V}^{-1}\mathbf{y}_b = \mathbf{0}$$

However, the vectors $\mathbf{x}_a,\boldsymbol{\lambda}_b,\mathbf{x}_b$ in these two equations are not all independent, so that eliminating $\mathbf{x}_b$ by the equation (6.5.73a) gives

$$[\partial\mathbf{F}/\partial t_b - (\mathbf{A} - \mathbf{G}\mathbf{C}^T\mathbf{V}^{-1}\mathbf{C})\mathbf{F}]\mathbf{x}_a + [\partial\mathbf{G}/\partial t_b - \mathbf{G}\mathbf{A}^T - \mathbf{B}\mathbf{W}\mathbf{B}^T - (\mathbf{A} - \mathbf{G}\mathbf{C}^T\mathbf{V}^{-1}\mathbf{C})\mathbf{G}]\boldsymbol{\lambda}_b$$
$$+ \partial\mathbf{r}_x/\partial t_b - \mathbf{G}\mathbf{C}^T\mathbf{V}^{-1}\mathbf{y}_b - (\mathbf{A} - \mathbf{G}\mathbf{C}^T\mathbf{V}^{-1}\mathbf{C})\mathbf{r}_x = \mathbf{0}$$

$$(-\partial\mathbf{Q}/\partial t_b + \mathbf{F}^T\mathbf{C}^T\mathbf{V}^{-1}\mathbf{C}\mathbf{F})\mathbf{x}_a + (\partial\mathbf{F}^T/\partial t_b - \mathbf{F}^T\mathbf{A}^T + \mathbf{F}^T\mathbf{C}^T\mathbf{V}^{-1}\mathbf{C}\mathbf{G})\boldsymbol{\lambda}_b$$
$$+ \partial\mathbf{r}_\lambda/\partial t_b - \mathbf{F}^T\mathbf{C}^T\mathbf{V}^{-1}\mathbf{y}_b + \mathbf{F}^T\mathbf{C}^T\mathbf{V}^{-1}\mathbf{C}\mathbf{r}_x = \mathbf{0}$$

These two equations are valid for arbitrary vectors $\mathbf{x}_a,\boldsymbol{\lambda}_b$. Hence, it results

$$\partial\mathbf{G}/\partial t_b = \mathbf{B}\mathbf{W}\mathbf{B}^T + \mathbf{G}\mathbf{A}^T + \mathbf{A}\mathbf{G} - \mathbf{G}\mathbf{C}^T\mathbf{V}^{-1}\mathbf{C}\mathbf{G} \tag{6.5.80a}$$

$$\partial\mathbf{Q}/\partial t_b = \mathbf{F}^T\mathbf{C}^T\mathbf{V}^{-1}\mathbf{C}\mathbf{F} \tag{6.5.80b}$$

$$\partial\mathbf{F}/\partial t_b = (\mathbf{A} - \mathbf{G}\mathbf{C}^T\mathbf{V}^{-1}\mathbf{C})\mathbf{F} \tag{6.5.80c}$$

$$\partial \mathbf{r}_x/\partial t_b = \mathbf{A}\mathbf{r}_x + \mathbf{G}\mathbf{C}^T\mathbf{V}^{-1}(\mathbf{y}_b - \mathbf{C}\mathbf{r}_x) + \mathbf{B}_u\mathbf{u}_b \quad (6.5.81)$$

$$\partial \mathbf{r}_\lambda/\partial t_b = \mathbf{F}^T\mathbf{C}^T\mathbf{V}^{-1}(\mathbf{y}_b - \mathbf{C}\mathbf{r}_x) \quad (6.5.82)$$

Equations (6.5.80a~c) for matrices $\mathbf{Q},\mathbf{G},\mathbf{F}$ are homogeneous but unrelated to the measurement $\mathbf{y}$. These equations apply to time-variant system matrices $\mathbf{A},\mathbf{B},\mathbf{C},\mathbf{V},\mathbf{W}$ too. The initial conditions have been given in equation (6.5.74). Linear equations (6.5.81)~(6.5.82) for the unknown vectors $\mathbf{r}_x$ and $\mathbf{r}_\lambda$ are the measurement $\mathbf{y}$ induced responses (two end forces), and the initial conditions are given also in equation (6.5.74). In the equations, $\mathbf{y}_b \equiv \mathbf{y}(t_b)$ and it appears linearly.

The derivation given above comes from partial differential with respect to the end-t_b. Fixing end-t_b and state vector $\mathbf{x}_b, \boldsymbol{\lambda}_b$, differentiating the equation (6.5.73) with respect to t_a gives

$$0 = \frac{\partial \mathbf{F}}{\partial t_a}\mathbf{x}_a + \mathbf{F}\frac{\partial \mathbf{x}_a}{\partial t_a} + \frac{\partial \mathbf{G}}{\partial t_a}\boldsymbol{\lambda}_a + \frac{\partial \mathbf{r}_x}{\partial t_a}$$

$$\frac{\partial \boldsymbol{\lambda}_a}{\partial t_a} = -\frac{\partial \mathbf{E}}{\partial t_a}\mathbf{x}_a - \mathbf{Q}\frac{\partial \mathbf{x}_a}{\partial t_a} + \frac{\partial \mathbf{F}^T}{\partial t_a}\boldsymbol{\lambda}_b + \frac{\partial \mathbf{r}_\lambda}{\partial t_a}$$

Using the equations (6.5.47b) and (6.5.47a') at $t = t_a$ and substituting into the above equation obtains

$$(\partial \mathbf{F}/\partial t_a + \mathbf{F}\mathbf{A})\mathbf{x}_a + \mathbf{F}\mathbf{B}\mathbf{W}\mathbf{B}^T\boldsymbol{\lambda}_a + (\partial \mathbf{G}/\partial t_a)\boldsymbol{\lambda}_b + \partial \mathbf{r}_x/\partial t_a = \mathbf{0}$$

$$-(\partial \mathbf{E}/\partial t_a + \mathbf{C}^T\mathbf{V}^{-1}\mathbf{C} + \mathbf{Q}\mathbf{A})\mathbf{x}_a + (\mathbf{A}^T - \mathbf{Q}\mathbf{B}\mathbf{W}\mathbf{B}^T)\boldsymbol{\lambda}_a + (\partial \mathbf{F}^T/\partial t_a)\boldsymbol{\lambda}_b$$
$$+ \partial \mathbf{r}_\lambda/\partial t_a + \mathbf{C}^T\mathbf{V}^{-1}\mathbf{y}_a - \mathbf{Q}\mathbf{B}_u\mathbf{u}_a = \mathbf{0}$$

However, the vectors $\mathbf{x}_a, \boldsymbol{\lambda}_a, \boldsymbol{\lambda}_b$ are not all independent. Using equation (6.5.73b) eliminates $\boldsymbol{\lambda}_a$ so that the remaining vectors $\mathbf{x}_a, \boldsymbol{\lambda}_b$ are independent, which derives

$$\left(\partial \mathbf{F}/\partial t_a + \mathbf{F}\mathbf{A} - \mathbf{F}\mathbf{B}\mathbf{W}\mathbf{B}^T\mathbf{Q}\right)\mathbf{x}_a + \left(\partial \mathbf{G}/\partial t_a + \mathbf{F}\mathbf{B}\mathbf{W}\mathbf{B}^T\mathbf{F}^T\right)\boldsymbol{\lambda}_b$$
$$+ \partial \mathbf{r}_x/\partial t_a + \mathbf{F}\mathbf{B}\mathbf{W}\mathbf{B}^T\mathbf{r}_\lambda = \mathbf{0}$$

$$\begin{bmatrix} \partial \mathbf{E}/\partial t_a + \mathbf{C}^T\mathbf{V}^{-1}\mathbf{C} + \mathbf{Q}\mathbf{A} \\ + (\mathbf{A}^T - \mathbf{Q}\mathbf{B}\mathbf{W}\mathbf{B}^T)\mathbf{Q} \end{bmatrix}\mathbf{x}_a - \left[\partial \mathbf{F}^T/\partial t_b + (\mathbf{A}^T - \mathbf{Q}\mathbf{B}\mathbf{W}\mathbf{B}^T)\mathbf{F}^T\right]\boldsymbol{\lambda}_b$$
$$- \partial \mathbf{r}_\lambda/\partial t_a - \mathbf{C}^T\mathbf{V}^{-1}\mathbf{y}_a + \mathbf{Q}\mathbf{B}_u\mathbf{u}_a - (\mathbf{A}^T - \mathbf{Q}\mathbf{B}\mathbf{W}\mathbf{B}^T)\mathbf{r}_\lambda = \mathbf{0}$$

Because of the independence of vectors $\mathbf{x}_a, \boldsymbol{\lambda}_b$, it derives

$$\partial \mathbf{F}/\partial t_a = -\mathbf{F}\cdot(\mathbf{A} - \mathbf{B}\mathbf{W}\mathbf{B}^T\mathbf{Q}) \quad (6.5.83)$$

$$\partial \mathbf{G}/\partial t_a = -\mathbf{F}\mathbf{B}\mathbf{W}\mathbf{B}^T\mathbf{F}^T \quad (6.5.84)$$

$$\partial \mathbf{Q}/\partial t_a = -\mathbf{C}^T\mathbf{V}^{-1}\mathbf{C} - \mathbf{Q}\mathbf{A} - \mathbf{A}^T\mathbf{Q} + \mathbf{Q}\mathbf{B}\mathbf{W}\mathbf{B}^T\mathbf{Q} \quad (6.5.85)$$

$$\partial \mathbf{r}_x/\partial t_a = -\mathbf{F}\mathbf{B}\mathbf{W}\mathbf{B}^T\mathbf{r}_\lambda \quad (6.5.86)$$

$$\partial \mathbf{r}_\lambda/\partial t_a = -(\mathbf{A}^T - \mathbf{Q}\mathbf{B}\mathbf{W}\mathbf{B}^T)\mathbf{r}_\lambda - \mathbf{C}^T\mathbf{V}^{-1}\mathbf{y}_a + \mathbf{Q}\mathbf{B}_u\mathbf{u}_a \quad (6.5.87)$$

For these equations the integration is reverse to the time coordinate with the initial conditions being

$$\left.\begin{array}{l}\mathbf{F}(t_b,t_b)=\mathbf{I}_n,\ \mathbf{G}(t_b,t_b)=\mathbf{Q}(t_b,t_b)=\mathbf{0},\\ \mathbf{r}_x(t_b,t_b)=\mathbf{0},\qquad \mathbf{r}_\lambda(t_b,t_b)=\mathbf{0}\end{array}\right\},\quad \text{when}\ t_a\to t_b \tag{6.5.88}$$

It is seen that the equation set (6.5.83)~(6.5.85) for $\mathbf{Q},\mathbf{G},\mathbf{F}$ is again homogeneous and the solution matrices are independent on $\mathbf{y}$. The measurement $\mathbf{y}$ influences only the vectors $\mathbf{r}_\lambda$ and $\mathbf{r}_x$, and the differential equation (6.5.81) for $\mathbf{r}_x$ is the same form to the equation for mean value $\hat{\mathbf{x}}$.

The equation (6.5.80) and (6.5.85) are the positive and reverse direction Riccati differential equations. For time-invariant system, the precise integration of Riccati differential equations will be given in the next sections.

Introducing interval mixed energy is quite useful for the solution of Riccati equations. It is seen that (6.5.80) is just the Riccati differential equation, the same as equation (6.5.50). The difference exists only on the initial condition. Let $t_a=t_0$, $t_b=t$ then because of $\mathbf{G}(0)=\mathbf{0}$, i.e. $\mathbf{G}(t_0)=\mathbf{0}$, which is in contrast to the condition of $\mathbf{P}(t_0)=\mathbf{P}_0$, hence the matrix $\mathbf{G}$ is different from the matrix $\mathbf{P}$. Nevertheless, the differential equations are the same, both the matrices $\mathbf{G}$ and $\mathbf{P}$ must be interrelated. For linear differential equation, the different initial conditions can be solved by means of the superposition principle. That is the addition of the initial value induced solution of the homogeneous equation and the non-homogeneous solution with null initial condition. But the Riccati differential equation is non-linear and such simple superposition is invalid.

The combined equation (6.5.77) of two intervals supplies the method transforming matrix $\mathbf{G}$ to solution matrix $\mathbf{P}$. Imagine that there is an infinitesimal fictitious interval at the end $t_0=0$ with the mixed energy interval matrices

$$\mathbf{Q}_1=\mathbf{0},\quad \mathbf{F}_1=\mathbf{I},\quad \mathbf{G}_1=\mathbf{P}_0 \tag{6.5.89}$$

be regarded as interval 1, and the interval $(0,t)$ with the mixed energy matrices $\mathbf{Q},\mathbf{G},\mathbf{F}$ be regarded as interval 2, then using the interval combination equation (6.5.77b) gives

$$\begin{aligned}&\mathbf{P}(t)=\mathbf{G}+\mathbf{F}(\mathbf{P}_0^{-1}+\mathbf{Q})^{-1}\mathbf{F}^T\\ &\mathbf{F}_c=\mathbf{F}(\mathbf{I}_n+\mathbf{P}_0\mathbf{Q})^{-1}\\ &\mathbf{Q}_c=(\mathbf{Q}^{-1}+\mathbf{P}_0)^{-1}\end{aligned} \tag{6.5.90}$$

Because $\mathbf{G}(0)=\mathbf{0}$, $\mathbf{Q}(0)=\mathbf{0}$, $\mathbf{F}(0)\to\mathbf{I}_n$ when $t\to 0$, it is easily verified $\mathbf{P}(0)\to\mathbf{P}_0$, so that the matrix $\mathbf{P}(t)$ satisfies the initial condition. The interval combination equation (6.5.77) does not influence the matrices $\mathbf{P},\mathbf{Q}_c,\mathbf{F}_c(t)$ to satisfy the differential equation (6.5.80). So the Riccati differential equation is still satisfied. Therefore the equation (6.5.90) derived solution matrix $\mathbf{P}(t)$ is the solution of differential equation (6.5.50). The physical interpretation is given in the next section.

Condition (6.5.74) corresponds to that at the end-t_b (or at end-t_a) there has not a concentrated fictitious interval as equation (6.5.89), hence the condition (6.5.74) is called the natural initial condition for the interval matrices.

§6.5.4.3, Physical interpretation of the solution of Riccati equation

Before description of precise integration for the solution matrix $\mathbf{P}(t)$ of Riccati differential equation, the investigation of the physical meaning of this matrix and the variational form is beneficial. The importance of matrix $\mathbf{P}(t)$ is that it is the variance matrix of the filter vector $\hat{\mathbf{x}}(t)$ of state $\mathbf{x}(t)$. From variational formulation, the role of the matrix is found as follows. After introducing the Lagrange multiplier for the index J of equation (6.5.44), the variational equation (6.5.45') is derived to be the functional $J_e(\mathbf{x},\boldsymbol{\lambda})$ of two kinds of independent variables $\mathbf{x},\boldsymbol{\lambda}$. The value of J_e equals to the original index J, but the arguments are different. The initial condition (6.5.43a) can also be satisfied by the variational principle, let (the special case of $\mathbf{u}=0$ is considered)

$$\begin{aligned} J_e = \int_0^t [\boldsymbol{\lambda}^T\dot{\mathbf{x}} - \boldsymbol{\lambda}^T\mathbf{A}\mathbf{x} - \boldsymbol{\lambda}^T(\mathbf{B}\mathbf{W}\mathbf{B}^T)\boldsymbol{\lambda}/2 + (\mathbf{y}-\mathbf{C}\mathbf{x})^T\mathbf{V}^{-1}(\mathbf{y}-\mathbf{C}\mathbf{x})/2]\mathrm{d}\tau \\ + (\mathbf{x}_0-\hat{\mathbf{x}}_0)\mathbf{P}_0^{-1}(\mathbf{x}_0-\hat{\mathbf{x}}_0)/2 \end{aligned} \tag{6.5.45''}$$

Carrying out the variational derivation and using integration by parts gives

$$\begin{aligned} \delta J_e = \int_0^t [\delta\boldsymbol{\lambda}^T(\dot{\mathbf{x}} - \mathbf{A}\mathbf{x} - \mathbf{B}\mathbf{W}\mathbf{B}^T\boldsymbol{\lambda}) - \delta\mathbf{x}^T(\dot{\boldsymbol{\lambda}} + \mathbf{A}^T\boldsymbol{\lambda} - \mathbf{C}^T\mathbf{V}^{-1}\mathbf{C}\mathbf{x} + \mathbf{C}^T\mathbf{V}^{-1}\mathbf{y})]\mathrm{d}\tau \\ + \boldsymbol{\lambda}^T(t)\cdot\delta\mathbf{x}(t) - \delta\mathbf{x}_0^T\cdot[\boldsymbol{\lambda}_0 - \mathbf{P}_0^{-1}(\mathbf{x}_0-\hat{\mathbf{x}}_0)] = 0 \end{aligned}$$

Let the dual equations (6.5.47a,b) and initial condition (6.5.43a) be satisfied, then

$$\delta J_e = \boldsymbol{\lambda}^T(t)\cdot\delta\mathbf{x}(t)$$

Note that the solution of the equation (6.5.47a,b) and initial condition (6.5.43a) is the mean value of state vector, but at the end $\tau=t$ the vector $\boldsymbol{\lambda}(t)$ can be arbitrary, hence the solution form is

$$\mathbf{x}(t) = \hat{\mathbf{x}}(t) + \mathbf{P}(t)\boldsymbol{\lambda}(t) \tag{6.5.48}$$

where the mean value vector $\hat{\mathbf{x}}(t)$ (filter solution) and the variance matrix $\mathbf{P}(t)$ are solutions of equations (6.5.49) and (6.5.50), respectively, and are deterministic values. So $\delta\mathbf{x}(t) = \mathbf{P}(t)\delta\boldsymbol{\lambda}(t)$ and then

$$\delta J_e = \delta[\boldsymbol{\lambda}^T(t)\mathbf{P}(t)\boldsymbol{\lambda}(t)/2] = \delta[(\mathbf{x}-\hat{\mathbf{x}})^T\mathbf{P}^{-1}(\mathbf{x}-\hat{\mathbf{x}})/2] = \delta[\boldsymbol{\lambda}^T\cdot(\mathbf{x}-\hat{\mathbf{x}})/2]$$

From this equation the physical interpretation of $\mathbf{P}$ is the ***flexibility matrix***. It can be modeled as a spring system, for which $\hat{\mathbf{x}}$ is the equilibrium point and $(\mathbf{x}-\hat{\mathbf{x}})$ is the deviation of displacement vector, and $\boldsymbol{\lambda} = \mathbf{P}^{-1}(\mathbf{x}-\hat{\mathbf{x}})$ is the force vector. Let the n unit vectors $\boldsymbol{\lambda}_1 = \{1,0,0,\cdots\}^T$, $\boldsymbol{\lambda}_2 = \{0,1,0,\cdots\}^T,\cdots$, and $\boldsymbol{\lambda}_n = \{0,0,\cdots,1\}^T$ be applied to the spring system in turn, then the n response deviations $(\mathbf{x}_1-\hat{\mathbf{x}})$, $(\mathbf{x}_2-\hat{\mathbf{x}}),\cdots$, and $(\mathbf{x}_n-\hat{\mathbf{x}})$ are the column vectors of the matrix $\mathbf{P}$, respectively. Hence, the matrix $\mathbf{P}(t)$ is interpreted as the ***filter flexibility matrix*** of the interval $[0,t)$ at the end t.

The method for determining the variance matrix is, finding the flexibility matrix of the homogeneous system, which gives the variance matrix. In chapter 3 when describing the least square problem, it has been found that the ***variance matrix***

is a flexibility matrix.

§6.5.5, Precise integration of the Riccati differential equation

For the linear equation (6.4.35) of prediction, the precise integration is given in section 6.4.4, which can be used under the time-invariant system. Precise integration method for Riccati differential equation is also given for a time-invariant system. Equation (6.5.90) has proposed the reduced method that the mixed energy matrices $\mathbf{G}(t),\mathbf{F}(t),\mathbf{Q}(t)$ corresponding to the null initial condition can be solved first, and then the ***solution matrix*** $\mathbf{P}(t)$ ***can be obtained via equation*** (6.5.90). Hence only the precise integration method for matrix $\mathbf{G}$ is needed.

When the time step η is selected, the whole interval is subdivided as in equation (6.5.68). The fundamental step has duration η, for which the mixed energy matrices $\mathbf{Q}(\eta),\mathbf{G}(\eta),\mathbf{F}(\eta)$ need to be computed. Previously these interval matrices are functions of the two ends t_a, t_b, however, for time-invariant system, which depends only on the interval length

$$\eta = t_b - t_a \tag{6.5.91}$$

but not relates to the initial point t_a, hence $\mathbf{Q}(t_a,t_b)=\mathbf{Q}(t_b-t_a)=\mathbf{Q}(\eta)$, etc. The equations (6.5.80)~(6.5.82) and (6.5.83)~(6.5.87) can be combined rewritten as

$$\mathrm{d}\mathbf{F}/\mathrm{d}\tau = (\mathbf{A}-\mathbf{G}\mathbf{C}^T\mathbf{V}^{-1}\mathbf{C})\mathbf{F} \qquad = \mathbf{F}(\mathbf{A}-\mathbf{B}\mathbf{W}\mathbf{B}^T\mathbf{Q}) \tag{6.5.92a,b}$$

$$\mathrm{d}\mathbf{G}/\mathrm{d}\tau = \mathbf{B}\mathbf{W}\mathbf{B}^T+\mathbf{G}\mathbf{A}^T+\mathbf{A}\mathbf{G}-\mathbf{G}\mathbf{C}^T\mathbf{V}^{-1}\mathbf{C}\mathbf{G} = \mathbf{F}\mathbf{B}\mathbf{W}\mathbf{B}^T\mathbf{F}^T \tag{6.5.93a,b}$$

$$\mathrm{d}\mathbf{Q}/\mathrm{d}\tau = \mathbf{F}^T\mathbf{C}^T\mathbf{V}^{-1}\mathbf{C}\mathbf{F} = \mathbf{C}^T\mathbf{V}^{-1}\mathbf{C}+\mathbf{Q}\mathbf{A}+\mathbf{A}^T\mathbf{Q}-\mathbf{Q}\mathbf{B}\mathbf{W}\mathbf{B}^T\mathbf{Q} \tag{6.5.94a,b}$$

When the measurement $\mathbf{y}$ and control $\mathbf{u}$ are the **special case** of constant vectors, the equations are

$$\mathrm{d}\mathbf{r}_x/\mathrm{d}\tau = (\mathbf{A}-\mathbf{G}\mathbf{C}^T\mathbf{V}^{-1}\mathbf{C})\mathbf{r}_x-\mathbf{G}\mathbf{C}^T\mathbf{V}^{-1}\mathbf{y}+\mathbf{B}_u\mathbf{u}_b \qquad = \mathbf{F}\mathbf{B}\mathbf{W}\mathbf{B}^T\mathbf{r}_\lambda \tag{6.5.95a,b}$$

$$\mathrm{d}\mathbf{r}_\lambda/\mathrm{d}\tau = \mathbf{F}^T\mathbf{C}^T\mathbf{V}^{-1}(\mathbf{y}-\mathbf{C}\mathbf{r}_x) = (\mathbf{A}^T-\mathbf{Q}\mathbf{B}\mathbf{W}\mathbf{B}^T)\mathbf{r}_\lambda+\mathbf{C}^T\mathbf{V}^{-1}\mathbf{y}-\mathbf{Q}\mathbf{B}_u\mathbf{u}_a \tag{6.5.96a,b}$$

These equations have two versions, which correspond to those derived from the ends t_a or t_b, respectively. Both versions are compatible to each other, which is ensured from the associative rule of interval combinations.

However, the measurement $\mathbf{y}$ is under stochastic process disturbance, so that it is not constant valued, hence the equation (6.5.95) and (6.5.96) can only be applied to special case. Special care must be taken at this point.

Equation set (6.5.92)~(6.5.94) are non-linear ODEs, so that doing precise integration must uses the structure of the problem itself. Comparing the precise integration of the matrix exponentiation, the first crux is the 2^N algorithm, because the matrix exponential function applies the additional theorem. Presently the interval combination algorithm can be used instead, from which the 2^N algorithm can be proposed. Let

$$\tau = \eta/2^N, \quad N = 20, \quad 2^N = 1048576 \tag{6.5.97}$$

where τ is extremely small. The 2^N algorithm needs an initialization interval and this extremely small interval τ can be used as the initial one. In case of matrix exponentiation, the Taylor series expansion can be used. For present case, there are the set of differential equations (6.5.92~94), from which the matrices $\mathbf{Q}(\tau), \mathbf{G}(\tau), \mathbf{F}(\tau)$ should be generated and should be computed with the relative error being beyond the computer double precision. Taylor series expansion method can still be applied presently and keep up to the fourth order τ^4 terms. Let

$$\mathbf{Q}(\tau) \approx \mathbf{e}_1\tau + \mathbf{e}_2\tau^2 + \mathbf{e}_3\tau^3 + \mathbf{e}_4\tau^4 \tag{6.5.98}$$

$$\mathbf{G}(\tau) \approx \mathbf{g}_1\tau + \mathbf{g}_2\tau^2 + \mathbf{g}_3\tau^3 + \mathbf{g}_4\tau^4 \tag{6.5.99}$$

$$\mathbf{F}(\tau) \approx \mathbf{I}_n + \mathbf{F}'(\tau), \quad \mathbf{F}'(\tau) \approx \mathbf{f}_1\tau + \mathbf{f}_2\tau^2 + \mathbf{f}_3\tau^3 + \mathbf{f}_4\tau^4 \tag{6.5.100}$$

where $\mathbf{e}_i, \mathbf{g}_i, \mathbf{f}_i (i = 1,2,3,4)$ need to be determined. The term $\mathbf{f}_1$ should not be confused with the previous one. Substituting the Taylor expansions (6.5.98~100) into (6.5.92a)~(6.5.96a), carrying out the multiplication and the coefficients of the various powers of τ equals zero, which gives

$$\begin{aligned}
&\mathbf{e}_1 = \mathbf{C}^T\mathbf{V}^{-1}\mathbf{C}, \quad \mathbf{g}_1 = \mathbf{B}\mathbf{W}\mathbf{B}^T, \quad \mathbf{f}_1 = \mathbf{A} \\
&\mathbf{e}_2 = (\mathbf{f}_1^T\mathbf{e}_1 + \mathbf{e}_1\mathbf{f}_1)/2, \quad \mathbf{g}_2 = (\mathbf{A}\mathbf{g}_1 + \mathbf{g}_1\mathbf{A}^T)/2, \quad \mathbf{f}_2 = (\mathbf{A}^2 - \mathbf{g}_1\mathbf{e}_1)/2 \\
&\mathbf{e}_3 = (\mathbf{f}_2^T\mathbf{e}_1 + \mathbf{e}_1\mathbf{f}_2 + \mathbf{f}_1^T\mathbf{e}_1\mathbf{f}_1)/3, \quad \mathbf{g}_3 = (\mathbf{A}\mathbf{g}_2 + \mathbf{g}_2\mathbf{A}^T - \mathbf{g}_1\mathbf{e}_1\mathbf{g}_1)/3 \\
&\mathbf{f}_3 = (\mathbf{A}\mathbf{f}_2 - \mathbf{g}_2\mathbf{e}_1 - \mathbf{g}_1\mathbf{e}_1\mathbf{f}_1)/3, \quad \mathbf{e}_4 = (\mathbf{f}_3^T\mathbf{e}_1 + \mathbf{e}_1\mathbf{f}_3 + \mathbf{f}_2^T\mathbf{e}_1\mathbf{f}_1 + \mathbf{f}_1^T\mathbf{e}_1\mathbf{f}_2)/4, \\
&\mathbf{g}_4 = (\mathbf{A}\mathbf{g}_3 + \mathbf{g}_3\mathbf{A}^T - \mathbf{g}_1\mathbf{e}_1\mathbf{g}_2 - \mathbf{g}_2\mathbf{e}_1\mathbf{g}_1)/4, \\
&\mathbf{f}_4 = (\mathbf{A}\mathbf{f}_3 - \mathbf{g}_3\mathbf{e}_1 - \mathbf{g}_2\mathbf{e}_1\mathbf{f}_1 - \mathbf{g}_1\mathbf{e}_1\mathbf{f}_2)/4
\end{aligned} \tag{6.5.101}$$

These coefficient matrices need only be computed one after another, with no iterative solutions. Note that $\mathbf{e}_i, \mathbf{g}_i, \mathbf{f}_i, (i = 1 \sim 4)$ are all $n \times n$ dimensioned and also $\mathbf{e}_i^T = \mathbf{e}_i$, $\mathbf{g}_i^T = \mathbf{g}_i$.

After computed these coefficient matrices and then substituting into (6.5.98~100), the mixed energy matrices $\mathbf{Q}(\tau), \mathbf{G}(\tau), \mathbf{F}'(\tau)$ are obtained. Because the interval length τ is extremely small, so the numerical result is precise up to the computer real word precision. This interval of length τ can be used as the initial interval of the 2^N algorithm. All the equation derivations before (6.5.96) are exact, except the Taylor expansions (6.5.98~100) truncate beyond τ^4. The first term of truncation is τ^5, which is of ratio τ^4 to the first term in the expansion. Because of $\tau^4 = (\eta/1048576)^4 \approx \eta^4 \cdot 10^{-24}$, this multiplier of relative error has been beyond the double precision 10^{-16} of real word, so this step of approximation reaches the full computer precision.

Having calculated the mixed energy expression of the interval τ, the equations (6.5.77a~c) can be recurrently executed N times, where $\mathbf{Q}, \mathbf{G}, \mathbf{F}$ are the matrices of mixed energy representation of equal length $(2^i\tau)$ interval. When the N times loop is finished, the matrices $\mathbf{Q},\mathbf{G},\mathbf{F}$ turn to be the mixed energy matrices $\mathbf{Q}(\eta),\mathbf{G}(\eta),\mathbf{F}(\eta)$ of the given length η interval, such algorithm is termed as 'interval doubling'. However, the interval-doubling algorithm was found numerically unreliable, see [128] chap.7. To solve this numerical problem, special attention must be taken in the execution of equation (6.5.77c), that ***the addition of*** $\mathbf{I}_n+\mathbf{F}'$ ***must not be executed in equation*** (6.5.100). This is the ***second crux of precise integration***. Because, when τ is extremely small, $\mathbf{F}'$ is also extremely small, the addition will seriously hurt the numerical precision because of round-off error. The cause of interval-doubling algorithm (*i.e.* 2^N algorithm) being considered unreliable comes from such numerical ill-conditioning. The similar situation appears also in the matrix exponential function. Hence, the equation (6.5.77) should be updated as

$$\mathbf{Q}_c=\mathbf{Q}+(\mathbf{I}+\mathbf{F}')^T(\mathbf{Q}^{-1}+\mathbf{G})^{-1}(\mathbf{I}+\mathbf{F}') \tag{6.5.102a}$$

$$\mathbf{G}_c=\mathbf{G}+(\mathbf{I}+\mathbf{F}')(\mathbf{G}^{-1}+\mathbf{Q})^{-1}(\mathbf{I}+\mathbf{F}')^T \tag{6.5.102b}$$

$$\begin{aligned}\mathbf{F}_c'=&(\mathbf{F}'-\mathbf{G}\mathbf{Q}/2)(\mathbf{I}+\mathbf{G}\mathbf{Q})^{-1}\\&+(\mathbf{I}+\mathbf{G}\mathbf{Q})^{-1}(\mathbf{F}'-\mathbf{G}\mathbf{Q}/2)+\mathbf{F}'(\mathbf{I}+\mathbf{G}\mathbf{Q})^{-1}\mathbf{F}'\end{aligned} \tag{6.5.102c}$$

That is, to keep track of always the incremental of the matrix $\mathbf{F}=\mathbf{I}_n+\mathbf{F}'$, then the numerical ill-condition problem is solved in the computation. These equations apply to the combination of two small equal length intervals.

Until now, the equations for the precise integration have been available, so that the algorithm can be given in meta language as follows:

[Give the matrices $\mathbf{A},\mathbf{B},\mathbf{C},\mathbf{W},\mathbf{V}$, and $\mathbf{P}_0$, select step size η, and $t_f=k_f\eta$]
[Calculate $\mathbf{C}^T\mathbf{V}^{-1}\mathbf{C}$, $\mathbf{B}\mathbf{W}\mathbf{B}^T$; let $N=20,\tau=\eta/2^N$]
[According to (6.5.98-101) calculate the matrices $\mathbf{Q}(\tau),\mathbf{G}(\tau),\mathbf{F}'(\tau)$]
for $(iter=0;\ iter<N;\ iter++)$ {
Comment: precise computation in the η interval
    [According to (6.5.102a~c) compute $\mathbf{Q}_c,\mathbf{G}_c,\mathbf{F}_c'$]
    [let $\mathbf{Q}=\mathbf{Q}_c;\mathbf{G}=\mathbf{G}_c;\mathbf{F}'=\mathbf{F}_c';$]
}
[$\mathbf{F}=\mathbf{I}+\mathbf{F}';$] Comment: The matrices $\mathbf{Q}(\eta),\mathbf{G}(\eta),\mathbf{F}(\eta)$ are obtained.
[$\mathbf{Q}_2=\mathbf{Q};\ \mathbf{G}_2=\mathbf{G};\ \mathbf{F}_2=\mathbf{F};\ \mathbf{G}_1=\mathbf{P}_0;\ \mathbf{Q}_1=\mathbf{0};\ \mathbf{F}_1=\mathbf{I};$]
Comment: initialization.
for $(k=0;\ k<k_f;\ k++)$ {Comment: stepwise forward
    [According to (6.5.77b), (6.5.77c) compute $\mathbf{G}_c$];
    [$\mathbf{G}_1=\mathbf{G}_c$] Comment: $\mathbf{G}_c$ is the matrix $\mathbf{P}(k\eta)$
} (6.5.103)

The above algorithm is the precise integration solution of Riccati differential equation for finite horizon (duration) $[0,t_f)$. Certainly, η is a relatively small time-step. Because the initialization of stepwise forward integration of the matrix $\mathbf{P}$ has been given the initial value $\mathbf{P}_0$, so that for each stepwise forward, the matrix $\mathbf{G}_c$ has been the solution of Riccati equation (6.5.50).

Kalman-Bucy filtering requires real time computation, however the computation of matrix $\mathbf{P}$ is unrelated to the real time measurement $\mathbf{y}$. Hence the matrix $\mathbf{P}$ can be computed and stored beforehand. The real time computation is then only for the solution of the linear time-variant differential equation (6.5.49).

The non-homogeneous vector terms $\mathbf{r}_x$ and $\mathbf{r}_\lambda$ can also be precisely integrated. However, the measurement vector $\mathbf{y}$ does not know beforehand, hence the effect of $\mathbf{y}$ can only be computed at the ***real time***. But some fundamental computations can still be executed beforehand, that the ***basis vectors*** of the measurement vector $\mathbf{y}$ can also be computed and stored with the matrices $\mathbf{Q},\mathbf{G},\mathbf{F}$ and $\mathbf{P}$ ***off-line***. The formulation and algorithm will be given in sections 6.5.7-8 for the integration of filter differential equation (6.5.49).

Algorithm (6.5.103) is given for the computation of finite horizon transient process. Sometimes, control theory needs to consider the infinite horizon filtering. In such case, the symmetric and positive definite solution matrix $\mathbf{P}_\infty$ of the algebraic Riccati equation (6.5.52) is needed, which can be obtained by taking the limit of $t_f \to \infty$ in algorithm (6.5.103). Because equation (6.5.77b) determines that $\mathbf{G}_c$ can only be continuously increasing so that the positive definiteness is ensured.

Hence, after computed the matrices $\mathbf{Q}(\eta),\mathbf{G}(\eta),\mathbf{F}(\eta)$ of the fundamental time interval η, the iterative computation is:

[$\mathbf{Q}(\eta),\mathbf{G}(\eta),\mathbf{F}(\eta)$ are obtained from the former part of (6.5.103)]
[Let $\mathbf{Q}_c=\mathbf{Q};\mathbf{G}_c=\mathbf{G};\mathbf{F}_c=\mathbf{F};$]
while $\left(\|\mathbf{F}_c\|>\varepsilon\right)$ $\{[\mathbf{Q}_1=\mathbf{Q}_2=\mathbf{Q}_c;\mathbf{G}_1=\mathbf{G}_2=\mathbf{G}_c;\mathbf{F}_1=\mathbf{F}_2=\mathbf{F}_c;]$
[According to (6.5.77a), (6.5.77b) and (6.5.77c) compute $\mathbf{Q}_c,\mathbf{G}_c,\mathbf{F}_c$]
}
[$\mathbf{P}_\infty=\mathbf{G}_c;$] Comment: For controllable and observable system, convergence is ensured. (6.5.104)

The gain matrix is computed after the ARE solution $\mathbf{P}_\infty$ is obtained

$$\mathbf{K}_\infty=\mathbf{P}_\infty\mathbf{C}^T\mathbf{R}^{-1} \tag{6.5.105}$$

Hence the precise computation for the matrix $\mathbf{P}_\infty$ is very important.

The matrix $\mathbf{P}_\infty$ satisfies the ARE (6.5.52), since no differentiation there, hence the precision can be verified easily. Computing

$$\mathbf{BWB}^T+\mathbf{AP}_\infty+\mathbf{P}_\infty\mathbf{A} \quad \text{and} \quad \mathbf{P}_\infty\mathbf{C}^T\mathbf{V}^{-1}\mathbf{CP}_\infty \tag{6.5.106}$$

and comparing both the matrices element by element, the coincident significant digits give the precision. Because $\mathbf{P}_\infty$ is obtained from equation (6.5.104) by iteration method, but not revised using the ARE, hence this comparison verifies also the reliability of $\mathbf{E}(\eta), \mathbf{G}(\eta), \mathbf{F}(\eta)$ computed. Below a numerical example is used to demonstrate.

Example 6.9, A one dimensional dynamic equation

$$\dot{x} = -ax + w(t), \quad \hat{x}_0 = 0, \quad P_0 = \sigma_0^2$$

where w is a Gaussian white noise with zero mean and variance σ_w^2. The measurement is

$$y(t) = x(t) + v(t)$$

where v is also a Gaussian white noise with zero mean and variance σ_v^2, the w, v and x_0 are all independent processes. The variance function of $x(t)$ is required.

Solution: $n = 1$ -D, with data $A = -a,\ B = 1,\ C = 1,\ \mathrm{W} = \sigma_w^2,\ V = \sigma_v^2$. It gives $CV^{-1}C = 1/\sigma_v^2$ and $BWB^T = \sigma_w^2$, hence the equation (6.5.50) becomes

$$\dot{p} = \sigma_w^2 - 2aP - P^2/\sigma_v^2, \quad P(0) = \sigma_0^2$$

This 1-D Riccati differential equation can be solved analytically. It derives to

$$\mathrm{d}t = \sigma_v^2 \mathrm{d}P / \left(\sigma_v^2\sigma_w^2 - 2a\sigma_v^2 P - P^2\right)$$

The quadratic term of denominator can be factorized, and integration gives

$$(P(t) - p_1)/(P(t) - p_2) = Ce^{-2\mu t}, \quad \mu = (a^2 + \sigma_w^2/\sigma_v^2)^{1/2}$$

$$p_{1,2} = -a\sigma_v^2 \pm \sigma_v\sqrt{a^2\sigma_v^2 + \sigma_w^2} = \sigma_v^2 \cdot (-a \pm \mu)$$

Substituting the initial condition gives

$$P(t) = (p_1 - p_2 ce^{-2\mu t})/(1 - ce^{-2\mu t}), \quad c = (\sigma_0^2 - p_1)/(\sigma_0^2 - p_2)$$

$$P_\infty = p_1 = -a\sigma_v^2 + \sigma_v\sqrt{a^2\sigma_v^2 + \sigma_w^2}, \quad \text{when } t \to \infty$$

The gain matrix is $k(t) = P(t)/\sigma_v^2,\ k_\infty = p_1/\sigma_v^2$.

Let the data be given as $\sigma_w = 0.8;\ \sigma_v = 0.2;\ \sigma_0 = 0.1;\ a = 0.8$. Selecting $\eta = 0.05$, the numerical solution of Riccati differential equation is listed in table 6.1.

Because of the non-linearity of Riccati differential equation, purely analytical solution rarely appears for multi-dimensional problem. For comparing the precision of precise integration result, this problem is also computed by the precise integration algorithm (6.5.103) and compared with the numerical result obtained from the analytical solution. In the ten significant digits listed in the table, the two algorithms give completely the same numerical results.

Table 6.1, The numerical solution of Riccati differential equation of example 6.9
$A = -0.8;\ B = 0.8;\ C = 5.0;\ \mathrm{W} = 1.0;\ \mathrm{V} = 1.0;\ P_0 = 0.01;\ \eta = 0.05$

T	0	0.05	0.10	0.15	0.20
Analytical	0.01	0.039142055	0.063584009	0.082872336	0.097374807
Pr. Integ.	0.01	0.039142055	0.063584009	0.082872336	0.097374807

continue

0.25	0.3	0.4	0.5	∞
0.107887105	0.1153066324	0.1239579232	0.1279397834	0.1311686244
0.107887105	0.1153066324	0.1239579232	0.1279397834	

Hence, the characteristic of highly precise result of the precise integration method is verified for this example. ##

More numerical result is unnecessary, because there are examples in structural mechanics, see chapter 5.

§6.5.6, Analytical solution of Riccati equation based on eigen-solutions

The analytical solution of Riccati differential equation can also be found based on all the eigen-solutions, and the method has been given in chapter 5. The crux is repeated here. Go back to the dual equation (6.5.47), the corresponding Hamilton matrix is

$$\mathbf{H}=\begin{bmatrix}\mathbf{A} & \mathbf{B}_w\mathbf{W}\mathbf{B}_w^T\\ \mathbf{C}^T\mathbf{V}^{-1}\mathbf{C} & -\mathbf{A}^T\end{bmatrix} \tag{6.5.107}$$

Composing the ***complete state vector*** $\mathbf{v}$ (because $\mathbf{x}$ has been nominated the term ***state vector***), and the homogeneous dual equation is combined as

$$\mathbf{v}=\{\mathbf{x}^T \quad \boldsymbol{\lambda}^T\}^T, \quad \dot{\mathbf{v}}=\mathbf{H}\mathbf{v} \tag{6.5.108}$$

From the corresponding eigen-equation $\mathbf{H}\boldsymbol{\psi}=\mu\boldsymbol{\psi}$ solves the eigen-matrix $\boldsymbol{\Psi}$

$$\mathbf{H}\boldsymbol{\Psi}=\boldsymbol{\Psi}\begin{bmatrix}\mathrm{diag}(\mu_i) & 0\\ 0 & -\mathrm{diag}(\mu_i)\end{bmatrix}, \quad \boldsymbol{\Psi}=\begin{bmatrix}\mathbf{X}_\alpha & \mathbf{X}_\beta\\ \mathbf{N}_\alpha & \mathbf{N}_\beta\end{bmatrix}\begin{matrix}n\\ n\end{matrix} \tag{6.5.109}$$

Composing the matrix $\mathbf{M}(\eta)$ and taking matrix inversion for it

$$\mathbf{M}(\eta)=\begin{bmatrix}\mathbf{X}_\alpha & \mathbf{X}_\beta\mathrm{diag}(e^{\mu_i\eta})\\ \mathbf{N}_\alpha\mathrm{diag}(e^{\mu_i\eta}) & \mathbf{N}_\beta\end{bmatrix}\begin{matrix}n\\ n\end{matrix} \tag{6.5.110}$$

$$\mathbf{M}^{-1}=\begin{bmatrix}\mathbf{A}_1 & \mathbf{A}_2\\ \mathbf{B}_1 & \mathbf{B}_2\end{bmatrix} \tag{6.5.111}$$

and

$$\mathbf{D}_{c\eta}\underset{\mathrm{def}}{=}\mathrm{diag}(e^{\mu_i\eta})\underset{\mathrm{def}}{=}\mathrm{diag}(e^{\mu_1\eta},e^{\mu_2\eta},\cdots,e^{\mu_n\eta})=\exp[\mathrm{diag}(\mu_i\eta)] \tag{6.5.112}$$

is a diagonal matrix. Each column of the matrix $\mathbf{M}(\eta)$ gives the two end-conditions of basis solution of the interval η. Using the matrix inversion lemma (6.5.30) gives

$$\left.\begin{aligned}&\mathbf{A}_1=(\mathbf{X}_\alpha-\mathbf{X}_\beta\mathbf{D}_{c\eta}\mathbf{N}_\beta^{-1}\mathbf{N}_\alpha\mathbf{D}_{c\eta})^{-1}, \quad \mathbf{A}_2=-\mathbf{X}_\alpha^{-1}\mathbf{X}_\beta\mathbf{D}_{c\eta}\mathbf{B}_2\\ &\mathbf{B}_2=(\mathbf{N}_\beta-\mathbf{N}_\alpha\mathbf{D}_{c\eta}\mathbf{X}_\alpha^{-1}\mathbf{X}_\beta\mathbf{D}_{c\eta})^{-1}, \quad \mathbf{B}_1=-\mathbf{N}_\beta^{-1}\mathbf{N}_\alpha\mathbf{D}_{c\eta}\mathbf{A}_1\end{aligned}\right\} \tag{6.5.113}$$

where η is the length of typical interval $\eta=t_b-t_a$. From which the interval mixed energy matrices are derived as

$$\begin{aligned}&\mathbf{Q}(\eta)=-(\mathbf{N}_\alpha\mathbf{A}_1+\mathbf{N}_\beta\mathbf{D}_{c\eta}\mathbf{B}_1), \quad \mathbf{F}(\eta)=\mathbf{X}_\alpha\mathbf{D}_{c\eta}\mathbf{A}_1+\mathbf{X}_\beta\mathbf{B}_1\\ &\mathbf{G}(\eta)=\mathbf{X}_\alpha\mathbf{D}_{c\eta}\mathbf{A}_2+\mathbf{X}_\beta\mathbf{B}_2, \quad \mathbf{F}^T(\eta)=\mathbf{N}_\alpha\mathbf{A}_2+\mathbf{N}_\beta\mathbf{D}_{c\eta}\mathbf{B}_2\end{aligned} \tag{6.5.114a~d}$$

The symmetry of matrices $\mathbf{Q}$ and $\mathbf{G}$, and that $\mathbf{F}$ and $\mathbf{F}^T$ being mutually transpose to each other can be proved from the above expressions, and that the differential equations (6.5.92~94) are all satisfied. Hence the analytical solution of interval mixed energy matrices based on the eigen-solutions is obtained. The derivation is given in section 5.8, and the detail is neglected.

Let $\eta = t - t_0$, then

$$\mathbf{P}(t) = \mathbf{G}(\eta) + \mathbf{F}(\eta)(\mathbf{I} + \mathbf{P}_0\mathbf{Q}(\eta))^{-1}\mathbf{P}_0\mathbf{F}^T(\eta) \tag{6.5.115}$$

which gives the solution of differential equation (6.5.50). When computing the above equation, the matrix $\mathbf{F}_c$ should also be computed simultaneously

$$\mathbf{F}_c(t) = \mathbf{F}(\eta)[\mathbf{I} + \mathbf{P}_0\mathbf{Q}(\eta)]^{-1}, \ \mathbf{Q}_c = (\mathbf{Q}^{-1} + \mathbf{P}_0)^{-1}, \quad \eta = t - t_0 \tag{6.5.116}$$

which is useful later in the precise computation of the filter vector $\hat{\mathbf{x}}(t)$.

A special case of $\eta \to \infty$ is considered next, which has the characteristics

$$\lim_{\eta\to\infty} \mathbf{D}_{c\eta} \to \mathbf{0}$$

Hence

$$\mathbf{A}_1 \to \mathbf{X}_\alpha^{-1}, \quad \mathbf{B}_2 \to \mathbf{N}_\beta^{-1}, \quad \mathbf{A}_2 \to \mathbf{0}, \quad \mathbf{B}_1 \to \mathbf{0}, \quad \text{when } \eta \to \infty$$

therefore

$$\mathbf{P}_\infty \to \mathbf{X}_\beta\mathbf{N}_\beta^{-1}, \quad \mathbf{S}_\infty = \mathbf{Q}_\infty \to -\mathbf{N}_\alpha\mathbf{X}_\alpha^{-1} \tag{6.5.117}$$

which gives the solution of ***algebraic Riccati equation***, see [129,130].

Analytical solution is certainly the best, but the present case needs to find all the eigen-solutions of matrix $\mathbf{H}$. The problem is that the Jordan normal form may appear, for which the eigen-solutions are numerically unstable. Precise integration method has no such problem that it is perfect also for Jordan normal form to appear. ***The eigen-solution based analytical method should combine with the precise integration method to solve the Riccati differential equation and the filter equation*** [103].

§6.5.7, Solution of single step filter equation [131]

The above two sections concentrate on the solution of Riccati differential equation (6.5.50). Another critical problem is the solution of filter differential equation (6.5.49). It should be emphasized that the filter equation needs ***real-time*** solution, which is a critical requirement and so careful investigation is needed. Examining the filter equation

$$\dot{\hat{\mathbf{x}}} = \mathbf{A}\hat{\mathbf{x}} - \mathbf{P}(t)\mathbf{C}^T\mathbf{V}^{-1}\mathbf{C}\hat{\mathbf{x}} + \mathbf{P}(t)\mathbf{C}^T\mathbf{V}^{-1}\mathbf{y} + \mathbf{B}_u\mathbf{u}, \qquad \hat{\mathbf{x}}(t_0) = \hat{\mathbf{x}}_0 \tag{6.5.49}$$

where the vectors $\mathbf{y}, \mathbf{u}$ cannot be determined beforehand but must be ***measured and computed at the real time.*** Examining further finds that the equation (6.5.49) is linear with respect to the filtered vector $\hat{\mathbf{x}}(t)$. Even if $\mathbf{A}$ is time invariant, but the term of $\mathbf{P}(t)$ implies that the filter equation is still a time-variant differential equation. According to the theory of ODE, solving the time variant equation should first solve the homogeneous equation

$$\dot{\hat{\mathbf{x}}}(t) = [\mathbf{A} - \mathbf{P}(t)\mathbf{C}^T\mathbf{V}^{-1}\mathbf{C}]\hat{\mathbf{x}}, \quad \hat{\mathbf{x}}(t_0) = \hat{\mathbf{x}}_0 \tag{6.5.118}$$

or $$\dot{\boldsymbol{\Phi}}=[\mathbf{A}-\mathbf{P}(t)\mathbf{C}^T\mathbf{V}^{-1}\mathbf{C}]\boldsymbol{\Phi},\quad \boldsymbol{\Phi}(t_0,t_0)=\mathbf{I}_n,\quad \hat{\mathbf{x}}(t)=\boldsymbol{\Phi}\,\hat{\mathbf{x}}_0 \tag{6.5.119}$$

The problem is to solve $\boldsymbol{\Phi}(t,t_0)$ in (6.5.119). In fact, this solution of time variant equation has really been computed, which is just the matrix $\mathbf{F}_c$ in equation (6.5.90), *i.e.*

$$\boldsymbol{\Phi}(t,t_0)=\mathbf{F}(\eta)[\mathbf{I}_n+\mathbf{P}_0\mathbf{Q}(\eta)]^{-1},\quad \eta=t-t_0 \tag{6.5.120}$$

The verification is as follows. First the initial condition, substituting into equation (6.5.74) verifies that the initial condition (6.5.119) is satisfied. Next to verify the differential equation, because the identity $\mathrm{d}\mathbf{X}^{-1}/\mathrm{d}t=-\mathbf{X}^{-1}\dot{\mathbf{X}}\mathbf{X}^{-1}$, where $\mathbf{X}$ is an arbitrary matrix. Using equation (6.5.92) gives

$$\begin{aligned}\dot{\boldsymbol{\Phi}}&=\dot{\mathbf{F}}(\mathbf{I}+\mathbf{P}_0\mathbf{Q})^{-1}-\mathbf{F}(\mathbf{I}+\mathbf{P}_0\mathbf{Q})^{-1}\mathbf{P}_0\dot{\mathbf{Q}}(\mathbf{I}+\mathbf{P}_0\mathbf{Q})^{-1}\\&=(\mathbf{A}-\mathbf{G}\mathbf{C}^T\mathbf{R}^{-1}\mathbf{C})\mathbf{F}(\mathbf{I}+\mathbf{P}_0\mathbf{Q})^{-1}-\mathbf{F}(\mathbf{I}+\mathbf{P}_0\mathbf{Q})^{-1}\mathbf{P}_0\mathbf{F}^T\mathbf{C}^T\mathbf{V}^{-1}\mathbf{C}\mathbf{F}(\mathbf{I}+\mathbf{P}_0\mathbf{Q})^{-1}\\&=[\mathbf{A}-(\mathbf{G}+\mathbf{F}(\mathbf{I}+\mathbf{P}_0\mathbf{Q})^{-1}\mathbf{P}_0\mathbf{F}^T)\mathbf{C}^T\mathbf{V}^{-1}\mathbf{C}]\boldsymbol{\Phi}=(\mathbf{A}-\mathbf{P}\mathbf{C}^T\mathbf{V}^{-1}\mathbf{C})\boldsymbol{\Phi}\end{aligned}$$

That the differential equation (6.5.119) satisfies too. According to the uniqueness theorem of ODE [36], the solution found is unique. Therefore, the solution of the time variant differential equation (6.5.119) is simply found. It should be pointed out, that the solution of homogeneous equation is independent on the values of measurement $\mathbf{y}$ and control $\mathbf{u}$, hence can be computed ***off-line*** beforehand. It is pointed out again, that the off-line computations should be executed and stored beforehand, in order to reduce the real time computation to lowest level. This is a fundamental principle of algorithm design.

Having solved the complete solution of the homogeneous differential equation, the variant coefficient method is used to find the solution of inhomogeneous equation and the solution is given in Duhamel integration form

$$\hat{\mathbf{x}}(t)=\boldsymbol{\Phi}(t,t_0)\cdot\left\{\int_{t_0}^{t}\boldsymbol{\Phi}^{-1}(\tau,t_0)[\mathbf{P}(\tau)\mathbf{C}^T\mathbf{V}^{-1}\mathbf{y}+\mathbf{B}_u\mathbf{u}]\mathrm{d}\tau+\hat{\mathbf{x}}_0\right\} \tag{6.5.121}$$

However, this integration should be executed effectively and precisely because it must be executed at the ***real-time***. The integration can be given in stepwise progressing form, for which the equation (6.5.121) is derived as that the integration begins from the arbitrary time station t_k. Let the impulse response matrix function of the linear system

$$\dot{\mathbf{x}}(t)=\mathbf{A}(t)\mathbf{x},\quad \mathbf{x}(t_0)=\mathbf{x}_0$$

be $$\dot{\boldsymbol{\Phi}}(t,t_0)=\mathbf{A}(t)\boldsymbol{\Phi},\quad \boldsymbol{\Phi}(t_0,t_0)=\mathbf{I} \tag{6.5.122}$$

for which the identity

$$\boldsymbol{\Phi}(t,t_0)=\boldsymbol{\Phi}(t,t_1)\,\boldsymbol{\Phi}(t_1,t_0),\quad t>t_1>t_0 \tag{6.5.123}$$

is proved simply that since $\mathbf{x}_1=\boldsymbol{\Phi}(t_1,t_0)\mathbf{x}_0$ and $\mathbf{x}(t)=\boldsymbol{\Phi}(t,t_1)\mathbf{x}_1$, therefore

$$\mathbf{x}(t)=\boldsymbol{\Phi}(t,t_0)\mathbf{x}_0=\boldsymbol{\Phi}(t,t_1)\cdot\boldsymbol{\Phi}(t_1,t_0)\mathbf{x}_0$$

Since $\mathbf{x}_0$ is an arbitrary n-dimensional vector, hence (6.5.123) must be valid.

The equation (6.5.121) is to be revised now. Suppose the integration has reached t_k that the vector $\hat{\mathbf{x}}_k$ is computed as

$$\hat{\mathbf{x}}_k=\boldsymbol{\Phi}(t_k,t_0)\hat{\mathbf{x}}_0+\boldsymbol{\Phi}(t_k,t_0)\int_{t_0}^{t_k}\boldsymbol{\Phi}^{-1}(\tau,t_0)[\mathbf{P}(\tau)\mathbf{C}^T\mathbf{V}^{-1}\mathbf{y}+\mathbf{B}_u\mathbf{u}]\mathrm{d}\tau$$

It requires to integrate the $\hat{\mathbf{x}}_{k+1}=\hat{\mathbf{x}}(t_{k+1})$ at the next time station t_{k+1}. The derivation is given as

$$\hat{\mathbf{x}}(t_{k+1})=\mathbf{\Phi}(t_{k+1},t_0)\cdot\left[\hat{\mathbf{x}}_0+\int_{t_0}^{t_{k+1}}\mathbf{\Phi}^{-1}(\tau,t_0)[\mathbf{P}(\tau)\mathbf{C}^T\mathbf{V}^{-1}\mathbf{y}+\mathbf{B}_u\mathbf{u}]\mathrm{d}\tau\right]$$
$$=\mathbf{\Phi}(t_{k+1},t_k)\mathbf{\Phi}(t_k,t_0)\left[\hat{\mathbf{x}}_0+\left(\int_{t_0}^{t_k}+\int_{t_k}^{t_{k+1}}\right)\mathbf{\Phi}^{-1}(\tau,t_0)[\mathbf{P}(\tau)\mathbf{C}^T\mathbf{V}^{-1}\mathbf{y}+\mathbf{B}_u\mathbf{u}]\mathrm{d}\tau\right]$$
$$=\mathbf{\Phi}(t_{k+1},t_k)\cdot\left[\hat{\mathbf{x}}_k+\int_{t_k}^{t_{k+1}}\mathbf{\Phi}(t_k,t_0)\left[\mathbf{\Phi}(\tau,t_k)\mathbf{\Phi}(t_k,t_0)\right]^{-1}[\mathbf{P}(\tau)\mathbf{C}^T\mathbf{V}^{-1}\mathbf{y}+\mathbf{B}_u\mathbf{u}]\mathrm{d}\tau\right]$$

so that

$$\hat{\mathbf{x}}_{k+1}=\mathbf{\Phi}(t_{k+1},t_k)\cdot\hat{\mathbf{x}}_k+\mathbf{\Phi}(t_{k+1},t_k)\int_{t_k}^{t_{k+1}}\mathbf{\Phi}^{-1}(\tau,t_k)[\mathbf{P}(\tau)\mathbf{C}^T\mathbf{V}^{-1}\mathbf{y}+\mathbf{B}_u\mathbf{u}]\mathrm{d}\tau \qquad (6.5.124)$$

which can be interpreted as that, the integration begins from t_k with the initial value $\hat{\mathbf{x}}_k$. The integration is one time step, and the situation becomes one step forward. The numerical integration is better to use the above equation (6.5.124), and the matrix $\mathbf{\Phi}(t_{k+1},t_k)$ is needed. The single step mixed energy matrices $\mathbf{F}(\eta),\mathbf{G}(\eta),\mathbf{Q}(\eta)$ has been obtained for step size $\eta=t_{k+1}-t_k$ and also the matrices $\mathbf{P}(t_k)$ and $\mathbf{P}(t_{k+1})$ are also obtained off-line. Thus

$$\mathbf{\Phi}(t_{k+1},t_k)=\mathbf{F}(\eta)[\mathbf{I}_n+\mathbf{P}(t_k)\mathbf{Q}(\eta)]^{-1},\quad \eta=t_{k+1}-t_k \qquad (6.5.125)$$

which is just the matrix $\mathbf{F}_c$ in equation (6.5.90) or (6.5.116), and can be termed as the ***single step transition matrix of the filter equation***. Until here, the derivations are all exact. But equation (6.5.124) has a fixed end integration, which can be computed only with some numerical approximation. Further investigation is needed.

The simplest method is to use the integration rules, such as trapezoidal approximation etc., but such method has not used enough the knowledge of system characteristics.

Although for arbitrary vectors $\mathbf{y}(\tau),\mathbf{u}(\tau)$ the precise integration method cannot apply, however, if these vectors can be ***linearly interpolated in the interval*** (t_k,t_{k+1})***, then equation (6.5.124) can still be integrated precisely.*** Similarly, ***when the matrix*** $\mathbf{H}$ ***has no Jordan form, the eigen-solution based analytical method can also be applied.*** Equation (6.5.120) explains that the homogeneous equation (6.5.119) can be solved firstly using the matrix $\mathbf{G}(t)$ instead of the matrix $\mathbf{P}(t)$, thereafter using the transformation (6.5.120) to obtain the solution of equation (6.5.119). To verify the filter equation (6.5.49), let $\eta=t-t_a$ and the equations to be solved first are

$$\dot{\mathbf{r}}_x(t_a,t)=[\mathbf{A}-\mathbf{G}(\eta)\mathbf{C}^T\mathbf{V}^{-1}\mathbf{C}]\mathbf{r}_x+\mathbf{G}(\eta)\mathbf{C}^T\mathbf{V}^{-1}\mathbf{y}+\mathbf{B}_u\mathbf{u},\quad \mathbf{r}_x(t_a,t_a)=\mathbf{0} \qquad (6.5.126a)$$

$$\dot{\mathbf{r}}_\lambda(t_a,t)=\mathbf{F}^T\mathbf{C}^T\mathbf{V}^{-1}(\mathbf{y}-\mathbf{C}\mathbf{r}_x),\qquad \mathbf{r}_\lambda(t_a,t_a)=\mathbf{0} \qquad (6.5.126b)$$

which are the equations (6.5.81~82). Thereafter execute the transformation

$$\mathbf{r}_{xp}(t_a,t)=\mathbf{r}_x(t_a,t)+\mathbf{F}(\eta)[\mathbf{I}_n+\mathbf{P}(t_a)\mathbf{Q}(\eta)]^{-1}\mathbf{P}(t_a)\mathbf{r}_\lambda(t_a,t) \qquad (6.5.127)$$

where t_a can be selected as t_k and then let $t=t_{k+1}$. From (6.5.126b) gives

$$\mathbf{r}_\lambda(t_k,t_{k+1})=\int_{t_k}^{t_{k+1}}\mathbf{F}^{\mathrm{T}}\mathbf{C}^{\mathrm{T}}\mathbf{V}^{-1}(\mathbf{y}-\mathbf{C}\mathbf{r}_x)\mathrm{d}t$$

therefore the equation (6.5.124) can be given as

$$\hat{\mathbf{x}}_{k+1}=\mathbf{\Phi}(t_{k+1},t_k)\cdot\hat{\mathbf{x}}_k+\mathbf{r}_{xp}(t_k,t_{k+1}) \tag{6.5.124'}$$

Verification is needed that $\mathbf{r}_{xp}(t_{\mathrm{a}},t)$ of (6.5.127) does satisfy the differential equation (6.5.49), which is given as

$$\dot{\mathbf{r}}_{xp}=\dot{\mathbf{r}}_x+\dot{\mathbf{F}}[\mathbf{I}_n+\mathbf{P}(t_{\mathrm{a}})\mathbf{Q}]^{-1}\mathbf{P}(t_{\mathrm{a}})\mathbf{r}_\lambda+\mathbf{F}[\mathbf{I}_n+\mathbf{P}(t_0)\mathbf{Q}]^{-1}\mathbf{P}(t_{\mathrm{a}})\dot{\mathbf{r}}_\lambda$$
$$-\mathbf{F}[\mathbf{I}_n+\mathbf{P}(t_{\mathrm{a}})\mathbf{Q}]^{-1}\mathbf{P}(t_{\mathrm{a}})\dot{\mathbf{Q}}[\mathbf{I}+\mathbf{P}(t_{\mathrm{a}})\mathbf{Q}]^{-1}\mathbf{P}(t_{\mathrm{a}})\mathbf{r}_\lambda$$

using the differential equations for $\dot{\mathbf{F}}$, $\dot{\mathbf{Q}}$ and $\dot{\mathbf{r}}_x$, it is verified as

$$\begin{aligned}\dot{\mathbf{r}}_{xp}&=(\mathbf{A}-\mathbf{G}\mathbf{C}^{\mathrm{T}}\mathbf{V}^{-1}\mathbf{C})\mathbf{r}_x+(\mathbf{A}-\mathbf{G}\mathbf{C}^{\mathrm{T}}\mathbf{V}^{-1}\mathbf{C})\mathbf{F}[\mathbf{I}+\mathbf{P}(t_{\mathrm{a}})\mathbf{Q}]^{-1}\mathbf{P}(t_{\mathrm{a}})\mathbf{r}_\lambda+\mathbf{B}_u\mathbf{u}\\
&\quad-\mathbf{F}[\mathbf{I}_n+\mathbf{P}(t_{\mathrm{a}})\mathbf{Q}]^{-1}\mathbf{P}(t_{\mathrm{a}})\mathbf{F}^{\mathrm{T}}\mathbf{C}^{\mathrm{T}}\mathbf{V}^{-1}\mathbf{C}\mathbf{F}[\mathbf{I}_n+\mathbf{P}(t_{\mathrm{a}})\mathbf{Q}]^{-1}\mathbf{P}(t_{\mathrm{a}})\mathbf{r}_\lambda+\mathbf{G}\mathbf{C}^{\mathrm{T}}\mathbf{V}^{-1}\mathbf{y}\\
&=[\mathbf{A}-(\mathbf{G}+\mathbf{F}[\mathbf{I}_n+\mathbf{P}(t_{\mathrm{a}})\mathbf{Q}]^{-1}\mathbf{P}(t_{\mathrm{a}})\mathbf{F}^{\mathrm{T}})\mathbf{C}^{\mathrm{T}}\mathbf{V}^{-1}\mathbf{C}]\mathbf{r}_x\\
&\quad+[\mathbf{G}+\mathbf{F}[\mathbf{I}_n+\mathbf{P}(t_{\mathrm{a}})\mathbf{Q}]^{-1}\mathbf{P}(t_{\mathrm{a}})\mathbf{F}^{\mathrm{T}}]\mathbf{C}^{\mathrm{T}}\mathbf{V}^{-1}\mathbf{y}+\mathbf{B}_u\mathbf{u}\\
&\quad+[\mathbf{A}-(\mathbf{G}+\mathbf{F}[\mathbf{I}_n+\mathbf{P}(t_{\mathrm{a}})\mathbf{Q}]^{-1}\mathbf{P}(t_{\mathrm{a}})\mathbf{F}^{\mathrm{T}})\mathbf{C}^{\mathrm{T}}\mathbf{V}^{-1}\mathbf{C}]\mathbf{F}[\mathbf{I}_n+\mathbf{P}(t_{\mathrm{a}})\mathbf{Q}]^{-1}\mathbf{P}(t_{\mathrm{a}})\mathbf{r}_\lambda\\
&=[\mathbf{A}-\mathbf{P}(t)\mathbf{C}^{\mathrm{T}}\mathbf{V}^{-1}\mathbf{C}]\left(\mathbf{r}_x+\mathbf{F}[\mathbf{I}_n+\mathbf{P}(t_{\mathrm{a}})\mathbf{Q}]^{-1}\mathbf{P}(t_{\mathrm{a}})\mathbf{r}_\lambda\right)+\mathbf{P}(t)\mathbf{C}^{\mathrm{T}}\mathbf{V}^{-1}\mathbf{y}+\mathbf{B}_u\mathbf{u}\end{aligned}$$

hence

$$\dot{\mathbf{r}}_{xp}=[\mathbf{A}-\mathbf{P}(t)\mathbf{C}^{\mathrm{T}}\mathbf{V}^{-1}\mathbf{C}]\mathbf{r}_{xp}+\mathbf{P}(t)\mathbf{C}^{\mathrm{T}}\mathbf{V}^{-1}\mathbf{y}+\mathbf{B}_u\mathbf{u} \tag{6.5.127'}$$

which is just the equation (6.5.49). The initial condition $\mathbf{r}_{xp}(t_{\mathrm{a}},t_{\mathrm{a}})=\mathbf{0}$ is ensured from (6.5.127) by the initial conditions in (6.5.126).

Comparing the equations (6.5.124) and (6.5.124') determines that $\mathbf{r}_{xp}$ is the integration term in (6.5.124). The verification is that, differentiating this term with respect to t_{k+1} knows that it satisfies the differential equation (6.5.49), and the initial condition when $t_{k+1}\to t_k$ is $\mathbf{0}$, which determines that this term is just $\mathbf{r}_{xp}$. Equation (6.5.127) explains that the computation of $\mathbf{r}_{xp}$ can be given as ***first computing*** $\mathbf{r}_x(t_k,t_{k+1})$ ***and*** $\mathbf{r}_\lambda(t_k,t_{k+1})$, ***then using the transformation*** (6.5.127) ***to obtain*** $\mathbf{r}_{xp}$. The computation of $\mathbf{\Phi}(t_{k+1},t_k)$ is also, first to compute the matrix $\mathbf{F}(\eta)$ with zero initial condition, thereafter to compute by the transformation (6.5.125).

The problem is, therefore, reduced to ***find the single step integration of*** $\mathbf{r}_{xp}$.

§6.5.7.1, Analytical single step integration for the filter equation

Consider the equation set of (6.5.80) and (6.5.81), that the equations for $\mathbf{r}_x$ and $\mathbf{r}_\lambda$ are of the same class to the equations for $\mathbf{F},\mathbf{G},\mathbf{Q}$. The interval combination equation (6.5.77) and (6.5.78) explains that the differential equation (6.5.126a,b) for $\mathbf{r}_x$ and $\mathbf{r}_\lambda$ can also be solved by the precise integration method. After the precise integration method for $\mathbf{Q}(\eta),\mathbf{F}(\eta),\mathbf{G}(\eta)$ is given in section 6.5.5, the analytical solution method is given in section 6.5.6. This fact implies that $\mathbf{r}_x$ and $\mathbf{r}_\lambda$ can also be integrated by the analytical method, which is given below.

Analytical integration method uses still the eigen-solution expansion. Rewriting the dual equation (6.5.47) as

$$\dot{\mathbf{v}} = \mathbf{H}\mathbf{v} + \mathbf{f}_1, \quad \mathbf{f}_1 = \begin{Bmatrix} \mathbf{B}_u\mathbf{u} \\ \mathbf{C}^{\mathrm{T}}\mathbf{V}^{-1}\mathbf{y} \end{Bmatrix}, \quad \mathbf{v} \underset{\mathrm{def}}{=} \begin{Bmatrix} \mathbf{x} \\ \boldsymbol{\lambda} \end{Bmatrix} \tag{6.5.108'}$$

of which the homogeneous equation solution corresponds to the computation of matrices $\mathbf{F}(\eta), \mathbf{G}(\eta), \mathbf{Q}(\eta)$, hence the inhomogeneous equation solution should be proposed further. The two end zero conditions are written as

$$\mathbf{x}(t_k) = \mathbf{0} \quad \text{and} \quad \boldsymbol{\lambda}(t_{k+1}) = \mathbf{0}.$$

Note that, the vectors $\mathbf{y}$ and $\mathbf{u}$ can only be given at the real time. No exact analytical expression is available in the interval (t_k, t_{k+1}) for arbitrary vectors of $\mathbf{y}$ and $\mathbf{u}$. Only at the grid point t_k and t_{k+1} the value of vectors $\mathbf{y}_k, \mathbf{u}_k, \mathbf{y}_{k+1}, \mathbf{u}_{k+1}$ or at the past time such as $\mathbf{y}_{k-1}, \mathbf{u}_{k-1}$ or $\dot{\mathbf{y}}_k, \dot{\mathbf{u}}_k$ etc. are available. Within the time interval (t_k, t_{k+1}) only interpolation function can be supplied. The interpolation usually uses simple functions, such as linear, quadratic functions etc. The vectors $\mathbf{y}$ and $\mathbf{u}$ are q- and m-dimensional vectors, respectively, their values cannot be given beforehand, however, they can be ***composed of*** q***- and*** m***-dimensional basis vector functions***, respectively. These basis vectors can be selected as the column vectors of $\mathbf{I}_q$ and $\mathbf{I}_m$ unit matrices. Therefore the inhomogeneous 'external force' vectors in the equation (6.5.108') should be extended as

$$\mathbf{f} = \begin{bmatrix} \mathbf{B}_u & \mathbf{0} \\ \mathbf{0} & \mathbf{C}^{\mathrm{T}}\mathbf{V}^{-1} \end{bmatrix} \begin{matrix} n \\ n \end{matrix} \quad (\text{columns: } m,\ q) \quad \text{and} \quad \mathbf{f}_1 = \mathbf{f} \cdot \begin{Bmatrix} \mathbf{u} \\ \mathbf{y} \end{Bmatrix} \tag{6.5.128a}$$

where $\mathbf{y}$ and $\mathbf{u}$ are interpolation functions in the interval (t_k, t_{k+1}). ***The integration of all the*** $(m+q)$ ***columns of the matrix*** $\mathbf{f}$ ***used as the inhomogeneous force terms in (6.5.108') can be executed off-line.*** Then at the system running time, the real computation needs only to execute matrix multiplication. Further within the interval (t_k, t_{k+1}), the interpolation for $\mathbf{u}$ and $\mathbf{y}$ can be

$$\mathbf{u}(\tau) = \mathbf{u}_0 + \tau\mathbf{u}_1 + \tau^2\mathbf{u}_2, \quad \mathbf{y}(\tau) = \mathbf{y}_0 + \tau\mathbf{y}_1 + \tau^2\mathbf{y}_2, \quad \tau = t - t_k \tag{6.5.128b}$$

where $\mathbf{u}_0, \mathbf{u}_1, \mathbf{u}_2$ and $\mathbf{y}_0, \mathbf{y}_1, \mathbf{y}_2$ are coefficient vectors to be determined according to the control and measurement data given at the real time. However, the ***matrix*** $\mathbf{f}$ ***is independent on the real time data, so that the*** $\mathbf{f}$ ***related computation can be executed off-line beforehand.*** Expanding the columns of matrix $\mathbf{f}$ with the eigenvectors of matrix $\mathbf{H}$

$$\mathbf{f} = \begin{bmatrix} \mathbf{X}_\alpha \\ \mathbf{N}_\alpha \end{bmatrix} \mathbf{f}_a + \begin{bmatrix} \mathbf{X}_\beta \\ \mathbf{N}_\beta \end{bmatrix} \mathbf{f}_b = \mathbf{\Psi} \begin{bmatrix} \mathbf{f}_a \\ \mathbf{f}_b \end{bmatrix} \begin{matrix} n \\ n \end{matrix} \quad (\text{columns: } (m+q)) \tag{6.5.129}$$

where $\mathbf{f}_a, \mathbf{f}_b$ are $(m+q)\times n$ coefficient matrices to be determined. Since $\mathbf{\Psi}$ is a symplectic matrix, left multiplying the above equation with $\mathbf{\Psi}^T\mathbf{J}$ gives

$$\begin{bmatrix} \mathbf{f}_a \\ \mathbf{f}_b \end{bmatrix} = -\mathbf{J}\mathbf{\Psi}^T\mathbf{J}\mathbf{f} = \mathbf{J}\mathbf{\Psi}^T\begin{bmatrix} \mathbf{0} & \mathbf{C}^{\mathrm{T}}\mathbf{V}^{-1} \\ \mathbf{B}_u & \mathbf{0} \end{bmatrix} \tag{6.5.130}$$

Thus $\mathbf{f}_a, \mathbf{f}_b$ can be computed ***off-line***, that $\mathbf{f}_a, \mathbf{f}_b$ are known matrices at the running time. Substituting the above equations into (6.5.108') derives the differential equation for τ

$$\dot{\mathbf{v}}(\tau) = \mathbf{H}\mathbf{v} + \mathbf{\Psi}\begin{bmatrix} \mathbf{f}_a \\ \mathbf{f}_b \end{bmatrix}\left(\begin{Bmatrix} \mathbf{u}_0 \\ \mathbf{y}_0 \end{Bmatrix} + \tau\begin{Bmatrix} \mathbf{u}_1 \\ \mathbf{y}_1 \end{Bmatrix} + \tau^2\begin{Bmatrix} \mathbf{u}_2 \\ \mathbf{y}_2 \end{Bmatrix}\right), \quad \begin{matrix} \mathbf{x}_0 = \mathbf{0} \\ \boldsymbol{\lambda}(\eta) = \mathbf{0} \end{matrix} \tag{6.5.131}$$

The solution of $\mathbf{v}(\tau)$ can also be expanded with the eigenvectors

$$\mathbf{v}(\tau) = \mathbf{\Psi}\begin{Bmatrix} \mathbf{a}(\tau) \\ \mathbf{b}(\tau) \end{Bmatrix}, \quad \begin{matrix} \mathbf{a}(\tau) \underset{\text{def}}{=} \{a_1(\tau), a_2(\tau), \cdots, a_n(\tau)\}^{\mathrm{T}} \\ \mathbf{b}(\tau) \underset{\text{def}}{=} \{b_1(\tau), b_2(\tau), \cdots, b_n(\tau)\}^{\mathrm{T}} \end{matrix} \tag{6.5.132}$$

Thus the equations for the components of vectors $\mathbf{a}(\tau)$ and $\mathbf{b}(\tau)$ [$\mathbf{f}_{ai}$ is the i-th row of $\mathbf{f}_a$]

$$\dot{a}_i(\tau) = \mu_i a_i(\tau) + \mathbf{f}_{ai}\left(\begin{Bmatrix} \mathbf{u}_0 \\ \mathbf{y}_0 \end{Bmatrix} + \tau\begin{Bmatrix} \mathbf{u}_1 \\ \mathbf{y}_1 \end{Bmatrix} + \tau^2\begin{Bmatrix} \mathbf{u}_2 \\ \mathbf{y}_2 \end{Bmatrix}\right) \tag{6.5.133a}$$

$$\dot{b}_i(\tau) = -\mu_i b_i(\tau) + \mathbf{f}_{bi}\left(\begin{Bmatrix} \mathbf{u}_0 \\ \mathbf{y}_0 \end{Bmatrix} + \tau\begin{Bmatrix} \mathbf{u}_1 \\ \mathbf{y}_1 \end{Bmatrix} + \tau^2\begin{Bmatrix} \mathbf{u}_2 \\ \mathbf{y}_2 \end{Bmatrix}\right) \tag{6.5.133b}$$

Integration of these equations needs to introduce the following functions

$$\int_0^\eta e^{\mu_i(\eta-\tau)}\mathrm{d}\tau = (e^{\mu_i\eta} - 1)/\mu_i \quad \underset{\text{def}}{=} \mathrm{e}_0(\mu_i)$$

$$\int_0^\eta e^{\mu_i(\eta-\tau)}\tau\mathrm{d}\tau = -\eta/\mu_i + (e^{\mu_i\eta} - 1)/\mu_i^2 \quad \underset{\text{def}}{=} \mathrm{e}_1(\mu_i)$$

$$\int_0^\eta e^{\mu_i(\eta-\tau)}\tau^2\mathrm{d}\tau = -\eta^2/\mu_i - 2\eta/\mu_i^2 + 2(e^{\mu_i\eta} - 1)/\mu_i^3 \quad \underset{\text{def}}{=} \mathrm{e}_2(\mu_i)$$

Note that, the functions $\mathrm{e}_1, \mathrm{e}_2, \mathrm{e}_3$ are not bold and are different to $\mathbf{e}_i$ in equation (6.5.98). Hence

$$a_i(\eta) = \mathbf{f}_{ai}\left[\begin{Bmatrix} \mathbf{u}_0 \\ \mathbf{y}_0 \end{Bmatrix}\mathrm{e}_0(\mu_i) + \begin{Bmatrix} \mathbf{u}_1 \\ \mathbf{y}_1 \end{Bmatrix}\mathrm{e}_1(\mu_i) + \begin{Bmatrix} \mathbf{u}_2 \\ \mathbf{y}_2 \end{Bmatrix}\mathrm{e}_2(\mu_i)\right] + a_i(0)e^{\mu_i\eta}$$

$$b_i(\eta) = \mathbf{f}_{bi}\left[\begin{Bmatrix} \mathbf{u}_0 \\ \mathbf{y}_0 \end{Bmatrix}\mathrm{e}_0(-\mu_i) + \begin{Bmatrix} \mathbf{u}_1 \\ \mathbf{y}_1 \end{Bmatrix}\mathrm{e}_1(-\mu_i) + \begin{Bmatrix} \mathbf{u}_2 \\ \mathbf{y}_2 \end{Bmatrix}\mathrm{e}_2(-\mu_i)\right] + b_i(0)e^{-\mu_i\eta}$$

where $\mathbf{f}_{ai}, \mathbf{f}_{bi}$ are row vectors picked from the i-th row of the matrices $\mathbf{f}_a, \mathbf{f}_b$, respectively. Let $\mathbf{a}_0 = \{a_1(0), \cdots, a_n(0)\}^T$ and $\mathbf{b}_0 = \{b_1(0), \cdots, b_n(0)\}^T$, the above equations have the form

$$\mathbf{a}(\eta) = \mathrm{diag}[\mathrm{e}_0(\mu_i)]\mathbf{f}_a\begin{Bmatrix} \mathbf{u}_0 \\ \mathbf{y}_0 \end{Bmatrix} + \mathrm{diag}[\mathrm{e}_1(\mu_i)]\mathbf{f}_a\begin{Bmatrix} \mathbf{u}_1 \\ \mathbf{y}_1 \end{Bmatrix} + \mathrm{diag}[\mathrm{e}_2(\mu_i)]\mathbf{f}_a\begin{Bmatrix} \mathbf{u}_2 \\ \mathbf{y}_2 \end{Bmatrix} + \mathbf{D}_{c\eta}\mathbf{a}_0$$

$$\mathbf{b}(\eta) = \text{diag}[\text{e}_0(-\mu_i)]\mathbf{f}_b\begin{Bmatrix}\mathbf{u}_0\\\mathbf{y}_0\end{Bmatrix} + \text{diag}[\text{e}_1(-\mu_i)]\mathbf{f}_b\begin{Bmatrix}\mathbf{u}_1\\\mathbf{y}_1\end{Bmatrix} + \text{diag}[\text{e}_2(-\mu_i)]\mathbf{f}_b\begin{Bmatrix}\mathbf{u}_2\\\mathbf{y}_2\end{Bmatrix} + \mathbf{D}_{c\eta}^{-1}\mathbf{b}_0 \tag{6.5.134}$$

where $\mathbf{a}_0$ and $\mathbf{b}_0$ are vectors to be determined. According to the two end boundary conditions of (6.5.131) derives

$$\mathbf{X}_\alpha\mathbf{a}_0 + \mathbf{X}_\beta\mathbf{D}_{c\eta}\mathbf{b}'_0 = \mathbf{0}\ , \qquad \text{where} \quad \mathbf{b}'_0 = \mathbf{D}_{c\eta}^{-1}\mathbf{b}_0 \tag{6.5.135a}$$

$$\begin{aligned}\mathbf{N}_\alpha\mathbf{D}_{c\eta}\mathbf{a}_0 + \mathbf{N}_\beta\mathbf{b}'_0 = &-\left(\mathbf{N}_\alpha\text{diag}[\text{e}_0(\mu_i)]\mathbf{f}_a + \mathbf{N}_\beta\text{diag}[\text{e}_0(-\mu_i)]\mathbf{f}_b\right)\begin{Bmatrix}\mathbf{u}_0\\\mathbf{y}_0\end{Bmatrix}\\ &-\left(\mathbf{N}_\alpha\text{diag}[\text{e}_1(\mu_i)]\mathbf{f}_a + \mathbf{N}_\beta\text{diag}[\text{e}_1(-\mu_i)]\mathbf{f}_b\right)\begin{Bmatrix}\mathbf{u}_1\\\mathbf{y}_1\end{Bmatrix}\\ &-\left(\mathbf{N}_\alpha\text{diag}[\text{e}_2(\mu_i)]\mathbf{f}_a + \mathbf{N}_\beta\text{diag}[\text{e}_2(-\mu_i)]\mathbf{f}_b\right)\begin{Bmatrix}\mathbf{u}_2\\\mathbf{y}_2\end{Bmatrix}\end{aligned} \tag{6.5.135b}$$

From the two equations $\mathbf{a}_0, \mathbf{b}_0$ are solved, then the solution $\mathbf{x}$ and $\boldsymbol{\lambda}$ of the inhomogeneous equation (6.5.108') is obtained.

The two vectors $\mathbf{r}_x(t_k, t_{k+1}), \mathbf{r}_\lambda(t_k, t_{k+1})$ in equation (6.5.127) are asserted that

$$\mathbf{r}_x(t_k, t_{k+1}) = \mathbf{x}(\eta), \qquad \mathbf{r}_\lambda(t_k, t_{k+1}) = \boldsymbol{\lambda}(0) \tag{6.5.136a,b}$$

where the vectors $\mathbf{x}$ and $\boldsymbol{\lambda}$ are the components of the ***complete state vector*** $\mathbf{v}$ in equation (6.5.108'). To prove the assertion, rewriting the equation (6.5.108') in dual equation form

$$\dot{\mathbf{x}}(\tau) = \mathbf{A}\mathbf{x} + \mathbf{B}_w\mathbf{W}\mathbf{B}_w^T\boldsymbol{\lambda} + \mathbf{B}_u\mathbf{u}\ \ , \quad \mathbf{x}(\tau = 0) = \mathbf{0} \tag{6.5.137a}$$

$$\dot{\boldsymbol{\lambda}}(\tau) = \mathbf{C}^T\mathbf{V}^{-1}\mathbf{C}\mathbf{x} - \mathbf{A}^T\boldsymbol{\lambda} - \mathbf{C}^T\mathbf{V}^{-1}\mathbf{y}\ \ , \quad \boldsymbol{\lambda}(\eta) = \mathbf{0} \tag{6.5.137b}$$

where $\tau = t - t_k$. Let

$$\mathbf{x}(\tau) = \mathbf{G}(\tau)\boldsymbol{\lambda}(\tau) + \mathbf{r}(\tau) \tag{6.5.137c}$$

where $\mathbf{G}(\tau)$ satisfies the Riccati equation

$$\dot{\mathbf{G}}(\tau) = \mathbf{B}_w\mathbf{W}\mathbf{B}_w^T + \mathbf{A}\mathbf{G} + \mathbf{G}\mathbf{A}^T - \mathbf{G}\mathbf{C}^T\mathbf{V}^{-1}\mathbf{C}\mathbf{G}, \quad \mathbf{G}(0) = \mathbf{0}.$$

Substituting (6.5.137c) into equation (6.5.137a,b) gives

$$\begin{aligned}&\dot{\mathbf{G}}\boldsymbol{\lambda} + \mathbf{G}\dot{\boldsymbol{\lambda}} + \dot{\mathbf{r}} = \mathbf{A}\mathbf{G}\boldsymbol{\lambda} + \mathbf{A}\mathbf{r} + \mathbf{B}_w\mathbf{W}\mathbf{B}_w^T\boldsymbol{\lambda} + \mathbf{B}_u\mathbf{u}\\ &\dot{\boldsymbol{\lambda}} = \mathbf{C}^T\mathbf{V}^{-1}\mathbf{C}\mathbf{G}\boldsymbol{\lambda} + \mathbf{C}^T\mathbf{V}^{-1}\mathbf{C}\mathbf{r} - \mathbf{A}^T\boldsymbol{\lambda} - \mathbf{C}^T\mathbf{V}^{-1}\mathbf{y}, \quad \boldsymbol{\lambda}(\eta) = \mathbf{0}\end{aligned} \tag{6.5.138a}$$

Eliminating $\dot{\boldsymbol{\lambda}}$ and since $\mathbf{G}$ satisfies the Riccati differential equation, so

$$\dot{\mathbf{r}}(\tau) = [\mathbf{A} - \mathbf{G}\mathbf{C}^T\mathbf{V}^{-1}\mathbf{C}]\mathbf{r} + \mathbf{G}\mathbf{C}^T\mathbf{V}^{-1}\mathbf{y} + \mathbf{B}_u\mathbf{u}, \quad \mathbf{r}(0) = \mathbf{0} \tag{6.5.138b}$$

This differential equation is the same as (6.5.126a), so $\mathbf{r}(\tau)$ is just $\mathbf{r}_x(t_a, t)$ in (6.5.126a), and $t_a = t_k$. Since (6.5.137c) and the boundary condition $\boldsymbol{\lambda}(\eta) = \mathbf{0}$, so the assertion (6.5.136a) is proved. The differential equation for $\mathbf{r}(\tau)$ has been separated with $\boldsymbol{\lambda}$.

Taking transpose of the equation (6.5.80c) gives $\dot{\mathbf{F}}^T = \mathbf{F}^T(\mathbf{A}^T - \mathbf{C}^T\mathbf{V}^{-1}\mathbf{C}\mathbf{G})$, and using equations (6.5.138b), (6.5.137b,c) gives

$$d(\mathbf{F}^T\boldsymbol{\lambda})/d\tau = -\mathbf{F}^T\mathbf{C}^T\mathbf{V}^{-1}(\mathbf{y} - \mathbf{C}\mathbf{r}) \quad , \quad \mathbf{F}^T\boldsymbol{\lambda}(\eta) = \mathbf{0}$$

Since $\mathbf{F}(0) = \mathbf{I}_n$, integration gives

$$\boldsymbol{\lambda}(0) = \mathbf{F}^T(0)\boldsymbol{\lambda}(0) = \int_0^\eta \mathbf{F}^T\mathbf{C}^T\mathbf{V}^{-1}(\mathbf{y} - \mathbf{C}\mathbf{r})d\tau$$

which is just the integration of equation (6.5.126b), the assertion (6.5.136b) is thus proved. ##

Numerical computation is investigated further.

§6.5.7.2, Analytical equation for single step integration

Based on the assertion (6.5.136a,b), the equation must be given for computation, and the ***off-line*** and ***on-line*** computations must be distinguished. The coefficient matrix of the left-hand side in equation (6.5.135) is the sub-matrices of $\mathbf{M}(\eta)$ of equation (6.5.110), whose inverse matrix is given in equation (6.5.111). Hence the solutions of $\mathbf{a}_0$ and $\mathbf{b}'_0$ are

$$\begin{aligned} \mathbf{a}_0 &= \mathbf{L}_{01}\begin{Bmatrix}\mathbf{u}_0\\ \mathbf{y}_0\end{Bmatrix} + \mathbf{L}_{11}\begin{Bmatrix}\mathbf{u}_1\\ \mathbf{y}_1\end{Bmatrix} + \mathbf{L}_{21}\begin{Bmatrix}\mathbf{u}_2\\ \mathbf{y}_2\end{Bmatrix} \\ \mathbf{b}'_0 &= \mathbf{L}_{02}\begin{Bmatrix}\mathbf{u}_0\\ \mathbf{y}_0\end{Bmatrix} + \mathbf{L}_{12}\begin{Bmatrix}\mathbf{u}_1\\ \mathbf{y}_1\end{Bmatrix} + \mathbf{L}_{22}\begin{Bmatrix}\mathbf{u}_2\\ \mathbf{y}_2\end{Bmatrix} \end{aligned} \qquad (6.5.139\text{a,b})$$

where

$$\begin{aligned}
\mathbf{L}_{01} &= -\mathbf{A}_2\left[\mathbf{N}_\alpha \mathrm{diag}[\mathrm{e}_0(\mu_i)]\mathbf{f}_a + \mathbf{N}_\beta \mathrm{diag}[\mathrm{e}_0(-\mu_i)]\mathbf{f}_b\right] \\
\mathbf{L}_{11} &= -\mathbf{A}_2\left[\mathbf{N}_\alpha \mathrm{diag}[\mathrm{e}_1(\mu_i)]\mathbf{f}_a + \mathbf{N}_\beta \mathrm{diag}[\mathrm{e}_1(-\mu_i)]\mathbf{f}_b\right] \\
\mathbf{L}_{21} &= -\mathbf{A}_2\left[\mathbf{N}_\alpha \mathrm{diag}[\mathrm{e}_2(\mu_i)]\mathbf{f}_a + \mathbf{N}_\beta \mathrm{diag}[\mathrm{e}_2(-\mu_i)]\mathbf{f}_b\right] \\
\mathbf{L}_{02} &= -\mathbf{B}_2\left[\mathbf{N}_\alpha \mathrm{diag}[\mathrm{e}_0(\mu_i)]\mathbf{f}_a + \mathbf{N}_\beta \mathrm{diag}[\mathrm{e}_0(-\mu_i)]\mathbf{f}_b\right] \\
\mathbf{L}_{12} &= -\mathbf{B}_2\left[\mathbf{N}_\alpha \mathrm{diag}[\mathrm{e}_1(\mu_i)]\mathbf{f}_a + \mathbf{N}_\beta \mathrm{diag}[\mathrm{e}_1(-\mu_i)]\mathbf{f}_b\right] \\
\mathbf{L}_{22} &= -\mathbf{B}_2\left[\mathbf{N}_\alpha \mathrm{diag}[\mathrm{e}_2(\mu_i)]\mathbf{f}_a + \mathbf{N}_\beta \mathrm{diag}[\mathrm{e}_2(-\mu_i)]\mathbf{f}_b\right]
\end{aligned}$$

The matrices $\mathbf{L}$ are all $n\times(m+q)$ dimensioned, and are independent on the measurement and control vectors, so that they can be computed ***off-line*** beforehand. The matrices $\mathbf{A}_2$ and $\mathbf{B}_2$ are given in equation (6.5.113) and can be computed ***off-line*** too. According to (6.5.134) it derives

$$\begin{aligned} \mathbf{a}(\eta) &= (\mathbf{D}_{c\eta}\mathbf{L}_{01} + \mathrm{diag}[\mathrm{e}_0(\mu_i)]\mathbf{f}_a)\begin{Bmatrix}\mathbf{u}_0\\ \mathbf{y}_0\end{Bmatrix} \\ &+ (\mathbf{D}_{c\eta}\mathbf{L}_{11} + \mathrm{diag}[\mathrm{e}_1(\mu_i)]\mathbf{f}_a)\begin{Bmatrix}\mathbf{u}_1\\ \mathbf{y}_1\end{Bmatrix} + (\mathbf{D}_{c\eta}\mathbf{L}_{21} + \mathrm{diag}[\mathrm{e}_2(\mu_i)]\mathbf{f}_a)\begin{Bmatrix}\mathbf{u}_2\\ \mathbf{y}_2\end{Bmatrix} \end{aligned} \qquad (6.5.140\text{a})$$

$$\mathbf{b}(\eta) = (\mathbf{L}_{02} + \mathrm{diag}[\mathrm{e}_0(-\mu_i)]\mathbf{f}_b)\begin{Bmatrix}\mathbf{u}_0\\ \mathbf{y}_0\end{Bmatrix} + (\mathbf{L}_{12} + \mathrm{diag}[\mathrm{e}_1(-\mu_i)]\mathbf{f}_b)\begin{Bmatrix}\mathbf{u}_1\\ \mathbf{y}_1\end{Bmatrix} + (\mathbf{L}_{22} + \mathrm{diag}[\mathrm{e}_2(-\mu_i)]\mathbf{f}_b)\begin{Bmatrix}\mathbf{u}_2\\ \mathbf{y}_2\end{Bmatrix} \quad (6.5.140b)$$

The matrices in the parenthesis of the above equations are all $n\times(m+q)$-dimensioned, and can be computed ***off-line*** for a given step size η. According to equation (6.5.136) it gives

$$\mathbf{r}_x(t_k,t_{k+1}) = \mathbf{X}_\alpha\mathbf{a}(\eta) + \mathbf{X}_\beta\mathbf{b}(\eta) = \mathbf{M}_{x0}\begin{Bmatrix}\mathbf{u}_0\\ \mathbf{y}_0\end{Bmatrix} + \mathbf{M}_{x1}\begin{Bmatrix}\mathbf{u}_1\\ \mathbf{y}_1\end{Bmatrix} + \mathbf{M}_{x2}\begin{Bmatrix}\mathbf{u}_2\\ \mathbf{y}_2\end{Bmatrix}$$

$$\begin{aligned}
\mathbf{M}_{x0} &= \left[\mathbf{X}_\alpha(\mathbf{D}_{c\eta}\mathbf{L}_{01} + \mathrm{diag}[\mathrm{e}_0(\mu_i)]\mathbf{f}_a) + \mathbf{X}_\beta(\mathbf{L}_{02} + \mathrm{diag}[\mathrm{e}_0(-\mu_i)]\mathbf{f}_b)\right]\\
\mathbf{M}_{x1} &= \left[\mathbf{X}_\alpha(\mathbf{D}_{c\eta}\mathbf{L}_{11} + \mathrm{diag}[\mathrm{e}_1(\mu_i)]\mathbf{f}_a) + \mathbf{X}_\beta(\mathbf{L}_{12} + \mathrm{diag}[\mathrm{e}_1(-\mu_i)]\mathbf{f}_b)\right] \quad (6.5.141a)\\
\mathbf{M}_{x2} &= \left[\mathbf{X}_\alpha(\mathbf{D}_{c\eta}\mathbf{L}_{21} + \mathrm{diag}[\mathrm{e}_2(\mu_i)]\mathbf{f}_a) + \mathbf{X}_\beta(\mathbf{L}_{22} + \mathrm{diag}[\mathrm{e}_2(-\mu_i)]\mathbf{f}_b)\right]
\end{aligned}$$

$$\mathbf{r}_\lambda(t_k,t_{k+1}) = \mathbf{N}_\alpha\mathbf{a}_0 + \mathbf{N}_\beta\mathbf{D}_{c\eta}\mathbf{b}_0' = \mathbf{M}_{\lambda 0}\begin{Bmatrix}\mathbf{u}_0\\ \mathbf{y}_0\end{Bmatrix} + \mathbf{M}_{\lambda 1}\begin{Bmatrix}\mathbf{u}_1\\ \mathbf{y}_1\end{Bmatrix} + \mathbf{M}_{\lambda 2}\begin{Bmatrix}\mathbf{u}_2\\ \mathbf{y}_2\end{Bmatrix},$$

$$\mathbf{M}_{\lambda 0} = (\mathbf{N}_\alpha\mathbf{L}_{01} + \mathbf{N}_\beta\mathbf{D}_{c\eta}\mathbf{L}_{02}) \;,\quad \mathbf{M}_{\lambda 1} = (\mathbf{N}_\alpha\mathbf{L}_{11} + \mathbf{N}_\beta\mathbf{D}_{c\eta}\mathbf{L}_{12}) \quad (6.5.141b)$$

$$\mathbf{M}_{\lambda 2} = (\mathbf{N}_\alpha\mathbf{L}_{21} + \mathbf{N}_\beta\mathbf{D}_{c\eta}\mathbf{L}_{22})$$

Substituting these into equation (6.5.127) gives

$$\mathbf{r}_{xp}(t_k,t_{k+1}) = \mathbf{M}_{p0}\begin{Bmatrix}\mathbf{u}_0\\ \mathbf{y}_0\end{Bmatrix} + \mathbf{M}_{p1}\begin{Bmatrix}\mathbf{u}_1\\ \mathbf{y}_1\end{Bmatrix} + \mathbf{M}_{p2}\begin{Bmatrix}\mathbf{u}_2\\ \mathbf{y}_2\end{Bmatrix} \quad (6.5.142)$$

$$\mathbf{M}_{p0} = \mathbf{M}_{x0} + \mathbf{\Phi}_{k+1,k}\mathbf{P}_k\mathbf{M}_{\lambda 0}\,,\qquad \mathbf{M}_{p1} = \mathbf{M}_{x1} + \mathbf{\Phi}_{k+1,k}\mathbf{P}_k\mathbf{M}_{\lambda 1}$$

$$\mathbf{M}_{p2} = \mathbf{M}_{x2} + \mathbf{\Phi}_{k+1,k}\mathbf{P}_k\mathbf{M}_{\lambda 2}\,,\qquad \mathbf{\Phi}_{k+1,k} = \mathbf{\Phi}(t_{k+1},t_k),\qquad \mathbf{P}_k = \mathbf{P}(t_k)$$

Equation (6.5.124) becomes

$$\hat{\mathbf{x}}_{k+1} = \mathbf{\Phi}(t_{k+1},t_k)\hat{\mathbf{x}}_k + \mathbf{M}_{p0}\begin{Bmatrix}\mathbf{u}_0\\ \mathbf{y}_0\end{Bmatrix} + \mathbf{M}_{p1}\begin{Bmatrix}\mathbf{u}_1\\ \mathbf{y}_1\end{Bmatrix} + \mathbf{M}_{p2}\begin{Bmatrix}\mathbf{u}_2\\ \mathbf{y}_2\end{Bmatrix} \quad (6.5.124")$$

This equation can be used for ***real time*** integration. The matrices $\mathbf{\Phi}(t_{k+1},t_k)$ and $\mathbf{M}_{pi}(i=0,1,2)$ can be computed ***off-line***, because they are independent on $\mathbf{y}$ and $\mathbf{u}$. The matrices $\mathbf{M}_{pi}$ may depend on the station number k. Because of the distinction of off- and on-line computations, the ***real time computation is limited only the equation*** (6.5.124") that the expense is greatly reduced. The real time operations are a $n\times n$ matrix multiplying a n vector and a $n\times(m+q)$ matrix multiplying a $(m+q)$ vector for three times, totally $n^2+3\times n\times(m+q)$ multiplication.

The eigen-solution expansion method has numerical problem when the Jordan normal form appears. However, the precise integration method does not affected by the possible Jordan normal form and still gives precise numerical results. The precise integration method is given in the next section.

§6.5.7.3, Taylor expansion of precise stepwise filtering

The superposition principle applies to linear system. For time-invariant system each time step η has identical characteristics, which implies that the mixed energy matrices $\mathbf{Q}(\eta),\mathbf{G}(\eta),\mathbf{F}(\eta)$ are unified. The difference only appears on different measurement $\mathbf{y}$ and control input $\mathbf{u}$. The measurement value $\mathbf{y}$ is assumed linear within the η interval. According to the superposition principle, if for both arbitrary constant $\mathbf{y}$ and linear measurement $\mathbf{y}=\mathbf{y}_1\cdot(t-t_a)$ the precise integration of system response have been carried out, then for arbitrary linear distributed measurement $\mathbf{y}$ in the time step η the system response can be obtained by the superposition principle. More clearly, let the left end be $t_a=0$ and denote the q-dimensional measurement vector $\mathbf{y}_0$. Let $\mathbf{Y}_0=\mathbf{I}_q$, its q column vectors are the basis of constant measurement vectors, that any vector $\mathbf{y}_0$ can be composed of these basis vectors. To model the linear measurement $\mathbf{y}=\mathbf{y}_1\cdot(t-t_a)$, the q columns of the matrix $\mathbf{Y}_1=\mathbf{I}_q\cdot(t-t_a)$ are used as the basis vectors. Any measurement in the time interval η can be linearly composed of these $q+q$ basis vectors. The similar basis vectors also apply to the control vector $\mathbf{u}$.

The precise integration method further subdivides the single step η into a large number of sub-intervals, as

$$\tau_e=\eta/2^N, \quad \text{such as } N=20, \quad 2^N=1048576 \text{ sub-intervals}$$

with very short step size τ_e. Introducing τ_e here is to distinguish the variable τ. For each η interval, the measurement $\mathbf{y}$ and control input $\mathbf{u}$ can be superimposed with the columns of the matrices

$$\mathbf{Y}_0=\mathbf{I}_q,\ \mathbf{Y}_1=\mathbf{I}_q\cdot(t-t_a), \quad \text{and} \quad \mathbf{U}_0=\mathbf{I}_m,\ \mathbf{U}_1=\mathbf{I}_m\cdot(t-t_a) \tag{6.5.143}$$

totally $2q$ measurement vector ***basis*** and $2m$ control input vector ***basis***. These basis vectors are measurement independent. At the points t_a and $t_a+\eta$ the vectors are denoted as $\mathbf{y}_0,\mathbf{u}_0$, and $\mathbf{y}_1,\mathbf{u}_1$, respectively, then the composition is

$$\mathbf{y}(\tau)=\mathbf{I}_q\mathbf{y}_0+\mathbf{I}_q(\mathbf{y}_1-\mathbf{y}_0)\tau/\eta \quad \text{and} \quad \mathbf{u}(\tau)=\mathbf{I}_m\mathbf{u}_0+\mathbf{I}_m(\mathbf{u}_1-\mathbf{u}_0)\tau/\eta \tag{6.5.144}$$

Therefore, corresponding to these $2q$ and $2m$ basis vectors in equation (6.5.143) solve the integration and denoted as

$$\mathbf{Y}_0=\mathbf{I}_q: \text{integration } \mathbf{R}_{xy}^{(0)}(t_a,t_b)=\mathbf{R}_{xy}^{(0)}(\tau),\quad \mathbf{R}_{\lambda y}^{(0)}(t_a,t_b)=\mathbf{R}_{\lambda y}^{(0)}(\tau) \tag{6.5.145a}$$

$$\mathbf{U}_0=\mathbf{I}_m: \text{integration } \mathbf{R}_{xu}^{(0)}(t_a,t_b)=\mathbf{R}_{xu}^{(0)}(\tau),\quad \mathbf{R}_{\lambda u}^{(0)}(t_a,t_b)=\mathbf{R}_{\lambda u}^{(0)}(\tau) \tag{6.5.145b}$$

$$\mathbf{Y}_1=\tau\mathbf{I}_q: \text{integration } \mathbf{R}_{xy}^{(1)}(0,\tau),\quad \mathbf{R}_{\lambda y}^{(1)}(0,\tau) \tag{6.5.145c}$$

$$\mathbf{U}_1=\tau\mathbf{I}_m: \text{integration } \mathbf{R}_{xu}^{(1)}(0,\tau),\quad \mathbf{R}_{\lambda u}^{(1)}(0,\tau) \tag{6.5.145d}$$

where $\tau=t-t_a$. Note that the basis matrices $\mathbf{R}_{xy}^{(0)},\mathbf{R}_{\lambda y}^{(0)}$ depend only on the interval length but not relate to the starting point, so that they are written with one argument τ, however the basis matrices $\mathbf{R}_{xy}^{(1)}, \mathbf{R}_{\lambda y}^{(1)}$ depend on two end times so that they must have two arguments. The integration matrices of $\mathbf{R}_{xy}^{(0)}, \mathbf{R}_{\lambda y}^{(0)}$ and $\mathbf{R}_{xy}^{(1)}, \mathbf{R}_{\lambda y}^{(1)}$, called as ***inhomogeneous basis matrices*** of $\mathbf{y}$, are $n \times q$ dimensioned and satisfying the differential equation $(i = 0,1)$

$$\partial \mathbf{R}_{xy}^{(i)} / \partial \tau = \mathbf{A}\mathbf{R}_{xy}^{(i)} + \mathbf{G}\mathbf{C}^{\mathrm{T}}\mathbf{V}^{-1}(\mathbf{Y}_i - \mathbf{C}\mathbf{R}_{xy}^{(i)}), \ \mathbf{R}_{xy}^{(i)} = \mathbf{0} \quad \text{when } \tau \to +0 \tag{6.5.146a}$$

$$\partial \mathbf{R}_{\lambda y}^{(i)} / \partial \tau = \mathbf{F}^{\mathrm{T}}\mathbf{C}^{\mathrm{T}}\mathbf{V}^{-1}(\mathbf{Y}_i - \mathbf{C}\mathbf{R}_{xy}^{(i)}), \qquad \mathbf{R}_{\lambda y}^{(i)} = \mathbf{0} \quad \text{when } \tau \to +0 \tag{6.5.146b}$$

where the initial conditions for $\mathbf{G}(\tau), \mathbf{F}(\tau)$ are $\mathbf{G}(0) = 0, \mathbf{F}(0) = \mathbf{I}_n$, *i.e.* t_{a} is the starting point of integration. Comparing equation (6.5.146) with (6.5.127'), the differences between $\mathbf{G}$ and $\mathbf{P}$, $\mathbf{F}$ and $\mathbf{F}_p$ are recognized. These differences can be compensated later, for which the method is still using the equation (6.5.127).

Similarly, the matrices $\mathbf{R}_{xu}^{(0)}, \mathbf{R}_{\lambda u}^{(0)}$ and $\mathbf{R}_{xu}^{(1)}, \mathbf{R}_{\lambda u}^{(1)}$, called as ***inhomogeneous basis matrices*** of $\mathbf{u}$, are $n \times m$ -dimensioned, satisfying the equations $(i = 0,1)$

$$\partial \mathbf{R}_{xu}^{(i)} / \partial \tau = (\mathbf{A} - \mathbf{G}\mathbf{C}^{\mathrm{T}}\mathbf{V}^{-1}\mathbf{C})\mathbf{R}_{xu}^{(i)} + \mathbf{B}_u \mathbf{U}_i, \ \mathbf{R}_{xu}^{(i)} = \mathbf{0} \ \text{ when } \tau \to +0 \tag{6.5.146c}$$

$$\partial \mathbf{R}_{\lambda u}^{(i)} / \partial \tau = -\mathbf{F}^{\mathrm{T}}\mathbf{C}^{\mathrm{T}}\mathbf{V}^{-1}\mathbf{C}\mathbf{R}_{xu}^{(i)}, \qquad \mathbf{R}_{\lambda u}^{(i)} = \mathbf{0} \quad \text{when } \tau \to +0 \tag{6.5.146d}$$

These equation are the extension of equations (6.5.81~82).

The Taylor series expansion of $\mathbf{Q}(\tau), \mathbf{G}(\tau), \mathbf{F}(\tau)$ is the same as (6.5.98~101) and the precise integration is also the same. The integration for constant basis of $\mathbf{Y}_0$ is examined first below, the Taylor series expansion of the corresponding matrices $\mathbf{R}_{xy}^{(0)}(\tau), \mathbf{R}_{\lambda y}^{(0)}(\tau)$ are given as

$$\mathbf{R}_{xy}^{(0)}(\tau) \approx \boldsymbol{\rho}_{xy01}\tau + \boldsymbol{\rho}_{xy02}\tau^2 + \boldsymbol{\rho}_{xy03}\tau^3 + \boldsymbol{\rho}_{xy04}\tau^4 \tag{6.5.147a}$$

$$\mathbf{R}_{\lambda y}^{(0)}(\tau) \approx \boldsymbol{\rho}_{\lambda y01}\tau + \boldsymbol{\rho}_{\lambda y02}\tau^2 + \boldsymbol{\rho}_{\lambda y03}\tau^3 + \boldsymbol{\rho}_{\lambda y04}\tau^4 \tag{6.5.147b}$$

The matrices $\mathbf{R}_{xy}^{(0)}$, $\mathbf{R}_{\lambda y}^{(0)}$ and the coefficient matrices $\boldsymbol{\rho}_{xy0i}$, $\boldsymbol{\rho}_{\lambda y0i}$ are all $n \times q$ dimensioned. The differential equations (6.5.146a,b) are

$$\mathrm{d}\mathbf{R}_{xy}^{(0)} / \mathrm{d}\tau = (\mathbf{A} - \mathbf{G}\mathbf{C}^{\mathrm{T}}\mathbf{V}^{-1}\mathbf{C})\mathbf{R}_{xy}^{(0)} + \mathbf{G}\mathbf{C}^{\mathrm{T}}\mathbf{R}^{-1}, \ \mathbf{R}_{xy}^{(0)}(0) = \mathbf{0}$$

$$\mathrm{d}\mathbf{R}_{\lambda y}^{(0)} / \mathrm{d}\tau = \mathbf{F}^{\mathrm{T}}\mathbf{C}^{\mathrm{T}}\mathbf{V}^{-1}(\mathbf{I}_q - \mathbf{C}\mathbf{R}_{xy}^{(0)}), \qquad \mathbf{R}_{\lambda y}^{(0)}(0) = \mathbf{0}$$

Substituting (6.5.147a,b) and (6.5.98~100) into the above equation and comparing the powers of τ gives

$$\begin{aligned}
&\boldsymbol{\rho}_{xy01} = \mathbf{0}, \ \boldsymbol{\rho}_{\lambda y01} = \mathbf{C}^T\mathbf{V}^{-1}; \ \boldsymbol{\rho}_{xy02} = \mathbf{g}_1\mathbf{C}^T\mathbf{V}^{-1}/2, \ \boldsymbol{\rho}_{\lambda y02} = \mathbf{f}_1^T\mathbf{C}^T\mathbf{V}^{-1}/2; \\
&\boldsymbol{\rho}_{xy03} = (\mathbf{A}\boldsymbol{\rho}_{xy02} + \mathbf{g}_2\mathbf{C}^T\mathbf{V}^{-1})/3, \ \boldsymbol{\rho}_{\lambda y03} = (\mathbf{f}_2^T\mathbf{C}^T\mathbf{V}^{-1} - \mathbf{e}_1\boldsymbol{\rho}_{xy02})/3; \\
&\boldsymbol{\rho}_{xy04} = (\mathbf{A}\boldsymbol{\rho}_{xy03} - \mathbf{g}_1\mathbf{e}_1\boldsymbol{\rho}_{xy02} + \mathbf{g}_3\mathbf{C}^T\mathbf{V}^{-1})/4 \\
&\boldsymbol{\rho}_{\lambda y04} = (\mathbf{f}_3^T\mathbf{C}^T\mathbf{V}^{-1} - \mathbf{f}_1^T\mathbf{e}_1\boldsymbol{\rho}_{xy02} - \mathbf{e}_1\boldsymbol{\rho}_{xy03})/4
\end{aligned} \tag{6.5.148a}$$

where $\mathbf{e}_1 = \mathbf{C}^T\mathbf{V}^{-1}\mathbf{C}$. These coefficient matrices can be computed one by one directly with no iteration, and then substituting into (6.5.147a,b) gives $\mathbf{R}_{xy}^{(0)}(\tau)$ and $\mathbf{R}_{\lambda y}^{(0)}(\tau)$. Because the relative error of truncated terms in the expansion (6.5.147) have been of the order of $O(\tau^4)$, which has been beyond the error of double precision real word of the computer today.

The derivation for the constant basis of $\mathbf{U}_0$ is similar, the corresponding Taylor series expansion of $\mathbf{R}_{xu}^{(0)}(\tau)$ and $\mathbf{R}_{\lambda u}^{(0)}(\tau)$ are

$$\mathbf{R}_{xu}^{(0)}(\tau) \approx \boldsymbol{\rho}_{xu01}\tau + \boldsymbol{\rho}_{xu02}\tau^2 + \boldsymbol{\rho}_{xu03}\tau^3 + \boldsymbol{\rho}_{xu04}\tau^4 \tag{6.5.147c}$$

$$\mathbf{R}_{\lambda u}^{(0)}(\tau) \approx \boldsymbol{\rho}_{\lambda u01}\tau + \boldsymbol{\rho}_{\lambda u02}\tau^2 + \boldsymbol{\rho}_{\lambda u03}\tau^3 + \boldsymbol{\rho}_{\lambda u04}\tau^4 \tag{6.5.147d}$$

The $\mathbf{R}_{xu}^{(0)}(\tau)$, $\mathbf{R}_{\lambda u}^{(0)}(\tau)$ and their coefficients are all $n \times m$ matrices, and the differential equations are, respectively

$$\mathrm{d}\mathbf{R}_{xu}^{(0)}/\mathrm{d}\tau = (\mathbf{A} - \mathbf{G}\mathbf{e}_1)\mathbf{R}_{xu}^{(0)} + \mathbf{B}_u, \quad \mathbf{R}_{xu}^{(0)}(0) = \mathbf{0}$$

$$\mathrm{d}\mathbf{R}_{\lambda u}^{(0)}/\mathrm{d}\tau = -\mathbf{F}^{\mathrm{T}}\mathbf{e}_1\mathbf{R}_{xu}^{(0)}, \quad \mathbf{R}_{\lambda u}^{(0)}(0) = \mathbf{0}$$

Substituting the equations (6.5.147c,d) and (6.5.98−100) into the above equations and comparing the coefficient of powers of τ derives

$$\begin{aligned}
&\boldsymbol{\rho}_{xu01} = \mathbf{B}_u, \ \boldsymbol{\rho}_{\lambda u01} = \mathbf{0}; \ \boldsymbol{\rho}_{xu02} = \mathbf{A}\boldsymbol{\rho}_{xu01}/2, \ \boldsymbol{\rho}_{\lambda u02} = -\mathbf{e}_1\boldsymbol{\rho}_{xu01}/2; \\
&\boldsymbol{\rho}_{xu03} = (\mathbf{A}\boldsymbol{\rho}_{xu02} - \mathbf{g}_1\mathbf{e}_1\boldsymbol{\rho}_{xu01})/3, \ \boldsymbol{\rho}_{\lambda u03} = -(\mathbf{e}_1\boldsymbol{\rho}_{xu02} + \mathbf{f}_1^{\mathrm{T}}\mathbf{e}_1\boldsymbol{\rho}_{xu01})/3; \\
&\boldsymbol{\rho}_{xu04} = (\mathbf{A}\boldsymbol{\rho}_{xu03} - \mathbf{g}_1\mathbf{e}_1\boldsymbol{\rho}_{xu02} - \mathbf{g}_2\mathbf{e}_1\boldsymbol{\rho}_{xu01})/4 \\
&\boldsymbol{\rho}_{\lambda u04} = -(\mathbf{e}\boldsymbol{\rho}_{xu03} + \mathbf{f}_1^{T}\mathbf{e}_1\boldsymbol{\rho}_{xu02} + \mathbf{f}_2^{\mathrm{T}}\mathbf{e}_1\boldsymbol{\rho}_{xu01})/4
\end{aligned} \tag{6.5.148b}$$

Directly computing these coefficient matrices and then substituting back into (6.5.147c,d) obtains $\mathbf{R}_{xu}^{(0)}(\tau)$ and $\mathbf{R}_{\lambda u}^{(0)}(\tau)$. The relative error has been beyond the double precision real word on the computer.

The above Taylor expansions apply to the constant part of $\mathbf{y}$ and $\mathbf{u}$ in the equations (6.5.145a,b). The linear distributed inhomogeneous terms need to consider the τ term equations (6.5.145c,d). The Taylor series expansion for $\mathbf{R}_{xy}^{(1)}(0,\tau)$ and $\mathbf{R}_{\lambda y}^{(1)}(0,\tau)$ are, respectively

$$\mathbf{R}_{xy}^{(1)}(0,\tau) \approx \boldsymbol{\rho}_{xy11}\tau + \boldsymbol{\rho}_{xy12}\tau^2 + \boldsymbol{\rho}_{xy13}\tau^3 + \boldsymbol{\rho}_{xy14}\tau^4 \tag{6.5.149a}$$

$$\mathbf{R}_{\lambda y}^{(1)}(0,\tau) \approx \boldsymbol{\rho}_{\lambda y11}\tau + \boldsymbol{\rho}_{\lambda y12}\tau^2 + \boldsymbol{\rho}_{\lambda y13}\tau^3 + \boldsymbol{\rho}_{\lambda y14}\tau^4 \tag{6.5.149b}$$

The differential equations are, respectively

$$\partial\mathbf{R}_{xy}^{(1)}(0,\tau)/\partial\tau = (\mathbf{A} - \mathbf{G}\mathbf{e}_1)\mathbf{R}_{xy}^{(1)} + \mathbf{G}\mathbf{C}^{\mathrm{T}}\mathbf{V}^{-1}\tau, \quad \mathbf{R}_{xy}^{(1)}(0,0) = \mathbf{0}$$

$$\partial\mathbf{R}_{\lambda y}^{(1)}(0,\tau)/\partial\tau = -\mathbf{F}^{\mathrm{T}}\mathbf{e}_1\mathbf{R}_{xy}^{(1)} + \mathbf{F}^{\mathrm{T}}\mathbf{C}\mathbf{V}^{-1}\tau, \qquad \mathbf{R}_{\lambda y}^{(1)}(0,0) = \mathbf{0}$$

Substituting with the Taylor expansions and comparing the powers of τ derives

$$\begin{aligned}
&\boldsymbol{\rho}_{xy11} = \mathbf{0}, \ \boldsymbol{\rho}_{\lambda y11} = \mathbf{0}; \ \boldsymbol{\rho}_{xy12} = \mathbf{0}, \ \boldsymbol{\rho}_{\lambda y12} = \mathbf{C}^{\mathrm{T}}\mathbf{V}^{-1}/2; \\
&\boldsymbol{\rho}_{xy13} = \mathbf{g}_1\mathbf{C}^{\mathrm{T}}\mathbf{V}^{-1}/3, \ \boldsymbol{\rho}_{\lambda y13} = \mathbf{f}_1^{\mathrm{T}}\mathbf{C}^{\mathrm{T}}\mathbf{V}^{-1}/3; \\
&\boldsymbol{\rho}_{xy14} = (\mathbf{A}\boldsymbol{\rho}_{xy13} + \mathbf{g}_2\mathbf{C}^{\mathrm{T}}\mathbf{V}^{-1})/4, \ \boldsymbol{\rho}_{\lambda y14} = (\mathbf{f}_2^{\mathrm{T}}\mathbf{C}^{\mathrm{T}}\mathbf{V}^{-1} - \mathbf{e}_1\boldsymbol{\rho}_{xy13})/4
\end{aligned} \tag{6.5.150a}$$

Directly computing the coefficient matrices from (6.5.150a) and substituting into (6.5.149a,b) the matrices $\mathbf{R}_{xy}^{(1)}(0,\tau)$ and $\mathbf{R}_{\lambda y}^{(1)}(0,\tau)$ are obtained precisely.

There is also the linear term of $\mathbf{u}$ to be computed. The Taylor series expansions are

$$\mathbf{R}_{xu}^{(1)}(0,\tau) \approx \boldsymbol{\rho}_{xu11}\tau + \boldsymbol{\rho}_{xu12}\tau^2 + \boldsymbol{\rho}_{xu13}\tau^3 + \boldsymbol{\rho}_{xu14}\tau^4 \tag{6.5.149c}$$

$$\mathbf{R}_{\lambda u}^{(1)}(0,\tau) \approx \boldsymbol{\rho}_{\lambda u11}\tau + \boldsymbol{\rho}_{\lambda u12}\tau^2 + \boldsymbol{\rho}_{\lambda u13}\tau^3 + \boldsymbol{\rho}_{\lambda u14}\tau^4 \tag{6.5.149d}$$

The differential equations are, respectively

$$\partial\mathbf{R}_{xu}^{(1)}(0,\tau)/\partial\tau = (\mathbf{A} - \mathbf{G}\mathbf{e}_1)\mathbf{R}_{xu}^{(1)} + \tau\mathbf{B}_u, \quad \mathbf{R}_{xu}^{(1)}(0,0) = 0$$

$$\partial\mathbf{R}_{\lambda u}^{(1)}/\partial\tau = -\mathbf{F}^{\mathrm{T}}\mathbf{e}_1\mathbf{R}_{xu}^{(1)}(0,\tau), \qquad \mathbf{R}_{\lambda u}^{(1)}(0,0) = 0$$

Substituting the expansions (6.5.149c~d) into the differential equations and comparing the powers of τ derives

$$\boldsymbol{\rho}_{xu11} = \mathbf{0},\ \boldsymbol{\rho}_{\lambda u11} = \mathbf{0}; \quad \boldsymbol{\rho}_{xu12} = \mathbf{B}_u/2,\ \boldsymbol{\rho}_{\lambda u12} = \mathbf{0};$$

$$\boldsymbol{\rho}_{xu13} = \mathbf{A}\boldsymbol{\rho}_{xu12}/3; \quad \boldsymbol{\rho}_{xu14} = (\mathbf{A}\boldsymbol{\rho}_{xu13} - \mathbf{g}_1\mathbf{e}_1\boldsymbol{\rho}_{xu12})/4; \tag{6.5.150b}$$

$$\boldsymbol{\rho}_{\lambda u13} = -\mathbf{e}_1\boldsymbol{\rho}_{xu12}/3; \quad \boldsymbol{\rho}_{\lambda u14} = -(\mathbf{e}_1\boldsymbol{\rho}_{xu13} + \mathbf{f}_1^T\mathbf{e}_1\boldsymbol{\rho}_{xu12})/4$$

These coefficient matrices can be directly computed one after another, and then $\mathbf{R}_{xy}^{(1)}(0,\tau)$ and $\mathbf{R}_{\lambda y}^{(1)}(0,\tau)$ are computed from (6.5.149c,d). The relative error has been beyond the computer precision.

Note that the precise integration computes only to the linear term here, however, the quadratic term can also be computed if required.

§6.5.7.4, Interval combination within the η interval

The precise integration subdivides the η step further into 2^N extremely small intervals with length $\tau = \eta/2^N$ (the subscript of τ_e is taken off here). The above derivation is only for the step size τ interval, and **N** times interval combination is necessary to recover the original length η. That kind of computational methodology for the mixed energy matrices $\mathbf{Q}(\eta), \mathbf{G}(\eta), \mathbf{F}(\eta)$ has been described in some detail below the equation (6.5.97). Here, the precise integration for $\mathbf{R}_x$ and $\mathbf{R}_\lambda$ is described.

For inhomogeneous terms (measurement and control) the interval combination equations are (6.5.78a,b). But these equations are used for one loading case. To consider the various possible cases for the measurement $\mathbf{y}$ as well as for the control input $\mathbf{u}$, the basis are given in equations (6.5.145a~d), respectively. The corresponding inhomogeneous basis are integrated to be $\mathbf{R}_x$ and $\mathbf{R}_\lambda$. Then the contiguous interval combination should consider the q-basis (for $\mathbf{y}$) or consider the m-basis (for $\mathbf{u}$) simultaneously. Hence the equation (6.5.78a,b) should be updated as

$$\mathbf{R}_{xc} = \mathbf{R}_{x2} + \mathbf{F}_2(\mathbf{I}_n + \mathbf{G}_1\mathbf{Q}_2)^{-1}(\mathbf{R}_{x1} + \mathbf{G}_1\mathbf{R}_{\lambda 2}) \tag{6.5.151a}$$

$$\mathbf{R}_{\lambda c} = \mathbf{R}_{\lambda 1} + \mathbf{F}_1^{\mathrm{T}}(\mathbf{I}_n + \mathbf{Q}_2\mathbf{G}_1)^{-1}(\mathbf{R}_{\lambda 2} - \mathbf{E}_2\mathbf{R}_{x1}) \tag{6.5.151b}$$

where the subscripts 1 and 2 represent left and right intervals, respectively, and the matrix $\mathbf{R}$ represents $n \times q$ matrix (for $\mathbf{y}$) or $n \times m$ matrix (for $\mathbf{u}$), respectively. Based on the equations (6.5.151a,b), in the single interval $(0,\eta)$, the basis matrices combination algorithm for $\mathbf{y}$ and for $\mathbf{u}$ are executed.

If within the interval length η there are only constant valued $\mathbf{y}$ and $\mathbf{u}$, then only the matrices $\mathbf{R}_{xy}^{(0)}(\tau),\mathbf{R}_{\lambda y}^{(0)}(\tau)$ and $\mathbf{R}_{xu}^{(0)}(\tau),\mathbf{R}_{\lambda u}^{(0)}(\tau)$ are required, and these matrices can be used to compute $\mathbf{R}_{xy}^{(0)}(2\tau),\mathbf{R}_{\lambda u}^{(0)}(2\tau)$ and $\mathbf{R}_{xu}^{(0)}(2\tau),\mathbf{R}_{\lambda u}^{(0)}(2\tau)$ by the equations (6.5.151a,b), which means the length of combined interval is doubled. After N times of iteration, the interval length is recovered to be η, and the matrices $\mathbf{R}_{xy}^{(0)}(\eta),\mathbf{R}_{\lambda u}^{(0)}(\eta)$ and $\mathbf{R}_{xu}^{(0)}(\eta),\mathbf{R}_{\lambda u}^{(0)}(\eta)$ are obtained. But there are further the linearly varying parts of $\mathbf{y}$ and $\mathbf{u}$, so further investigations are needed, as follows.

Based on $\mathbf{R}_{xy}^{(1)}(0,\tau),\mathbf{R}_{\lambda y}^{(1)}(0,\tau)$, etc., compute $\mathbf{R}_{xy}^{(1)}(0,2\tau),\mathbf{R}_{\lambda y}^{(1)}(0,2\tau)$

Based on $\mathbf{R}_{xu}^{(1)}(0,\tau),\mathbf{R}_{\lambda u}^{(1)}(0,\tau)$, etc., compute $\mathbf{R}_{xu}^{(1)}(0,2\tau),\mathbf{R}_{\lambda u}^{(1)}(0,2\tau)$

Within the interval η, both $\mathbf{y}$ and $\mathbf{u}$ are assumed linearly varying, the interval $(0,2\tau)$ is within $(0,\eta)$ so that they are linear too. The interval $(0,2\tau)$ is combined contiguously by $(0,\tau)$ and $(\tau,2\tau)$. The distribution $t\mathbf{I}$ in the interval $(0,2\tau)$, where t is the running argument, corresponding to the combination of $t\mathbf{I}$ distribution in the interval $(0,\tau)$ and both the constant distribution $\tau\mathbf{I}$ in the interval $(\tau,2\tau)$ in addition to the linear distribution $(t-\tau)\mathbf{I}$ in the interval $(\tau,2\tau)$, see figure 6.7. To express with the interval combination equation:

$$(0,\tau)\ \text{linear distribution:}\ \mathbf{R}_{x1}=\mathbf{R}_{x}^{(1)}(0,\tau),\mathbf{R}_{\lambda 1}=\mathbf{R}_{\lambda}^{(1)}(0,\tau) \tag{6.5.152}$$

$$(\tau,2\tau)\ \text{constant distribution:}\ \mathbf{R}_{x2}=\tau\cdot\mathbf{R}_{x}^{(0)}(\tau)+\mathbf{R}_{x}^{(1)}(0,\tau),\quad \text{plus}$$

$$(\tau,2\tau)\ \text{linear distribution:}\ \mathbf{R}_{\lambda 2}=\tau\cdot\mathbf{R}_{\lambda}^{(0)}(\tau)+\mathbf{R}_{\lambda}^{(1)}(0,\tau) \tag{6.5.153}$$

The combination of the two intervals $(0,\tau)$ and $(\tau,2\tau)$ gives the linear distribution of

$$(0,2\tau)\ \text{linear distribution:}\quad \mathbf{R}_{x}^{(1)}(0,2\tau),\mathbf{R}_{\lambda}^{(1)}(0,2\tau)$$

the interval combination equations are (6.5.151a,b). Thus, the equations for the N recursive interval combinations for forming the linear distribution in the interval $(0,\eta)$ have been available. The computation for the matrices $\mathbf{R}_{xy}^{(0)}(\eta),\mathbf{R}_{xu}^{(0)}(\eta)$, $\mathbf{R}_{\lambda y}^{(0)}(\eta),\mathbf{R}_{\lambda u}^{(0)}(\eta)$ and $\mathbf{R}_{xy}^{(1)}(0,\eta),\mathbf{R}_{xu}^{(1)}(0,\eta)$, $\mathbf{R}_{\lambda y}^{(1)}(0,\eta),\mathbf{R}_{\lambda u}^{(1)}(0,\eta)$, or simply denoted $\mathbf{R}_x,\mathbf{R}_\lambda$, are independent on the real time measurement $\mathbf{y}$ and control input $\mathbf{u}$. Therefore, the computation of $\mathbf{R}_x,\mathbf{R}_\lambda$ can be executed and stored into the database at the same time with $\mathbf{Q}(\eta),\mathbf{G}(\eta),\mathbf{F}(\eta)$ *off-line* beforehand, and can be reloaded from the database at real-time

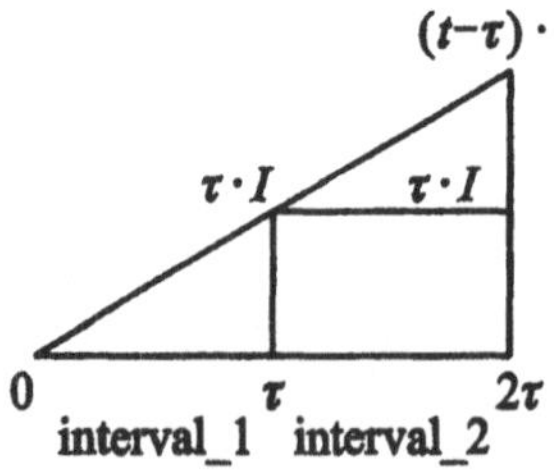

Figure 6.7, *Interval* $(0,2\tau)$ *subdivision*

performance. The interval combination equations for $\mathbf{Q},\mathbf{G},\mathbf{F}$ are (6.5.102a~c), and the interval combination equations for $\mathbf{R}_x,\mathbf{R}_\lambda$ are (6.5.151a,b). Presently it is the combination of two intervals with equal length τ, so that the equations are

$$\mathbf{R}_{xc}=\mathbf{R}_{x2}+\mathbf{F}(\tau)(\mathbf{I}_n+\mathbf{G}(\tau)\mathbf{Q}(\tau))^{-1}(\mathbf{R}_{x1}+\mathbf{G}(\tau)\mathbf{R}_{\lambda 2}) \tag{6.5.154a}$$

$$\mathbf{R}_{\lambda c}=\mathbf{R}_{\lambda 2}+\mathbf{F}^{\mathrm{T}}(\tau)(\mathbf{I}_n+\mathbf{Q}(\tau)\mathbf{G}(\tau))^{-1}(\mathbf{R}_{\lambda 2}-\mathbf{Q}(\tau)\mathbf{R}_{x1}) \tag{6.5.154b}$$

The algorithm for the matrices of the fundamental interval η is given as follows:

Comment: The 2^N algorithm to generate the basis matrices $\mathbf{R}_{xy}^{(0)}(\eta),\mathbf{R}_{xu}^{(0)}(\eta)$, $\mathbf{R}_{\lambda y}^{(0)}(\eta),\mathbf{R}_{\lambda u}^{(0)}(\eta)$ of constant distribution and the basis matrices $\mathbf{R}_{xy}^{(1)}(0,\eta),\mathbf{R}_{xu}^{(1)}(0,\eta)$, $\mathbf{R}_{\lambda y}^{(1)}(0,\eta),\mathbf{R}_{\lambda u}^{(1)}(0,\eta)$ of linear distribution, respectively, in the fundamental interval $(0,\eta)$ with the matrices $\mathbf{F}(\eta),\mathbf{Q}(\eta),\mathbf{G}(\eta)$ simultaneously.

[Original data: n,m,q,l and matrices $\mathbf{A}_{n\times n},\mathbf{B}_1^{n\times l},\mathbf{B}_u^{n\times m},\mathbf{C}^{q\times n},\mathbf{W}^{l\times l},\mathbf{V}^{q\times q}$]

[Select the time step η, let $\tau=\eta/2^N$, $N=20$]

[According to (6.5.98~101) generate $\mathbf{Q}(\tau),\mathbf{G}(\tau),\mathbf{F}'(\tau)$; and $\mathbf{e}_1=\mathbf{C}^{\mathrm{T}}\mathbf{V}^{-1}\mathbf{C}$]

[According to (6.5.147~150) generate: $\mathbf{R}_{xy}^{(0)}(\eta),\mathbf{R}_{xu}^{(0)}(\eta),\mathbf{R}_{\lambda y}^{(0)}(\eta),\mathbf{R}_{\lambda u}^{(0)}(\eta)$; and

$\mathbf{R}_{xy}^{(1)}(0,\eta),\mathbf{R}_{xu}^{(1)}(0,\eta),\mathbf{R}_{\lambda y}^{(1)}(0,\eta),\mathbf{R}_{\lambda u}^{(1)}(0,\eta)$]

for (*iter*=0; *iter*<*N*; *iter*++) {

[Let $\mathbf{R}_{x1}=\mathbf{R}_{xy}^{(1)}(0,\tau);\mathbf{R}_{\lambda 1}=\mathbf{R}_{\lambda y}^{(1)}(0,\tau);$

$\mathbf{R}_{x2}=\tau\cdot\mathbf{R}_{xy}^{(0)}(\tau)+\mathbf{R}_{x1};\mathbf{R}_{\lambda 2}=\tau\cdot\mathbf{R}_{\lambda y}^{(0)}(\tau)+\mathbf{R}_{\lambda 1};$]

[Using (6.5.154a,b) compute $\mathbf{R}_{xc},\mathbf{R}_{\lambda c}$;]

[Let $\mathbf{R}_{xy}^{(1)}(0,2\tau)=\mathbf{R}_{xc}$; $\mathbf{R}_{\lambda y}^{(1)}(0,2\tau)=\mathbf{R}_{\lambda c}$;]

[Let $\mathbf{R}_{x1}=\mathbf{R}_{xu}^{(1)}(0,\tau)$; $\mathbf{R}_{\lambda 1}=\mathbf{R}_{\lambda u}^{(1)}(0,\tau)$;]

[$\mathbf{R}_{x2}=\tau\cdot\mathbf{R}_{xu}^{(0)}(\tau)+\mathbf{R}_{x1}$; $\mathbf{R}_{\lambda 2}=\tau\cdot\mathbf{R}_{\lambda u}^{(0)}(\tau)+\mathbf{R}_{\lambda 1}$;]

[Using (6.5.154a,b) compute $\mathbf{R}_{xc},\mathbf{R}_{\lambda c}$;]

[Let $\mathbf{R}_{xu}^{(1)}(0,2\tau)=\mathbf{R}_{xc}$; $\mathbf{R}_{\lambda u}^{(1)}(0,2\tau)=\mathbf{R}_{\lambda c}$;]

Comment: Above for linear distributed $\mathbf{y}$ and $\mathbf{u}$ (6.5.152~153), then interval combination. Below algorithm is for constant distributed basis matrices.

[$\mathbf{R}_{x1}=\mathbf{R}_{xy}^{(0)}(\tau);\mathbf{R}_{\lambda 1}=\mathbf{R}_{\lambda y}^{(0)}(\tau);\mathbf{R}_{x2}=\mathbf{R}_{x1};\mathbf{R}_{\lambda 2}=\mathbf{R}_{\lambda 1};$]

[Using (6.5.154a,b) compute $\mathbf{R}_{xc},\mathbf{R}_{\lambda c}$;]

[Let $\mathbf{R}_{xy}^{(0)}(2\tau)=\mathbf{R}_{xc};\mathbf{R}_{\lambda y}^{(0)}(2\tau)=\mathbf{R}_{\lambda c}$;]

[$\mathbf{R}_{x1}=\mathbf{R}_{xu}^{(0)}(\tau);\mathbf{R}_{\lambda 1}=\mathbf{R}_{\lambda u}^{(0)}(\tau);\mathbf{R}_{x2}=\mathbf{R}_{x1};\mathbf{R}_{\lambda 2}=\mathbf{R}_{\lambda 1};$]

[Using (6.5.154a,b) compute $\mathbf{R}_{xc},\mathbf{R}_{\lambda c}$;]

[Let $\mathbf{R}_{xu}^{(0)}(2\tau)=\mathbf{R}_{xc};\mathbf{R}_{\lambda y}^{(0)}(2\tau)=\mathbf{R}_{\lambda c}$;]

Comment: The load basis matrices $\mathbf{R}_x,\mathbf{R}_\lambda$ are computed. Below for updating $\mathbf{Q}(\tau),\mathbf{G}(\tau),\mathbf{F}'(\tau)$

[Using (6.5.102a~c) compute $\mathbf{Q}_c,\mathbf{G}_c,\mathbf{F}'_c$;]

[Let $\mathbf{Q}(2\tau)=\mathbf{Q}_c;\mathbf{G}(2\tau)=\mathbf{G}_c;\mathbf{F}'(2\tau)=\mathbf{F}'_c$;]

$\tau=2*\tau$;

} $\mathbf{F}(\eta)=\mathbf{I}+\mathbf{F}'$; (6.5.155)

Comment: After the iteration, $\mathbf{R}_{xy}^{(0)}(\eta),\mathbf{R}_{\lambda y}^{(0)}(\eta),\mathbf{R}_{xu}^{(0)}(\eta),\mathbf{R}_{\lambda u}^{(0)}(\eta)$ and $\mathbf{R}_{xy}^{(1)}(0,\eta)$ $\mathbf{R}_{\lambda y}^{(1)}(0,\eta)$, $\mathbf{R}_{xu}^{(1)}(0,\eta),\mathbf{R}_{\lambda u}^{(1)}(0,\eta)$ and also $\mathbf{Q}(\eta),\mathbf{G}(\eta),\mathbf{F}(\eta)$ are all generated. These matrices are independent on the starting point t_k. The computation is ***off-line***.

The above precise integration computation can be used when the Jordan normal form appears. The analytical solution should be used in combination with the precise integration approach, as that given in [103] for the solution of Riccati differential equations. Distinguishing off-line with on-line is quite important for real time control problems.

§6.5.8, Integration of filter equation for the whole interval

The algorithm (6.5.155) given above are the computation of a fundamental $(0,\eta)$ interval. Because of that within the fundamental interval η, the vectors $\mathbf{y}$ and $\mathbf{u}$ are considered linearly distributed. Hence the linear basis computations in the η interval are performed ***off-line*** in order to reduce the real time computation as little as possible. However, the algorithm is only for the length η time step, this fundamental basis computation is unified for all the time steps. The computation for whole the time interval $(0,t_f)$ has some other parts needed to be performed off-line, i.e. the transformation of $\mathbf{P}_k$ at the time station t_k. According to (6.5.124), the time instant t_k can be regarded as the starting point of integration. Each step integrating one η length, with the initial condition

$$\mathbf{x}(t)=\hat{\mathbf{x}}_k+\mathbf{P}_k\boldsymbol{\lambda}(t), \quad \text{when } t=t_k \tag{6.5.156}$$

where $\hat{\mathbf{x}}_k$ is the filtered state at the last time step, and $\mathbf{P}_k=\mathbf{P}(t_k)$ is the solution of Riccati differential equation, and is solved off-line before.

After the single step η integration of $\mathbf{Q}(\eta),\mathbf{G}(\eta),\mathbf{F}(\eta)$, then treating the matrix $\mathbf{P}_k$ as the initial variance matrix '$\mathbf{P}_0$', using (6.5.90) to obtain the matrix $\mathbf{F}_c(\eta)$, which is just the matrix

$$\mathbf{F}_{pk}(\eta)=\boldsymbol{\Phi}(t_{k+1},t_k)=\mathbf{F}(\eta)(\mathbf{I}_n+\mathbf{P}_k\mathbf{Q}(\eta))^{-1} \tag{6.5.156}$$

The basis matrices $\mathbf{R}_{xy}^{(i)}(0,\tau),\mathbf{R}_{\lambda y}^{(i)}(0,\tau)$ and $\mathbf{R}_{xu}^{(i)}(0,\tau),\mathbf{R}_{\lambda u}^{(i)}(0,\tau)$ computed above satisfying the differential equations (6.5.146a~d), comparing which with (6.5.127') there are the differences of $\mathbf{G},\mathbf{F}$ with $\mathbf{P},\mathbf{F}_p$, which should be amended. The method is again using the transformation (6.5.127), where the t_k is regarded as the starting point, therefore the equations are revised as from t_k to t_{k+1} that

$$\begin{aligned}
\mathbf{R}_{xyp}^{(0)}(\eta) &= \mathbf{R}_{xy}^{(0)}(\eta)+\mathbf{F}(\eta)(\mathbf{P}_k^{-1}+\mathbf{Q}(\eta))^{-1}\cdot\mathbf{R}_{\lambda y}^{(0)}(\eta)\\
\mathbf{R}_{xup}^{(0)}(\eta) &= \mathbf{R}_{xu}^{(0)}(\eta)+\mathbf{F}(\eta)(\mathbf{P}_k^{-1}+\mathbf{Q}(\eta))^{-1}\cdot\mathbf{R}_{\lambda u}^{(0)}(\eta)\\
\mathbf{R}_{xyp}^{(1)}(0,\eta) &= \mathbf{R}_{xy}^{(1)}(0,\eta)+\mathbf{F}(\eta)(\mathbf{P}_k^{-1}+\mathbf{Q}(\eta))^{-1}\cdot\mathbf{R}_{\lambda y}^{(1)}(0,\eta)\\
\mathbf{R}_{xup}^{(1)}(0,\eta) &= \mathbf{R}_{xu}^{(0)}(0,\eta)+\mathbf{F}(\eta)(\mathbf{P}_k^{-1}+\mathbf{Q}(\eta))^{-1}\cdot\mathbf{R}_{\lambda u}^{(1)}(0,\eta)
\end{aligned}\tag{6.5.157}$$

Here again to explain, that the basis matrices $\mathbf{R}^{(0)}$ correspond to the constant load, hence it depends only on the interval length, one argument. The matrices $\mathbf{R}^{(1)}$ depend also on the start point, so that two arguments are needed. The subscript p represents that the matrix $\mathbf{P}$ transformation has been performed. The matrix $\mathbf{R}_{\lambda p}$ has been unrelated to filtering but it is useful later for smoothing computation. Using equation (6.6.15b) gives the equation for $\mathbf{R}_{\lambda p}$.

The algorithm (6.5.155) generated basis matrices depend only on η but not relate to t_k, however, $\mathbf{R}_{xyp}^{(i)}$ and $\mathbf{R}_{xup}^{(i)}$ depend also on t_k hence they should be stored for each time step. Therefore at t_k, two $n\times n$ matrices $\mathbf{P}_k,\mathbf{F}_{pk}$, two $n\times q$ matrices $\mathbf{R}_{xyp}^{(0)}(\eta),\mathbf{R}_{xyp}^{(1)}(\eta)$ and two $n\times m$ matrices $\mathbf{R}_{xup}^{(0)}(\eta),\mathbf{R}_{xup}^{(1)}(0,\eta)$ must be stored. The ***off-line*** algorithm is given as:

[Using (6.5.155) compute the matrices $\mathbf{Q}(\eta),\mathbf{G}(\eta),\mathbf{F}'(\eta)$, $\mathbf{R}_{xy}^{(0)}(\eta),\mathbf{R}_{\lambda y}^{(0)}(\eta)$, $\mathbf{R}_{xu}^{(0)}(\eta)$, $\mathbf{R}_{\lambda u}^{(0)}(\eta)$; and $\mathbf{R}_{xy}^{(1)}(0,\eta),\mathbf{R}_{\lambda y}^{(1)}(0,\eta),\mathbf{R}_{xu}^{(1)}(0,\eta),\mathbf{R}_{\lambda u}^{(1)}(0,\eta)$ of the fundamental interval η]
Comment: execute once
[Let $\mathbf{P}=\mathbf{P}_0$;] Comment: $\mathbf{P}_0$ input here
[Let $\mathbf{Q}_2=\mathbf{Q}(\eta)$; $\mathbf{G}_2=\mathbf{G}(\eta)$; $\mathbf{F}_2=\mathbf{F}(\eta)$;]
for ($k=0$; $k<k_f$; $k++$) {
  [Compute $\mathbf{T}_1=(\mathbf{I}_n+\mathbf{P}\mathbf{Q}_2)^{-1}$; $\mathbf{T}_2=\mathbf{T}_1\times\mathbf{P}$;]
  [$\mathbf{P}=\mathbf{G}_2+\mathbf{F}_2\times\mathbf{T}_2\times\mathbf{F}_2^{\mathrm{T}}$; matrix $\mathbf{P}$ is stored as $\mathbf{P}_{k+1}$ of station $(k+1)$]
  [$\mathbf{F}_p=\mathbf{F}_2\times\mathbf{T}_1$; Store $\mathbf{F}_p$ at the station $(k+1)$]
  [$\mathbf{R}_{xyp}^{(0)}=\mathbf{R}_{xy}^{(0)}+\mathbf{F}_2\times\mathbf{T}_2\times\mathbf{R}_{\lambda y}^{(0)}$; Store $\mathbf{R}_{xyp}^{(0)}$ at the station $(k+1)$]
  [$\mathbf{R}_{xup}^{(0)}=\mathbf{R}_{xu}^{(0)}+\mathbf{F}_2\times\mathbf{T}_2\times\mathbf{R}_{\lambda u}^{(0)}$; Store $\mathbf{R}_{xup}^{(0)}$ at the station $(k+1)$]
  [$\mathbf{R}_{xyp}^{(1)}=\mathbf{R}_{xy}^{(1)}+\mathbf{F}_2\times\mathbf{T}_2\times\mathbf{R}_{\lambda y}^{(1)}$; Store $\mathbf{R}_{xyp}^{(1)}$ at the station $(k+1)$]
  [$\mathbf{R}_{xup}^{(1)}=\mathbf{R}_{xu}^{(1)}+\mathbf{F}_2\times\mathbf{T}_2\times\mathbf{R}_{\lambda u}^{(1)}$; Store $\mathbf{R}_{xup}^{(1)}$ at the station $(k+1)$]
}
Comment: The algorithm is unrelated to measurement and control input, hence ***off-line***. (6.5.158a)

After the ***off-line*** computation is consummated, the ***real time*** computation is

[$k=0$; *i.e.* $t=0$ is the starting point]
[Input $\hat{\mathbf{x}}_0$, the measurement $\mathbf{y}_0$ and control $\mathbf{u}_0$; and stored in $\hat{\mathbf{x}}_-,\mathbf{y}_-,\mathbf{u}_-$;]
for ($k=1$; $k \le k_f$; $k++$) { Comment: Time step integration.
  [Load in $\mathbf{F}_p, \mathbf{R}_{xyp}^{(0)}, \mathbf{R}_{xup}^{(0)}, \mathbf{R}_{xyp}^{(1)}, \mathbf{R}_{xup}^{(1)}$ from station k ;]
  [Read measurement $\mathbf{y}_k$ and control $\mathbf{u}_k$]*
  [$\hat{\mathbf{x}}_k = \mathbf{F}_p\hat{\mathbf{x}}_{k-1} + \mathbf{R}_{xyp}^{(0)} * \mathbf{y}_{k-1} + \mathbf{R}_{xyp}^{(1)} * (\mathbf{y}_k - \mathbf{y}_{k-1})/\eta + \mathbf{R}_{xup}^{(0)}\mathbf{u}_{k-1}$
    $+ \mathbf{R}_{xup}^{(1)}(\mathbf{u}_k - \mathbf{u}_{k-1})/\eta$;]
  [$\hat{\mathbf{x}}_- = \hat{\mathbf{x}}_k$; $\mathbf{y}_- = \mathbf{y}_k$; $\mathbf{u}_- = \mathbf{u}_k$;]**
} Comment: $\hat{\mathbf{x}}_-$ is $\hat{\mathbf{x}}_{k-1}$, similarly for $\mathbf{y}_-,\mathbf{u}_-$ (6.5.158b)

The above real time algorithm has not taken the factor of control system into account. The line marked * means that it requires revision. Because $\mathbf{u}_k$ requires the value of $\hat{\mathbf{x}}_k$ to calculate. Also the operation of the actuator of feedback control should also be considered. The simplest way is using constant value in the interval (t_k, t_{k+1}), which corresponds to that the line marked with * takes the value $\mathbf{u}_k = \mathbf{u}_{k-1}$. The line marked with ** means that the $\mathbf{u}_- = \mathbf{u}_k$ instruction should use the separation principle of LQG, $\mathbf{u}_- = \mathbf{R}^{-1}\mathbf{B}_2^{\mathrm{T}}\mathbf{S}_k\hat{\mathbf{x}}_k$, where $\mathbf{S}_k$ is the solution of the Riccati differential equation of optimal control in the future time interval.

The algorithm (6.5.158a,b) is given based on the precise integration method, but there are the equation (6.5.142) and (6.5.124") derived from the analytical method. The comparison between these two approaches gives the corresponding relationship. Two approaches can be used interchangeably.

First, the equations of precise integration derive only for linear interpolation, but the analytical approach gives quadratic interpolation in (6.5.128b). Hence, the comparison should go back to linear, *i.e.* regard $\mathbf{y}_2, \mathbf{u}_2$ being zero. So that the step forward equation used in (6.5.158b) is

$$\hat{\mathbf{x}}_{k+1} = \mathbf{F}_{pk}\hat{\mathbf{x}}_k + \begin{bmatrix} \mathbf{R}_{xup}^{(0)} \\ \mathbf{R}_{xyp}^{(0)} \end{bmatrix} \begin{Bmatrix} \mathbf{u}_k \\ \mathbf{y}_k \end{Bmatrix} + \begin{bmatrix} \mathbf{R}_{xup}^{(1)} \\ \mathbf{R}_{xyp}^{(1)} \end{bmatrix} \begin{Bmatrix} (\mathbf{u}_{k+1} - \mathbf{u}_k)/\eta \\ (\mathbf{y}_{k+1} - \mathbf{y}_k)/\eta \end{Bmatrix} \tag{6.5.159}$$

Comparing this equation with (6.5.124"), then $\mathbf{F}_{pk}$ is the matrix $\boldsymbol{\Phi}$ in equation (6.5.119); and the matrices in (6.5.142) correspond to

$$\mathbf{M}_{p0} = \begin{bmatrix} \mathbf{R}_{xup}^{(0)} \\ \mathbf{R}_{xyp}^{(0)} \end{bmatrix}, \quad \mathbf{M}_{p1} = \begin{bmatrix} \mathbf{R}_{xup}^{(1)} \\ \mathbf{R}_{xyp}^{(1)} \end{bmatrix} \tag{6.5.160}$$

Hence the result from analytical method is easy to use in computations. Note that the matrices in equation (6.5.158) of interval (t_k, t_{k+1}) are stored at the station $k+1$.

§6.5.8.1, Numerical example

The Riccati differential equation is non-linear, only one-dimensional problem has purely analytical solution. The filtering problem has also the time variant differential equation to solve, in order to estimate the state vector $\mathbf{x}$ under the measured vector $\mathbf{y}$. The practical measured $\mathbf{y}$ is under the random disturbance, but as the numerical example, the vector $\mathbf{y}$ is assumed a constant vector, so as to carry out the numerical result of the vector $\hat{\mathbf{x}}$ by means of precise integration. Its numerical result can be compared with the analytical solution of one-dimensional problem. In the case of multi-dimensional problem, the comparison can be given for the numerical results obtained for differential step sizes.

Example 6.10, Give dimension $n = 1$, and $A = -0.8; W = 1.0; V = 1.0;$ $B = 0.8; C = 5.0; P_0 = 0.01;$ the solution $\mathbf{P}(t)$ of Riccati differential equation, and $\mathbf{F}_p(t)$, $\mathbf{R}_{xyp}^{(0)}(t)$ are required. Further, assuming $\hat{x}_0 = 0.0; y(t) = 0.5$ compute the filter solution $\hat{x}(t)$.

Solution: Select the step size as $\eta = 0.05$. The analytical solution is available for one-dimensional problem. Both the numerical results from precise integration method and from analytical solution completely coincides to each other, and is given in the table 6.3 below

Table 6.3, solutions of Riccati equation and related matrices of example 6.10

t=	0.0	0.05	0.10	0.15	0.20	0.25
P*1E3	10.0	39.142	63.584	82.872	97.375	107.887
Fp*1E3		931.327	900.562	876.284	858.030	844.798
Rx0*1E3		6.036	12.312	17.264	20.988	23.687
$\hat{x}$ *1E3	0.0	3.018	8.874	16.408	24.572	32.602

continue

t=	0.30	0.35	0.40	0.45	0.50	0.55
P*1E3	115.307	120.445	123.958	126.338	127.940	129.014
Fp*1E3	835.460	828.992	824.570	821.575	819.558	818.206
Rx0*1E3	25.592	26.911	27.813	28.424	28.835	29.111
$\hat{x}$ *1E3	40.033	46.643	52.367	57.235	61.325	64.732

continue

t=	0.60	0.65	0.70	0.75	0.80
P*1E3	129.733	130.212	130.532	130.745	130.887
Fp*1E3	817.302	816.698	816.296	816.028	815.849
Rx0*1E3	29.296	29.419	29.501	29.555	29.592
$\hat{x}$ *1E3	67.553	69.880	71.793	73.363	74.649

When selecting different step sizes for precise integration, such as $\eta = 0.025$ and $\eta = 0.1$ etc., the numerical results are still the same, so that the precision of precise integration method is verified perfectly for one-dimensional problem. ##

A multi-dimensional example is supplied further.

Example 6.11: With dimension n=4 and the data of matrices are given as

$$\mathbf{A} = \begin{bmatrix} 0 & 0 & -2 & 0.5 \\ 0 & 0 & 1 & -.5 \\ 1 & 0 & 0 & 0 \\ 0 & 1 & 0 & 0 \end{bmatrix}, \quad \mathbf{BWB}^T = \begin{bmatrix} 2 & -1 & 0 & 0 \\ -1 & 1 & 0 & 0 \\ 0 & 0 & 1 & 0 \\ 0 & 0 & 0 & 2 \end{bmatrix}, \quad \begin{matrix} \mathbf{C} = [0,\ 0,\ 0,\ 0.5],\ \mathbf{V} = [1] \\ \mathbf{P}_0 = \mathrm{diag}[0.1,\ 0.1,\ 0.1,\ 0.1] \\ \hat{\mathbf{x}}_0 = \mathbf{0}; \quad \mathbf{y} = 2.0\ ; \end{matrix}$$

The measured vector $\mathbf{y}$ ought to be given at each step, which should be near by the motion under the dynamic equation. For simplicity, it is assumed always $\mathbf{y} = 2.0$, and it is pointed out that the deviation of measurement data is quite large.

Solution: For comparison purpose, step sizes of η=0.2 and η=0.8 are selected in the computation, that the numerical results of different step-sizes can be compared. The numerical results of the matrix $\mathbf{P}(t)$ at time t=0.8, 1.6, 2.4, 3.2 and $\mathbf{P}_\infty$ for the two step sizes are completely the same as given by

$$\mathbf{P}(0.8) = \begin{bmatrix} 1.78125 & -.85855 & -.02226 & -.01341 \\ -.85855 & 0.92609 & 0.04335 & 0.6718 \\ -.02226 & 0.04335 & 0.86167 & 0.02591 \\ -.01341 & 0.6718 & 0.02591 & 1.55834 \end{bmatrix}, \quad \mathbf{P}(1.6) = \begin{bmatrix} 3.29278 & -1.6199 & -.01481 & -.17031 \\ -1.6199 & 1.66723 & 0.04221 & 0.23289 \\ -.01481 & 0.04221 & 1.67061 & -.01755 \\ -.17031 & 0.23289 & -.01755 & 2.5098 \end{bmatrix}$$

$$\mathbf{P}(2.4) = \begin{bmatrix} 4.77971 & -2.3283 & -.09309 & -.32622 \\ -2.3283 & 2.31155 & 0.14642 & 0.45497 \\ -.09309 & 0.14642 & 2.41387 & -.11841 \\ -.32622 & 0.45497 & -.11841 & 3.0679 \end{bmatrix}, \quad \mathbf{P}(3.2) = \begin{bmatrix} 6.28667 & -3.0539 & -.09534 & -.61718 \\ -3.0539 & 2.90016 & 0.25705 & 0.71555 \\ -.09534 & 0.25705 & 3.03551 & -.23121 \\ -.61718 & 0.71555 & -.23121 & 3.4600 \end{bmatrix}$$

The matrices $\mathbf{R}_{xup}^{(0)}, \mathbf{R}_{xyp}^{(0)}$ and $\mathbf{R}_{xup}^{(1)}, \mathbf{R}_{xyp}^{(1)}$ are needed stored for each time steps, and the data is too much, so that only the matrix $\mathbf{R}_{xyp}^{(0)}$ is given at part of time stations.

$\mathbf{R}_{xyp}^{(0)} \times 1E3 = \{88.8350,\ \ -6.62121,\ \ -340.201,\ \ 1268.603\}^T$, t: 12.0 ~ 12.8

$\mathbf{R}_{xyp}^{(0)} \times 1E3 = \{87.8004,\ \ -5.55990,\ \ -333.450,\ \ 1266.010\}^T$, t: 11.2 ~ 12.0

...

$\mathbf{R}_{xyp}^{(0)} \times 1E3 = \{42.9259,\ \ -33.72541,\ \ 14.42738,\ \ 309.9350\}^T$, t: 0.0 ~ 0.8

The above are the off-line results. Below are the filter results, but under the assumed measured data of $\mathbf{y} = 2.0$, which is not a real one and unreasonable, but just for showing the computation.

$\mathbf{x}(t) \times 1E3 =$

t=0.0	0.0000	0.0000	0.0000	0.0000
t=0.8	85.85	-67.45	28.85	619.87
t=1.6	333.60	-362.72	201.25	1634.55
t=2.4	470.88	-750.97	516.03	2301.42
t=3.2	248.06	-1028.42	775.63	2582.10
t=4.0	-151.59	-1228.95	730.80	2668.09
t=4.8	-294.40	-1535.29	420.01	2622.05
t=5.6	-59.86	-1969.45	114.57	2436.04
t=6.4	293.97	-2369.86	0.2215	2163.16
......				
t=12.8	565.24	2699.35	-167.06	1663.80

When computing for different step-sizes, which give again the same results. ##

As mentioned, that the measured data y is unreasonable, and can only be used for comparison of different step-sizes.

After all, the Kalman-Bucy filtering is a fundamental part of control and signal processing etc. and numerical computation is indispensable. Traditionally, the computation is classified as the off-line computation of Riccati differential equation, and the on-line computation of filter equation. The former needs to solve a non-linear differential equation, and the latter is to solve a time-variant differential equation. Both problems have difficulties in numerical solution. Based on the time-invariant behavior of the system, the precise integration method proposed in this book solves the Riccati differential equation precisely first, and then the $\hat{\mathbf{x}}$ precise integration of the filter equation is also solved based on the linear interpolation within the time steps of length η. Especially, the computation is sub-divided into ***off-line*** and ***on-line***, that the ***real time computation is reduced to its minimum***, which is extremely important.

The high precision of the precise integration method is quite beneficial to various applications.

§6.6, Optimal smoothing and computations

Three types of estimation, namely ***Prediction, Filtering, and Smoothing***, are introduced in section 6.3. Prediction and filtering have been described in sections 6.4 and 6.5, respectively, and the computations are given by the precise integration method or the eigen-solution based analytical approach. For prediction, there is no measurement to check with, and can only base on the system model to estimate the state of system in the future. The filtering estimates the state at the present time, except the knowledge of mathematical model of the system, there is measurement data until present time to check with, which means that the measurement data used does not object to ***causality*** for filtering. Smoothing state estimation uses all the measured data before and after the time of estimation, which does not fulfill the causality condition. Smoothing is the state estimation for past time. The classification is expressed simply as

Smoothing $\rightarrow$ past; Filtering $\rightarrow$ present; Prediction $\rightarrow$ future.

Smoothing uses all the measured data, so that its estimation for the state at past time is much reliable. Filtering usually applies to the real time estimation and control. The prediction has no data to check its estimation, so that its possible deviation is largest among the three kinds of estimations.

Smoothing itself has three kinds of estimations. The first kind of smoothing is called as ***fixed interval smoothing*** and can be denoted as $\mathbf{x}(t_e \mid t_f)$, the meaning is to estimate the state $\mathbf{x}(t_e \mid t_f)$ at the given time instant t_e with the measured data $\mathbf{y}$ of whole the time interval $[0, t_f)$. Because the finish time t_f is a fixed instant, so that it is called as the ***fixed time interval smoothing*** or ***fixed interval smoothing***. The second type of smoothing is called as ***fixed point smoothing*** and is denoted as $\mathbf{x}(t_e \mid t\ )$, where t_e is a fixed instant but t is continuously progressing, which means to estimate the state vector $\mathbf{x}$ at the fixed time instant t_e using all the measured data until present time t. The third kind of smoothing is called as the ***fixed delay smoothing***, and is written as $\mathbf{x}(t_e \mid t_e + k_f\eta)$, where η is a reasonable small time interval and k_f is a fixed number of intervals. Which means that the measurement data has been progressed to the time instant $t_e + k_f\eta$, but the estimated state $\mathbf{x}$ is at the instant t_e. The time t_e progresses steadily, and the time difference is taken always a constant $k_f\eta$, so that it is called as fixed delay smoothing.

The fixed interval smoothing applies to the case that the in-situ recorded data is brought back to the laboratory, then analyzes the state vector in the whole time interval.

The fixed point smoothing is useful for identifying the state vector at a key instant. Such as for satellite launching, the satellite free motion begins at a definite time, which is critical. Using all the measured data before and after the free motion to estimate the state at the beginning point of free motion, this is the fixed point smoothing.

The fixed delay smoothing can be used for communication systems. After long distance transmission, the dispatched signal reaches the receiver, there are a number of disturbances along the way of transmission including the environmental disturbances such as atmospheric noises etc. The received signal has been mixed with these disturbances. If the estimation bases only on the measurements until time t_e, then the result is filtering. However, if a small time delay $t - t_e = k_f\eta$ is acceptable, where t is the current time, then the time delay improves the quality of estimation.

Smoothing does not consider the causality relation, which is characterized by that the measured data used for estimation involves before and after the estimation instant t_e. The difference between smoothing and filtering is that the analysis interval extends beyond the estimated time instant t_e. For filter analysis, the estimated time is at the end of the analyzed interval, but for smoothing the estimated time is **within** the analyzed time interval. Therefore, ***the interval mixed energy method can also be applied to smoothing computation***, which supplies the **unified** approach to the three kinds of smoothing and also filtering. The method of interval mixed energy is quite different to direct numerical integration, and can use the precise integration method or the eigen-solution based analytical method to develop a whole set of algorithms.

§6.6.1, Optimal smoothing of continuous-time linear system

The filtering of linear system described in last section pays attention to the state estimation at the end point of time interval, which is determined from the ***causality*** requirement. However, smoothing does not care about the causality that the estimated point t_e locates within the time interval $[t_0, t_f)$, see figure 6.8. The ***causality*** condition characterizes the difference between smoothing and filtering.

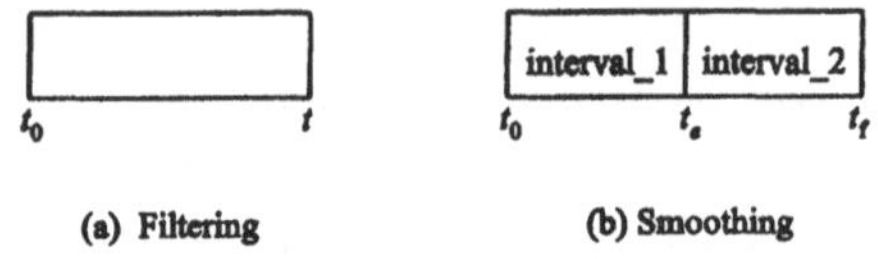

Figure 6.8, *Sketches for filtering and smoothing*

The case of correlated dynamic noise and measurement noise is considered in filtering problem in last section, for which there are only more derivations and complicated equations. The solution methodology is fundamentally the same to the case of uncorrelated dynamic and measurement noises. Hence, only the case of uncorrelated dynamic and measurement Gaussian white noises $\mathbf{w}(t)$ and $\mathrm{v}(t)$ is considered below for smoothing, *i.e.* assuming (6.5.42a~c) is valid. The fundamental equations are

$$\dot{\mathbf{x}} = \mathbf{A}\mathbf{x} + \mathbf{B}_w\mathbf{w} + \mathbf{B}_u\mathbf{u} \tag{6.6.1}$$

$$\mathbf{y} = \mathbf{C}\mathbf{x} + \mathbf{v} \tag{6.6.2}$$

The initial condition is given as

$$\mathbf{x}(0) = \mathbf{x}_0 = \hat{\mathbf{x}}_0 + \mathbf{P}_0\boldsymbol{\lambda}_0 \tag{6.6.3}$$

where $\hat{\mathbf{x}}_0$ is a given vector and $\mathbf{P}_0$ is a given initial variance matrix, symmetric and non-negative definite, $\mathbf{P}_0 = \mathrm{E}[(\mathbf{x}_0 - \hat{\mathbf{x}}_0),(\mathbf{x}_0 - \hat{\mathbf{x}}_0)^T]$. The vector $\hat{\mathbf{x}}_0$ is initially the filtering of the vector $\mathbf{x}_0$ at time $t_0 = 0$, but as time evolves more measurement data $\mathbf{y}$ are available, to estimate the mean value of the vector $\mathbf{x}_0$ at $t_0 = 0$ based on these data implies the objection of causality. Therefore, the estimation of $\mathbf{x}_0$ becomes smoothing mean value $\bar{\mathbf{x}}_0$, which is different from $\hat{\mathbf{x}}_0$ and is reflected on that the vector $\boldsymbol{\lambda}(0)$ is not zero. However, $\bar{\mathbf{x}}_0$ is only a mean vector, smoothing has also its variance matrix $\overline{\mathbf{P}}_0$. The smoothing variance matrix $\overline{\mathbf{P}}_0$ is smaller than the filtering variance matrix $\mathbf{P}_0$ because there are more measured data for smoothing with comparison to filtering.

Hence, smoothing estimation bases also on the measurement $\mathbf{y}$ to estimate $\bar{\mathbf{x}}$ such that the index J functional is minimized, but the integration bound should be extended to t_f

$$J = \int_0^{t_f} [\mathbf{w}^{\mathrm{T}}\mathbf{W}^{-1}\mathbf{w} + \mathbf{v}^{\mathrm{T}}\mathbf{V}^{-1}\mathbf{v}]\mathrm{d}t/2 + (\mathbf{x}(0) - \hat{\mathbf{x}}_0)^{\mathrm{T}}\mathbf{P}_0^{-1}(\mathbf{x}(0) - \hat{\mathbf{x}}_0)/2, \quad \min_{\mathbf{x}} J \tag{6.6.4}$$

Substituting equation (6.6.2) into the index functional gives

$$J = \int_0^{t_f} [\mathbf{w}^{\mathrm{T}}\mathbf{W}^{-1}\mathbf{w} + (\mathbf{y} - \mathbf{C}\mathbf{x})^{\mathrm{T}}\mathbf{V}^{-1}(\mathbf{y} - \mathbf{C}\mathbf{x})]\mathrm{d}t/2 + (\mathbf{x}(0) - \hat{\mathbf{x}}_0)^{\mathrm{T}}\mathbf{P}_0^{-1}(\mathbf{x}(0) - \hat{\mathbf{x}}_0)/2, \qquad \min_{\mathbf{x}} J$$

The above minimization is under the constraint of dynamic equation (6.6.1). Introducing the Lagrange multiplier vector $\boldsymbol{\lambda}(t)$ derives the extended functional J_e

$$J_e = \int_0^{t_f} \begin{bmatrix} \boldsymbol{\lambda}^{\mathrm{T}}(\dot{\mathbf{x}} - \mathbf{A}\mathbf{x} - \mathbf{B}_w\mathbf{w} - \mathbf{B}_u\mathbf{u}) + \mathbf{w}^{\mathrm{T}}\mathbf{W}^{-1}\mathbf{w}/2 \\ + (\mathbf{y} - \mathbf{C}\mathbf{x})^{\mathrm{T}}\mathbf{V}^{-1}(\mathbf{y} - \mathbf{C}\mathbf{x})/2 \end{bmatrix} \mathrm{d}t + (\mathbf{x}(0) - \hat{\mathbf{x}}_0)^{\mathrm{T}}\mathbf{P}_0^{-1}(\mathbf{x}(0) - \hat{\mathbf{x}}_0)/2, \quad \min_{\mathbf{x}} \max_{\boldsymbol{\lambda}} \min_{\mathbf{w}} J_e, \ \text{ or } \ \delta J_e = 0 \tag{6.6.5}$$

The minimization of $\mathbf{w}$ is first performed, which derives $-\mathbf{B}_w^T\boldsymbol{\lambda} + \mathbf{W}^{-1}\mathbf{w} = \mathbf{0}$, then

$$\mathbf{w} = \mathbf{W}\mathbf{B}_w^T\boldsymbol{\lambda} \tag{6.6.6}$$

Substituting the above relation into J_e derives the variational principle with two kinds of variables

$$\mathbf{J}_e = \int_0^{t_f} \left[\boldsymbol{\lambda}^T(\dot{\mathbf{x}} - \mathbf{A}\mathbf{x} - \mathbf{B}_u\mathbf{u}) + \left((\mathbf{y} - \mathbf{C}\mathbf{x})^T\mathbf{V}^{-1}(\mathbf{y} - \mathbf{C}\mathbf{x}) - \boldsymbol{\lambda}^T\mathbf{B}_w\mathbf{W}\mathbf{B}_w^T\boldsymbol{\lambda}\right)/2\right]\mathrm{d}t$$
$$+ (\mathbf{x}(0) - \hat{\mathbf{x}}_0)^T\mathbf{P}_0^{-1}(\mathbf{x}(0) - \hat{\mathbf{x}}_0)/2, \qquad \min_{\mathbf{x}} \max_{\boldsymbol{\lambda}} J_e \tag{6.6.7}$$

where the measurement $\mathbf{y}$ is given vector and does not vary. Expanding the variational principle derives

$$\dot{\mathbf{x}} = \mathbf{A}\mathbf{x} + \mathbf{B}_w\mathbf{W}\mathbf{B}_w^T\boldsymbol{\lambda} + \mathbf{f}_1, \quad \mathbf{f}_1 = \mathbf{B}_u\mathbf{u} \tag{6.6.8a}$$

$$\dot{\boldsymbol{\lambda}} = \mathbf{C}^T\mathbf{V}^{-1}\mathbf{C}\mathbf{x} - \mathbf{A}^T\boldsymbol{\lambda} + \mathbf{f}_2, \quad \mathbf{f}_2 = -\mathbf{C}^T\mathbf{V}^{-1}\mathbf{y} \tag{6.6.8b}$$

The two equations are dual to each other. The initial condition has been given in equation (6.6.3), which is also the natural boundary condition of the variational principle (6.6.7). As the boundary condition at $t = t_f$, the natural boundary condition derived from (6.6.7) is

$$\boldsymbol{\lambda}_f = \mathbf{0}, \text{ as } t = t_f \tag{6.6.9}$$

The dual unknowns $\mathbf{x}, \boldsymbol{\lambda}$ are still state and co-state vectors. Vector $\mathbf{y}$ is the measured data, which is deterministic for a sample process, hence no variation. The vectors $\mathbf{v}, \mathbf{w}$ in the functional (6.6.4) are considered the Gaussian white noises, and hence the dual vectors $\mathbf{x}, \boldsymbol{\lambda}$ in (6.6.7) are also Gaussian distributed stochastic processes. A Gaussian process can always be expressed as the sum of a smoothing mean vector plus a zero mean stochastic vector process. However, smoothing computation uses measurement $\mathbf{y}$ in whole time interval $[0, t_f)$ to estimate the state vector of some past time instant, hence the mean value is certainly different to the filter mean value. For example, $\hat{\mathbf{x}}_0$ is the initial condition of state vector mean value, in filtering analysis, this vector will not be revised later, but smoothing computation requires to revise its mean value based on the measured data in the later time interval $[0, t_f)$. Hence, there is the smoothing mean value $\bar{\mathbf{x}}_0 = \mathbf{x}(0 \mid t_f)$, which makes use of all the measurement $\mathbf{y}$ in $[0, t_f)$ to estimate $\mathbf{x}_0$. The dual vector $\mathbf{x}, \boldsymbol{\lambda}$ solved from the variational principle (6.6.7) of $\mathbf{J}_e$, *i.e.* the solution of the dual equations (6.6.8a~b) and the respective boundary conditions is the smoothing mean vector.

Smoothing is not limited only on the revision of the mean value of state vector, but also that the variance matrix $\mathbf{P}_0$ of filtering should also be revised. Since there is more measured data used in the estimation, the result is more reliable and the variance of smoothing is smaller.

As mentioned above, the vectors $\mathbf{x}, \boldsymbol{\lambda}$ are stochastic processes. Taking average value to the equations (6.6.8a,b) derives $\mathbf{x}, \boldsymbol{\lambda}$ to be deterministic vectors $\bar{\mathbf{x}}(t \mid t_f), \bar{\boldsymbol{\lambda}}(t \mid t_f)$. When $t = t_f$ $\bar{\mathbf{x}}(t_f \mid t_f) = \hat{\mathbf{x}}(t_f)$, which means that the front of smoothing is just filtering, and also $\bar{\boldsymbol{\lambda}}(t_f \mid t_f) = \mathbf{0}$ from (6.6.9). The filtering is always at the front and $\bar{\boldsymbol{\lambda}} = \hat{\boldsymbol{\lambda}}_f = \mathbf{0}$. But internal to the interval, the smoothing co-state $\bar{\boldsymbol{\lambda}}(t \mid t_f)$ is not a null vector, when $t < t_f$.

In this section, no complex conjugate is used, hence a bar above represents smoothing value, which makes no confusion.

§6.6.2, Interval mixed energy and differential equations for smoothing

The filtering vector is at the front t of the interval $[0,t)$, where time t evolves steadily. The smoothing considered time instant t_e is an internal point within the interval $[0,t_f)$, that the measurement $\mathbf{y}$ has reached t_f, but the state vector $\mathbf{x}(t_e \mid t_f)$ at t_e is estimated. Smoothing estimated state vector is within the interval, which is quite different to filtering for which the frontier state vector $\mathbf{x}(t \mid t)$ is to estimate.

Equation (6.5.69) introduced the interval mixed energy $V(\mathbf{x}_a, \boldsymbol{\lambda}_b)$, where (t_a, t_b) is a time interval, and the boundary conditions are given in (6.5.74). The quadratic form of $V(\mathbf{x}_a, \boldsymbol{\lambda}_b)$ is given by (6.5.72), where the interval matrices $\mathbf{Q}(t_a,t_b)$, $\mathbf{G}(t_a,t_b)$, $\mathbf{F}(t_a,t_b)$ and vectors $\mathbf{r}_x(t_a,t_b)$, $\mathbf{r}_\lambda(t_a,t_b)$ satisfy the differential equations (6.5.80-82) and (6.5.83-87). The differential equation (6.5.80a) coincides with the Riccati differential equation (6.5.50) but the boundary condition is different to that in (6.5.50). The differential equations (6.5.81) and (6.5.49) are also similar, but the initial condition is different and the matrix $\mathbf{G}$ in the differential equation is again different to the matrix $\mathbf{P}$ in (6.5.49).

In filtering analysis, the precise integration of mixed energy matrices $\mathbf{F}(\eta), \mathbf{G}(\eta), \mathbf{Q}(\eta)$ of the fundamental interval η is first carried out by the precise integration method. Then the factor of initial condition matrix $\mathbf{P}_0$ is considered by the revision equation (6.5.90). For the filter equation (6.5.49), the same method can also be applied. When the Hamilton matrix $\mathbf{H}$ does not appear Jordan normal form, these computations can use the eigen-solution based analytical method.

In addition to filtering, the smoothing computation has the interval (t_e, t_f) past the estimated point t_e, for which the interval mixed energy method can still be applied. In smoothing analysis, the fact that the two ends t_a and t_b of interval can be arbitrarily selected is quite useful. Now the method of equation (6.5.89~90) is further investigated. Let the variational principle (6.5.45') be rewritten as

$$\delta\left\{\int_{t_0}^{t} [\boldsymbol{\lambda}^T(\dot{\mathbf{x}} - \mathbf{B}_u\mathbf{u}) - \mathrm{H}(\mathbf{x},\boldsymbol{\lambda}) - \mathbf{x}^T\mathbf{C}^T\mathbf{V}^{-1}\mathbf{y}]\mathrm{d}\tau + (\mathbf{x}_0 - \hat{\mathbf{x}}_0)^{\mathrm{T}}\mathbf{P}_0^{-1}(\mathbf{x}_0 - \hat{\mathbf{x}}_0)/2\right\} = 0$$

where $\mathrm{H}(\mathbf{x},\boldsymbol{\lambda})$ is given in (6.5.70) and the last term represents the inhomogeneous boundary condition at t_0. According to the definition of interval mixed energy (6.5.69), the variational principle can be written as

$$\delta\left\{\boldsymbol{\lambda}^T(t)\mathbf{x}(t) - V(\mathbf{x}_0, \boldsymbol{\lambda}(t)) + (\mathbf{x}_0 - \hat{\mathbf{x}}_0)^T\mathbf{P}_0^{-1}(\mathbf{x}_0 - \hat{\mathbf{x}}_0)/2\right\} = 0$$

The last term can be rewritten as

$$(\mathbf{x}_0 - \hat{\mathbf{x}}_0)^{\mathrm{T}}\mathbf{P}_0^{-1}(\mathbf{x}_0 - \hat{\mathbf{x}}_0)/2 = \max_{\boldsymbol{\lambda}_0}\left[\boldsymbol{\lambda}_0^{\mathrm{T}}(\mathbf{x}_0 - \hat{\mathbf{x}}_0) - \boldsymbol{\lambda}_0^{\mathrm{T}}\mathbf{P}_0\boldsymbol{\lambda}_0/2\right]$$

Substituting into the variational principle and note that the $\boldsymbol{\lambda}_0$ maximization can been involved in the variational operation, so that

$$\delta\left\{\boldsymbol{\lambda}^T\mathbf{x} - V(\mathbf{x}_0, \boldsymbol{\lambda}) + \boldsymbol{\lambda}_0^T\mathbf{x}_0 - \boldsymbol{\lambda}_0^T\mathbf{I}_n\hat{\mathbf{x}}_0 - \boldsymbol{\lambda}_0^T\mathbf{P}_0\boldsymbol{\lambda}_0/2\right\} = 0$$

where the last two terms represents a virtual 'interval', *i.e.* at the t_0 end the mixed energy is given as

$$V_0(\hat{\mathbf{x}}_0,\boldsymbol{\lambda}_0)=\boldsymbol{\lambda}_0^T\mathbf{I}_n\hat{\mathbf{x}}_0+\boldsymbol{\lambda}_0^T\mathbf{P}_0\boldsymbol{\lambda}_0/2$$

It corresponds to that the initial condition at the $t_0=0$ end is treated as a virtual 'interval'

$$\mathbf{Q}_0=\mathbf{0},\quad \mathbf{F}_0=\mathbf{I}_n,\quad \mathbf{G}_0=\mathbf{P}_0;\qquad \mathbf{r}_x(0)=\mathbf{0},\quad \mathbf{r}_\lambda(0)=\mathbf{0} \tag{6.6.10}$$

or given as $[\mathbf{Q},\mathbf{G},\mathbf{F}]_0=[\mathbf{0},\mathbf{P}_0,\mathbf{I}_n]$. Based on the associative rule of interval combination $\backsim$, the integration of matrix differential equation (6.5.80) for $\mathbf{Q},\mathbf{G},\mathbf{F}$ can be first performed under the initial condition (6.5.74), then the mixed energy (6.6.10) is treated as interval 1 and (t_0,t) is treated as interval 2, executing the interval combination algorithm (6.5.77a~c) gives

$$\mathbf{P}(t)=\mathbf{G}+\mathbf{F}(\mathbf{P}_0^{-1}+\mathbf{Q})^{-1}\mathbf{F}^{\mathrm{T}} \tag{6.6.11a}$$

$$\mathbf{F}_p(t)=\mathbf{F}(\mathbf{I}_n+\mathbf{P}_0\mathbf{Q})^{-1} \tag{6.6.11b}$$

$$\mathbf{Q}_p(t)=\mathbf{Q}_0+(\mathbf{Q}^{-1}+\mathbf{P}_0)^{-1} \tag{6.6.11c}$$

or written as

$$[\mathbf{Q}_p(t),\mathbf{P}(t),\mathbf{F}_p(t)]=[\mathbf{0},\mathbf{P}_0,\mathbf{I}_n]\backsim[\mathbf{Q}(t),\mathbf{G}(t),\mathbf{F}(t)] \tag{6.6.11'}$$

Subscript p means that the initial variance matrix being $\mathbf{P}_0$ is considered.

To verify the differential equation satisfied by the matrix $\mathbf{P}(t)$, since only the matrices $\mathbf{Q}(t),\mathbf{G}(t),\mathbf{F}(t)$ depend on time t, they satisfy the differential equation (6.5.80). Direct differentiating verifies

$$\begin{aligned}\dot{\mathbf{P}}(t)&=\dot{\mathbf{G}}+\dot{\mathbf{F}}(\mathbf{P}_0^{-1}+\mathbf{Q})^{-1}\mathbf{F}^T+\mathbf{F}(\mathbf{P}_0^{-1}+\mathbf{Q})^{-1}\dot{\mathbf{F}}^T-\mathbf{F}(\mathbf{P}_0^{-1}+\mathbf{Q})^{-1}\dot{\mathbf{Q}}(\mathbf{P}_0^{-1}+\mathbf{Q})^{-1}\mathbf{F}^T\\
&=\mathbf{B}_w\mathbf{W}\mathbf{B}_w^T+\mathbf{G}\mathbf{A}^T+\mathbf{A}\mathbf{G}-\mathbf{G}\mathbf{C}^T\mathbf{V}^{-1}\mathbf{C}\mathbf{G}\\
&\quad+\mathbf{F}(\mathbf{P}_0^{-1}+\mathbf{Q})^{-1}\mathbf{F}^T(\mathbf{A}-\mathbf{G}\mathbf{C}^{\mathrm{T}}\mathbf{V}^{-1}\mathbf{C})^T+(\mathbf{A}-\mathbf{G}\mathbf{C}^T\mathbf{V}^{-1}\mathbf{C})\mathbf{F}(\mathbf{P}_0^{-1}+\mathbf{Q})^{-1}\mathbf{F}^T\\
&\quad-\mathbf{F}(\mathbf{I}+\mathbf{P}_0\mathbf{Q})^{-1}\mathbf{P}_0\mathbf{F}^T\mathbf{C}^T\mathbf{V}^{-1}\mathbf{C}\mathbf{F}\mathbf{P}_0(\mathbf{I}+\mathbf{P}_0\mathbf{Q})^{-T}\mathbf{F}^T\\
&=\mathbf{B}_w\mathbf{W}\mathbf{B}_w^T+\left[\mathbf{G}+\mathbf{F}(\mathbf{P}_0^{-1}+\mathbf{Q})^{-1}\mathbf{F}^T\right]\mathbf{A}^T+\mathbf{A}\left[\mathbf{G}+\mathbf{F}(\mathbf{P}_0^{-1}+\mathbf{Q})^{-1}\mathbf{F}^T\right]\\
&\quad-\left[\mathbf{G}+\mathbf{F}(\mathbf{P}_0^{-1}+\mathbf{Q})^{-1}\mathbf{F}^T\right]\mathbf{C}^T\mathbf{V}^{-1}\mathbf{C}\left[\mathbf{G}+\mathbf{F}(\mathbf{P}_0^{-1}+\mathbf{Q})^{-1}\mathbf{F}^T\right]\end{aligned}$$

so that

$$\dot{\mathbf{P}}(t)=\mathbf{B}_w\mathbf{W}\mathbf{B}_w^T+\mathbf{P}\mathbf{A}^T+\mathbf{A}\mathbf{P}-\mathbf{P}\mathbf{C}^T\mathbf{V}^{-1}\mathbf{C}\mathbf{P} \tag{6.6.12}$$

and the initial condition is verified as

$$\mathbf{P}(0)=\mathbf{0}+\mathbf{I}(\mathbf{P}_0^{-1}+\mathbf{0})^{-1}\mathbf{I}=\mathbf{P}_0$$

Based on the uniqueness theorem of ODE, the interval combination algorithm (6.6.11a) computed matrix $\mathbf{P}(t)$ is the solution matrix of the filtering Riccati equation (6.5.50). For $\mathbf{F}_p(t)$

$$\begin{aligned}\dot{\mathbf{F}}_p(t)&=\dot{\mathbf{F}}(\mathbf{I}+\mathbf{P}_0\mathbf{Q})^{-1}-\mathbf{F}(\mathbf{I}+\mathbf{P}_0\mathbf{Q})^{-1}\mathbf{P}_0\dot{\mathbf{Q}}(\mathbf{I}+\mathbf{P}_0\mathbf{Q})^{-1}\\
&=(\mathbf{A}-\mathbf{G}\mathbf{C}^T\mathbf{V}^{-1}\mathbf{C})\mathbf{F}(\mathbf{I}+\mathbf{P}_0\mathbf{Q})^{-1}-\mathbf{F}(\mathbf{I}+\mathbf{P}_0\mathbf{Q})^{-1}\mathbf{P}_0\mathbf{F}^T\mathbf{C}^T\mathbf{V}^{-1}\mathbf{C}\mathbf{F}(\mathbf{I}+\mathbf{P}_0\mathbf{Q})^{-1}\\
&=\left[\mathbf{A}-(\mathbf{G}+\mathbf{F}(\mathbf{P}_0^{-1}+\mathbf{Q})^{-1}\mathbf{F}^T)\mathbf{C}^T\mathbf{V}^{-1}\mathbf{C}\right]\mathbf{F}_p=(\mathbf{A}-\mathbf{P}\mathbf{C}^T\mathbf{V}^{-1}\mathbf{C})\mathbf{F}_p\end{aligned}$$

so that the differential equation and initial condition are

$$\dot{\mathbf{F}}_p=(\mathbf{A}-\mathbf{P}\mathbf{C}^T\mathbf{V}^{-1}\mathbf{C})\mathbf{F}_p\;,\quad \mathbf{F}_p(0)=\mathbf{I} \tag{6.6.13}$$

For $\mathbf{Q}_p(t)$, it gives

$$\dot{\mathbf{Q}}_p(t) = \mathbf{F}_p^T\mathbf{C}^T\mathbf{V}^{-1}\mathbf{C}\mathbf{F}_p \ , \ \mathbf{Q}_p(0) = \mathbf{Q}_0(=\mathbf{0}) \tag{6.6.14}$$

All of the above verification explains that the initial condition can be solved by the transformation (6.6.11) after the interval $(t_0 = 0, t)$ mixed energy matrices $\mathbf{Q}(t), \mathbf{G}(t), \mathbf{F}(t)$ with the initial condition (6.5.74) are computed. All these derivation is very similar to that given in last section, *i.e.* a unified approach.

This method corresponds to the associative rule of interval combination operation.

Similarly for vectors $\mathbf{r}_{xp}$ and $\mathbf{r}_{\lambda p}$, the interval combination equation (6.5.78a,b) derives

$$\mathbf{r}_{xp} = \mathbf{r}_x + \mathbf{F}(\mathbf{P}_0^{-1} + \mathbf{Q})^{-1}\mathbf{r}_\lambda \tag{6.6.15a}$$

$$\mathbf{r}_{\lambda p} = (\mathbf{I} + \mathbf{Q}\mathbf{P}_0)^{-1}\mathbf{r}_\lambda \tag{6.6.15b}$$

The differential equation for $\mathbf{r}_{xp}(0,t)$ can be derived from (6.5.80~82) as

$$\begin{aligned}
\dot{\mathbf{r}}_{xp} &= \dot{\mathbf{r}}_x + \mathbf{F}(\mathbf{P}_0^{-1} + \mathbf{Q})^{-1}\mathbf{r}_\lambda - \dot{\mathbf{F}}(\mathbf{P}_0^{-1} + \mathbf{Q})^{-1}\dot{\mathbf{Q}}(\mathbf{P}_0^{-1} + \mathbf{Q})^{-1}\mathbf{r}_\lambda + \mathbf{F}(\mathbf{P}_0^{-1} + \mathbf{Q})^{-1}\dot{\mathbf{r}}_\lambda \\
&= (\mathbf{A} - \mathbf{G}\mathbf{C}^T\mathbf{V}^{-1}\mathbf{C})\mathbf{r}_x + \mathbf{G}\mathbf{C}^T\mathbf{V}^{-1}\mathbf{y} + \mathbf{B}_u\mathbf{u} + (\mathbf{A} - \mathbf{G}\mathbf{C}^T\mathbf{V}^{-1}\mathbf{C})\mathbf{F}(\mathbf{P}_0^{-1} + \mathbf{Q})^{-1}\mathbf{r}_\lambda \\
&\quad - \mathbf{F}(\mathbf{P}_0^{-1} + \mathbf{Q})^{-1}\mathbf{F}^T\mathbf{C}^T\mathbf{V}^{-1}\mathbf{C}\mathbf{F}(\mathbf{P}_0^{-1} + \mathbf{Q})^{-1}\mathbf{r}_\lambda + \mathbf{F}(\mathbf{P}_0^{-1} + \mathbf{Q})^{-1}\mathbf{F}^T\mathbf{C}^T\mathbf{V}^{-1}(\mathbf{y} - \mathbf{C}\mathbf{r}_x)
\end{aligned}$$

So

$$\dot{\mathbf{r}}_{xp} = (\mathbf{A} - \mathbf{P}\mathbf{C}^T\mathbf{V}^{-1}\mathbf{C})\mathbf{r}_{xp} + \mathbf{P}\mathbf{C}^T\mathbf{V}^{-1}\mathbf{y} + \mathbf{B}_u\mathbf{u} \ , \ \mathbf{r}_{xp}(0) = \mathbf{0} \tag{6.6.16a}$$

Similarly

$$\dot{\mathbf{r}}_{\lambda p} = \mathbf{F}_p^T\mathbf{C}^T\mathbf{V}^{-1}(\mathbf{y} - \mathbf{C}\mathbf{r}_{xp}) \ , \ \mathbf{r}_{\lambda p}(0) = \mathbf{0} \tag{6.6.16b}$$

The equations derived above explain that the combination of the initial condition of 'virtual' interval and the interval (t_0, t) gives the mixed energy matrices and vectors [equations (6.6.11a~c) and (6.6.15a~b)]

$$\mathbf{Q}_p(t),\ \mathbf{P}(t),\ \mathbf{F}_p(t); \qquad \mathbf{r}_{xp}(t),\ \mathbf{r}_{\lambda p}(t) \tag{6.6.17}$$

which satisfy the differential equations (6.6.12~16). These equations are similar to (6.5.80~82) in turn. The combined interval is $[t_0, t)$ and the mixed energy is

$$V_p(\hat{\mathbf{x}}_0, \boldsymbol{\lambda}) = \boldsymbol{\lambda}^T\mathbf{F}_p\hat{\mathbf{x}}_0 + \boldsymbol{\lambda}^T\mathbf{P}\boldsymbol{\lambda}/2 - \hat{\mathbf{x}}_0\mathbf{Q}_p\hat{\mathbf{x}}_0/2 + \boldsymbol{\lambda}^T\mathbf{r}_{xp} + \hat{\mathbf{x}}_0^T\mathbf{r}_{\lambda p} \tag{6.6.18}$$

From equation $\mathbf{x} = \partial V_p/\partial\boldsymbol{\lambda}$ gives

$$\mathbf{x} = \mathbf{F}_p\hat{\mathbf{x}}_0 + \mathbf{P}\boldsymbol{\lambda} + \mathbf{r}_{xp} \tag{6.6.19}$$

The set of equations applies to arbitrary time t in the whole interval $0 \le t \le t_f$. As mentioned above, the filter solution implies that the time t is a free end of the interval $[t_0 = 0, t)$, *i.e.* $\boldsymbol{\lambda}(t) = \mathbf{0}$. If an appropriate vector $\boldsymbol{\lambda}(t_e) = \bar{\boldsymbol{\lambda}}(t_e)$ is selected, then $\mathbf{x}(t_e) = \bar{\mathbf{x}}(t_e)$ and it gives the smoothing solution, that is the difference of smoothing to filtering.

Firstly, it is to point out that the filter vector is

$$\hat{\mathbf{x}}(t) = \mathbf{F}_p(t)\hat{\mathbf{x}}_0 + \mathbf{r}_{xp}(t) \tag{6.6.20}$$

which is verified as follows. The initial condition at $t = 0$ is obtain based on the (6.6.16a) and (6.6.13) that $\hat{\mathbf{x}}(0) = \hat{\mathbf{x}}_0$. Then differentiating gives

$$\dot{\hat{\mathbf{x}}}(t) = (\mathbf{A} - \mathbf{P}\mathbf{C}^T\mathbf{V}^{-1}\mathbf{C})\mathbf{F}_p\hat{\mathbf{x}}_0 + (\mathbf{A} - \mathbf{P}\mathbf{C}^T\mathbf{V}^{-1}\mathbf{C})\mathbf{r}_{xp} + \mathbf{P}\mathbf{C}^T\mathbf{V}^{-1}\mathbf{y} + \mathbf{B}_u\mathbf{u}$$

$$= (\mathbf{A} - \mathbf{P}\mathbf{C}^T\mathbf{V}^{-1}\mathbf{C})\hat{\mathbf{x}} + \mathbf{P}\mathbf{C}^T\mathbf{V}^{-1}\mathbf{y} + \mathbf{B}_u\mathbf{u}$$

Comparing with equation (6.5.49) determines that $\hat{\mathbf{x}}$ is the filtering vector. Hence, equation (6.6.19) can be rewritten as

$$\boldsymbol{\xi}(t) = \mathbf{P}\bar{\boldsymbol{\lambda}}(t) \quad , \quad \boldsymbol{\xi} = \bar{\mathbf{x}} - \hat{\mathbf{x}} \tag{6.6.21}$$

where the vector $\boldsymbol{\xi}(t)$ is the revision of smoothing to the filtering $\hat{\mathbf{x}}$. The vector $\boldsymbol{\xi}(t)$ is not a null vector within the interval, only at the end point $t = t_f$, $\boldsymbol{\xi}(t_f) = \mathbf{0}$, where the smoothing solution reduces to be filtering. The dual differential equation of the dual vectors $\boldsymbol{\xi}, \bar{\boldsymbol{\lambda}}$ can be obtained by combining (6.6.8) and (6.5.49)

$$\dot{\boldsymbol{\xi}}(t) = \mathbf{A}\boldsymbol{\xi} + \mathbf{B}_w\mathbf{W}\mathbf{B}_w^T\bar{\boldsymbol{\lambda}} - \mathbf{P}\mathbf{C}^T\mathbf{V}^{-1}(\mathbf{y} - \mathbf{C}\hat{\mathbf{x}}), \text{ with } \boldsymbol{\xi}(t_f) = \mathbf{0} \tag{6.6.22}$$

$$\dot{\bar{\boldsymbol{\lambda}}}(t) = \mathbf{C}^T\mathbf{V}^{-1}\mathbf{C}\boldsymbol{\xi} - \mathbf{A}^T\bar{\boldsymbol{\lambda}} - \mathbf{C}^T\mathbf{V}^{-1}(\mathbf{y} - \mathbf{C}\hat{\mathbf{x}}), \text{ with } \bar{\boldsymbol{\lambda}}(t_f) = \mathbf{0} \tag{6.6.23}$$

At the end point $t = t_f$, the zero boundary conditions are for the mean value. Substituting (6.6.21) into (6.6.23) gives the differential equation

$$\dot{\bar{\boldsymbol{\lambda}}} = (-\mathbf{A}^T + \mathbf{C}^T\mathbf{V}^{-1}\mathbf{C}\mathbf{P})\bar{\boldsymbol{\lambda}} - \mathbf{C}^T\mathbf{V}^{-1}(\mathbf{y} - \mathbf{C}\hat{\mathbf{x}}), \quad \bar{\boldsymbol{\lambda}}(t_f) = \mathbf{0} \tag{6.6.24}$$

For a deterministic measurement vector $\mathbf{y}$, the vector $\bar{\boldsymbol{\lambda}}$ solved from the integration of equation (6.6.24) gives the deterministic smoothing co-state mean value.

However, application needs to solve the state vector $\bar{\mathbf{x}}(t)$ directly. Because of equation (6.6.21)

$$\bar{\boldsymbol{\lambda}}(t) = \mathbf{P}^{-1}(\bar{\mathbf{x}} - \hat{\mathbf{x}}) \tag{6.6.21'}$$

where the over-bar represents smoothing mean value. Substituting into equation (6.6.8a) gives

$$\dot{\bar{\mathbf{x}}}(t) = \mathbf{A}\bar{\mathbf{x}} + \mathbf{B}_w\mathbf{W}\mathbf{B}_w^T\mathbf{P}^{-1}(\bar{\mathbf{x}} - \hat{\mathbf{x}}), \text{ with } \bar{\mathbf{x}}(t_f) = \hat{\mathbf{x}}(t_f) \tag{6.6.25}$$

where $\hat{\mathbf{x}}(t)$ is the filtering solution, which is regarded as a known function in smoothing solution. The factor $\mathbf{B}_w\mathbf{W}\mathbf{B}_w^T\mathbf{P}^{-1}$ is the gain matrix of smoothing, where $\mathbf{P}(t)$ is the variance matrix of filtering and is regarded also as a known function. Note that the precise integration solution of $\hat{\mathbf{x}}(t)$ and $\mathbf{P}(t)$ has been described in detail in section 6.5, so that only the solution of (6.6.25) is given in detail below.

Differential equation (6.6.25) is time variant. The integration for a general time variant system is a problem. The usual FDM integration is not precise enough. Using the method of ***back substitution equation*** (see section 5.9) of internal vectors to solve the differential equation has a number of benefits. If the corresponding Hamilton matrix $\mathbf{H}$ does not appear Jordan normal form in eigen-solutions, then the analytical method can also apply. In the filtering and wave-guide problems, the analytical method is quite effective in solving these problems.

Differential equation for smoothing variance matrix is given in next section.

§6.6.3, Mean value and variance of smoothing

The dual differential equation of smoothing is derived in the last section, which is useful for theoretical development, however, used in numerical integration directly is inappropriate. Currently, the numerical solution of ODEs generally uses FDM, which is error prone. Algorithm with interval combination solution is far better, because such method in combination with the precise integration method solves the differential equation up to the computer precision.

Interval combination equations have been derived in section 6.5.4.1. Let $t_a = t_0$, $t_b = t$ and $t_c = t_f$, and the interval-1 being selected as $[t_0, t)$ then the matrices and vectors in equation (6.6.17) are $\mathbf{Q}_1, \mathbf{P}, \mathbf{F}_1, \mathbf{r}_{x1}, \mathbf{r}_{\lambda 1}$. The interval-2 is selected as (t, t_f) and the boundary condition is $\bar{\boldsymbol{\lambda}}(t_f) = \mathbf{0}$ with the matrices and vectors denoted as $\mathbf{Q}_2, \mathbf{G}_2, \mathbf{F}_2, \mathbf{r}_{x2}, \mathbf{r}_{\lambda 2}$. According to the back substitution equation (6.5.76a,b), it derives

$$\bar{\mathbf{x}} = (\mathbf{I} + \mathbf{P}\mathbf{Q}_2)^{-1}(\mathbf{F}_1\hat{\mathbf{x}}_0 + \mathbf{r}_{x1} + \mathbf{P}\mathbf{r}_{\lambda 2}) \tag{6.6.26}$$

$$\bar{\boldsymbol{\lambda}} = (\mathbf{I} + \mathbf{Q}_2\mathbf{P})^{-1} \cdot [-\mathbf{Q}_2(\mathbf{F}_1\hat{\mathbf{x}}_0 + \mathbf{r}_{x1}) + \mathbf{r}_{\lambda 2}] \tag{6.6.27}$$

then using equation (6.6.20)

$$\bar{\mathbf{x}} = (\mathbf{I} + \mathbf{P}\mathbf{Q}_2)^{-1}(\hat{\mathbf{x}} + \mathbf{P}\mathbf{r}_{\lambda 2}) \tag{6.6.26'}$$

$$\bar{\boldsymbol{\lambda}} = (\mathbf{I} + \mathbf{Q}_2\mathbf{P})^{-1}(-\mathbf{Q}_2\hat{\mathbf{x}} + \mathbf{r}_{\lambda 2}) \tag{6.6.27'}$$

where $\bar{\mathbf{x}}, \bar{\boldsymbol{\lambda}}$ are the vectors of smoothing solution, and $\hat{\mathbf{x}}$ is the filter vector.

The vectors $\bar{\mathbf{x}}, \bar{\boldsymbol{\lambda}}$ are dual to each other, so that they are of same importance, but application inclines to the state vector $\bar{\mathbf{x}}$. The physical meaning of vector $\bar{\boldsymbol{\lambda}}$ corresponds to the internal force between the intervals-1 and 2 in structural mechanics. Rewrite the equation (6.6.26') as

$$\bar{\mathbf{x}} = (\mathbf{P}^{-1} + \mathbf{Q}_2)^{-1}(\mathbf{P}^{-1}\hat{\mathbf{x}} + \mathbf{r}_{\lambda 2}) \tag{6.6.26''}$$

from which and considering the structural mechanics interpretation, a flexibility matrix $\mathbf{P}_s$ is introduced as

$$\mathbf{P}_s = (\mathbf{P}^{-1} + \mathbf{Q}_2)^{-1} \tag{6.6.28}$$

which is the variance matrix $\mathbf{P}_s(t) = \bar{\mathbf{P}}(t)$ of smoothing solution of state vector $\bar{\mathbf{x}}(t)$. Before verification, the physical interpretation is clarified first. Matrix $\mathbf{P}(t)$ is the solution of Riccati differential equation of interval $[t_0, t)$ with the initial condition $\mathbf{P}(0) = \mathbf{P}_0$, see equation (6.5.50), interpreted in structural mechanics, the flexibility matrix of the interval $[t_0, t)$ at the end t. Matrix $\mathbf{Q}_2(t, t_f)$ is from the mixed energy matrix of the interval (t, t_f) and as $t \to t_f$ the natural boundary condition (6.5.74) is satisfied. The physical interpretation of $\mathbf{Q}_2(t, t_f)$ in structural mechanics is the stiffness matrix of the interval (t, t_f) at the left end t, and $\mathbf{r}_{\lambda 2}$ is the corresponding force vector of the interval (t, t_f). Therefore, the physical interpretation of the matrix $\mathbf{P}_s$ in structural mechanics is clear, that it is the internal flexibility matrix of the whole interval $[t_0, t_f)$ at the station t, which is interpreted in turn as follows.

That the parenthesis in equation (6.6.28) gives the stiffness matrix, which is the sum of the right end stiffness matrix $\mathbf{P}^{-1}$ of the interval $[t_0,t)$, *i.e.* the inverse of flexibility matrix $\mathbf{P}$, and the left end stiffness matrix $\mathbf{Q}_2$ of interval (t,t_f). Then $\mathbf{P}_s$ is the inverse of a stiffness matrix, a flexibility matrix internal to the whole interval $[t_0,t_f)$ at t. Note that a flexibility matrix corresponds to the variance matrix in least square analysis. The quadratic index is just a least square.

The differential equation of $\mathbf{P}_s = \mathbf{P}_s(t \mid t_f)$ is derived as follows. Let $\mathbf{K}_s = \mathbf{P}_s^{-1} = \mathbf{P}^{-1} + \mathbf{Q}_2$, then

$$\dot{\mathbf{P}}_s = -\mathbf{P}_s\dot{\mathbf{K}}_s\mathbf{P}_s = -\mathbf{P}_s(\dot{\mathbf{Q}}_2 - \mathbf{P}^{-1}\dot{\mathbf{P}}\mathbf{P}^{-1})\mathbf{P}_s$$

Substituted with $\dot{\mathbf{Q}}_2 = \partial\mathbf{Q}_2/\partial t_a = -\mathbf{C}^T\mathbf{V}^{-1}\mathbf{C} - \mathbf{Q}_2\mathbf{A} - \mathbf{A}^T\mathbf{Q}_2 + \mathbf{Q}_2\mathbf{B}_1\mathbf{W}\mathbf{B}_1^T\mathbf{Q}_2$ and (6.6.12), and the derivation is as follows

$$\begin{aligned}\dot{\mathbf{P}}_s &= \mathbf{P}_s[\mathbf{C}^T\mathbf{V}^{-1}\mathbf{C} + \mathbf{Q}_2\mathbf{A} + \mathbf{A}^T\mathbf{Q}_2 - \mathbf{Q}_2\mathbf{B}_w\mathbf{W}\mathbf{B}_w^T\mathbf{Q}_2 \\ &\quad + \mathbf{P}^{-1}(\mathbf{B}_w\mathbf{W}\mathbf{B}_w^T + \mathbf{P}\mathbf{A}^T + \mathbf{A}\mathbf{P} - \mathbf{P}\mathbf{C}^T\mathbf{V}^{-1}\mathbf{C}\mathbf{P})\mathbf{P}^{-1}]\mathbf{P}_s \\ &= \mathbf{P}_s[(\mathbf{P}^{-1} + \mathbf{Q}_2)\mathbf{A} + \mathbf{A}^T(\mathbf{P}^{-1} + \mathbf{Q}_2) - \mathbf{Q}_2\mathbf{B}_w\mathbf{W}\mathbf{B}_w^T\mathbf{Q}_2 + \mathbf{P}^{-1}\mathbf{B}_w\mathbf{W}\mathbf{B}_w^T\mathbf{P}^{-1}]\mathbf{P}_s \\ &= \mathbf{A}\mathbf{P}_s + \mathbf{P}_s\mathbf{A}^T + \mathbf{P}_s[\mathbf{P}^{-1}\mathbf{B}_w\mathbf{W}\mathbf{B}_w^T(\mathbf{P}^{-1} + \mathbf{Q}_2) - (\mathbf{P}^{-1} + \mathbf{Q}_2)\mathbf{B}_w\mathbf{W}\mathbf{B}_w^T\mathbf{Q}_2]\mathbf{P}_s\end{aligned}$$

substituting $\mathbf{P}_s^{-1} = \mathbf{P}^{-1} + \mathbf{Q}_2$, the derivation continues as

$$\begin{aligned}\dot{\mathbf{P}}_s(t \mid t_f) &= \mathbf{A}\mathbf{P}_s + \mathbf{P}_s\mathbf{A}^T + \mathbf{P}_s\mathbf{P}^{-1}\mathbf{B}_w\mathbf{W}\mathbf{B}_w^T - \mathbf{B}_w\mathbf{W}\mathbf{B}_w^T\mathbf{Q}_2\mathbf{P}_s \\ &= \mathbf{P}_s(\mathbf{A}^T + \mathbf{P}^{-1}\mathbf{B}_w\mathbf{W}\mathbf{B}_w^T) + (\mathbf{A} + \mathbf{B}_w\mathbf{W}\mathbf{B}_w^T\mathbf{Q}_2)\mathbf{P}_s \\ &= \mathbf{P}_s(\mathbf{A} + \mathbf{B}_w\mathbf{W}\mathbf{B}_w^T\mathbf{P}^{-1})^T + (\mathbf{A} + \mathbf{B}_w\mathbf{W}\mathbf{B}_w^T\mathbf{P}^{-1})\mathbf{P}_s - \mathbf{B}_w\mathbf{W}\mathbf{B}_w^T(\mathbf{Q}_2 + \mathbf{P}^{-1})\mathbf{P}_s\end{aligned}$$

The differential equation is thus derived as

$$\dot{\mathbf{P}}_s(t \mid t_f) = \mathbf{P}_s(\mathbf{A} + \mathbf{B}_w\mathbf{W}\mathbf{B}_w^T\mathbf{P}^{-1})^T + (\mathbf{A} + \mathbf{B}_w\mathbf{W}\mathbf{B}_w^T\mathbf{P}^{-1})\mathbf{P}_s - \mathbf{B}_w\mathbf{W}\mathbf{B}_w^T \tag{6.6.29}$$

which is the differential equation of $\mathbf{P}_s$. The boundary condition is obtained as that from (6.6.28), $\mathbf{Q}_2 \to \mathbf{0}$ when $t \to t_f$, so that

$$\mathbf{P}_s(t \to t_f \mid t_f) = \mathbf{P}(t_f) \tag{6.6.30}$$

This explains that the boundary condition of differential equation (6.6.29) is at t_f, where the smoothing variance $\mathbf{P}_s$ equals to the filtering variance matrix $\mathbf{P}(t_f)$, for which the algorithm has been given in the last section.

Therefore, the differential equation (6.6.29) and boundary condition (6.6.30) for the matrix $\mathbf{P}_s$ is derived. Based on the uniqueness theorem of ODE, the solution is unique. Although, the derivation to reach the differential equation is different from the traditional approach, but the resulted equations are the same. The present derivation gives the equation (6.6.28) for $\mathbf{P}_s$, which can be used for computation. From (6.6.30), the integration is along the reverse direction from t_f back to t, and the end condition gives symmetric positive definite matrix. Looking from the differential equation (6.6.29) it is seen that $\mathbf{P}_s$ must be symmetric at arbitrary time $t < t_f$. According to equation (6.6.28) the matrix $\mathbf{Q}_2$ is necessary, which can be computed by the precise integration method too. Matrix $\mathbf{Q}_2$ plays a key role for LQ optimal control and is also symmetric and positive definite, so that the $\mathbf{P}_s$ computed by (6.6.28) must be symmetric and positive definite, and holds the inequality

$$\mathbf{x}^T(\mathbf{P}-\mathbf{P}_s)\mathbf{x}>0 \tag{6.6.31}$$

which explains that the smoothing variance matrix $\mathbf{P}_s(t)=\overline{\mathbf{P}}(t)$ is smaller than filtering, which coincides to the theoretical consideration.

Equation (6.6.28) means that it is unnecessary to integrate the differential equation (6.6.29) in smoothing computation, but simply to use the filter matrix $\mathbf{P}$ and the matrix $\mathbf{Q}_2$ then compute by equation (6.6.28). For time invariant system the matrix $\mathbf{Q}_2$ can be computed by means of mixed energy matrices $\mathbf{Q}(\eta),\mathbf{G}(\eta),\mathbf{F}(\eta)$ of fundamental interval η, which is also obtained in computing matrix $\mathbf{P}$. So equation (6.6.28) proposes a different way for the computation of variance matrix by means of the precise integration method.

Before continuing description, the new term of ***innovation*** should be clarified. Certainly the state vectors $\hat{\mathbf{x}},\overline{\mathbf{x}}$ and the output vector $\mathbf{z}$ of (6.5.6) generated from them are of most concern. However, in the index of variational principle (6.5.44) or the discrete-time version (6.5.12), the disturbance vector $\mathbf{v}$ is of concern. For a sample of measurement vector $\mathbf{y}$, after the index is minimized and the vectors $\hat{\mathbf{x}}$ (filtering) and $\overline{\mathbf{x}}$ (smoothing) are obtained, then the disturbance $\hat{\mathbf{v}}$ is found as

$$\hat{\mathbf{v}}=\mathbf{y}-\mathbf{C}\hat{\mathbf{x}} \tag{6.6.32}$$

termed as "***Innovation***". The dynamic and measurement noises are modeled by zero mean and mutually independent white noises and expressed by the equations (6.5.42a~c). The innovation vector $\hat{\mathbf{v}}(t)$ is the unbiased estimation of $\mathbf{v}$ based on the sampling $\mathbf{y}$. For a given sample $\mathbf{y}$, the innovation vector $\hat{\mathbf{v}}(t)$ is deterministic, however, if treating the measurement $\mathbf{y}$ as assemble, then the innovation assemble of $\hat{\mathbf{v}}$ is also a white noise stochastic process, whose mean value is a null vector and the variance matrix is $\mathbf{V}$, see (6.5.42b).

For $\mathbf{w}$, there is also filtering estimation, but filter requires $\hat{\boldsymbol{\lambda}}=\mathbf{0}$ hence $\hat{\mathbf{w}}=\mathbf{0}$. The reason is clear that the filter uses no measurement later than the current time, therefore giving zero mean value. For smoothing, the dual vector $\overline{\boldsymbol{\lambda}}$ is estimated not null, so that $\overline{\mathbf{w}}$ is also estimated correspondingly. For measurement noise, there is also smoothing vector $\overline{\mathbf{v}}$, which is obtained from the smoothing estimation of $\overline{\mathbf{x}}$. Having computed the smoothing $\overline{\mathbf{x}}$ and $\overline{\boldsymbol{\lambda}}$, the smoothing estimation of noises $\overline{\mathbf{v}}$ and $\overline{\mathbf{w}}$ is not so much demanded.

Using equation (6.6.28) computing smoothing variance matrix $\mathbf{P}_s(t)$ is easy, which is independent on the sample $\mathbf{y}$, but depends on the system model. But the vectors $\overline{\mathbf{x}}$ and $\overline{\boldsymbol{\lambda}}$ computation depend on the measurement sampling vector $\mathbf{y}$, and is given implicitly via the vectors $\hat{\mathbf{x}},\mathbf{r}_{\lambda 2}$ in the back substitution equations (6.6.26'~27'). Vector $\hat{\mathbf{x}}$ is filter solution, and the equation for vector $\mathbf{r}_{\lambda 2}$ is given in filter analysis equation (6.5.126b), (6.5.136b).

The precise integration algorithm given before is applicable for time-invariant systems; the time-variant system computation should be further investigated, but the computation for time-invariant system is still a fundamental step, and is the main topic investigated below.

§6.6.3.1, Precise integration for single time step

The three kinds of smoothing, *i.e.* fixed interval smoothing, fixed point smoothing and fixed delay smoothing, have been described before. The corresponding algorithms need to be developed based on the interval mixed energy matrices and the back substitution equations. The interval mixed energy method is a unified approach, which can be used according to the different requirements of three smoothing algorithms.

Based on the above equations, the filtering computation must be performed before the smoothing computation. Comparing to the filtering algorithm, the smoothing needs further the algorithm of back substitution equations (6.6.26') and (6.6.27'). The filtering can disregard the used data of measurement vector, because no back substitution is needed; but the smoothing must refer to the used data. Hence the smoothing computation needs to store more data than filtering.

The smoothing variance matrix $\mathbf{P}_s$ can be computed via equation (6.6.28), which is obviously independent on the measurement $\mathbf{y}$, hence it can be computed and stored beforehand, which is useful in the back substitution. The matrices $\mathbf{P}$ and $\mathbf{Q}_2$ are the prerequisite of equation (6.6.28) and are obtained before.

The vectors $\hat{\mathbf{x}}$ and $\mathbf{r}_{\lambda 2}$ depend on the measurement $\mathbf{y}$. How to reduce the computational expense is an important problem such as for fixed delay smoothing, and will be discussed later. Similar situation also appears for the control vector $\mathbf{u}$.

Time integration needs uniform time steps, described as $t_k = k\eta, k = 0,1,\cdots,k_f$, with step size η. The situation is the same as filtering. The superposition principle applies to the linear system analysis, so that for each time step η the system mixed energy matrices $\mathbf{Q}(\eta),\mathbf{G}(\eta),\mathbf{F}(\eta)$ are all the same. The sole difference is the different values of measurement vectors $\mathbf{y}$. Assuming the measurement value is linear with respect to the time interval η. If precise integration has been executed for constant basis $\mathbf{y} = \mathbf{I}_q$ and linearly distributed basis vectors $\mathbf{y} = \mathbf{I}_q \cdot (t - t_a)$, then according to the superposition principle any linearly distributed measurement can be obtained from these basis vectors via the superposition principle. For control vector $\mathbf{u}$ the computation is similar. Situation is the same as in filter computation, see sections 6.5.7. Following the same method and algorithm (6.5.155) the single step η mixed energy matrices $\mathbf{Q}(\eta),\mathbf{F}(\eta),\mathbf{G}(\eta)$, constant basis matrices $\mathbf{R}_{xy}^{(0)}(\eta),\mathbf{R}_{xu}^{(0)}(\eta)$, $\mathbf{R}_{\lambda y}^{(0)}(\eta)$, $\mathbf{R}_{\lambda u}^{(0)}(\eta)$ and the linear basis matrices $\mathbf{R}_{xy}^{(1)}(0,\eta)$, $\mathbf{R}_{xu}^{(1)}(0,\eta)$, $\mathbf{R}_{\lambda y}^{(1)}(0,\eta),\mathbf{R}_{\lambda u}^{(1)}(0,\eta)$ can all be computed. These matrices are independent on the measurement sample vector $\mathbf{y}$ and to the control input $\mathbf{u}$, and can be used for all the intervals of η length. Any linear measurement $\mathbf{y}$ and control $\mathbf{u}$ are analyzed in the η time step, so that the integration with step length η is easy.

§6.6.4, Three kinds of smoothing algorithms

The interval mixed energy method proposes a unified approach to the fixed interval, fixed point and fixed delay smoothing.

§6.6.4.1, Fixed interval smoothing

Let the given interval be discretized as

$$t_0 = 0,\ t_1 = \eta, \cdots,\ t_k = k\eta, \cdots,\ t_f = k_f\eta \tag{6.6.33}$$

and $\mathbf{y}_k, \mathbf{u}_k (k = 0,1,\cdots,k_f)$ are given, the computation of smoothing mean value and the variance matrix $\bar{\mathbf{x}}_k, \mathbf{P}_k (k = 0,\cdots,k_f)$ are needed, for which the matrices of step η are considered computed as described in previous section.

According to equation (6.6.26~28), the matrices $\mathbf{P}_k, \mathbf{Q}_{(k_f - k)}$ should be first computed for all the stations k, thereafter the smoothing variance matrix $\mathbf{P}_{sk}$ is computed and stored. The filtering vector $\hat{\mathbf{x}}_k$ is computed too. Because the causality is of no concern for smoothing, hence the algorithm equation (6.6.26") is used. Remaining the term $\mathbf{r}_{\lambda 2}$ should be considered, which is induced from the external force of interval (t_k, t_f), for which the computation can still use the interval combination equations (6.5.77~78).

Using the associative rule of the interval combination operation $\backsim$, the computation of $\mathbf{r}_{\lambda 2}$ can be executed along the reverse direction. Suppose the force vectors $\mathbf{r}_x, \mathbf{r}_\lambda (t_{k+1}, t_f)$ have been obtained for the $k+1$ station and they can be treated as the vectors $\mathbf{r}_{x2}$ and $\mathbf{r}_{\lambda 2}$ in the equation (6.5.78). The force vectors $\mathbf{r}_x, \mathbf{r}_\lambda$ of the single interval (t_k, t_{k+1}) are regarded as $\mathbf{r}_{x1}$ and $\mathbf{r}_{\lambda 1}$, and can be computed by the equation (6.5.141) analytically, or based on the precise integration (6.5.155) computed matrices

$$\begin{aligned}\mathbf{r}_x(t_k, t_{k+1}) = {} & \mathbf{R}_{xu}^{(0)} \cdot \mathbf{u}_k + \mathbf{R}_{xy}^{(0)} \cdot \mathbf{y}_k \\ & + \mathbf{R}_{xu}^{(1)} \cdot (\mathbf{u}_{k+1} - \mathbf{u}_k)/\eta + \mathbf{R}_{xy}^{(1)} \cdot (\mathbf{y}_{k+1} - \mathbf{y}_k)/\eta\end{aligned} \tag{6.6.34a}$$

$$\begin{aligned}\mathbf{r}_\lambda(t_k, t_{k+1}) = {} & \mathbf{R}_{\lambda u}^{(0)} \cdot \mathbf{u}_k + \mathbf{R}_{\lambda y}^{(0)} \cdot \mathbf{y}_k \\ & + \mathbf{R}_{\lambda u}^{(1)} \cdot (\mathbf{u}_{k+1} - \mathbf{u}_k)/\eta + \mathbf{R}_{\lambda y}^{(1)} \cdot (\mathbf{y}_{k+1} - \mathbf{y}_k)/\eta\end{aligned} \tag{6.6.34b}$$

In using equation (6.5.78), matrices $\mathbf{Q}_1, \mathbf{F}_1, \mathbf{G}_1$ are the computed $\mathbf{Q}(\eta), \mathbf{F}(\eta), \mathbf{G}(\eta)$, and $\mathbf{Q}_2, \mathbf{F}_2, \mathbf{G}_2$ are the (t_{k+1}, t_f) interval matrices $\mathbf{Q}(\Delta t), \mathbf{F}(\Delta t), \mathbf{G}(\Delta t)$ of $\Delta t = (k_f - (k+1))\eta$

$$(t_k, t_{k+1}) \backsim (t_{k+1}, t_f) = (t_k, t_f)$$

Thus, based on equations (6.5.77~78) the $\mathbf{Q}, \mathbf{F}, \mathbf{G}$, $\mathbf{r}_x(t_k, t_f)$ and $\mathbf{r}_\lambda(t_k, t_f)$ of the interval (t_k, t_f) are obtained, which forms the reverse direction recurrence. Then the right-hand side terms in the equation (6.6.26') are all obtained. Detail is neglected.

§6.6.4.2, Fixed point smoothing

For fixed point smoothing, the interval combination $\backsim$ of equations (6.5.77~78) is still applied. The fixed instant t_e means the interval-1, $[t_0,t_e)$, being fixed, and the interval-2, (t_e,t), where t is steadily increasing. The computation for interval-1, $[t_0,t_e)$, generates first $\mathbf{Q},\mathbf{F},\mathbf{G}$ and $\mathbf{r}_x,\mathbf{r}_\lambda$ of the open interval (t_0,t_e), then the initial condition at t_0 is accounted for by equations (6.6.11) and (6.6.15). The computation of the inhomogeneous term of each unit interval η is solved using (6.6.34). As the successive computation for the open interval (t_0,t_e) can use interval combination $\backsim$ recursively.

When the length (t_e,t) is long enough, the respective matrix $\mathbf{F}(t_e,t)$ tends to be a null matrix, and the matrices $\mathbf{Q},\mathbf{G}$ tend to their limit $\mathbf{Q}_\infty,\mathbf{G}_\infty$, the smoothing value and its variance matrix at t_e converge to the limit too, and the iteration is stopped. Detail is neglected.

§6.6.4.3, Fixed delay smoothing

Fixed delay (time lag) smoothing computation is very close to filtering, because it often needs real time computation too. The mean value of such smoothing has no fixed limit value. Computing can still use equations (6.5.77~78), *i.e.* the interval combination $\backsim$. The present time t steadily progresses, with the smoothing instant t_e progresses with the same speed to keep the interval-2 length (t_e,t) unchanged, and the interval-1 is $[t_0,t_e)$, whose length continuously increases. Although the length of interval-2 (t_e,t) does not change, but the inhomogeneous term induced from the measurement and control vectors $\mathbf{y},\mathbf{u}$ changes, because of time progressing.

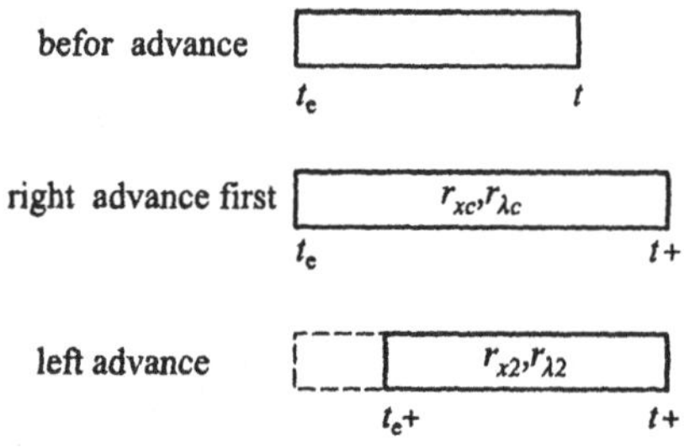

Figure 6.9, *Attach and remove one step*

The computation for the interval-1, $[t_0,t_e)$, uses still the interval combination $\backsim$. Each time step increases a unit interval $(t_e,t_e+\eta)$, then t_e progress. As interval-2, (t_e,t), its mixed energy matrices $\mathbf{Q},\mathbf{F},\mathbf{G}$ keep unchanged but the inhomogeneous force terms $\mathbf{r}_x$ and $\mathbf{r}_\lambda$ changed, because a unit interval is attached from the right but a unit interval is detached from the left, respectively, figure 6.9. The equation for the attachment of a unit interval $(t,t+\eta)$ from the right has been given in equation (6.5.78), but the equation for detaching a unit interval from the left is needed. The order of execution can be first attaching from the right end then detaching from the left end. Let the matrices $\mathbf{Q}_2,\mathbf{G}_2,\mathbf{F}_2$ be the interval matrices of (t_e,t), and $\mathbf{Q}_1,\mathbf{F}_1,\mathbf{G}_1$ denote the interval mixed energy matrices $\mathbf{Q}(\eta),\mathbf{F}(\eta),\mathbf{G}(\eta)$ of unit length

η. Let $\mathbf{E}_c, \mathbf{F}_c, \mathbf{G}_c, \mathbf{r}_{xc}, \mathbf{r}_{\lambda c}$ denote the interval matrices and vectors after combination from the right unit interval. It gives

$$\mathbf{r}_{\lambda c} = \mathbf{r}_{\lambda 1} + \mathbf{F}_1^T (\mathbf{I}_n + \mathbf{Q}_2 \mathbf{G}_1)^{-1} (\mathbf{r}_{\lambda 2} - \mathbf{Q}_2 \mathbf{r}_{x1}) \tag{6.5.78a}$$

$$\mathbf{r}_{xc} = \mathbf{r}_{x2} + \mathbf{F}_2 (\mathbf{I}_n + \mathbf{G}_1 \mathbf{Q}_2)^{-1} (\mathbf{r}_{x1} + \mathbf{G}_1 \mathbf{r}_{\lambda 2}) \tag{6.5.78b}$$

Next, detaching a unit interval from the left, for which the vectors $\mathbf{r}_{x1}$ and $\mathbf{r}_{\lambda 1}$ are known, so that the vectors $\mathbf{r}_{x2}$ and $\mathbf{r}_{\lambda 2}$ need to be solved, for which the vector $\mathbf{r}_{\lambda 2}$ can be solved first, then solving $\mathbf{r}_{x2}$ is easy via equation (6.5.78b). From equation (6.5.78a) the vector $\mathbf{r}_{x2}$ is solved as

$$\mathbf{r}_{\lambda 2} = \mathbf{Q}_2 \mathbf{r}_{x1} + (\mathbf{I}_n + \mathbf{Q}_2 \mathbf{G}_1) \mathbf{F}_1^{-T} (\mathbf{r}_{\lambda c} - \mathbf{r}_{\lambda 1}) \tag{6.6.35a}$$

and

$$\mathbf{r}_{x2} = \mathbf{r}_{xc} - \mathbf{F}_2 (\mathbf{I}_n + \mathbf{G}_1 \mathbf{Q}_2)^{-1} (\mathbf{r}_{x1} + \mathbf{G}_1 \mathbf{r}_{\lambda 2}) \tag{6.6.35b}$$

The computation for fixed time delay smoothing is not difficult either. The smoothing variance matrix $\mathbf{P}_s$ can be computed by equation (6.6.28). Detail is neglected.

§6.7, Optimal control

The whole time interval $[t_0, t_f]$ is subdivided into the past time interval $[t_0, t)$ of filtering and the future time interval $(t, t_f]$ of control by the present time t, where $t_0 \le t \le t_f$, see figure 6.4. The past time interval has been the history, for which the analysis can only be the state vector recognition, which means using filtering analysis to obtain mean value $\hat{\mathbf{x}}(t)$ and variance matrix $\mathrm{P}(t)$, or further the system parameter identification. In section 6.5, the filter theory and computation has been described in some detail.

The future time interval is controllable. The linear optimal control theory is based on minimization of a quadratic index functional under the constraint of dynamic equation, which is a criterion for optimization. Therefore the problem is described with the ***linear dynamic equation*** and also with the ***quadratic functional index***, so that it is termed as ***Linear Quadratic*** (LQ) optimal control problem. The LQ control analysis is for the future time interval $(t, t_f]$, hence no measurement at all, but there is the output vector $\mathbf{z}(t)$. The performance of state vector in the future time interval is completely under the governing of dynamic equation but with no measurement data to check with. The control vector $\mathbf{u}(\tau)$ is selected in the analysis according to the state vector $\mathbf{x}(\tau)$ in interval $t < \tau \le t_f$.

Although the filter problem in the past time interval and the LQ optimal control in the future time interval are for two different time intervals. However, the analysis for the two contiguous time intervals is the composition of the whole control analysis, that the filter and LQ control are connected at the present time t. For LQ control the initial condition is

$$\mathbf{x}(\tau) = \hat{\mathbf{x}}(t), \quad \text{when } \tau = t \tag{6.7.1}$$

which means that the ***filtering mean-value of the state vector is treated as the initial value of the LQ control***, which is the ***state vector continuity*** for the connection of two intervals. Based on LQ control analysis of future time interval, the control vector $\mathbf{u}(\tau)$ is determined based on the initial condition (6.7.1). Especially, the control vector $\mathbf{u}(t)$ at the present time is obtained. This $\mathbf{u}(t)$ is considered the control vector of the whole time interval $[t_0, t_f]$ at the present time. Hence, determining the LQ control theory and computation for the future time interval $(t, t_f]$ is necessary. This model gives the characteristics of **LQG (Linear Quadratic Gaussian)** optimal control theory, with the connection (6.7.1) of the two intervals, and is termed the ***separation principle***.

§6.7.1, Theory of LQ optimal control for the future time interval

Let the linear dynamic equation be given as

$$\dot{\mathbf{x}}(\tau) = \mathbf{A}(\tau)\mathbf{x}(\tau) + \mathbf{B}_u(\tau)\mathbf{u}(\tau) + \mathbf{B}_w(\tau)\mathbf{w}(\tau) \tag{6.7.2}$$

The initial condition has been given in equation (6.7.1) and output is the p-dimensional vector

$$\mathbf{z}(\tau) = \mathbf{C}_z(\tau)\mathbf{x}(\tau) + \mathbf{D}_u(\tau)\mathbf{u}(\tau) \tag{6.7.3}$$

where the system matrices $\mathbf{A}, \mathbf{B}_u, \mathbf{B}_1, \mathbf{C}_z, \mathbf{D}_u$ can be functions of time that the theory applies to time-variant system. In computation, however, these matrices are usually considered as time-invariant, so that the precise integration method or the analytical method can be applied.

The control vector $\mathbf{u}(\tau)$ in dynamic equation can be selected arbitrarily. The criterion of selection is the minimization of the ***quadratic functional index*** J defined as the least square of

$$J = \int_t^{t_f} (\mathbf{z}^T\mathbf{z}/2)\mathrm{d}\tau + \mathbf{x}_f^T\mathbf{S}_f\mathbf{x}_f/2\,, \qquad \mathbf{x}_f \underset{\text{def}}{=} \mathbf{x}(t_f) \tag{6.7.4a}$$

where $\mathbf{S}_f$ is a given ***symmetric and non-negative definite matrix***. Substituting the expression (6.7.3) into J, and note that using appropriate linear transformation, the output equation can be derived as

$$\mathbf{C}_z^T\mathbf{D}_u = 0, \qquad \mathbf{D}_u^T\mathbf{D}_u = \mathbf{I}_m$$

Therefore

$$J = \int_t^{t_f} (\mathbf{x}^T\mathbf{C}_z^T\mathbf{C}_z\mathbf{x}/2 + \mathbf{u}^T\mathbf{u}/2)\mathrm{d}\tau + \mathbf{x}_f^T\mathbf{S}_f\mathbf{x}_f/2\,, \quad \min_{\mathbf{u}} J \tag{6.7.4b}$$

which is a ***conditional minimization***, the condition is the dynamic equation (6.7.2). Introducing the Lagrange multiplier function $\boldsymbol{\lambda}(\tau)$ (n-dimensional vector) derives the variational principle

$$J_A = \int_t^{t_f} [\boldsymbol{\lambda}^T(\dot{\mathbf{x}} - \mathbf{A}\mathbf{x} - \mathbf{B}_u\mathbf{u} - \mathbf{B}_w\mathbf{w}) + \mathbf{x}^T\mathbf{C}_z^T\mathbf{C}_z\mathbf{x}/2 + \mathbf{u}^T\mathbf{u}/2]\mathrm{d}\tau + \mathbf{x}_f^T\mathbf{S}_f\mathbf{x}_f/2\,,$$

$$\delta J_A = 0$$

where J_A is the extended index functional, and the variation has been free from constraint with three kinds of independent variables $\mathbf{x}, \boldsymbol{\lambda}, \mathbf{u}$. Because there is no time derivative relating $\mathbf{u}$ in the integrand of functional J_A, so that the minimization with respect to $\mathbf{u}$ can be performed first (the Pontriagin minimization principle), which gives

$$\mathbf{u} = \mathbf{B}_u^T \boldsymbol{\lambda} \tag{6.7.5}$$

Substituting back into the expression of J_A gives

$$J_A = \int_t^{t_f} [\boldsymbol{\lambda}^T \dot{\mathbf{x}} - \mathrm{H}(\mathbf{x},\boldsymbol{\lambda}) - \boldsymbol{\lambda}^T \mathbf{B}_w \mathbf{w}] \mathrm{d}\tau + \mathbf{x}_f^T \mathbf{S}_f \mathbf{x}_f / 2, \quad \min_{\mathbf{x}} \max_{\boldsymbol{\lambda}} J_A \tag{6.7.6}$$

$$\mathrm{H}(\mathbf{x},\boldsymbol{\lambda}) = \boldsymbol{\lambda}^T \mathbf{A}\mathbf{x} + \boldsymbol{\lambda}^T \mathbf{B}_u \mathbf{B}_u^T \boldsymbol{\lambda}/2 - \mathbf{x}^T \mathbf{C}_z^T \mathbf{C}_z \mathbf{x}/2 \tag{6.7.7}$$

Certainly the initial condition (6.7.1) is to be satisfied. There are only two kinds of variable dual to each other, *i.e.* $\mathbf{x},\boldsymbol{\lambda}$, in the variational functional, which has been typical variational principle of a Hamilton system. Therefore the problem is ***analogous to structural mechanics*** given in chapter 5. Especially, the dual equations and boundary conditions derived from the variational principle (6.7.6) give a two point boundary value problem.

$$\dot{\mathbf{x}}(\tau) = \mathbf{A}\mathbf{x} + \mathbf{B}_u \mathbf{B}_u^T \boldsymbol{\lambda} + \mathbf{B}_w \mathbf{w}, \quad \mathbf{x}(t) = \hat{\mathbf{x}}(t) \tag{6.7.8a}$$

$$\dot{\boldsymbol{\lambda}}(\tau) = \mathbf{C}_z^T \mathbf{C}_z \mathbf{x} - \mathbf{A}^T \boldsymbol{\lambda}, \quad \boldsymbol{\lambda}(t_f) = -\mathbf{S}_f \mathbf{x}_f \tag{6.7.8b}$$

where the filtered vector $\hat{\mathbf{x}}(t)$ is considered given. This is a set of non-homogeneous dual equations, where $\mathbf{w}$ is a non-homogeneous term, a ***zero-mean white noise***. In H_∞ control theory however, the white noise $\mathbf{w}$ is not zero-mean, which is the ***contrast between the two theories***. Taking mathematical expectation gives the homogeneous equation as

$$\dot{\mathbf{x}}_a(\tau) = \mathbf{A}\mathbf{x}_a + \mathbf{B}_u \mathbf{B}_u^T \boldsymbol{\lambda}_a, \quad \mathbf{x}_a(t) = \hat{\mathbf{x}}(t) \tag{6.7.8'a}$$

$$\dot{\boldsymbol{\lambda}}_a(\tau) = \mathbf{C}_z^T \mathbf{C}_z \mathbf{x}_a - \mathbf{A}^T \boldsymbol{\lambda}_a, \quad \boldsymbol{\lambda}_a(t_f) = -\mathbf{S}_f \mathbf{x}_{fa} \tag{6.7.8'b}$$

Before solving the dual equations (6.7.8'a,b), the analogy theory between structural mechanics and optimal control is presented first. The variational principle of the homogeneous dual equations (6.7.8'a,b) is

$$J_A = \int_t^{t_f} [\boldsymbol{\lambda}^T \dot{\mathbf{x}} - \mathrm{H}(\mathbf{x},\boldsymbol{\lambda})] \mathrm{d}\tau + \mathbf{x}_f^T \mathbf{S}_f \mathbf{x}_f / 2, \quad \min_{\mathbf{x}} \max_{\boldsymbol{\lambda}} J_A \tag{6.7.6a}$$

On the other hand, from structural mechanics side, the dual equations, the variational principle and the Hamilton function are (5.2.10~12). Comparison shows that both sides are the same problem mathematically. The term of $\mathbf{S}_f$ corresponds to that in the variational principle (5.2.12), (5.2.13) a term of deformation energy $\mathbf{x}_f^T \mathbf{S}_f \mathbf{x}_f/2$ at the end-z_f is appended, the physical meaning of which is an elastic support at z_f. The correspondence relationship of both sides can be seen as

Structural mechanics	LQ optimal control
Displacement, internal force $\mathbf{q},\mathbf{p}$	Dual vectors $\mathbf{x},\boldsymbol{\lambda}$
Longitudinal coordinate z	time coordinate τ
Interval $(z_0, z_f]$	time interval $(t, t_f]$
Mixed energy $\mathrm{H}(\mathbf{q},\mathbf{p})$	Hamilton function $\mathrm{H}(\mathbf{x},\boldsymbol{\lambda})$
$\mathbf{A},\mathbf{B},\mathbf{D}$	$\mathbf{A},\mathbf{C}_z^T\mathbf{C}_z,\mathbf{B}_u\mathbf{B}_u^T$

Elastic support matrix $\mathbf{S}_f$	End index matrix $\mathbf{S}_f$
Compatibility equation (5.2.10a)	Dynamic equation (6.7.8a')
Equilibrium equation (5.2.10b)	Dual equation (6.7.8b')
Deformation energy, Action function	Index J, (6.7.4)
Etc.	

The above comparison gives the ***analogy relationship between structural mechanics and optimal control***, which is useful for both sides.

The solution of the dual equations can use the form of boundary condition in (6.7.8'b), let

$$\boldsymbol{\lambda}_a(\tau) = -\mathbf{S}(\tau)\mathbf{x}_a(\tau) \tag{6.7.9}$$

Substituting into (6.7.8'a,b) and eliminating $\dot{\mathbf{x}}_a$ gives

$$(\dot{\mathbf{S}} + \mathbf{C}_z^T\mathbf{C}_z + \mathbf{A}^T\mathbf{S} + \mathbf{S}\mathbf{A} - \mathbf{S}\mathbf{B}_u\mathbf{B}_u^T\mathbf{S})\cdot\mathbf{x}_a = \mathbf{0}$$

Because $\mathbf{x}_a(\tau)$ is arbitrarily selected vector, so that

$$\dot{\mathbf{S}}(\tau) = -\mathbf{C}_z^T\mathbf{C}_z - \mathbf{A}^T\mathbf{S} - \mathbf{S}\mathbf{A} + \mathbf{S}\mathbf{B}_u\mathbf{B}_u^T\mathbf{S}, \quad \mathbf{S}(t_f) = \mathbf{S}_f \tag{6.7.10}$$

It is again a matrix Riccati differential equation for the finite time interval $(t, t_f]$, and the boundary condition is given at t_f where $\mathbf{S}_f$ is a symmetric and non-negative definite matrix, that the integration should be backward for time τ. Because $\mathbf{S}_f$ is symmetric, the right hand side of (6.7.10) is also symmetric, so that $\mathbf{S}(\tau)$ is a symmetric matrix. It can also be proved that ***if*** $(\mathbf{A}, \mathbf{B}_u)$ ***is controllable, and*** $(\mathbf{A}, \mathbf{C}_z)$ ***is observable, then*** $\mathbf{S}(\tau)$ ***is a positive definite symmetric matrix.*** See section 6.7.2.2.

When $(t_f - t)$ is quite long and tends to infinity, the time invariant system has the limit

$$\mathbf{S}_\infty = \lim_{\tau\to\infty} \mathbf{S}(\tau)$$

This matrix satisfies the Algebraic Riccati Equation, (ARE)

$$-\mathbf{C}_z^T\mathbf{C}_z - \mathbf{A}^T\mathbf{S}_\infty - \mathbf{S}_\infty\mathbf{A} + \mathbf{S}_\infty\mathbf{B}_u\mathbf{B}_u^T\mathbf{S}_\infty = \mathbf{0} \tag{6.7.11}$$

Such situation is similar to the Riccati equation in Kalman-Bucy filter.

The importance of the solution matrix of Riccati equation is that substituting equation (6.7.9) into (6.7.5) gives

$$\mathbf{u}(\tau) = -\mathbf{B}_u^T\mathbf{S}(\tau)\mathbf{x}_a(\tau) = -\mathbf{K}(\tau)\mathbf{x}_a(\tau), \text{ where } \mathbf{K}(\tau) = \mathbf{B}_u^T\mathbf{S}(\tau) \tag{6.7.12}$$

This equation gives the ***feedback control vector***, and from the equation (6.7.8a') derives

$$\dot{\mathbf{x}}_a(\tau) = [\mathbf{A} - \mathbf{B}_u\mathbf{B}_u^T\mathbf{S}(\tau)]\mathbf{x}_a, \quad \mathbf{x}_a(t) = \hat{\mathbf{x}}(t) \tag{6.7.13}$$

It is the differential equation for state mean-value. $\mathbf{K}(\tau)$ is called the ***gain matrix.***

Equation (6.7.13) is a ***linear homogeneous time-variant*** differential equation, so that its solution is proportional to the initial value $\hat{\mathbf{x}}(t)$. The solution $\mathbf{x}_a(\tau)$ is a mean-value. If the disturbance is $\mathbf{w} = \mathbf{0}$ in the dynamic equation (6.7.2), then the solution is only a mean-value. Mean-value is deterministic, hence the calculus is as usual, but if the stochastic disturbance $\mathbf{w}(\tau)$ is considered, the state vector is also a stochastic process, then the mean square calculus is necessary. The control input is

$$\mathbf{u}(\tau) = -\mathbf{K}(\tau)\mathbf{x}(\tau), \quad \mathbf{K}(\tau) = \mathbf{B}_u^T\mathbf{S}(\tau) \tag{6.7.12'}$$

where the gain matrix $\mathbf{K}(\tau)$ is still deterministic. Substituting into (6.7.2) gives

$$\dot{\mathbf{x}}(\tau) = [\mathbf{A} - \mathbf{B}_u\mathbf{B}_u^T\mathbf{S}(\tau)]\mathbf{x} + \mathbf{B}_w\mathbf{w} \tag{6.7.13'}$$

Therefore the differential equation for the mean-value state vector $\mathbf{x}_a(\tau)$ is its homogeneous equation. Since $\mathbf{w}(\tau)$ is a white noise disturbance, then $\mathbf{x}(t)$ is a stochastic process, of which the ***mean-value*** $\mathbf{x}_a(\tau)$ ***is deterministic.*** Although the equation (6.7.13) is a set of time-variant differential equations, but it can be solved by the precise integration method, see section 6.7.3.2. In case of ***robust control,*** $\mathbf{w}(\tau)$ is ***not a zero-mean white noise,*** then the selection of $\mathbf{w}(\tau)$ becomes a critical point.

In LQG applications, only the mean-value is used, hence the solutions of Riccati differential equation (6.7.10) and the state mean value differential equation (6.7.13), respectively, are the concentration. Before numerical solution, it is important to consider the stability of the system.

§6.7.2, Stability analysis

According to the LQ optimal control theory, the differential equation (6.7.13) for mean-value of state vector is derived above. The stability problem for the differential equation follows. Differential equation (6.7.13) is a time-variant one, so that the method of finding the eigenvalues to check their real parts all being negative no longer applies. Using the second method of Lyapunov is a reasonable choice. A positive definite Lyapunov function is necessary to select and to check that it always decreases with time.

Stability is the characteristic of a linear system itself. Hence in stability analysis the external force can be given as zero $\mathbf{w} = \mathbf{0}$. Then the governing equations become a homogeneous set (6.7.8'a,b), which comes from (6.7.6) by letting $\mathbf{w} = \mathbf{0}$. Numerically, $J = J_{\mathrm{A}}$, so that

$$J(t) = J_{\mathrm{A}}(t) = \mathbf{x}_{\mathrm{f}}^T\mathbf{S}_{\mathrm{f}}\mathbf{x}_{\mathrm{f}}/2 + \int_t^{t_{\mathrm{f}}}[\boldsymbol{\lambda}^T\dot{\mathbf{x}} - \boldsymbol{\lambda}^T\mathbf{A}\mathbf{x} - \boldsymbol{\lambda}^T\mathbf{B}_u\mathbf{B}_u^T\boldsymbol{\lambda}/2 + \mathbf{x}^T\mathbf{C}_z^T\mathbf{C}_z\mathbf{x}/2]\mathrm{d}\tau$$

$$= \mathbf{x}_{\mathrm{f}}^T\mathbf{S}_{\mathrm{f}}\mathbf{x}_{\mathrm{f}}/2 + \int_t^{t_{\mathrm{f}}}[\boldsymbol{\lambda}^T\mathbf{B}_u\mathbf{B}_u^T\boldsymbol{\lambda}/2 + \mathbf{x}^T\mathbf{C}_z^T\mathbf{C}_z\mathbf{x}/2]\mathrm{d}\tau \quad \text{Comment: Using (6.7.8'a)}$$

$$= \mathbf{x}_{\mathrm{f}}^T\mathbf{S}_{\mathrm{f}}\mathbf{x}_{\mathrm{f}}/2 + \int_t^{t_{\mathrm{f}}}[\boldsymbol{\lambda}^T\dot{\mathbf{x}} - \boldsymbol{\lambda}^T\mathbf{A}\mathbf{x} + \mathbf{x}^T\dot{\boldsymbol{\lambda}} + \mathbf{x}^T\mathbf{A}^T\boldsymbol{\lambda}]\mathrm{d}\tau/2 \quad \text{Comment: (6.7.8')}$$

$$= -\boldsymbol{\lambda}^T(t)\mathbf{x}(t)/2 = \mathbf{x}^T(t)\mathbf{S}(t)\mathbf{x}(t)/2 \tag{6.7.14}$$

where the equation (6.7.9) is used and $\mathbf{x}_a$ is just the $\mathbf{x}$. The index J is non-negative, so that the matrix $\mathbf{S}$ is certainly symmetric and non-negative definite.

The statement of $\mathbf{S}(t)$ being symmetric and non-negative definite is not enough, it requires to prove that ***if*** ($\mathbf{A},\mathbf{B}_u$) ***is controllable and*** ($\mathbf{A},\mathbf{C}_z$) ***is observable then*** $\mathbf{S}(t)$ ***must be a symmetric and positive definite matrix.*** This statement will be proved below.

§6.7.2.1, Gram matrices of controllability and observability

When considering controllability the output is unrelated. Hence the assumption $\mathbf{C}_z = \mathbf{0}$ is made and the homogeneous dual equations are reduced to be [Comparing (6.4.29)]

$$\dot{\mathbf{x}} = \mathbf{A}\mathbf{x} + \mathbf{B}_u\mathbf{B}_u^T\boldsymbol{\lambda}, \qquad \dot{\boldsymbol{\lambda}} = -\mathbf{A}^T\boldsymbol{\lambda} \tag{6.7.15}$$

The initial condition is $\mathbf{x}(0) = \mathbf{x}_0$, where $\mathbf{x}_0$ is an arbitrary vector. For a given time instant t, $t > t_0$, $\boldsymbol{\lambda}$ is solved as

$$\boldsymbol{\lambda}(\tau) = \exp[\mathbf{A}^T \cdot (t-\tau)]\boldsymbol{\lambda}(t) \tag{6.7.16}$$

Therefore the state vector is computed as

$$\begin{aligned}\mathbf{x}(t) &= \exp[\mathbf{A}\cdot(t-t_0)]\cdot\mathbf{x}_0 + \int_{t_0}^{t} \exp[\mathbf{A}\cdot(t-\tau)]\mathbf{B}_u\mathbf{B}_u^T\boldsymbol{\lambda}(\tau)\mathrm{d}\tau \\ &= \exp[\mathbf{A}\cdot(t-t_0)]\cdot\mathbf{x}_0 + \int_{t_0}^{t} \exp[\mathbf{A}\cdot(t-\tau)]\mathbf{B}_u\mathbf{B}_u^T\exp[\mathbf{A}^T\cdot(t-\tau)]\mathrm{d}\tau\cdot\boldsymbol{\lambda}(t) \\ &= \exp[\mathbf{A}\cdot(t-t_0)]\cdot\mathbf{x}_0 + \mathbf{W}_c(t,t_0)\cdot\boldsymbol{\lambda}(t)\end{aligned}$$

where $$\mathbf{W}_c(t,t_0) \underset{\text{def}}{=} \int_{t_0}^{t} \exp[\mathbf{A}\cdot(t-\tau)]\mathbf{B}_u\mathbf{B}_u^T\exp[\mathbf{A}^T\cdot(t-\tau)]\mathrm{d}\tau$$

For time-invariant system, $\mathbf{W}_c(t,t_0) = \mathbf{W}_c(t-t_0)$ depends only on the time difference, which is just the ***controllability Gram matrix*** (6.1.38b). The positive definiteness of this Gram matrix means that initiating from $\mathbf{x}_0$ to reach an arbitrary state $\mathbf{x}(t)$ can always be realized with the control $\mathbf{u} = \mathbf{B}_u^T\boldsymbol{\lambda}$. Because $\mathbf{W}_c$ is assumed positive definite, so that $\boldsymbol{\lambda}(t)$ can be solved form the above equation. The above derivation gives the equation (6.1.37b).

The derivation of observability is also required. Observability is the dual to controllability. When considering observability, the control vector is unrelated, hence let $\mathbf{B}_u = \mathbf{0}$, then

$$\dot{\mathbf{x}}_2 = \mathbf{A}\mathbf{x}_2, \quad \dot{\boldsymbol{\lambda}}_2 = -\mathbf{A}^T\boldsymbol{\lambda}_2 + \mathbf{C}_z^T\mathbf{C}_z\mathbf{x}_2 \tag{6.7.17}$$

Here for distinction with controllability analysis, a subscript 2 is put on. First solves

$$\mathbf{x}_2(\tau) = \exp[\mathbf{A}\cdot(\tau-t_0)]\cdot\mathbf{x}_0 \tag{6.7.18}$$

Then solve

$$\begin{aligned}\boldsymbol{\lambda}_2(t_0) &= \exp[\mathbf{A}\cdot(t-t_0)]\cdot\boldsymbol{\lambda}_2(t) + \int_{t_0}^{t} \exp[\mathbf{A}^T\cdot(\tau-t_0)]\mathbf{C}_z^T\mathbf{C}_z\mathbf{x}_2(\tau)\mathrm{d}\tau \\ &= \exp[\mathbf{A}\cdot(t-t_0)]\cdot\boldsymbol{\lambda}_2(t) + \int_{t_0}^{t} \exp[\mathbf{A}^T\cdot(\tau-t_0)]\mathbf{C}_z^T\mathbf{C}_z\exp[\mathbf{A}\cdot(\tau-t_0)]\mathrm{d}\tau\cdot\mathbf{x}_0 \\ &= \exp[\mathbf{A}^T\cdot(t-t_0)]\cdot\boldsymbol{\lambda}_2(t) - \mathbf{W}_o(t,t_0)\cdot\mathbf{x}_0\end{aligned}$$

where $$\mathbf{W}_o(t,t_0) \underset{\text{def}}{=} \int_{t_0}^{t} \exp[\mathbf{A}^T\cdot(\tau-t_0)]\mathbf{C}_z^T\mathbf{C}_z\exp[\mathbf{A}\cdot(\tau-t_0)]\mathrm{d}\tau$$

For time-invariant system, $\mathbf{W}_o(t,t_0) = \mathbf{W}_o(t-t_0)$ depends only on time difference, which is just the ***observability Gram matrix***, see equation (6.1.38b). The positive definiteness determines that $\mathbf{x}_0$ is solved as

$$\mathbf{x}_0 = \mathbf{W}_o^{-1} \cdot \left(\exp[\mathbf{A}^T \cdot (t - t_0)]\boldsymbol{\lambda}_2(t) - \boldsymbol{\lambda}_2(t_0)\right)$$

But $\boldsymbol{\lambda}_2$ is not measurable, the measured value can only be $\mathbf{z} = \mathbf{C}_z\mathbf{x}_2$. From the equation

$$\dot{\boldsymbol{\lambda}}_2 = -\mathbf{A}^T\boldsymbol{\lambda}_2 + \mathbf{C}_z^T\mathbf{z}$$

solves

$$\boldsymbol{\lambda}_2(t_0) = \exp[\mathbf{A}^T \cdot (t - t_0)]\boldsymbol{\lambda}_2(t) - \int_{t_0}^{t} \exp[\mathbf{A}^T \cdot (\tau - t_0)]\mathbf{C}_z^T\mathbf{z}(\tau)\mathrm{d}\tau$$

therefore
$$\mathbf{x}_0 = \mathbf{W}_o^{-1} \cdot \int_{t_0}^{t} \exp[\mathbf{A}^T \cdot (\tau - t_0)]\mathbf{C}_z^T\mathbf{z}(\tau)\mathrm{d}\tau \tag{6.7.19}$$

which explains that $\mathbf{x}_0$ is solved by means of the measured $\mathbf{z}(\tau)$, *i.e.* observable.

The homogeneous equation (6.7.8') corresponds to the variational principle

$$J_{\mathrm{A}} = \int_t^{t_f} [\boldsymbol{\lambda}^T\dot{\mathbf{x}} - \mathrm{H}(\mathbf{x}, \boldsymbol{\lambda})]\mathrm{d}\tau + \mathbf{x}_f^T\mathbf{S}_f\mathbf{x}_f/2 , \quad \delta J_{\mathrm{A}} = 0 \tag{6.7.6'}$$

where H is given by (6.7.7). The equation set for controllability (6.7.15) and the set for observability (6.7.17) correspond, respectively, to the Hamilton functions $\mathrm{H}_c(\mathbf{x}, \boldsymbol{\lambda}) = \boldsymbol{\lambda}^T\mathbf{A}\mathbf{x} + \boldsymbol{\lambda}^T\mathbf{B}_u\mathbf{B}_u^T\boldsymbol{\lambda}/2$ and $\mathrm{H}_o(\mathbf{x}, \boldsymbol{\lambda}) = \boldsymbol{\lambda}^T\mathbf{A}\mathbf{x} - \mathbf{x}^{\mathrm{T}}\mathbf{C}_z^T\mathbf{C}_z\mathbf{x}/2$.

Verify directly that the controllability and observability Gram matrices ($n \times n$ matrices)

$$\mathbf{W}_c(t) = \int_0^t e^{\mathbf{A}\tau}\mathbf{B}_u\mathbf{B}_u^T e^{\mathbf{A}^T\tau}\mathrm{d}\tau \quad \text{and} \quad \mathbf{W}_o(t) = \int_0^t e^{\mathbf{A}^T\tau}\mathbf{C}_z^T\mathbf{C}_z e^{\mathbf{A}\tau}\mathrm{d}\tau$$

satisfy, respectively, the Lyapunov differential equations [cf. Equation (6.4.65)]

$$\dot{\mathbf{W}}_c(t) = \mathbf{A}\mathbf{W}_c + \mathbf{W}_c\mathbf{A}^T + \mathbf{B}_u\mathbf{B}_u^T \quad \text{and} \quad \dot{\mathbf{W}}_o(t) = \mathbf{W}_o\mathbf{A} + \mathbf{A}^T\mathbf{W}_o + \mathbf{C}_z^T\mathbf{C}_z$$

For $\mathbf{W}_c(t)$ the verification uses the identity $f(t) \equiv \int_0^t (\mathrm{d}/\mathrm{d}\tau)f(\tau)\mathrm{d}\tau + f(0)$ giving

$$\begin{aligned}\dot{\mathbf{W}}_c(t) &= e^{\mathbf{A}t}\mathbf{B}_u\mathbf{B}_u^T e^{\mathbf{A}^Tt} = \int_0^t(\mathrm{d}/\mathrm{d}\tau)[e^{\mathbf{A}\tau}\mathbf{B}_u\mathbf{B}_u^T e^{\mathbf{A}^T\tau}]\mathrm{d}\tau + \mathbf{B}_u\mathbf{B}_u^T \\ &= \mathbf{A}\mathbf{W}_c + \mathbf{W}_c\mathbf{A}^T + \mathbf{B}_u\mathbf{B}_u^T\end{aligned}$$

§6.7.2.2, Positive definiteness of the Riccati matrix

After the controllability and observability Gram matrices $\mathbf{W}_c$ and $\mathbf{W}_o$ are derived, the ***positive definiteness of the matrix*** $\mathbf{S}(t)$ ***in LQ control and the matrix*** $\mathbf{P}(t)$ ***in Kalman-Bucy filter*** can be proved ***under the condition of controllability and observabiltity*** of the system.

The matrix $\mathbf{S}(t)$ is examined first. Equation (6.7.14) gives the relation between the matrix $\mathbf{S}(t)$ and the index functional $J(t)$. The integrand in index $J(t)$ is composed of two terms $\mathbf{u}$ and $\mathbf{C}_z\mathbf{x}$, see equation (6.7.4). If the two terms cannot identically be zero simultaneously, then $\mathbf{S}(t)$ is ensured a decreasing function, and because $\mathbf{S}(t_f)$ is non-negative definite so that $\mathbf{S}(t)$ is positive definite. Assuming $\mathbf{u}(\tau) = \mathbf{0}$, *i.e.* $\mathbf{B}_u^T\boldsymbol{\lambda} = \mathbf{0}$, it needs to verify that under this condition the output vector $\mathbf{z} = \mathbf{C}_z\mathbf{x}$ may not be zero. Because $\mathbf{u} = \mathbf{0}$, so the homogeneous dual equations are

(6.7.17), which corresponds to the case of observability analysis. Then, according to equation (6.7.18) it solves $\mathbf{x}(\tau)=\exp[\mathbf{A}\cdot(\tau-t)]\cdot\mathbf{x}(t)$.

Substituting back into the integration of $J(t)$ gives

$$\mathbf{x}^T(t)\int_t^{t_f}\exp[\mathbf{A}^T(\tau-t)]\mathbf{C}_z^T\mathbf{C}_z\exp[\mathbf{A}(\tau-t)]\mathrm{d}\tau\cdot\mathbf{x}(t)/2=\mathbf{x}^T(t)\mathbf{W}_o(t_f,t)\mathbf{x}(t)/2$$

where the integration in the above equation gives the observability Gram matrix. According to the observability condition of the system, it is positive definite. Hence as $\mathbf{x}(t)$ is not identically zero, $J(t)>0$ is always valid, *i.e.* $\mathbf{S}(t)$ is positive definite, also is a decreasing function of time t. ##

Let $\mathbf{S}_f$ be a null matrix, then $\mathbf{S}(t)$ is $\mathbf{Q}(t)$, thus the matrix $\mathbf{Q}(t)$ is also positive definite. ##

The Kalman-Bucy filter and LQ optimal control are dual problems to each other. The positive definiteness of the matrix $\mathbf{P}(t)$ is proved below, the conditions are the controllability of $(\mathbf{A},\mathbf{B}_w)$ and the observability of $(\mathbf{A},\mathbf{C}_y)$. For simplicity, $\mathbf{B}_w,\mathbf{C}_y$ are written as $\mathbf{B},\mathbf{C}$ below.

Let us begin with the index function $J(t)$ of equation (6.5.44). For the stability analysis or for the computation of variance matrix $\mathbf{P}(t)$, it is appropriate to assume $\mathbf{y}=\mathbf{0},\mathbf{u}=\mathbf{0},\hat{\mathbf{x}}_0=\mathbf{0}$, hence $\hat{\mathbf{x}}(t)\equiv\mathbf{0}$. For convenience, the noise $\mathbf{w}$ can be normalized to be unit intensity and the index function is given as

$$J(t)=\mathbf{x}_0^T\mathbf{P}_0^{-1}\mathbf{x}_0/2+\int_{t_0}^t[\mathbf{w}^T\mathbf{w}+\mathbf{x}^T\mathbf{C}^T\mathbf{V}^{-1}\mathbf{C}\mathbf{x}]\mathrm{d}\tau/2$$

Introducing the Lagrange multiplier vector $\boldsymbol{\lambda}(\tau)$, that the extended index $J_A(t)$ has the same numerical value with the index $J(t)$, so that

$$\begin{aligned}J(t)=J_A(t)&=\int_{t_0}^t[\boldsymbol{\lambda}^T\dot{\mathbf{x}}-\boldsymbol{\lambda}^T\mathbf{A}\mathbf{x}-\boldsymbol{\lambda}^T\mathbf{B}\mathbf{B}\boldsymbol{\lambda}/2+\mathbf{x}^T\mathbf{C}^T\mathbf{V}^{-1}\mathbf{C}\mathbf{x}/2]\mathrm{d}\tau\\&+\mathbf{x}_0^T\mathbf{P}_0^{-1}\mathbf{x}_0/2=\int_{t_0}^t[\boldsymbol{\lambda}^T\mathbf{B}\mathbf{B}^T\boldsymbol{\lambda}/2+\mathbf{x}^T\mathbf{C}^T\mathbf{V}^{-1}\mathbf{C}\mathbf{x}/2]\mathrm{d}\tau+\mathbf{x}_0^T\mathbf{P}_0^{-1}\mathbf{x}_0/2\end{aligned}\tag{6.7.20'}$$

Using the dual equations gives

$$\begin{aligned}J(t)&=\mathbf{x}_0^T\mathbf{P}_0^{-1}\mathbf{x}_0/2+\int_{t_0}^t[\boldsymbol{\lambda}^T\dot{\mathbf{x}}-\boldsymbol{\lambda}^T\mathbf{A}\mathbf{x}+\mathbf{x}^T\dot{\boldsymbol{\lambda}}+\mathbf{x}^T\mathbf{A}^T\boldsymbol{\lambda}]\mathrm{d}\tau/2\\&=\mathbf{x}_0^T\mathbf{P}_0^{-1}\mathbf{x}_0/2+\int_{t_0}^t[\boldsymbol{\lambda}^T\dot{\mathbf{x}}+\mathbf{x}^T\dot{\boldsymbol{\lambda}}]\mathrm{d}\tau/2=\mathbf{x}^T(t)\mathbf{P}^{-1}(t)\mathbf{x}(t)/2\end{aligned}\tag{6.7.20}$$

where the relation $\mathbf{x}(t)=\mathbf{P}(t)\boldsymbol{\lambda}(t)$ is used. Equation (6.7.20') determines that the index $J(t)$ is non-negative and non-decreasing function of time t. However, the further requirement is to prove that $J(t)$ is a positive definite and increasing function of time t, which determines that $\mathbf{P}(t)$ is positive definite.

If $\mathbf{C}\mathbf{x}(\tau)$ is not zero, then $J(t)$ has been ensured the above behavior. Assume $\mathbf{C}\mathbf{x}(\tau)=\mathbf{0}$, then the homogeneous dual equations become

$$\dot{\mathbf{x}}(\tau)=\mathbf{A}\mathbf{x}+\mathbf{B}\mathbf{B}^T\boldsymbol{\lambda}\,,\qquad\dot{\boldsymbol{\lambda}}(\tau)=-\mathbf{A}^T\boldsymbol{\lambda}\tag{6.7.21}$$

From the latter equation solves $\boldsymbol{\lambda}$ as

$$\boldsymbol{\lambda}(\tau)=\exp[-\mathbf{A}^T\cdot(\tau-t_0)]\cdot\boldsymbol{\lambda}_0$$

where $\boldsymbol{\lambda}_0$ is the initial disturbance. Substituting into the integration of equation (6.7.20') gives

$$\int_{t_0}^{t} \boldsymbol{\lambda}^T \mathbf{B}\mathbf{B}^T \boldsymbol{\lambda} \mathrm{d}\tau = \int_{t_0}^{t} \boldsymbol{\lambda}_0^T \exp[-\mathbf{A}(\tau - t_0)]\mathbf{B}\mathbf{B}^T \exp[-\mathbf{A}^T(\tau - t_0)]\boldsymbol{\lambda}_0 \mathrm{d}\tau$$
$$= \boldsymbol{\lambda}^T(t)\mathbf{W}_c(t, t_0)\boldsymbol{\lambda}(t)$$

where $\mathbf{W}_c$ is the controllability Gram matrix and is ensured positive definite, and then $\mathbf{P}(t)$ is positive definite. ##

Similarly, as described in section 6.5, $\mathbf{G}(t)$ is also a positive definite matrix.

The positive definiteness of the solution matrices of Riccati differential equations explains the close relationship between the ***positive definiteness of index*** with the ***controllability and observability***. It is clarified as follows. Let us examine the interval mixed energy and variational principle from structural mechanics side. For interval (t_a, t_b), there is the integration term U in the extended functional J_A

$$U(t_a, t_b) = \int_{t_a}^{t_b} [\boldsymbol{\lambda}^T \dot{\mathbf{x}} - \boldsymbol{\lambda}^T \mathbf{A}\mathbf{x} - \boldsymbol{\lambda}^T \mathbf{B}\mathbf{B}^T \boldsymbol{\lambda}/2 + \mathbf{x}^T \mathbf{C}^T \mathbf{C}\mathbf{x}/2] \mathrm{d}\tau$$
$$= \int_{t_a}^{t_b} [\boldsymbol{\lambda}^T \mathbf{B}\mathbf{B}^T \boldsymbol{\lambda}/2 + \mathbf{x}^T \mathbf{C}^T \mathbf{C}\mathbf{x}/2] \mathrm{d}\tau \tag{6.7.22}$$

where $\mathbf{W} = \mathbf{I}, \mathbf{V} = \mathbf{I}$ are taken without loss of generality. The integrand is obviously non-negative. Furthermore, the controllability states that if the state vector makes $\mathbf{C}\mathbf{x}(t) \equiv \mathbf{0}$, then $U(t_a, t_b)$ is still positive definite, and the observability states that if $\mathbf{B}^T \boldsymbol{\lambda}(t) \equiv \mathbf{0}$, then $U(t_a, t_b)$ is positive definite also, so that $U(t_a, t_b)$ ***is a positive definite functional.*** The variation $\delta U = 0$ derived dual equations are homogeneous

$$\dot{\mathbf{x}} = \mathbf{A}\mathbf{x} + \mathbf{B}\mathbf{B}^T \boldsymbol{\lambda} \tag{6.7.23a}$$

$$\dot{\boldsymbol{\lambda}} = \mathbf{C}^T \mathbf{C}\mathbf{x} - \mathbf{A}^T \boldsymbol{\lambda} \tag{6.7.23b}$$

Solution of the dual equations needs the two point boundary value conditions, which can be given as

$$\mathbf{x}(t_a) = \mathbf{x}_a, \quad \mathbf{x}(t_b) = \mathbf{x}_b \tag{6.7.23c}$$

The integration term $U(t_a, t_b)$ depends on the states at the two ends, so that it can be written as $U(t_a, \mathbf{x}_a; t_b, \mathbf{x}_b)$, which is the action function in analytical dynamics. For simplicity it can be written as $U(t_a, t_b)$. In structural mechanics, $U(t_a, t_b)$ is the ***deformation energy*** of the interval (t_a, t_b), and "***controllability and observability***" ensures the ***positive definiteness of the deformation energy.*** Therefore, it is clarified that the controllability and observability can be interpreted as the ***positive definiteness of the index*** $J_A(t_a, t_b)$ ***for any time interval*** (t_a, t_b), which corresponds to the deformation energy in structural mechanics. The controllability and observability conditions described in section 6.1 apply only to the time-invariant system. For ***time-variant system*** the condition can be extended to the ***positive definiteness of the index function*** for arbitrary time interval.

The definition of mixed energy $V(\mathbf{x}_a, \boldsymbol{\lambda}_b)$ is [Compare equation (6.5.69)]

$$V(\mathbf{x}_a, \boldsymbol{\lambda}_b) = \boldsymbol{\lambda}_b^T \mathbf{x}_b - \int_{t_0}^{t} [\boldsymbol{\lambda}^T \dot{\mathbf{x}} - \boldsymbol{\lambda}^T \mathbf{A}\mathbf{x} - \boldsymbol{\lambda}^T \mathbf{B}\mathbf{B}^T \boldsymbol{\lambda}/2 + \mathbf{x}^T \mathbf{C}^T \mathbf{C}\mathbf{x}/2] \mathrm{d}\tau$$
$$= \boldsymbol{\lambda}_b^T \mathbf{x}_b - U(t_a, t_b) \tag{6.7.24a}$$

which is a homogeneous quadratic form and its general form is

$$V(\mathbf{x}_a,\boldsymbol{\lambda}_b)=\boldsymbol{\lambda}_b^T\mathbf{F}\mathbf{x}_a+\boldsymbol{\lambda}_b^T\mathbf{G}\boldsymbol{\lambda}_b/2-\mathbf{x}_a^T\mathbf{Q}\mathbf{x}_a/2 \tag{6.7.24b}$$

Variational derivation for (6.7.24a) gives

$$\delta V(\mathbf{x}_a,\boldsymbol{\lambda}_b)=\boldsymbol{\lambda}_a^T\delta\mathbf{x}_a+\mathbf{x}_b^T\delta\boldsymbol{\lambda}_b\ ,\ \ \boldsymbol{\lambda}_a=\partial V/\partial\mathbf{x}_a,\ \ \mathbf{x}_b=\partial V/\partial\boldsymbol{\lambda}_b \tag{6.7.25}$$

hence

$$\mathbf{x}_b=\mathbf{F}\mathbf{x}_a+\mathbf{G}\boldsymbol{\lambda}_b,\quad \boldsymbol{\lambda}_a=-\mathbf{Q}\mathbf{x}_a+\mathbf{F}^T\boldsymbol{\lambda}_b \tag{6.7.26}$$

The quadratic form of $U(\mathbf{x}_a,\mathbf{x}_b)$ can be derived from the mixed energy quadratic form (6.7.24b)

$$U(\mathbf{x}_a,\mathbf{x}_b)=\mathbf{x}_a^T\mathbf{K}_{aa}\mathbf{x}_a/2+\mathbf{x}_b^T\mathbf{K}_{bb}\mathbf{x}_b/2+\mathbf{x}_b^T\mathbf{K}_{ba}\mathbf{x}_a[=\boldsymbol{\lambda}_b^T\mathbf{x}_b-V(\mathbf{x}_a,\boldsymbol{\lambda}_b)] \tag{6.7.27}$$

The corresponding relation is given as [Compare equations (5.7.13), (5.7.26)]

$$\mathbf{K}_{bb}=\mathbf{G}^{-1},\quad \mathbf{K}_{aa}=\mathbf{Q}+\mathbf{F}^T\mathbf{G}^{-1}\mathbf{F},\quad \mathbf{K}_{ba}=-\mathbf{G}^{-1}\mathbf{F} \tag{6.7.28a}$$

$$\mathbf{G}=\mathbf{K}_{bb}^{-1},\ \ \mathbf{Q}=\mathbf{K}_{aa}-\mathbf{K}_{aa}\mathbf{K}_{bb}^{-1}\mathbf{K}_{ba},\ \ \mathbf{F}=-\mathbf{K}_{bb}^{-1}\mathbf{K}_{ba},\ \ \mathbf{K}_{ab}=\mathbf{K}_{ba}^T \tag{6.7.28b}$$

The controllability and observability ensure the positive definiteness of the potential energy quadratic form $U(\mathbf{x}_a,\mathbf{x}_b)$. Certainly the matrix $\mathbf{K}_{bb}$ is positive definite and so is $\mathbf{G}$. Based on the identity

$$\begin{bmatrix}\mathbf{K}_{aa} & \mathbf{K}_{ab}\\ \mathbf{K}_{ba} & \mathbf{K}_{bb}\end{bmatrix}\equiv\begin{bmatrix}\mathbf{I}_n & \mathbf{K}_{ab}\mathbf{K}_{bb}^{-1}\\ \mathbf{0} & \mathbf{I}_n\end{bmatrix}\times\begin{bmatrix}\mathbf{K}_{aa}-\mathbf{K}_{ab}\mathbf{K}_{bb}^{-1}\mathbf{K}_{ba} & \mathbf{0}\\ \mathbf{0} & \mathbf{K}_{bb}\end{bmatrix}\times\begin{bmatrix}\mathbf{I}_n & \mathbf{0}\\ \mathbf{K}_{bb}^{-1}\mathbf{K}_{ba} & \mathbf{I}_n\end{bmatrix}$$

the matrix $\mathbf{Q}$ is also positive definite. This conclusion is valid for arbitrary interval (t_a,t_b). Except the positive definiteness mentioned above, from the interval combination algorithm (6.5.77a,b), the matrices $\mathbf{Q}$ and $\mathbf{G}$ are the increasing matrices of the interval length $\eta=t_b-t_a$. A further problem is the upper bound of matrices $\mathbf{Q}$ and $\mathbf{G}$.

The matrix $\mathbf{G}$ is considered first. The interval combination algorithm (6.5.77) is given in the mixed energy representation. Correspondingly, there is the interval combination algorithm (5.7.5) in potential energy representation, so that it is seen that the inverse matrix of $\mathbf{G}$ is $\mathbf{K}_{bb}$, which is decreasing with the increasing of η. Hence there is upper bound for the matrix $\mathbf{G}$. For the matrix $\mathbf{Q}$, because of (5.7.5), the matrix $\mathbf{K}_{aa}$ decreases with the increasing of η. From equation (6.7.28b) it is seen that $\mathbf{Q}$ is always less than $\mathbf{K}_{aa}$, hence $\mathbf{Q}$ has upper bound too.

The problem of the limit of matrix $\mathbf{F}$ as $\eta\to\infty$ remains to be solved. From the interval combination algorithm (6.5.77a,b) it is seen, that as η tends very large the matrices $\mathbf{G}$ and $\mathbf{Q}$ almost reach their limits, and if the matrix $\mathbf{F}$ does not tends zero, then the matrices $\mathbf{G}$ and $\mathbf{Q}$ will further increase a finite value, a contradiction. Hence it must be $\mathbf{F}(\eta)\to 0$ as $\eta\to\infty$.

From structure mechanics side, the conclusion is that the controllability and observability requires the positive definiteness of the potential energy, which ***excludes the existence of rigid body motion.*** On the other hand, the positive definiteness of flexibility matrix excludes the internal force with no deformation, *i.e.* rigid body forces.

§6.7.2.3, Stability analysis based on Lyapunov second method

The stability problem of the differential equation (6.7.13) for the mean value of state vector of LQ control is analyzed first. The analysis is meaningful also for finite duration problem. Because the differential equation (6.7.13) is time variant, the method of checking the real part of all the eigenvalues no longer applies, so that the Lyapunov's second method is used for the stability analysis. The key step for the Lyapunov second method is to find a positive definite Lyapunov function of the state vector $L(\mathbf{x}_a)$. The index function $J(t)$ is readily selected as the Lyapunov function

$$L(\mathbf{x}_a) = J(t) = \mathbf{x}_a^T(t)\mathbf{S}(t)\mathbf{x}_a(t)/2$$

Its positive definiteness has been proved. Its time derivative is

$$-(\mathbf{x}_a^T\mathbf{C}_z^T\mathbf{C}_z\mathbf{x}_a/2 + \mathbf{u}^T\mathbf{u}/2) = \dot{L}(\mathbf{x}_a) < 0$$

which is easily seen from the definition of $J(t)$. The positive definiteness of the observable Gram matrix determines that the derivative decreases continuously until t_f, where $L(\mathbf{x}_a) = \mathbf{x}_f^T\mathbf{S}_f\mathbf{x}_f/2$, and the stability conclusion is drawn from the Lyapunov theory. ##

The filtering problem derives the differential equation (6.5.49), for which the stability needs verification too. Because the matrix $\mathbf{P}(t)$ appears in the coefficient, it is time-invariant differential equation again. The stability analysis can also use the Lyapunov second method. The stability analysis of linear differential equation can assume the non-homogeneous term equals zero

$$\dot{\hat{\mathbf{x}}}(t) = \mathbf{A}\hat{\mathbf{x}} - \mathbf{P}(t)\mathbf{C}^T\mathbf{C}\hat{\mathbf{x}}, \quad \hat{\mathbf{x}}(0) = \hat{\mathbf{x}}_0 \tag{6.7.29}$$

where $\mathbf{P}(t)$ is the solution of equation (6.5.50). The two equations are derived from the homogeneous dual equations of (6.5.47)

$$\dot{\mathbf{x}}(\tau) = \mathbf{A}\mathbf{x} + \mathbf{B}\mathbf{B}^T\boldsymbol{\lambda}, \quad \mathbf{P}_0^{-1}\cdot(\mathbf{x}(0) - \hat{\mathbf{x}}_0) = \boldsymbol{\lambda}(0) \tag{6.7.30a}$$

$$\dot{\boldsymbol{\lambda}}(\tau) = \mathbf{C}^T\mathbf{C}\mathbf{x} - \mathbf{A}^T\boldsymbol{\lambda}, \qquad \boldsymbol{\lambda}(t) = \mathbf{0} \tag{6.7.30b}$$

where $\hat{\mathbf{x}}_0$ is the initial mean vector, and $\mathbf{P}_0$ is the initial variance. The vector $\hat{\mathbf{x}}_0$ is the initial deviation, and the stability is induced from the arbitrary initial vector. The solution of the homogeneous dual equations can use the substitution

$$\mathbf{x}(t) = \hat{\mathbf{x}}(t) + \mathbf{P}(t)\boldsymbol{\lambda}(t) \tag{6.7.31}$$

transforming to the equations (6.7.29) and (6.5.50). The variational principle of the dual equations is

$$\begin{aligned} J_A(t) = &\int_{t_0}^{t} [\boldsymbol{\lambda}^T\dot{\mathbf{x}} - \boldsymbol{\lambda}^T\mathbf{A}\mathbf{x} - \boldsymbol{\lambda}^T\mathbf{B}\mathbf{B}^T\boldsymbol{\lambda}/2 + \mathbf{x}^T\mathbf{C}^T\mathbf{C}\mathbf{x}/2]\mathrm{d}\tau \\ &+ (\mathbf{x}(0) - \hat{\mathbf{x}}_0)^T\mathbf{P}_0^{-1}(\mathbf{x}(0) - \hat{\mathbf{x}}_0)/2, \qquad \delta J_A = 0 \end{aligned} \tag{6.7.32}$$

Note that numerically $J_A(t) = J(t)$. Using the equation (6.7.30a) gives

$$J_A(t) = (\mathbf{x}(0) - \hat{\mathbf{x}}_0)^T\mathbf{P}_0^{-1}(\mathbf{x}(0) - \hat{\mathbf{x}}_0)/2 + \int_{t_0}^{t} [\boldsymbol{\lambda}^T\mathbf{B}\mathbf{B}^T\boldsymbol{\lambda}/2 + \mathbf{x}^T\mathbf{C}^T\mathbf{C}\mathbf{x}/2]\mathrm{d}\tau \tag{6.7.32a}$$

The equation (6.7.30a) corresponds to the compatibility equation in structural mechanics, so the above integration is potential energy. Integration by parts from the extended index (6.7.32), and let the equation (6.7.30b) satisfied, using two end boundary conditions gives the complementary energy as

$$J_A(t) = (\mathbf{x}(0) - \hat{\mathbf{x}}_0)^T \mathbf{P}_0^{-1}(\mathbf{x}(0) - \hat{\mathbf{x}}_0)/2 - \boldsymbol{\lambda}^T(0)\mathbf{x}_0 + \boldsymbol{\lambda}^T(t)\mathbf{x}(t)$$
$$- \int_{t_0}^{t} [\boldsymbol{\lambda}^T \mathbf{B}\mathbf{B}^T \boldsymbol{\lambda}/2 + \mathbf{x}^T \mathbf{C}^T \mathbf{C}\mathbf{x}/2] d\tau$$
$$= \hat{\mathbf{x}}_0^T \mathbf{P}_0^{-1} \hat{\mathbf{x}}_0/2 - \mathbf{x}_0^T \mathbf{P}_0^{-1} \mathbf{x}_0/2 - \int_{t_0}^{t} [\boldsymbol{\lambda}^T \mathbf{B}\mathbf{B}^T \boldsymbol{\lambda}/2 + \mathbf{x}^T \mathbf{C}^T \mathbf{C}\mathbf{x}/2] d\tau \quad (6.7.32b)$$

The above two formulations of $J_A(t)$ are numerically equal. Adding the two formulae together gives

$$J_A(t) = -\boldsymbol{\lambda}_0^T \hat{\mathbf{x}}_0/2 = -(\mathbf{x}_0 - \hat{\mathbf{x}}_0)^T \mathbf{P}_0^{-1} \hat{\mathbf{x}}_0/2 , \quad \mathbf{x}_0 \underset{\text{def}}{=} \mathbf{x}(0) \quad (6.7.33)$$

and $$U(t_0, t) = \int_{t_0}^{t} [\boldsymbol{\lambda}^T \mathbf{B}\mathbf{B}^T \boldsymbol{\lambda} + \mathbf{x}^T \mathbf{C}^T \mathbf{C}\mathbf{x}] d\tau/2 = -\boldsymbol{\lambda}_0^T \mathbf{x}_0/2$$

where the vectors $\mathbf{x}_0, \boldsymbol{\lambda}_0$ are all smoothing solution. From equation (6.7.32b) know that $U(t_0, t)$ is the upper bounded.

The natural boundary condition of filtering at the present time t is $\boldsymbol{\lambda}(t) = \mathbf{0}$. From the definition of mixed energy gives

$$J_A(t) = (\mathbf{x}_0 - \hat{\mathbf{x}}_0)^T \mathbf{P}_0^{-1}(\mathbf{x}_0 - \hat{\mathbf{x}}_0)/2 - V(\mathbf{x}_0, \boldsymbol{\lambda}(t))$$
$$= (\mathbf{x}_0 - \hat{\mathbf{x}}_0)^T \mathbf{P}_0^{-1}(\mathbf{x}_0 - \hat{\mathbf{x}}_0)/2 + \mathbf{x}_0^T \mathbf{Q}(\eta)\mathbf{x}_0/2 , \quad \eta = t - t_0$$

This mixed energy corresponds to the interval (t_0, t), where the latter equal sign uses the substitution $\boldsymbol{\lambda}(t) = \mathbf{0}$. The variation of $\mathbf{x}_0$ should minimize $J(t)$, which gives the smoothing state estimation

$$\bar{\mathbf{x}}_0 = (\mathbf{I} + \mathbf{P}_0 \mathbf{Q}(\eta))^{-1} \hat{\mathbf{x}}_0 \quad (6.7.34)$$

From the above equation

$$\hat{\mathbf{x}}(t) = \mathbf{F}\bar{\mathbf{x}}_0 + \mathbf{G}\boldsymbol{\lambda}(t) = \mathbf{F}(\eta) \cdot [\mathbf{I} + \mathbf{P}_0 \mathbf{Q}(\eta)]^{-1} \hat{\mathbf{x}}_0$$

Because $\mathbf{F}(\eta) \to \mathbf{0}$ when $\eta \to \infty$, so that $\hat{\mathbf{x}}(t) \to \mathbf{0}$ as $\eta \to \infty$. So that the system is asymptotically stable.

The stability can also be proved by the Lyapunov second method. Note $U(t_0, t)$ is a function of the two end displacement vectors, which means $U(\bar{\mathbf{x}}_0, \hat{\mathbf{x}}(t))$ having the upper bound U_p, hence the Lyapunov function can be selected as

$$L(\mathbf{x}) = U_p - U(t_0, t) \quad (6.7.35)$$

The system asymptotic stability is proved based on the Lyapunov second theorem. ##

§6.7.3, Precise computation of LQ control

The fundamental behavior has been described in the last section, below is for the precise computations. From the experience of filter computation and that in chapter 5, the matrices of mixed energy quadratic form should be computed first. For LQ control, only $\mathbf{F}(\eta), \mathbf{G}(\eta), \mathbf{Q}(\eta)$ is required, where η is a reasonable interval length. Two methods can be used for the computation, namely:

a) Precise integration method

It does not need to solve all the eigen-solutions of the Hamilton matrix, but some CPU expense is required for matrix computations.

b) Analytical method bases on eigen-solutions

This method needs to solve all the eigen-solutions of the Hamilton matrix then according to the equation (6.5.114a~d) in section 6.5.6 the matrices $\mathbf{Q}(\eta),\mathbf{G}(\eta)$ and $\mathbf{F}(\eta)$ are obtained. The Hamilton matrix is asymmetric, Jordan normal form may possibly appear, and in this case numerical instability emerges for the eigen-solutions. The same problem appears also in exponential matrix computation. Hence, when Jordan form does not appear, the analytical method is beneficial; however, when the eigenvalues nearly appear duplicate roots, precise integration is necessary. The numerical stable behavior of precise integration method is quite attractive.

In the problem of system identification, the Hamilton matrix is due to change, the eigen-solution expansion method is attractive.

After the step size η is selected, the precise integration or eigen-solution method can be used. So the matrices $\mathbf{Q}(\eta),\mathbf{G}(\eta),\mathbf{F}(\eta)$ can be regarded computed as given in chapter 5 or in section 6.5. The subsequent computations are

a) Solve the Riccati differential equation (6.7.10), and

b) Solve the mean state vector from equation (6.7.13).

§6.7.3.1, Precise solution of Riccati differential equation

In section 6.5.4 the interval mixed energy is described in detail. The boundary condition of differential equation (6.7.10) is given at the end $t=t_{\mathrm{f}}$, so that the integration for t is along the reverse direction. The reverse differential equations (6.5.83~85) of interval matrices with $\mathbf{V}=\mathbf{I}_q,\mathbf{W}=\mathbf{I}_m$ derive

$$\partial\mathbf{F}/\partial t_a=-\mathbf{F}(\mathbf{A}-\mathbf{B}_u\mathbf{B}_u^T\mathbf{Q}),\ \text{when } t_a=t_b,\ \mathbf{F}=\mathbf{I}_n \tag{6.7.36a}$$

$$\partial\mathbf{G}/\partial t_a=-\mathbf{F}\mathbf{B}_u\mathbf{B}_u^T\mathbf{F}^T,\ \text{when } t_a=t_b,\ \mathbf{G}=\mathbf{0} \tag{6.7.36b}$$

$$\partial\mathbf{Q}/\partial t_a=-\mathbf{C}_z^T\mathbf{C}_z-\mathbf{Q}\mathbf{A}-\mathbf{A}^T\mathbf{Q}+\mathbf{Q}\mathbf{B}_u\mathbf{B}_u^T\mathbf{Q},\ \text{when } t_a=t_b,\mathbf{Q}=\mathbf{0} \tag{6.7.36c}$$

Time-invariant system changes the PDE to be ODE but the integration is still along reverse direction. Comparing (6.7.10) with (6.7.36a), the differential equation is the same and $t_{\mathrm{b}}=t_{\mathrm{f}}$, but the boundary conditions are different. Hence $\mathbf{Q}(t,t_{\mathrm{f}})$ is not $\mathbf{S}(t)$, the matrix $\mathbf{Q}(t,t_{\mathrm{f}})$ corresponds to zero end condition, *i.e.* corresponds to an open interval (t,t_{f}). The variational principle (6.7.6') for homogeneous equations is

$$\begin{aligned}V(\mathbf{x}(t),\boldsymbol{\lambda}_{\mathrm{f}})&=\boldsymbol{\lambda}_{\mathrm{f}}^T\mathbf{x}_{\mathrm{f}}-\int_t^{t_{\mathrm{f}}}[\boldsymbol{\lambda}^T\dot{\mathbf{x}}-\mathrm{H}(\mathbf{x},\boldsymbol{\lambda})]\mathrm{d}\tau\\&=-\mathbf{x}^T(t)\mathbf{Q}\mathbf{x}(t)/2+\boldsymbol{\lambda}_{\mathrm{f}}^T\mathbf{F}\mathbf{x}(t)+\boldsymbol{\lambda}_{\mathrm{f}}^T\mathbf{G}\boldsymbol{\lambda}_{\mathrm{f}}/2\end{aligned} \tag{6.7.37a}$$

which is the mixed energy, and relates to the extended index

$$J_{\mathrm{A}}=\mathbf{x}_{\mathrm{f}}^T\mathbf{S}_{\mathrm{f}}\mathbf{x}_{\mathrm{f}}/2+\boldsymbol{\lambda}_{\mathrm{f}}^T\mathbf{x}_{\mathrm{f}}-V(\mathbf{x}(t),\boldsymbol{\lambda}_{\mathrm{f}}),\quad \max_{\boldsymbol{\lambda}_{\mathrm{f}}}\min_{\mathbf{x}_{\mathrm{f}}} J_{\mathrm{A}} \tag{6.7.37b}$$

It is derived as

$$J_{\mathrm{A}}=\mathbf{x}^T(t)[\mathbf{Q}+\mathbf{F}^T(\mathbf{S}_{\mathrm{f}}^{-1}+\mathbf{G})^{-1}\mathbf{F}]\mathbf{x}(t)/2$$

Comparing with equation (6.7.14) gives

$$\mathbf{S}(t)=\mathbf{Q}+\mathbf{F}^T(\mathbf{S}_{\mathrm{f}}^{-1}+\mathbf{G})^{-1}\mathbf{F} \tag{6.7.38}$$

where $\mathbf{Q},\mathbf{G},\mathbf{F}$ are the mixed energy sub-matrices of interval (t,t_{f}). This equation explains that the end condition $\mathbf{S}_{\mathrm{f}}$ can be disregarded first. Reverse integrating the interval mixed energy for t first, and then computing according to equation (6.7.38) gives the solution matrix of Riccati equation.

A simple interpretation of equation (6.7.38) is that it can be regarded as the result of interval combination. Comparing with equation (6.5.77a), that the matrices $\mathbf{Q},\mathbf{G},\mathbf{F}$ of interval (t,t_{f}) are treated as the interval-1, which is combined with a fictitious interval-2 with the matrices $(\mathbf{S}_{\mathrm{f}},\mathbf{I},\mathbf{0})$, therefore the matrix $\mathbf{S}(t)$ is just the matrix $\mathbf{Q}_c$ at the left-hand side. The legitimacy of using this interval combination algorithm is again based on the association rule of interval combination operation.

The precise solution of Riccati differential equation (6.7.10) is described until here. Based on the description above, it is easy to do programming on the computer. This algorithm is similar to the solution of the respective Riccati differential equation of Kalman-Bucy filtering. Also, when the interval duration of integration is long enough, the solution of the differential equation tends to be the solution matrix $\mathbf{S}_\infty$ of the corresponding algebraic Riccati equation.

The importance of the solution $\mathbf{S}(t)$ is that the feedback control vector $\mathbf{u}$ can be determined according to equation (6.7.12). This equation requires the state vector, so that the LQ control theory is based on the whole state feedback. However, the whole state vector $\mathbf{x}(t)$ is not available at the current time t, so that $\mathbf{x}(t)$ is substituted by the filtered vector $\hat{\mathbf{x}}(t)$ computed from the past interval. Later in section 6.7.4 such kind method of measurement feedback, which is called as the ***separation principle***, is introduced.

§6.7.3.2, Integration of state differential equation

Under LQ control theory, the state vector $\mathbf{x}(\tau)$ in the future time interval should be solved from the differential equation (6.7.13), which is a time variant set of ODE. A general time variant ODE is very difficult to find its analytical solution or to solve by precise integration method, so that some approximation must be made in finding the solution. However, the ODE (6.7.13) is derived from a time invariant system, hence the precise integration method can still be used to solve the state vector precisely.

In the above, the interval mixed energy is given in equation (6.7.37), where the matrix $\mathbf{F}$ satisfies the differential equation (6.7.36a). Note that the time variant term at the right hand side is $\mathbf{Q}(t_{\mathrm{a}})$ but not $\mathbf{S}(t_{\mathrm{a}})$, that the different equation is different. To amend this disagreement, the virtual concentrated interval at the end time t_{f} used as the interval-2 is combined with the interval mixed energy of the interval (τ,t_{f}), which is used as the interval-1, it gives

$$\mathbf{F}_c(\tau)=(\mathbf{I}+\mathbf{G}(\tau,t_{\mathrm{f}})\mathbf{S}_{\mathrm{f}})^{-1}\mathbf{F}(\tau,t_{\mathrm{f}}) \qquad t\le\tau\le t_{\mathrm{f}} \tag{6.7.39}$$

Now the differential equation satisfied by the matrix $\mathbf{F}_c(\tau)$ is to be determined. Certainly the interval matrices of $\mathbf{F},\mathbf{G},\mathbf{Q}$ satisfy equation (6.7.36), where t_{a} is τ. Using the inverse matrix differential $\mathrm{d}(\mathbf{X}^{-1})/\mathrm{d}\tau = -\mathbf{X}^{-1}\dot{\mathbf{X}}\mathbf{X}^{-1}$ derives

$$\begin{aligned} \mathrm{d}\mathbf{F}_c/\mathrm{d}\tau &= -(\mathbf{I}+\mathbf{G}\mathbf{S}_{\mathrm{f}})^{-1}(\mathrm{d}\mathbf{G}/\mathrm{d}\tau)\mathbf{S}_{\mathrm{f}}(\mathbf{I}+\mathbf{G}\mathbf{S}_{\mathrm{f}})^{-1}\mathbf{F}+(\mathbf{I}+\mathbf{G}\mathbf{S}_{\mathrm{f}})^{-1}(\mathrm{d}\mathbf{F}/\mathrm{d}\tau) \\ &= (\mathbf{I}+\mathbf{G}\mathbf{S}_{\mathrm{f}})^{-1}\mathbf{F}\mathbf{B}_u\mathbf{B}_u^T\mathbf{F}^T(\mathbf{S}_{\mathrm{f}}^{-1}+\mathbf{G})^{-1}\mathbf{F}-(\mathbf{I}+\mathbf{G}\mathbf{S}_{\mathrm{f}})^{-1}\mathbf{F}(\mathbf{A}-\mathbf{B}_u\mathbf{B}_u^T\mathbf{Q}) \\ &= -(\mathbf{I}+\mathbf{G}\mathbf{S}_{\mathrm{f}})^{-1}\mathbf{F}\cdot[\mathbf{A}-\mathbf{B}_u\mathbf{B}_u^T(\mathbf{Q}+\mathbf{F}^T(\mathbf{S}_{\mathrm{f}}^{-1}+\mathbf{G})^{-1}\mathbf{F})] \\ &= -\mathbf{F}_c\times[\mathbf{A}-\mathbf{B}_u\mathbf{B}_u^T\mathbf{S}(\tau)] \end{aligned} \tag{6.7.40}$$

Therefore it derives that

$$\mathrm{d}(\mathbf{F}_c^{-1})/\mathrm{d}\tau = [\mathbf{A}-\mathbf{B}_u\mathbf{B}_u^T\mathbf{S}(\tau)]\cdot\mathbf{F}_c^{-1} \tag{6.7.40'}$$

Note that $\mathbf{F}_c(t_{\mathrm{f}})=\mathbf{I}_n$, this differential equation is the same as (6.7.13), but $\mathbf{F}_c^{-1}$ is a matrix. To transform to the vector $\mathbf{x}_{\mathrm{a}}(t)$ and satisfying the initial condition, let

$$\mathbf{x}_{\mathrm{a}}(\tau)=\mathbf{F}_c^{-1}(\tau)\times\xi(t), \quad \xi(t)=\mathbf{F}_c(t)\times\hat{\mathbf{x}}(t) \tag{6.7.41}$$

where τ is the argument $t\le\tau\le t_{\mathrm{f}}$. Hence the vector $\mathbf{x}_{\mathrm{a}}(\tau)$ as a function of τ satisfies the equation (6.7.13), because the matrix $\mathbf{F}_c^{-1}$ satisfies (6.7.40'), and the initial condition in (6.7.13) can also be fulfilled.

The equation is quite simple, but directly used in computation may still have problem, which is because when both t,τ are large, the matrix $\mathbf{F}(\tau)$ and then the matrix $\mathbf{F}_c(\tau)$ tend to be null matrices. And these matrices tend to be singular as $\tau\to\infty$, which causes ill-conditioning problem in numerical computation. To avoid such problem, the strategy of stepwise progressing can be used. Because the matrix $\mathbf{S}$ has been all computed at the grid time points, it needs to compute the state vector $\mathbf{x}_{\mathrm{a}}$ at $\tau=t+\eta$, which needs the matrices $\mathbf{F}_c(t)$ and $\mathbf{F}_c(t+\eta)$. Regard the time instant $t+\eta$ as the current end time t_{f}, and the matrix $\mathbf{S}(t+\eta)$ as the current $\mathbf{S}_{\mathrm{f}}$. Using equation (6.7.39), the two matrices $\mathbf{F}_c$ at time instants $\tau=t$ and $\tau=t+\eta$ are computed as

$$\tau=t:\ \mathbf{F}_c(t)=[\mathbf{I}+\mathbf{G}(t,t+\eta)\mathbf{S}(t+\eta)]^{-1}\mathbf{F}(t,t+\eta)=[\mathbf{I}+\mathbf{G}(\eta)\mathbf{S}(t+\eta)]^{-1}\mathbf{F}(\eta)$$

$$\tau=t+\eta:\ \mathbf{F}_c(t+\eta)=[\mathbf{I}+\mathbf{G}(t+\eta,t+\eta)\mathbf{S}(t+\eta)]^{-1}\mathbf{F}(t+\eta,t+\eta)=\mathbf{I}$$

Therefore from (6.7.41), the integration of interval $t\sim t+\eta$ is given as

$$\mathbf{x}_{\mathrm{a}}(t+\eta)=\mathbf{\Phi}_c(t+\eta,t)\cdot\mathbf{x}_{\mathrm{a}}(t),\ \ \mathbf{\Phi}_c(t+\eta,t)=[\mathbf{I}+\mathbf{G}(\eta)\mathbf{S}(t+\eta)]^{-1}\mathbf{F}(\eta) \tag{6.7.42}$$

Hence the state vector precise integration needs also the computation of interval mixed energy. The computational steps are given as

1) Compute $\mathbf{F}(\eta),\mathbf{G}(\eta),\mathbf{Q}(\eta)$ of single step, and the matrices $\mathbf{S}(t)$ at all grid points.
2) Compute the matrices $\mathbf{\Phi}_c(t+\eta,t)$ according to equation (6.7.42) at all the grid points.
3) Integration of state vector ODE, which is only the matrix-vector multiplication in (6.7.42).

Therefore the integration of state differential equation is reduced to the interval matrices computation and matrix-vector multiplication, which gives the precise results approaching the full computer precision.

§6.7.4, Measurement feedback optimal control

In applications, the state vector is not available precisely, which can be estimated by various means. Usually the LQG (Linear Quadratic Gaussian) theory is applied to the measurement feedback control. Simply looking from the LQ control theory, the control rule (6.7.12) uses the state vector $\mathbf{x}(t)$ to feedback, however the state vector has not been measured precisely. Hence, at the present time instant t, the state vector estimation $\hat{\mathbf{x}}(t)$ of Kalman-Bucy filter is used in place of the state vector $\mathbf{x}(t)$ to compute the feedback control vector $\mathbf{u}(t)$, in order to realize the optimal control based on the measurement data. The computation of control vector $\mathbf{u}(t)$ requires instantaneous response, which needs ***real time*** computation and is a key step in LQG optimal control.

In figure 6.4, the whole time interval $[t_0, t_f]$ is subdivided as the combination of the past time interval $[t_0, t)$ and the future time interval $(t, t_f]$ with the current time instant be denoted as t. The state space control analysis requires solving the control vector $\mathbf{u} = -\mathrm{K}\mathbf{y}$, where $\mathbf{y}$ denotes the measurement before the present time instant t, and K denotes a causal operator, which uses the past measured data $\mathbf{y}$ to compute the current control vector. In future time interval, the LQ control theory is used so as to minimize the least square index functional. At the present time, the connection of past and future intervals, the state vector uses the optimal filter state vector $\hat{\mathbf{x}}(t)$ in place of the state vector, which gives the equation

$$\mathbf{u}(t) = -\mathbf{B}_u^T \mathbf{S}(t) \hat{\mathbf{x}}(t) \tag{6.7.12'}$$

Note, both the computations of LQ control in the future time interval and the Kalman-Bucy filtering in the past time interval, respectively, are carried out separately. Their connection is, at the present time using the equation (6.7.12') to give the control law, where the substitution of $\hat{\mathbf{x}}(t)$ connects the past interval with the future interval. Such analysis strategy is termed as the ***separation principle***.

After the LQ control and the Kalman-Bucy filter problems are solved, using the separation principle combining these two results gives the LQG control theory. The previous sections have described the Kalman-Bucy filter computations in some detail, and have also described the LQ control theory. Combining the results of both the intervals with equation (6.7.12') gives the computation of LQG control.

§6.8, Robust control

The LQG (Linear Quadratic Gaussian) methodology of measurement feedback optimal control is described in some detail, based on combining the Kalman-Bucy filtering and LQ control state feedback theory using the separation principle. The LQG theory is very nice and composes the main progresses of control theory in twenty more years of 196* to 197*. However, the robustness of a control system is a fundamental requirement.

The control system engineer must be assured that a design functions properly before committing to implementation. Such assurance can be obtained by analyzing control system stability and performance with respect to a range of plant models, which is expected to encompass the actual plant. This type of analysis is termed robustness analysis.

The analysis of robustness requires that the discrepancy between the mathematical model of the plant and the actual plant be quantified. Since a perfect mathematical model of the plant is not available, so that a set of mathematical models is defined which includes the actual plant dynamics.

This model is specified with a nominal plant and a set of perturbations termed admissible perturbations. The admissible perturbations are typically bounded, where the bound is dependent on the uncertainty in the model, [132].

The full state-feedback LQ controllers and the Kalman-Bucy filters, considered separately have impressive robust properties. But it is shown that this robustness may be destroyed when combining the LQ control and Kalman-Bucy filter to become a LQG controller. In general, there is no guarantee on the robustness of the LQG optimal control, and the robustness of each design should be carefully verified.

The dynamic equation is fundamental for a linear system, where the $n \times n$ plant matrix $\mathbf{A}$ is regarded completely given in the analysis. However, the mathematical modeling is constructed based on a number of assumptions in order to yield a manageably simple model. Hence a perfect mathematical model of the plant is not available. In applications, the given plant matrix $\mathbf{A}$ should be considered having some kinds of error $\mathbf{A}_\Delta$. A controller that functions adequately for all admissible perturbations is termed ***robust***. The real plant matrix $\mathbf{A}_r$ can then be considered as

$$\mathbf{A}_r = \mathbf{A} \pm \mathbf{A}_\Delta \tag{6.8.1}$$

where $\mathbf{A}$ is a nominal plant matrix from the mathematical model and $\mathbf{A}_\Delta$ is the possible deviation of the matrix, its value is not known and so is considered random. The matrix $\mathbf{A}_\Delta$ is considered with zero-mean value and the deviation is not very large, *i.e.* the norm $\|\mathbf{A}_\Delta\|$ being limited. Robust control is to analyze the system stability and performance under the uncertainty of plant matrix $\mathbf{A}_\Delta$ and/or other model uncertainties, which is very important for applications. The H_∞ linear optimal control theory is developed since 1981, and is still investigated intensively. See [50,132-137].

There are a number of robust control analyses based on various theories, however, only the state space based approach, the H_∞ linear optimal control is considered below. Similar methodology is used as in the LQG theory, that the finite time interval $[t_0, t_f]$ of control problem is subdivided into the past interval $[t_0, t)$ and the future time interval $(t,t_f]$ subdivided by the present time instant t. For the past time interval $[t_0,t)$ the robust filtering analysis is needed, while in the future time interval the robust state feedback optimal control $(t,t_f]$ is to be executed. Afterwards, the results of the two time intervals require to be combined together at the present time instant t, so that the feedback control vector $\mathbf{u}(t)$ is generated. Although the problems of the two time intervals are different, but there exists still a unified variational principle [94], that not only the equations of individual intervals can be derived but also the connection condition at the present time t can be generated form the unified variational principle.

A general linear plant description is given by the differential equations

Dynamic: $\dot{\mathbf{x}} = \mathbf{A}\mathbf{x} + \mathbf{B}_w\mathbf{w} + \mathbf{B}_u\mathbf{u}$

Output: $\mathbf{z} = \mathbf{C}_z\mathbf{x} + \mathbf{D}_{11}\mathbf{w} + \mathbf{D}_{12}\mathbf{u}$

Measurement: $\mathbf{y} = \mathbf{C}_y\mathbf{x} + \mathbf{D}_{21}\mathbf{w} + \mathbf{D}_{22}\mathbf{u}$

The solution of such form of general plant should first simplify the plant by the LFT (***Linear Fractional Transformation***), that the existence of the terms $\mathbf{D}_{11}$ and $\mathbf{D}_{22}$ complicates the calculations and formulas, distracting attention from the central ideas. The LFTs carry the simplified system fundamental equations being of the form

Dynamic equation: $\dot{\mathbf{x}} = \mathbf{A}\mathbf{x} + \mathbf{B}_w\mathbf{w} + \mathbf{B}_u\mathbf{u}$ (6.8.2)

Output equation: $\mathbf{z} = \mathbf{C}_z\mathbf{x} \qquad + \mathbf{D}_{12}\mathbf{u}$ (6.8.3)

Measurement equation: $\mathbf{y} = \mathbf{C}_y\mathbf{x} + \mathbf{D}_{21}\mathbf{w}$ (6.8.4)

Where $\mathbf{x}$ denotes the state, $\mathbf{y}$ the measurement, $\mathbf{z}$ the output, $\mathbf{u}$ the control, and $\mathbf{w}$ the input disturbance white noise vectors, with dimensions n,q,p,m,l, respectively, where $p \geq m, q \geq l$, and the conditions

$$\mathbf{D}_{12}^T\mathbf{D}_{12} = \mathbf{I}_m, \quad \mathbf{D}_{21}\mathbf{D}_{21}^T = \mathbf{I}_q \tag{6.8.5}$$

are satisfied, see [50,132~135], otherwise $\mathbf{D}_{12}^T\mathbf{D}_{12} = \mathbf{0}$, $\mathbf{D}_{21}\mathbf{D}_{21}^T = \mathbf{0}$. The form of plant equations (6.8.2~5) can be called as the norminal model. Therefore all the subsequent robust analysis of H_∞ linear optimal control will be based on such norminal model.

In LQG theory, the input disturbance vector $\mathbf{w}$ is considered as a given zero-mean white noise and the variance matrix $\mathbf{W}$ is given. Such model of external disturbance assumes that the disturbance is neutral, *i.e.* it cannot consider that the white noise $\mathbf{w}$ may be selected to maximize the system deviation.

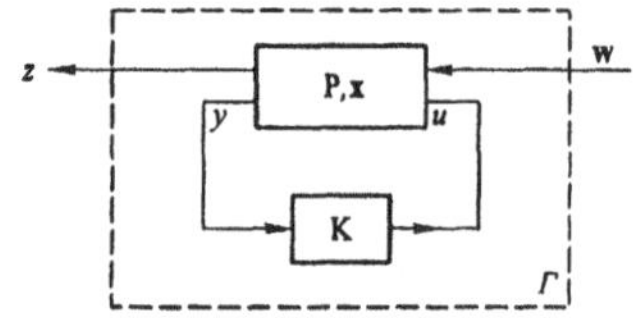

(a) The nominal system Γ

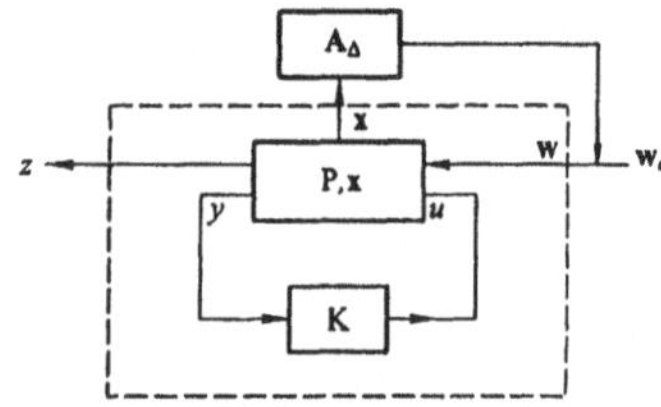

(b) $\mathbf{A}_\Delta$ is combined into w

Figure 6.10, (a) *The nominal system*; (b) $\mathbf{A}_\Delta$ *is merged into* $\mathbf{w}$

However, the robust control needs to consider **all possible variations of the plant matrix** $\mathbf{A}_\Delta$, so that the most unfavorable variation of $\mathbf{A}_\Delta$ must be selected. If takes the effect of the term $\mathbf{A}_\Delta$ into consideration, the dynamic equation will become

$$\dot{\mathbf{x}} = \mathbf{A}\mathbf{x} + (\mathbf{B}_w\mathbf{w} + \mathbf{A}_\Delta\mathbf{x}) + \mathbf{B}_u\mathbf{u}$$

where $\mathbf{A}_\Delta$ is somewhat arbitrary except with limited norm $\|\mathbf{A}_\Delta\|$. Let the term $\mathbf{A}_\Delta\mathbf{x}$ be combined into $\mathbf{B}_w\mathbf{w}$, which is represented by the parenthesis in the above equation. Figure 6.10(a) gives the system block diagram of equations (6.8.2~5), which is represented with the block P, and K is a causal operator for computing the control vector $\mathbf{u}$. In figure 6.10(b), the factor of $\mathbf{A}_\Delta$ has been merged into the disturbance noise $\mathbf{w}$. Because of the merge of this $\mathbf{A}_\Delta$ uncertain term, the neutral characteristic of the random input disturbance $\mathbf{w}$ is changed. The external disturbance $\mathbf{w}$ is thought of neutral originally in the LQG control theory. However, since the $\mathbf{A}_\Delta$ term is merged in for robust control, which is only an uncertain but time-invariant factor, so that the disturbance $\mathbf{w}$ is not a zero-mean neutral vector but should be biased to deteriorate the system performance. The model deviation $\mathbf{A}_\Delta$ should take robust behavior into consideration that the ***most unfavorable selection to the system response must be considered***. Because of the norm $\|\mathbf{A}_\Delta\|$ is limited, which means that the norm of $\mathbf{w}$ is also limited.

The disturbance $\mathbf{w}$ and the control $\mathbf{u}$ have their completely contrast characteristics. The selection of the control vector $\mathbf{u}$ is the right of the system designer, for whom the selection criterion is to minimize the index of system performance. On the other hand, the disturbance $\mathbf{w}$ is out of control of the designer that the robust consideration needs to maximize the index of the system, which means that the system performance should consider any unfavorable possibility. The selection of $\mathbf{w}$ is no longer a zero-mean white noise for H_∞ robust control, but a ***white noise with non-zero mean***, a covariance stationary stochastic process, that the non-zero mean biases to ***maximize the index***. Note that the mean value of a stochastic process is deterministic. The distinction of the LQG control versus the robust H_∞ control is just laid on that $\mathbf{w}$ is considered a ***zero mean white noise for LQG control***, but is considered ***non-zero mean white noise for*** H_∞ ***control***.

In H_∞ control, such completely contrast characteristics of the control $\mathbf{u}$ and the disturbance $\mathbf{w}$ are quite typical in **game theory**. The present situation corresponds to the case of ***zero-sum game between two-players***. 'Zero-sum' means that the payoff function $J(\mathbf{u},\mathbf{w})$ to the player-A implies the payoff function $-J(\mathbf{u},\mathbf{w})$ to the player-B, see [138]. The connection between the game theory and robust control can be found in [50,136].

The above discussion is only interpretation and reasoning. Mathematical theory is necessary for deeper understanding and serious description. The analysis will be given for the ***future time interval***, for the ***past time interval***, and for their ***connection at the present time instant*** t, one after another. The mathematical analysis for two individual time intervals, the H_∞ ***control*** and H_∞ ***filter***, and further connecting the two intervals at the present time instant t, to form the H_∞

measurement feedback control of the complete time interval $[t_0, t_f]$ will all be given in the following variational formulation

$$\gamma^2 = \|\Gamma\| = \max_{\mathbf{w}}(\|\Gamma\mathbf{w}\|/\|\mathbf{w}\|), \quad \text{where } \|\mathbf{w}\| = \int(\mathbf{w}^T\mathbf{w}/2)\mathrm{d}\tau, \ [\mathbf{W} = \mathbf{I}] \tag{6.8.6}$$

The norm of the disturbance $\mathbf{w}$ has been clarified, but for the initial value problem of filtering there is the contribution from the disturbance of initial condition. The norm of the system response $\Gamma\mathbf{w}$ should also be clarified. Since the system response depends both on the intervals of control and on filtering, and for these two intervals the norm will be clarified at the problem description. Obviously, the system response depends on both the optimal selection of the control vector $\mathbf{u}$ for future time interval, and on the causal operator K of filter for past time interval. After all, the system response $\Gamma\mathbf{w}$ is represented by the output vector $\mathbf{z}$, hence $\|\Gamma\mathbf{w}\|$ can also be written as $\|\Gamma_{zw}\|$ to explain the factor of input-output.

The variational expression (6.8.6) is homogeneous for the disturbance $\mathbf{w}$ both at the denominator and numerator. Such expression corresponds to that in the figure 6.10(b) the external disturbance $\mathbf{w}_e$ is zero, which means that the disturbance $\mathbf{w}$ is caused by $\mathbf{A}_\Delta$. Hence the critical norm γ_{cr}^2 calculated according to equation (6.8.6) is practically an upper bound. This upper bound γ_{cr}^2 means that the system is only excited by the disturbance $\mathbf{w}$, which is solely due to the possible $\mathbf{A}_\Delta$ with no any other external disturbance $\mathbf{w}_e$. As the parameter γ^2 exceeds this upper bound γ_{cr}^2, the system will not be unstable due to the excitation of disturbance $\mathbf{w}$. In other words, in case of $\gamma_{\mathrm{cr}}^2 < \gamma^2$, the system can withstand the external disturbance $\mathbf{w}_e$, which implies that the system is stable. On the contrary, if the parameter γ^2 does not exceed this upper bound, then the system will generate non-zero solution even there is no external excitation $\mathbf{w}_e$, which means that the system is unstable. This upper bound γ_{cr}^2 of γ^2 means the upper bound of unstable. Otherwise, it can be said that γ_{cr}^2 is the lower bound of stability. Finding this bound γ_{cr}^2 is a key step of analysis that, as $\gamma_{\mathrm{cr}}^2 < \gamma^2$ the system is stable.

Let us begin with the robust control analysis for future time interval.

§6.8.1, *Analysis of* H_∞ *state feedback robust control* [92]

State feedback means that the measurement equation (6.8.4) is not required in the analysis. The feedback control can only be used to the future time interval $(t, t_f]$. The system input is $\mathbf{w}$ and system output is given by $\mathbf{z}$. The selection of control $\mathbf{u}$ should make the output vector $\mathbf{z}$ minimized. The output $\mathbf{z}$ is measured by the norm

$$\|\mathbf{z}\| = \int_t^{t_f} \mathbf{z}^T\mathbf{z}\mathrm{d}\tau/2 + \mathbf{x}_f^T\mathbf{S}_f\mathbf{x}_f/2, \quad \min_{\mathbf{u}}\|\mathbf{z}\| \tag{6.8.7}$$

where the closed interval end is used at the end t_f, which is represented by the latter term in the above equation and is seen in section 6.7.

The output $\mathbf{z}$ is excited by the disturbance $\mathbf{w}$. The norm of $\mathbf{w}$ is

$$\|\mathbf{w}\| = \int_t^{t_f} (\mathbf{w}^T\mathbf{w}/2)\mathrm{d}\tau, \text{ i.e. } [\mathbf{W} = \mathbf{I}]$$

as given in equation (6.8.6). According to the idea of robust control as discussed above, the excitation $\mathbf{w}$ should consider the most unfavorable selection, that the norm of output $\|\mathbf{z}\|$ must be maximized, so that

$$\gamma^2 = \|\mathbf{z}\| / \|\mathbf{w}\|, \qquad \max_{\mathbf{w}} \min_{\mathbf{u}} \gamma^2 = \gamma_{cr}^2 \tag{6.8.8}$$

which completes the formulation of mathematical model. The parameter γ^2 is called as ***Induced norm*** of the state feedback system Γ. The induced norm is extended from the norms $\|\mathbf{w}\|$ and $\|\mathbf{z}\|$ of the quadratic integrable functional space. This sort of norm induced by the 2-norm in L_2 space gives the H_∞ norm in control theory, which stands for the ***Hardy-space*** in functional analysis. Describing in the frequency domain, the system transfer function $\mathbf{G}(s)$ is analytic in the open-right-half plane, which means that the system is stable, see [50].

The extreme condition of the induced norm in equation (6.8.8) is a variational problem. Because the ***dynamic equation and the output equation*** need to be satisfied beforehand, hence the variational principle is ***conditional***. The equation (6.8.8) is rewritten as

$$J_c = \int_t^{t_f} \left(\mathbf{z}^T\mathbf{z}/2 - \gamma^2\mathbf{w}^T\mathbf{w}/2\right)\mathrm{d}\tau + \mathbf{x}_f^T\mathbf{S}_f\mathbf{x}_f/2, \quad \max_{\mathbf{w}} \min_{\mathbf{u}} J_c \tag{6.8.8'}$$

The constraint conditions are equation (6.8.2) and (6.8.3) and the initial condition

$$\mathbf{x}(\tau) = \bar{\mathbf{x}}(t), \quad \text{when} \quad \tau \to t \tag{6.8.9}$$

Substituting (6.8.3) into (6.8.8') and introducing the Lagrange parameter vector for the equation (6.8.2) gives

$$J_{cA} = \int_t^{t_f} \begin{bmatrix} \boldsymbol{\lambda}^T(\dot{\mathbf{x}} - \mathbf{A}\mathbf{x} - \mathbf{B}_w\mathbf{w} - \mathbf{B}_2\mathbf{u}) - \gamma^2\mathbf{w}^T\mathbf{w}/2 \\ +\mathbf{u}^T\mathbf{u}/2 + \mathbf{u}^T\mathbf{D}_{12}^T\mathbf{C}_z\mathbf{x} + \mathbf{x}^T\mathbf{C}_z^T\mathbf{C}_z\mathbf{x}/2 \end{bmatrix} \mathrm{d}\tau + \mathbf{x}_f^T\mathbf{S}_f\mathbf{x}_f/2, \tag{6.8.10'}$$

$$\delta J_{cA} = 0$$

J_{cA} is a functional of extended index, inside which there are four kinds of independent variables. Performing the maximization with respect to $\mathbf{w}$ and minimization with respect to the control vector $\mathbf{u}$ gives

$$\mathbf{w} = -\gamma^{-2}\mathbf{B}_w^T\boldsymbol{\lambda}, \quad \mathbf{u} = \mathbf{B}_u^T\boldsymbol{\lambda} - \mathbf{D}_{12}^T\mathbf{C}_z\mathbf{x} \tag{6.8.10''}$$

$$J_{cA} = \int_t^{t_f} \begin{bmatrix} \boldsymbol{\lambda}^T(\dot{\mathbf{x}} - \tilde{\mathbf{A}}\mathbf{x}) + \mathbf{x}^T\tilde{\mathbf{C}}^T\tilde{\mathbf{C}}\mathbf{x}/2 \\ -\boldsymbol{\lambda}^T(\mathbf{B}_u\mathbf{B}_u^T - \gamma^{-2}\mathbf{B}_w\mathbf{B}_w^T)\boldsymbol{\lambda}/2 \end{bmatrix} \mathrm{d}\tau + \mathbf{x}_f^T\mathbf{S}_f\mathbf{x}_f/2, \tag{6.8.10}$$

$$\tilde{\mathbf{A}} = \mathbf{A} - \mathbf{B}_u\mathbf{D}_{12}^T\mathbf{C}_z, \quad \tilde{\mathbf{C}}^T\tilde{\mathbf{C}} = \mathbf{C}_z^T(\mathbf{I}_p - \mathbf{D}_{12}\mathbf{D}_{12}^T)\mathbf{C}_z, \quad \delta J_{cA} = 0$$

where the dual vectors $\mathbf{x}$ and $\boldsymbol{\lambda}$ are the two kinds of independent variables and γ^{-2} is the parameter to be determined.

Performing the variational operation of the stationary value for J_{cA} gives

$$\dot{\mathbf{x}}(\tau) = \tilde{\mathbf{A}}\mathbf{x} + (\mathbf{B}_u\mathbf{B}_u^T - \gamma^{-2}\mathbf{B}_w\mathbf{B}_w^T)\boldsymbol{\lambda} \tag{6.8.11a}$$

$$\dot{\boldsymbol{\lambda}}(\tau) = \tilde{\mathbf{C}}^T\tilde{\mathbf{C}}\mathbf{x} - \tilde{\mathbf{A}}^T\boldsymbol{\lambda} \tag{6.8.11b}$$

The natural boundary condition derived at the end $\tau = t_f$ is

$$\boldsymbol{\lambda}_f = -\mathbf{S}_f\mathbf{x}_f \tag{6.8.12}$$

The solution of the dual equation (6.8.11a,b) uses the usual method, introducing

$$\boldsymbol{\lambda}(\tau) = -\mathbf{X}(\tau)\cdot\mathbf{x}(\tau) \tag{6.8.13}$$

then the dual differential equations are derived to the matrix Riccati differential equation and the state differential equation with boundary conditions

$$\dot{\mathbf{X}}(\tau) = -\tilde{\mathbf{C}}^T\tilde{\mathbf{C}} - \tilde{\mathbf{A}}^T\mathbf{X} - \mathbf{X}\tilde{\mathbf{A}} + \mathbf{X}(\mathbf{B}_u\mathbf{B}_u^T - \gamma^{-2}\mathbf{B}_w\mathbf{B}_w^T)\mathbf{X}, \quad \mathbf{X}(t_f) = \mathbf{S}_f \tag{6.8.14}$$

$$\dot{\mathbf{x}}(\tau) = [\tilde{\mathbf{A}} - (\mathbf{B}_u\mathbf{B}_u^T - \gamma^{-2}\mathbf{B}_w\mathbf{B}_w^T)\mathbf{X}(\tau)]\mathbf{x}, \quad \mathbf{x}(\tau = t) = \bar{\mathbf{x}}(t) \tag{6.8.15}$$

The equation (6.8.14) can compare with (6.7.10), and the equation (6.8.15) can compare with (6.7.13). The terms with γ^{-2} are additional, which is the characteristic of H_∞ robust control. The computation of γ_{cr}^{-2} is very important. In figure 6.10(b) let $\mathbf{w}_e = \mathbf{0}$, the term $\mathbf{B}_w\mathbf{w}$ becomes all from the uncertainty of $\mathbf{A}_\Delta$, *i.e.* the self-disturbance induced from the deviation of the plant matrix $\mathbf{A}$. The problem becomes the stability analysis under self-excitation. The pre-conditions of analysis are that $(\mathbf{A}, \mathbf{C}_z)$ is observable and $(\mathbf{A}, \mathbf{B}_u)$ is controllable. When $\gamma^{-2} \to 0$, the problem reduces to be LQ control, which means the factor of disturbance $\mathbf{A}_\Delta$ disappeared, that no model error certainly reduces to the LQ control. On the other hand, larger γ^{-2} means the effect of $\mathbf{A}_\Delta$ increased, but γ^{-2} cannot indefinitely increase that as γ^{-2} increases to its critical value γ_{cr}^{-2}, the Riccati differential equation (6.8.14) will not exist solution in the interval $(t, t_f]$, which means that the system becomes unstable. So that the computation of γ_{cr}^{-2} is necessary, and the precise integration method is quite useful still in this case.

The application of precise integration for the solution of Riccati differential equation has been given in section 6.5 in detail. However, there is further the parameter γ^{-2} to be selected in H_∞ robust control. Thus in the case of LQ control, under the condition of $\gamma^{-2} = 0$ the solution of Riccati needs only once, but in the case H_∞ control, the solution of the Riccati differential equation requires to be carried out for various values of parameter γ^{-2} iteratively.

Based on the analogy relationship between structural mechanics and optimal control, the statement of the ***induced norm*** (6.8.8) corresponds to the ***eigenvalue in structural mechanics***, *i.e.* the Euler's structural stability force or the fundamental frequency of structural vibration [92], is shown first below. Then the behavior of the critical parameter γ_{cr}^{-2} is made clearer. Based on this understanding, the precise integration computation is proposed for the optimal identification of the critical parameter γ_{cr}^{-2} of robust H_∞ control. Extended W-W algorithm [41] in eigenvalue computation in structural mechanics can be invoked in combination with the precise integration of the matrix Riccati differential equation to give the precise computation of $\gamma_{\rm cr}^{-2}$. Note that the Euler stability is for the finite interval $(t, t_{\rm f}]$, that this kind of stability is no longer the asymptotic stability as usually described in system stability theory. Note, in structural engineering, the Euler stability critical value $\gamma_{\rm cr}^{-2}$ is inappropriate for practical problems, which is a common understanding.

Similarly, the critical parameter $\gamma_{\rm cr}^{-2}$ can only supply a bound for control system design. In applications, only the ***sub-optimal parameter***

$$\gamma^{-2} < \gamma_{\rm cr}^{-2} \tag{6.8.16}$$

is used in system design. For sub-optimal parameter, the equations (6.8.14) and (6.8.15) still need to be solved, and the precise integration method can also be used for this two differential equations. For the case of Jordan normal form does not appear, the analytical method is applicable too, otherwise the precise integration method can be used.

The bound (6.8.16) of parameter γ^{-2} is extremely important, and the bound $\gamma_{\rm cr}^{-2}$ is just the ***Rayleigh quotient*** [92,93], and is the main concern in the text below.

§6.8.1.1, Extended Rayleigh quotient

The variational principle (6.8.10) can be rewritten in the form

$$\delta(\Pi_1 - \gamma^{-2}\Pi_2) = 0 \tag{6.8.17}$$

where

$$\Pi_1 = \int_t^{t_{\rm f}} [-\boldsymbol{\lambda}^T\dot{\mathbf{x}} + \boldsymbol{\lambda}^T\widetilde{\mathbf{A}}\mathbf{x} + \boldsymbol{\lambda}^T\mathbf{B}_u\mathbf{B}_u^T\boldsymbol{\lambda}/2 - \mathbf{x}^T\widetilde{\mathbf{C}}^T\widetilde{\mathbf{C}}\mathbf{x}/2]\mathrm{d}\tau - \mathbf{x}_{\rm f}^T\mathbf{S}_{\rm f}\mathbf{x}_{\rm f}/2 \tag{6.8.18a}$$

$$\Pi_2 = \int_t^{t_{\rm f}} [\boldsymbol{\lambda}^T\mathbf{B}_w\mathbf{B}_w^T\boldsymbol{\lambda}/2]\mathrm{d}\tau \tag{6.8.18b}$$

This problem is not necessarily always solvable. When the initial condition is given as $\mathbf{x}(\tau = t) = \mathbf{0}$, for an arbitrarily selected parameter γ^{-2}, it has only the trivial solution $\mathbf{x}(\tau) = \mathbf{0}$. Only when the parameter holds a special value $\gamma^{-2} = \gamma_{\rm cr}^{-2}$, the variational problem (6.8.17) has non-trivial solution. In such case the variational equation (6.8.17) can also be written as

$$\gamma^{-2} = \Pi_1/\Pi_2 \quad , \quad \gamma_{\rm cr}^{-2} = \min_{\boldsymbol{\lambda}} \max_{\mathbf{x}}(\Pi_1/\Pi_2) \tag{6.8.19}$$

This is a Rayleigh quotient. The traditional Rayleigh quotient is for one kind of variable, but the variational principle (6.8.19) has two kinds of independent variables $\mathbf{x}$ and $\boldsymbol{\lambda}$, hence it is an ***extended Rayleigh quotient*** [41].

From equation (6.8.18b), the functional Π_2 is non-negative and it must be positive at the real solution. Then examining the functional Π_1, because Π_2 is independent on $\mathbf{x}$, hence in Π_1 the maximization with respect to the vector $\mathbf{x}$ can be performed first, it derives to

$$\Pi_1 = [-\boldsymbol{\lambda}^T\mathbf{x}]_t^{t_f} + \int_t^{t_f} \begin{bmatrix} \mathbf{x}^T(\dot{\boldsymbol{\lambda}} + \tilde{\mathbf{A}}^T\boldsymbol{\lambda} - \tilde{\mathbf{C}}^T\tilde{\mathbf{C}}\mathbf{x}) \\ +\boldsymbol{\lambda}^T\mathbf{B}_u\mathbf{B}_u^T\boldsymbol{\lambda}/2 + \mathbf{x}^T\tilde{\mathbf{C}}^T\tilde{\mathbf{C}}\mathbf{x}/2 \end{bmatrix} \mathrm{d}\tau - \mathbf{x}_f^T\mathbf{S}_f\mathbf{x}_f/2$$

$$= \int_t^{t_f} \left[\boldsymbol{\lambda}^T\mathbf{B}_u\mathbf{B}_u^T\boldsymbol{\lambda}/2 + \mathbf{x}^T\tilde{\mathbf{C}}^T\tilde{\mathbf{C}}\mathbf{x}/2\right]\mathrm{d}\tau + \mathbf{x}_f^T\mathbf{S}_f\mathbf{x}_f/2$$

where both the initial condition $\mathbf{x}(t) = \mathbf{0}$ and the natural boundary condition at the finish time (6.8.12) as well as the differential equation (6.8.11b) are satisfied. The above equation also verifies the positive definiteness of Π_1.

Further, if the matrix $\tilde{\mathbf{C}}^T\tilde{\mathbf{C}}$ is positive definite, then from (6.8.11b) derives

$$\mathbf{x} = (\tilde{\mathbf{C}}^T\tilde{\mathbf{C}})^{-1}(\dot{\boldsymbol{\lambda}} + \tilde{\mathbf{A}}^T\boldsymbol{\lambda}) \tag{6.8.20}$$

Substituting back into the integral expression of Π_1 gives

$$\Pi_1 = \int_t^{t_f} \begin{bmatrix} \dot{\boldsymbol{\lambda}}^T(\tilde{\mathbf{C}}^T\tilde{\mathbf{C}})^{-1}\dot{\boldsymbol{\lambda}}/2 + \dot{\boldsymbol{\lambda}}^T\tilde{\mathbf{A}}^T\boldsymbol{\lambda} \\ +\boldsymbol{\lambda}^T(\mathbf{B}_u\mathbf{B}_u^T + \tilde{\mathbf{A}}(\tilde{\mathbf{C}}^T\tilde{\mathbf{C}})^{-1}\tilde{\mathbf{A}}^T)\boldsymbol{\lambda}/2 \end{bmatrix} \mathrm{d}\tau + \boldsymbol{\lambda}_f^T\mathbf{S}_f^{-1}\boldsymbol{\lambda}_f/2 \tag{6.8.21}$$

where $\mathbf{x}_f$ is also substituted with the condition (6.8.12). Therefore the extended Rayleigh quotient variational principle (6.8.19) becomes with one kind of variables $\boldsymbol{\lambda}$. This form of variational principle is typically the usual Rayleigh quotient. However, for a general control system the matrix $\tilde{\mathbf{C}}$ is $q \times n$ dimensional and $\tilde{\mathbf{C}}^T\tilde{\mathbf{C}}$ cannot ensure positive definiteness, hence the extended Rayleigh quotient functional (6.8.19) is necessary. The conditions of $(\mathbf{A},\mathbf{B}_u)$ being controllable and $(\mathbf{A},\mathbf{C}_z)$ being observable ensure that the extended Rayleigh quotient is positive definite.

Rayleigh quotient is fundamental to structural stability and vibration, that it represents the Euler critical load or natural vibration frequency. The important statement is that the ***controllability and observability ensure that the analogy corresponding strain energy is positive definite in structural mechanics***. Therefore the eigenvalue must be positive. Theory of H_∞ control required parameter γ_{cr}^{-2} is the smallest eigenvalue, that only the γ^{-2} satisfying (6.8.16) is applicable for the control system. This situation is just the same to the Euler load P_{cr} of structural stability, that the structure can withstand only the external load P less than P_{cr}. The analogy of both sides is beneficial for understanding.

One problem should be pointed out here for the future time interval, the running time τ varies from the initial time t to t_f, but the initial time t is also running from t_0 to t_f. The eigenvalue γ_{cr}^{-2} is thus a function of the interval length $t_f - t$, and the lowest minimum eigenvalue γ_{cr}^{-2} appears at $t = t_0$. Therefore, in computing γ_{cr}^{-2}, the initial time should use t_0.

For the eigenvalue γ_{cr}^{-2}, the dual differential equations (6.8.11a~b) with boundary conditions (6.8.12) and $\mathbf{x}(t_0) = \mathbf{0}$ exist a nontrivial solution. This is a typical feature of Rayleigh quotient eigen-solution. For a given parameter $\gamma_\#^{-2}$, if $\gamma_\#^{-2} < \gamma_{cr}^{-2}$ then for arbitrary initial time t in the interval $(t_0, t_f]$, the solution of

Riccati equation (6.8.14) exists, but as $\gamma_{\#}^{-2} = \gamma_{\text{cr}}^{-2}$ then the solution matrix $\mathbf{X}(\tau)$ tends to be singular but its element value tends to infinity at the $\tau = t_0$ end. When $\gamma_{\#}^{-2} > \gamma_{\text{cr}}^{-2}$, then the solution does not exist for the whole interval $(t_0, t_f]$. However, as the initial instant t is close to t_f, the solution of Riccati equation must exist. Then the idea of ***conjugate point*** [50] is mentioned here, as $\gamma_{\#}^{-2} > \gamma_{\text{cr}}^{-2}$ there is a conjugate time $t_{\#}$, $t_0 < t_{\#} < t_f$, such that solution matrix of the Riccati equation tends to be singular at the $t = t_{\#}$ end.

The conjugate point applies only to the fundamental eigenvalue of the Rayleigh quotient, because it uses the solution of Riccati differential equation. For a Rayleigh quotient, there are a number of eigen-solutions, which is quite useful for perturbation analysis or for sub-structural modal synthesis etc. Thus conjugate point analysis is not enough for finding all the eigen-solutions for a given parameter $\gamma_{\#}^{-2} > \gamma_{\text{cr}}^{-2}$. The interval mixed energy and eigenvalue count method can be used to solve such problem.

§6.8.1.2, Interval mixed energy

The interval of the variational principle of control is $(t, t_f]$. For the sake of solving the Riccati differential equation by means of precise integration method, the interval mixed energy is introduced. For two time instants t_a and t_b, for which $t \le t_a < t_b \le t_f$, the time interval (t_a, t_b) is composed. If the state vector $\mathbf{x}_a = \mathbf{x}(t_a)$ is given at t_a and the dual vector $\boldsymbol{\lambda}_b = \boldsymbol{\lambda}(t_b)$ is given at t_b, then the vectors $\mathbf{x}(\tau), \boldsymbol{\lambda}(\tau)$ in the time interval are completely determined. The interval mixed energy is defined as

$$V(\mathbf{x}_a, \boldsymbol{\lambda}_b) = \boldsymbol{\lambda}_b^T \mathbf{x}_b - \int_{t_a}^{t_b} [\boldsymbol{\lambda}^T \dot{\mathbf{x}} - \mathrm{H}(\mathbf{x}, \boldsymbol{\lambda})] \mathrm{d}\tau \tag{6.8.22}$$

with

$$\mathrm{H}(\mathbf{x}, \boldsymbol{\lambda}) = \boldsymbol{\lambda}^T \tilde{\mathbf{A}} \mathbf{x} + \boldsymbol{\lambda}^T (\mathbf{B}_u \mathbf{B}_u^T - \gamma^{-2} \mathbf{B}_w \mathbf{B}_w^T) \boldsymbol{\lambda} / 2 - \mathbf{x}^T \tilde{\mathbf{C}}^T \tilde{\mathbf{C}} \mathbf{x} / 2 \tag{6.8.23}$$

Obviously, the mixed energy $V(\mathbf{x}_a, \boldsymbol{\lambda}_b)$ is a quadratic form of $\mathbf{x}_a$ and $\boldsymbol{\lambda}_b$, and the general form is

$$V(\mathbf{x}_a, \boldsymbol{\lambda}_b) = \boldsymbol{\lambda}_b^T \mathbf{F} \mathbf{x}_a + \mathbf{x}_a^T \mathbf{Q} \mathbf{x}_a / 2 - \boldsymbol{\lambda}_b^T \mathbf{G} \boldsymbol{\lambda}_b / 2 \tag{6.8.24}$$

where $\mathbf{Q}(t_a, t_b)$, $\mathbf{G}(t_a, t_b)$ and $\mathbf{F}(t_a, t_b)$ are $n \times n$ matrices, and $\mathbf{Q}^T = \mathbf{Q}$, $\mathbf{G}^T = \mathbf{G}$.

Two contiguous time intervals can be combined to form a longer time interval, see figure 6.5. The combination equations have been given in (6.5.77)

$$\mathbf{Q}_c = \mathbf{Q}_1 + \mathbf{F}_1^T (\mathbf{Q}_2^{-1} + \mathbf{G}_1)^{-1} \mathbf{F}_1 \tag{6.8.25}$$

$$\mathbf{G}_c = \mathbf{G}_2 + \mathbf{F}_2 (\mathbf{G}_1^{-1} + \mathbf{Q}_2)^{-1} \mathbf{F}_2^T \tag{6.8.26}$$

$$\mathbf{F}_c = \mathbf{F}_2 (\mathbf{I} + \mathbf{G}_1 \mathbf{Q}_2)^{-1} \mathbf{F}_1 \tag{6.8.27}$$

These equations can be used recursively. The matrices $\mathbf{Q}, \mathbf{G}, \mathbf{F}$ represent only the characteristics of the interval at the two ends, however the internal characteristic of the interval has not been determined. The internal eigenvalue γ_{cr}^{-2} count is to be supplied now, according to structural mechanics the eigenvalue count is necessary which can be supplied by means of the W-W algorithm, see chapter 2. However, the original W-W algorithm was derived for the case of given displacement boundary condition at the two ends, which corresponds to the variational principle with one kind of variables. Note that the original form of Rayleigh quotient is also for one kind of variables.

However, mixed energy is for the variational functional of dual vectors at the two ends, hence the ***extended W-W algorithm*** applies. These problems have been described in chapter 2. For present case, the vector $\boldsymbol{\lambda}_b$ should be regarded as the 'displacement' and the vector $\mathbf{x}_a$ is regarded as the 'internal force', the dual vectors. This analogy is the reverse direction of extended W-W algorithm presented in chapter 2. Reverse direction means that the left end is given the internal force and the right end of the interval is given the displacement. The problem is proposed as that for given parameter $\gamma_\#^{-2} = \omega_\#^2$, the inner eigenvalue $\omega^2 < \omega_\#^2$ count of the interval (t_a, t_b), with the two end vectors being given as $\mathbf{x}_a = \mathbf{0}$ and $\boldsymbol{\lambda}_b = \mathbf{0}$, is denoted as $J_R(\omega_\#^2)$, *i.e.* the number of eigenvalues of $\gamma_{cr}^{-2} < \gamma_\#^{-2}$. The count J_R is certainly a function of t_a and t_b, but for convenience it is not written as $J_R(\omega_\#^2, t_a, t_b)$.

Let $J_{R1}(\omega_\#^2)$ denote the eigenvalue count of interval-1 under the condition $\omega^2 < \omega_\#^2$, similarly for $J_{R2}(\omega_\#^2)$ and $J_{Rc}(\omega_\#^2)$, see figure 6.11. The eigenvalue count equation for interval combination is given as

$$J_{Rc}(\omega_\#^2) = J_{R1}(\omega_\#^2) + J_{R2}(\omega_\#^2) - s\{\mathbf{Q}_2\} + s\{\mathbf{G}_1 + \mathbf{Q}_2^{-1}\} \tag{6.8.28}$$

where $s\{\mathbf{M}\}$ represents that the number of negative entries in the diagonal matrix $\mathbf{D}$, which comes from the triangle factorization $\mathbf{M} = \mathbf{LDL}^T$ of the symmetric matrix $\mathbf{M}$.

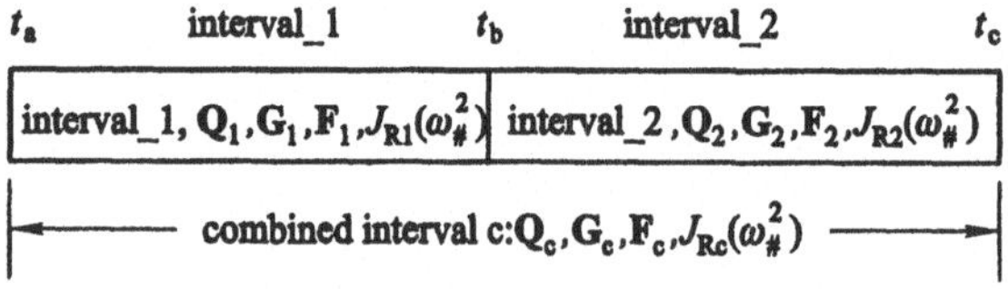

Figure 6.11, *Eigenvalue count for interval combination*

Therefore, the interval mixed energy representation for eigenvalue problem should be extended as $(\mathbf{Q}, \mathbf{G}, \mathbf{F}, J_R)$, that there are three matrices and an eigenvalue count, all of them are the functions of the two end-time instants t_a, t_b and the parameter $\gamma_\#^{-2} = \omega_\#^2$. The LQ optimal control or Kalman-Bucy filter problems correspond to the selection of $\gamma_\#^{-2} = \omega_\#^2 = 0$, so that the count J_R is definitely zero and is not mentioned in the mixed energy representation for LQ control or Kalman-Bucy filter. In H_∞ control theory, the eigenvalue count is a key parameter, so that it must be involved in the mixed energy representation. However, the interval mixed energy matrices are still written as $\mathbf{Q}(t_a,t_b)$ etc. that the parameter $\gamma_\#^{-2}$ usually does not express explicitly.

The above derivation applies also to time variant system. In the numerical computation for time invariant systems, the mixed energy representation $\mathbf{Q},\mathbf{G},\mathbf{F},J_R$ depends only on the interval length $\tau = t_b - t_a$, so that they can be written as

$$\mathbf{Q} = \mathbf{Q}(\tau),\quad \mathbf{G} = \mathbf{G}(\tau),\quad \mathbf{F} = \mathbf{F}(\tau),\ \text{where } \tau = t_b - t_a \tag{6.8.29}$$

They satisfy the differential equations below.

$$\begin{aligned} \mathrm{d}\mathbf{Q}/\mathrm{d}\tau &= \mathbf{F}^T\mathbf{C}^T\mathbf{C}\mathbf{F} \\ &= \mathbf{C}^T\mathbf{C} + \mathbf{A}^T\mathbf{Q} + \mathbf{Q}\mathbf{A} - \mathbf{Q}(\mathbf{B}_u\mathbf{B}_u^T - \gamma_\#^{-2}\mathbf{B}_w\mathbf{B}_w^T)\mathbf{Q} \end{aligned} \tag{6.8.30a}$$

$$\begin{aligned} \mathrm{d}\mathbf{G}/\mathrm{d}\tau &= (\mathbf{B}_u\mathbf{B}_u^T - \gamma_\#^{-2}\mathbf{B}_w\mathbf{B}_w^T) + \mathbf{A}\mathbf{G} + \mathbf{G}\mathbf{A}^T - \mathbf{G}\mathbf{C}^T\mathbf{C}\mathbf{G} \\ &= \mathbf{F}(\mathbf{B}_u\mathbf{B}_u^T - \gamma_\#^{-2}\mathbf{B}_w\mathbf{B}_w^T)\mathbf{F}^T \end{aligned} \tag{6.8.30b}$$

$$\mathrm{d}\mathbf{F}/\mathrm{d}\tau = \mathbf{F}\left[\mathbf{A} - (\mathbf{B}_u\mathbf{B}_u^T - \gamma_\#^{-2}\mathbf{B}_w\mathbf{B}_w^T)\mathbf{Q}\right] \quad = (\mathbf{A} - \mathbf{G}\mathbf{C}^T\mathbf{C})\mathbf{F} \tag{6.8.30c}$$

where the matrices $\tilde{\mathbf{A}},\tilde{\mathbf{C}}$ are written as $\mathbf{A},\mathbf{C}$ etc. The dependence of matrices $\mathbf{Q},\mathbf{G},\mathbf{F}$ on the original system matrices $\mathbf{A},\mathbf{B}_u,\mathbf{B}_w,\mathbf{C}$ can be seen from the above equations. The initial conditions of these differential equations are

$$\mathbf{Q} = \mathbf{0},\quad \mathbf{G} = \mathbf{0},\quad \mathbf{F} = \mathbf{I}_n,\quad J_R = 0,\quad \text{when } \tau = 0 \tag{6.8.31}$$

The derivation of these differential equations is almost the same as those in section 6.5. If select $t_b = t_f$, the differential equation (6.8.30a) is the same as the Riccati differential equation (6.8.14), since $\mathrm{d}/\mathrm{d}\tau = -\partial/\partial t_a$. But the boundary condition (6.8.31) is different to that given in equation (6.8.14). Similar to the method in section 6.5, the matrices $\mathbf{Q},\mathbf{G},\mathbf{F}$ and the eigenvalue count J_R for the grid points of the open time interval (t_0,t_f) can be computed first, then compute all the grid points t in the whole interval $(t_0,t_f]$ by executing

$$\mathbf{S} = \mathbf{Q} + \mathbf{F}^T(\mathbf{S}_f^{-1} + \mathbf{G})^{-1}\mathbf{F} \tag{6.8.32}$$

which gives the solution matrix of Riccati differential equation.

But these computed grid points are discrete, solely based on the matrices $\mathbf{S}$ of these discrete points are not enough to make sure that if there has been no any conjugate point within the whole interval $(t_0,t_f]$. It cannot make sure the number of eigenvalues of the Rayleigh quotient (6.8.19). To find the number of eigenvalues, one can compute the quadruple $(\mathbf{Q},\mathbf{G},\mathbf{F},J_R)$ for the open time interval (t_0,t_f) first, then compute the eigenvalue count of the whole time interval $(t_0,t_f]$ by the equation

$$J_{Rf} = J_R - s\{\mathbf{S}_f\} + s\{\mathbf{S}_f^{-1} + \mathbf{G}\} \tag{6.8.33}$$

If $J_{Rf} = 0$ then $\gamma_\#^{-2}$ is a sub-optimal parameter, *i.e.* within the whole time interval $(t_0,t_f]$ there has no singular point. Based on this criterion, a searching method, such as the bisection method, for the parameter $\gamma_\#^{-2}$, is proposed, so as to solve the eigenvalue γ_{cr}^{-2} to arbitrary assigned precision. Such method is the same to the application of the W-W algorithm in the elastic stability or natural vibration problems in structural mechanics.

§6.8.1.3, Precise integration

The integration of the governing equation of H_∞ control is similar to that in LQ control. For a given parameter $\gamma_{\#}^{-2}$, the solution of the non-linear Riccati differential equation (6.8.14) is solved first, then to solve the state differential equation (6.8.15). Numerical integration needs a given time step η, which should be a reasonable step size. When η is selected, the very small time step size τ is selected again (we have seen a number of times) as

$$\tau = \eta / 2^N \quad \text{Such as selecting } N = 20, \; 2^N = 1048576 \tag{6.8.34}$$

For such a small time step τ, using the Taylor series expansion gives very precise numerical result.

The precise integration method uses the time interval mixed energy combination in the integration algorithm. First, an interval mixed energy of an initial interval τ is necessary. The set (6.8.30) gives the differential equation with initial condition (6.8.31). Because size τ is very small, so the power series expansion is applied and truncated after τ^4, which gives

$$\mathbf{F}'(\tau) = \mathbf{f}_1\tau + \mathbf{f}_2\tau^2 + \mathbf{f}_3\tau^3 + \mathbf{f}_4\tau^4 + O(\tau^5), \quad \mathbf{F}(\tau) = \mathbf{I} + \mathbf{F}'(\tau) \tag{6.8.35a}$$

$$\mathbf{G}(\tau) = \mathbf{g}_1\tau + \mathbf{g}_2\tau^2 + \mathbf{g}_3\tau^3 + \mathbf{g}_4\tau^4 + O(\tau^5) \tag{6.8.35b}$$

$$\mathbf{Q}(\tau) = \mathbf{e}_1\tau + \mathbf{e}_2\tau^2 + \mathbf{e}_3\tau^3 + \mathbf{e}_4\tau^4 + O(\tau^5) \tag{6.8.35c}$$

These series expansions have satisfied the initial condition (6.8.31). Substituting into the differential equations (6.8.30) and comparing the coefficients of various powers of τ derives

$$\begin{aligned}
&\mathbf{e}_1 = \mathbf{C}^T\mathbf{C}, \qquad \mathbf{g}_1 = (\mathbf{B}_u\mathbf{B}_u^T - \gamma_{\#}^{-2}\mathbf{B}_1\mathbf{B}_1^T), \quad \mathbf{f}_1 = \mathbf{A} \\
&\mathbf{e}_2 = (\mathbf{f}_1^T\mathbf{e}_1 + \mathbf{e}_1\mathbf{f}_1)/2, \quad \mathbf{g}_2 = (\mathbf{A}\mathbf{g}_1 + \mathbf{g}_1\mathbf{A}^T)/2, \quad \mathbf{f}_2 = (\mathbf{A}\mathbf{f}_1 - \mathbf{g}_1\mathbf{e}_1)/2 \\
&\mathbf{e}_3 = (\mathbf{f}_2^T\mathbf{e}_1 + \mathbf{e}_1\mathbf{f}_2 + \mathbf{f}_1^T\mathbf{e}_1\mathbf{f}_1)/3, \quad \mathbf{g}_3 = (\mathbf{A}\mathbf{g}_2 + \mathbf{g}_2\mathbf{A}^T - \mathbf{g}_1\mathbf{e}_1\mathbf{g}_1)/3 \\
&\mathbf{f}_3 = (\mathbf{A}\mathbf{f}_2 - \mathbf{g}_2\mathbf{e}_1 - \mathbf{g}_1\mathbf{e}_1\mathbf{f}_1)/3, \quad \mathbf{e}_4 = (\mathbf{f}_3^T\mathbf{e}_1 + \mathbf{e}_1\mathbf{f}_3 + \mathbf{f}_2^T\mathbf{e}_1\mathbf{f}_1 + \mathbf{f}_1^T\mathbf{e}_1\mathbf{f}_2)/4 \\
&\mathbf{g}_4 = (\mathbf{g}_3\mathbf{A}^T + \mathbf{A}\mathbf{g}_3 - \mathbf{g}_2\mathbf{e}_1\mathbf{g}_1 - \mathbf{g}_1\mathbf{e}_1\mathbf{g}_2)/4, \\
&\mathbf{f}_4 = (\mathbf{A}\mathbf{f}_3 - \mathbf{g}_3\mathbf{e}_1 - \mathbf{g}_2\mathbf{e}_1\mathbf{f}_1 - \mathbf{g}_1\mathbf{e}_1\mathbf{f}_2)/4
\end{aligned} \tag{6.8.36}$$

The computation of these equations can be executed successively with no iteration, where $\mathbf{e}_i, \mathbf{g}_i, \mathbf{f}_i, (i = 1 \sim 4)$ are all $n \times n$ matrices, and $\mathbf{e}_i^T = \mathbf{e}_i, \mathbf{g}_i^T = \mathbf{g}_i$.

The matrices $\mathbf{Q}(\tau), \mathbf{G}(\tau), \mathbf{F}'(\tau)$ computed from equations (6.8.35) and (6.8.36) correspond to $J_R(\tau) = 0$, because τ is extremely small. The mixed energy expression has been obtained for the interval τ and it can be used as the initial interval for the 2^N combination algorithm.

The equations derived before are all exact, except the Taylor expansion (6.8.35) truncated after the τ^4 term, for which the ratio of the error term to the first term is of the order of $O(\tau^4)$. Because of $\tau^4 = (\eta/2^N) \approx \eta^4 \cdot 10^{-24}$, the relative error has been beyond the double precision of real number of 10^{-16}. So this approximation has really reached the double precision of computer real word.

Having determined the mixed energy of initial time interval τ, the interval combination algorithm (6.5.77) and (6.8.28) for two equal length intervals should be executed N times so as to obtain the mixed energy representation $(\mathbf{Q},\mathbf{G},\mathbf{F},J_R)$ of the time interval η. Special attention is needed that in the combination algorithm (6.8.27), the addition $\mathbf{I}+\mathbf{F}'$ in (6.8.35a) should not be executed. Because τ is very small, so is $\mathbf{F}'$ that the addition will be seriously decreasing the computational precision. To free from such problem the equations

$$\mathbf{Q}_c = \mathbf{Q} + (\mathbf{I}+\mathbf{F}')^T(\mathbf{Q}^{-1}+\mathbf{G})^{-1}(\mathbf{I}+\mathbf{F}') \tag{6.8.37a}$$

$$\mathbf{G}_c = \mathbf{G} + (\mathbf{I}+\mathbf{F}')(\mathbf{G}^{-1}+\mathbf{Q})^{-1}(\mathbf{I}+\mathbf{F}')^T \tag{6.8.37b}$$

$$\mathbf{F}'_c = \mathbf{F}'(\mathbf{I}+\mathbf{G}\mathbf{Q})^{-1} + (\mathbf{I}+\mathbf{G}\mathbf{Q})^{-1}\mathbf{F}' + \mathbf{F}'(\mathbf{I}+\mathbf{G}\mathbf{Q})^{-1}\mathbf{F}' \tag{6.8.37c}$$

are used instead of the three equations in (6.5.77). Note that the equations (6.8.37a~c) are used for equal length small intervals.

Up to here, the equations for the precise integration have been available. In LQ control theory, the solution matrix of Riccati differential equation is always positive definite. For finite time interval $(t_0,t_f]$, after the execution of equations (6.8.32~33), the condition $J_{Rc}=0$ means that $\gamma_{\#}^{-2}$ is a sub-optimal parameter. For infinite time interval problem, if the appearing of condition $J_R>0$ means that the parameter $\gamma_{\#}^{-2}$ is too large and is not sub-optimal.

If $\gamma_{\#}^{-2}=0$ is selected, the problem reduces to be LQ control. For a controllable and observable system the matrices $\mathbf{Q},\mathbf{G}$ must be positive definite, hence monitoring the count J_R is unnecessary. But when $\gamma_{\#}^{-2}>0$, the matrix $\mathbf{G}$ may be indefinite, hence monitoring the count J_R is necessary until the last step of (6.8.33). Since $\gamma_{\#}^{-2}>0$, the matrix $\mathbf{G}$ may still be indefinite, for which $J_R>0$. The algorithm is given below.

§6.8.1.4, Algorithm

The algorithm for optimal parameter γ_{cr}^{-2} is the bisection iterative solution for the given parameter $\gamma_{\#}^{-2}$. The iteration algorithm is proposed as follows. The first part is precise integration algorithm for the fundamental interval η

[Give the dimension n, the matrices $\mathbf{A},\mathbf{B}_w,\mathbf{B}_u,\mathbf{C},\mathbf{D}$ and $\mathbf{S}_f$ at boundary, the time interval $(0,t_f]$, selecting a parameter $\gamma_{\#}^{-2}$]

[Computing $\mathbf{C}^T\mathbf{C}$ and $(\mathbf{B}_u\mathbf{B}_u^T - \gamma_{\#}^{-2}\mathbf{B}_w\mathbf{B}_w^T)$, selecting step size η. Let $N=20$ and $\tau=\eta/2^N$]

[According to (6.8.35) and (6.8.36) compute $\mathbf{Q}(\tau),\mathbf{G}(\tau),\mathbf{F}'(\tau)$; and $J_R = 0$;]
for ($iter = 0$; $iter < N$; $iter$ ++) {
    [Triangular factorize $\mathbf{G}$ and $(\mathbf{Q}+\mathbf{G}^{-1})$; J_{Rc} is obtained from (6.8.33);]
    [According to (6.8.37) computes $\mathbf{Q}_c,\mathbf{G}_c,\mathbf{F}_c'$]
    [Let $\mathbf{Q}_c,\mathbf{G}_c,\mathbf{F}_c'$ and $J_R = J_{Rc}$;]
}
[Let $\mathbf{F} = \mathbf{F}_c = \mathbf{I} + \mathbf{F}'$;]
Comment: $(\mathbf{Q},\mathbf{G},\mathbf{F},J_R)$ has been of the interval η (6.8.38)

Continued, two cases of infinite time interval and finite time interval are considered separately. For infinite time interval, the algorithm is

[Select an initial parameter $\gamma_\#^{-2}$]
while (the precision of $\gamma_\#^{-2}$ is not enough) {
    while ($\|\mathbf{F}_c\| > \varepsilon$) {
        [Let $\mathbf{Q}_1 = \mathbf{Q}_2 = \mathbf{Q}_c;\mathbf{G}_1 = \mathbf{G}_2 = \mathbf{G}_c;\mathbf{F}_1 = \mathbf{F}_2 = \mathbf{F}_c;J_{R1} = J_{R2} = J_{Rc};$]
        [From (6.5.77) compute $\mathbf{Q}_c,\mathbf{G}_c,\mathbf{F}_c$ and from (6.8.28) compute J_{Rc} ;]
        if ($J_{Rc} > 0$) { $\gamma_\#^{-2}$ is too large; Break;}
    }
    if ($J_R == 0$) [$\gamma_\#^{-2}$ is H_∞ sub-optimal, increase $\gamma_\#^{-2}$] else [reduce $\gamma_\#^{-2}$]
    [Change the revision size of $\gamma_\#^{-2}$, which gives the precision]
} (6.8.39)

For case of finite time interval, let the grid be $t_0 = 0,\cdots,t_k = k\eta,\cdots,t_f = k_f\eta$. Assuming $k_f = 2^{N_1}$, where N_1 is a given number such as 4 or 5, then after (6.8.38) executes

[Select an initial parameter $\gamma_\#^{-2}$]
while (the precision of $\gamma_\#^{-2}$ is not enough) {
    for ($iter = 0$; $iter < N_1$; $iter$ ++) {
        [Let $\mathbf{Q}_1 = \mathbf{Q}_2 = \mathbf{Q}_c;\mathbf{G}_1 = \mathbf{G}_2 = \mathbf{G}_c;\mathbf{F}_1 = \mathbf{F}_2 = \mathbf{F}_c;J_{R1} = J_{R2} = J_{Rc};$]
        [Based on (6.5.77) compute $\mathbf{Q}_c,\mathbf{G}_c,\mathbf{F}_c$ and from (6.8.28) compute J_{Rc}]
    }
    [Let $\mathbf{Q} = \mathbf{Q}_c;\mathbf{G} = \mathbf{G}_c;\mathbf{F} = \mathbf{F}_c;J_R = J_{Rc};$ execute (6.8.32), (6.8.33);]
    if ($J_R == 0$) [$\gamma_\#^{-2}$ is H_∞ sub-optimal, increase $\gamma_\#^{-2}$] else [reduce $\gamma_\#^{-2}$]
    [Change the size of revision of $\gamma_\#^{-2}$, which gives the precision]
} (6.8.40)

Such as using the iteration method of bisection for $\gamma_{\#}^{-2}$, until enough precision is reached. That is the upper and lower bound iteration method for the least eigenvalue of the Rayleigh quotient.

To show the effectiveness of the algorithm, two numerical examples are given below. The lower bound of the least eigenvalue will be selected as the sub-optimal parameter. In the algorithm, very close upper and lower bounds are given for the least eigenvalue γ_{cr}^{-2} up to six significant digits. The eigenvalue γ_{cr}^{-2} is called the optimal parameter of H_∞ control, and it is often written as γ_{opt}^{-2}. Correspondingly, the solution matrix of Riccati differential equation has the characteristic of tending to be singular at the end point with very large value. Such ill-conditioned solution matrix with its element increased indefinitely is unrealistic to generate the gain matrix and the control vector. The H_∞ theory emphasizes the robustness of system, the H_∞ control optimality has concentrated only on the eigenvalue and the computation corresponds to the direct problem in structural mechanics.

From system design consideration, the output related matrices $\mathbf{C}_z, \mathbf{D}_{12}, \mathbf{C}_y, \mathbf{D}_{21}$ and the input related matrices $\mathbf{B}_w, \mathbf{B}_u$ etc. all have parameters to be selected, which correspond to the problems of system synthesis and optimization. This consideration corresponds to the problem of structural optimization in structural mechanics.

It should be emphasized that the Euler stability critical value is only a bound, which cannot be used in practice. When facing the stability problem in structural engineering, the allowable loading factor must be reduced from the critical value with a large factor.

Example 6.12: A one-dimensional problem is proposed in order to compare with the analytical solution.

Let $n=1$, $\mathbf{A}=0.8$; $\mathbf{B}_w=0.8$; $\mathbf{B}_u=3.0$; $\mathbf{C}=0.8$; $\mathbf{S}_f=0.01$, and the time interval is given as $(0,t_f)=(0,0.8)$. The optimal parameter γ_{cr}^{-2} and the corresponding solution of the Riccati differential equation are required.

Solution: Substituting the original data into the Riccati differential equation gives

$$\dot{\mathbf{X}}(t)=-0.64-1.6\mathbf{X}+(9.0-0.64\gamma_{\#}^{-2})\mathbf{X}^2, \quad \mathbf{X}(0.8)=0.01;$$

where $\mathbf{X}(t)$ is a scalar function, for which the analytical solution is easy to find.

Using precise integration method in combination with the extended W-W algorithm and using the bisection method searching the optimal parameter γ_{cr}^{-2}, after 20 times iterations gives

Lower bound=19.198742; *Upper bound*=19.198751;

From which, the lower bound parameter is used as the optimal parameter, which gives γ_{cr}^{-2}=19.198742. For this γ_{cr}^{-2}, the numerical values of the analytical solution and the precise integration solution of the Riccati differential equation are listed in the following table 6.3.

Table 6.3, The numerical results of an 1-D example. Comment: **16-div** and **8-div** represent the precise integration results of 16 or 8 uniformly subdivided intervals, respectively.

Time=	0.75	0.70	0.65	0.60	0.55	0.50
Analytic	.0442861	.0819794	.1239372	.1712935	.2255860	.2889585
16-Div.	.0442861	.0819794	.1239372	.1712935	.2255860	.2889585
8-Div.		.0819794		.1712935		.2889585

Continue

Time=	0.45	0.40	0.35	0.30	0.25	0.20
Analytic	.3645008	.4568356	.5732203	.7257400	.9361363	1.247880
16-Div.	.3645008	.4568356	.5732203	.7257400	.9361363	1.247880
8-Div.		.4568356		.7257400		1.247880

Continue

Time=	0.15	0.10	0.05	0.0
Analytic	1.762386	2.783875	5.833399	**1576124**
16-Div.	1.762386	2.783875	5.833399	**1562431**
8-Div.		2.783875		

It is found in the table, that the numerical results of analytical and precise integration solutions are the same. Only the last significant digit, the eight-th, may have difference of 1, which can be interpreted coming from rounding error. The left end of the interval, 0.0 is a singular point, only at this point the numerical results of precise integration and analytical solutions have some difference. However, the solution matrix should tend to infinity at the singular point, but because the parameter $\gamma_{\rm cr}^{-2}$ is selected less than the exact value an extremely small value, so that the solution becomes a very large value, but this value itself is again imprecise. ##

Example 6.13: Selecting n=4 with the matrices

$$\mathbf{A}=\begin{bmatrix}-0.08 & -3.0 & 0.0 & 2.0\\ 0.04 & -1.4 & 10.0 & 0.0\\ -0.01 & -4.0 & -2.8 & 0.0\\ 0.0 & 0.0 & 1.0 & 0.0\end{bmatrix};\ \mathbf{B}_w=\begin{bmatrix}0.0\\ 0.0\\ 0.2\\ .001\end{bmatrix},\ \mathbf{B}_u=\begin{bmatrix}0.0\\ 0.0\\ -0.30\\ 0.0\end{bmatrix},$$

$$\mathbf{C}=\mathrm{diag}\begin{bmatrix}0.5 & 0.5 & 1.0 & 0.0\end{bmatrix}.$$

The solutions of infinite interval and finite interval are required.

Solution:

a) For infinite time interval, using the precise integration method in combination of the extended W-W algorithm and bisection searching the optimal parameter $\gamma_{\rm cr}^{-2}$ gives

Lower bound=2.17081; *Upper bound*=2.17082

The lower bound is used as $\gamma_{\rm cr}^{-2}$, correspondingly the solution matrix $\mathbf{S}_\infty$, as anticipated, is almost a singular matrix with very large elements.

b) For finite time interval $t_{\rm f}=4.5$ and $\mathbf{S}_{\rm f}=\mathrm{diag}\begin{bmatrix}0.1 & 0.1 & 0.1 & 0.1\end{bmatrix}$. The bounds of $\gamma_{\rm cr}^{-2}$ are

Lower bound=60.54692; *Upper bound*=60.54697;

The lower bound is used as $\gamma_{\rm cr}^{-2}$, and the corresponding solution matrix of Riccati equation will not be listed here to save space. As anticipated, the solution matrix $\mathbf{S}(0)$ at the end point is almost singular with very large numerical value elements. ##

§6.8.2, H_∞ *filtering*

The optimal filtering for the state vector is a necessary part of an optimal control system. As in LQG, the H_∞ robust control theory is composed of both the H_∞ filtering and the H_∞ state feedback control. The two theories are mutually dual problems. The control theory needs state feedback, which assumes that the whole state vector is measured precisely so that the state estimation is unnecessary, but it is almost impossible. Therefore usually the measurement feedback requires the filter to supply the state estimation, as that in LQG theory. In fact, the performance of a real system is always under some kind of dynamic random disturbance and the measurement is again under measuring noise. Hence the H_∞ filtering computation is one of the central part of the H_∞ robust control theory. The control is for the future time interval, and the filtering is for the analysis of past time interval. The two parts can be solved separately, and such kind of treatment is termed as the ***separation theory***. The future and past intervals link together at the present time instant t, and the linking condition gives the H_∞ feedback-control rule and also gives the boundary condition for the past interval state estimation. So the H_∞ filtering analysis is necessary.

Filter analysis is necessary for control analysis, but the application of filtering is not only for control. Hence robust filter itself is also important. The state space description of filter problem is

$$\dot{\mathbf{x}} = \mathbf{A}\mathbf{x} + \mathbf{B}_w\mathbf{w} + \mathbf{B}_u\mathbf{u} \tag{6.8.2}$$

$$\mathbf{z} = \mathbf{C}_z\mathbf{x} \qquad + \mathbf{D}_{12}\mathbf{u} \tag{6.8.3}$$

$$\mathbf{y} = \mathbf{C}_y\mathbf{x} + \mathbf{D}_{21}\mathbf{w} \tag{6.8.4}$$

which applies to the past time interval $[0,t)$. The vector $\mathbf{w}$ is a Gaussian stochastic white noise input, l-dimensioned, but for H_∞ filter the disturbance $\mathbf{w}$ is no longer a zero-mean vector, which involves both the dynamic and measurement noises with

$$\text{var}[\mathbf{w}(t), \mathbf{w}(t+\tau)] = \mathbf{W}\delta(\tau), \quad \text{with } \mathbf{W} = \mathbf{I} \tag{6.8.41}$$

The vector $\mathbf{u}$ is the control of past interval, and is only a given vector in filter analysis m-dimensioned. Because of the noise input, the state vector $\mathbf{x}(\tau)$ is also a Gaussian stochastic process n-dimensioned non-stationary vector with the initial vector $\mathbf{x}(0)$ being of mean value $\hat{\mathbf{x}}_0$ and variance matrix $\mathbf{P}_0$. $\mathbf{y}(\tau)$ is a q-dimensional measurement vector, and $\mathbf{z}$ is a p-dimensional output vector to be used for estimation. The matrices $\mathbf{A}, \mathbf{B}_w, \mathbf{B}_u, \mathbf{C}_y, \mathbf{C}_z, \mathbf{D}_{12}, \mathbf{D}_{21}$ have appropriate dimensions and are given matrices. The vectors $\mathbf{x}(0), \mathbf{w}(\tau)$ are uncorrelated to each other. $(\mathbf{A}, \mathbf{B}_1)$ is controllable and $(\mathbf{A}, \mathbf{C}_y)$ is observable. The linking condition at the present time $\tau = t$ can be derived from the variational principle of H_∞ control as a natural boundary condition.

The H_∞ filtering problem requires to find the linear causal operator F so that the estimation of the output vector

$$\hat{\mathbf{z}}(t) = \mathrm{F}(\mathbf{y}, \mathbf{u}) \tag{6.8.42}$$

makes the norm of the error $(\hat{\mathbf{z}}-\mathbf{C}_z\mathbf{x}-\mathbf{D}_{12}\mathbf{u})$ satisfying the following condition

$$\int_0^t(\hat{\mathbf{z}}-\mathbf{C}_z\mathbf{x}-\mathbf{D}_{12}\mathbf{u})^T(\hat{\mathbf{z}}-\mathbf{C}_z\mathbf{x}-\mathbf{D}_{12}\mathbf{u})\mathrm{d}\tau/2<\gamma^2\|\mathbf{w}\| \tag{6.8.43}$$

where γ^2 is a given parameter. Obviously, this condition will be fulfilled for a large parameter γ^2, but it must have a lower bound γ_{cr}^2, whose reciprocal $\gamma_{\mathrm{cr}}^{-2}$ will be shown an eigenvalue of a Rayleigh quotient [93].

For infinite horizon $t_{\mathrm{f}}\to\infty$, the present time t may also tend to infinity, in such case the proposition becomes the H_∞ filter for the infinite horizon (interval) $[0,\infty)$ problem.

The robust filter needs to consider the possible error of the plant matrix $\mathbf{A}$ of the dynamic equation. The uncertainty of the plant matrix $\mathbf{A}$ brings the error term $\mathbf{A}_\Delta\mathbf{x}$, where the matrix $\mathbf{A}_\Delta$ should consider the most unfavorable possibility. The error term $\mathbf{A}_\Delta\mathbf{x}$ is treated as the disturbance input $\mathbf{w}$ of the dynamic equation. The fundamental consideration of the Kalman-Bucy filter is to select the state vector $\hat{\mathbf{x}}(\tau)$ so as to minimize the norm of the zero-mean white noise $\mathbf{w}$, *i.e.* $\min_{\mathbf{x}}\|\mathbf{w}\|$, thereafter to compute the output vector $\hat{\mathbf{z}}$ with equation (6.8.3). Because of $\mathbf{w}$ engaged the term $\mathbf{A}_\Delta\mathbf{x}$ in H_∞ theory, and the fundamental statement of robust filter must consider the most unfavorable system uncertainty $\mathbf{A}_\Delta\mathbf{x}$. The former statement of simply minimizing the norm of the white noise $\mathbf{w}$ is no longer appropriate. That the Kalman filter treats the white noise $\mathbf{w}$ only as neutral, that the most unfavorable characteristic of $\mathbf{w}$ to the output $\hat{\mathbf{z}}$ has not been taken into account. For robust control, the selection of the disturbance $\mathbf{w}$ is not zero-mean and need to maximize the induced norm of the H_∞ filtering. Rewriting the equation (6.8.43) as

$$\max_{\mathbf{w}}\|(\hat{\mathbf{z}}-\mathbf{C}_z\mathbf{x}-\mathbf{D}_{12}\mathbf{u})\|/\|\mathbf{w}\|<\gamma^2 \tag{6.8.44}$$

The left-hand side can be interpreted as to ***maximize*** as much output deviation energy per unit disturbance energy of $\mathbf{w}$. Certainly $\mathbf{w}$ satisfies equation (6.8.41), and is still white noise.

The equation (6.8.44) has considered the unfavorable selection of $\mathbf{w}$. On the contrary, the selection of the linear and causal operator F in equation (6.8.42) must ***minimize*** the estimation error. Therefore the equation (6.8.44) should further be rewritten as

$$\max_{\mathbf{w}}\min_{\mathrm{F}}\|(\hat{\mathbf{z}}-\mathbf{C}_z\mathbf{x}-\mathbf{D}_{12}\mathbf{u})\|/\|\mathbf{w}\|=\gamma_{\mathrm{cr}}^2<\gamma^2 \tag{6.8.45}$$

The above equation shows the saddle point characteristic of the critical value γ_{cr}^2, which corresponds again to the situation of ***two players zero-sum game*** theory. Note the estimation $\hat{\mathbf{z}}$ itself has included the factor of the selection of operator F. Note also that the parameter γ_{cr}^2 here corresponds to the H_∞ filter, which having the same sign but is different to the γ_{cr}^2 of H_∞ state feedback control.

The norm of the unfavorable disturbance is thus expressed as

$$2\|\mathbf{w}\|=\int_0^t\mathbf{w}^T\mathbf{w}\mathrm{d}\tau+(\mathbf{x}(0)-\hat{\mathbf{x}}_0)^T\mathbf{P}_0^{-1}(\mathbf{x}(0)-\hat{\mathbf{x}}_0) \tag{6.8.46}$$

In taking maxi-minimization, the vectors $\mathbf{x}, \mathbf{w}, \mathbf{u}, \mathbf{z}, \mathbf{y}$ should satisfy the equations (6.8.2~4). The equation (6.8.45) expresses two folds of considerations, the first is to find the lower bound γ_{cr}^{2} of the allowable (sub-optimal) parameter γ^{-2}, and the second is to find the filter output $\hat{\mathbf{z}}$ under the given sub-optimal parameter $\gamma^{-2} < \gamma_{cr}^{-2}$, which needs finding the filtered state $\hat{\mathbf{x}}(t)$ and the variance matrix $\mathbf{P}(t)$. The filtered vector $\hat{\mathbf{x}}(t)$ and the variance matrix $\mathbf{P}(t)$ are given at the present time. If the vector $\hat{\mathbf{x}}(\tau)$ and the variance matrix $\mathbf{P}(\tau)$ for $\tau < t$ are required, then these results are the smoothing quantities.

Comparing with the Euler stability critical loading in structural engineering is meaningful. Firstly, the computation of the critical axial loading factor P_{cr} is necessary. But this factor P_{cr} cannot be used directly in engineering practice, that when the loading factor approaches P_{cr} only a small external disturbance will induce a very large deformation for the structure. Because of which, the second problem is to determine an appropriate critical loading factor $P < P_{cr}$.

The above two folds are different problems, but the equations are of the same sort. The eigenvalue problem corresponds to the axial load with no external force in structural mechanics. For H_∞ filter, the governing equation for eigenvalue problem is homogeneous with no regard to the inhomogeneous measurement value, and the unknown value is eigenvalue γ_{cr}^{-2}. The filtering problem considers the measurement value $\mathbf{y}(\tau)$, which corresponds to the external force in structural analysis with the parameter γ^{-2} given. The selection of the parameter γ^{-2} must satisfy the condition $\gamma^{-2} < \gamma_{cr}^{-2}$, which resembles the Euler critical load in structural engineering, and the parameter γ^{-2} is called as sub-optimal. Note that the equation (6.8.43) is an inequality, for which the solution is not unique, there are a number of solution of the vector $\hat{\mathbf{z}}$ satisfy the inequality. Rewriting the condition as the ***mini-maximization of the index*** J_p

$$J_p = \int_0^t [\mathbf{w}^T \mathbf{w}/2 - \gamma^{-2}(\hat{\mathbf{z}} - \mathbf{C}_z\mathbf{x} - \mathbf{D}_{12}\mathbf{u})^T(\hat{\mathbf{z}} - \mathbf{C}_z\mathbf{x} - \mathbf{D}_{12}\mathbf{u})/2]\mathrm{d}\tau \\ + (\mathbf{x}(0) - \hat{\mathbf{x}}_0)\mathbf{P}_0^{-1}(\mathbf{x}(0) - \hat{\mathbf{x}}_0)/2, \qquad \min_{\mathbf{w}} \max_{\mathrm{F}} J_p \tag{6.8.47}$$

then the filtered vector $\hat{\mathbf{z}}$ obtained is called as the ***central solution***. In fact, the vector $\hat{\mathbf{z}}$ is determined from equation (6.8.42), where $\mathbf{y}, \mathbf{u}$ are given vectors and F is a linear operator. The operator F depends only on the system matrices $\mathbf{A}, \mathbf{B}_w, \mathbf{B}_u, \mathbf{C}_y, \mathbf{C}_z, \mathbf{D}_{12}, \mathbf{D}_{21}$ but not depends on the disturbance $\mathbf{w}$ directly. Taking mean value to the equation (6.8.3) gives

$$\hat{\mathbf{z}} = \mathbf{C}_z\hat{\mathbf{x}} + \mathbf{D}_{12}\mathbf{u} \tag{6.8.48}$$

which means that the estimated state vector $\hat{\mathbf{x}}$ determines the estimation of vector $\hat{\mathbf{z}}$.

The variational principle (6.8.47) is conditional, and the conditions are equations (6.8.2) and (6.8.4). Introducing the Lagrange multipliers, the n-dimensional vector $\boldsymbol{\lambda}(\tau)$ and the q-dimensional vector $\boldsymbol{\rho}(\tau)$ derives the extended index

$$J_{pA}=\int_0^t\begin{bmatrix}\mathbf{w}^T\mathbf{w}/2-\gamma^{-2}(\hat{\mathbf{z}}-\mathbf{C}_z\mathbf{x}-\mathbf{D}_{12}\mathbf{u})^T(\hat{\mathbf{z}}-\mathbf{C}_z\mathbf{x}-\mathbf{D}_{12}\mathbf{u})/2\\+\boldsymbol{\rho}^T(\mathbf{y}-\mathbf{C}_y\mathbf{x}-\mathbf{D}_{21}\mathbf{w})+\boldsymbol{\lambda}^T(\dot{\mathbf{x}}-\mathbf{A}\mathbf{x}-\mathbf{B}_w\mathbf{w}-\mathbf{B}_u\mathbf{u})\end{bmatrix}\mathrm{d}\tau \tag{6.8.49}$$

$$+(\mathbf{x}(0)-\hat{\mathbf{x}}_0)^{\mathrm{T}}\mathbf{P}_0^{-1}(\mathbf{x}(0)-\hat{\mathbf{x}}_0)/2$$

Taking minimization with respect to $\mathbf{w}$ first gives

$$\mathbf{w}=\mathbf{B}_w^T\boldsymbol{\lambda}+\mathbf{D}_{21}^T\boldsymbol{\rho}$$

Substituting back into (6.8.49) and using equation (6.8.5) gives

$$J_{pA}=\int_0^t\begin{bmatrix}\boldsymbol{\lambda}^T(\dot{\mathbf{x}}-\mathbf{A}\mathbf{x})-\boldsymbol{\lambda}^T\mathbf{B}_w\mathbf{B}_w^T\boldsymbol{\lambda}/2+\boldsymbol{\rho}^T(\mathbf{C}_y\mathbf{x}+\mathbf{D}_{21}\mathbf{B}_w^T\boldsymbol{\lambda}-\mathbf{y})-\boldsymbol{\rho}^T\boldsymbol{\rho}/2\\-\boldsymbol{\lambda}^T\mathbf{B}_u\mathbf{u}-\gamma^{-2}(\hat{\mathbf{z}}-\mathbf{C}_z\mathbf{x}-\mathbf{D}_{12}\mathbf{u})^T(\hat{\mathbf{z}}-\mathbf{C}_z\mathbf{x}-\mathbf{D}_{12}\mathbf{u})/2\end{bmatrix}\mathrm{d}\tau$$

$$+(\mathbf{x}(0)-\hat{\mathbf{x}}_0)^{\mathrm{T}}\mathbf{P}_0^{-1}(\mathbf{x}(0)-\hat{\mathbf{x}}_0)/2$$

Then taking minimization with respect to $\boldsymbol{\rho}$ gives

$$\boldsymbol{\rho}=-\mathbf{D}_{21}\mathbf{B}_w^T\boldsymbol{\lambda}-\mathbf{C}_y\mathbf{x}+\mathbf{y}$$

$$J_{pA}=\int_0^t\begin{bmatrix}\boldsymbol{\lambda}^T(\dot{\mathbf{x}}-\bar{\mathbf{A}}\mathbf{x})-\boldsymbol{\lambda}^T\mathbf{B}\mathbf{B}^T\boldsymbol{\lambda}/2+\mathbf{x}^T(\mathbf{C}_y^T\mathbf{C}_y-\gamma^{-2}\mathbf{C}_z^T\mathbf{C}_z)\mathbf{x}/2\\-\boldsymbol{\lambda}^T(\mathbf{B}_u\mathbf{u}+\mathbf{B}_w\mathbf{D}_{21}^T\mathbf{y})-\mathbf{x}^T\left(\mathbf{C}_y^T\mathbf{y}-\gamma^{-2}\mathbf{C}_z^T(\hat{\mathbf{z}}-\mathbf{D}_{12}\mathbf{u})\right)\end{bmatrix}\mathrm{d}\tau \tag{6.8.50}$$

$$+(\mathbf{x}(0)-\hat{\mathbf{x}}_0)^{\mathrm{T}}\mathbf{P}_0^{-1}(\mathbf{x}(0)-\hat{\mathbf{x}}_0)/2,\qquad \delta J_{pA}=0$$

where

$$\bar{\mathbf{A}}=\mathbf{A}-\mathbf{B}_w\mathbf{D}_{21}^T\mathbf{C}_y,\quad \mathbf{B}\mathbf{B}^T=\mathbf{B}_w(\mathbf{I}_l-\mathbf{D}_{21}^T\mathbf{D}_{21})\mathbf{B}_w^T \tag{6.8.51}$$

The variational principle (6.8.50) has two kinds of independent variables $\mathbf{x}(\tau),\boldsymbol{\lambda}(\tau)$, both being n-dimensioned vectors and dual to each other. Performing the variational derivation gives the dual differential equations

$$\dot{\mathbf{x}}=\bar{\mathbf{A}}\mathbf{x}+\mathbf{B}\mathbf{B}^T\boldsymbol{\lambda}+\mathbf{B}_u\mathbf{u}+\mathbf{B}_W\mathbf{D}_{21}^T\mathbf{y} \tag{6.8.52a}$$

$$\dot{\boldsymbol{\lambda}}=(\mathbf{C}_y^T\mathbf{C}_y-\gamma^{-2}\mathbf{C}_z^T\mathbf{C}_z)\mathbf{x}-\bar{\mathbf{A}}^T\boldsymbol{\lambda}-\mathbf{C}_y^T\mathbf{y}+\gamma^{-2}\mathbf{C}_z^T(\hat{\mathbf{z}}-\mathbf{D}_{12}\mathbf{u}) \tag{6.8.52b}$$

The boundary condition at $\tau=0$ is

$$\mathbf{x}(0)=\hat{\mathbf{x}}_0+\mathbf{P}_0\boldsymbol{\lambda}(0),\quad \text{when } \tau=0 \tag{6.8.53}$$

The boundary condition at the other end $\tau=t$ needs further investigation. If the connection condition with the future time interval $(t,t_{\mathrm{f}}]$ at the present time $\tau=t$ is disregarded, then the natural boundary condition of J_{pA} at $\tau=t$ is

$$\boldsymbol{\lambda}(\tau)=\mathbf{0},\quad \text{when } \tau=t \tag{6.8.54}$$

Using this completely free natural boundary condition at the present time $\tau=t$, the resulting vector is filter, the H_∞ filter vector.

The application of filtering is not necessarily limited only to the control problems, hence the computation of H_∞ filter is even more important. Control is one of the important applications of filtering, and in this case there is the connection to future time interval at the present time instant $\tau=t$. No matter what form of boundary condition is used at $\tau=t$, there must be a vector denoted as

$$\boldsymbol{\lambda}(\tau\to t)=\bar{\boldsymbol{\lambda}}(t) \tag{6.8.55}$$

So the problem becomes to find the equation for this vector $\bar{\lambda}(t)$. The feature of H_∞ control problem is the connection to the future time interval, so that the equation for $\bar{\lambda}(t)$ should be derived according to this connection condition, which will be given later. Leaving the vector $\bar{\lambda}(t)$ to be determined, the equations (6.8.52~53) should be satisfied first and the variational index J_{pA} becomes

$$\delta J_{pA} = [\delta \mathbf{x}(t)]^T \bar{\boldsymbol{\lambda}}(t) \tag{6.8.56}$$

From what follows, the solution of the ***free boundary condition*** (6.8.54) is termed as the ***filter solution***, whereas the other boundary condition with $\bar{\lambda}(t)$ being not a zero vector is termed as ***estimation solution***, which looked likes smoothing but it does not offence the causality so it is not smoothing. Evidently filtering is a special case $\boldsymbol{\lambda}(t) = \mathbf{0}$ of estimation solution.

The H_∞ filter solution is described first, that except the boundary condition at $\tau = t$ the equations for estimation solution are the same as filter solution. Therefore the majority of solution procedure for the filter solution is also useful for the estimation solution.

§6.8.2.1, Solution of the dual equations

Comparing the equations (6.8.52a,b) and (6.5.47a,b) knows that both set are practically the same. Both are first order inhomogeneous differential equations with two end boundary conditions. Both are filter problems so that the boundary conditions coincide to each other, with the same solution method. The inhomogeneous term can be solved after the homogeneous equation is solved. Similar to that in section 6.5 let

$$\mathbf{x}(\tau) = \hat{\mathbf{x}}(\tau) + \mathbf{P}(\tau)\boldsymbol{\lambda}(\tau) \tag{6.8.57}$$

Substituting into equations (6.8.52a,b) gives

$$\dot{\mathbf{x}} = \dot{\hat{\mathbf{x}}} + \dot{\mathbf{P}}\boldsymbol{\lambda} + \mathbf{P}\dot{\boldsymbol{\lambda}} = \bar{\mathbf{A}}\hat{\mathbf{x}} + \bar{\mathbf{A}}\mathbf{P}\boldsymbol{\lambda} + \mathbf{B}\mathbf{B}^T\boldsymbol{\lambda} + \mathbf{B}_u\mathbf{u} + \mathbf{B}_w\mathbf{D}_{12}\mathbf{y}$$

$$\dot{\boldsymbol{\lambda}} = (\mathbf{C}_y^T\mathbf{C}_y - \gamma^{-2}\mathbf{C}_z^T\mathbf{C}_z)(\hat{\mathbf{x}} + \mathbf{P}\boldsymbol{\lambda}) - \bar{\mathbf{A}}^T\boldsymbol{\lambda} - \mathbf{C}_y^T\mathbf{y} + \gamma^{-2}\mathbf{C}_z^T(\hat{\mathbf{z}} - \mathbf{D}_{12}\mathbf{u})$$

Eliminating $\dot{\boldsymbol{\lambda}}$ in the above equations derives the Riccati differential equation

$$\dot{\mathbf{P}}(\tau) = \mathbf{B}\mathbf{B}^T + \bar{\mathbf{A}}\mathbf{P} + \mathbf{P}\bar{\mathbf{A}}^T - \mathbf{P}(\mathbf{C}_y^T\mathbf{C}_y - \gamma^{-2}\mathbf{C}_z^T\mathbf{C}_z)\mathbf{P}\,, \quad \mathbf{P}(0) = \mathbf{P}_0 \tag{6.8.58}$$

and the mean value differential equation for the filter

$$\dot{\hat{\mathbf{x}}}(\tau) = \bar{\mathbf{A}}\hat{\mathbf{x}} - \mathbf{P}(\tau)(\mathbf{C}_y^T\mathbf{C}_y - \gamma^{-2}\mathbf{C}_z^T\mathbf{C}_z)\hat{\mathbf{x}}$$
$$+ \mathbf{P}(\tau) \times [\mathbf{C}_y^T\mathbf{y} + \gamma^{-2}\mathbf{C}_z^T(\hat{\mathbf{z}} - \mathbf{D}_{12}\mathbf{u})] + \mathbf{B}_u\mathbf{u} + \mathbf{B}_w\mathbf{D}_{12}\mathbf{y}$$

The matrix $\mathbf{P}(\tau)$ should first be solved from the Riccati differential equation. Then, substituting the matrix $\mathbf{P}(\tau)$ into the filter equation solves the filter vector $\hat{\mathbf{x}}(t)$. It is easily seen that the matrix $\mathbf{P}(\tau)$ is independent on the measurement $\mathbf{y}$ and the control input $\mathbf{u}$, *i.e.* independent on the in-homogenous terms. Comparing to the Kalman filter the γ^{-2} term is appended. Presently the matrix $(\mathbf{C}_y^T\mathbf{C}_y - \gamma^{-2}\mathbf{C}_z^T\mathbf{C}_z)$ is no longer a non-negative definite matrix. However, γ^{-2} cannot be too large, otherwise $\mathbf{P}(\tau)$ will be unsolvable. The critical value γ_{cr}^{-2} can be found from the existence condition of the solution, which will be given in the next section.

The vector $\hat{\mathbf{z}}$ in the filter equation should be substituted with equation (6.8.48), which gives

$$\dot{\hat{\mathbf{x}}}(\tau) = \overline{\mathbf{A}}\hat{\mathbf{x}} - \mathbf{P}(\tau)\mathbf{C}_y^T(\mathbf{y} - \mathbf{C}_y\hat{\mathbf{x}}) + \mathbf{B}_u\mathbf{u} + \mathbf{B}_w\mathbf{D}_{12}\mathbf{y} \tag{6.8.59}$$

The precise integration method for this equation is very similar to the solution of Kalman-Bucy filter equation given in the section 6.5.7, which has been described in detail. Hence, the problem is reduced to the determination of the critical parameter γ_{cr}^{-2} for the solvability condition of the Riccati differential equation (6.8.58) and afterwards, to the precise solution of (6.8.58) with the given sub-optimal parameter $\gamma^{-2} < \gamma_{cr}^{-2}$.

§6.8.2.2, Extended Rayleigh quotient

To solve the Riccati differential equation (6.8.58), the first step is the determination of γ_{cr}^{-2}. The equation (6.8.58) can only be solved when $\gamma^{-2} < \gamma_{cr}^{-2}$. Note that the equation is derived from the dual equations with no regard to the non-homogeneous term, hence the critical value γ_{cr}^{-2} can be found under the assumption $\hat{\mathbf{x}}_0 = \mathbf{0}, \mathbf{u} = \mathbf{0}, \mathbf{y} = \mathbf{0}, \hat{\mathbf{z}} = \mathbf{0}$. The corresponding variational principle (6.8.50) becomes

$$J_A = \int_0^t [\boldsymbol{\lambda}^T\dot{\mathbf{x}} - \mathrm{H}(\mathbf{x},\boldsymbol{\lambda})]\mathrm{d}\tau + (\mathbf{x}(0))^T\mathbf{P}_0^{-1}\mathbf{x}(0)/2, \quad \delta J_A = 0 \tag{6.8.60}$$

$$\mathrm{H}(\mathbf{x},\boldsymbol{\lambda}) = \boldsymbol{\lambda}^T\overline{\mathbf{A}}\mathbf{x} - \mathbf{x}^T(\mathbf{C}_y^T\mathbf{C}_y - \gamma^{-2}\mathbf{C}_z^T\mathbf{C}_z)\mathbf{x}/2 + \boldsymbol{\lambda}^T\mathbf{B}\mathbf{B}^T\boldsymbol{\lambda}/2 \tag{6.8.61}$$

From this variational principle derives the homogeneous dual equations of the equations (6.8.52a,b)

$$\dot{\mathbf{x}} = \overline{\mathbf{A}}\mathbf{x} + \mathbf{B}\mathbf{B}^T\boldsymbol{\lambda}, \quad \dot{\boldsymbol{\lambda}} = (\mathbf{C}_y^T\mathbf{C}_y - \gamma^{-2}\mathbf{C}_z^T\mathbf{C}_z)\mathbf{x} - \overline{\mathbf{A}}^T\boldsymbol{\lambda} \tag{6.8.62a,b}$$

and the initial and end boundary conditions

$$\boldsymbol{\lambda}_0 = \mathbf{P}_0^{-1}\mathbf{x}_0, \quad \boldsymbol{\lambda}_f = \boldsymbol{\lambda}(t_f) = \mathbf{0}$$

where $\mathbf{x}_0, \boldsymbol{\lambda}_0$ are $\mathbf{x}(0), \boldsymbol{\lambda}(0)$, and the end time is t_f, the reason is explained as follows. Because the present time t increases from $t_0 = 0$ to t_f, and γ_{cr}^{-2} will reach its minimum value at $t = t_f$, so that the end time of the past interval be given as t_f is appropriate. The variational principle (6.8.60) can be rewritten as

$$\delta(\Pi_1 - \gamma_{cr}^{-2}\Pi_2) = 0 \tag{6.8.63}$$

$$\Pi_1 = \int_0^{t_f} [\boldsymbol{\lambda}^T\dot{\mathbf{x}} - \boldsymbol{\lambda}^T\overline{\mathbf{A}}\mathbf{x} + \mathbf{x}^T\mathbf{C}_y^T\mathbf{C}_y\mathbf{x}/2 - \boldsymbol{\lambda}^T\mathbf{B}\mathbf{B}^T\boldsymbol{\lambda}/2]\mathrm{d}\tau + \mathbf{x}_0^T\mathbf{P}_0^{-1}\mathbf{x}_0/2$$

$$\Pi_2 = \int_0^{t_f} [\mathbf{x}^T\mathbf{C}_z^T\mathbf{C}_z\mathbf{x}/2]\mathrm{d}\tau$$

where $\mathbf{x}(\tau), \boldsymbol{\lambda}(\tau)$ are independently varied vectors. Obviously Π_2 is non-negative. If the selection of $\mathbf{x}(\tau), \boldsymbol{\lambda}(\tau)$ satisfies the equation (6.8.62a), then

$$\Pi_1 = \int_0^{t_f} [\mathbf{x}^T \mathbf{C}_y^T \mathbf{C}_y \mathbf{x}/2 + \boldsymbol{\lambda}^T \mathbf{B}\mathbf{B}^T \boldsymbol{\lambda}/2] \mathrm{d}\tau + \mathbf{x}_0^T \mathbf{P}_0^{-1} \mathbf{x}_0/2 > 0$$

The larger sign is ensured from both the controllability and observability conditions. The variational equation (6.8.63) can be rewritten in the form of the ***extended Rayleigh quotient*** with two kinds of variables

$$\gamma_{\mathrm{cr}}^{-2} = \min_{\mathbf{x}} \max_{\boldsymbol{\lambda}} (\Pi_1/\Pi_2) \tag{6.8.64}$$

In the variational equations (6.8.63~64), the parameter γ^{-2} has been appended with the subscript $_{\mathrm{cr}}$, since otherwise if $\gamma^{-2} < \gamma_{\mathrm{cr}}^{-2}$, the original problem can have only trivial solution $\mathbf{x} = \boldsymbol{\lambda} = \mathbf{0}$ because $\hat{\mathbf{x}}_0 = \mathbf{0}, \mathbf{u} = \mathbf{0}, \mathbf{y} = \mathbf{0}, \hat{\mathbf{z}} = \mathbf{0}$. As a matter of fact, it is an eigenvalue problem and the behavior of the extended Rayleigh quotient is the same as the eigenvalue problem of a self-adjoint operator in mathematical physics.

The H_∞ filter problem needs only the smallest eigenvalue, which is the same as the H_∞ control problem.

For convenience of understanding, a special case of full rank matrix $\mathbf{B}\mathbf{B}^T$ is considered, *i.e.* positive definite. Hence equation (6.8.62a) can be used to maximize Π_1, which derives

$$\boldsymbol{\lambda} = (\mathbf{B}\mathbf{B}^T)^{-1}(\dot{\mathbf{x}} - \overline{\mathbf{A}}\mathbf{x})$$

hence

$$\Pi_1 = \int_0^{t_f} [\mathbf{x}^T \mathbf{C}_y^T \mathbf{C}_y \mathbf{x}/2 + (\dot{\mathbf{x}} - \overline{\mathbf{A}}\mathbf{x})^T (\mathbf{B}\mathbf{B}^T)^{-1} (\dot{\mathbf{x}} - \overline{\mathbf{A}}\mathbf{x})/2] \mathrm{d}\tau + \mathbf{x}_0^T \mathbf{P}_0^{-1} \mathbf{x}_0/2$$

$$= \int_0^{t_f} [\dot{\mathbf{x}}^T \mathbf{K}_{22} \dot{\mathbf{x}} - \dot{\mathbf{x}}^T \mathbf{K}_{21} \mathbf{x} - \mathbf{x}^T \mathbf{K}_{21}^T \dot{\mathbf{x}} + \mathbf{x}^T \mathbf{K}_{11} \mathbf{x}] \mathrm{d}\tau/2 + \mathbf{x}_0^T \mathbf{P}_0^{-1} \mathbf{x}_0/2 > 0$$

where

$$\mathbf{K}_{22} = (\mathbf{B}\mathbf{B}^T)^{-1}, \quad \mathbf{K}_{21} = (\mathbf{B}\mathbf{B}^T)^{-1}\overline{\mathbf{A}}, \quad \mathbf{K}_{11} = \mathbf{C}_y^T \mathbf{C}_y + \overline{\mathbf{A}}^T (\mathbf{B}\mathbf{B}^T)^{-1} \overline{\mathbf{A}}$$

This form of functional Π_1 has only one kind of variables $\mathbf{x}$. The one kind of variables variational principle is popular in structural mechanics, that in the analysis of structural stability or of structural vibration natural vibration both the Rayleigh quotients have such form. In structural mechanics, the two kinds of variables mixed energy variational principle can be derived from the one kind of variable system as derived in section 5.7. The potential energy expression is the same as the above expression. The extended W-W algorithm for the eigenvalue count of two kinds of variables and mixed energy can also be derived from the W-W algorithm of one kind of variables as described in section 2.2. The eigenvalue $\gamma_{\mathrm{cr}}^{-2}$ computation is certainly very important, that using the precise integration method in combination with the mixed variable W-W algorithm, the eigenvalue $\gamma_{\mathrm{cr}}^{-2}$ can be found up to arbitrarily assigned precision.

Both the critical parameters of future time interval H_∞ control and of past time interval H_∞ filtering are all extended Rayleigh quotient, that they are resemble to each other. In fact, both problems are dual to each other as was seen in the case of LQG problem.

§6.8.2.3, Interval mixed energy

The idea of interval mixed energy has been appeared several times for various problems, so that it is only briefly described here. Let $\gamma_{\#}^{-2}$ denote the trial parameter of the critical parameter $\gamma_{\rm cr}^{-2}$. Let the interval be denoted as $(t_{\rm a},t_{\rm b})$, that the instants $t_{\rm a},t_{\rm b}$ satisfy $0 \le t_{\rm a} < t_{\rm b} \le t_{\rm f}$. Let the two end boundary conditions of the interval be given as that the state vector $\mathbf{x}_{\rm a}$ and the co-state vector $\boldsymbol{\lambda}_{\rm b}$ are given at the two ends $t_{\rm a}$ and $t_{\rm b}$, respectively. Therefore, the dual vectors $\mathbf{x}(\tau),\boldsymbol{\lambda}(\tau)$ within the interval are determined. The ***mixed energy*** of the interval $(t_{\rm a},t_{\rm b})$ is defined as

$$V(\mathbf{x}_{\rm a},\boldsymbol{\lambda}_{\rm b}) = \boldsymbol{\lambda}_{\rm b}^T\mathbf{x}_{\rm b} - \int_{t_{\rm a}}^{t_{\rm b}}[\boldsymbol{\lambda}^T\dot{\mathbf{x}} - \mathrm{H}(\mathbf{x},\boldsymbol{\lambda})]\mathrm{d}\tau \tag{6.8.65}$$

$$\mathrm{H}(\mathbf{x},\boldsymbol{\lambda}) = \boldsymbol{\lambda}^T\overline{\mathbf{A}}\mathbf{x} - \mathbf{x}^T(\mathbf{C}_y^T\mathbf{C}_y - \gamma_{\#}^{-2}\mathbf{C}_z^T\mathbf{C}_z)\mathbf{x}/2 + \boldsymbol{\lambda}^T\mathbf{B}\mathbf{B}^T\boldsymbol{\lambda}/2$$

Obviously, the mixed energy is a homogeneous quadratic form of the vectors $\mathbf{x}_{\rm a}$ and $\boldsymbol{\lambda}_{\rm b}$, for which the general expression is

$$V(\mathbf{x}_{\rm a},\boldsymbol{\lambda}_{\rm b}) = \boldsymbol{\lambda}_{\rm b}^T\mathbf{F}\mathbf{x}_{\rm a} + \mathbf{x}_{\rm a}^T\mathbf{Q}\mathbf{x}_{\rm a}/2 - \boldsymbol{\lambda}_{\rm b}^T\mathbf{G}\boldsymbol{\lambda}_{\rm b}/2 \tag{6.8.66}$$

where $\mathbf{Q}(t_{\rm a},t_{\rm b})$, $\mathbf{G}(t_{\rm a},t_{\rm b})$ and $\mathbf{F}(t_{\rm a},t_{\rm b})$ are all $n\times n$ matrices, and also $\mathbf{Q}^T = \mathbf{Q}$, $\mathbf{G}^T = \mathbf{G}$.

Two contiguous intervals can be combined into one longer interval, for which the elimination and combination equations have been given in equation (6.5.77) and are not repeated here. The interval combination procedure can be invoked recursively. However, the matrices $\mathbf{Q},\mathbf{G},\mathbf{F}$ express only the characteristic of the interval at the two ends, with no knowledge to the internal behavior of the interval. Presently, the eigenvalue $\gamma_{\rm cr}^{-2}$ is of concern. According to the W-W algorithm for structural mechanics the eigenvalue count is necessary, see chapter 2. The original form of the W-W algorithm applies only to the given two end displacement boundary condition, which corresponds to one kind of variables potential energy (dynamic stiffness) variational principle in structural mechanics. The basic form of Rayleigh quotient is also for one kind of variables. According to the analogy between structural mechanics and optimal control, the original form corresponds to the boundary condition of given state vectors $\mathbf{x}_{\rm a},\mathbf{x}_{\rm b}$ at the two ends $t_{\rm a},t_{\rm b}$, respectively, which does not fit the requirement of the present problem. Transform the dynamic stiffness potential energy of the vectors $\mathbf{x}_{\rm a},\mathbf{x}_{\rm b}$ into the mixed energy of mixed variable vectors $\mathbf{x}_{\rm a}$ and $\boldsymbol{\lambda}_{\rm b}$, correspondingly the original W-W algorithm should also be transformed into the extended W-W algorithm [41]. The extended W-W algorithm supplies the eigenvalue count of the mixed energy representation and is used as the supplement to the mixed energy matrices $\mathbf{Q},\mathbf{G},\mathbf{F}$. Comparing to structural mechanics, the vectors $\mathbf{x}_{\rm a}$, $\boldsymbol{\lambda}_{\rm b}$ correspond to displacement and internal force vectors, respectively. The eigenvalue count equations of extended W-W algorithm for interval combination can be transplanted from structural mechanics as (see section 2.2)

$$J_{mc}(\gamma_{\#}^{-2}) = J_{m1}(\gamma_{\#}^{-2}) + J_{m2}(\gamma_{\#}^{-2}) - s\{\mathbf{G}_2\} + s\{\mathbf{G}_1^{-1} + \mathbf{Q}_2\} \tag{6.8.67}$$

where $\gamma_{\#}^{-2}$ is the given trial parameter for eigenvalue, $J_m(\gamma_{\#}^{-2})$ represents the number of eigenvalues, which are less than $\gamma_{\#}^{-2}$, of the interval (t_a, t_b) with the two end conditions $\mathbf{x}_a = \mathbf{0}, \boldsymbol{\lambda}_b = \mathbf{0}$. Certainly, the count of eigenvalues is also a function of t_a and t_b, $J_m(\gamma_{\#}^{-2}, t_a, t_b)$, but for simplicity it is still written as $J_m(\gamma_{\#}^{-2})$. The sign $s\{\mathbf{M}\}$ represents the number of negative entries in the diagonal matrix $\mathbf{D}$, where the symmetric matrix $\mathbf{M}$ is triangular factorized as $\mathbf{M} = \mathbf{LDL}^T$.

Therefore, the interval mixed energy representation for eigenvalue problem is extended as $(\mathbf{Q}, \mathbf{G}, \mathbf{F}, J_m)$, a quadruple composed of three matrices and an eigenvalue count. All of them are functions of the two end time instants t_a, t_b and the given parameter $\gamma_{\#}^{-2}$. However, usually the parameter $\gamma_{\#}^{-2}$ is not written explicitly, such as still written as $\mathbf{Q}(t_a, t_b)$, etc. The LQ control or Kalman-Bucy filter theories correspond to the selection $\gamma_{\#}^{-2} = 0$, so that the eigenvalue count J_m always equals zero, hence in these cases the eigenvalue count is unnecessary in the mixed energy representation. However, in H_∞ theory the eigenvalue is a key parameter, so that the mixed energy representation must add the eigenvalue count and becomes of the form $(\mathbf{Q}, \mathbf{G}, \mathbf{F}, J_m)$.

The above derivation applies also to time variant system. When the system is time invariant, the matrices and count $\mathbf{Q}, \mathbf{G}, \mathbf{F}, J_m$ depend only on the interval length $\tau = t_b - t_a$, *i.e.*

$$\mathbf{Q} = \mathbf{Q}(\tau), \quad \mathbf{G} = \mathbf{G}(\tau), \quad \mathbf{F} = \mathbf{F}(\tau), \quad \text{where} \quad \tau = t_b - t_a \tag{6.8.68}$$

They satisfy the differential equations

$$\begin{aligned} \mathrm{d}\mathbf{Q}/\mathrm{d}\tau &= \mathbf{F}^T(\mathbf{C}_y^T\mathbf{C}_y - \gamma_{\#}^{-2}\mathbf{C}_z^T\mathbf{C}_z)\mathbf{F} \\ &= (\mathbf{C}_y^T\mathbf{C}_y - \gamma_{\#}^{-2}\mathbf{C}_z^T\mathbf{C}_z) + \overline{\mathbf{A}}^T\mathbf{Q} + \mathbf{Q}\overline{\mathbf{A}} - \mathbf{QBB}^T\mathbf{Q} \end{aligned} \tag{6.8.69a}$$

$$\begin{aligned} \mathrm{d}\mathbf{G}/\mathrm{d}\tau &= \mathbf{BB}^T + \overline{\mathbf{A}}\mathbf{G} + \mathbf{G}\overline{\mathbf{A}}^T - \mathbf{G}(\mathbf{C}_y^T\mathbf{C}_y - \gamma_{\#}^{-2}\mathbf{C}_z^T\mathbf{C}_z)\mathbf{G} \\ &= \mathbf{FBB}^T\mathbf{F}^T \end{aligned} \tag{6.8.69b}$$

$$\mathrm{d}\mathbf{F}/\mathrm{d}\tau = \mathbf{F}[\overline{\mathbf{A}} - \mathbf{BB}^T\mathbf{Q}] \quad = [\overline{\mathbf{A}} - \mathbf{G}(\mathbf{C}_y^T\mathbf{C}_y - \gamma_{\#}^{-2}\mathbf{C}_z^T\mathbf{C}_z)]\mathbf{F} \tag{6.8.69c}$$

These equations give the relation between the system matrices and the interval matrices $\mathbf{Q}, \mathbf{G}, \mathbf{F}$. The initial conditions are

$$\mathbf{Q} = \mathbf{0}, \quad \mathbf{G} = \mathbf{0}, \quad \mathbf{F} = \mathbf{I}_n, \quad J_m = 0, \quad \text{when } \tau = 0 \tag{6.8.70}$$

The derivation of these equations is almost the same as comparing to that in section 6.5. If selecting $t_a = t_0 = 0$, then the differential equation (6.8.69b) is the same as the Riccati differential equation in (6.8.58), but the boundary condition (6.8.70) is different. However, for an arbitrary point $t_b \in (0, t_f)$ after found the matrices $\mathbf{Q}, \mathbf{G}, \mathbf{F}$, execute the equation

$$\mathbf{P}(t_b) = \mathbf{G} + \mathbf{F}(\mathbf{P}_0^{-1} + \mathbf{Q})^{-1}\mathbf{F}^T \tag{6.8.71}$$

the solution of (6.8.58) is obtained. But the grid points are discrete, that computing the matrices only at these discrete points cannot make sure if there has no conjugate point within the whole time interval $[t_0, t_f)$. To make sure if the solution of the Riccati differential equation has any conjugate point, the eigenvalue count must be computed as

$$J_{mp} = J_m - s\{\mathbf{P}_0\} + s\{\mathbf{P}_0^{-1} + \mathbf{Q}\} \tag{6.8.72}$$

The criterion $J_{mp} = 0$ means that $\gamma_{\#}^{-2}$ is a H_∞ sub-optimal parameter. Based on this criterion, the searching method can be used for $\gamma_{\#}^{-2}$, such as using the bisection method, then the real eigenvalue γ_{cr}^{-2} can be found up to arbitrary precision. The methodology is the same as using the W-W algorithm in eigenvalue problems in structural stability analysis or in structural natural vibration.

§6.8.2.4, Precise integration method

Direct integration for the Riccati differential equation faces the non-linear problem. The usual numerical integration method usually applies finite difference approximation, which is error prone and has also the problem of numerical instability. The precise integration method in combination with the W-W algorithm can solve the Riccati differential equation and its eigenvalue with the numerical results up to the computer precision.

Numerical integration need a time step size η, which is a reasonable value not too large. Follow the given step size η, a very small time step τ is selected, see (6.8.34). Because the interval elimination and combination require an initial interval with its interval matrices be determined very precisely (up to the computer precision), that $\tau = \eta / 2^N$ is used as the initial interval. For such a small time interval τ, the power series expansion has very high precision. The series expansion can select the same as equation (6.8.35), for which the error has been beyond the precision of the real number of computer double precision.

The series expansion (6.8.35) has satisfied the initial condition (6.8.70). Substituting into the equations (6.8.69a~c) and comparing the coefficients of various powers of τ successively gives

$$\begin{aligned}
&\mathbf{e}_1 = (\mathbf{C}_y^T\mathbf{C}_y - \gamma_{\#}^{-2}\mathbf{C}_z^T\mathbf{C}_z),\ \ \mathbf{g}_1 = \mathbf{B}\mathbf{B}^T,\ \ \mathbf{f}_1 = \overline{\mathbf{A}} \\
&\mathbf{e}_2 = (\mathbf{f}_1^T\mathbf{e}_1 + \mathbf{e}_1\mathbf{f}_1)/2,\ \ \mathbf{g}_2 = (\overline{\mathbf{A}}\mathbf{g}_1 + \mathbf{g}_1\overline{\mathbf{A}}^T)/2,\ \ \mathbf{f}_2 = (\overline{\mathbf{A}}\mathbf{f}_1 - \mathbf{g}_1\mathbf{e}_1)/2 \\
&\mathbf{e}_3 = (\mathbf{f}_2^T\mathbf{e}_1 + \mathbf{e}_1\mathbf{f}_2 + \mathbf{f}_1^T\mathbf{e}_1\mathbf{f}_1)/3,\ \ \mathbf{g}_3 = (\overline{\mathbf{A}}\mathbf{g}_2 + \mathbf{g}_2\overline{\mathbf{A}}^T - \mathbf{g}_1\mathbf{e}_1\mathbf{g}_1)/3 \\
&\mathbf{f}_3 = (\overline{\mathbf{A}}\mathbf{f}_2 - \mathbf{g}_2\mathbf{e}_1 - \mathbf{g}_1\mathbf{e}_1\mathbf{f}_1)/3,\ \ \mathbf{e}_4 = (\mathbf{f}_3^T\mathbf{e}_1 + \mathbf{e}_1\mathbf{f}_3 + \mathbf{f}_2^T\mathbf{e}_1\mathbf{f}_1 + \mathbf{f}_1^T\mathbf{e}_1\mathbf{f}_2)/4 \\
&\mathbf{g}_4 = (\mathbf{g}_3\overline{\mathbf{A}}^T + \overline{\mathbf{A}}\mathbf{g}_3 - \mathbf{g}_2\mathbf{e}_1\mathbf{g}_1 - \mathbf{g}_1\mathbf{e}_1\mathbf{g}_2)/4, \\
&\mathbf{f}_4 = (\overline{\mathbf{A}}\mathbf{f}_3 - \mathbf{g}_3\mathbf{e}_1 - \mathbf{g}_2\mathbf{e}_1\mathbf{f}_1 - \mathbf{g}_1\mathbf{e}_1\mathbf{f}_2)/4
\end{aligned} \tag{6.8.73}$$

where $\mathbf{e}_i, \mathbf{g}_i, \mathbf{f}_i, (i = 1 \sim 4)$ are all $n \times n$ matrices, and $\mathbf{e}_i^T = \mathbf{e}_i, \mathbf{g}_i^T = \mathbf{g}_i$. These equations require only computing successively with no iteration. After computed the matrices $\mathbf{Q}(\tau), \mathbf{G}(\tau), \mathbf{F}'(\tau)$ according to the equation (6.8.35) and (6.8.73), the eigenvalue count $J_m(\tau) = 0$ is applied since τ is selected extremely small and then the quadruple of mixed energy representation of interval τ is composed. It can be used as the initial time interval for the 2^N algorithm of interval combination.

So far, all the derivations are exact except the series expansion (6.8.35) is truncated after the term τ^4. The relative error of the truncation to the first term is $O(\tau^4)$. Since $\tau^4 = (\eta/2^N) \approx \eta^4 \cdot 10^{-24}$, the relative error has been beyond the significant decimal digits 10^{-16} of the double precision real number of the computer, so that practically this series expansion has no net effect to the computer arithmetic.

Based on the mixed energy representation of the initial time interval τ, the interval combination algorithm should be executed N times to obtain the mixed energy representation $(\mathbf{Q}, \mathbf{G}, \mathbf{F}, J_m)$ of time interval η. Special attention must be taken on that the matrix addition $\mathbf{I} + \mathbf{F}'$ in the combination equation (6.8.35a) should not be executed. The reason is that $\mathbf{F}'$ is very small when τ is very small and the addition will seriously drop the arithmetic precision due to round-off operations. Using equations (6.8.37a~c) instead of (6.5.77a~c) can avoid such round-off error of arithmetic precision.

Until here, the equations for precise integration have been completed. Control theory requires the solution matrix of the Riccati differential equation being always symmetric and positive definite (or at least non-negative definite). For finite horizon (time interval), this requires the criterion $J_{mp} = 0$ after the execution of equations (6.8.71~72), in order to ensure that $\gamma_{\#}^{-2}$ is a sub-optimal parameter, and in such case the parameter $\gamma_{\#}^{-2}$ can be increased. Otherwise if $J_m > 0$, then the parameter is too large, and should be decreased, then compute again. For the computation of infinite horizon problem, the appearance of $J_m > 0$ at any stage means that the parameter $\gamma_{\#}^{-2}$ is too large and should be decreased.

§6.8.2.5, Algorithm

The algorithm for the optimal parameter γ_{cr}^{-2} (eigenvalue) can be the bisection searching for the given parameter $\gamma_{\#}^{-2}$ based on the W-W algorithm. For a given parameter $\gamma_{\#}^{-2}$, the computation is the sub-optimal algorithm. The searching of the eigenvalue γ_{cr}^{-2} can be only the bisection algorithm, which is so popular and is avoided.

First, the precise integration algorithm for the fundamental interval η is given as

[Given n, and matrices $\mathbf{A}, \mathbf{B}_1, \mathbf{C}_y, \mathbf{C}_z, \mathbf{D}_{12}, \mathbf{D}_{21}$ and initial variance matrix $\mathbf{P}_0$]

[Give the time interval $[0, t_f)$. Select a trial parameter $\gamma_{\#}^{-2}$]

[From (6.8.51) compute the matrices $\overline{\mathbf{A}}, \mathbf{BB}^T$ and $(\mathbf{C}_y^T\mathbf{C}_y - \gamma_{\#}^{-2}\mathbf{C}_z^T\mathbf{C}_z)$;]

[Select a fundamental time step η. Let $N = 20$; and $\tau = \eta/2^N$;]

[According to (6.8.35), (6.8.73) computing $\mathbf{Q}(\tau), \mathbf{G}(\tau), \mathbf{F}'(\tau)$; let $J_m = 0$;]

for ($iter = 0$; $iter < N$; $iter++$) {

[Triangular factorizing the matrices $\mathbf{G}$ and $(\mathbf{Q}+\mathbf{G}^{-1})$, compute J_{mc} according to (6.8.67);]

[According to (6.8.37) compute $\mathbf{Q}_c, \mathbf{G}_c, \mathbf{F}_c'$;]

[Let $\mathbf{Q}_c, \mathbf{G}_c, \mathbf{F}_c'$ $J_m = J_{mc}$;]

}

[Let $\mathbf{F} = \mathbf{F}_c = \mathbf{I} + \mathbf{F}'$;]

Comment: $(\mathbf{Q}, \mathbf{G}, \mathbf{F}, J_m)$ has been for the interval η (6.8.74)

Below, two cases of infinite horizon and finite horizon should be distinguished. For the case of infinite horizon, the algorithm is

while $(\|\mathbf{F}_c\| > \varepsilon)$ {

[Let $\mathbf{Q}_1 = \mathbf{Q}_2 = \mathbf{Q}_c; \mathbf{G}_1 = \mathbf{G}_2 = \mathbf{G}_c; \mathbf{F}_1 = \mathbf{F}_2 = \mathbf{F}_c; J_{m1} = J_{m2} = J_{mc};$]

[Using (6.5.77) compute $\mathbf{Q}_c, \mathbf{G}_c, \mathbf{F}_c$, and compute J_{mc} based on (6.8.67)]

if $(J_{mc} > 0)$ { $\gamma_{\#}^{-2}$ is too large; Break;}

} Comment: If $\gamma_{\#}^{-2}$ is too large, decrease $\gamma_{\#}^{-2}$; or $\gamma_{\#}^{-2}$ is H_∞ sub-optimal, increase $\gamma_{\#}^{-2}$; (6.8.75)

For case of finite horizon, the mesh is $t_0 = 0, \cdots, t_k = k\eta, \cdots t_f = k_f \eta$. After the computation of (6.8.74), execute

[$\mathbf{Q}_2 = \mathbf{Q}_c; \mathbf{G}_2 = \mathbf{G}_c; \mathbf{F}_2 = \mathbf{F}_c; J_{m2} = 0;$ $\mathbf{Q}_1 = \mathbf{0}; \mathbf{G}_1 = \mathbf{P}_0; \mathbf{F}_1 = \mathbf{I}; J_{m1} = 0;$]

for ($iter = 1$; $iter \le k_f$; $iter++$) {

Comment: $\mathbf{Q}_2, \mathbf{G}_2, \mathbf{F}_2, J_{m2}$ have not changed

[Using (6.5.77) compute $\mathbf{Q}_c, \mathbf{G}_c, \mathbf{F}_c$ and compute J_{mc} based on (6.8.67)]

[Let $\mathbf{Q}_1 = \mathbf{Q}_c; \mathbf{G}_1 = \mathbf{G}_c; \mathbf{F}_1 = \mathbf{F}_c; J_{m1} = J_{mc};$]

if ($J_{mc} > 0$) Break;

}

if $(J_{mc} == 0)$ { $\gamma_{\#}^{-2}$ is lower bound of γ_{cr}^{-2}, sub-optimal and $\gamma_{\#}^{-2}$ can be increased; if the precision is enough, stop.}

else { $\gamma_{\#}^{-2}$ is too large, the upper bound of γ_{cr}^{-2}; reduce $\gamma_{\#}^{-2}$;} (6.8.76)

For the parameter $\gamma_{\#}^{-2}$, iterates using such as the bisection method until satisfied. This is the least eigenvalue iteration of Rayleigh quotient, which gives the lower and upper bounds.

In the next section, some numerical example will be given for the measurement feedback H_∞ optimal control problem, which certainly involves the computation of H_∞ filtering. Hence no numerical example is given here.

§6.8.3, Synthesis variational principle for measurement feedback control [94]

The whole time interval $[0,t_f]$ is composed of the ***past*** $[0,t)$ and the ***future*** $(t,t_f]$ two ***time intervals***, connected at the ***present time instant*** t. ***Synthesizing*** means the application of the results of the two intervals and the connection condition at the present time instant t in the variational principle for the whole time interval. The connection condition is the ***continuity condition of the state vector*** that the initial condition of the state vector of the future time interval equals the state estimation vector $\bar{\mathbf{x}}(t)$ of the past interval at its end time t. The state estimation value $\bar{\mathbf{x}}(t)$ is not necessary the filter value $\hat{\mathbf{x}}(t)$, which gives the difference between LQG and H_∞ theory and is determined from the variational principle (6.8.6). It is emphasized that the filter value is obtained from the boundary condition $\boldsymbol{\lambda}(t) = \mathbf{0}$, but the estimation state value $\bar{\mathbf{x}}(t)$ does not use such boundary condition. The boundary condition is derived from the variational principle as a natural boundary (connection) condition, based on the state vector continuity condition at the present time t. According to the proposition of TPBVP, there should be n boundary conditions for each of both the connected time intervals at the present time t, respectively, so that $2n$ conditions should be given. Considering from the whole time interval, there must be $2n$ connection conditions at t. The ***continuity of the state vector*** $\mathbf{x}(t)$ proposed n connection conditions, and the other n conditions will be derived as the ***natural connection condition*** of the variational principle (6.8.6).

Internal to the future time interval and to the past time interval, there is no connection, the only connection of the two intervals is at the present time t. Hence from equation (6.8.6), the integral of the future time interval derives still J_c of equation (6.8.8'). Introducing the dual vector $\boldsymbol{\lambda}(\tau)$ as given in section 6.8.1, it is derived that the index variation of the future time interval is

$$\delta J_c = -\boldsymbol{\lambda}^T(t) \cdot \delta\mathbf{x}(t)$$

Substituting the equation (6.8.13) gives

$$\delta J_c = \mathbf{x}^T(t)\mathbf{X}(t) \cdot \delta\mathbf{x}(t)$$

Because of the continuity condition (6.8.9) for the state vector $\mathbf{x}(t)$, it gives

$$\delta J_c = \bar{\mathbf{x}}^T(t)\mathbf{X}(t) \cdot \delta\mathbf{x}(t) \tag{6.8.77}$$

Note that the variation of state vector $\delta\mathbf{x}$ is different from the estimated state vector $\bar{\mathbf{x}}(t)$ of the past time interval.

In the variational principle (6.8.6), the integral of the past time interval gives still the functional J_p in equation (6.8.47), the derivation of which is independent on the future time interval except at the present time t. Obviously, the functional J_c and J_p can arbitrarily multiply individual multipliers to them, respectively, without any consequence to the derived equations within the two time intervals. Because the functional J_c and J_p are for different time intervals, any of their ***linear combination*** can still derive the differential equations of the two intervals separately. Only at the time instant t, the connecting point, the linear combination influences the resulted equations derived there. Hence an appropriate linear combination is to find.

Let us go back to the variational principle (6.8.6) to find the appropriate linear combination, that the expression of norm of the disturbance $\mathbf{w}$ is the same for the future and for the past intervals. Hence the linear combination of the norm of J_c and J_p is required to keep the unified form of the norm of disturbance $\mathbf{w}$. Comparing the $\mathbf{w}$ contributions in the functional (6.8.8') and (6.8.47) determines that the functional J of the whole time interval should be composed as

$$J = J_p - \gamma^{-2} J_c \tag{6.8.78}$$

That the norm contributed by the disturbance $\mathbf{w}$ of the past and future intervals has the same form as given in (6.8.6). Based on this functional J, the differential equations derived in the past and future time intervals are the same as given in the two previous sections, so that only the connection equation at the time instant t is to be derived further. According to the variational equations (6.8.77) and (6.8.56) derives

$$\delta J = \delta J_p - \gamma^{-2}\delta J_c = [\delta \mathbf{x}(t)]^T[\bar{\boldsymbol{\lambda}}(t) - \gamma^{-2}\mathbf{X}(t)\bar{\mathbf{x}}(t)] = 0$$

where because $\delta\mathbf{x}(t)$ is arbitrarily varied, hence

$$\bar{\boldsymbol{\lambda}}(t) = \gamma^{-2}\mathbf{X}(t)\bar{\mathbf{x}}(t) \tag{6.8.79}$$

The vectors $\bar{\mathbf{x}}(t), \bar{\boldsymbol{\lambda}}(t)$ above are the estimation solutions of the past time interval, and the equation (6.8.79) is the another boundary condition at $\tau = t - 0$ for the estimation in past interval. Because the equation is derived from the functional of the whole time interval, so it is called as the ***natural connection condition***, the another n connection conditions. The derivation of this natural connection condition is based on the ***continuity of the state vector*** $\mathbf{x}$. However, the two intervals have their own individual dual vectors, respectively, and are independent on each other before connection. Therefore at the two sides of the present time t the two vectors of $\boldsymbol{\lambda}$ have no continuity condition.

Comparing the condition (6.8.79) with the natural boundary condition (6.8.54) of a filter, the difference is obvious. Therefore, the connection condition of H_∞ measurement feedback control should be strictly distinguished with the boundary condition (6.8.54) of H_∞ filter theory. The past interval analysis should use the natural connection condition as the boundary condition at the time t for H_∞ measurement feedback control. Because it is different to pure filtering so the term ***estimation*** is used. The LQG theory is the limiting case of H_∞ theory with $\gamma^{-2} = 0$, that the connection condition (6.8.79) reduces to (6.8.54) as $\gamma^{-2} \to 0$.

Substituting $\tau = t$ into the equation (6.8.57), it is seen that when the boundary condition (6.8.54) is used, the state vector gives naturally the filter value. But the state vector continuity condition determines that vectors $\bar{\mathbf{x}}(t), \bar{\boldsymbol{\lambda}}(t)$ are used, which gives

$$\bar{\mathbf{x}}(t) = \hat{\mathbf{x}}(t) + \mathbf{P}(t)\bar{\boldsymbol{\lambda}}(t) \tag{6.8.80}$$

The simultaneous equation (6.8.79) and (6.8.80) can be used to solve the unknown dual vectors $\bar{\mathbf{x}}(t)$ and $\bar{\boldsymbol{\lambda}}(t)$, and it obtains

$$\bar{\mathbf{x}}(t) = [\mathbf{I} - \gamma^{-2}\mathbf{P}(t)\mathbf{X}(t)]^{-1}\hat{\mathbf{x}}(t) \tag{6.8.81}$$

This equation gives the relation between the filter vector $\hat{\mathbf{x}}(t)$ and the estimation vector $\bar{\mathbf{x}}(t)$. Hence the H_∞ filter state vector $\hat{\mathbf{x}}(t)$ and the variance matrix $\mathbf{P}(t)$ for the past time interval, and the H_∞ state feedback optimal control solution matrix $\mathbf{X}(t)$ of the Riccati differential equation should first be computed, thereafter the H_∞ ***measurement feedback state estimation vector*** $\bar{\mathbf{x}}(t)$ can be computed by equation (6.8.81). Based on $\bar{\mathbf{x}}(t)$, the feedback control vector can be computed from equations (6.8.10'') and (6.8.13)

$$\mathbf{u}(t) = -[\mathbf{B}_u^T\mathbf{X}(t) - \mathbf{D}_{12}^T\mathbf{C}_z]\bar{\mathbf{x}}(t) \tag{6.8.82}$$

The feedback control vector should be computed at the real-time, so that fast computation of the control vector is required. All the computation, which can be executed ***off-line***, should be carried out beforehand and stored, and be retrieved at the running time of the control system. The state vector estimation $\bar{\mathbf{x}}(t)$ depends on the real time measurement so that it can only be computed at the real time. The matrices $\mathbf{P}(t)$ and $\mathbf{X}(t)$ do not depend on the measurement $\mathbf{y}(\tau)$ nor the control $\mathbf{u}(\tau)$, so that they can be computed ***off-line***. In the equation (6.8.81) only the filter vector $\hat{\mathbf{x}}(t)$ requires real-time computation. Therefore the filter computation is extremely important, for which the precise integration has been described in detail in section 6.5.

§6.8.3.1, Rayleigh quotient

The critical parameters $\gamma_{\rm cr}^{-2}$ of the H_∞ filter and the H_∞ state feedback control are described in the sections 6.8.1 and 6.8.2, respectively. These two critical parameters are different and are denoted as $\gamma_{\rm cr,f}^{-2}$ and $\gamma_{\rm cr,c}^{-2}$, respectively. According to the variational equation (6.8.6), the measurement feedback control system has a unified critical parameter $\gamma_{\rm cr}^{-2}$, since the equation (6.8.81) explains that the factor $[\mathbf{I} - \gamma^{-2}\mathbf{P}(t)\mathbf{X}(t)]$ should be non-singular everywhere [134], which is again a condition to the critical parameter $\gamma_{\rm cr}^{-2}$. In previous sections the conditions $\gamma^{-2} < \gamma_{\rm cr,c}^{-2}$ and $\gamma^{-2} < \gamma_{\rm cr,f}^{-2}$ have been derived for the H_∞ state feedback control and for the H_∞ filtering, respectively. The control and filter critical values $\gamma_{\rm cr,c}^{-2}$ and $\gamma_{\rm cr,f}^{-2}$, respectively, are certainly two bounds to the critical parameter $\gamma_{\rm cr}^{-2}$ of the measurement feedback control system of the whole time interval $[0, t_{\rm f}]$. The equation (6.8.81) proposes the third bound in addition to the previous two. This condition can be rewritten as

$$\det[\mathbf{P}^{-1}(t) - \gamma^{-2}\mathbf{X}(t)] > 0 \tag{6.8.83}$$

This equation has the same form of the eigenvalue problem of structural vibration problem, that the matrix $\mathbf{P}^{-1}(t) \sim \mathbf{K}$ resembles the stiffness matrix of a structure, and $\mathbf{X}(t) \sim \mathbf{M}$ resembles the mass matrix of the structure. As the time t is changing, equation (6.8.83) proposes an infinite number of eigenvalue problems. The eigenvalue problem can be written in the typical ***Rayleigh quotient*** form

$$\gamma^{-2} \le \min_{\mathbf{x}}[\mathbf{x}^T\mathbf{P}^{-1}(t)\mathbf{x}/\mathbf{x}^T\mathbf{X}(t)\mathbf{x}] \tag{6.8.84}$$

where the time t is only a parameter. Hence the measurement feedback critical parameter γ_{cr}^{-2} can be written as

$$\gamma^{-2} \le \gamma_{\text{cr}}^{-2} = \inf_t\{\min_{\mathbf{x}}[\mathbf{x}^T\mathbf{P}^{-1}(t)\mathbf{x}/\mathbf{x}^T\mathbf{X}(t)\mathbf{x}]\} \tag{6.8.85}$$

The conditions $\gamma^{-2} < \gamma_{\text{cr,f}}^{-2}$ and $\gamma^{-2} < \gamma_{\text{cr,c}}^{-2}$ are two bounds of H_∞ filtering and H_∞ state feedback control, respectively, exactly two conditions. The above third condition is for arbitrary time t. Solvable conditions of the matrices $\mathbf{X}(t),\mathbf{P}(t)$ imply that the former two conditions have been satisfied, so that the third condition is based on the former two conditions. The $n \times n$ matrices $\mathbf{X}(t)$ and $\mathbf{P}(t)$ are both positive definite and symmetric. The computation of ***Rayleigh quotient of two positive definite symmetric matrices*** is a very popular problem, that a number of standard algorithms are available. However, this computation must be executed for ***all the time grid points***, so that the third condition is by no means one condition but quite a number of conditions.

§6.8.3.2, Variance matrix of the state estimation vector

It has been shown that the matrix $\mathbf{P}(t)$ is the variance of the filter vector $\hat{\mathbf{x}}(t)$, however, the estimated vector $\bar{\mathbf{x}}(t)$ is used now, whose variance matrix is also of concern. For which, the analogy relationship to structural mechanics can also be used. According to the analogy, the variance matrix corresponds to the flexibility matrix in structural mechanics. Then the estimation state vector $\bar{\mathbf{x}}(t)$ can be generated as follows: let all measurement vector $\mathbf{y} = \mathbf{0}$ with the initial state $\hat{\mathbf{x}}_0 = \mathbf{0}$ also, which corresponds to zero external forces. Then let the unit forces $\mathbf{e}_i, i = 1,...,n$ one after another acts at the time t sequentially, where $\mathbf{e}_i$ is the i-th column vector of the unit matrix $\mathbf{I}_n$. Under this unit force load, the displacement vector (*i.e.* the state vector) solution is obtained as

$$\delta J = \delta J_{\text{p}} + \gamma^{-2}\delta J_{\text{c}} = \delta\mathbf{x}^{\text{T}}(t)\big(-\boldsymbol{\lambda}(t) + \gamma^{-2}\mathbf{X}(t)\mathbf{x}(t) + \mathbf{e}_i\big) = 0\,, \quad i = 1,...,n$$

Note that because $\mathbf{y} = \mathbf{0}$ so that $\hat{\mathbf{x}}(t) = \mathbf{0}$ in the equation (6.8.80) for all $i = 1,...,n$. Therefore, it solves $\boldsymbol{\lambda}(t) = \mathbf{P}^{-1}(t)\mathbf{x}(t)$, substituting which into the above equation gives

$$\mathbf{x}_i(t) = \mathbf{x}(t) = (\mathbf{P}^{-1}(t) - \gamma^{-2}\mathbf{X}(t))^{-1}\mathbf{e}_i\,, \quad \text{for all } \; i = 1,...,n$$

These $\mathbf{x}_i(t)$ are the displacement vectors under the unit force vector $\mathbf{e}_i$. Using displacement vectors $\mathbf{x}_i(t)$ as the i-th columns composes the variance matrix $\mathbf{Z}(t)$ as

$$\mathbf{Z}(t) = [\mathbf{P}^{-1}(t) - \gamma^{-2}\mathbf{X}(t)]^{-1} \tag{6.8.86}$$

§6.8.3.3, Algorithm

The precise integration computations for the H_∞ state feedback control and for H_∞ filtering are described above. The condition (6.8.84), however, cannot be checked for all the continuous-time t, but can only be checked at the discrete time grid points. Usually, regular mesh is used in the computation, the whole time interval $[0, t_\mathrm{f}]$ is subdivided into N_int fundamental small length $\eta = t_\mathrm{f} / N_\mathrm{int}$ intervals. For a given parameter γ^{-2}, the precise integration method can be used to compute the interval mixed energy matrices and the respective eigenvalue count for the fundamental interval of length η. Such computations, certainly, should be carried out both for H_∞ state feedback optimal control and for H_∞ filter. Based on the method of interval elimination and combination, the Riccati matrices $\mathbf{X}_k$ and $\mathbf{P}_k$ for $k = 0,1,\ldots,N_\mathrm{int}$ of both the H_∞ state feedback optimal control and the H_∞ filtering can be computed for all the grid points. Using the mixed energy extended W-W algorithm, the trial parameter $\gamma_\#^{-2}$ is ensured to be sub-optimal for both the H_∞ control and H_∞ filter, respectively.

However, the parameter $\gamma_\#^{-2}$ needs further verification to meet the condition of (6.8.84), which is the verification for all the grid points.

The computation of the measurement feedback H_∞ optimal control has also two stages. The first stage is determination of the critical parameter γ_cr^{-2} for measurement feedback H_∞ optimal control. Afterwards, for a selected sub-optimal parameter $\gamma^{-2} < \gamma_\mathrm{cr}^{-2}$ carry out the precise integration computations. The latter stage computation is almost the same as the computation for a LQG problem, so that the detail is omitted. The important part is the determination of critical parameter γ_cr^{-2} for the measurement feedback H_∞ optimal control. Obviously, $\gamma_\mathrm{cr}^{-2} \le \gamma_\mathrm{cr,c}^{-2}$ and $\gamma_\mathrm{cr}^{-2} \le \gamma_\mathrm{cr,f}^{-2}$, so that after the parameters of $\gamma_\mathrm{cr,c}^{-2}$ and $\gamma_\mathrm{cr,f}^{-2}$ are computed, the searching range of the critical parameter γ_cr^{-2} is finite. The algorithm can be described as

[Give dimension n ; system matrices $\mathbf{A}, \mathbf{B}_w, \mathbf{C}_y, \mathbf{C}_z, \mathbf{D}_{12}, \mathbf{D}_{21}$ and end matrices $\mathbf{P}_0, \mathbf{S}_\mathrm{f}$]

[Give time interval $[0, t_\mathrm{f}]$. Selecting a parameter $\gamma_\#^{-2}$;]

while ($\gamma_\#^{-2}$ is not precise enough) {

[Compute state feedback H_∞ control solution, checking if $\gamma_{\#}^{-2}$ is sub-optimal, solve $\mathbf{X}(t)$;]

if ($\gamma_{\#}^{-2}$ is not sub-optimal) {reduce $\gamma_{\#}^{-2}$; Break;}

[Compute H_∞ filter solution, checking if $\gamma_{\#}^{-2}$ is sub-optimal, then solve $\mathbf{P}(t)$;]

if ($\gamma_{\#}^{-2}$ is not sub-optimal) {reduce $\gamma_{\#}^{-2}$; Break;}

for ($k = 0; k \le N_{\text{int}}; k++$) {

[Solve the Rayleigh quotient (6.8.84); checking if $\gamma_{\#}^{-2}$ is less than the Rayleigh quotient;]

if ($\gamma_{\#}^{-2}$ is too large) Break;

}

if ($\gamma_{\#}^{-2}$ is too large) {reduce $\gamma_{\#}^{-2}$; Break;}

[$\gamma_{\#}^{-2}$ is sub-optimal; increasing $\gamma_{\#}^{-2}$;]

} Comment: end of loop, $\gamma_{\#}^{-2}$ is the measurement feedback H_∞ critical parameter γ_{cr}^{-2} (6.8.87)

Two numerical examples are given below, in order to see the relation among various critical parameters. It must be $\gamma_{\text{cr}}^{-2} \le \gamma_{\text{cr,c}}^{-2}$ and $\gamma_{\text{cr}}^{-2} \le \gamma_{\text{cr,p}}^{-2}$. But there is no definite relation between the parameters $\gamma_{\text{cr,c}}^{-2}$ and $\gamma_{\text{cr,p}}^{-2}$.

Example 6.14, A one-dimensional example gives the comparison between the results of precise integration and analytical solution. Suppose
$n = 1$, $\mathbf{A} = 0.8$; $\mathbf{B}_1 = 0.8$; $\mathbf{B}_2 = 2.0$; $\mathbf{C}_y = 3.0$; $\mathbf{C}_z = 0.8$, $\mathbf{S}_f = 0.01$, $\mathbf{P}_0 = 0.01$; $\mathbf{D}_{12} = \mathbf{0}$, $\mathbf{D}_{21} = \mathbf{0}$. The time interval is $(0, t_f) = (0, 0.8)$. The problem is to find the measurement feedback H_∞ control optimal parameter γ_{cr}^{-2}, and the Riccati matrices $\mathbf{X}(t), \mathbf{P}(t)$ and $\mathbf{Z}(t)$.

For one-dimensional problem, the differential equation can be solved analytically. Substituting the original data into the Riccati differential equations gives

$$\dot{\mathbf{X}} = -0.64 - 1.6\mathbf{X} + (9.0 - 0.64\gamma^{-2})\mathbf{X}^2, \quad \mathbf{X}(0.8) = 0.01;$$

$$\dot{\mathbf{P}} = 0.64 + 1.6\mathbf{P} - (4.0 - 0.64\gamma^{-2})\mathbf{P}^2, \quad \mathbf{P}(0) = 0.01;$$

where $\mathbf{X}(t)$ and $\mathbf{P}(t)$ are scalar functions and analytical solutions can be found. The numerical results below are the solutions from both the precise integration and analytical method. The two different methods give almost completely the same result, so that only one set of numerical result is listed.

For H_∞ state feedback control, the critical parameter is $\gamma_{\text{cr,c}}^{-2} = 11.38624$

For H_∞ filtering, the critical parameter is $\gamma_{\text{cr,p}}^{-2} = 19.19874$

However, equation (6.8.86) gives the measurement feedback H_∞ control critical parameter $\gamma_{\text{cr}}^{-2} = 7.623978$. For this lower bound critical parameter the three variance matrices $\mathbf{X}(t), \mathbf{P}(t)$ and $\mathbf{Z}(t)$ are all scalars and are listed in table 6.4. Because the critical parameter γ_{cr}^{-2} is less than both $\gamma_{cr,c}^{-2}$ and $\gamma_{cr,p}^{-2}$, so the two Riccati differential equations have no singular point. However, the measurement feedback variance matrix $\mathbf{Z}(t)$ has a singular point nearby the internal point $t \approx 0.30$. It can be seen that $\mathbf{Z}(0.30)$ is very large in table 6.4.

Table 6.4, The variance matrices of Example 6.14.

Time	0.0	0.05	0.10	0.15	0.20
X(*t*)=	1.430932	1.216512	1.038753	0.889047	0.761284
P(*t*)=	0.01	0.043976	0.080115	0.117988	0.157065
Z(*t*)=	0.011225	0.074266	0.219171	0.589149	1.776944

continue

0.25	0.30	0.35	0.40	0.45	0.50
0.651009	0.554896	0.470413	0.395595	0.328899	0.269092
0.196743	0.236378	0.275331	0.313005	0.348884	0.382551
8.368041	**577614.2**	21.938593	5.592052	2.787334	1.777840

continue

0.55	0.60	0.65	0.70	0.75	0.80
0.215179	0.166347	0.121927	0.081362	0.044185	0.01
0.413708	0.442176	0.467883	0.490852	0.511181	0.529024
1.287594	1.006730	0.828009	0.705731	0.617516	0.551258

##

Example 6.15, A multi-dimensional problem is illustrated, for which the precise integration method reaches almost the computer precision. It can also use the analytical method based on the eigen-solutions. Both methods give the same result. The original data is

$$n = 5, \quad t_f = 4.0$$

$$\mathbf{A} = \begin{bmatrix} -.08 & -3.0 & 0 & 2.0 & 0.5 \\ 0.04 & -1.4 & 10.0 & 0 & 0 \\ -.01 & -4.0 & -2.8 & 0 & 0 \\ 0 & 0 & 1.0 & 0 & 0.3 \\ 0.2 & 0 & 0 & 0 & 1.0 \end{bmatrix}; \quad \mathbf{B}_w = \begin{bmatrix} 0 & 0 & 0 \\ 0 & 0 & 0 \\ 0.2 & 0 & 0.3 \\ 0 & 0.4 & 0 \\ 0.5 & 0 & 0 \end{bmatrix}; \quad \mathbf{B}_u = \begin{bmatrix} 0 \\ -0.1 \\ -3.3 \\ 0 \\ 1.1 \end{bmatrix};$$

$$\mathbf{C}_y = \begin{bmatrix} 1.0 & 0.0 & 0.0 & 0.0 & 0.0 \\ 0.0 & 0.0 & 0.0 & 4.0 & 0.0 \end{bmatrix}, \quad \mathbf{C}_z = \begin{bmatrix} 0.5 & 0 & 0 & 0 & -.5 \\ 0 & 0 & 0.8 & 0 & 0 \\ 0 & 0 & 0 & 0.5 & 0 \end{bmatrix}, \quad \begin{array}{l} \mathbf{D}_{12} = \mathbf{0}, \ \mathbf{D}_{21} = \mathbf{0} \\ \mathbf{S}_f = \mathbf{I} \times 1E-3 \\ \mathbf{P}_0 = \mathbf{I} \times 1E-3 \end{array}$$

The precise integration method gives the eigenvalues as below.

The state feedback H_∞ control critical parameter: $\gamma_{cr,c}^{-2} = 0.546001$

The H_∞ filter critical parameter: $\gamma_{cr,p}^{-2} = 1.499859$

However, (6.8.85) gives the measurement feedback critical parameter $\gamma_{cr}^{-2} = 0.1611565$, which is far apart from $\gamma_{cr,c}^{-2}$ and $\gamma_{cr,p}^{-2}$. The variance matrices $\mathbf{X}(t), \mathbf{P}(t)$ and $\mathbf{Z}(t)$ can also be computed, but not listed for saving space. It should be pointed, nearby the point $t = 2.0$ the variance matrix $\mathbf{Z}(t)$ is almost singular. As a matter of fact, for an optimal parameter γ_{cr}^{-2} there must be a time point t in the domain $[0, t_f]$, where the matrix $\mathbf{Z}(t)$ is singular. In application, the optimal parameter γ_{cr}^{-2} makes the matrix $\mathbf{Z}(t)$ tends to be singular somewhere, so that it is not applicable. The similar situation appears also in the structural engineering of Euler stability load, which cannot be used in the engineering application either. ##

One point must be emphasized that in determining the optimal parameter γ_{cr}^{-2}, any internal time station can be a marginal point. So that solving only the ARE and skipped the internal time point is inappropriate, that the solution of the transient of Riccati differential equation is necessary.

So far the state space based theory of linear optimal control is described in some detail above. Readers can find that the analogy theory between structural mechanics and optimal control is quite helpful in understanding.

Concluding remarks:

The optimal control theory, structural vibration and wave propagation etc. are described under a ***unified theoretical framework***. They are closely related to that in analytical dynamics. The reader can find that the methodology used for these different problems is parallel, the differential equations, variational principles and others, are of the same form. To these subjects described in this book, if one studied one subject among them, then it will be easier for him to understand the others. The control theory is shown developed very close to that in applied mechanics, that parallel studying both the subjects is not very difficult, since the mathematical theory is the same. However, the linear control theory is only a fundamental part of the control theory, there are further problems of parameter identification, adaptive control, de-centralized control theory etc. Further researches are definitely necessary.

The problems solved in this book are all finite dimensional however it by no means to say such methodology can only solve finite dimensional problems. Theory of elasticity is a typical problem of infinite dimensional problem, by using the duality system approach described in the book [23], the fundamental equations are derived to a linear system form, and then the method of separation of variables is applied. The solution methodology becomes rational, rather than the try and error style semi-inverse approach. This sort of developments will be given in another book.

This book describes mainly for linear system, however, the duality methodology can be used not only for linear system, but also for non-linear system, see [140].

The methodology described in the present book can not only be used for teaching but also be used for further research.

一阴一阳之谓道

--易经，系辞

References

[1] R. Courant & D. Hilbert, *Methods of mathematical physics*. Vol.I and Vol.II, Interscience publishers Wiley, N.Y., 1953.

[2] S.P. Timoshenco & J.N. Goodier, *Theory of elasticity*. McGraw-Hill, 1951.

[3] S.P. Timoshenco & S. Woinowsky-Krieger, *Theory of plates & Shells*. McGraw-Hill, NY, 1959.

[4] S.P. Timoshenco & D. Young, *Vibration problems in engineering, 3rd Ed.* Van Nostrand, 1955.

[5] S.P. Timoshenco & J.M. Gere, *Theory of elastic stability*. McGraw-Hill, NY, 1961.

[6] S.P Timoshenco, *Advanced strength of material*. Van Nostrand, NY, 1958.

[7] O.C. Zienkiewicz & R. Taylor, *The finite element method*, 4-th ed. McGraw-Hill, NY, 1989.

[8] K.J. Bathe & E.L. Wilson, *Numerical methods for finite element analysis*, Prentice-Hall, NJ, 1977.

[9] F.E. Kadestensor and D.H. Norrie: *Finite element handbook*, McGraw-Hill, NY, 1987.

[10] E.T. Whittaker, *A treatise on the analytical dynamics*. Cambridge Univ. Press, 4-th ed. 1952.

[11] H. Goldstein, *Classical mechanics*, 2nd ed. Addison-Wesley, 1980.

[12] D.T. Greenwood, *Classical dynamics*. Dover, NY, 1997.

[13] L. Landau & E. Lifshitz, *Mechanics,* 3rd Edition, Pergamon, Oxford, 1976.

[14] J.V. Jose and E.J. Saletan, *Classical dynamics: A contemporary approach*. Cambridge University Press, 1999.

[15] H. Weyl: *The classical groups; Their Invariants and representations*. University press, Princeton, 1939.

[16] H. Kwakernaak & R. Sivan, *Linear optimal control systems*, Wiley-Interscience, NY, 1972.

[17] R. Stengel, *Stochastic optimal control*, Wiley, NY, 1986.

[18] D.Z. Zheng, *Linear system theory*. Tsing-Hua University Press, 1990. (Chinese)

[19] B.D.O. Anderson & J.B. Moore, *Optimal control: Linear quadratic methods*, Prentice-Hall, NJ, 1990.

[20] W.X. Zhong & X.X. Zhong, "Computational structural mechanics, optimal control and semi-analytical method for PDE", *Computers & Structures*, vol.**37**(6):993-1004,1990.

[21] W.X. Zhong, J.H. Lin & C.H Qiu, "Computational structural mechanics and optimal control—The simulation of substructural chain theory to linear quadratic optimal control problems", *Intern. J. Num. Meth. Eng.* **33**:197-211, 1992.

[22] W.X. Zhong, H.J. OuYang & Z.C. Deng, *Computational structural mechanics and Optimal control*. Dalian University of Technology Press, 1993. (Chinese)

[23] W.X. Zhong, *A new systematic methodology for theory of elasticity*. Dalian University of Technology Press, 1995. (Chinese)

[24] D.J. Inman, *Vibration, with control measurement and stability*. Prentice-Hall, N.J., 1989.

[25] L. Meirovitch, *Dynamics and Control of structures*. J. Wiley & Sons, 1992.

[26] Zhong Wan-xie and Yang Zai-shi: "On the computation of the main eigen-pairs of the continuous-time linear quadratic control problem", *Applied Math. and Mech.*, vol.**12**(1):49-54, 1991.
[27] W.X. Zhong, and F.W. Williams: 'A precise time step integration method', *Proc. Inst. Mech. Engr.* vol.**208**:427-430, 1994.
[28] E. Angel & R. Bellman, *Dynamic programming and partial differential equations*, Academic press, NY, 1972.
[29] C.B. Moler & C.F. Van Loan, "Nineteen dubious ways to compute the exponential of a matrix", *SIAM Review*, **20**:801-836, 1978.
[30] G.H. Golub & C.F. Van Loan, *Matrix computation.* Johns Hopkins Univ. Press, 1983.
[31] J.H. Lin, W.P. Shen & F.W. Williams, "Accurate high-speed computation of non-stationary random seismic responses", *Engineering structures*, vol.**19**(7):586 -593, 1997.
[32] W.X. Zhong, J.P. Zhu and X.X. Zhong, "A precise time integration algorithm for non-linear systems", *Proc. WCCM-3*,vol.1,pp.12-17, 1994.
[33] Y. Liu, *The precise integration of Hamilton system under parametric excitation.* Doctoral thesis, Shanghai Jiao-Tong University, 1996. (Chinese)
[34] X.D. Kong, *The precise integration for ODEs and its applications in multi-body dynamics.* Doctoral thesis, Dalian University of Technology, 1998. (Chinese)
[35] T.D. Burton, *Introduction to dynamic systems analysis.* McGraw-Hill, NY, 1994.
[36] D. Zwillinger, *Handbook of differential equations*, 2nd ed., McGraw-Hill, 1992.
[37] A.H. Nayfeh, *Perturbation methods*, J. Wiley and Sons, 1973.
[38] E.J. Hinch, *Perturbation methods*, Cambridge Univ. Press, 1991.
[39] G. Strang, *Introduction to applied mathematics*, Wellesley Cambridge press, Massachusett, 1986.
[40] W.H. Wittrick and F.W. Williams, "A general algorithm for computing natural frequencies of elastic structures". *Quart. J. Mech. Appl. Math.*, **24**, 263-284, 1971.
[41] W.X. Zhong, F.W. Williams and P.N. Bennett, "Extension of the Wittrick-Williams algorithm to mixed variable systems", *J. Vib. & Acous. Trans ASME*, **119**:334-340, 1997.
[42] J.H. Wilkinson, C. Reinsch, *Handbook of automatic computation*, vol.2, Linear algebra, Springer_verlag, 1971.
[43] W.H. Press *et.al*, *Numerical Recipes in C*, Cambridge Univ. Press, 1992.
[44] Crossley, "Vibration of multiple degrees of freedom system", *Handbook of engineering mechanics*, ed. W. Flugge, McGraw-Hill, 1962
[45] W.X. Zhong & J.H. Lin, "Whiplash effect of high buildings caused by earthquakes", *Journal of Vibration and Shock*, 1985, issue(2):1-6. (Chinese)
[46] A.Y.T. Leung, *Dynamic stiffness & sub-structures*. Springer, London, 1993.
[47] M.C. Pease, *Methods of matrix algebra.* Academic press, NY, 1965.
[48] P. Dennery, A. Krzywicki, *Mathematics for physicists*, Dover, 1995.
[49] W.X. Zhong & J.H. Lin, "Adjoint subspace iteration method for large scale eigenvalue problem of asymmetric real matrices", *Computational structural mechanics and its applications*, vol.7(4):1-10, 1989.
[50] M. Green & D.J.N. Limebeer, *Linear robust control.* Prentice_Hall, NJ, 1995.

[51] W. Zhang, *Theoretical basis of rotor dynamics*. Science press, Beijing, 1990. (Chinese)

[52] B.E. Yang, "Eigenvalue inclusion principles for discrete gyroscopic systems", *J. Appl. Mech.,* **59**: 278-283,1992.

[53] C.F. Van Loan, "A symplectic method for approximating all the eigenvalues of a Hamiltonian matrix", *Linear algebra and its application*, 233~251, 1984.

[54] H.G. Chetaev, *Stability of Motion.* Defense Press, Beijing, 1955.

[55] W.X. Zhong, J.H. Lin, J.P. Zhu, "Computation of gyroscopic systems and symplectic eigensolutions of skew-symmetric matrices", *Computers & structures*, vol. **52**(5):999-1009, 1994.

[56] W.X. Zhong, X.X. Zhong, "On the computation of anti-symmetric matrices", *Proc. Of EPMESC-4*, vol.2:1309-1316(late papers), Jul.30~Aug.2, 1992. Also, *J. Mathematical research and exposition*, vol.**15**(1):123-128, (Chinese) 1995.

[57] Z. Kang, W.X. Zhong, "Numerical study on parametric resonance of cables in the cable stayed bridge", *China Civil Engineering Journal*, Vol.**31**(4):14~22, 1998.

[58] W. Feller, *An introduction to probability theory and its applications*. Wiley, NY, 1950.

[59] M. Loeve, *Probability theory*. Springer, NY, 1978.

[60] N.U. Prabhu, *Stochastic processes—basic theory and its applications.* Macmillan, NY, 1965.

[61] D. Lamberton & B. Lapeyre, *Introduction to stochastic calculus applied to finance*. Chapman & Hall, London, 1996.

[62] W.Q. Zhu, *Random vibration*. Science Press, Beijing, 1992. (Chinese)

[63] N.C. Nigam, *Introduction to random analysis*, MIT press, Cambridge, 1983.

[64] R.W. Clough, J. Penzien, *Dynamics of structures*, McGraw-Hill, NY, 1993.

[65] D.E. Newland, *An Introduction to random vibration and spectral analysis*. Longmans, 1984.

[66] A.D. Kiureghian, A. Neuenhofer, "Response spectrum method for multi-support seismic excitations", *Earthquake engineering and structural dynamics*, vol.**21**: 713-740, 1992.

[67] E.H. Zavoni, E.H. Venmark, "Seismic random vibrarion analysis of multi-support structural systems", *ASCE J. Eng. Mech.*, **120**(5):1107-28, 1994.

[68] A.D. Kiureghian, A. Neuenhofer, "A discussion on above [3]", *ASCE J. Eng. Mech.*, vol.**121**(9):1037, 1995. H.Z. Ernesto, E.H. Venmark, "Closure on the discussion", *ASCE J. Eng. Mech.*, vol.**121**(9):1038, 1995.

[69] B. Oksendal, *Stochastic differential equations*, 4th ed. Springer-Verlag, Berlin, 1995.

[70] J.H. Lin, "A deterministic algorithm of stochastic seismic responses", *Chinese Journal of Earthquake Engineering Vibration,* vol.**5**(1), 89-93,1985. (Chinese)

[71] J.H. Lin, W.P. Shen, H.M. Song & D.K. Sun, "High-precision integration of mixed type for analysis of non-stationary random responses", *Journal of Vibration Engineering,* vol.**8**(2):127-135, 1995. (Chinese)

[72] J.H. Lin, Y.H. Zhang, D.K. Sun & Y. Sun, "Fast and precise computation of structural responses to non-uniformly modulated evolutionary random excitations", *Chinese Journal of Computational Mechanics*, vol.**14**(1):2-8, 1997.

[73] J.H. Lin, W.S. Zhang, F.W. Williams, "Pseudo-excitation algorithm for

nonstationary random seismic responses", *Engineering structures*, vol.**16**:270-276, 1994.

[74] J.H. Lin, D.K. Sun, Y. Sun, F.W. Williams, "Structural responses to non-uniformly modulated evolutionary random seismic excitations", *Communications in numerical methods in engineering*, vol.**13**:605-616, 1997.

[75] J.H. Lin, W.S. Zhang, J.J. Li, "Structural responses to arbitrarily coherent stationary random excitations", *Computers & structures*, vol.**44**(3):683-687, 1994.

[76] J.H. Lin, W.X. Zhong, W.S. Zhang & D.K. Sun: "High efficiency computation of the variances of structural evolutionary random responses", *Vibration & Shock*, vol.**7**(4):209-216, 2000.

[77] M.B. Priestly, "Evolutionary spectra and non-stationary process", *J. Royal Statis. Soc.* Ser.B, vol.**27**:204, 1965.

[78] W.X. Zhong, "Review of a high-efficiency algorithm series for structural random responses", *Progress in natural sciences*, vol.**6**(3):257-268, 1996.

[79] Y.L. Xu, D.K. Sun, J.M. Ko & J.H. Lin, "Buffeting analysis of long span bridges: A new algorithm", *Computers & Structures*, vol.**68**:303-313, 1998.

[80] A.H. Liang, X.L. Du & H.Q. Chen, "Random seismic response analysis of arch dams based on non-stationary random earthquake motion field", *SUILI XUEBAO(Journal of Hydraulics)*, vol.**6**:21-25, 1999. (Chinese)

[81] G.X. Chen, J.F. Xie & K.X. Zhang, "Two-dimensional stochastic analysis of the aseismic characteristics of earth dams", *Earthquake engineering and engineering vibration*, vol.**14**(3):81-89, 1994. (Chinese)

[82] L.C. Fan, J.J. Wang & W. Chen, "Response characteristics of long-span cable-stayed bridges under non-uniform seismic action", *Chinese Journal of Computational Mechanics*, vol.**18**,(3):358~363, 2001. (Chinese)

[83] J. Connor, *Wave, current and wind loads*. Dept. Civil Eng., MIT, Cambridge, Massachusetts, 1979.

[84] J.H. Lin, F.W. Williams, W.S. Zhang, "A new approach to multiphase-excitation stochastic seismic response", *Microcomputers in civil engineering*, vol.**8**:283-290, 1993.

[85] J.H. Lin, Y.H. Zhang, Y. Zhao: "Seismic analysis methods of long-span structures and recent advances", *Advances of mechanics*, vol.**31**(3):350-360, 2001. (Chinese)

[86] C.W.S. To, "Response statics of discretized structures to non-stationary random excitation", *J. sound & vibration*, vol.**105**(2):217-231, 1986.

[87] Zhong Wan-xie, Lin Jia-hao and Qiu Chun-hang: 'On the theory of sub-structural chain and the eigen-problems----The simulation to linear quadratic control problems', *Proc. of the 2nd world congress on computational mechanics (Extended Abstract)*, p.772, Stuttgart, 1990.

[88] Zhong Wan-xie: 'On the simulation of computational mechanics to optimal control and its applications', *Proc. of the EPMESC-3 Conference*, Macao, 1990.

[89] W.X. Zhong & X.X. Zhong, "Elliptic partial differential equation and optimal control", *Numerical methods for PDE*, vol.**8**(2):149-169, 1992.

[90] W.X. Zhong, Z.S. Yang, "Partial differential equations and Hamiltonian system", *Computational mechanics in structural engineering*, pp32-48, Elsevier, 1992.

[91] W.X. Zhong & X.X. Zhong, "On the transverse eigenvector solution of the

elliptic partial differential equation in the prismatic domain", *Journ. on Numerical methods and computer applications*, vol.**13**(2):107~118, 1992. (Chinese)

[92] W.X. Zhong, W.P. Howson & F.W. Williams, " H_∞ control state feedback and Rayleigh quotient", *Computer Methods in applied mechanics and engineering*, vol.**191**:489-501, 2001.

[93] W.X. Zhong, F.W. Williams, " H_∞ filtering with secure eigenvalue calculation and precise integration", *Int. J. Numer. Meth. Engng.*, Vol.**46**, 1017-1030, 1999.

[94] W.X. Zhong, "Variational method and computation for H_∞ control ", *Applied Math. & Mech.*, vol.**21**(12), 2000, also *Proc. APCOM-4,* Singapore, 1999.

[95] W.X. Zhong, F.W. Williams, "Physical interpretation of the symplectic orthogonality of the eigen-solutions of a Hamiltonian or symplectic matrix", *Computers & Structures*, vol.**49**(4):749-750, 1993.

[96] W.X. Zhong, "Reciprocal theorem and symplectic orthogonality", *Acta Mechanica Sinica*, vol.**24**(4):432-437,1992. (Chinese)

[97] Zhong Wanxie, Zhong Xiangxiang, "The Differential Equations of the Interval Mixed Energy Matrices of LQ Control and Its Applications", *Acta Automatica,* Vol.18, No.3, 325-331, (Chinese) 1992.

[98] Zhong Wanxie, "The Precise Integration for Matrix Riccati Equations", *Journal of Computational Structural Mechanics and Applications*, vol.**11**(2):113-119, 1994, (Chinese).

[99] W.X. Zhong, "The method of precise integration of finite strip and wave guide problems", *Proc. Intern. Conf. on Computational Method in Struct. and Geotech. Eng.*, pp.50-60, 1994, HongKong, Eds. P.K.K. Lee, L.G. Tham, Y.K. Cheung.

[100] W.X. Zhong and J.P. Zhu, "Precise time integration for the matrix Riccati equation", *J. Num. Method & Comp. Appl.*, Vol.**17**(1):26-35,1996.

[101] W.X. Zhong, "Precise integration of eigen-waves for layered-media", *Proc. EPMESC-5*, vol.2:1209-1220, 1995, Macao.

[102] W.X. Zhong, Y.K. Cheung and Y. Li: "The precise finite strip method", *Computers & structures*, **69**:779-783, 1998.

[103] W.X. Zhong: "Combined method for the solution of asymmetric Riccati differential equations", *Computer Methods in Appl. Mech. & Eng.* Vol.**191**:93-102, 2001.

[104] W.X. Zhong: "The dynamic programming method for the loading of influence function of highway bridges", *The Cognition and Practice of Mathematics*, 1978.

[105] K.F. Graff, *Wave motion in elastic solid*, Oxford: Clarendon press, 1975.

[106] J.D. Achenbach, *Wave propagation in elastic solids*, North-Holland, 1973.

[107] J.F. Doyle, *Wave propagation in structures*, Springer, NY, 1989.

[108] W.M. Ewing, W.S. Jardetzky, F. Press, *Elastic waves in layered media*,McGraw-Hill,NY, 1957.

[109] L.M. Brekhovskikh, *Waves in layered media*. Academic press, NY, 1980.

[110] B.L.N. Kennett, *Seismic wave propagation in stratified media*, Cambridge univ. Press, 1983.

[111] K. Aki, P.G. Richards, *Quantitative seismology*, W.H. Freeman and Company, San Francisco, 1980.

[112] S.A. Rizzi, J.F. Doyle, "Spectral analysis of wave motion in plane solids", *Trans ASME J. vib. & acoust.*, vol.**114**:133-140, 1992.

[113] D.J. Mead, "A general theory of Harmonic wave propagation in linear periodic systems with multiple coupling", *J. sound & vib.*, vol.**27**:235-260, 1973.

[114] W.X. Zhong, F.W. Williams, "Wave propagation for repetitive structures and symplectic mathematics", *Proc. Inst. Mech. Engrs., part C*, vol.**206**:371-379, 1992.

[115] W.X. Zhong, F.W. Williams, "The eigensolutions of wave propagation for repetitive structures", *Structural engineering and mechanics*, vol.**1**(1):47-60, 1993.

[116] F.W. Williams, W.X. Zhong and P.N. Bennett, "Computation of the eigenvalues of wave propagation in periodic substructural systems", *J. Vib. & Acous. Trans ASME*, **115**:422-426, 1993.

[117] W.X. Zhong, F.W. Williams, "On the direct solution of wave propagation for repetitive structures", *J sound & vib.*, vol.**181**(3):485-501, 1995.

[118] W.X. Zhong & F.W. Williams, "On the localization of the vibration mode of sub-structural chain-type structure", *Proc. Inst. Mech. Engrs., part C*, vol.**205**(4): 281-288, 1991.

[119] C. Kittel: *Introduction to solid-state physics,* J. Wiley & Sons, NY, 1986.

[120] F.J. Taylor: *Principles of signals and systems*, McGraw-Hill, NY, 1994.

[121] P.A. Cook: *Nonlinear dynamical systems*, Prentice-Hall, NJ, 1994.

[122] W. Hahn: *Stability of motion.* Springer, Berlin, 1967.

[123] W. Zhong: "Numerical solution of Lyapunov differential equation", *Proc. of Conference for Modern Math. and Mech. (MMM-7),* pp.511-520, Shanghai Univ. press, 1997.

[124] J.H. Lin, W.X. Zhong, W.S. Zhang and D.K. Sun: "High efficiency computation of the variances of structural evolutionary random responses", *Shock & Vibration*, vol.7(4):209-216, 2000.

[125] *Kalman Filtering: Theory and Application*, Edited by H.W. Sorenson, IEEE press, NY, 1985.

[126] F.Y. Wang: "The derivation of Kalman filter and interpolation of quadratic optimal control", *Journal of Zhejiang university*, vol.**23**(2):193-204, 1989.

[127] G.S. Sidhu and G.J. Bierman: "Integration free interval doubling for Riccati equations", *IEEE Trans Automat. control*, vol.**22**, 1977.

[128] S. Bittanti, A. Laub and Willem, *The Riccati Rquation,* Springer, 1991.

[129] W.F. Arnold, A.J. Laub: "Generalized eigenproblem algorithms and software for algebraic Riccati equations", *Proc. IEEE*, vol.**72**, 1984.

[130] T. Pappas, A.J. Laub & N.R. Sandell: "On the numerical solution of the discrete-time algebraic Riccati equation", *IEEE Trans-AC,* vol.**25**(4):631-641, 1980.

[131] W.X. Zhong: "The precise integration of Kalman-Bucy filter", *Journal of Dalian university of Technology*, vol.**39**(2), 1999.

[132] J.B. Burl, *Linear optimal control*, Addison-Wesley, CA, 1999.

[133] B. M. Chen: H_∞ *control and its Applications*. Springer, 1998

[134] J.C. Doyle, K. Glover, P.P. Khargonekar & B.A. Francis, "State space solution to standard H_2 and H_∞ control problems", *IEEE Trans-AC*, vol.**34**:831-847,

1989.

[135] K.M. Zhou, J.C. Doyle & K. Grove, *Robust and optimal control*, Prentice-Hall, NJ, 1996.

[136] T. Basar & P. Bernhard, H_∞-*optimal control and related minimax design problems—a dynamic game approach*, Birkhauser, Boston, 1995.

[137] M. Bala Subrahmanyam, *Finite Horizon* H_∞ *and related control problems*. Birkhauser, Boston, 1995.

[138] M. Dresher: *The mathematics of games of strategy-theory and applications*. Dover, N.Y. 1981.

[139] N. Kalouptsidis: *Signal processing systems—Theory and design*. J. Wiley & Sons, NY, 1997.

[140] J.E. Marsden, T.S. Ratiu: *Introduction mechanics and symmetry*. 2nd ed. Springer, 1999.

[141] Zhong Wanxie, Zhong Xiangxiang: 'On the adjoint simplectic inverse substitution method for main eigensolutions of a large Hamiltonian matrix.' *Journ. of System Eng*. vol.**1**, pp.41-50, 1991.

Index